W9-BQZ-523

ANNUAL REVIEW OF FLUID MECHANICS

ANNUAL REVIEW OF FLUID MECHANICS

VOLUME 19, 1987

JOHN L. LUMLEY, *Co-Editor*
Cornell University

MILTON VAN DYKE, *Co-Editor*
Stanford University

HELEN L. REED, *Associate Editor*
Arizona State University

ANNUAL REVIEWS INC. 4319 EL CAMINO WAY P.O. BOX 10139 PALO ALTO, CALIFORNIA 94303-0897 USA

ANNUAL REVIEWS INC.
Palo Alto, California, USA

International Standard Serial Number: 0066-4189
International Standard Book Number: 0-8243-0719-4
Library of Congress Catalog Card Number: 74-80866

TYPESET BY A.U.P. TYPESETTERS (GLASGOW) LTD., SCOTLAND
PRINTED AND BOUND IN THE UNITED STATES OF AMERICA

SOME RELATED ARTICLES IN OTHER *ANNUAL REVIEWS*

From the *Annual Review of Astronomy and Astrophysics*, Volume 24 (1986)

The Physics of Supernova Explosions, S. E. Woosley and Thomas A. Weaver

From the *Annual Review of Earth and Planetary Sciences*, Volume 14 (1986)

El Niño, Mark A. Cane

Coastal Processes and the Development of Shoreline Erosion, Paul D. Komar and Robert A. Holman

From the *Annual Review of Physical Chemistry*, Volume 37 (1986)

Dynamics in Polyatomic Fluids: A Kinetic Theory Approach, Robert G. Cole and Glenn T. Evans

Thermodynamic Behavior of Fluids Near the Critical Point, J. V. Sengers and J. M. H. Levelt-Sengers

Annual Review of Fluid Mechanics
Volume 19, 1987

CONTENTS

ANNUAL REVIEWS INC. is a nonprofit scientific publisher established to promote the advancement of the sciences. Beginning in 1932 with the *Annual Review of Biochemistry*, the Company has pursued as its principal function the publication of high quality, reasonably priced *Annual Review* volumes. The volumes are organized by Editors and Editorial Committees who invite qualified authors to contribute critical articles reviewing significant developments within each major discipline. The Editor-in-Chief invites those interested in serving as future Editorial Committee members to communicate directly with him. Annual Reviews Inc. is administered by a Board of Directors, whose members serve without compensation.

ca. 1938

L. Pauling

Ann. Rev. Fluid Mech. 1987. 19: 1–25

LUDWIG PRANDTL AND HIS KAISER-WILHELM-INSTITUT

K. Oswatitsch

Institut für Strömungslehre und Wärmeübertragung, Technische Universität, Wien, Austria

K. Wieghardt

Institut für Schiffbau, Universität Hamburg, Hamburg, West Germany

1. INTRODUCTION

To describe the work of Ludwig Prandtl, we have had to restrict ourselves to activities under his direct influence and responsibility, i.e. his own publications (Prandtl 1961), and those of his immediate coworkers and pupils as long as they stayed in Göttingen. Some of these individuals later on became well known for their own work elsewhere in Germany or in various other countries, especially after the war. To cite Flügge-Lotz & Flügge (1973), "The seeds sown by Prandtl have sprouted in many places, and there are now many 'second-growth' Göttingers who do not even know that they are." Even with the restriction to Göttingen papers, we could not strive for completeness, although we have tried to depict the variety of Prandtl's interests. In particular, we have had to omit all the extensive work done at the *Aerodynamische Versuchsanstalt* (AVA), directed by A. Betz since 1937, although there too worked many members of the Göttingen school who have won for themselves a name in aerodynamics. A short history of the AVA from 1907 to 1982 is given by Wuest (1982).

2. PRANDTL'S FIRST 30 YEARS

Ludwig Prandtl was born on 4 February 1875 in Freising, north of Munich, as son of a professor at the Agricultural Central School there.

0066–4189/87/0115–0001$02.00

To get a feeling for this time, a few events of the year 1875 may be recalled: Hans Christian Andersen and Eduard Mörike died, Thomas Mann and Albert Schweitzer were born, Mark Twain's *The Adventures of Tom Sawyer* came out, there were first performances of Bizet's *Carmen* at Paris and of Tchaikovsky's piano Concerto No. 1 at Boston. Last but not least, London's sewerage system was completed (Grun 1975).

As to science: at the same time a young student, M. Planck (1858–1947), asked Ph. von Jolly (1809–84) whether he should study physics or rather music, in which he was also interested. The answer he received was that, in general, physics is a completed science, although there are little niches here and there where further work might be useful. Yet, as we know now, Planck's quantum theory came in 1900, Einstein's relativity in 1905, and, which is of interest here, Prandtl's boundary layer in 1904 after the first trial flight of Zeppelin in 1900 and the first successful flight of a powered airplane by Orville and Wilbur Wright in 1903.

Starting in 1894, Prandtl studied mechanical engineering at the *Technische Hochschule* (TH) in Munich and became an assistant of A. Föppl (1854–1924), who was also his doctoral father. His thesis was "On Tilting Phenomena, an Example of Unstable Elastic Equilibrium" (Prandtl 1900), and his graduation to Dr. phil. took place at the University of Munich on 29 January 1900. (The degrees Dipl.-Ing. and Dr.-Ing. had only just been introduced at the TH in 1899.)

He then worked for less than two years at a factory in Nuremberg (later called MAN). One of his tasks there was to improve the suction of sawdust from a wood-cutting machine; he failed because he used a diffuser with too large an opening angle. However, this little mishap haunted him for years, even after he had left the firm, until at last he "invented" boundary layers and separation in 1904 (Prandtl 1904a).

Nevertheless, before that, he was appointed professor at the TH in Hanover on 1 October 1901 at the age of 26. This might have been the summit of his career, but actually it was just the beginning. For in the meantime the famous mathematician Felix Klein (1849–1925) at Göttingen University had become interested in Prandtl. For many years to come he played a very important role in the life of Prandtl, who simply said of him, "He was my fate" and kept in his office a large portrait of Klein. To understand this influence of Klein, some peculiarities of the German system of higher education—especially for engineers—in the nineteenth century should be outlined first (Manegold 1970).

A neat example of the deep roots of the "inferiority complex" of many engineers is to be seen already at the very beginning of engineering science. Probably the first chair in this field was established at Leyden University in 1600, a chair for fortification, on the recommendation of the Quartermaster-General Simon Stevin, with the stipulation that the lectures be given in Dutch because later on these students would have to deal mainly with humble workers. Only after exactly 100 years, in 1700, was this professor allowed to teach in Latin, just like all his other colleagues.

Later on, German philosophers like Schelling and Schleiermacher of the early nineteenth century still had a following in Prandtl's lifetime. They maintained that if universities had to cultivate natural sciences besides philosophy at all, then it must be done only for their own sake without any practical or even technical purpose—no bread science! As Goethe said to Eckermann in 1829, "Whereas we Germans are drudging with the solution of philosophical problems, the English laugh at us and win the world with their common sense." Klein's opinion was that the real task of natural science is not to explain nature, a goal that would never be reached in the last resort, but rather to master nature.

Moreover, the Industrial Revolution had come to Germany somewhat belatedly. It is true that polytechnical schools or institutes were founded after the model of Monge's École Polytechnique in Paris (1794), first in Prague (1806),[1] Vienna, and Graz (1815), and then in Karlsruhe (1832) and many other German-speaking states. For some decades the Viennese school had the best reputation, and even young men of noble birth did not feel ashamed of studying there. Only at the end of the century did all these schools become THs and so, at least officially, of the same rank as the older universities. On the other hand, many THs in Central Europe and Scandinavia had the same structure as the German ones, and there the language mostly used for publications on engineering sciences was German (e.g. Oseen 1927), becoming English after the Second World War.

At the end of the nineteenth century, Klein had already been fighting for many years, and in grand style, against the then common polarization between practical engineers and noble thinkers at universities. At first he had tried to reach a general close cooperation between THs and universities. Especially after his second visit to the United States during the world exhibition in Chicago (1893), he was much impressed by the superiority of American technology and of the pragmatic teaching at the Massachusetts Institute of Technology. His idea was—roughly—that the "front-line officers" for industry should come from THs and their "general staff" from universities. But both groups protested violently. The engineers again felt subordinated, and the war cry of the others was "Against materialism and Americanism!"

Hence, Klein pursued the less ambitious idea that at least his Göttingen

[1] The first German universities were Prague (1348) and Vienna (1365) after the Sorbonne in Paris (1255).

University should have a few chairs for technical applied sciences as well as for pure sciences—in their mutual interest. This plan was supported strongly in Berlin by *Ministerialdirektor* F. Althoff. (At this time all education was directed from the center in Berlin. The Kaiser himself signed the letter of appointment for each professor.) To find financial support, professors and interested directors of industry founded in 1898 the *Göttinger Vereinigung zur Förderung der angewandten Physik und Mathematik*. It was the first of its kind in Germany, with H. T. Böttinger, director of *Farbwerke Elberfeld*, and Klein as co-chairmen (Kraemer 1975).

As early as 1900, Klein had made inquiries about Prandtl, as shown in still existing letters from A. Föppl, A. Stodola, and the director of MAN (Rotta 1985). Meanwhile, Prandtl had become a full professor in Hanover, whereas Klein could offer him only an *Extraordinariat*, comparable to an associate professorship without faculty membership, although the difference in salary was to be paid by the *Göttinger Vereinigung*. But at last, in 1904, yet still before Prandtl's lecture in Heidelberg on boundary layers, Klein succeeded in enticing Prandtl to come to Göttingen against the wishes of A. Föppl, his doctoral father, who later became his father-in-law in 1909. In due course, in 1907, Prandtl became full professor again, and his chair for mechanics was the only one at a German university for almost a half-century. And so the quiet university town of Göttingen later became the center of aerodynamic research in Germany.

3. PRANDTL'S FIRST TWO DECADES AT GÖTTINGEN

In Hanover, Prandtl met the mathematician C. Runge (1856–1927), who sometimes was called the father of numerics. Soon they became friends, and together they went to Göttingen in 1904/5, where they combined their two institutes into one *Institut für angewandte Mathematik und Mechanik*.

It was the time of the Zeppelin and Parseval airships, and in 1906 the *Studiengesellschaft für Motorluftschiffahrt* was founded in Berlin. Prandtl proposed the erection of a test station corresponding to already existing ship model basins. In particular, he designed two wind tunnels, one of the open Eiffel type and the other a return-circuit tunnel, but still with closed test section. The latter tunnel was built in 1908 with a borrowed 30-HP motor in a tiny house on rented ground in Göttingen.

In those first years at Göttingen, Prandtl (1906) published the first estimate of the thickness of a shock wave. In his thesis, Magin (1908) showed by schlieren photographs of the flow in a Laval nozzle the limited upstream influence of supersonic flow, which was not taken for granted

80 years ago. Also, the supersonic Prandtl-Meyer flow around a corner was originated by Prandtl (1907) and Meyer (1908). H. Blasius (1907) calculated the first laminar boundary layer along a flat plate, and four years later Hiemenz (1911) did the same for a circular cylinder. Both works were dissertations. Prandtl (1910), who was not aware of an older paper by Reynolds (1874), showed again in mathematically improved form the analogy between resistance and heat transfer. Prandtl's pitot-static tube is described in Prandtl (1912).

Th. von Kármán (1881–1963) came to Göttingen in 1908. At first he finished his thesis on a buckling problem of his own (von Kármán 1909). Then he was employed by Prandtl to work on Zeppelin problems in the new wind tunnel. Just a year later he became *Privatdozent* at the university.

In some Central European universities one can apply, after the doctor's degree, for "habilitation" in order to become a *Privatdozent*. This requires a special habilitation paper and a colloquium before the faculty and results in permission—and then the duty—to teach and examine in a special field. This right is the so called *venia legendi*. The *Privatdozent* does not receive a salary; usually he earns his living as a scientific assistant or practicing physician, lawyer, and so on.[2] To become professor later on at one's own university or somewhere else it is, of course, helpful to be a *Privatdozent* at a well-known faculty and to have a certain repertoire of lectures.

In 1908 and 1909, F. W. Lanchester paid two short visits to Göttingen that greatly influenced the later development of wing theory. Almost 20 years later, when Prandtl delivered the Wilbur Wright Memorial Lecture in 1927, Lanchester remembered, "Professor Runge, who introduced us, acted as interpreter; Professor Prandtl and I could only smile at each other, for neither could speak the other's language" (Prandtl 1927).

The work at the test station became more and more important. Hence, the university took over the wind tunnel previously under the management of the aforementioned *Göttinger Vereinigung*. In 1911 they engaged A. Betz (1885–1968), who had just obtained his degree Dipl.-Ing. in naval architecture from the TH in Berlin. Prandtl and Betz complemented each other ideally. Betz was not only a scientist of high rank but also an efficient administrator—quite in contrast to Prandtl. Very soon he became the right hand of Prandtl, and in 1924 he was appointed vice-chairman of the institution that later became the *Aerodynamische Versuchsanstalt* (AVA), now the Göttingen center of the *Deutsche Forschungs- und Versuchs-Anstalt für Luft- und Raumfahrt* (DFVLR).

In 1911 the *Wissenschaftliche Vereinigung für Luftfahrt* was founded,

[2] For example, young Schopenhauer was a *Privatdozent*. Yet, since he offered his lectures at exactly the same hours as his great enemy Hegel, he attracted only a few students.

now called the *Deutsche Gesellschaft für Luft- und Raumfahrt* (DGLR). In an expertise for the establishment of the *Deutsche Versuchsanstalt für Luftfahrt* (DVL) in Berlin-Adlershof in 1912, Prandtl warned against the appointment of too many *Beamte*, i.e. officials with automatic tenure. Since one could not foretell the future of aviation, the director should be given free hand as much as possible to cope with yet unexpected developments.

Von Kármán's paper on his famous vortex street was presented at the Göttingen Academy (von Kármán 1911, 1912, von Kármán & Rubach 1912). In 1913 he became professor at the TH Aachen, where he worked until 1929, except for five years of war service in the Austro-Hungarian air force (where, e.g., R. von Mises and K. von Terzaghi were other famous members). Communication between Göttingen and the new center in Aachen remained close, of course.

Wind-tunnel tests at this time yielded many new insights. For example, the drag coefficient of a sphere found at Göttingen was about twice the value G. Eiffel (1832–1923) had found in Paris. Hence, Eiffel (1912) made more tests with three spheres of various sizes and at various speeds and discovered the then shocking dependence of the drag coefficient on Reynolds number. In searching for an explanation, Prandtl (1914) supposed that the French boundary layer had become turbulent. Hence, he forced the boundary layer of his own sphere to also become turbulent by a trip wire, which indeed decreased the drag drastically. Incidentally, Prandtl often regretted that Eiffel did not want to continue their correspondence after the war.

During the First World War a bigger wind tunnel (*Kanal* 1) was built with an open test section of 4 m^2 and 300 HP; it was the first one of the Göttingen type. Regular test series with a three-component balance were started in 1918. The older tunnel of 1908—now called *Kanal* 2—was rebuilt,[3] also with an open test section, on the grounds of the new institute, *Modellversuchsanstalt für Aerodynamik*, founded in 1915. Its proprietor was the *Kaiser-Wilhelm-Gesellschaft* (KWG)—now the *Max-Planck-Gesellschaft* (MPG)—which was founded in 1911. The land was donated by H. Th. Böttinger, and so the institute address was *Böttinger-Str*. 4-8.

Tests with airfoils and wings, together with the earlier conversations with Lanchester, stimulated Prandtl (1918, 1919a) to propound his wing theory. At the same time two theses came out: that of Munk (1918) on elliptic lift distribution as an optimum and that of Betz (1919a) on the lift of a rectangular wing. Betz (1919b) also proposed the first theory for

[3] Both tunnels were dismantled in 1948.

propellers with minimum energy loss, which Prandtl (1919b) presented to the academy with a supplement.

Another important thesis was that of K. Pohlhausen (1921) on the approximate calculation of laminar boundary layers by using the momentum equation that von Kármán (1921) had proposed at Aachen. It was the first of the now so-called integral methods.

Initiated by Prandtl, E. Pohlhausen (1921), the brother of K. Pohlhausen, wrote on the heat transfer on a plate with laminar boundary layer. Concerning this topic Prandtl always mentioned in his lectures and in his book that the "Prandtl number" had been formulated already by W. Nusselt (1909).

The thesis of Birnbaum (1924) deals with unsteady profile theory. Better known is his previously published paper (Birnbaum 1923) on the influence of camber on a profile in steady flow. Here, by request of Prandtl, work was finished that had come to a stop at the end of the war.

J. Ackeret (1927) reports in the *Handbuch der Physik*, Vol. 7, that Prandtl had given a formula expressing the compressibility effect on profile lift in subsonic flow in his mechanics seminar in 1922; this formula was later rediscovered by Glauert (1927). These were the first formulations of what is now known in more general form as the Prandtl-Glauert analogy [cf. Busemann (1928), where also test results of Göttingen and NACA are shown].

4. ORGANIZATION PROBLEMS IN SCIENCE

The Kaiser-Wilhelm-Institut (KWI) and the Aerodynamische Versuchsanstalt (AVA)

In 1907 Prandtl was offered a position at the TH Stuttgart, which he did not accept. Later, in 1920 and again in 1922, he was asked to accept the chair of his father-in-law, A. Föppl, at the TH Munich. But then the KWG proposed the foundation of a *KWI für Hydro- und Aerodynamik* at Göttingen with a regular budget, whereas the AVA essentially had to earn its own money.[4] Already in 1911, when the *KWG zur Förderung der Wissenschaften* was founded, Prandtl had written—again encouraged by Klein—a memorandum for such a KWI. But the KWG had previously promised to support the aforementioned DVL at Berlin-Adlershof,

[4] It might be mentioned here that most German research institutes such as, e.g., DVL (now DFVLR), AVA, and KWG (now MPG) are only registered associations (*eingetragene Vereine*). Insofar as they use public money, they are controlled by the state audit office. The individual KWIs (MPIs) are not corporate bodies, but the mother society [KWG (MPG)] is.

founded in 1912. And then came the First World War and inflation. But now, after the promise of KWG, Prandtl remained in Göttingen.

His *KWI für Strömungsforschung* came into existence in 1925. At the opening ceremony, the founder and president of KWG, Adolf von Harnack, explained: "With great care do we look for a good researcher. When we have found him we direct currents of good will to him, procure as much financial aid as possible, and leave it to him to do what he thinks right. We ourselves no longer exercise any influence on him. Our task is finished with his selection and the provision of means" (Betz 1957).

The new institute had, e.g., two high-pressure steel vessels, each with a volume of 10 m^3 for high subsonic and supersonic tests as well as cavitation experiments, and a rotating chamber of 3-m diameter and up to 60 rpm for the study of Coriolis forces in fluid flow. Unfortunately, most prospective investigators fell seasick in this closed room without a view of a fixed horizon as in a merry-go-round.

Prandtl directed his KWI with a staff of up to about 40 until 1945. His university institute for mechanics he handed over in 1934 to M. Schuler, well known for his work on gyroscopes. The name of the *Versuchsanstalt für Aerodynamik* was changed in 1919 to the *Aerodynamische Versuchsanstalt* (AVA). It was on the same grounds as the KWI and became an independent registered association in 1937 under its director, Betz, with over 700 employees during the war. Prandtl was chairman of the AVA directorate. Despite these organizational differences, there was always close contact between members of both institutions, sometimes even leading to marriage.

The research topic of the AVA was aerodynamics of airplanes and their propulsion in the widest sense. After the First World War, sometimes its main income was earned by investigations ordered by foreign countries (Wuest 1982). On the other hand, Prandtl at his KWI could freely plan farsighted research in various directions such as, e.g., meteorology.

Nowadays, when all activities are required to be of "relevance to society," whatever that means, the importance of such completely free research is often not appreciated. Yet commissions appointed to fix specific directions for future research work usually do not show the foresight and versatility of top experts. Without Prandtl and his KWI, Göttingen would never have emerged as a main center for flow research.

Foundation of the Zeitschrift für Angewandte Mathematik und Mechanik (ZAMM) and the Gesellschaft für Angewandte Mathematik und Mechanik (GAMM)

Prandtl delivered his famous lecture on flows with very small friction at the Third Mathematical Congress in Heidelberg in 1904. But there he found no response at all, with the only interested listener being Klein. At that time in Germany, such a lecture would have been appropriate to either the *Deutsche Mathematiker-Vereinigung* (DMV) founded in 1890 or the *Verein Deutscher Ingenieure* (VDI) founded in 1856.[5]

A special journal, the *Zeitschrift für Angewandte Mathematik und Mechanik* (ZAMM), was founded only after the First World War in 1921, with R. von Mises as editor and published by VDI-Verlag. The idea of establishing a corresponding society emerged first in correspondence between von Mises, Prandtl, and H. J. Reissner, also in 1921. A few sentences out of a letter from Prandtl to von Mises might be mentioned here because they are characteristic of his scientific credo (Gericke 1972):

> Against your name proposal (Society of Applied Mathematics and Mechanics) I will stick to mine (Society of Technical Mechanics) because I wish that the main object of the new society be rather the field of mechanics and its applications in civil and mechanical engineering. Applied mathematicians, i.e. those doing useful mathematics[6] for the main purpose of this society, are very welcome. What I would like to prevent is the dominance of mathematics and the mathematical treatment of problems. I think experiments ought to be stressed at least as much as theories.

Yet eventually, the *Gesellschaft für Angewandte Mathematik und Mechanik* (GAMM) was founded in 1923 at Marburg, where the first executive committee (Prandtl president, H. J. Reissner vice-president, von Mises secretary) and the first scientific committee (Emde, Finsterwalder, von Kármán, Knoblauch, and Trefftz) were elected.

The objective of GAMM is the cultivation and advancement of scientific work and international cooperation in the field of applied mathematics as well as in all branches of mechanics and physics that number among the foundations of engineering sciences. The Society pursues this goal chiefly by organizing scientific meetings (once a year).

For some unknown reason, GAMM has never applied for registration into the official *Vereinsregister* (register of associations)—perhaps simply because of the juristic inefficiency of its members. However, this semi-official status turned out to be helpful in 1933 for surviving and then in 1945 for reestablishing.

Already in 1922, von Kármán & Levi-Civita (1924) had initiated the first postwar international meeting of hydro- and aerodynamicists in Innsbruck, Austria. The first International Congress of Applied Mechanics in Delft in 1924 was initiated by C. B. Biezeno, J. M. Burgers, J. A.

[5] A time when, e.g., Prussia, Bavaria, and Hesse were still separate states, although members of a custom union since 1833.

[6] In 1931/32 Prandtl gave a lecture series in Göttingen with the provocative title "Intuitive and Useful Mathematics."

Schouten, and E. B. Wolff; there the idea of the International Union of Theoretical and Applied Mechanics was born.

5. THE FIRST TWELVE YEARS OF THE KWI

The young Swiss engineer J. Ackeret (1898–1981) came in 1921 to the AVA to continue his studies in flow research at his own cost. Yet, in 1922 he was given an assistantship when C. Wieselsberger (1887–1941) left for Japan. He proved extremely active and versatile; for example, he developed small electric motors with up to 30,000 rpm for model propellers (Ackeret 1924). When the planning for the new KWI began in 1922, Prandtl and Ackeret designed the supersonic and the cavitation tunnels, and he became *Abteilungsleiter* at the KWI. His work there can be described only sketchily here.

W. Tollmien (1900–68) became assistant to Prandtl in 1924. Prandtl & Tollmien (1924) generalized Ekman's spiral for the wind distribution above the rotating Earth to the much more realistic turbulent flow with a 1/7-power profile. This subject was later further developed in a thesis of Roux (1935). The thesis of Tollmien (1925) deals with an unsteady boundary layer, and that of Nikuradse (1925) with turbulent pipe flow.

Many other important developments took place in 1925. In that year the former naval architect Betz (1925) invented his wake-survey method to measure the drag of an airfoil in flight or in a wind tunnel. Yet, only in 1951 did the aeronautical engineer M. Tulin extend this method to ship models in a tank, mainly to separate viscous and wave resistance of a ship.

In 1925, Ackeret (1925) found his formulae for profile forces in supersonic flow.

Also in 1925, A. Busemann (b. 1901) came from Braunschweig, where he had received his Dr.-Ing. degree under O. Föppl. He built the rotating chamber, and Prandtl (1926a) gave a short report on first experiments; it was not much used, except in the thesis of Fabricius (1936) and by Stümke (1940).

Last but not least, the turbulent year of 1925 saw Prandtl (1925) propose the concept of the mixing length in turbulent flow, although he called it "stopping distance" (*Bremsweg*) in his report. The final name *Mischungsweg* appeared a year later in Prandtl (1926b). It gave rise to an exciting "race" with von Kármán (1930), who produced similar and sometimes identical results by assumptions based on dimensional analysis only. Later on, G. I. Taylor (1935, 1936) gave in his statistical theory of turbulence the fundamentals for a really sound physical theory; here, mixing lengths turn up again as length integrals of correlation functions. He also initiated the hot-wire technique for measuring velocity fluctuations.

However, this theory was complicated and not helpful for quick practical applications such as pipe flow or plate resistance. One important "practical" result of Taylor (1936), the dependence of sphere drag on turbulence of the oncoming flow, could also be derived by dimensional analysis only, following Taylor's arguments (Wieghardt 1940).

Ackeret (1926) published tests on boundary-layer suction. He also made cavitation tests in the new facility as shown in his article on cavitation in the *Handbuch der Experimentalphysik*, Vol. 4, Part 1 (Ackeret 1931). This tome gives in several treatises by various authors an excellent survey of fluid mechanics of that time.

Tollmien (1926) used the mixing length to calculate free turbulent flow. Betz (1927) gave the first extensive aerodynamics of the windmill, which has now become topical again.

In connection with the Prandtl-Glauert analogy, a paper by Busemann (1928) has been previously mentioned. In this paper the small tunnel for high subsonic and supersonic flow is described; it was the first tunnel of the blow-down type. It had a test section of only 6 cm^2 and was always meant as a prototype for a later, bigger tunnel. By request of Prandtl, the title of his paper had the supplement, "(With Regard to Airscrews)," just as Glauert was also thinking of this application only. It would have seemed hubris to think of planes flying at such speeds. At that time, only the Mach angle was used; the designation "Mach number" was introduced by Ackeret in his habilitation paper at ETH Zürich in 1929.

Ackeret (1927) wrote the first monograph on gasdynamics in the *Hand-*

Figure 1 Photo showing (left to right) J. Ackeret, L. Prandtl, A. Betz, and R. Seiferth, taken in 1923.

buch der Physik; it was his last paper written in Göttingen. Early in 1927 he left for a post in industry (Escher-Wyss), and then he became *ausserordentlicher* (associate) professor in 1931 and later, in 1934, full professor and director of the newly founded institute for aerodynamics at ETH Zürich. His successor at the KWI was Busemann, with whom he had cooperated intimately.

Another remarkable year was 1929. Prandtl & Busemann (1929) found the first characteristics method for supersonic flow, which—in a modern sense—is not approximate as stated in its title. Admittedly, it is a panel method that, however, tends toward the exact solution for small panels. With it, profile flows, oscillating free jets, and parallel-flow nozzles were constructed; these results appeared later in many publications.

In 1929, Busemann (1929a) published his shock-polar diagram and a short paper (1929b) on the pressure on a conical tip, which gave all flow essentials and the appropriate calculation method.

Also in 1929, the work of Tollmien (1929) on the instability of the laminar boundary layer along a plate was presented by Prandtl at the academy. Whereas the thesis of Tietjens (1923) had in fact yielded amplification of disturbances, Tollmien calculated for the first time a critical Reynolds number for the beginning of instability. Following this work, Tollmien traveled to the United States and spent three years (1930–33) at Caltech in Pasadena.

Figure 2 The Kaiser-Wilhelm-Institut für Strömungsforschung at Göttingen about 1937.

Besides all these original papers, experimental work in wind tunnels was continually in progress. The main results were collected and published by Prandtl & Betz (1921, 1923, 1927, 1932). Also at this time, Prandtl's lectures were published (Prandtl & Tietjens 1929, 1931).

Busemann did his habilitation in 1930 at Göttingen and became lecturer for fluid- and thermodynamics at the TH Dresden in 1931. Later on, in 1936, he became professor and director of a new institute for gasdynamics at the *Luftfahrtforschungsanstalt* in Braunschweig.

At the end of the twenties, H. Schlichting (1907–82) studied at Göttingen. In 1930 he wrote his thesis on free turbulence in a wake (Schlichting 1930) and joined the KWI.

Semiempirical integral methods for the calculation of turbulent boundary layers at a given pressure distribution were developed in two theses by Buri (1931) and Gruschwitz (1931). The latter was much used for some time.

Nikuradse (1932, 1933) published his test results on turbulent flow through smooth and rough pipes; in order to define a special but reproducible roughness, the so-called sand grain roughness was invented. For many technical applications these two papers proved to be very important and were widely acknowledged. Unfortunately, this increased his self-esteem to such a height that he tried to replace Prandtl as director after Hitler had come to power. It was, indeed, a dangerous attack, for Nikuradse knew at least one man high up in the Nazi regime, whereas neither Prandtl nor Betz ever became party members in spite of their important positions. Luckily Prandtl was victorious. Nikuradse had to leave KWI and—without Prandtl's guidance—he never again wrote a paper worth mentioning.

Prandtl & Schlichting (1934) used these results of pipe flow to compute the resistance of a rough plate. This was of great interest, e.g., for airplanes and ships, and in particular for the extrapolation of model tests.

Schlichting (1932a,b, 1933, 1935) extended Tollmien's stability theory in various directions, especially in later years, when he had left Göttingen in 1935 for a post in industry (Dornier) and then later (since 1939) as professor at TH Braunschweig. While at Göttingen he also wrote on the rectangular wing in supersonic flow (Schlichting 1936); it was the first paper on a finite supersonic wing. For this he used the acceleration potential introduced by Prandtl (1936a).

Tollmien (1935) proved that for inviscid flow to be unstable, Rayleigh's condition of an inflection point in the velocity profile of a boundary layer is not only necessary but also sufficient. He did his habilitation in 1936 and became the successor of Trefftz at TH Dresden in 1937.

W. Frössel (1936) reported on tests on pipe friction in subsonic and supersonic flow. A theoretical addition was given later in the thesis of Koppe (1947).

6. THE SEPARATION OF THE AVA AND THE LAST DECADE OF THE KAISER-WILHELM-INSTITUT

In 1937 the management of the AVA and KWI was separated because of the large expansion of AVA and its administration. Prandtl became chairman of the AVA directorate and Betz director of AVA, which had eight divisions or institutes. Cooperation with KWI was close, in particular with the institute for theoretical aerodynamics under Betz (deputy: F. Riegels) and with those for unsteady aerodynamics under W.-G. Küssner and for high-speed problems under O. Walchner. Most of the AVA institutes were on the same grounds as the KWI, yet there were two separate entrances. Luckily, control at the KWI entrance was less formal for its "golden horde."

In spite of the high prestige of Prandtl and Betz, Göttingen did not have the biggest wind tunnels in Germany. These were at the DVL in Berlin and at the *Luftfahrtforschungsanstalt* in Braunschweig, also of Göttingen type. The biggest supersonic tunnels, all of blow-down type, were in Braunschweig, Aachen, and Peenemünde. Hence, Betz and Walchner built such a tunnel from their own revenues.

At the much smaller KWI, the hierarchy was more of a family type. Prandtl's administrative deputy was H. Reichardt, who started in 1927 at AVA and came in 1930 to KWI; he worked mainly on turbulence and cavitation. A small bureau for design was led by W. Frössel, who also supervised the workshop. But, as he liked to point out, he was never officially appointed thereto. He had come to KWI as a locksmith apprentice, but then he performed tests as reported in his aforementioned paper of 1936. Despite his age, to become admitted to the university as a doctorand, he had to pass the high school examination. At last he took his doctoral degree with a thesis on lubrication (Frössel 1941), a field in which he carried on successful research.

In fluid mechanics, theoreticians can be useful of course, but left alone they would soon dry up. Hence, it seems appropriate to mention the nonscientific staff too. Prandtl's truly faithful secretary was Miss Eleonore von Seebach (called "Lorchen"), who also typed all of our papers. (At AVA, Miss Kreibohm worked as Betz' secretary for a half-century!) Bookkeeping was done by Mrs. Grüber and her help. There were also at least six girls using automatic calculating machines, sliderules, or graphical methods. In the workshop were several foremen, and the test work was done by a few but excellent technicians; for example, one specialized in hot-wire tests, another in photography and optical methods, and so on. Together with the two watchmen, Prandtl's crew had up to 40 people.

In the second half of the thirties, the KWI building and facilities were somewhat enlarged. At this time Dr.-Ing. F. Schultz-Grunow came to the institute because Prandtl wanted a representative of mechanical engineering; Prandtl liked to call himself a plain mechanical engineer. In 1937 he engaged Dr. phil. H. Görtler, a pure mathematician. K. Oswatitsch, who had done his doctorate in theoretical physics at Graz University, came in 1938, at first as a stipendiary of the *Deutsche Forschungsgemeinschaft* (German Research Association). The same year K. Wieghardt joined, although he had done his doctorate at Göttingen only with the grade "good." Dipl.-Ing. H. Schuh came in 1939 from Vienna.

Of Schultz-Grunow's papers we recall three (Schultz-Grunow 1939, 1940, 1942): The first one was his habilitation paper on inertia of turbulence, the second was on the outer law for the turbulent boundary layer past a plate, and the third concerned his characteristics method for one-dimensional wave propagation, for which examples of applications reappeared in many textbooks. Before this last paper came out, he accepted a chair at TH Aachen in 1941.

As for Görtler, for his first official task Prandtl gave him a differential equation, saying "I want a solution starting at zero linearly, reaching a maximum, and then coming down to an asymptote. Please, figure it out." But Görtler could not find such a solution. And at last, as a good mathematician, he even succeeded in constructing a rigorous nonexistence proof for such a solution! But, when he proudly showed this to Prandtl, the latter only said, "Oh, is that so? Well, it might be that this constant here also depends a little bit on Reynolds number." Besides this experience, Görtler was the only one who became acquainted with fluid mechanics through private lessons by Prandtl. In his habilitation paper (Görtler 1940), he showed the instability of flow along a concave wall that generates longitudinal Görtler vortices, analogous to Taylor vortices between two rotating cylinders (Taylor 1923). In 1944 Görtler took a position at the University of Freiburg.

At the end of the thirties, a series of theses on wing theory came out. Three of these used the new acceleration potential: Kinner (1937) for the circular wing, Krienes (1940) for the elliptic wing, and Schade (1940) for the oscillating circular wing. The rectangular wing in Wieghardt's (1939) thesis was represented by four horseshoe vortices calculated numerically. In a footnote there, the integral equation for a continuous vortex distribution is mentioned, which is usually ascribed to E. Reissner (1949); it can also be found in Mattioli (1939). Later on, Kinner worked in industry, and Krienes and Schade at AVA.

To Oswatitsch, Prandtl gave a special problem "so that something is done again in gasdynamics"; at that time, this was—compared with wing theory, turbulence, and boundary-layer theory—a field of minor interest. Already in the supersonic part of the small KWI tunnel, there had appeared inexplicable schlieren called "the ghost" by Busemann. At first, by deriving a formula for the growth of droplets with time, Oswatitsch (1942) found out that sufficient condensation of humidity was impossible during the short stay of the air in the nozzle. But then, inspired by a paper on the breakdown of a supersaturated thermodynamic state, he was able to give a detailed theory for the ghost. Here a chance cluster of molecules can serve as the kernel of a droplet. Meanwhile, such humidity effects were avoided by using silicagel desiccators in front of all blow-down tunnels.

Another problem was the pressure recovery of a ramjet at Mach number 3. At that time, no better recovery seemed obtainable than that behind a single normal shock. However, Oswatitsch (1944) developed a concept in which the air is compressed first by a conical tip in one or several oblique shocks to a lower supersonic flow and only then in a normal shock to subsonic flow leading to the combustion chamber. At that time this was a complicated diffuser for which the KWI tunnel was too small. But at last, his experiments in the AVA tunnel of Walchner were successful.

In Oswatitsch (1945) the analytic formulation for the drag as an integral over the entropy flux is given.

Wieghardt measured in 1942 the drag increase of a plate due to single roughnesses such as rivets, holes, etc.; this wartime report was reprinted in Wieghardt (1953). He also measured the temperature distribution due to a point or line heat source at the bottom of a turbulent boundary layer; the results might be used in meteorology when the much greater eddy viscosity of the atmosphere is allowed for. This report of 1944 came out in Wieghardt (1948a). In his habilitation paper of 1944 (Wieghardt 1948b), he uses the energy equation for boundary layers for a calculation method. Only after the war was it found out that Leibenson (1935) had published this equation much earlier.

Schuh refined the hot-wire technique and worked also on heat transfer. His publications came out mainly after the war, including his thesis (Schuh 1946a) and Schuh (1946b).

Prandtl himself became more and more interested in dynamic meteorology. He considered it to be a main field of application of future fluid dynamics, whereas Sir G. I. Taylor (1886–1975) had come from meteorology to fluid mechanics. This is evident from some papers and lectures, as, e.g., in Prandtl (1932, 1936b). There are preliminary sketches of what was worked out later in two theses: Kropatschek (1935), on a simplified model of general circulation, and Stümke (1940), on generation of cyclones by source effects with a few tests in the rotating chamber. Further papers of this series are on atmospheric lee waves behind mountains by Lyra (1940, 1943) and on wind deflection due to Coriolis forces above a mountain chain by Görtler (1941) or above a single mountain by Rothstein (1943). Both Stümke and Rothstein remained at KWI during the war.

In 1942 Prandtl's book, *Führer durch die Strömungslehre*, came out (Prandtl 1942a). It was also the third edition of his much smaller *Abriss der Strömungslehre*. The new title gave rise to some jokes, of course. But in German there is only one word for leader and guide; a *Führer durch London* is just a Baedeker. There were five editions and translations into English, French, Japanese, Polish, and Russian. After his death we carried on three further editions, the eighth in 1984. All of his articles in journals are reprinted in Prandtl (1961).

Initiated by the manifold and precise measurements of Reichardt (1941, 1942), Prandtl (1942b) proposed a theory for free turbulent flow such as jets and wakes. With it Görtler (1942) calculated some flows and found better experimental verification than previously, except for the edge regions.

In 1944 Prandtl listed seven problems "that should be investigated after the peace treaty." One of us still has a copy of this list with two pages of explanations. The problems are (*a*) influence of Richardson number on turbulence, (*b*) flow and heat transfer on a heated inclined plane for meteorological applications, (*c*) turbulence theory as in Prandtl (1945a), (*d*) generation of sand ripples in water or air, (*e*) boundary layer on rotating turbine blades, (*f*) generation of circulation by shaking a partly filled vessel, and (*g*) refined experimental study of turbulent correlation functions. On problems (*a*) and (*f*), he wrote two short papers (Prandtl 1944, 1949).

For problem (*c*) he advanced a transport equation for the turbulent kinetic energy when only one of the mean velocity gradients is predominant (Prandtl 1945a). This paper has indeed initiated almost all of modern turbulence modeling, especially after computers became available. But Boussinesq's exchange coefficient, nowadays called eddy viscosity, is still used as a scalar, in spite of Prandtl's warning that in general it should be a tensor.

Another paper (Prandtl 1945b) marks the beginning of three-dimensional boundary-layer theory. For 40 years, only two-dimensional boundary layers had been considered either in plane or axisymmetric flow!

Early in April 1945 American troops came to Göttingen, and some weeks later we belonged to the British occupation zone, about 10 km off the Russian zone. Previously the principals of the KGW and scientists from other KWIs—including celebrities like Hahn, Heisenberg, and Planck

—had come as guests to the grounds of KWI and AVA. Also, a huge mysterious machine had been stored there: the first Zuse computer.

For almost two years the British Ministry of Supply employed many KWI and AVA scientists for a special task. With Betz as editor, a review of all work done in Germany during the war in the field of aero- and hydrodynamics was written down on 7000 typed pages in German English: These were known as the Göttingen monographs of 1945/46. A similar, shorter review was done at the same time for the FIAT review of German sciences in the years 1939–46. These reviews were useful and enlightening—at least for the writers. Otherwise, a response did not seem to be forthcoming. On the other hand, many foreign colleagues had previously come for interviews, so that the main points of scientific progress in various countries had already been discussed, such as, e.g., sweptback wings (cf. Busemann 1971, Wuest 1982) or the experimental verification of instability theory by Schubauer & Skramstad (1947). Most of us saw von Kármán for the first time, in American uniform and with cigar.

Since AVA had belonged to the air ministry, it was dissolved. It was reestablished in 1953, now under the auspices of MPG. Betz retired in 1957 and Schlichting became director of AVA (Wuest 1982).

As it turned out, KWI had been under the ministry of education and so it could carry on. Only the name was changed—in 1948 it was renamed the *Max-Planck-Institut*. In 1946 Prandtl retired as institute director. He retained only a department of MPI; two others were set up for Betz and Tollmien, who had come back from Dresden.

Prandtl engaged the meteorologist E. Kleinschmidt, who wrote a paper on tropical cyclones (Kleinschmidt 1951) that is still cited often; he also wrote the chapter on meteorology for the sixth and seventh editions of Prandtl's book. Another paper still much used is Ludwieg (AVA) & Tillmann (KWI) (1949); it gives a general formula for the wall friction when the size and the form parameters of a turbulent velocity profile are known.

Oswatitsch left Göttingen in 1946, Wieghardt in 1949.

Tollmien became successor to Prandtl's chair when Prandtl retired from the university in 1947.

7. PRANDTL'S WORK IN SOLID MECHANICS

So far we have mentioned only papers by Prandtl or his staff that are in the field of fluid mechanics. However, those working in solid mechanics also consider Prandtl as their colleague. Hence, a short survey of some of his work in this field is added here.

Prandtl (1903, 1904b) advanced his soap-film analogy for the elastic stress of a twisted bar. For a fully plastic material, the analogy of the sand hill was found by Nadai (1923a).

Prandtl (1920) determined the pressures under which blunt edges of ductile metals yield, and in Prandtl (1921) he generalized the concept of the ideally plastic substance (without strain hardening; cf. also Prandtl 1924).

Prandtl (1928) described an apparently simple model for solid bodies to calculate hysteresis, the dependence of yield stress upon strain velocity, and how temperature changes a state from solid to fluid. Reuss (1930) in Budapest also incorporated the elastic strains in the plastic domain, although in a more formal general manner; hence, this Prandtl body is often called a Prandtl-Reuss body. In particular, this model yields the hyperbolic sine speed law of deformation for which Eyring (1936) gave a more molecular explanation. Hence, rheologists call it now the Prandtl-Eyring model. For the pipe flow of such a fluid, see Prandtl (1950).

Prandtl (1933) extended his solid-body model of 1928 to investigate the time-dependent fracture of a brittle solid.

There were at least two young scientists at Prandtl's mechanics institute who later became well known all over the world. The first was A. Nadai (1883–1963), who came from Hungary in 1919; he became *Privatdozent* (Nadai 1923b) and left Göttingen in 1927 for the United States. The second was W. Prager (1903–80), from Karlsruhe, who did his doctorate and habilitation (1927) at the TH Darmstadt; he then worked at Göttingen from 1929–33. When Hitler came to power he went to the University of Istanbul, where he held the chair for theoretical mechanics until he left in 1941 for Brown University in Providence, Rhode Island.

8. PRANDTL'S PERSONALITY

Besides those pupils of Prandtl we have mentioned so far, there were, of course, many more who later on held important positions in the field of fluid dynamics. Some more of the over 80 students who did their doctorate under Prandtl's supervision are the following: C. Wieselsberger (1922), who worked in Japan before he became successor to von Kármán in Aachen; C. Tietjens (1923), who spent many years in India; H. Blenk (1924), the first director of the *Forschungsanstalt* in Braunschweig; D. Küchemann (1938), who joined the AVA and after the war the Royal Aircraft Establishment at Farnborough.

Those who were later on department heads at AVA were J. Stüper (1932), F. Riegels (1938), H. Ludwieg (1939), and W. Wuest (1941). D. T.

Dumitrescu (1941) became professor at Bucharest. B. Dolaptschiew (1937, 1938) wrote his thesis at Göttingen but took his degree from Sofia University, where he also became professor after the war.

These special examples might create the wrong impression that Prandtl had influenced the job hunting of his disciples, but actually we never even heard a rumor of this sort. He was always outspoken yet strictly neutral, also to his coworkers. He almost always had time to speak with them, but he did not obtrude his ideas, since he expected them to work as independently as possible. By way of conversation he just dropped scientific nuggets, which were, of course, collected with great care. Naturally, one had to report some progress from time to time, otherwise one was told to find another post somewhere else; yet, plenty of time was conceded in these very few cases. Most of his staff he had got to know beforehand as doctoral candidates for about two years time.

He expected his students to attend his lectures, or else he would ask, "Oh, do you know all that?" But he was not a fascinating lecturer, simply because he was often thinking too far ahead. Young students want to hear that this is black and that is white. So often, only the older ones understood why he formulated so cautiously, thinking of some far away exceptions. And the lecturer at the Wednesday colloquium had to expect harsh questions by Prandtl or Betz as soon as he became vague or tiresome.

Prandtl's working method is perhaps best described by von Kármán (1954):

> Prandtl, an engineer by training, was endowed with rare vision for the understanding of physical phenomena and unusual ability in putting them into relatively simple mathematical form. His control of mathematical methods and tricks was limited; many of his collaborators and followers surpassed him in solving difficult mathematical problems. But his ability to establish systems of simplified equations which expressed the essential physical relations and dropped the nonessentials was unique, I believe, even compared with his great predecessors in the field of mechanics—men like Leonhard Euler and D'Alembert.

Sometimes Prandtl could even bypass mathematics, at least qualitatively. For example, after some months, hard-working Dolaptschiew (1937, 1938) had calculated paths along which the vortices of a Kármán street move after a special periodic disturbance. Full of pride, he brought his results to Prandtl. But instead of listening to Dolaptschiew's broken German, Prandtl turned over the papers and began to design the paths the vortices ought to take. And indeed, after a few minutes his drawing looked very much like the calculated one. Similarly, he often solved scientific problems by visualizing the physical processes described by mathematical equations (Prandtl 1948).

The invisible foundation of Prandtl's ability in theory making was certainly a very intimate knowledge of mechanics acquired by many close observations of more or less intriguing processes, either in the laboratory or else in daily life, as certified by many anecdotes. On his way home he would explain to his companion the standing waves on the rainwater approaching the gully, or the different sound reflection of the footfalls when walking along a paling or a wall. Once, in the United States, he performed slight knee-bendings at the top floor of a skyscraper to measure the natural frequency of the building. Only when he had found his result confirmed roughly by the formula for the fixed-free bar was he ready to enter the restaurant. As a last example, we offer the following personal anecdote: Besides writing a thesis his doctoral candidates also had to give a colloquium lecture on a different subject. For this important event, one of us (KW) had bought new shoes with rubber soles. Unfortunately, they made a horrible squeaking noise on the parquet of the auditorium. And indeed, very soon KW noticed that Prandtl gazed fascinatedly at his shoes—alas, it was quite obvious that he was just advancing an acoustic theory of this interesting phenomenon.

Besides science, Prandtl had, of course, other interests, particularly music. He had perfect pitch and played the piano, mostly classical music; sometimes, he played a waltz for dancing when he had invited young people to his home. His mode of life in quiet Göttingen was quite unpretentious; he lived in a flat within 20 minutes walking distance of his institute. The only luxury was a little house in the Alps for vacations. He did know, of course, that he was exceptionally gifted and productive. Görtler (1975) once was present when Prandtl opened a letter announcing a new high honor. Prandtl showed him the letter, saying "Well, they might have thought of me a bit earlier."

In his family life the most tragic year was 1941. First his wife died, then the husband of his elder daughter lost his life as a soldier in Russia, and their baby also died.

In 1950 he could still enjoy three remarkable jubilees: his seventy-fifth birthday, the fiftieth anniversary of his doctorate, and the twenty-fifth anniversary of his institute. But in spring 1952, illness prevented him from further work and led to his death on 15 August 1953. Everyone who met Prandtl will always remember the dignity and kindheartedness of this great man.

Literature Cited

Ackeret, J. 1924. Motoren zum Antrieb von kleinen Modell-Luftschrauben. *Z. Flugtech.* 15: 101–3

Ackeret, J. 1925. Luftkräfte an Flügeln, die mit grösserer als Schallgeschwindikeit bewegt werden. *Z. Flugtech. Motorluftsch.* 16: 72–74

Ackeret, J. 1926. Grenzschichtabsaugung. *Z. VDI* 70: 1153

Ackeret, J. 1927. Gasdynamik. In *Handbuch*

der Physik, 7: 289–342. Berlin: Springer-Verlag

Ackeret, J. 1931. Kavitation. In *Handbuch der Experimentalphysik*, 4(1): 463–86. Leipzig: Akad. Verlag

Betz, A. 1919a. *Beiträge zur Tragflächentheorie mit besonderer Berücksichtigung des einfachen rechteckigen Flügels*. Dissertation. Univ. Göttingen

Betz, A. 1919b. Schraubenpropeller mit geringstem Energieverlust. *Nachr. Ges. Wiss. Göttingen* 1919: 193–213

Betz, A. 1925. Ein Verfahren zur direkten Ermittlung des Profilwiderstands. *Z. Flugtech. Motorluftsch.* 16: 42

Betz, A. 1927. Die Windmühlen im Lichte neuerer Forschung. *Naturwissenschaften* 15: 905–14

Betz, A. 1957. Lehren einer fünfzigjährigen Strömungsforschung. 1st L. Prandtl Memorial Lecture. *Z. Flugwiss.* 5: 97–105

Birnbaum, W. 1923. Die tragende Wirbelfläche als Hilfsmittel zur Behandlung des ebenen Problems der Tragflügeltheorie. *ZAMM* 3: 290–97

Birnbaum, W. 1924. *Das ebene Problem des schlagenden Flügels*. Dissertation. Univ. Göttingen (1923). *ZAMM* 4: 277–92 (1924)

Blasius, H. 1907. *Grenzschichten in Flüssigkeiten kleiner Reibung*. Dissertation. Univ. Göttingen. *Z. Math. Phys.* 57: 1–37 (1908)

Blenk, H. 1924. *Der Eindecker als tragende Wirbelfläche*. Dissertation. Univ. Göttingen. *ZAMM* 5: 36–47 (1925)

Buri, A. 1931. *Eine Berechnungsgrundlage für die turbulente Grenzschicht bei beschleunigter und verzögerter Grundströmung*. Dissertation. ETH Zürich

Busemann, A. 1928. Profilmessungen bei Geschwindigkeiten nahe der Schallgeschwindigkeit. *Jahrb. Wiss. Ges. Luftfahrt*, Vol. 95

Busemann, A. 1929a. Verdichtungsstösse in ebenen Gasströmungen. In *Vorträge aus dem Gebiet der Aerodynamik, Aachen*, ed. A. Gilles, L. Hopf, T. von Kármán, p. 162. Berlin: Springer-Verlag

Busemann, A. 1929b. Drücke auf kegelförmige Spitzen bei Bewegung mit Überschallgeschwindigkeiten. *ZAMM* 9: 496–98

Busemann, A. 1971. Compressible flow in the thirties. *Ann. Rev. Fluid Mech.* 3: 1–10

Dolaptschiew, B. 1937. Über die Stabilität der Kármánschen Wirbelstrasse. *ZAMM* 17: 313–23

Dolaptschiew, B. 1938. Störungsbewegungen (Bahnen) der einzelnen Wirbel der Kármánschen Wirbelstrasse. *ZAMM* 18: 263–71

Dumitrescu, D. T. 1941. *Strömung an einer Luftblase im senkrechten Rohr*. Dissertation. Univ. Göttingen. *ZAMM* 23: 139–49 (1943)

Eiffel, G. 1912. Sur la résistance des sphères dans l'air en mouvement. *C. R. Acad. Sci. Paris* 155: 1597–1602

Eyring, H. J. 1936. Viscosity, plasticity and diffusion as examples of absolute reaction rates. *J. Chem. Phys.* 4: 283–89

Fabricius, W. 1936. Studien über inhomogene gekrümmte Strömungen. *Ing.-Arch.* 7: 410–27

Flügge-Lotz, I., Flügge, W. 1973. Ludwig Prandtl in the nineteen-thirties: reminiscences. *Ann. Rev. Fluid Mech.* 5: 1–8

Frössel, W. 1936. Strömungen in glatten und rauhen Rohren mit Über- und Unterschallgeschwindigkeit. *Forsch. Ing.-Wesen* 7: 75–84

Frössel, W. 1941. Berechnung der Reibung und Tragkraft eines endlich breiten Gleitschuhs auf ebener Gleitbahn. *ZAMM* 21: 321–40

Gericke, H., ed. 1972. *50 Jahre GAMM*, *Beih. Ing.-Arch.*, Vol. 41. 36 pp.

Glauert, H. 1927. The effect of compressibility on the lift of an aerofoil. *Proc. R. Soc. London Ser. A* 118: 113–19

Görtler, H. 1940. *Über eine dreidimensionale Instabilität laminarer Grenzschichten an konkaven Wänden*. Habilitation. Univ. Göttingen. *Nachr. Ges. Wiss. Göttingen* 1941: 1–26

Görtler, H. 1941. Einfluss der Bodentopographie auf Strömungen über der rotierenden Erde. *ZAMM* 21: 279–303

Görtler, H. 1942. Berechnung von Aufgaben der freien Turbulenz aufgrund eines neuen Näherungsansatzes. *ZAMM* 22: 244–54

Görtler, H. 1975. Ludwig Prandtl—Persönlichkeit und Wirken. *Z. Flugwiss.* 23: 153–62

Grun, B. 1975. *The Timetables of History*. London: Thames & Hudson. 661 pp.

Gruschwitz, E. 1931. *Die turbulente Reibungsschicht in ebener Strömung bei Druckabfall und Druckanstieg*. Dissertation. Univ. Göttingen (1932). *Ing.-Arch.* 2: 321–46 (1931)

Hiemenz, K. 1911. *Die Grenzschicht an einem in den gleichförmigen Flüssigkeitsstrom eingetauchten geraden Kreiszylinder*. Dissertation. Univ. Göttingen. *Dinglers Polytech. J.* 326: 321–24

Kinner, W. 1937. *Die kreisförmige Tragfläche auf potentialtheoretischer Grundlage*. Dissertation. Univ. Göttingen. *Ing.-Arch.* 8: 47–80

Kleinschmidt, E. 1951. Grundlagen der Theorie der tropischen Zyklone. *Arch. Meteorol. Geophys. Bioklimatol. Ser. A* 4: 53–72

Koppe, M. 1947. *Reibungseinfluss auf stationäre Rohrströmung bei hohen Geschwindigkeiten*. Dissertation. Univ. Göttingen

Kraemer, K. 1975. Geschichte der Gründung des Max-Planck-Instituts für Strömungsforschung in Göttingen. In *Max-Planck-Institut für Strömungsforschung Göttingen 1925–1975*, pp. 16–34. Göttingen: MPI für Strömungsforsch.

Krienes, K. 1940. *Die elliptische Tragfläche auf potentialtheoretischer Grundlage*. Dissertation. Univ. Göttingen. *ZAMM* 20: 65–88

Kropatschek, F. 1935. Die Mechanik der grossen Zirkulation der Atmosphäre. *Beitr. Phys. freien Atmos*. 22: 272–98

Küchemann, D. 1938. *Strömungsbewegung in einer Gasströmung mit Grenzschicht*. Dissertation. Univ. Göttingen. *ZAMM* 18: 207–22

Leibenson, L. S. 1935. The energy form of the integral condition in the theory of the boundary layer. *Dokl. Akad. Nauk SSSR* 2: 22–24

Ludwieg, H. 1939. *Über Potentialströmungen mit unstetigen Randbedingungen*. Dissertation. Univ. Göttingen

Ludwieg, H., Tillmann, W. 1949. Untersuchungen über die Wandschubspannung in turbulenten Reibungsschichten. *Ing.-Arch*. 17: 288–99

Lyra, G. 1940. Über den Einfluss von Bodenerhebungen auf die Strömung einer stabil geschichteten Atmosphäre. *Beitr. Phys. freien Atmos*. 26: 197–206

Lyra, G. 1943. Theorie der stationären Leewellenströmung in freier Atmosphäre. *ZAMM* 23: 1–28

Magin, E. 1908. Optische Untersuchung über den Ausfluss von Luft durch eine Lavaldüse. *Mitt. Forsch. Ing.-Wesen* 62: 1–29

Manegold, K.-H. 1970. *Universität, Technische Hochschule und Industrie*. Berlin: Duncker & Humblot

Mattioli, G. D. 1939. Theorie des dünnen Tragflügels bei beliebigen Umrissformen. *Ing.-Arch*. 10: 153–59

Meyer, T. 1908. Über zweidimensionale Bewegungsvorgänge in einem Gas, das mit Überschallgeschwindigkeit strömt. *Mitt. Forsch. Ing.-Wesen* 62: 31–67

Munk, M. 1918. *Isoperimetrische Aufgaben aus der Theorie des Fluges*. Dissertation. Univ. Göttingen

Nadai, A. 1923a. Der Beginn des Fliessvorganges in einem tordierten Stab. *ZAMM* 3: 442–54

Nadai, A. 1923b. *Über die Biegung durchlaufender Platten und der rechteckigen Platte mit freien Rändern, und Die Verbiegungen in einzelnen Punkten unterstützter kreisförmiger Platten*. Habilitation. Univ. Göttingen

Nikuradse, J. 1925. *Untersuchung über die Geschwindigkeitsverteilung in turbulenten Strömungen*. Dissertation. Univ. Göttingen. *VDI-Forsch. Arb. Heft 281* (1926)

Nikuradse, J. 1932. Gesetzmässigkeiten der turbulenten Strömung in glatten Rohren. *VDI-Forsch. Heft 356*

Nikuradse, J. 1933. Strömungsgesetze in rauhen Rohren. *VDI-Forsch. Heft 361*

Nusselt, W. 1909. Der Wärmeübergang in Rohrleitungen. *VDI-Z*. 53: 1750, 1808. Also *VDI-Forsch. Heft 89* (1910)

Oseen, C. W. 1927. *Hydrodynamik*. Leipzig: Akad. Verlag.

Oswatitsch, K. 1942. Kondensationserscheinungen in Überschalldüsen. *ZAMM* 22: 1–14

Oswatitsch, K. 1944. Der Druckwiedergewinn bei Geschossen mit Rückstossantrieb bei hohen Überschallgeschwindigkeiten. *KWI-Ber*. Also *NACA TM 1140*

Oswatitsch, K. 1945. Der Luftwiderstand als Integral des Entropiestromes. *Nachr. Wiss. Ges. Göttingen* 1945: 88–90

Pohlhausen, E. 1921. Der Wärmeaustausch zwischen festen Körpern und Flüssigkeiten mit kleiner Reibung und kleiner Wärmeleitung. *ZAMM* 1: 115–21

Pohlhausen, K. 1921. *Zur näherungsweisen Integration der Differentialgleichung der laminaren Grenzschicht*. Dissertation. Univ. Göttingen. *ZAMM* 1: 252–68

Prandtl, L. 1900. *Kipperscheinungen*. Dissertation. Univ. München

Prandtl, L. 1903. Zur Torsion von prismatischen Stäben. *Phys. Z*. 4: 758

Prandtl, L. 1904a. Über Flüssigkeitsbewegung bei sehr kleiner Reibung. *Int. Math.-Kongr., Heidelberg, 3rd, 1904*, pp. 484–91. Leipzig: Teubner (1905)

Prandtl, L. 1904b. Eine neue Darstellung von Torsionsspannungen bei prismatischen Stäben von beliebigem Querschnitt. *Jahresber. Dtsch. Math.-Ver*. 13: 31–36

Prandtl, L. 1906. Zur Theorie des Verdichtungsstosses. *Z. Ges. Turbinenwesen* 3: 241–45

Prandtl, L. 1907. Neue Untersuchungen über die strömende Bewegung der Gase und Dämpfe. *Phys. Z*. 8: 23–30

Prandtl, L. 1910. Eine Beziehung zwischen Wärmeaustausch und Strömungswiderstand der Flüssigkeiten. *Phys. Z*. 11: 1072–78

Prandtl, L. 1912. Erläuterungsbericht zu den "Regeln für Leistungsversuche an Ventilatoren." *VDI-Verlag Heft*

Prandtl, L. 1914. Der Luftwiderstand von Kugeln. *Nachr. Ges. Wiss. Göttingen* 1914: 177–90

Prandtl, L. 1918. Tragflügeltheorie, 1. Mitteilung. *Nachr. Ges. Wiss. Göttingen* 1918: 451–77. Also *NACA TN 9* (1920)
Prandtl, L. 1919a. Tragflügeltheorie, 2. Mitteilung. *Nachr. Ges. Wiss. Göttingen* 1919: 107–37
Prandtl, L. 1919b. Zusatz zu A. Betz: Schraubenpropeller mit geringstem Energieverlust. *Nachr. Ges. Wiss. Göttingen* 1919: 213–17
Prandtl, L. 1920. Über die Härte plastischer Körper. *Nachr. Ges. Wiss. Göttingen* 1920: 74–85
Prandtl, L. 1921. Über die Eindringungsfestigkeit (Härte) plastischer Baustoffe und die Festigkeit von Schneiden. *ZAMM* 1: 15–21
Prandtl, L. 1924. Elastisch bestimmte und elastisch unbestimmte Systeme. In *Festschrift zum 70. Geburtstag von A. Föppl*, pp. 52–61. Berlin: Springer-Verlag
Prandtl, L. 1925. Bericht über Untersuchungen zur ausgebildeten Turbulenz. *ZAMM* 5: 136–39
Prandtl, L. 1926a. Erste Erfahrungen mit dem rotierenden Laboratorium. *Naturwissenschaften* 14: 425–27
Prandtl, L. 1926b. Bericht über neuere Turbulenzforschung. In *Hydraulische Probleme*, pp. 1–13. Berlin: VDI-Verlag
Prandtl, L. 1927. The generation of vortices in fluids of small viscosity (Wilbur Wright Memorial Lecture 1927). *Proc. Aeronaut. Soc.* 31: 720–43. Also *Z. Flugtech. Motorluftsch.* 18: 489–96
Prandtl, L. 1928. Ein Gedankenmodell zur kinetischen Theorie der festen Körper. *ZAMM* 8: 85–106
Prandtl, L. 1932. Meteorologische Anwendung der Strömungslehre. *Beitr. Phys. freien Atmos.* 19: 188–202
Prandtl, L. 1933. Ein Gedankenmodell für den Zerreissvorgang spröder Körper. *ZAMM* 13: 129–33
Prandtl, L. 1936a. Beitrag zur Theorie der tragenden Fläche. *ZAMM* 16: 360–61
Prandtl, L. 1936b. Beiträge zur Mechanik der Atmosphäre. *Ber. Meteorol. Assoc. Int. Geodät. Geophys. Union, Edinburgh, 1936*, pp. 1–32. Paris: Dupont (1939)
Prandtl, L. 1942a. *Führer durch die Strömungslehre*. Braunschweig: Vieweg
Prandtl, L. 1942b. Bemerkungen zur Theorie der freien Turbulenz. *ZAMM* 22: 241–43
Prandtl, L. 1944. Neuere Erkenntnisse der meteorologischen Strömungslehre. *Schriftenr. Dtsch. Akad. Luftfahrtforsh.* 8: 157–79
Prandtl, L. 1945a. Über ein neues Formelsystem für die ausgebildete Turbulenz. *Nachr. Ges. Wiss. Göttingen* 1945: 6–19
Prandtl, L. 1945b. Über Reibungsschichten bei dreidimensionalen Strömungen. In *Festschrift A. Betz*, pp. 134–41
Prandtl, L. 1948. Mein Weg zu hydrodynamischen Theorien. *Phys. Bl.* 4: 89–92
Prandtl, L. 1949. Erzeugung von Zirkulation beim Schütteln von Gefässen. *ZAMM* 29: 8–9
Prandtl, L. 1950. Fliessgesetze normal-zäher Stoffe im Rohr. *ZAMM* 30: 169–74
Prandtl, L. 1961. *Gesammelte Abhandlungen*, ed. W. Tollmien, H. Schlichting, H. Görtler, Vols. 1–3. Berlin: Springer-Verlag
Prandtl, L., Betz, A. 1921–1932. *Ergebnisse der Aerodynamischen Versuchsanstalt zu Göttingen. 1. Lief. 1921, 2. Lief. 1923, 3. Lief. 1927, 4. Lief. 1932.* (2nd ed. 1935). München-Berlin: Oldenburg
Prandtl, L., Busemann, A. 1929. Näherungsverfahren zur zeichnerischen Ermittlung von ebenen Strömungen mit Überschallgeschwindigkeit. *Festschift zum 70. Geburtstag von A. Stodola*, pp. 499–509. Zürich: Füssli
Prandtl, L., Schlichting, H. 1934. Das Widerstandsgesetz rauher Platten. *Werft, Reederei, Hafen* 15: 1–4
Prandtl, L., Tietjens, C. 1929, 1931. *Hydro- und Aeromechanik*, 2 Vols. Berlin: Springer-Verlag
Prandtl, L., Tollmien, W. 1924. Die Windverteilung am Erdboden errechnet aus den Gesetzen der Rohrströmung. *Z. Geophys.* 1: 47–55
Reichardt, H. 1941. Über eine neue Theorie der Turbulenz. *ZAMM* 21: 257–64
Reichardt, H. 1942. Gesetzmässigkeiten der freien Turbulenz. *VDI-Forsch., Heft 414*
Reissner, E. 1949. Note on the theory of lifting surfaces. *Proc. Natl. Acad. Sci. USA* 35: 208–15
Reuss, A. 1930. Berücksichtigung der elastischen Formänderung in der Plastizitätstheorie. *ZAMM* 10: 266–74
Reynolds, O. 1874. On the extent and action of the heating surface for steam boilers. *Proc. Lit. Philos. Soc. Manchester* 14: 7–12
Riegels, F. 1938. Zur Kritik des Hele-Shaw-Versuchs. *ZAMM* 18: 95–106
Rothstein, W. 1943. Strömungen über Bodenerhebungen auf der rotierenden Erde. *ZAMM* 23: 72–80
Rotta, J. C. 1985. Die Berufung von Ludwig Prandtl nach Göttingen. *Luft- und Raumfahrt* 6: 53–56
Roux, L. 1935. Turbulente Windströmungen auf der rauhen Erdoberfläche. *Z. Geophys.* 11: 165–87
Schade, T. 1940. Theorie der schwingenden kreisförmigen Tragfläche auf potentialtheoretischer Grundlage. *Luftfahrt-Forsch.* 17: 387
Schlichting, H. 1930. *Über das ebene Wind-*

schattenprofil. Dissertation. Univ. Göttingen. *Ing.-Arch.* 1 : 533
Schlichting, H. 1932a. Über die Stabilitätstheorie der Couette-Strömung. *Ann. Phys.* 14 : 905–36
Schlichting, H. 1932b. Über die Entstehung der Turbulenz in einem rotierenden Zylinder. *Nachr. Ges. Wiss. Göttingen* 1932 : 160–98
Schlichting, H. 1933. Zur Entstehung der Turbulenz bei der Plattenströmung. *Nachr. Ges. Wiss. Göttingen* 1933 : 181–208
Schlichting, H. 1935. Amplitudenverteilung und Energiebilanz der kleinen Störungen bei der Plattengrenzschicht. *Nachr. Ges. Wiss. Göttingen* 1935 : 47–78
Schlichting, H. 1936. Tragflügeltheorie bei Überschallgeschwindigkeiten. *Luftfahrt-Forsch.* 13 : 320–35
Schubauer, G. B., Skramstad, H. K. 1947. Laminar boundary layer oscillations and transition on a flat plate. *NACA Rep. 909*
Schuh, H. 1946a. *Laminarer Wärmeübergang in Grenzschichten bei hohen Geschwindigkeiten.* Dissertation. Univ. Göttingen
Schuh, H. 1946b. Einige Probleme des Wärmeübergangs und der Diffusion bei Laminarströmung längs einer ebenen Platte. *Österr. Ing.-Arch.* 2 : 346–60
Schultz-Grunow, F. 1939. Über das Nachwirken der Turbulenz bei örtlich und zeitlich verzögerter Grenzschichtströmung. Habil. Göttingen. *Proc. 5th Int. Congr. Appl. Mech., Cambridge, Mass., 5th, 1938*, pp. 428–35. New York : Wiley
Schultz-Grunow, F. 1940. Neues Reibungswiderstandsgesetz für glatte Platten. *Luftfahrt-Forsch.* 17 : 239–46
Schultz-Grunow, F. 1942. Nichtstationäre eindimensionale Gasbewegung. *Forsch. Ing.-Wesen* 13 : 125
Stümke, H. 1940. Rotationssymmetrische Gleichgewichtsstörungen in einer isothermen Atmosphäre nebst einem Modellversuch mit rotierender Flüssigkeit. *Z. Geophys.* 16 : 127–49
Stüper, J. 1932. Der durch einen Freistrahl hindurchgesteckte Tragflügel. *Ing.-Arch.* 3 : 338–55
Taylor, G. I. 1923. Stability of a viscid liquid contained between two rotating cylinders. *Philos. Trans. R. Soc. London Ser. A* 223 : 289–343
Taylor, G. I. 1935, 1936. Statistical theory of turbulence. *Proc. R. Soc. London Ser. A* 151 : 421–78, 156 : 307–17
Tietjens, C. 1923. *Beiträge zum Turbulenzproblem.* Dissertation. Univ. Göttingen. *ZAMM* 5 : 200–17 (1925)
Tollmien, W. 1925. *Zeitliche Entwicklung der laminaren Grenzschicht am rotierenden Kreiszylinder*. Dissertation. Univ. Göttingen
Tollmien, W. 1926. Berechnung turbulenter Ausbreitungsvorgänge. *ZAMM* 6 : 468–78
Tollmien, W. 1929. Über die Entstehung der Turbulenz. *Nachr. Ges. Wiss. Göttingen* 1929 : 21–44
Tollmien, W. 1935. Ein allgemeines Kriterium der Instabilität laminarer Geschwindigkeitsverteilungen. *Nachr. Ges. Wiss. Göttingen* 1935 : 79–114
von Kármán, T. 1909. Untersuchungen über die Knickfestikeit. *Mitt. Forsch. Ing.-Wesen* 81 : 1–44
von Kármán, T. 1911, 1912. Über den Mechanismus des Widerstandes, den ein bewegter Körper in einer Flüssigkeit erfährt. *Nachr. Ges. Wiss. Göttingen* 1911 : 509–17, 1912 : 547–56
von Kármán, T. 1921. Über laminare und turbulente Reibung. *ZAMM* 1 : 233–52
von Kármán, T. 1930. Mechanische Ähnlichkeit und Turbulenz. *Nachr. Ges. Wiss. Göttingen* 1930 : 58. Also *Proc. Int. Congr. Tech. Mech., 3rd, Stockholm, 1930*, 1 : 85–93
von Kármán, T. 1954. *Aerodynamics : Selected Topics in the Light of their Historical Development.* Ithaca, NY : Cornell Univ. Press. 203 pp.
von Kármán, T., Levi-Civita, T., eds. 1924. *Vorträge aus dem Gebiete der Hydro- und Aerodynamik.* Berlin : Springer-Verlag
von Kármán, T., Rubach, H. 1912. Über den Mechanismus des Flüssigkeits- und Luftwiderstandes. *Phys. Z.* 13 : 49–59
Wieghardt, K. 1939. Über die Auftriebsverteilung des einfachen Rechteckflügels über die Tiefe. *ZAMM* 19 : 257–70
Wieghardt, K. 1940. Über die Wirkung der Turbulenz auf den Umschlagpunkt. *ZAMM* 20 : 58–59
Wieghardt, K. 1948a. Über Ausbreitungsvorgänge in turbulenten Reibungsschichten. *ZAMM* 28 : 346–55
Wieghardt, K. 1948b. Über einen Energiesatz zur Berechnung laminarer Grenzschichten. *Ing.-Arch.* 16 : 231–42
Wieghardt, K. 1953. Erhöhung des turbulenten Reibungswiderstands durch Oberflächenstörungen. *Schiffstechnik* 2 : 65–81
Wieselsberger, C. 1922. Zur Theorie des Tragflügels bei gekrümmter Flugbahn. *ZAMM* 2 : 325–40
Wuest, W. 1941. *Über die Entstehung von Wasserwellen durch Wind.* Dissertation. Univ. Göttingien. *ZAMM* 29 : 239–52 (1949)
Wuest, W. 1982. *Sie zähmten den Sturm, 75 Jahre AVA.* Göttingen : DFVLR-AVA, Göttinger Tageblatt

Ann. Rev. Fluid Mech. 1987. 19 : 27–52

CONFINED VORTICES IN FLOW MACHINERY

Marcel Escudier

Fluid Mechanics Research Group, Brown Boveri Research Center, 5405 Baden, Switzerland[1]

INTRODUCTION

Local recirculation or extensive regions of reversed flow, periodic fluctuations in pressure and velocity, and high levels of kinetic-energy dissipation are the principal characteristics of (sufficiently intense) confined vortex flows. The present article represents a necessarily descriptive rather than analytical discussion of swirling-flow behavior in various practical devices where these effects may be either the desired result of design or unavoidable, possibly unforeseen, side effects. One of the most widespread technical applications of swirl is for flame stabilization in furnaces and combustion chambers, whereby fuel and air are mixed in a zone of recirculating flow leading to a stable, compact flame. In this instance, discrete-frequency noise and vibration may be undesirable side effects. Vibration due to excess swirl also represents a problem in the draft tubes of water turbines, and the features of vortex flows can also adversely affect the performance of axial-radial outlet casings of axial turbines and compressors. The basic characteristics of fluidic vortex valves, which range in size from miniature fluidic devices to flood-dam control valves, are a consequence of the dissipative nature of vortex flows. Examples of other devices where the vortex character of the flow plays a central role include dust cyclones, hydrocyclones, and the Ranque-Hilsch refrigeration tube.

As a preliminary, we review some basic aspects of confined vortex flows—in particular, the concepts of supercritical and subcritical flow and vortex breakdown.

[1] Present address: Fluid Mechanics Department, Schlumberger Cambridge Research Limited, P.O. Box 153, Cambridge CB3 0HG, England.

0066–4189/87/0115–0027$02.00

BASIC ASPECTS OF CONFINED VORTEX FLOWS

In laboratory investigations of vortex flows in tubes, swirl may be imparted to the flow by adjustable guidevanes (Harvey 1962, Sarpkaya 1971, Faler & Leibovich 1977), by fixed swirlers similar to those employed in combustors, or by tangential inflow through a long slit (Escudier et al. 1980). Measurements in both the guidevane systems and the slit-tube arrangement reveal that the swirl velocity v has a variation with radius r well represented by

$$v = \frac{\Gamma}{2\pi r}(1 - e^{-r^2/\delta^2}) + \frac{1}{2}\omega r. \tag{1}$$

This equation is an empirical expression based upon the result (with $\omega = 0$) first derived by Burgers (1940) for a viscous line vortex of circulation Γ and uniform axial velocity $w = 2\alpha x$, where α is a constant, x the axial distance along the vortex, $\delta \equiv \sqrt{2\nu/\alpha}$ a measure of the radius of the viscous core, and ν the kinematic viscosity. The ω-term was introduced by Escudier et al. (1982) to account for the effects of Taylor-Görtler vortices that form on the tube wall, mix up the fluid outside the core, and result in a relatively uniform, stable distribution of vorticity. As illustrated by Figure 1*a*, the Taylor-Görtler vortices are especially strong for jet-driven vortex flow. No comparably simple and useful expression is available to describe the axial velocity, although Faler & Leibovich (1977) and others have suggested the use of an exponential function to represent the departure from uniformity.

The core produced in a guidevane system is generally much larger in radius than that produced in a slit tube because in the former case much of the vortical fluid that constitutes the core is continuously swept in from the endwall boundary layer, whereas in the latter case the core circulation results primarily from vorticity shed during the starting process. For a given flow geometry, the diameter of the vortex core decreases with increasing Reynolds number, in much the same way as does the thickness of a boundary layer, with a corresponding increase in the maximum swirl velocity. In this sense we can regard increases in both guidevane angle and Reynolds number as increases in swirl. The static pressure p falls monotonically to a minimum at the axis ($r = 0$) by virtue of the radial momentum equation for a cylindrical flow $dp/dr = \rho v^2/r$, where ρ is the fluid density. If no account were taken of changes in stagnation pressure p_0, it is clear that the axial velocity would exhibit a jetlike overshoot in the vortex core. This is qualitatively the behavior observed at low swirl, although the overshoot is reduced by the reduction in p_0 due to viscous effects in the core region. The laminar appearance of vortex cores, even at relatively high Reynolds numbers, is consistent with the stable character of the vortex core according to, e.g., the inviscid Rayleigh criterion for stability: $d(v^2r^2)/dr > 0$.

Since (1) is merely an empirical formula that adequately represents the mean swirl with suitable choices for the parameters Γ, δ, and ω, it is immaterial whether the flow is laminar or turbulent. Much more important is the criticality of the flow [i.e. whether the axial flow velocity everywhere exceeds the relative phase velocity of long, upstream-directed, longitudinal

A

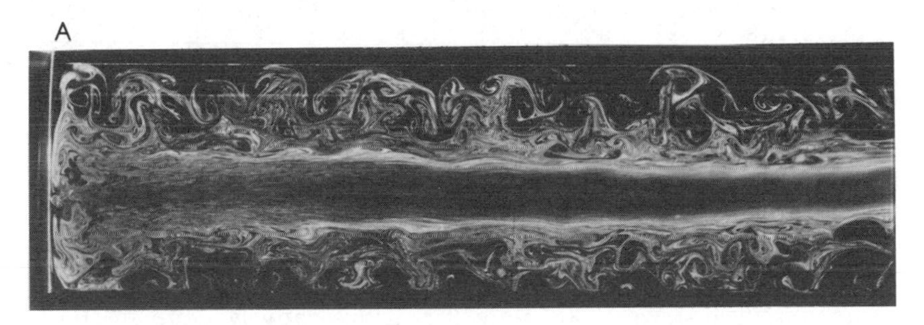

B

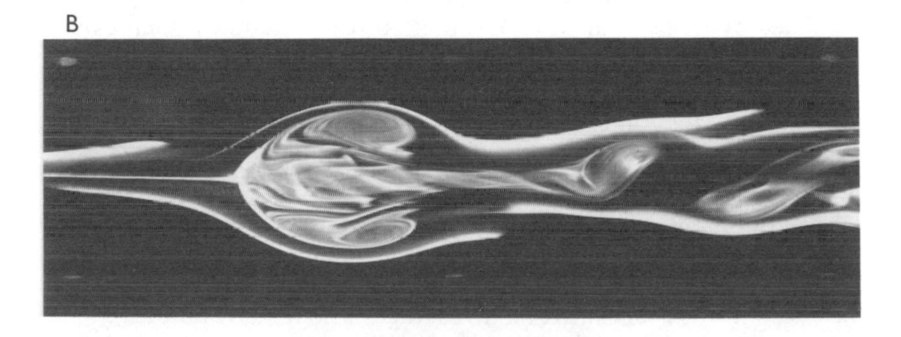

C

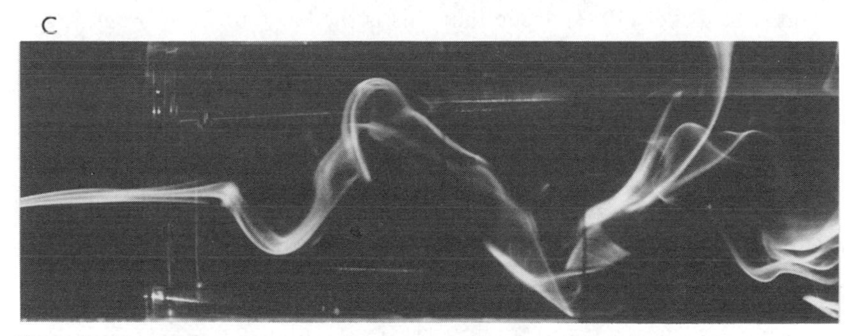

Figure 1 Vortex flow phenomena in axisymmetric tubes (flow left to right): (*a*) Taylor-Görtler vortices in a jet-driven flow; (*b*) axisymmetric breakdown produced by flow through guidevanes; (*c*) spiral breakdown produced in a jet-driven flow.

inertial waves (supercritical flow) or vice versa (subcritical flow)]. To determine the criticality quantitatively involves solving Benjamin's "critical" equation (Benjamin 1962), which requires (usually unavailable) accurate knowledge of the axial- and swirl-velocity distributions. A convenient rule of thumb, suggested by Squire (1962), is that for a flow to be supercritical, we must have $v/w < 1.0$ at the location of maximum swirl. In practice, it is rarely possible to determine analytically whether a real flow is supercritical or subcritical, and the best indicator is the nature of the flow itself.

Once the degree of swirl reaches a certain magnitude, the vortex structure at some axial location typically undergoes a complicated transition (vortex breakdown) to a new flow state. This topic remains controversial, and there is no general agreement regarding the physical mechanisms involved and the criteria for the occurrence of vortex breakdown: Reviews have been given in this series by Hall (1972) and Leibovich (1978). Generally accepted is that the transition is between a supercritical upstream state and a subcritical downstream state (i.e. with relatively more swirl). A recent hypothesis, proposed by Escudier & Keller (1983) and further elucidated by Keller et al. (1985), is that vortex breakdown is basically a two-stage transition, with the first between two different supercritical states, and the second to the final subcritical state. The intermediate state has the form of an inner stagnant core separated from an outer zone of irrotational fluid by a thin layer of rotational fluid emanating from the upstream core. This layer is subject to an intense shear stress, becomes unstable, and rolls up into a spiral: In effect, the vortex core takes on a spiral form precessing around the stagnation region. However, if the stagnant fluid (in the case of a liquid flow) is replaced by air, the viscous stress and instability mechanism are removed, there is no roll up of the rotational layer, and a stable axisymmetric bubble form of breakdown is obtained. Escudier & Keller (1983) argue that this is the basic form of breakdown, and that viscous effects (albeit omnipresent) are only a secondary contaminating influence. Although viscous action or a density decrease (as in combustion) may subsequently return the flow to supercritical, an important consequence if it remains subcritical is that the entire subcritical region, including the breakdown zone, must adjust to match downstream conditions. It is also the case that with increasing swirl, the breakdown region moves upstream until practically the entire flow downstream of the swirler is subcritical. Under carefully controlled conditions at moderate Reynolds numbers it is possible to produce a stable axisymmetric breakdown flow, as illustrated by Figure 1*b*. At the higher Reynolds numbers typical of most practical situations, the breakdown structure (for the same swirl) has the appearance of a precessing spiral, as in Figure 1*c*, even though the associated mean flow is much the same as for the axisymmetric form. The periodicity of many vortex flows appears to be a consequence of this precession. Vortex breakdown also appears to result in a loss of stability and, except at very low Reynolds numbers, is invariably accompanied by a transition to either turbulence or a higher turbulence level.

Vortex flows have not been neglected by computational fluid dynamicists, particularly with respect to combustion-chamber applications. Two difficulties arise that are particular to swirling flows: the influence of rotation on turbulence, and upstream influence in subcritical flows. Efforts to improve turbulence models for swirling flows continue, while the problem of upstream influence, which manifests itself in the specification of the downstream boundary conditions, is arguably more fundamental and has been largely ignored by workers in this area. Gosman et al. (1977) and Boysan & Swithenbank (1981) attempted to circumvent the latter problem by effectively extending their domain of computation sufficiently far downstream to where the flow would be supercritical. More typical is the work of Lilley & Rhode (1982) in which all variables are assumed to have zero axial gradient at the outlet (Neumann conditions); such an assumption does not appear to be adequate in situations where reversed flow occurs, since no information is provided regarding the nature of the fluid entering the calculation domain from downstream.

At this point we consider briefly the nondimensional parameters that are used to characterize vortex flows. We suppose that at some location there is a spiral inflow with circulation Γ and volume flow rate Q uniformly distributed along a length L. Dimensional considerations then show that a global description of the flow must include the dynamic parameters $\mathrm{Re} \equiv Q/L\nu$ and $\Omega \equiv \Gamma L/Q$, where Re is a Reynolds number and Ω a swirl parameter, or their equivalent. If at some radius along the length L the inflow has radial and tangential velocity components $V \cos \beta$ and $V \sin \beta$, respectively, then we have $\Omega = \tan \beta$. In a guidevane system, for example, β is related to the vane angle ϕ by $\tan \beta = \sin \phi/(\cos \phi - m/R_\mathrm{p})$, where m is the chordwise distance from the pivot point of a vane to its trailing edge, and R_p is the radius of the pivot-point pitch circle. In the case of flow into a circular tube of diameter D through a tangential slit of width t, it can be shown that $\Omega = \pi D/t$, whereas for an axial plane-vane swirler (typical of combustion systems) under the assumptions of no separation from the vanes, uniform total pressure, and zero radial velocity, we have

$$\Omega = \frac{(2\gamma - 1)}{\gamma(\gamma - 1)^{1/2}} \frac{(1 - s^{1-1/\gamma})}{(1 - s^{2-1/\gamma})},$$

where $\gamma = 1 + \tan^2 \phi$, and s is the ratio of the inner to outer swirler radius.

More elaborate swirl numbers are also used, and these are defined in terms of the fluxes of axial and angular momenta at some axial location:

$$A \equiv \int_F \rho uvr\, dF \quad \text{and} \quad M \equiv \int_F (\rho u^2 + p - p_R)\, dF,$$

where F is the flow cross section, $u(r)$ and $v(r)$ are the radial distributions of the axial and swirl velocities, and p_R is the static pressure at the outer radius R of the swirler. In draft-tube problems, $AR/\rho Q^2$ is sometimes employed, whereas in the combustion literature a swirl number $S \equiv A/MR$ is commonly introduced. Evaluation of the momentum-flux integrals to calculate a swirl number makes far greater demands than the evaluation of Ω and is ultimately no more informative. For example, if u is assumed to be uniform and v to correspond to a Rankine vortex with core radius δ, then we have $S = 2\alpha(1-\varepsilon/2)/(1-\alpha^2+2\alpha^2 \ln \varepsilon)$, where $\varepsilon \equiv \delta/R$ and $\alpha \equiv \Omega R/4L$. Although this expression emphasizes the importance of the core size, to be of practical value ε must be eliminated, for example, by assuming solid-body rotation ($\varepsilon = 1$), in which case there is a one-to-one correspondence between S and Ω.

SWIRL-STABILIZED COMBUSTION

Figure 2 shows a fixed-geometry swirl generator typical of those employed in the combustors of large industrial gas turbines for electric-power generation. Swirl is introduced into the primary airflow entering a combustion chamber to produce a localized zone of recirculating flow within which fuel and air are subject to intense turbulent mixing with a much longer residence time than would be the case without swirl. The result is a stable, compact flame. In certain applications (for example, where it is necessary to compensate for large changes in the airflow rate) an adjustable guidevane type of swirl register may be used. A fixed geometry swirler is obviously simpler, more robust, and adequate for situations where changes in flow rate are negligible, as is the case for synchronous-drive industrial gas turbines and also household burners.

An important feature of swirl-stabilized combustion is the different character of the burning and nonburning flow fields. Mass conservation requires that the reduction in gas density across the reaction zone be accompanied by a corresponding increase in axial velocity, whereas the swirl velocity is not directly or greatly affected. Thus, where a cold flow would be subcritical downstream of the recirculation zone, the corresponding flow with combustion is likely to return to supercritical. This difference explains the problems with backflow into the combustion chamber under nonburning conditions encountered by Altgeld et al.

(1983), and it is evident that great care must be exercised in transferring the results of nonburning experiments to the burning situation. The differences between vortex flows that return to supercritical and those that remain subcritical were demonstrated by Escudier & Keller (1985) with a flow geometry similar to that of a swirl-stabilized combustion chamber. In these experiments with water, acceleration of the flow was due only to boundary-layer growth. Some of the results are shown in Figure 3, in the form of streamline plots derived from laser-Doppler anemometer measurements of the mean axial velocity. In the supercritical case the influence of a contraction seven tube diameters downstream of the inlet is discernible only in the immediate vicinity of the exit, as it would be in the absence of swirl. With slightly higher swirl, where the acceleration is insufficient to return the flow to a supercritical condition, the contraction effect is felt throughout the entire flow field.

Precession of the spiral vortex structure after breakdown was identified by Syred & Beér (1972) as a possible source of discrete-frequency noise in combustion chambers (although not necessarily the most serious). Typical of vortex-flow oscillations is a linear variation of frequency with flow rate

Figure 2 A typical industrial combustion-chamber swirler (fixed geometry).

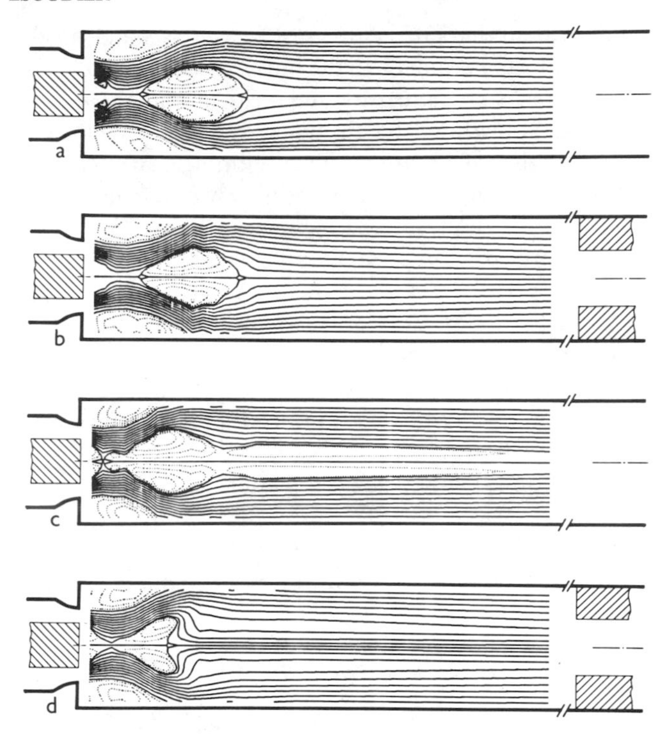

Figure 3 Influence of an exit contraction on swirling flow in a combustor geometry: (*a*) supercritical flow ($\Omega = 5.2$, Re = 10,600) with no exit contraction; (*b*) supercritical flow with exit contraction (54.5% diameter reduction); (*c*) subcritical flow ($\Omega = 22.4$, Re = 10,600) with no exit contraction; (*d*) subcritical flow with exit reduction.

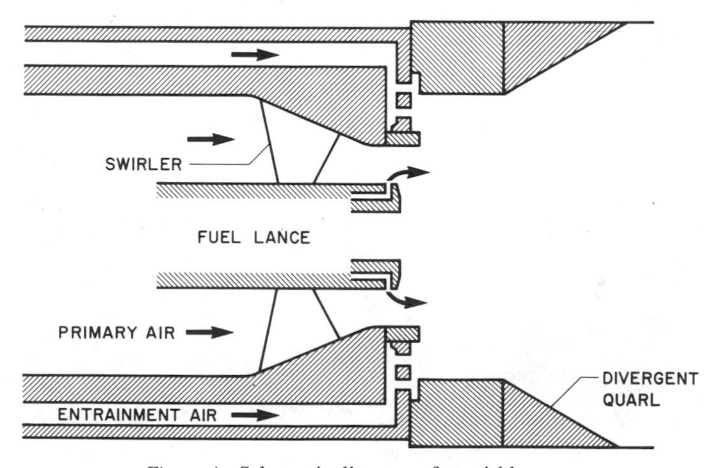

Figure 4 Schematic diagram of a swirl burner.

(constant Strouhal number), although this may be modified by the strong density variations in combustion flows and also by the proximity of the oscillation frequency to an eigenfrequency of the combustion chamber. In practice, a systematic development of the burner geometry, including the swirler, fuel-injector, contraction, quarl, and secondary air inlets (see Figure 4), is needed to arrive at a satisfactory arrangement for a given application with low noise, a compact flame, and acceptable pollutant levels.

VORTEX AMPLIFIERS

The geometry shown in Figure 5 is typical of a radial vortex valve: The proportions given are the optimum values found experimentally by Brombach (1972). The essential characteristic of a vortex valve is the increased resistance to flow produced by the introduction of swirl into the valve chamber, either by reversal of the flow direction (diode) or by the introduction of control flow (triode).

Two factors of importance in defining vortex-valve performance are the turndown ratio TR and the cutoff control pressure ratio PR (Wormley 1976). If Q_m is the maximum throughflow (zero control flow) and Q_c the control flow necessary to reduce the supply flow to zero, then $TR = Q_m/Q_c$, while $PR \equiv p_c/p_s$, where p_c is the control pressure corresponding to Q_c, and p_s is the (constant) supply pressure (p_c and p_s measured above ambient). Depending upon the geometry, the relationship between total flow rate and control pressure or control flow may be single-valued (proportional) or multivalued (bistable). Figure 6, based upon the work of Lawley & Price (1972), shows the static characteristics of a typical single-exit vortex triode. An ejector can be incorporated in the control-flow line to improve the turndown ratio for vortex valves operating with a control pressure significantly higher than p_s: With the low-pressure chamber of the ejector connected to the valve outlet, the quantity of pure control fluid can be considerably reduced. This suggestion was first made by Fitt (1974). Syred and his coworkers have also discussed various vortex valve-ejector combinations (Syred 1975, Owen & Syred 1982) as well as the problems of instabilities and a discontinuity in the characteristic of a vortex amplifier, which are again associated with vortex breakdown (MacGregor & Syred 1982, MacGregor et al. 1983).

The inherent reliability of vortex amplifiers has led to their widespread use in applications where maintenance is difficult and operating safety of vital importance: in systems handling or controlling hot or toxic (especially radioactive) materials (Grant 1975, Brodersen & Papadopoulos 1981, Etherington 1984), and in large hydraulic installations, where J. Giesecke,

H. Brombach, and others at the University of Stuttgart have made major contributions to the use of large vortex valves in sewage-treatment plants, flood-control reservoirs, and irrigation systems (e.g. Brombach & Neumayer 1977). Brombach (1984) has recently described a novel conical vortex valve, several hundred of which, in the size range 150–500 mm, have been in use in sewage systems for a number of years. This design has only a single-inlet port (tangential); swirl is limited at partial load by the existence of a free liquid surface in the valve chamber that is installed with its axis at an angle of about 45° to the vertical. Only when the valve is full

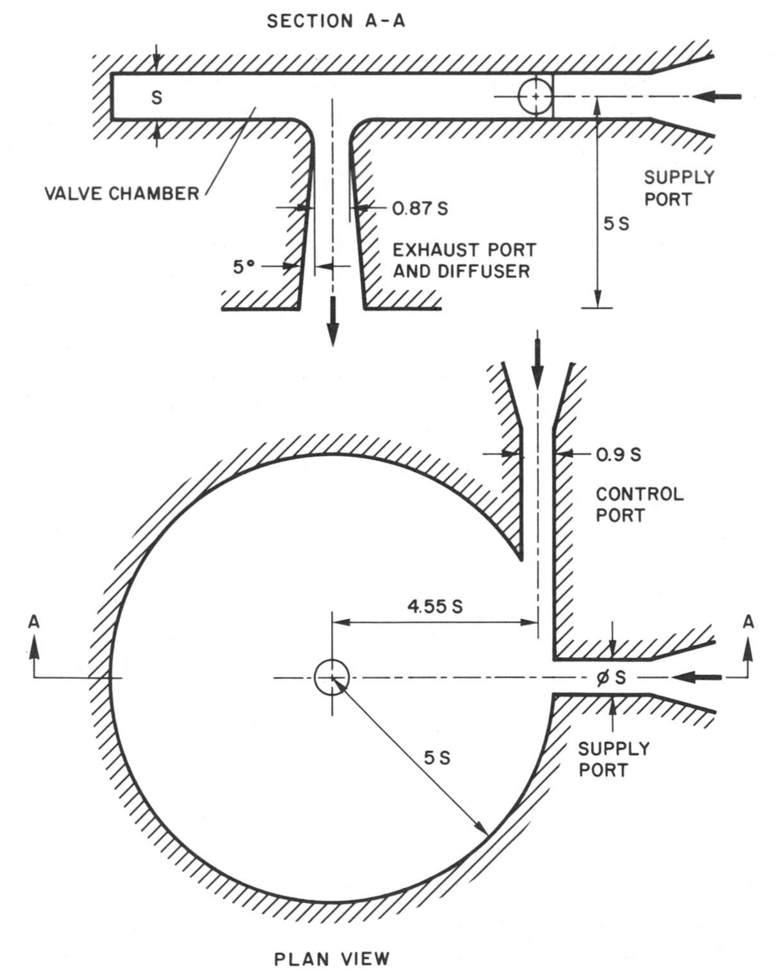

Figure 5 Optimum proportions for a conventional vortex-valve design (after Brombach 1972).

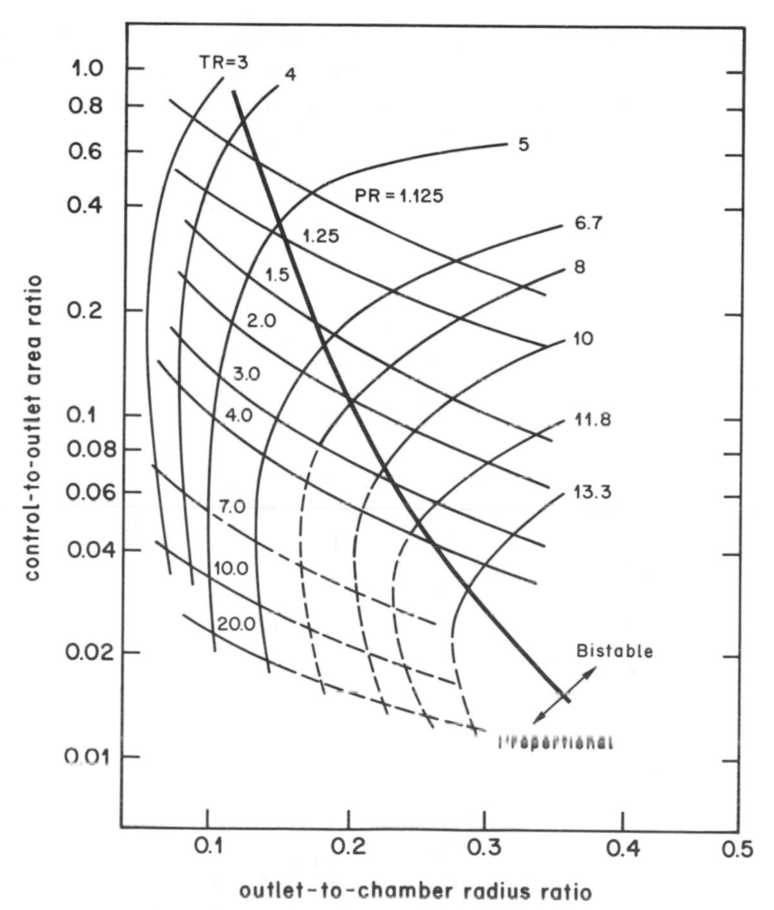

Figure 6 Vortex triode characteristics (after Lawley & Price 1972).

of water is intense vortex motion sustained. In an earlier article, Brombach (1981) described a variant of the conical valve design with an inlet diameter of 900 mm and maximum chamber diameter of 2.7 m for use in a flood-retention reservoir. Even larger, with an outlet diameter of 7 m, is the vortex diode installed at the Oraison hydroelectric power plant in France (Jeanpierre et al. 1966) as part of the relief bypass system through which the turbine flow is diverted in the event of a power failure.

TURBOMACHINE OUTLETS

A problem inherent in the design of many axial-flow turbomachines is that of collecting the axial throughflow and diverting it radially away from the machine's main axis. This is the case, for example, for a large turboset where steam flows through several turbines running on a common shaft and also in the valves that regulate the steam supply to the turbines. A commonly employed solution is the use of a ring-type exit chamber (or exhaust hood or ring diffuser) with annular inflow and radial outflow (see Figure 7). Little has been published about the design of such exhausts, although a great deal of proprietary information must exist and a comprehensive design guide is in preparation (Japikse 1987). The articles by Deich & Zaryankin (1972), Suter & Girsberger (1974), and Keller (1980) are probably indicative of industrial design practice whereby a turbine exhaust hood is treated (inadequately) as a combination of an annular diffuser and a plenum chamber. Perhaps even more indicative of the state

Figure 7 Photograph of an 88-MW axial compressor.

of the art are patent applications. One of the most significant in this area is that of Herzog & Kovacik (1964), who recognized that the key to improving the flow behavior of an exhaust hood (or scroll, as they term it) lies in weakening the vortex. More recent patent applications that have a similar aim include those of Garkuša & Dobrynin (1983) and Takamura (1983).

Escudier & Merkli (1977, 1979), motivated by a model study of steam-regulating valves, carried out an experimental investigation of the flow in the idealized exit-chamber model shown in Figure 8 in which the inlet-annulus width t could be varied. The results of this and a later study (Merkli 1978) revealed that the actual characteristics of exhaust hoods could be undesirably different from those intended by the designer. An intense vortex generated by the tangential inflow into a large chamber of near-circular cross section dominates the flow behavior. In addition to extremely high pressure loss, some configurations were found to give rise to periodic behavior, as illustrated by the Strouhal-number plots in Figure 9. The oscillations were strongest for $t/R = 0.31$ (typical of practical proportions, since the annulus cross section is then equal to that of the outlet pipe). The corresponding pressure-loss coefficient $\xi = 2\Delta p/\rho V^2$ (V is the bulk velocity in the annulus) was an order of magnitude greater than that for a comparable 90° bend. The characteristics of the ring-type exit-chamber model are entirely consistent with those of vortex diodes, vortex oscillators, and vortex whistles (Vonnegut 1954, Chanaud 1963).

A simple analysis of the mean flow in a ring-type exit chamber by Escudier (1979) led to reasonable agreement with experimental values of

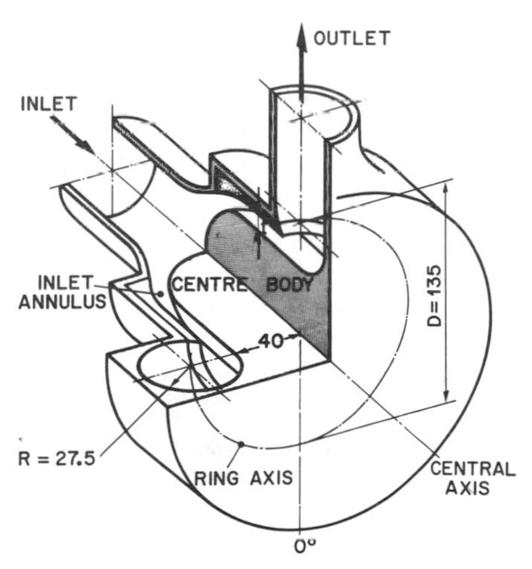

Figure 8 Idealized exit-chamber model (dimensions in millimeters).

ξ. The analysis is in the spirit of Borda and Carnot, in that the loss is assumed to result from dissipation of the axial and swirling kinetic energy of the flow. The flow structure in the ring was taken as a core of diameter δ with solid-body rotation and zero axial velocity surrounded by a potential vortex with uniform axial velocity. Following Binnie & Hookings (1948), Lewellen (1971), and others, it was assumed that the core size adjusts in such a way as to achieve maximum throughflow ($\partial\xi/\partial\gamma = 0$), although there appears to be no formal basis for this assumption. The final result can be expressed as

$$\xi = \frac{1}{4}\left(\frac{D}{D_e}\right)^4\left[\left(\frac{\gamma}{1-\gamma}\right)^2 + \frac{1}{F}(3-2\ln\gamma)\right],$$

where $F = 8(dt)^2/(DD_e)^2 = (1-\gamma)^3/\gamma^2$ and $\gamma = (\delta/D_e)^2$.

The oscillatory behavior in the model exit chamber was suppressible with a splitter plate in the exit pipe, but at the expense of a further 25% increase in ξ. Explicit recognition of the vortex influence enabled Merkli (1978) to achieve remarkable improvements in the performance of exhaust hoods. For example, with a square cross section of side D, the loss was cut by about 20%, and introduction of a circular collar with an optimized beaklike cross section (Figure 10) reduced ξ to 0.65 with complete absence

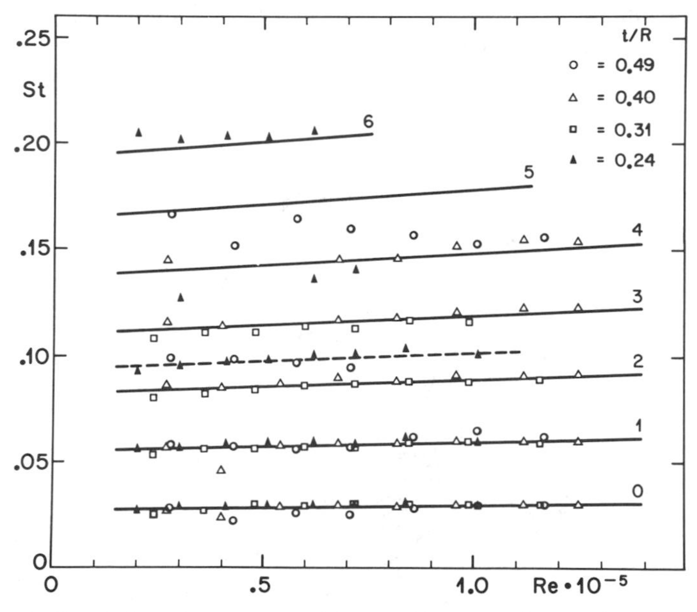

Figure 9 Strouhal-number variation with Reynolds number for model exit chamber.

Figure 10 Low-loss cross section for an exhaust chamber.

of periodic pressure fluctuations. This optimized design represents a compromise between vortex formation and blockage.

CYCLONE SEPARATORS

Cyclone separators and hydrocyclones represent one of the most widespread and oldest applications of confined-vortex phenomena. Since the efficiency is poor for dust particles smaller than about 5 μm diameter, cyclone dust separators are widely used as final collectors for large particles or as precleaners ahead of electrostatic precipitators and other collectors better suited to small particles. Present-day cyclones differ in detail but are basically similar in principle to the device patented by Morse (1886). Current practice, essentially established by ter Linden (1949) and Stairmand (1951), is exemplified by the reverse-flow cyclone shown in Figure 11. The basic components of a cyclone are usually a cylindrical-conical body with either a tangential inlet or an axial inlet with guidevanes to generate swirl; a cover plate with a central "vortex finder" extending some distance into the cyclone; and a dust collector at the cone vertex with a central "vortex breaker" to prevent dust being sucked back into the flow.

The behavior of dust in a cyclone varies strongly with particle size, which changes the balance of the inertia, gravity, and viscous forces acting on the particles. The majority of the coarsest particles (> 10 μm) entering a tangential-inlet cyclone migrate to the sidewall, where they are transported along a spiral path through the boundary layer to the collector (Figure 12). It is interesting to note that the now classical cone shape was already a part of Morse's (1886) design, long before it was shown by Taylor (1950) that the boundary layer generated by inviscid swirling flow in a conical channel has a component of velocity directed toward the vertex. Dust separation occurs when the boundary-layer fluid turns sharply back toward the exit. A certain amount of dust is transported through the boundary layer on the cover plate, and part of the purpose of the vortex finder (also incorporated into one of Morse's designs) is to prevent this dust from entering the outflow. Fine particles tend to closely follow the gas flow and become distributed throughout the cyclone interior by turbulent diffusion, so that a large fraction leaves with the "clean" gas, an effect that is partially counteracted by agglomeration if the concentration is high. In fact, Abrahamson et al. (1978) argue that small particles must agglomerate in the collection hopper if they are to remain collected.

A critical aspect of cyclone design is minimization of the overall pressure loss, and here the vortex finder again plays an important role: The sepa-

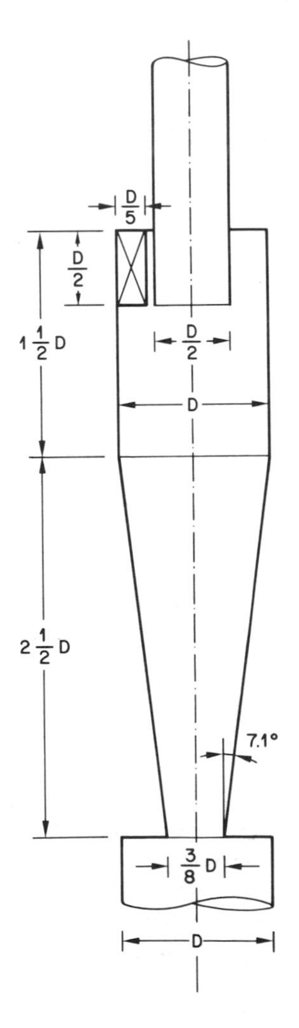

Figure 11 High-efficiency cyclone design (after Stairmand 1951).

ration efficiency is adversely affected if the exit diameter is too large, whereas the "vortex-valve" effect becomes unacceptably high if it is too small. Browne & Strauss (1978) report a 25% reduction in pressure loss by introducing guidevanes to remove swirl from the outlet flow [Ogawa (1984) reproduces part of a patent from 1901 showing essentially the same idea], whereas Boadway (1984) suggests designing the vortex finder as an axial-radial diffuser with a radial outlet as an alternative approach to decreasing the pressure loss in a hydrocyclone.

Extensive treatments of cyclone and hydrocyclone design are given by Rietema & Verver (1961), Brauer & Varma (1981), Gupta et al. (1984), and Ogawa (1984).

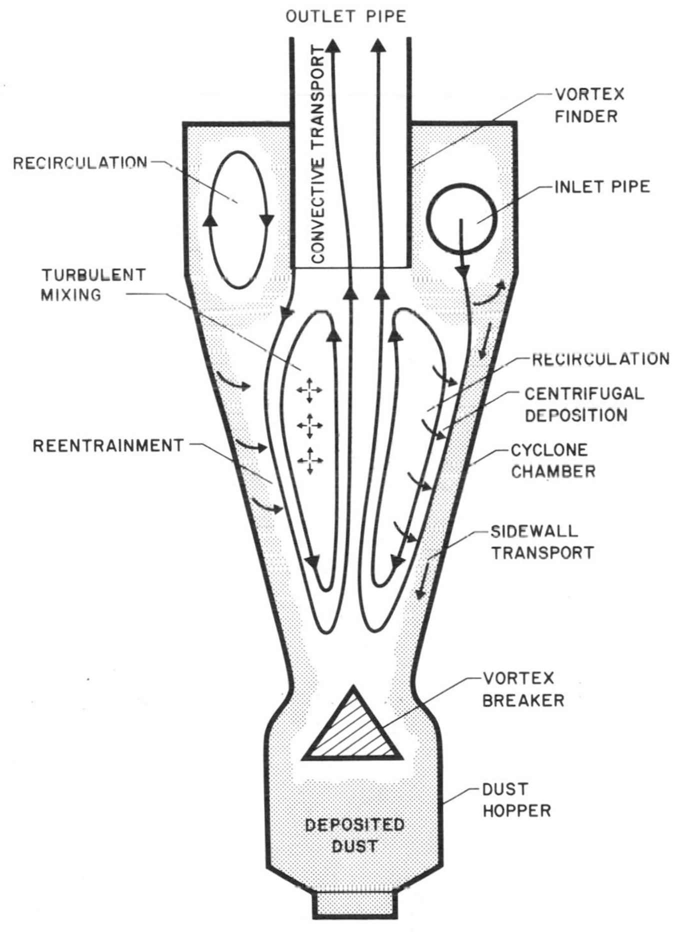

Figure 12 Dust-flow patterns in a reverse-flow cyclone.

DRAFT-TUBE SURGE

Falvey (1971) cites numerous references, the earliest published in 1924, to the problem of low-frequency (typically a few hertz) vibration experienced with Francis- and Kaplan (propeller)-type hydraulic turbines operating at partial load or overload. It has long been recognized that the vibration originates from surges caused by the intense vortex flow in the turbine draft tube downstream of the runner (Figure 13). The tests of Kubota & Matsui (1972) on a model turbine showed maximum surge amplitudes at about 60% discharge, and also that pressure fluctuations occurred throughout the entire turbine. At partial load a high level of residual swirl in the draft tube results from a mismatch between the swirl generated by the wicket gates (guidevanes) and the angular momentum extracted by the turbine runner. From field observations of full-scale machines, Rheingans (1940) deduced that the surge frequency was proportional to the turbine rotational speed, while Winkelhaus et al. (1946) were able to photograph the helical structure of the cavitated vortex core in a draft tube, a much more recent and excellent example of which is shown in Figure 14 (from Dörfler 1980). Particularly interesting is the published discussion of Rheingans' paper, which reveals how widespread the surge problem was at the time. Recent papers (e.g. Nishi et al. 1984, Dörfler 1985) suggest that the problem is extant.

Gerich & Raabe (1975) argued that the unsteady vortex flow in a draft

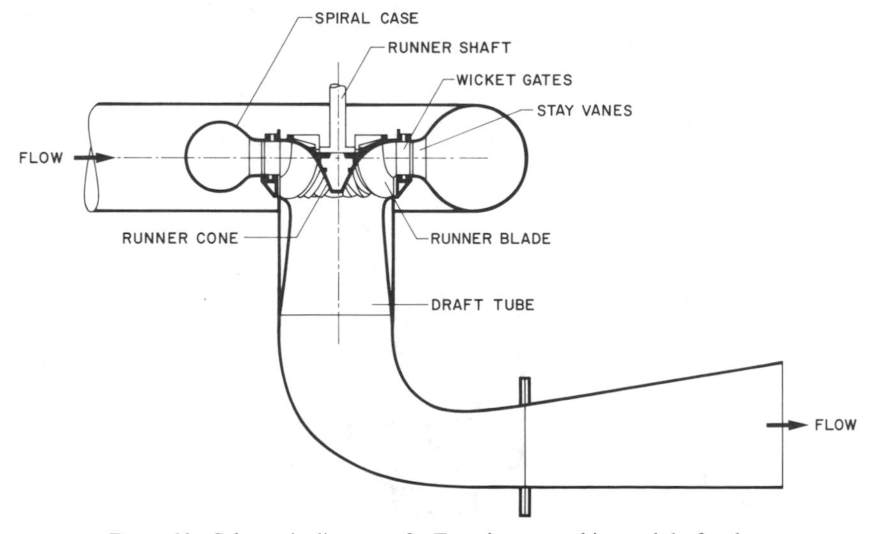

Figure 13 Schematic diagram of a Francis-type turbine and draft tube.

tube is associated with flow separation in the runner-blade channels. A more widely accepted viewpoint is that the phenomenon is a manifestation of vortex breakdown, a suggestion first made by Cassidy & Falvey (1970), who carried out a systematic investigation of the oscillating flow in swirling airflow using a guidevane arrangement with the aim of establishing guidelines for the surge characteristics of hydraulic turbines. Their results showed that above a critical swirl number, both the Strouhal number fR^3/Q and the pressure amplitude $\langle p' \rangle R^3/A$ were linearly dependent on the swirl number $AR/\rho Q^2$, where A is the flux of angular momentum and Q the volume flow rate.

The presence of a limited quantity of air or water vapor in the flow in a water turbine provides a degree of elasticity, termed cavitation compliance by Rubin (1966). Dörfler (1980) suggested that this elasticity leads to a form of resonance in the draft tube excited by the precessing inhomo-

Figure 14 Cavitating vortex flow in a draft tube (from Dörfler 1980).

geneous pressure field associated with the spiral vortex core, a viewpoint that is shared by Nishi et al. (1984). The surge intensity increases with the quantity of entrapped air or vapor to a maximum and then decreases again to a negligible level. Ulith (1968) indicates that the necessary amount of air to be injected for maximum attenuation is between 1 and 20‰ by volume, depending upon whether injection takes place upstream or downstream of the runner. The suppression mechanism is suggested to be as follows. The air injected enters the recirculation region, producing an essentially axisymmetric stable flow—a hollow (air) core surrounded by swirling water flow. In the sense of the vortex-breakdown literature, air injection changes the breakdown form from spiral to bubble.

The various techniques that have been employed to reduce or suppress surge are intended either to produce an axisymmetric draft-tube flow or to reduce swirl in the draft tube below the critical level. In addition to air injection, reducing the asymmetry can be achieved by introducing either a coaxial hollow cylinder or an extension of the runner cone into the draft tube. The swirl level can be reduced by incorporating longitudinal fins or splitter plates into the draft tube, or by injecting counterswirling fluid into the flow. Seybert et al. (1978) reviewed the general problem of surge suppression, with emphasis on the counterswirl approach, and Grein (1980) has given a general survey of the causes and prevention of vibrations in Francis turbines, with emphasis on the draft-tube surge problem. Palde (1972) showed that the shape of the draft tube has a significant influence on the surging characteristics of hydraulic turbines, a further aspect that is consistent with the work of Sarpkaya (1971) and others on vortex breakdown in flared tubes.

RANQUE-HILSCH TUBE

Ranque (1933) was the first to observe that the total temperature in an intense gas vortex decreases with approach to the axis, i.e. an energy separation takes place. Ranque's discovery excited little interest until Hilsch (1946) carried out the first systematic experiments aimed at optimizing Ranque's device for maximum efficiency. High-pressure air or other gas is introduced into the Ranque-Hilsch tube (Figure 15) tangentially at one end. Cold air flows out through a central orifice at the inlet end, and warm air from an annular gap at the other. The counterflow configuration as depicted here is not vital, as was already known to Ranque, and some workers (Bruun 1969, Kurosaka 1982) have investigated a uniflow arrangement. With inlet air at 11 bar and 20°C, Hilsch obtained temperature differences as large as 70°C and cold temperatures as low as −45°C for a cold flow rate of about 20% of the total throughflow. A much more extensive series of optimization experiments carried out by Westley (1955, 1957) showed that temperatures below $-40°C$ could be achieved at an inlet pressure of 7 bar, and that for optimum conditions the temperature drop was close to 50% of the isentropic value.

Although at the present time there is no possibility of analyzing the complex flow field in a Ranque-Hilsch tube in detail, including the effects of three dimensionality, compressibility, and turbulence, there have been a number of attempts to discuss limited aspects of the problem. Remarkably, questions are still raised regarding the basic mechanism of energy separation. Deissler & Perlmutter (1958, 1960) performed a quasi-two-dimensional steady-flow analysis of the mean velocity and temperature distributions in a vortex. They were able to achieve acceptable agreement with Hilsch's data and concluded that to account for the energy separation the flow must be turbulent and compressible. They also argued that the total temperature of a fluid element spiraling toward the axis is increased primarily by the kinetic- and pressure-energy contributions to the shear work in the core. Reynolds (1961) came to similar conclusions based upon order-of-magnitude estimates of the various terms in the energy equation. An extensive analysis of energy separation in various devices was given by Graham (1972).

In a recent article, Kurosaka (1982) argued that the energy separation in a vortex is a consequence of acoustic streaming associated with intense discrete-frequency sound produced by the precessing vortex core. On the basis of a second-order analysis, it is shown that acoustic streaming produces an excess of angular momentum in the outer region of the vortex and thereby an increase in the total temperature. Kurosaka also presents experiments on a uniflow vortex tube that reveal a drastic drop in separation efficiency if the vortex whistling is attenuated by tuned acoustic cavities radial to the vortex axis. Although the experiments show clearly that acoustic suppression has a major influence on the energy separation

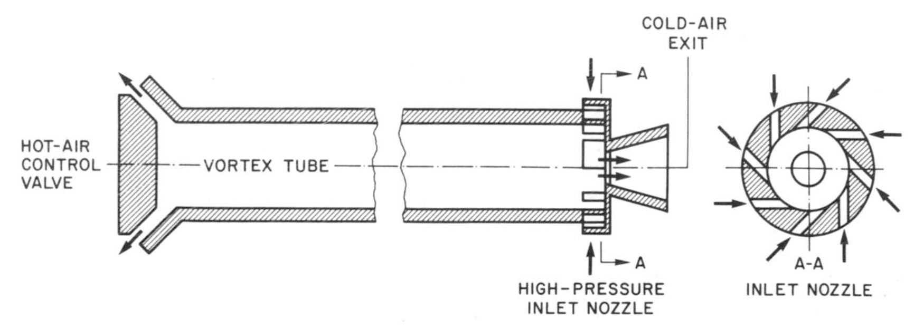

Figure 15 Schematic diagram of a Ranque-Hilsch tube (after Westley 1955).

process in his device, it is unlikely that the acoustic-streaming explanation is valid in general. Hilsch, Westley (1955), and Huub, for example, make no mention of whistling, and comment on the noise generated only in passing. McDuffie (1979), who also suggests a direct link between whistling and energy separation in a Ranque-Hilsch tube, identified high-intensity ultrasound (140 dBa, 31 kHz) at the exit of his tube and suggests that what Hilsch called a bubbling sound could have been related to ultrasonic whistling.

It is suggested here that the direct effect of Kurosaka's suppressors is to change the basic flow structure from an asymmetric precessing vortex core (spiral breakdown) to a symmetric annular flow (bubble breakdown), with an accompanying loss of discrete-frequency sound excitation (much as in the draft-tube surge problem) and also a redistribution of angular momentum. Since Kurosaka's device has a very low length/diameter ratio (about 3.1 compared with between 17 and 65 for Hilsch, and 66 for Westley), it is probable that at no location was there an intense vortex core, and thus that the breakdown structure occupied the entire tube. In the case of very long vortex tubes, viscous action may be sufficient to bring the flow back to supercritical conditions at the hot end, so that the hot (annular) valve acts merely as a throttle with limited influence on the upstream flow structure. Hilsch (1946), for example, remarks that the tube should be long enough for the rotating component of the gas motion to decrease sufficiently. Kurosaka's discussion leaves little doubt that energy separation occurred when the exit flow was subcritical: He mentions that radial insertion of a probe interfered with the measurements; a coaxial probe inserted from the exhaust also influenced whistling and energy separation; it was necessary to introduce an exit plug to prevent flow into the exhaust from the environment; and a finger(!) near the exhaust detected a change from a spiral outflow during whistling to an axial outflow when whistling was suppressed.

The application of the vortex tube that motivated Hilsch's investigation was gas liquification. A number of subsequent studies were concerned with the device as a fluidic refrigerator, and Kurosaka's experiments were carried out on a modified commercial Ranque-Hilsch tube designed for refrigeration applications. Other authors have investigated the possibility that vortex tubes could be used for the separation of isotopes or other gas mixtures, although the effect here appears to be a more straightforward centrifugal separation (Linderstrøm-Lang 1964). A related application, first suggested by Kerrebrock & Meghrablian (1961), was to use vortex containment for a gaseous fission rocket. The basic concept was for an annular gaseous cloud of fissioning fuel to transfer heat to a propellant such as hydrogen flowing radially inward through the fuel. Development work on such rockets appears to have ceased in the early 1970s, with the most recent references found being those of Lakshmikantha (1973) and Black et al. (1973).

Other less exotic uses proposed for the Ranque-Hilsch tube have been as a cooling system for electronic components in supersonic aircraft, with shock compression to provide a sufficiently high driving pressure (*Machine Design* 1961), and as a cooler for the drinking water of a locomotive crew (Stefanides 1962). Vonnegut (1950) investigated the possibility of using the cooling effect to compensate for the aerodynamic heating of airborne thermometers.

ACKNOWLEDGMENTS

The photographs in Figures 7 and 14 are reproduced by kind permission of Gebrüder Sulzer AG and Sulzer Escher Wyss AG, respectively. I am also indebted to Maria Zamfirescu for preparing the line drawings, Luděk Hlaváč for printing the photographs, and Birgit Nowatzek for typing and incorporating endless changes in the manuscript.

Literature Cited

Abrahamson, J., Martin, C. G., Wong, K. K. 1978. The physical mechanisms of dust collection in a cyclone. *Trans. Inst. Chem. Eng.* 56: 168–77

Altgeld, H., Jones, W. P., Wilhelmi, J. 1983. Velocity measurements in a confined swirl driven recirculating flow. *Exp. Fluids* 1: 73–78

Benjamin, T. B. 1962. Theory of the vortex breakdown phenomenon. *J. Fluid Mech.* 14: 593–628

Binnie, A. M., Hookings, G. A. 1948. Laboratory experiments on whirlpools. *Proc. R. Soc. London Ser. A* 194: 398–415

Black, D. L., Kyslinger, J. A., Ravets, J. M. 1983. The colloid core reactor rocket engine concept. *Trans. Am. Nucl. Soc.* 17: 6–7

Boadway, J. D. 1984. A hydrocyclone with recovery of velocity energy. *Proc. Int. Conf. Hydrocyclones, 2nd, Bath*, pp. 99–108

Boysan, F., Swithenbank, J. 1981. Numerical prediction of confined vortex flows. *Proc. Int. Conf. Numer. Methods in Laminar Turbul. Flow, 2nd, Venice*, pp. 425–37

Brauer, H., Varma, Y. B. G. 1981. *Air Pollution Control Equipment*. Berlin/Heidelberg/New York: Springer-Verlag. 388 pp.

Brodersen, R. K., Papadopoulos, J. G. 1981. Hot gas control system design and vortex valve tests. *IEEE Trans. Autom. Control* AC-26: 625–37

Brombach, H. 1972. Vortex devices in hydraulic engineering. *Proc. Cranfield Fluidics Conf., 5th, Uppsala*, pp. B1-1–B1-12

Brombach, H. 1981. Flood protection by vortex valves. *Trans. ASME, J. Dyn. Syst. Meas. Control* 103: 338–41

Brombach, H. 1984. Vortex flow controllers in sanitary engineering. *Trans. ASME, J. Dyn. Syst. Meas. Control* 106: 129–33

Brombach, H., Neumayer, H. 1977. Development of a very large radial vortex valve. *J. Fluid Control. Fluid. Q.* 9: 59–74

Browne, J. M., Strauss, W. 1978. Pressure drop reduction in cyclones. *Atmos. Environ.* 12: 1213–21

Bruun, H. H. 1969. Experimental investigation of the energy separation in vortex tubes. *J. Mech. Eng. Sci.* 11: 567–82

Burgers, J. M. 1940. Application of a model system to illustrate some points of the statistical theory of free turbulence. *Proc. K. Ned. Akad. Wet.* 43: 2–12

Cassidy, J. J., Falvey, H. T. 1970. Observations of unsteady flow arising after vortex breakdown. *J. Fluid Mech.* 41: 727–36

Chanaud, R. C. 1963. Experiments concerning the vortex whistle. *J. Acoust. Soc. Am.* 35: 953–60

Deich, M. Ye., Zaryankin, A. Ye. 1972. Gas dynamics of diffusers and exhaust ducts of turbomachines. *US Dept. Commer. FTD-MT-24-1450-71*. 467 pp. Transl. of *Gazodinamika Diffuzorov i Vykhlopnykh Patrubkov Turbomashin*. Moscow: Energiya. 384 pp. (1970)

Deissler, R. G., Perlmutter, M. 1958. An analysis of the energy separation in laminar and turbulent compressible vortex flows. *Proc. Heat Transfer Fluid Mech. Inst.*, pp. 40–53. Stanford, Calif: Stanford Univ. Press

Deissler, R., Perlmutter, M. 1960. Analysis of the flow and energy separation in a turbulent vortex. *Int. J. Heat Mass Transfer* 1: 173–91

Dörfler, P. 1980. Mathematical model of the pulsations in Francis turbines caused by the vortex core at part load. *Escher Wyss News* 1/2: 101–6

Dörfler, P. 1985. Francis turbine surge prediction and prevention. *Proc. Waterpower/'85, Las Vegas*, pp. 1–10

Escudier, M. P. 1979. Estimation of pressure loss in ring-type exit chambers. *Trans. ASME, J. Fluids Eng.* 101: 511–16

Escudier, M. P., Bornstein, J., Maxworthy, T. 1982. The dynamics of confined vortices. *Proc. R. Soc. London Ser. A* 382: 335–60

Escudier, M. P., Bornstein, J., Zehnder, N. 1980. Observations and LDA measurements of confined turbulent vortex flow. *J. Fluid Mech.* 98: 49–63

Escudier, M. P., Keller, J. 1983. Vortex breakdown: a two-stage transition. *AGARD CP No. 342*, Pap. 25

Escudier, M. P., Keller, J. J. 1985. Recirculation in swirling flow: a manifestation of vortex breakdown. *AIAA J.* 23: 111–16

Escudier, M. P., Merkli, P. 1977. Flow in ring-type inlet and outlet chambers. *Brown Boveri Res. Cent. Intern. Rep. KLR 77-105 B*. 123 pp.

Escudier, M. P., Merkli, P. 1979. Observations of the oscillatory behaviour of a confined ring vortex. *AIAA J.* 17: 253–60

Etherington, C. 1984. Power fluidics technology and its application in the nuclear industry. *Nucl. Energy* 23: 227–35

Faler, J. H., Leibovich, S. 1977. Disrupted states of vortex flow and vortex breakdown. *Phys. Fluids* 20: 1385–1400

Falvey, H. T. 1971. Draft tube surges. A review of present knowledge and an annotated bibliography. *US Bur. Reclam. Rep. REC-ERC-71-42*. 25 pp.

Fitt, P. W. 1974. A vortex valve and ejector combination for improved turn down ratios. *Proc. Cranfield Fluid. Conf., 6th, Cambridge*, pp. F1-1–F1-8

Garkuša, A. V., Dobrynin, V. E. 1983. Exhaust pipe of turbine. *US Patent No. 4,398,865*

Gerich, R., Raabe, J. 1975. Measurement of the unsteady and cavitating flow in a model Francis turbine of high specific speed. *Trans. ASME, J. Fluids Eng.* 97: 402–11

Gosman, A. D., Khalil, E. E., Whitelaw, J. H. 1977. The calculation of two-dimensional turbulent recirculating flows. In *Turbulent Shear Flows 1*, ed. F. Durst, B. E. Launder, F. W. Schmidt, J. H. Whitelaw, pp. 237–55. Berlin/Heidelberg/New York: Springer-Verlag. 415 pp.

Graham, P. A. 1972. A theoretical study of fluid dynamic energy separation. *Rep. No. TR-ES-721*, George Washington Univ., Washington, DC. 383 pp.

Grant, J. 1975. Power fluidics in nuclear engineering. *Atom* 225: 108–16

Grein, H. 1980. Vibration phenomena in Francis turbines: their causes and prevention. *Proc. IAHR Symp., 10th, Tokyo*

Gupta, A. K., Lilley, D. G., Syred, N. 1984. *Swirl Flows*. Kent, Engl: Abacus. 475 pp.

Hall, M. G. 1972. Vortex breakdown. *Ann. Rev. Fluid Mech.* 4: 195–218

Harvey, J. K. 1962. Some observations of the vortex breakdown phenomenon. *J. Fluid Mech.* 14: 585–92

Herzog, J., Kovacik, J. M. 1964. Exhaust scroll for turbomachine. *US Patent No. 3,120,374*

Hilsch, R. 1946. Die Expansion von Gasen im Zentrifugalfeld als Kälteprozess. *Z. Naturforsch.* 1: 208–14. Transl., 1947, in *Rev. Sci. Instrum.* 18: 108–13

Japikse, D. 1987. *Turbomachinery Volutes, Hoods, Collectors, and Industrial Inlets. Design Technol. Ser.*, Vol. 2. Norwich, Conn: Concepts ETI, Inc. In press

Jeanpierre, D., Lachal, A., van Thienen, N. 1966. La chambre d'eau de l'usine d'Oraison. *Houille Blanche* 7: 815–21

Keller, H. 1980. Design of non-rotating parts. In *Steam Turbines for Large Power Outputs. VKI Lecture Ser. No. 1980-06*, Chap. 3

Keller, J. J., Egli, W., Exley, J. 1985. Force- and loss-free transitions between flow states. *ZAMP* 36: 854–89

Kerrebrock, J. L., Meghrablian, R. V. 1961. Vortex containment for the gaseous-fission rocket. *J. Aerosp. Sci.* 28: 710–24

Kubota, T., Matsui, H. 1972. Cavitation characteristics of forced vortex core in the flow of the Francis turbine. *Fuji Electr. Rev.* 18: 102–8

Kurosaka, M. 1982. Acoustic streaming in swirling flow and the Ranque-Hilsch (vortex-tube) effect. *J. Fluid Mech.* 124: 139–72

Lakshmikantha, H. 1973. *Investigation of*

two-phase vortex flows with applications to a cavity nuclear rocket. PhD thesis. Mass. Inst. Technol., Cambridge. 321 pp.

Lawley, T. J., Price, D. C. 1972. Design of vortex fluid amplifiers with asymmetrical flow fields. *Trans. ASME, J. Dyn. Syst. Meas. Control* 94: 82–84

Leibovich, S. 1978. The structure of vortex breakdown. *Ann. Rev. Fluid Mech.* 10: 221–46

Lewellen, W. S. 1971. A review of confined vortex flows. *NASA CR-1772*. 219 pp.

Lilley, D. G., Rhode, D. I. 1982. A computer code for swirling turbulent axisymmetric recirculating flows in practical isothermal combustor geometries. *NASA CR 3442*. 130 pp.

Linderstrøm-Lang, C. U. 1964. Gas separation in the Ranque-Hilsch vortex tube. *Int. J. Heat Mass Transfer* 7: 1195–1206

MacGregor, S. A., Syred, N. 1982. Effect of outlet diffusers on vortex amplifier characteristics. *J. Fluid Control. Fluidics Q.* 14: 1–11

MacGregor, S. A., Syred, N., Markland, E. 1983. Instabilities associated with the outlet flow in vortex amplifiers. *J. Fluid Control. Fluidics Q.* 19: 29–37

Machine Design. 1961. Laboratory curiosity tries out for a cooling job. 33: 14

McDuffie, N. G. 1979. Resonance in the Ranque-Hilsch vortex tube. *ASME Pap. 79-HT-16*

Merkli, P. 1978. Experimental geometrical parameter study of turbomachine outlet chambers. *Brown Boveri Res. Cent. Intern. Rep. KLR 78-102 B*. 104 pp.

Morse (init. unknown). 1886. Staubsammler. *Ger. Pat. No. 39219*

Nishi, M., Matsunaga, S., Kubota, T., Senoo, Y. 1984. Surging characteristics of conical and elbow-type draft tubes. *Proc. IAHR Symp., 12th, Stirling*, pp. 272–83

Ogawa, A. 1984. *Separation of Particles From Air and Gases*, Vols. 1, 2. Boca Raton, Fla: CRC Press. 178 pp., 152 pp.

Owen, I., Syred, N. 1982. Upgrading vortex amplifier performance by matched ejectors. *J. Fluid Control. Fluidics Q.* 14: 29–37

Palde, U. J. 1972. Influence of draft tube shape on surging characteristics of reaction turbines. *US Bur. Reclam. Rep. REC-ERC-72-24*. 29 pp.

Ranque, G. 1933. Expériences sur la détente giratoire avec productions simultanées d'un échappement d'air chaud et d'un échappement d'air froid. *J. Phys. Radium* 4: 112S–15S

Reynolds, A. J. 1961. Energy flows in a vortex tube. *ZAMP* 12: 343–57

Rheingans, W. J. 1940. Power swings in hydroelectric power plants. *Trans. ASME* 62: 171–84

Rietema, K., Verver, C. G., eds. 1961. *Cyclones in Industry*. Amsterdam: Elsevier. 151 pp.

Rubin, S. 1966. Longitudinal instability of liquid rockets due to propulsion feedback (POGO). *J. Spacecr. Rockets* 3: 1188–95

Sarpkaya, T. 1971. On stationary and travelling vortex breakdowns. *J. Fluid Mech.* 45: 545–59

Seybert, T. A., Gearhart, W. S., Falvey, H. T. 1978. Studies of a method to prevent draft tube surge in pump-turbines. *Proc. ASCE-IAHR/AIHR-ASME Joint Symp. Des. Oper. of Fluid Mach., Fort Collins, Colo.*, 1: 151–66

Squire, H. B. 1962. Analysis of the "vortex breakdown" phenomenon. Part 1. In *Miszellanen der Angewandten Mechanik*, ed. M. Schäfer, pp. 306–12. Berlin: Akad.-Verlag. 332 pp.

Stairmand, C. J. 1951. The design and performance of cyclone separators. *Trans. Inst. Chem. Eng.* 29: 356–83

Stefanides, E. J. 1962. Ranque vortex tube cools locomotive drinking water. *Des. News* 1962 (Jan. 24): 14–15

Suter, P., Girsberger, R. 1974. Strömungstechnische Gestaltung des Austrittsstutzens von Axialmaschinen. In *Traupel-Festschrift*, pp. 305–50. Zürich: Juris-Verlag. 377 pp.

Syred, N. 1975. A review of the performance of thin cylindrical vortex devices with reference to power fluidics. *Proc. Inst. Meas. Control Conf. Process Control by Power Fluid. Direct Fluid. Control, Sheffield*, pp. 3-1–3-33

Syred, N., Beér, J. 1972. The damping of precessing vortex cores by combustion in swirl generators. *Astronaut. Acta* 17: 783–801

Takamura, T. 1983. Diffuser and exhaust gas collector arrangement. *US Patent No. 4,391,566*

Taylor, G. I. 1950. The boundary layer in the converging nozzle of a swirl atomizer. *Q. J. Mech. Appl. Math.* 3: 129–39

ter Linden, A. J. 1949. Investigations into cyclone dust collectors. *Proc. Inst. Mech. Eng.* 160: 233–40

Ulith, P. 1968. A contribution to influencing the part-load behavior of Francis turbines by aeration and σ-value. *Proc. IAHR Symp., Lausanne*, pp. B1/1–B1/12

Vonnegut, B. 1950. Vortex thermometer for measuring true air temperature and true air speeds in flight. *Rev. Sci. Instrum.* 21: 136–41

Vonnegut, B. 1954. A vortex whistle. *J. Acoust. Soc. Am.* 26: 18–20

Westley, R. 1955. Optimum design of a vortex tube for achieving large temperature drop ratios. *Coll. Aeronaut., Cranfield, Note No. 30*. 30 pp.

Westley, R. 1957. Vortex tube performance data sheets. *Coll. Aeronaut., Cranfield, Note No. 67* (supplement to *Note No. 30*). 37 pp.

Winkelhaus, L. E., Hebert, D. J., Wigle, D. A. 1946. Hydraulic model studies for the turbines at Grand Coulee Power Plant, Columbia Basin Project. *US Bur. Reclam. Hydraul. Lab. Rep. No. Hyd.-198*

Wormley, D. N. 1976. A review of vortex diode and triode static and dynamic design techniques. *J. Fluid Control. Fluid. Q.* 8: 85–112

Ann. Rev. Fluid Mech. 1987. 19: 53–74

TURBULENT SECONDARY FLOWS

Peter Bradshaw

Department of Aeronautics, Imperial College of Science and Technology, London SW7 2BY, England

1. INTRODUCTION

A more precise, but less concise, title for this article would have been "Turbulent flows with longitudinal (i.e. streamwise) mean vorticity." We review here the current knowledge of conventional three-dimensional ("3D") boundary layers and other thin shear layers, as well as the flows with embedded streamwise vortices comprised in the turbomachinery engineer's definition of secondary flow. Problems of interaction between the turbulent region and the nominally nonturbulent adjacent stream are not discussed.

It has been realized over the last few years that turbulence models originally developed for two-dimensional ("2D") flows may not give good results when extended, by plausible arguments, to 3D flows, even for simple boundary layers or other thin shear layers. This implies that even mild three dimensionality of the mean flow produces significant changes in turbulence structure parameters, since the empirical constants in calculation methods can or should be interpreted as structure parameters. Also, in recent years there has been an increase of interest in predicting embedded-vortex flows: These are nearly always present at the lateral boundaries of 3D thin shear layers—for instance, in flows near the root or tip of a wing or of a turbomachine blade. In both these types of 3D flow, streamwise mean vorticity is generated by lateral deflection or "skewing" of a preexisting shear layer.

The generation of skew-induced streamwise vorticity (Prandtl's first kind of secondary flow) is an essentially inviscid process, given a shear layer with spanwise vorticity: Viscous or turbulent stresses generally cause the streamwise mean vorticity to decay (strictly, to diffuse). The mechanism is the same whether the secondary vorticity is distributed throughout the

0066–4189/87/0115–0053$02.00

shear layer, as in a conventional 3D boundary layer, or concentrated into identifiable streamwise vortices. Of course, once the vorticity has appeared, it is diffused and (generally) reduced by Reynolds stresses and viscous stresses.

In turbulent flows in, say, straight noncircular ducts, streamwise mean vorticity can actually be generated by the Reynolds stresses. This is Prandtl's second kind of secondary flow, referred to here as stress-induced secondary flow. (It is different from *reduction* of skew-induced flow by Reynolds stresses.) This surprising phenomenon has no counterpart in laminar flow, and indeed it cannot be reproduced by any turbulence model that uses an isotropic eddy viscosity. However, stress-induced streamwise vorticity is generally much weaker than vorticity induced by skewing of initially spanwise vortex lines.

The test cases for the 1980–81 AFOSR-HTTM-Stanford conference on complex turbulent flows (Kline et al. 1982) contained only a few discrete-vortex flows, and for administrative reasons it was not possible to include 3D boundary layers on swept wings. Five years later, in spite of a good deal of work now in progress on a wide range of flows, there are still only a few experiments on 3D flows that include detailed measurements of all components of Reynolds stress, let alone the higher order statistics required in modern turbulence models. This review is as much a list of inadequacies as achievements.

We use rectangular Cartesian axes, neglecting metric terms for simplicity. We choose x roughly streamwise (e.g. along the centerline of an aircraft), y normal to the plane of a thin shear layer (e.g. normal to the wing surface), and z spanwise. (These are not streamline coordinates, which are used here only in special cases.) The exact transport equation for the x-wise mean vorticity,

$$U\frac{\partial\Omega_x}{\partial x}+V\frac{\partial\Omega_x}{\partial y}+W\frac{\partial\Omega_x}{\partial z}=\Omega_x\frac{\partial U}{\partial x}+\Omega_y\frac{\partial U}{\partial y}+\Omega_z\frac{\partial U}{\partial z}$$
$$+\left(\frac{\partial^2}{\partial y^2}-\frac{\partial^2}{\partial z^2}\right)(-\overline{vw})+\frac{\partial^2}{\partial y\partial z}(\overline{v^2}-\overline{w^2})+\nu\nabla^2\Omega_x, \tag{1}$$

helps to distinguish the mechanisms of vorticity generation mentioned above:

Mechanism (1)—"skew-induced" generation by quasi-inviscid deflection of existing mean vorticity, corresponding to the second and third terms on the right of Equation (1), found in laminar or turbulent flow;

Mechanism (2)—"strcss-induced" generation by turbulent (Reynolds) stresses, the relative sizes of the normal-stress and shear-stress contributions depending on the orientation of the y and z axes.

We also need to distinguish two types of secondary-flow patterns in the yz-plane:

Type (a)—"cross flow," as in a 3D boundary layer or other thin shear layer, in which the axial vorticity $\partial W/\partial y - \partial V/\partial z$ resides almost entirely in $\partial W/\partial y$.

Type (b)—identifiable streamwise vortices, with $\partial W/\partial y$ roughly equal to $\partial V/\partial z$.

Mechanism (1) can lead to flows of type (a) or type (b): So can mechanism (2), but the latter is normally identified only when it produces identifiable vortices.

An important difference between types (a) and (b) is that while positive streamwise acceleration always stretches and amplifies streamwise vorticity, via the first term on the right side of Equation (1), a shear layer with cross flow is merely thinned—V remaining small—whereas a true vortex "spins up." After infinite stretching, or at infinite Reynolds number, a 3D shear layer becomes a vortex sheet and a true vortex becomes a vortex line.

An obvious effect of streamwise mean vorticity on turbulent eddies is that they are rotated about a streamwise axis. In the case of type (a) secondary flow, the spanwise shear $\partial W/\partial y$ can only rotate the eddies through 90° at most, whereas type (b)—streamwise vortices—can rotate them indefinitely. The central question is, How do 3D effects, such as the rotation of eddies by streamwise mean vorticity, affect the structure parameters of the turbulence (i.e. the empirical constants of the turbulence models)? The main difficulty in experimental work is the need for statistics of the v and w velocity fluctuations [see Equation (1)]; the shear stress in the cross-flow plane, $-\rho\overline{vw}$, is particularly difficult to measure. It became clear at a recent European Mechanics Colloquium ("Euromech 202") that even the laser-Doppler velocimeter, which has been regarded as a very promising technique for complex-flow measurements, cannot at present be relied on to produce accurate vw statistics. The hot-wire technique requires either a three-wire probe, which has some calibration uncertainties, or the ill-conditioned subtraction of conventional cross-wire readings. Large-eddy simulations or full turbulence simulations (Rogallo & Moin 1984) are only just now becoming computationally and financially possible for 3D flows: They will be a useful source of quasi-experimental data in the next few years, but they have not yet had any impact on the problems discussed in this review.

The lecture notes of Bradshaw (1986) provide basic information about 3D turbulent flows omitted from the present review. I also omit topics not directly related to turbulence, notably the topology of surface streamlines and separation lines (Tobak & Peake 1982, Hornung & Perry 1984) and the treatment of viscous/inviscid interactions: In many practical cases, upstream influence (ellipticity) within the turbulent region is important. Again, the references quoted here are only a small selection of the total literature: A list of over 2000 references on turbulence, including 3D flows, is available on disk or tape from the author.

2. SKEW-INDUCED SECONDARY FLOW

2.1 *Three-Dimensional Boundary Layers*

This subject has been reviewed explicitly in the *Annual Review of Fluid Mechanics* by Eichelbrenner (1973) and Cousteix (1986) and has, of course, figured in many other articles in this series (e.g. Landweber & Patel 1979) and elsewhere. A recent review with a very extensive reference list is that of Cebeci (1984). These authors concentrated on computational matters: The conference proceedings edited by Fernholz & Krause (1982) is an introduction to current experimental work.

The simplest 3D flow occurs when an initially two-dimensional boundary layer, developing on a surface near the xz-plane, is deflected sideways by a pressure gradient $\partial p/\partial z$, which itself changes fairly slowly in the z direction, so that z-wise velocity gradients remain small compared with y-wise velocity gradients. The mean vorticity equation (1) shows that x-wise mean vorticity arises because of the "skewing" of the ω_z vortex lines by $\partial U/\partial z$ (third term on the right). Since ω_y is zero, at least initially, $\partial U/\partial z$ is equal to $\partial W/\partial x$, and thus the streamwise vorticity can also be said to arise from $\partial W/\partial x$. For small flow deflection angles, $\tan^{-1}(W/U)$, we have

$$U\frac{\partial \Omega_x}{\partial x} = \Omega_z\frac{\partial U}{\partial z} = \Omega_z\frac{\partial W}{\partial x}, \tag{2a}$$

$$\frac{1}{\Omega_z}\frac{\partial \Omega_x}{\partial x} = \frac{1}{U}\frac{\partial W}{\partial x}, \tag{2b}$$

$$\frac{d}{dx}\frac{\Omega_x}{\Omega_z} \doteq \frac{d}{dx}\frac{W}{U}. \tag{2c}$$

Equation (2c) is the Squire-Winter-Hawthorne (SWH) secondary-flow formula (e.g. Horlock & Lakshminarayana 1973), which simply states that vortex lines are skewed through an angle equal and opposite to that through which the flow has turned. Resolved with respect to axes along and normal to the *local* external-flow direction, this approximate formula yields the outer part of the triangular plot of W against U (Figure 1) popularized by Johnston (1960) but attributed by him to Gruschwitz

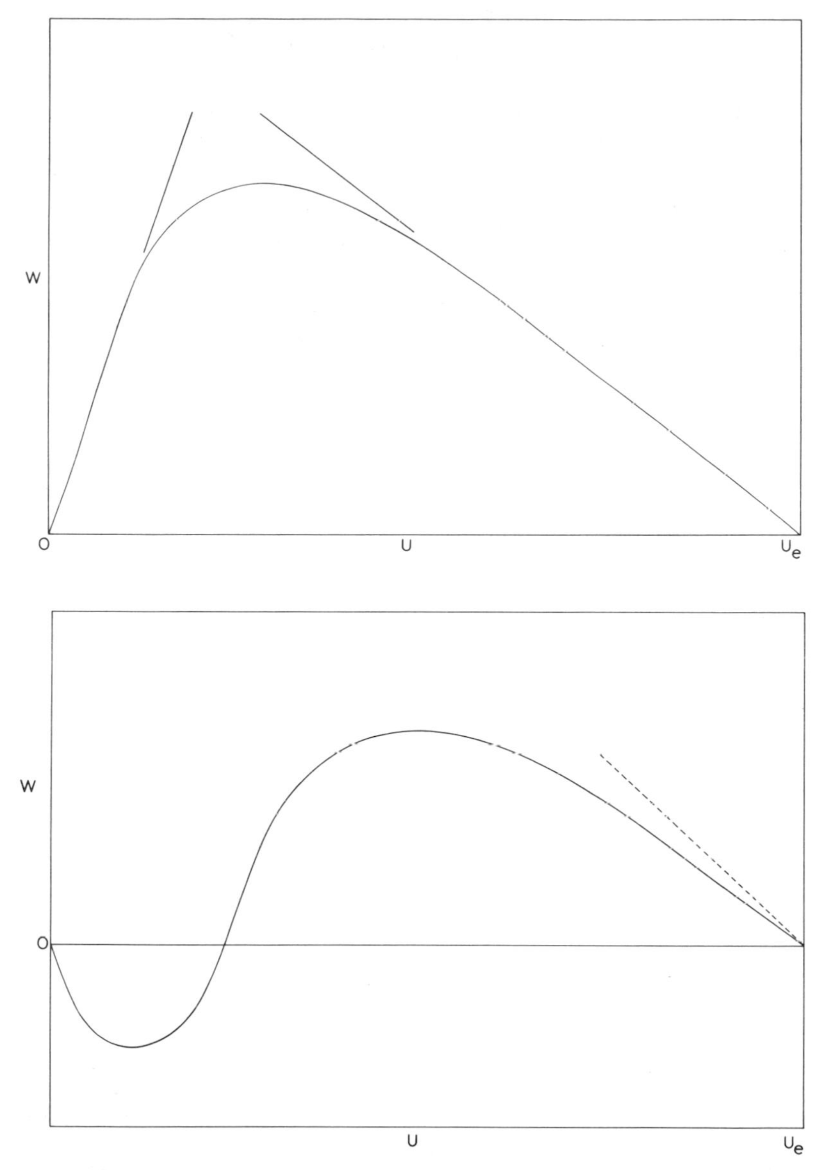

Figure 1 "Triangular" plots of mean-velocity profile (axes aligned with the local external-stream direction). (*Top*) standard profile; (*bottom*) "crossover" profile.

[1935; translated as Grushwitz (sic) 1984]. The inner leg of the triangular plot corresponds to an internal layer of negative longitudinal vorticity $\partial W/\partial y$, required to satisfy the no-slip condition for W at the surface.

The triangular plot illustrates the response of a 3D boundary layer to pressure gradient. In 2D flow, the mean vorticity (approximately $-\partial U/\partial y$), the total pressure, and the Reynolds stresses on a given streamline are unaffected by streamwise pressure gradient as such and therefore change only slowly in the outer part of the flow. However, an internal layer grows out from the surface as the effect of the no-slip boundary condition propagates out through the flow. At the surface, the total pressure equals the static pressure, and there is thus a rise in total pressure near the wall (Figure 2) corresponding to positive $\partial\tau/\partial y$. In the 3D case, consider the boundary layer on an infinite (untapered) swept wing, on which the isobars will be parallel to the generators so that streamwise and spanwise pressure gradients are linked: This simplifies thought but not the physics. The internal layer now has to satisfy the no-slip boundary condition on W as well as on U. The total pressure is again nearly constant on a streamline outside the internal layer (e.g. Johnston 1970), and the SWH formula is a good approximation for the cross flow. Especially in boundary layers that encounter a 3D pressure gradient after developing in two dimensions, the outer part of the triangular plot is usually an admirably straight line, and

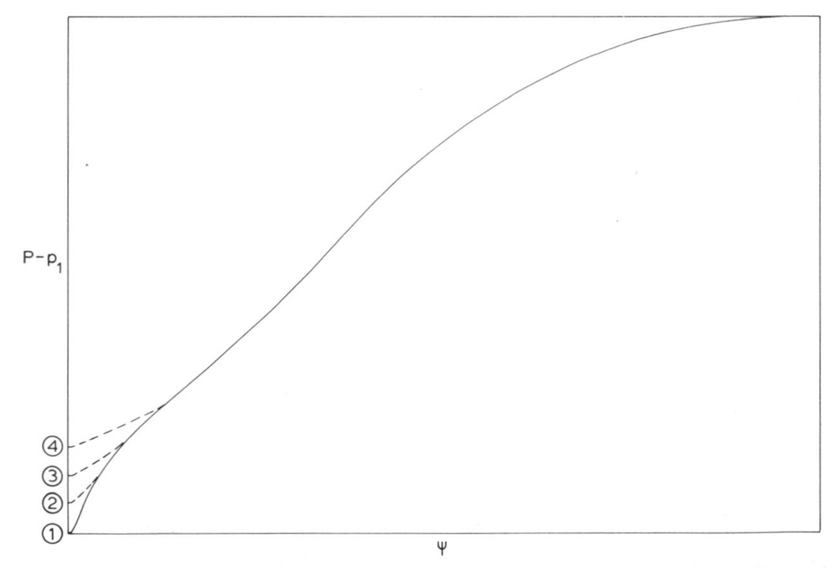

Figure 2 Total pressure versus stream function. (1) Initial station; (2, 3, 4) increasing static pressure.

even departures from the SWH value of dW/dU are small. These small discrepancies are attributed partly to inaccuracy of the simple formula for large turning angles and large velocity defects, as well as, of course, partly to the effects of Reynolds stresses—so a second-order version of the SWH formula would not be worthwhile.

If the spanwise pressure gradient changes sign after some downstream distance, the slow-moving fluid near the surface will again be deflected more strongly than the outer layer, and so-called "crossover" profiles, in which W changes sign, may occur (Figure 1). No new physics is involved, but it is difficult to fit crossover profiles by means of a simple family of shapes, and this complicates calculations by "integral" (generalized Galerkin) methods (Cousteix 1986). These methods are still popular for 3D flow calculations, because they require much less computer time and storage than fully 3D finite-difference methods, but give adequate results for nonseparated flow over wings.

Since, by definition, velocity gradients normal to the surface in a 3D boundary layer are considerably larger than gradients in the x or z directions, only the thin shear layer ("boundary-layer") form of the mean-motion equations is required for discussion of the effects of pressure gradient (although the thin-shear-layer approximation may not be accurate enough for many real-life problems, especially in turbomachinery):

$$\frac{DU}{Dt} \equiv U\frac{\partial U}{\partial x} + V\frac{\partial U}{\partial y} + W\frac{\partial U}{\partial z} = -\frac{1}{\rho}\frac{\partial p}{\partial x} - \frac{\partial \overline{uv}}{\partial y} + \nu\frac{\partial^2 U}{\partial y^2}, \tag{3}$$

$$\frac{DW}{Dt} \equiv U\frac{\partial W}{\partial x} + V\frac{\partial W}{\partial y} + W\frac{\partial W}{\partial z} = -\frac{1}{\rho}\frac{\partial p}{\partial z} - \frac{\partial \overline{vw}}{\partial y} + \nu\frac{\partial^2 W}{\partial y^2}, \tag{4}$$

$$\frac{\partial U}{\partial x} + \frac{\partial V}{\partial y} + \frac{\partial W}{\partial z} = 0. \tag{5}$$

The U-component motion is controlled by $\partial\overline{uv}/\partial y$, as in 2D flow, whereas the W-component motion is controlled by $\partial\overline{vw}/\partial y$. It has been pointed out by several authors that $\overline{uw}$ can be considerably larger than $\overline{vw}$, so that the term $\partial\overline{uw}/\partial x$, omitted from Equation (4), may sometimes be significant. The thin-shear-layer equations are invalid if z-wise gradients are not small, as in embedded vortex flows. The Reynolds-stress transport equations for $\overline{uv}$ and $\overline{vw}$ (again to the thin-shear-layer approximation and neglecting viscous terms that are insignificant outside the viscous sublayer) are

$$\frac{D(-\overline{uv})}{Dt} = \overline{v^2}\frac{\partial U}{\partial y} - \overline{\frac{p'}{\rho}\left(\frac{\partial u}{\partial y} + \frac{\partial v}{\partial x}\right)} + \frac{\partial}{\partial y}\left(\frac{\overline{p'u}}{\rho} + \overline{uv^2}\right), \tag{6}$$

$$\frac{D(-\overline{vw})}{Dt} = \overline{v^2}\underset{(a)}{\frac{\partial W}{\partial y}} - \underset{(b)}{\overline{\frac{p'}{\rho}\left(\frac{\partial w}{\partial y}+\frac{\partial v}{\partial z}\right)}} + \underset{(c)}{\frac{\partial}{\partial y}\left(\frac{\overline{p'w}}{\rho}+\overline{vw^2}\right)}. \tag{7}$$

We see that in each case the "generation" term (a), representing the creation of the Reynolds stress by interaction of existing turbulence with the mean-velocity gradients, is simply $\overline{v^2}$ times the appropriate mean-velocity gradient.

To discuss the behavior of the Reynolds stresses, we consider the simple case of an initially 2D boundary layer ($W = 0$, $\overline{vw} = 0$) on, say, an infinite swept wing. The SWH analysis implies that changes to $\partial U/\partial y$ caused by moderate lateral deflection of the flow are small, so the generation term in the $\overline{uv}$ equation will change only slowly. The change in $\partial W/\partial y$ in the outer layer roughly follows the SWH formula, so that the generation term (a) in the $\overline{vw}$ equation (7) increases roughly proportional to flow deflection angle, which in turn is very roughly proportional to distance downstream from the start of pressure gradient. Since $\overline{vw}$ is initially zero, Equation (7), giving the rate of change of $\overline{vw}$ along a mean streamline, implies that $\overline{vw}$ is very roughly of order (deflection angle)^2. In short, the "generation" terms, at least, suggest that both shear stresses, $-\rho\overline{uv}$ and $-\rho\overline{vw}$, respond rather slowly to pressure gradient. However, the response of vw is not *negligible* to the boundary-layer approximation, and it is therefore unlike the response of uv on a given streamline outside the "internal layer" in a 2D flow. In general, we must expect the other terms in the Reynolds-stress transport equations, namely the pressure-strain "redistribution" terms (b) and the turbulent-transport terms (c), to react at the same rate as the generation terms. Indeed the "rapid" part of the pressure-strain redistribution term in the $\overline{vw}$ equation, which involves $\partial W/\partial y$, can and should be regarded as an immediate response to the distortion of the turbulence by $\partial W/\partial y$. The representation of the pressure-strain terms by plausible combinations of the Reynolds stresses and mean-velocity gradients is the central problem of turbulence modeling. We do not discuss general principles of turbulence modeling here: The reviews by Lakshminarayana (1985) and Cousteix (1986) concentrate on modeling for 3D flows, whereas the books by Rodi (1980) and Bradshaw et al. (1981) are more general introductions.

This discussion of the effects of pressure gradient in a 3D boundary layer is ostensibly only qualitative, but the SWH formula gives an adequate quantitative description of the many laboratory or real-life flows in which a strong transverse pressure gradient is applied to an initially 2D boundary layer. This implies that empirical Reynolds-stress models are seriously tested only within the internal layer, and the influence of Reynolds-stress perturbations in the outer layer may be so small that any model would serve if it did not grossly overestimate the stresses.

2.2 *The Inner Layer and the "Law of the Wall"*

Most current turbulence models for 2D flow reduce to the mixing-length formula

$$-\overline{uv} = l^2(\partial U/\partial y)^2 \tag{8}$$

in a "constant-stress" layer close to the surface, where the shear stress is equal to the wall value and thus provides a velocity scale $u_\tau = (\tau_w/\rho)^{1/2}$, with the length scale being provided by the distance from the wall itself. The mixing-length formula and the law of the wall can be obtained by dimensional analysis alone, without invoking any phenomenological model, correct or otherwise. The mixing-length formula seems to be adequate for the inner layer of all but the most strongly perturbed 2D flows. In three dimensions, most models again reduce, in the inner layer, to a logical extension of the mixing-length formula, which predicts that the shear-stress magnitude is

$$[(-\overline{uv})^2+(-\overline{vw})^2]^{1/2} = l^2[(\partial U/\partial y)^2+(\partial W/\partial y)^2], \tag{9}$$

the shear-stress "vector," with components $(\overline{uv}, \overline{vw})$ supposedly being in the same direction as the velocity gradient vector $(\partial U/\partial y, \partial W/\partial y)$.

A detailed review of law-of-the-wall modifications for 3D flow was given by Pierce et al. (1983), who distinguish "scalar" models (which provide formulae for the variation of velocity magnitude with distance from the surface but do not show how the direction of the velocity varies) and more complex methods that consider the velocity as a vector. Goldberg & Reshotko (1984) discussed asymptotic matching of the inner layer, the outer layer, and an intermediate layer. Perry & Joubert (1965) pointed out that if the 3D mixing-length formula [Equation (9)] is integrated with respect to y, assuming that the shear stress and velocity gradient are in the same direction, the velocity that appears in 2D flow is replaced by the arc distance around the triangular plot (Figure 1). Van den Berg (1975) produced what is probably still the most methodical discussion of the law of the wall, improving on Perry & Joubert's analysis by estimating the mean-flow acceleration, and hence the shear-stress gradients, from a first approximation to the velocity profile. He found that with the x and z axes aligned along and normal to the direction of the surface shear-stress vector, the streamwise velocity component in the sublayer varies linearly with y^+, with a small parabolic correction due to stress gradient that would be present even in 2D flow, whereas the "spanwise" velocity in the sublayer varies as y^{+2}. In both cases, the coefficients of the terms in y^{+2} are the shear-stress gradients estimated as above. If we assume that both sublayer formulas hold out to $y^+ = 11$ and then join the "mixing-length" formulas, then we obtain Equations (19) and (20) of van den Berg's paper as a law of the wall for U and W. In brief, van den Berg's analysis provides a semiempirical estimate of the cross flow at the edge of the sublayer. Unlike the analysis of Townsend (1961) from which it descends, van den Berg's treatment neglects turbulent transport in the y direction, which would lead to a difference between the directions of the shear stress and of the mean-velocity gradient even in the "log law" region. The measurements of Johnston (1970) in a flow with large stress gradients suggest that this may be an important effect.

Some 2D Reynolds-stress models have been extended into the viscous sublayer by adding extra viscosity-dependent terms or by allowing the empirical constants to depend on a local eddy Reynolds number (e.g. Jones & Launder 1972, Patel et al. 1985, So & Yoo 1986). Detailed turbulence measurements in the viscous sublayer are scarce, and the models have to be adjusted to yield the right mean-velocity profile. This reviewer is not aware of any explicit inclusion of 3D effects in the models by making their constants depend on the cross flow. It appears that any improvement on van den Berg's correlation of the sublayer cross flow must await more experimental data.

2.3 *The Outer Layer*

In the outer layer, two serious effects of three dimensionality on the turbulence structure have been identified. First, the direction of the shear-stress vector $(\overline{uv}, \overline{vw})$ lags behind the direction of the "velocity-gradient vector" $(\partial U/\partial y, \partial W/\partial y)$—which remains normal to the mean vorticity vector—as the latter is skewed. This effect is not reproduced by isotropic eddy-viscosity models, and it is underestimated even by current Reynolds-stress transport models adapted from 2D methods—that is, the lag is not entirely the effect of mean-transport terms, but it implies changes in turbulence structure, evidently in the main pressure-strain "redistribution" terms in Equations (6) and (7).

Secondly, the *magnitude* of the shear stress and the ratio of shear stress to turbulent energy are often smaller than predicted by current models, which again implies a change (an increase) in the pressure-strain terms. The most detailed evidence for the shear-stress behavior is provided by the experiments on a simulated 35° "infinite" swept wing by van den Berg et al. (1975), with the turbulence measurements being given by Elsenaar & Boelsma (1974). Bradshaw & Pontikos (1985) made more detailed turbulence measurements in a similar configuration. The last-named experiment stops some distance short of separation, but the structural changes are nearly as strong as in the separating boundary layer of van den Berg et al.: This implies that difficulties in calculating the latter flow cannot be wholly attributed to significant interaction between the boundary layer and the external stream, or to the neglect of pressure gradients normal to the surface. Johnston (1970) found very large lag effects, and a decrease in the stress/energy ratio from a rather high initial value, in the separating boundary layer upstream of a swept, forward-facing step. Similar effects can be seen, less clearly, in several experiments on flow around surface-mounted obstacles (e.g. Anderson & Eaton 1986): Müller (1982) finds no great change in the stress/energy ratio, but streamline divergence was probably significant in his flow. It must be said again that *mean-flow* predictions of boundary layers with large pressure gradient are not a good test of the turbulence model in the outer layer.

Rotta (1979) has suggested an improvement to the conventional method of modeling the "rapid" part of the pressure-strain term (the part depending on the mean-velocity gradient) in 3D flow. If the pressure-strain terms (*b*) in Equations (6) and (7) are formally evaluated from the Poisson equation for the fluctuating pressure, the factors that weight the two components of mean-velocity gradient arise from spatial correlations that are in general anisotropic—in particular, they are asymmetrical about the x and z axes. In other words, if we follow Rotta and model these terms as $K_{xx}\overline{v^2}\partial U/\partial y$ and $K_{zz}\overline{v^2}\partial W/\partial y$, the coefficients K_{xx} and K_{zz} defined by these expressions will differ because of the asymmetry of the correlations. Rotta defines

$$T = (1 - K_{zz})/(1 - K_{xx}) \tag{10}$$

and shows that for a particular eddy-viscosity method (the k, L method, closely related to the k, ε method), the ratio of the eddy viscosity for vw to the eddy viscosity for uv is equal to T. In the experiments commonly used as test cases for calculation methods, 3D pressure gradients are applied to initially 2D boundary layers, so that $\overline{vw}$ grows only slowly, and the measured eddy-viscosity ratio (in streamline coordinates) may be 0.5 or less. Abid & Schmitt (1984) have shown that Rotta's correction, incorporated into several turbulence models (of which the most successful was an algebraic stress model—see Section 3.4), produces great improvements in the prediction of shear-stress direction. T is necessarily adjusted to optimize results, in the hope that its value is nearly universal: The velocity correlations on which it depends have not been measured.

Rotta's analysis identifies a key point in turbulence modeling for 3D flows. Unfortunately, the definition of the two eddy-viscosity components, of the correlation "asymmetry," and of T itself must be referred to arbitrarily chosen x and z axes. Choosing the direction of the shear stress as the x-axis in defining T gives a trivial result; choosing the direction of the mean-velocity gradient causes the correction to have a negligible effect to begin with, and then to run away once there is a significant difference between the direction of the resultant shear stress and that of the mean-velocity gradient; and choosing the mean-velocity direction, as was done by Rotta, violates the principle of Galilean (translational) invariance. In a skewed flow obeying the SWH formula, the angle between the mean-velocity vector and the velocity-gradient vector is just twice the total turning angle, so the effect of T is related to the turning angle. If the flow ceases to skew and relaxes to 2D conditions in a new flow direction, T should have no effect. This suggests that the effect on turbulence structure represented by T is a "fading-memory" integral of the rate of skewing. The work of Anderson & Eaton (1986) on a series of wedge-shaped obstacles, giving different total turning angles and turning rates, should provide useful information.

In principle, Rotta's correction to the pressure-strain redistribution model could improve the prediction of resultant shear-stress magnitude as well, but Rotta's original results show that the decrease in the ratio of shear-stress magnitude to turbulent energy is greatly underestimated unless the cross flow is very large. One difficulty is that positive and negative cross-flow angles should give the same effect, and this implies that the correction must vary as (cross-flow angles)^2 for small cross flow, whereas the structural changes observed in experiments do seem to vary almost linearly with cross-flow magnitude.

2.4 *Three-Dimensional Wakes*

Cousteix & Pailhas (1983) have measured the mean velocities and all the Reynolds stresses in the boundary layer and wake of a 22.5° swept wing, with an upper-surface cross-flow angle of roughly 45° at the trailing edge that produces a strongly asymmetrical wake. Detailed results given by Cousteix et al. (1980) show that the ratio of shear-stress magnitude to turbulent energy increases from about 0.12 near the trailing edge to over 0.2 at one chord downstream, in both cases ignoring values near the wake centerline, where the resultant shear stress necessarily changes sign. The best agreement between their data and a transport-equation method using Rotta's correction was obtained with $T = 1$ (i.e. no correction).

3. VORTEX FLOWS

If the lateral deflection that produces longitudinal vorticity extends for only a small *spanwise* distance, then the longitudinal vorticity becomes concentrated into a vortex. We first discuss trailing vortices in the inviscid flow behind a body, as an introduction, and then the more complicated cases in which the vortex remains embedded in a shear layer.

3.1 *Isolated Vortices*

In most cases, such as that of the vortex from the leading edge of a slender wing, the shear layer that feeds the vortex is sufficiently thin to be idealized as a "vortex sheet." The spiral ingestion of the sheet into the vortex itself is complicated, beautiful (e.g. Gad-el-Hak & Blackwelder 1985), and largely inviscid—in the sense that viscous or turbulent thickening of the vortex sheet does not greatly affect the process. Quasi-inviscid vortex rollup was reviewed by Smith (1986). Here, we discuss the turbulence processes once a concentrated vortex has been formed. Reynolds-stress measurements in isolated vortices are rare: Phillips & Graham (1984) present detailed data suitable for developing and testing calculation methods.

An obvious feature of flow visualizations of vortices, natural and artificial, is a core in which the time-averaged motion is nominally solid-body rotation and in which turbulent mixing is very much reduced. A confusing result is that the dye or smoke used to visualize the flow as a whole may never be mixed into the core: Its absence does not mean that it has been "centrifuged" out, and smoke introduced into the core will certainly stay there, while low-density tracer fluid (heated air) remains outside (Cutler & Bradshaw 1986). The most spectacular evidence of suppression of turbulence mixing is the "wave-guide" property of the vortex core, in which quite large axial velocities can be sustained over long distances (see Figure 5 of Bradshaw 1973).

Now in the measurements of Phillips & Graham and of Cutler & Bradshaw, there are significant Reynolds stresses within the core. Since turbulent mixing is suppressed, the core must be wandering in the yz-plane, thus smearing out the mean-velocity pattern and contributing to the Reynolds-averaged rate of momentum transfer. Longitudinal waves may contribute to $\overline{u^2}$. As in other cases where the eye can distinguish between turbulence and unsteadiness, a Reynolds-stress model that works well in truly turbulent flow may perform poorly (Majumdar & Rodi 1985). Clearly, vortex-core waves and wandering may depend on the initial conditions or on directional fluctuations in the surrounding flow.

3.2 *Junction Flows*

The classical example of generation of concentrated vortices from an oncoming boundary layer is the "horseshoe" vortex around the front of a tall obstacle—a wing, a turbomachine blade, or a building—in a boundary layer (Baker 1980). Again the process of generation is quasi-inviscid

(Morton 1984), with the size and strength of the vortex depending on the leading-edge shape. (If the leading edge is sharp, the concentrated vortex may be close enough to the surface to be rapidly diffused by viscous or Reynolds stresses.) In practice, a complete Navier-Stokes calculation is more convenient than an inviscid calculation embedded in a shear-layer calculation. Briley & McDonald (1982) treated wing/body junctions with swept leading edges in a general nonorthogonal coordinate system that resolved the viscous sublayer. Their predictions did not agree with measurements well downstream of the leading edge: The turbulence model was fairly crude, but false-diffusion errors, notoriously large in recirculating flows, may also have contributed.

Another region of rapid change in wing/body junction flows is at the trailing edge, where the two legs of the horseshoe vortex meet. Flow visualization on the body (e.g. Young 1977, Harsh & Pierce 1985) shows a strong "fishtail" divergence of the streamlines. The size of the turning angles alone indicates that the effect is confined to the slow-moving fluid in and near the viscous sublayer, but the fan can spread out to a distance at least as large as the oncoming boundary-layer thickness. The explanation appears to be the sudden release of the spanwise component of surface shear stress at the wing trailing edge, which allows the two vortices to propel fluid toward the body with consequent divergence near the body surface. This would be a good test of the viscous sublayer model: Apart from this, it is a reminder that surface-flow patterns in three dimensions can be grossly unrepresentative of the behavior of the main part of the boundary layer, let alone that of the external flow.

At high lift, the leg of the horseshoe vortex on the upper ("suction") surface may move out onto the wing or blade, whereas the pressure-surface vortex is deflected downward and, in a turbomachine compressor, commonly reaches the suction-surface junction of the next blade with large effects on machine performance. Moore & Ransmayr (1984), in one of a series of papers on horseshoe vortices in turbomachines, show—at least in the case of typical turbine blades—that the leading-edge shape does not affect the horseshoe vortex or the losses very much; however, Mehta (1984) showed that vortex strength increased with nose bluntness in the nonlifting case. The two results are not necessarily incompatible, because the net circulation of the longitudinal vortex system, due to lift, can in principle greatly exceed the circulation in either leg of the horseshoe vortex as a result of the blade thickness.

One of the most detailed investigations of turbine-blade junction flow is that of Langston et al. (1977; see also Langston 1980), but full turbulence measurements would involve an extremely large amount of work; hence, we must turn to simpler flows to provide test cases for embedded-vortex turbulence models.

Turbulence measurements in a highly simplified wing/body junction were reported by Shabaka & Bradshaw (1981). Other less detailed measurements include those of McMahon et al. (1981; see also Kubendran et al. 1984) and of Nakamura et al. (1982). Nakayama & Rahai (1984) made measurements *behind* the junction between a flat-plate body and a thin, flat-plate nonlifting "wing" whose leading edge was at the front of the body. They found two pairs of (stress-induced) vortices in each corner at the trailing edge. Kornilov & Kharitonov (1984) made measurements in a similar thin-plate rig but with the wing leading edge downstream of that of the body so that a (weak) horseshoe vortex was generated. They found that the vortex leg in each streamwise wing/body corner was gradually replaced by a pair of stress-induced vortices, as found by Nakayama & Rahai.

3.3 *Embedded Vortices*

Single vortices or vortex pairs on substantially flat surfaces, as distinct from streamwise corners, appear in a wide variety of strengths in a wide variety of engineering flows. If the vortices are weak, the velocity scales of the turbulence will be those of the host boundary layer, but if the mean vorticity in the vortex is not small compared with that in the boundary layer, the vortex will tend to impose its own scales.

Isolated embedded vortices can be produced by "vortex generators"—miniature rectangular or half-delta-shaped wings protruding from the main surface—that are frequently used to promote mixing and thus delay separation. Also, any pair of vortices with the "common flow" between them directed toward a solid surface will tend to drift apart under the induced velocity field of the vortex images below the surface. Shabaka et al. (1985) reported detailed turbulence measurements in a weak single vortex that entered a wind-tunnel floor boundary layer from a vortex generator mounted on the floor of the wind-tunnel settling chamber (thus reducing the effect of the total-pressure deficit in the vortex-generator wake). Better mean-velocity measurements are presented by Westphal et al. (1986) using more modern data-logging equipment and a stronger vortex, but their Reynolds-stress measurements exclude $\overline{vw}$: They report the effects of streamwise pressure gradient and point out that although the vortex generally grows in rough proportion to the undisturbed boundary-layer thickness, its cross section becomes noticeably flatter with increasing x, even in zero pressure gradient. Eibeck & Eaton (1985) investigated the effect of a streamwise vortex on heat transfer from a uniformly heated surface: They conclude that a constant Reynolds analogy factor adequately correlates the data, except perhaps near the peak in surface shear stress.

Computations of isolated embedded vortices were made by Liandrat et al. (1986), who compared the measurements of Shabaka et al. with a variety of turbulence models. Even a simple one-equation model, similar to that used by Briley & McDonald (1982), satisfactorily predicted the main features of the flow, provided that the "boundary-layer" thickness used as the turbulence length scale was taken as the local thickness of the shear layer (thus greatly increasing the length scale in the neighborhood of the vortex). However, Reynolds-stress transport models were required to give adequate predictions of normal stresses, and the secondary shear stresses were still underpredicted. Presumably the main need, as in the 3D boundary layers discussed above, is for improved modeling of the pressure-strain term, although the behavior of the triple-product turbulent-transport terms is itself extremely complicated and plays an important part in the Reynolds-stress balance, especially near the vortex perimeter.

Liandrat et al. also compare their predictions with the measurements of Mehta & Bradshaw (1986) on an embedded vortex pair with the "common flow" upward, so that the vortices remain just above (and extract fluid from) the boundary layer. In this case, only one-equation model predictions were reported, and discrepancies were unsatisfactorily large.

In general, it appears that even the most refined turbulence models do not give adequate predictions of the cross-stream intensities and secondary shear stresses that control the diffusion of streamwise vorticity. Fortunately, this diffusion is extremely slow, so that large percentage errors in its prediction may have comparatively little effect on overall flow properties like spanwise-average skin friction and heat-transfer rate. One should not assume that a model that is satisfactory for 3D boundary layers will be able to cope with the more complicated vortex flows.

3.4 *Flow in Curved Ducts*

The prototype flow is that through a curved elbow in a circular pipe or—easier to discuss—a square duct. If the flow is nearly fully developed initially, a vortex pair that virtually fills the duct will develop, with the common flow toward the center of curvature: If the boundary layers on the duct walls are thin, cross flow eventually leads to a vortex pair at the middle of the inside wall. The most recent detailed measurements in turbulent flows are by Chang et al. (1983) in a square duct and Azzola & Humphrey (1984) in a pipe. An earlier study on a curved square duct by Humphrey et al. (1981) did not include measurements of the secondary shear stresses, and there appear to be no measurements of triple products as yet. In addition to the skew-induced vortex pair, bend flows suffer from subsidiary effects that are likely to cause further changes in the turbulence structure. Longitudinal streamline curvature tends to stabilize the turbulence near the inner wall and destabilize that near the outer wall. Also, the flow over the sidewalls is strongly skewed and is therefore likely to suffer the same changes in eddy structure as the 3D boundary layers discussed above.

Iacovides & Launder (1985) report successful computations for flow in a circular-section bend using an algebraic stress model (ASM). As the ASM (e.g. Rodi 1980) is simpler than a full transport model but can in principle provide the anisotropic eddy viscosity essential in 3D flows, some discussion is warranted. In the ASM, rates of mean and turbulent transport of each Reynolds stress are assumed to be proportional to the Reynolds stress being transported: This assumption is plausible but not rigorous. The transport rates can therefore be deduced by full modeling of one Reynolds-stress transport equation, usually that for the turbulent kinetic energy. Pressure-strain terms must be modeled directly in each transport equation. The ASM approximation necessarily causes increasing errors as transport terms become large, so ASM methods are probably not serious contenders for 3D flows whose mean strain rates change rapidly in the stream direction. However, they should be more acceptable for streamwise-vortex flows if cyclic changes of flow properties along the spiral streamlines are not too large.

4. JETS IN CROSS-FLOW

The "bent-over" jet emerging from a wall into a cross stream has been studied by many investigators with reference to jet-lift aircraft and to transpiration cooling. Crabb et al. (1981) give a useful review of previous work and present combined laser-Doppler and hot-wire measurements for two ratios of jet velocity to cross-flow velocity. Their measurements include all three mean-velocity components and all Reynolds stresses except vw, but the flow does not seem to be accurately symmetrical about the center plane. In one of a series of papers, Andreopoulos & Rodi (1984) present hot-wire measurements of all three velocity components and all six Reynolds stresses for jet to cross-flow velocity ratios up to 2. They comment that $\overline{vw}$ is closely related to the gradients $\partial V/\partial x$ and $\partial W/\partial y$ and could probably be simulated in general by an eddy-viscosity model. Andreopoulos & Rodi measured triple products and deduced a turbulent-energy balance in the center plane only. Triple hot-wire probes were used for some of the measurements, including $\overline{vw}$ and the triple products. A recent experiment on a cross-stream jet, providing useful insights but not detailed turbulence measurements, is that of Broadwell & Breidenthal (1984). Shayesteh et al. (1985) present hot-wire measurements of all mean-velocity components, Reynolds stresses, and relevant triple products in a jet that emerges from one wall of a wind tunnel and impinges on the opposite wall. Limited pulsed-wire measurements (not including $\overline{vw}$) were made in the recirculating flow. Related computations of impinging jets in a cross stream were reported by Childs & Nixon (1985) using a two-equation (k, ε) turbulence model. Results were not entirely satisfactory, and later work has shown that even algebraic stress models cannot deal with this rapidly changing flow.

5. STRESS-INDUCED SECONDARY FLOWS

In any turbulent flow that is not exactly two dimensional or exactly axisymmetric, Reynolds-stress gradients can actually produce vorticity. The classical example is the straight rectangular duct investigated by Nikuradse in 1930 (see Schlichting 1979). Contrarotating pairs of vortices appear in each corner, deforming the axial-velocity contours. A very comprehensive review was prepared by Gessner (1979) for the 1980–81 Stanford meeting. He reported that none of the square-duct measurements then available included the yz-plane shear stress $\overline{vw}$, which is of course an essential part of the process of vorticity generation [Equation (1)]. The measurements of Aly et al. (1978) in an equilateral triangular duct include all the Reynolds stresses except $\overline{vw}$: Apparently, the only $\overline{vw}$ measurements in straight noncircular ducts are those of Hooper & Wood (1984) in flow in a parallel rod bundle—a very complicated cross section. Again, there seem to be no systematic triple-product measurements.

Adequate agreement with experiment in rectangular-duct flows has been obtained with a wide range of turbulence models, almost always involving an ad hoc modification to allow for constraint by *two* adjacent walls. This correction is necessarily nonunique, even if it reduces to an established flat-wall proximity correction in 2D flows: Thus the model can be adjusted arbitrarily to optimize agreement with the data set. Of course, a more rational approach cannot be expected without measurements of the terms in the Reynolds-stress transport equations.

Undoubtedly the most spectacular stress-induced secondary flows are found in 3D free jets and wall jets. The free jet from (say) a rectangular nozzle of large but finite aspect ratio will become approximately circular in cross section as it spreads, but then it will continue to change shape: Thus the jet from a horizontally oriented nozzle becomes noticeably vertically oriented before finally relaxing to a circular shape very far downstream. The reason is that turbulent stresses drive the mean flow into approximate axisymmetry while still themselves far from axisymmetric in distribution. A turbulence model that predicts stress-induced secondary flows in a duct reasonably well should also predict the jet flow acceptably, but reminders of the lack of close connection between the mean flow and the turbulence stresses are always salutary. The classical experiment on jets from rectangular nozzles is that of Sforza et al. (1966): There seem to be no modern turbulence measurements.

The 3D wall jet (Launder & Rodi 1983), produced when a jet from (say) a circular nozzle blows tangential to a flat surface, spreads normal to the surface at about the same rate as a 2D wall jet, but the spanwise spreading rate is very much larger, with an included angle of at least 45° between the outer boundaries of the jet. Launder & Rodi suggest that skew-induced vorticity is mainly responsible, but calculations using an isotropic eddy-viscosity model (not capable of producing stress-induced vorticity) predicted a growth-rate ratio of only about 2 : 1, one third of the correct value. It is therefore likely that stress-induced vorticity plays a large part. Now $\overline{w^2}-\overline{v^2}$, which is small in most free-shear flows, is large near a solid surface because v is reduced to zero at the surface [that is, $\partial(\overline{w^2}-\overline{v^2})/\partial y$ is negative, presumably most strongly so near the center of the wall jet, and $\partial^2(\overline{w^2}-\overline{v^2})/\partial y \partial z$ is positive in the positive-z half of the jet, leading to generation of negative ω_x as required].

6. CONCLUSIONS AND OUTLOOK

Most reviews of turbulence end, as they always have, with the statement that we cannot calculate all flows of engineering interest to engineering accuracy. However, the best modern methods allow *almost* all flows to be calculated to higher accuracy than the best-informed guess, which means that the methods are genuinely useful even if they cannot replace experiments. Flows with strong skew-induced streamwise vorticity or flows dominated by stress-induced vorticity are particularly challenging, and the main conclusion of the present review is that we lack basic physical understanding of the effect of mean-flow three dimensionality on turbulence structure. As always, the main question is the behavior of the pressure-strain term in the Reynolds-stress transport equations. If models based directly on these equations fail to reproduce, say, the decline in shear-stress magnitude in a boundary layer with cross flow, it seems unlikely that simpler models will do better.

This reviewer's opinion is that engineering calculations will have to be done by Reynolds-averaged methods for the foreseeable future, but that computer simulations of eddy motion can and will provide the detailed statistics—above all, the pressure-fluctuation statistics—that cannot be adequately measured. The step from two-dimensional to three-dimensional mean flow is as difficult in simulation as in experiment, but it is to be hoped that the input to turbulence modeling from simulations, which has already begun in two-dimensional flows, will soon be extended to three dimensions.

ACKNOWLEDGMENTS

I am grateful to Dr. A. D. Cutler and Dr. V. Baskaran for detailed comments on a draft of this paper, and to Prof. K. D. Papailiou for a very helpful review of the turbomachine problem. Many of the authors cited kindly gave advance details of their work or helpful comments on the draft.

Literature Cited

Abid, R., Schmitt, R. 1984. Critical examination of turbulence models for a separated three-dimensional boundary layer. *Rech. Aerosp.* 1984-6: 1–17

Aly, A. M. M., Trupp, A. C., Gerrard, A. D. 1978. Measurements and prediction of fully developed turbulent flow in an equilateral triangular duct. *J. Fluid Mech.* 85: 57–83

Anderson, S. D., Eaton, J. K. 1986. Experimental study of a pressure-driven, three-dimensional turbulent boundary layer. *AIAA Pap. 86-0211*

Andreopoulos, J., Rodi, W. 1984. Experimental investigation of jets in a crossflow. *J. Fluid Mech.* 138: 93–127

Azzola, J., Humphrey, J. A. C. 1984. Developing turbulent flow in a 180° curved pipe and its downstream tangent. *Rep. LBL-17681*, Lawrence Berkeley Lab., Calif.

Baker, C. J. 1980. The turbulent horseshoe vortex. *J. Wind Eng. Ind. Aerodyn.* 6: 9–23

Bradshaw, P. 1973. Effects of streamline curvature on turbulent flow. *AGARDograph 169, AD-768 316*

Bradshaw, P. 1986. Physics and modelling of three-dimensional boundary layers. In *Computations of Three-Dimensional Boundary Layers Including Separation, Lect. Course 1985–7*, ed. J. Cousteix. Brussels: Von Kármán Inst.

Bradshaw, P., Pontikos, N. S. 1985. Measurement in the turbulent boundary layer on an "infinite" swept wing. *J. Fluid Mech.* 159: 105–30

Bradshaw, P., Cebeci, T., Whitelaw, J. H. 1981. *Engineering Calculation Methods for Turbulent Flow.* London: Academic. 331 pp.

Briley, W. R., McDonald, H. 1982. Computations of turbulent horseshoe vortex flow past swept and unswept leading edges. *Rep. R-82-92001-F*, Sci. Res. Associates, Glastonbury, Conn.

Broadwell, J. E., Breidenthal, R. E. 1984. Structure and mixing of a transverse jet in incompressible flow. *J. Fluid Mech.* 148: 405–12

Cebeci, T. 1984. Problems and opportunities with three-dimensional boundary layers. *AGARD Rep. 719*, Pap. No. 6

Chang, S. M., Humphrey, J. A. C., Johnson, R. W., Launder, B. E. 1983. *Turbulent momentum and heat transport in flow through a 180° bend of square cross-section.* Presented at Int. Symp. Turbulent Shear Flow, 4th, Karlsruhe

Childs, R. E., Nixon, D. 1985. Simulation of impinging turbulent jets. *AIAA Pap. 85-0047*

Cousteix, J. 1986. Three-dimensional and unsteady boundary-layer computations. *Ann. Rev. Fluid Mech.* 18: 173–96

Cousteix, J., Pailhas, G. 1983. Three-dimensional wake of a swept wing. In *Structure of Complex Turbulent Flow*, ed. R. Dumas, L. Fulachier, pp. 208–18. Berlin: Springer-Verlag

Cousteix, J., Aupoix, B., Pailhas, G. 1980. Synthèse de résultants théoriques et expérimentaux sur les couches limites et sillages turbulents tridimensionnels. *ONERA Note Tech. 1980-4*

Crabb, D., Durao, D. F. G., Whitelaw, J. H. 1981. A round jet normal to a crossflow. *Trans. ASME, J. Fluids Eng.* 103: 142–53

Cutler, A. D., Bradshaw, P. 1986. The interaction between a strong longitudinal vortex and a boundary layer. *AIAA Pap. 86-1071*

Eibeck, P. A., Eaton, J. K. 1985. Heat-transfer effects of a longitudinal vortex imbedded in a turbulent boundary layer. *Rep.*, Dept. Mech. Eng., Stanford Univ., Calif.

Eichelbrenner, E. A. 1973. Three-dimensional boundary layers. *Ann. Rev. Fluid Mech.* 5: 339–60

Elsenaar, A., Boelsma, S. H. 1974. Measurement of the Reynolds stress tensor in a three-dimensional turbulent boundary layer under infinite swept wing conditions. *NLR TR 70495 U*

Fernholz, H. H., Krause, E., eds. 1982. *Three-Dimensional Turbulent Boundary Layers*. Berlin: Springer-Verlag. 389 pp.

Gad-el-Hak, M., Blackwelder, R. F. 1985. The discrete vortices from a delta wing. *AIAA J.* 23: 961–62

Gessner, F. 1979. Corner flow data evaluation. *Rep.*, Dept. Mech. Eng., Univ. Wash., Seattle (see also Kline et al. 1982)

Goldberg, U., Reshotko, E. 1984. Scaling and modelling of three-dimensional, pressure driven turbulent boundary layers. *AIAA J.* 22: 914–20

Gruschwitz, E. 1935. Turbulente Reibungsschichten mit Sekundärströmung. *Ing.-Arch.* 6: 355–65

Gruschwitz, E. 1984. Turbulent boundary layers with secondary flow. *NASA TM 77494* (see preceding entry)

Harsh, M. D., Pierce, F. J. 1985. An experimental investigation of a turbulent junction vortex. *Rep. VPI-E-85-4*, Va. Polytech. Inst. State Univ., Blacksburg

Hooper, J. D., Wood, D. H. 1984. Fully developed rod bundle flow over a large range of Reynolds number. *Nucl. Eng. Des.* 83: 31–46

Horlock, J. H., Lakshminarayana, B. 1973. Secondary flows: theory, experiment, and application in turbomachinery aerodynamics. *Ann. Rev. Fluid Mech.* 5: 247–80

Hornung, H. G., Perry, A. E. 1984. Some aspects of three-dimensional separation. *Z. Flugwiss.* 8: 77–87, 155–60

Humphrey, J. A. C., Whitelaw, J. H., Yee, G. 1981. Turbulent flow in a square duct with strong curvature. *J. Fluid Mech.* 103: 443–63

Iacovides, H., Launder, B. E. 1985. ASM predictions of turbulent momentum and heat transfer in coils and U-bends. *Proc. Int. Conf. Numer. Methods in Laminar and Turbul. Flow, 4th, Swansea.* In press

Johnston, J. P. 1960. On three-dimensional turbulent boundary layers generated by secondary flow. *Trans. ASME* 82D: 233–48

Johnston, J. P. 1970. Measurements in a three-dimensional turbulent boundary layer induced by a swept, forward-facing step. *J. Fluid Mech.* 42: 823–44

Jones, W. P., Launder, B. E. 1972. The prediction and laminarization with a two-equation model of turbulence. *Int. J. Heat Mass Transfer* 15: 301–14

Kleine, S. J., Cantwell, B., Lilley, G. M., eds. 1982. *Proc. AFOSR-IFP-Stanford Conf. Computation of Complex Turbul. Flows.* Stanford, Calif: Thermosci. Div., Stanford Univ. 1551 pp.

Kornilov, V. I., Kharitonov, A. M. 1984. Investigation of the structure of turbulent flows in streamwise asymmetric corners. *Exp. Fluids* 2: 205–12

Kubendran, L. R., McMahon, H., Hubbartt, J. 1984. Interference drag in a simulated wing-fuselage juncture. *NASA-CR-3811*

Lakshminarayana, B. 1985. Turbulence modeling for complex flows. *AIAA Pap. 85-1652*

Landweber, L., Patel, V. C. 1979. Ship boundary layers. *Ann. Rev. Fluid Mech.* 11: 173–205

Langston, L. S. 1980. Crossflows in a turbine cascade passage. *Trans. ASME, J. Eng. Power* 102: 866–74

Langston, L. S., Nice, M. L., Hooper, R. M. 1977. Three-dimensional flow within a turbine cascade passage. *Trans. ASME, J. Eng. Power* 99: 21–28

Launder, B. E., Rodi, W. 1983. The turbulent wall jet—measurements and modeling. *Ann. Rev. Fluid Mech.* 15: 429–59

Liandrat, J., Aupoix, B., Cousteix, J. 1986. Calculation of longitudinal vortices imbedded in a turbulent boundary layer. In *Turbulent Shear Flows 5*. Berlin: Springer-Verlag. In press

McMahon, H., Hubbartt, J., Kubendran, L. R. 1981. Mean velocities and Reynolds stresses in a juncture flow. *NASA-CR-164029*

Majumdar, S., Rodi, W. 1985. *Numerical calculations of turbulent flow past circular cylinders.* Presented at Symp. Numer. and Phys. Aspects of Aerodyn. Flows, 3rd, Long Beach, Calif.

Mehta, R. D. 1984. Effect of wing nose shape on the flow in a wing/body junction. *Aeronaut. J.* 88: 456–60

Mehta, R. D., Bradshaw, P. 1986. Longitudinal vortices imbedded in turbulent boundary layers. Part II. Vortex pair with "common flow" upwards. Submitted for publication

Moore, J., Ransmayr, A. 1984. Flow in a

turbine cascade, part 1—losses and leading edge effects. *Trans. ASME, J. Eng. Gas Turbines Power* 106: 400–8

Morton, B. R. 1984. The generation and decay of vorticity. *Geophys. Astrophys. Fluid Dyn.* 28: 277–308

Müller, U. 1982. Measurement of the Reynolds stresses and the mean-flow field in a three-dimensional pressure-driven boundary layer. *J. Fluid Mech.* 119: 121–53

Nakamura, I., Miyota, M., Kushida, T., Kagiya, Y. 1982. Some measurements in the intermittent region of a turbulent boundary layer along a corner. See Fernholz & Krause 1982, pp. 199–209

Nakayama, A., Rahai, H. R. 1984. Measurement of turbulent flow behind a flat plate mounted normal to the wall. *AIAA J.* 22: 1817–19

Patel, V. C., Rodi, W., Scheurer, G. 1985. Turbulence models for near-wall and low Reynolds number flows: a review. *AIAA J.* 23: 1308–19

Perry, A. E., Joubert, P. N. 1965. A three-dimensional turbulent boundary layer. *J. Fluid Mech.* 22: 285–304

Phillips, W. R. C., Graham, J. A. H. 1984. Reynolds-stress measurements in a turbulent trailing vortex. *J. Fluid Mech.* 147: 353–71

Pierce, F. J., McAllister, J. E., Tennant, M. H. 1983. A review of near-wall similarity models in three-dimensional turbulent boundary layers. *Trans. ASME, J. Fluids Eng.* 105: 251–69

Rodi, W. 1980. *Turbulence Models and Their Applications in Hydraulics.* Delft: Int. Assoc. Hydraul. Res.

Rogallo, R. S., Moin, P. 1984. Numerical simulation of turbulent flows. *Ann. Rev. Fluid Mech.* 16: 99–137

Rotta, J. C. 1979. A family of turbulence models for three-dimensional boundary layers. In *Turbulent Shear Flows 1*, ed. F. Durst, B. E. Launder, F. W. Schmidt, J. H. Whitelaw, pp. 267–78. Berlin: Springer-Verlag

Schlichting, H. 1979. *Boundary Layer Theory*. New York: McGraw-Hill. 817 pp.

Sforza, P. M., Steiger, M. H., Trentacoste, N. 1966. Studies on three-dimensional viscous jets. *AIAA J.* 4: 800–6

Shabaka, I. M. M. A., Bradshaw, P. 1981. Turbulent flow measurements in an idealized wing/body junction. *AIAA J.* 19: 131–32

Shabaka, I. M. M. A., Mehta, R. D., Bradshaw, P. 1985. Longitudinal vortices imbedded in turbulent boundary layers. Part 1. Single vortex. *J. Fluid Mech.* 155: 37–57

Shayesteh, M. V., Shabaka, I. M. M. A., Bradshaw, P. 1985. Turbulence structure of a three-dimensional impinging jet in a cross stream. *AIAA Pap. 85-0044*

Smith, J. H. B. 1986. Vortex flows in aerodynamics. *Ann. Rev. Fluid Mech.* 18: 221–42

So, R. M. C., Yoo, G. J. 1986. A full Reynolds-stress closure for low Reynolds number turbulent flows. *NASA CR-3994*

Tobak, M., Peake, D. J. 1982. Topology of three-dimensional separated flows. *Ann. Rev. Fluid Mech.* 14: 61–85

Townsend, A. A. 1961. Equilibrium layers and wall turbulence. *J. Fluid Mech.* 11: 97–120

van den Berg, B. 1975. A three-dimensional law of the wall for turbulent shear flows. *J. Fluid Mech.* 70: 149–60

van den Berg, B., Elsenaar, A., Lindhout, J. P. F., Wesseling, P. 1975. Measurements in an incompressible three-dimensional turbulent boundary layer, under infinite swept-wing conditions, and comparison with theory. *J. Fluid Mech.* 70: 127–48

Westphal, R. V., Eaton, J. K., Pauley, W. R. 1986. Interaction between a vortex and a turbulent boundary layer in a streamwise pressure gradient. In *Turbulent Shear Flows 5*. Berlin: Springer-Verlag. In press

Young, A. D. 1977. Some special boundary layer problems. *Z. Flugwiss.* 1: 401–14

Ann. Rev. Fluid Mech. 1987. 19 : 75–97

UPSTREAM BLOCKING AND AIRFLOW OVER MOUNTAINS

Peter G. Baines

CSIRO, Division of Atmospheric Research, Private Bag No. 1, Mordialloc, Victoria 3195, Australia

INTRODUCTION

The reality of upstream blocking in stratified flows has been recognized for many years. If a stratified flow with Brunt-Väisälä frequency N ($N^2 = -g/\rho \; d\rho/dz$, where ρ is the fluid density, g the acceleration due to gravity, and z the vertical coordinate) is set in motion with mean velocity U over a (two-dimensional) obstacle of height h, then naive energy arguments (and common sense) indicate that if Nh/U is sufficiently large, fluid near the ground would be blocked on the upstream side and not flow over the obstacle. Casual observations and "folklore" have long indicated that this phenomenon is common near mountain ranges in the atmosphere. However, the nature and mechanics of how it occurs have only recently become clear. It is now known that upstream blocking in large-Reynolds-number flows propagates as a wave phenomenon, generated by nonlinear effects over the topography. These waves may be linear or nonlinear depending on circumstances, and they propagate primarily as "columnar" motions, meaning that they permanently alter the density and horizontal velocity profiles as they pass through the fluid ahead of the obstacle. Blocking occurs when these changes reach sufficient amplitude. Since they alter the upstream conditions, the understanding of these upstream disturbances caused by the obstacle is a prerequisite for calculating the steady-state flow over an obstacle, regardless of the other details of the flow. These effects generally depend on the topography being approximately two-dimensional (2D) with sufficiently large height. They are common in geophysical situations such as fjords, estuaries, and in the atmosphere.

Since blocking is primarily a two-dimensional stratified phenomenon, in this review we exclude the effects of rotation and are concerned with

0066–4189/87/0115–0075$02.00

topography that is at least nearly two dimensional. The literature on stratified flow over topography is quite large, but most of the earlier studies were focused on downstream phenomena such as lee waves and windstorms, rather than upstream effects. This article is primarily concerned with the latter effects, and downstream-flow properties are only discussed insofar as they relate to upstream phenomena. From this viewpoint, the study of the subject began with the pioneering work of Long (1954, 1955). Since then, the state of the subject has been reviewed by Long (1972) and, for laboratory experiments, by Baines & Davies (1980).

The character of the flow will depend on the mean density stratification of the fluid, and here there are two main considerations. Firstly, the stratification may take the form of a number of homogeneous layers, or the density may vary continuously with height. In the latter case a layered model can be used as an approximation, although many layers may be needed. Secondly, the fluid depth may be finite or infinite. In the finite-depth case the stratified fluid is bounded above by a rigid horizontal boundary or an infinitely deep homogeneous layer, so that all upward-propagating energy is reflected downward; the vertical spectrum of linear internal waves consists of discrete modes. In the infinite-depth case, wave energy may propagate upward out of the region of interest without any downward reflection. This may be achieved by a fluid that is effectively infinitely deep or that has a region which absorbs and dissipates internal wave energy above some sufficiently high level. The vertical spectrum of internal wave energy is continuous, with no downward energy propagation. The behavior of finite- and infinite-depth systems is quite different in general. In particular, finite-depth systems contain an additional parameter—the total depth of the stratified fluid. Furthermore, the linearized solutions (for flow over obstacles with small h) become singular for layered and finite-depth systems when the speed of an internal wave mode is zero relative to the topography, whereas this does not occur for infinite-depth systems without trapped modes.

Both finite- and infinite-depth continuously stratified systems may contain a critical layer, which in the present context implies a level in the flow where the (initial) mean velocity of the fluid is zero relative to the topography. Critical layers introduce considerable complications, and in order to focus on the essentials of upstream blocking we assume that they are initially absent in the flows discussed here. However, topographic disturbances may themselves produce *local* critical levels.

A simple criterion for upstream blocking can be obtained from the following energy argument, due to Sheppard (1956). One may liken stratified fluid approaching an obstacle to balls being rolled uphill. Relative to neutral stratification, an approaching fluid particle must overcome a potential-energy deficit due to the stratification if it is to surmount the barrier. For continuously stratified fluid, a fluid particle with velocity U will not have sufficient kinetic energy to reach a height h if

$$\tfrac{1}{2}\rho U^2 < g \int_0^h (h-z)\left(-\frac{d\rho}{dz}\right) dz.$$

In the particular case where U and the Brunt-Väisälä frequency N are constant with height, this relation gives the criterion for blocking as

$$\frac{Nh}{U} > 1.$$

Laboratory experiments with 3D axisymmetric and near-axisymmetric obstacles (Hunt & Snyder 1980, Snyder et al. 1985) show that this criterion agrees closely with observations taken on the centerline. However, this agreement must be regarded as almost a coincidence, since the theoretical derivation ignores the effects of neighboring fluid particles through the pressure term. For two-dimensional topography this energy argument is not consistent with observations, and the value of Nh/U required for blocking is closer to 2, as shown below.

The most common dimensionless number in this topic is the Froude (pronounced "Frood") number F. Unfortunately, this name is used for different quantities in different circumstances by different people. In the flow of homogeneous fluid with a free surface, F is defined to be $U/(gl)^{1/2}$, where l may be an obstacle length L, the fluid depth D, or (conceivably but rarely) an obstacle height h. So defined, F may represent any one of three parameters, and these have very different physical significance. The first [$U/(gL)^{1/2}$] was used extensively by William Froude and relates to wave drag. The second has been commonly termed the Froude number since the work of Moritz Weber (Rouse & Ince 1957) and is the ratio of a fluid speed to a linear wave speed. The use of "Froude number" for both terms must be regarded as accepted terminology. For the case of continuous stratification with constant N, we have the corresponding parameters U/NL, U/ND and U/Nh. All three (plus their squares and reciprocals and suitable constant multiples) have been termed the "Froude number" by various authors. This proliferation of the term has caused unnecessary confusion because, again, these three parameters have very different physical significance: The first relates to internal wave drag, the second is the ratio of a fluid speed to a wave speed, and the third relates to nonlinear wave steepening and upstream blocking. By analogy with free-surface flows it may (regrettably) be regarded as accepted practice to term U/NL and U/ND Froude numbers, but there seems to be little sense[1] or justification for using the same appellation for U/Nh (although the present author is as guilty of this in the past as anyone else). I suggest that it is more appropriate to write this number as Nh/U, and we leave it nameless with no symbol in this article. A suitable name might be "Nhu."

In the following sections we consider the nature of the blocking phenomenon in systems of increasing complexity. We begin with a single homogeneous layer with a free surface, then proceed to multilayer systems, and finally discuss continuously stratified systems of finite and infinite depth. Most theoretical studies have assumed that the obstacle has a long horizontal length scale, so that the flow is mostly hydrostatic (apart from certain situations mentioned below); this provides a substantial simplification of the equations, and the flows calculated should at least be representative of the character of flow over shorter obstacles, because the essential nonlinearities are retained.

SINGLE LAYER

We consider the flow of a single layer over a long (slowly varying) obstacle, so that the flow is mostly in hydrostatic balance. We also note that the equations governing hydrostatic flow of a single layer are the same as those for hydrostatic flow of a two-layer system with an infinitely deep inert upper layer, if g is replaced by $g' = g(\rho_1 - \rho_2)/\rho_1$.

This single-layer system provides examples of the two main types of nonlinear disturbances produced by topography in finite-depth flows. The first of these is the *hydraulic jump*, which is the end result of a steepening process due to nonlinear advection. For many purposes these jumps may be regarded as traveling discontinuities that do not change their shape or properties with time; their detailed structure will depend on a balance between nonlinear steepening and a combination of linear dispersion, dissipation, and wave breaking. The second type of disturbance is the *rarefaction*—a term borrowed from gasdynamics, but here the word implies that the *disturbance* is being rarefied, rather than the fluid density. This type occurs when the trailing part of the disturbance travels more slowly than the leading part (conversely to the hydraulic-jump case), so that nonlinear advection causes the disturbances to become progressively more stretched out as time passes. Both of these types of disturbance are important for stratified flows over obstacles in general.

The effects of two-dimensional topography on a single layer have been

[1] This point will be discussed in more detail in the monograph "Topographic effects in stratified flows" by the author.

investigated by Long (1954, 1970, 1972—theory and experiments) and independently by Houghton & Kasahara (1968—theory and numerical experiments). Their results have been summarized in a unified form in Baines & Davies (1980). If a fluid layer of depth d_0 is impulsively set into motion with velocity u_0 in the presence of an obstacle of maximum height h, the resulting flow may be characterized by two dimensionless parameters—a Froude number $F_0 = u_0/(gd_0)^{1/2}$ and $H = h/d_0$. From the equations of momentum and mass conservation, one may infer that the final steady state depends on F_0 and H, as shown in Figure 1. The equations for the various curves are

$$\mathrm{A'B'+A'E'}: H = 1-\frac{3}{2}F_0^{2/3}+\frac{1}{2}F_0^2,$$

$$\mathrm{A'F'}: H = \frac{8(F_0^2+1)^{3/2}+1}{16F_0^2}-\frac{1}{4}-\frac{3}{2}F_0,$$

$$\mathrm{B'C'}: F_0 = (H-1)\left(\frac{1+H}{2H}\right)^{1/2}.$$

To the left of curve F′A′B′, where the flow is either supercritical ($F_0 > 1$) or subcritical ($F_0 < 1$), the flow upstream and downstream is the same as the initial undisturbed flow (apart from transients), and the flow over the obstacle is given by the Bernoulli equation. To the right of B′C′ the obstacle height is sufficiently large to completely block the flow. When the flow is partially or totally blocked, a hydraulic jump propagates upstream to infinity, reducing the incident mass flux and altering the upstream fluid

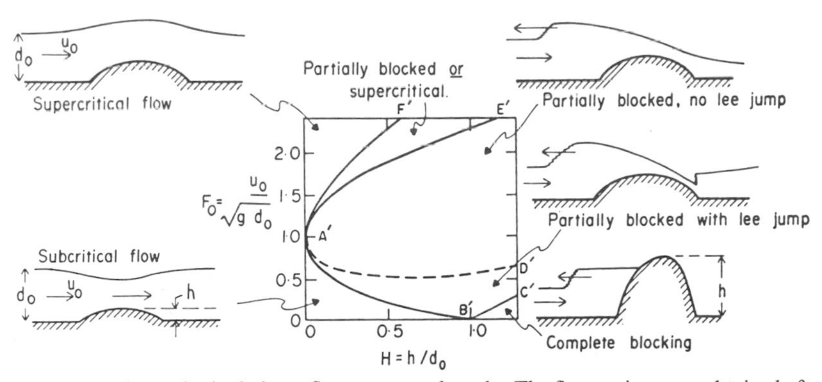

Figure 1 Hydrostatic single-layer flow over an obstacle: The flow regimes are obtained after an impulsive start from rest for various values of F_0 and H, where h is the maximum obstacle height and u_0 is the speed of the obstacle relative to the initial undisturbed stream, which has depth d_0.

velocity u and layer thickness d. An equation relating jump speed to the change in conditions across the jump may be derived and used to obtain the properties of the overall flow. In the partially blocked case, flow over the obstacle crest is controlled by the local condition $F = u/(gd)^{1/2} = 1$. On the downstream side, a hydraulic jump may be attached to the obstacle (below A′D′) or swept downstream (above A′D′): farther downstream, a rarefaction (simple wave) disturbance connects the flow to the original undisturbed state. In the region E′A′F′ the flow may be either partially blocked or supercritical, depending on the initial conditions, so that a hysteresis phenomenon exists in this system. The existence of these double equilibria has been verified numerically by Pratt (1983) and experimentally by Baines (1984).

When two long obstacles are present in two-dimensional flow, the hydrostatic long-wave model may not be applicable. If the steady-state flow for a single obstacle is everywhere subcritical or supercritical, the steady-state flow pattern for each of two long obstacles of the same height will be the same as that for a single obstacle. However, if upstream blocking occurs, the long-wave theory may yield no sensible answer; in these cases nonlinear wave trains are observed in the region between the obstacles (Pratt 1984). Various flow regimes obtained experimentally for a range of heights of two obstacles are shown in Figure 2. Apart from possible wave breaking, the observed flows were all completely steady. This phenomenon may be interpreted, at least in part, with the theory of Benjamin & Lighthill (1954). We define the mass flux Q, energy R, and momentum flux S of a uniform stream of velocity u_1 and depth d_1, taking density as unity, by $Q = u_1 d_1$, $R = \frac{1}{2}u_1^2 + gd_1$, $S = u_1^2 d_1 + \frac{1}{2}gd_1^2$. Then if R_c and S_c denote the values of R and S for a *critical* stream ($F = u_1/\sqrt{gd_1} = 1$) of given volume flux Q, the possible values of R and S for steady flows on this stream are given in Figure 3. The upper boundary of the cusp represents subcritical uniform stream flows ($F < 1$), the lower boundary represents supercritical uniform flows ($F > 1$) and solitary waves, and the region in between represents

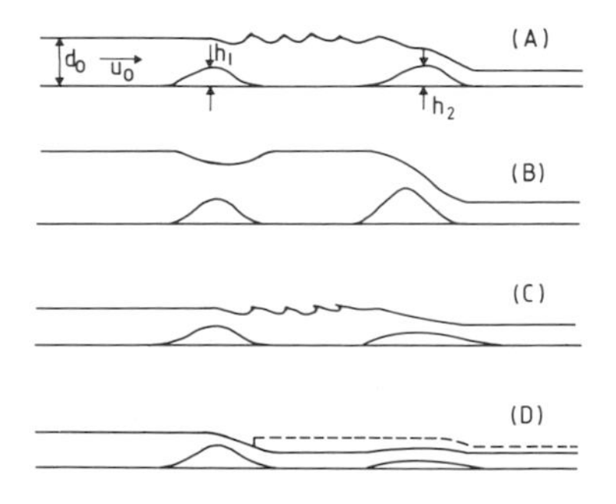

Figure 2 Sketches of experimentally found flow regimes for a single layer over two obstacles (from Pratt 1984). (*a*) $h_1 \approx h_2$; laminar lee waves between obstacles. (*b*) $h_1 < h_2$; long-wave subcritical flow between obstacles. (*c*) $h_1 > h_2$; breaking lee waves. (*d*) $h_1 \gg h_2$; long-wave supercritical flow between obstacles (solid line) or containing hydraulic jump (dashed line).

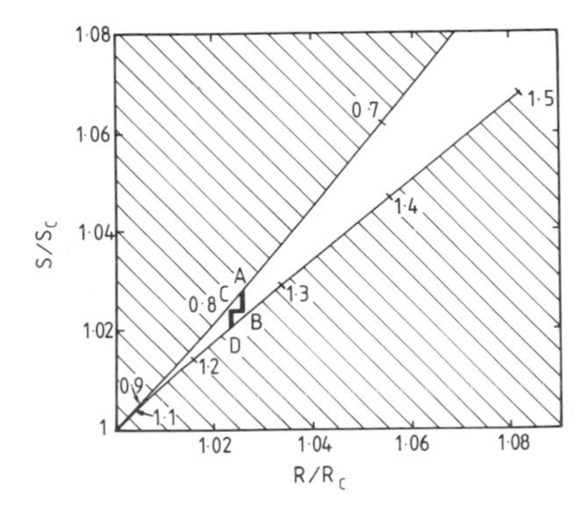

Figure 3 The energy density (R)–momentum flux (S) diagram for possible steady states of a single layer with given mass flux Q. For critical flow, we have $R = R_c$, $S = S_c$. Numbers on the cusp denote values of F (from Benjamin & Lighthill 1954).

flow with cnoidal wave trains. Flow over an obstacle causes a decrease in S equal to the (inviscid) drag force; hence, in passing over an obstacle, the point on the diagram representing the stream flow moves downward from the upper branch of the cusp. For a single obstacle it may reach the lower branch, but, with a second obstacle, in the cases of interest it will only traverse part of this gap (AB in Figure 3), giving a cnoidal wave train downstream of the first obstacle. If these waves are large enough to break, then a decrease in R will result (BC), and then a further decrease in S (CD) at the second obstacle. The details of these changes in S may be dependent on the spacing between the obstacles and their shape; the phenomenon needs further study. If several obstacles are present in the flow, we may expect a succession of such zigzags in the R-S plane, so that R and S decrease toward their minimum values R_c and S_c and the downstream flow tends toward criticality. On the other hand, if any one obstacle blocks the flow, it is blocked everywhere.

For three-dimensional (3D) topography (for example, a 3D barrier in a channel) the flow will be totally blocked if and only if the barrier is higher than the 2D blocking height (given in Figure 1) continuously across the channel. Also, if the channel is narrow relative to the longitudinal length scale, the flow may be controlled by a critical condition that depends on the topographic height profile at the "minimum gap" (rather than a single height); we discuss this point in a broader framework below.

TWO OR MORE LAYERS

The upstream effects of two-layer flow have been investigated numerically by Houghton & Isaacson (1970) and experimentally by Long (1954, 1974) and the author (Baines 1984). The last paper gives a comprehensive description of the various flow types that occur with two immiscible fluids when the flow is commenced from a state of rest, so that the velocities of the two layers are initially equal. The experiments have been carried out with moderately long obstacles (with length comparable to the depth), and the observations have been satisfactorily compared with results from a hydraulic two-layer model (using mass and momentum equations for each layer). The observed upstream disturbances may take one of three forms, as follows. (*a*) A hydraulic jump (Figure 4*a*), similar in character to those observed in single-layer flows. The jump is undular at small amplitudes; at large amplitudes the interface becomes turbulent at and on the lee side of the crests due to Kelvin-Helmholtz instability. (*b*) A limiting bore plus a rarefaction (Figure 4*b*). As the amplitude of a bore and the downstream lower-layer depth are increased, the effect of the upper-layer thickness becomes more important; the speed of the bore tends to a maximum value at a particular amplitude, and the energy loss across the bore decreases to zero at (or very near) this same point. This bore of maximum amplitude and zero dissipation is termed a "limiting bore," and it consists of a monotonic increase in the lower-layer depth, which still propagates without changing shape. If the downstream lower-layer depth is forced to increase further, this must result in a rarefaction that propagates more slowly than the bore. (*c*) A pure rarefaction (Figure 4*c*). If the lower-layer depth is initially greater than or equal to a value that is approximately half the total depth (and depends on ρ_2/ρ_1), an increase in the lower-layer depth is propagated as a rarefaction only.

When hydraulic jumps are present, the hydraulic model requires a relationship between the jump speed and the conditions upstream and downstream of it. In order to obtain this relationship for multilayered

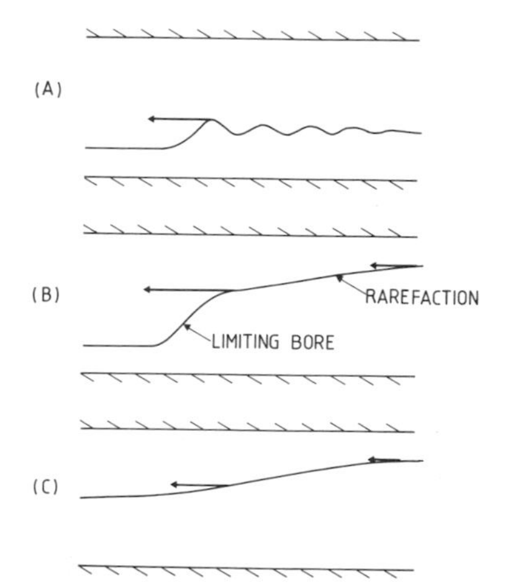

Figure 4 Examples of the types of non-linear disturbances in two-layer flow. (*a*) A hydraulic jump, which propagates at constant speed without changing shape. (*b*) A limiting bore and rarefaction; the limiting bore is a bore of maximum amplitude that propagates at constant speed without changing shape, and for the following rarefaction, the leading part propagates faster than the trailing part. (*c*) A pure rarefaction. The arrows represent relative propagation speeds of the interface height.

flows, an assumption about the flow within the jump is required. One assumption that meets the requirements is that the flow within the jump is hydrostatic, and this is equivalent to the assumptions used by Yih & Guha (1955), Houghton & Isaacson (1970), Long (1970, 1974), Su (1976), and Baines (1984). However, it is obviously not strictly correct, and Chu & Baddour (1977) and Wood & Simpson (1984) have suggested that for two-layer systems, it may be replaced by an assumption of conservation of energy in the contracting layer in the jump. In cases where the two criteria have been compared with observations (Wood & Simpson 1984, Baines 1984), the difference between them is small and the comparisons are inconclusive, and hence the question of the most appropriate assumption is still open.

We now consider the results for flow between rigid upper and lower boundaries with $(\rho_1 - \rho_2)/\rho_1 \ll 1$, starting from a state of rest. The resulting flow may be specified by three parameters F_0, H, and r, where

$$F_0 = \frac{u_0}{c_0}, \qquad c_0^2 = \frac{g(\rho_1 - \rho_2)}{\dfrac{\rho_1}{d_{10}} + \dfrac{\rho_2}{d_{20}}}, \qquad H = \frac{h}{D}, \qquad r = \frac{d_{10}}{D},$$

where u_0 is the initial fluid velocity relative to the topography, ρ_1, d_{10} and ρ_2, d_{20} denote the density and initial thickness of the lower and upper layers, respectively, h is the maximum height of the obstacle, and the total depth $D = d_{10} + d_{20}$. Figure 5 shows the model results in terms of F_0, H for $r = 0.1$, 0.5. For $r = 0.1$ the diagram is very similar to Figure 1 for a single layer when $F_0 \lesssim 1.4$. However, when $F_0 > 1.4$ the upstream disturbance may be sufficiently large for the flow to become critical immediately upstream of the obstacle (the dashed line in Figure 5*a*); this marks an upper limit to the magnitude of the upstream disturbance, which does not increase further if H is increased. On part of this curve the upstream bore has reached its maximum amplitude, and a small-amplitude rarefaction follows it. Flow states with upstream bores in the two-state (hysteresis) region may not be realizable experimentally because of interfacial friction (Baines 1984). For $r = 0.5$, on the other hand, no upstream jumps occur, and the only upstream disturbances are of the rarefaction type. As r increases from 0.1 to 0.5, the F_0-H diagram evolves continuously from Figure 5*a* to Figure 5*b*.

Mathematical analyses of the nonlinear region near resonance ($F_0 \sim 1$) have recently been carried out by Grimshaw & Smyth (1986) and by W. K. Melville & K. R. Helfrich (private communication). These studies enable the fluid response for fairly long obstacles with small H to be calculated as the solution of a forced Korteweg–de Vries (KdV) equation;

an extended KdV (EKdV) equation incorporating cubic nonlinearities is required to model two-layer effects, such as limiting bores. An example of the results obtained from the KdV equation is shown in Figure 6. Results are in qualitative agreement with laboratory observations for small r, and Melville & Helfrich obtained reasonable detailed quantitative agreement with the EKdV equation for larger r in some cases.

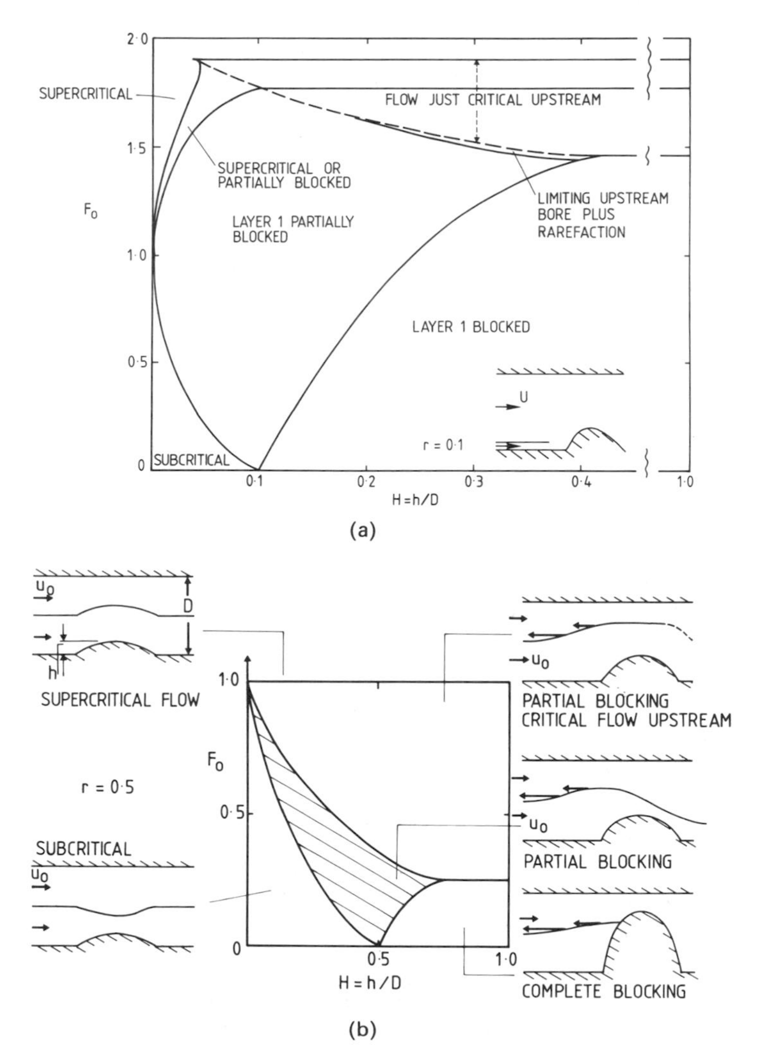

Figure 5 Flow-regime diagrams in terms of F_0, H for two-layer flows: (*a*) $r = 0.1$; upstream disturbances are mostly hydraulic jumps (cf Figure 1). (*b*) $r = 0.5$; upstream disturbances are all rarefactions.

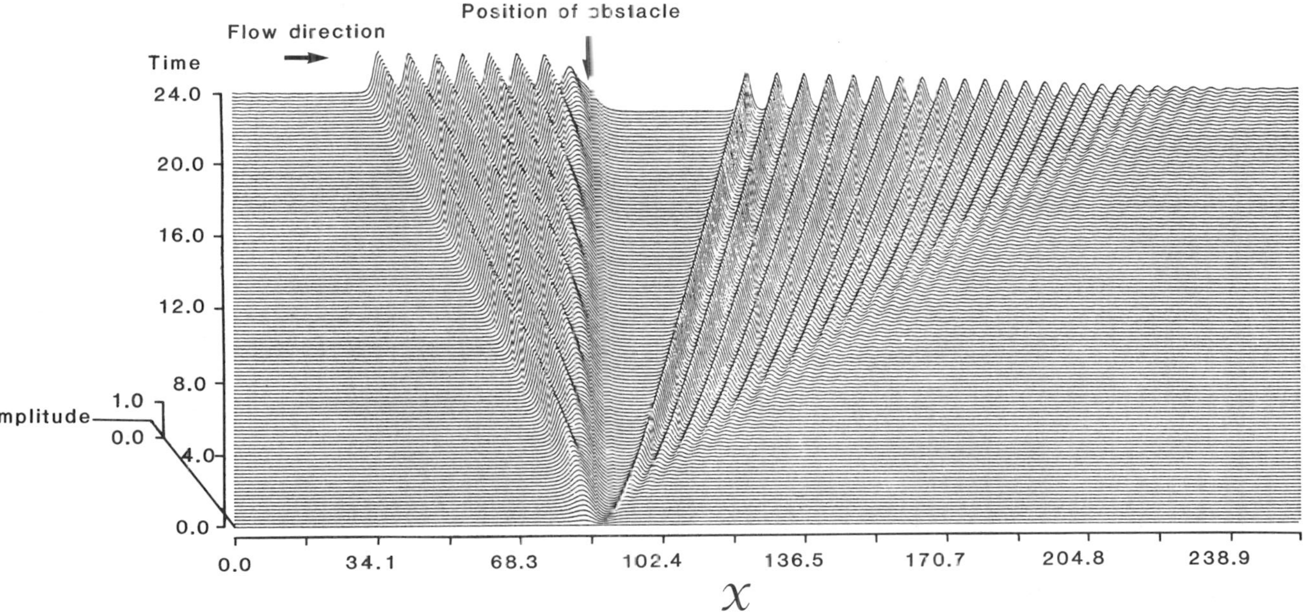

Figure 6 An example of a solution of the forced Korteweg–de Vries equation at resonance, showing an upstream (undamped) undular bore that parallels the two-layer experiments with a thin lower layer ($r \ll 1$). For larger r, cubic nonlinearities must be included to obtain agreement with observations (from Grimshaw & Smyth 1986).

The hydraulic model may be extended to systems with more than two layers (and hence with more than one internal mode) by the following procedure (P. G. Baines, submitted for publication, 1986). It may be shown (Benton 1954, Lee & Su 1977) that at the crest of an obstacle, either the horizontal gradients of all interfaces must vanish or else *one* mode must be critical there (i.e. its propagation speed relative to the topography must be zero). If one particular mode is critical at the crest and the obstacle height is increased by a small amount, the flow may adjust to retain this critical condition by sending a small-amplitude columnar disturbance upstream that has the structure of the critical mode. This disturbance will have the character of a jump if $dc/da > 0$ and a rarefaction if $dc/da < 0$, where c is the linear wave speed propagating against the upstream flow and a is the amplitude of the *preceding* columnar disturbances. By these means, it is possible to construct the F_0-H diagram for any number of layers, although the procedure becomes more difficult as H increases and the upstream disturbances become more complex. In particular, criteria for blocking of the lowest layer may be obtained. Figure 7 shows the F_0-H diagram for three layers between rigid boundaries, originally of equal thickness and with equal density increments. Up to the point where blocking of the lowest layer begins, the upstream disturbances are pure rarefactions. Treatment of the flow with a blocked layer present is

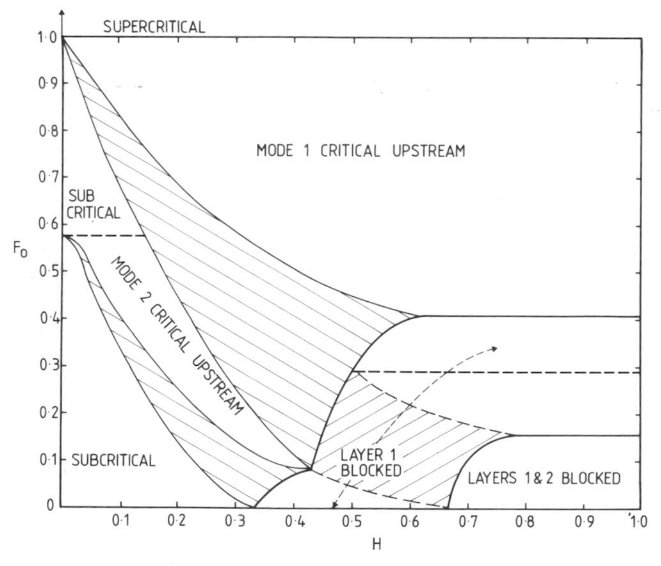

Figure 7 F_0-H regime diagram for three-layer flow, with layer velocities and thicknesses initially equal. In the shaded regions the upstream disturbance of the appropriate mode increases in amplitude as H increases.

more complicated, and the details are given in P. G. Baines & F. Guest (submitted for publication, 1986).

For hydraulic flows (long obstacles) governed by a critical condition at $dh/dx = 0$, in many cases (such as the one- and two-layer systems described above) the upstream flow may be determined independently of the downstream flow, although this will not be true in general. Downstream flows are more complicated and are less well understood. Smith (1976) observed lee waves behind short obstacles in subcritical two-layer flows and found that the wave amplitudes were substantially larger than those predicted by linear theory. For long obstacles in two-layer flow with $r \ll 0.5$, when upstream bores are present the lee-side flow may contain a stationary jump or the jump may be swept downstream. When the flow is critical just upstream, it must be supercritical over the obstacle and then adjust to downstream conditions by a sudden descent of the interface on the lee side to another supercritical state through what is sometimes called a "hydraulic drop" (Baines 1984). These flows may be affected by lee-side flow separation, which is conspicuously present in some cases when the upstream flow is supercritical. Lawrence (1985) has made a detailed study of downstream flow features using miscible fluids and a large flume that permits the flow to exist in a steady state for long periods. In particular, the nature of mixing processes and their dependence on various features of flows with a downstream hydraulic jump have been explored; mixing was observed to be due primarily to Kelvin-Helmholtz billows in the region of maximum shear upstream of the hydraulic jump, rather than to processes within the jump itself. Armi (1986) has made an experimental study of two-layer flow through horizontal contractions.

There do not appear to have been any relevant studies of layered flows with three-dimensional barriers, or with two or more barriers.

CONTINUOUSLY STRATIFIED FLOW—FINITE DEPTH

One might expect that the upstream phenomena present in layered flows would have their analogues in continuously stratified fluid. This is in fact the case, although the subject has developed quite independently. Virtually all reported studies of upstream effects in continuously stratified fluids have used approximately uniform stratification, i.e. fluid with constant Brunt-Väisälä frequency N. In a fluid of depth D, mean velocity U, and obstacle height h, we have the dimensionless parameters

$$F_0 = \frac{\pi U}{ND}, \qquad H = \frac{h}{D}, \qquad \frac{Nh}{U} = \frac{\pi H}{F_0}.$$

Here F_0 (sometimes written as $1/K$) is a Froude number based on the lowest internal wave mode. Linear theory (with small h) does not predict steady upstream disturbances unless $F_0 < 1$ and the topography is semi-infinite [or effectively semi-infinite (Wong & Kao 1970)]. In this case, upstream columnar motions of $O(h)$ are obtained as linear "transients"; these are not in fact transient, because the obstacle has no downstream end, and so they constitute a steady upstream disturbance. For obstacles of finite length, weakly nonlinear theories by Benjamin (1970) (single layer), Keady (1971) (two layer) and McIntyre (1972) (constant N) predict an $O(h^2)$ columnar motion upstream of and related to the downstream lee-wave train; for the constant-N case, these effects are numerically very small and have been looked for experimentally without success (Baines 1977). Solutions to the linear equations, in fact, become singular when the phase (and group) velocity of long waves for some internal mode is zero relative to the topography. For constant N this implies $F_0 = 1/n$, where n is an integer. This resonance causes nonlinear terms to become significant over the obstacle, even for small H, and it is this process that causes the steady-state upstream disturbances. For stratification with constant N the nonlinear steepening effects are extremely small, so that upstream disturbances propagate as linear waves (though their generation over the obstacle is nonlinear), even for moderate amplitudes (provided that the background state is not significantly altered). Consequently, only modes that are subcritical ($c_n = ND/n\pi > U$) can propagate upstream.

The first observations of upstream effects in continuously stratified fluids were made by Long (1955), who observed upstream jets and blocking close to the obstacle when $F_0 < 1$ and Nh/U was sufficiently large. Wei et al. (1975) noticed that these upstream disturbances propagated far upstream as unattenuated columnar linear modes; the obstacles used in their experiments were steep sided, and Wei et al. regarded these upstream effects as consequences of lee-side separation and a turbulent wake. The present author (Baines 1977, 1979a,b) observed these columnar modes and upstream blocking for smooth streamlined obstacles and described their properties for various values of F_0 and H. For small Nh/U, linear lee-wave theory describes the steady-state flow quite well, except near the points of resonance ($1/F_0 \lesssim n$) (Baines 1979a). For $1/n+1 < F_0 < 1/n$, as H increases, a critical value is reached beyond which steady upstream columnar motions of mode n are observed, and this height is zero for $F_0 = 1/n$.

The analysis of Grimshaw & Smyth (1986) generalizes the forced Korteweg–de Vries equation for two-layer flow near resonance to arbitrary finite-depth flows near resonance ($F_0 \sim 1/n$); the coefficients are dependent on the mean velocity profile and stratification. Comparisons between this model and experiments with continuous stratification have yet to be made.

As Nh/U increases, the upstream disturbances in the laboratory experiments are observed to increase in amplitude until upstream blocking occurs. If $F_0 < 0.5$ this occurs for $Nh/U \gtrsim 2$, for obstacles of witch of Agnesi shape. The onset of upstream blocking is manifested as a layer of fluid of finite thickness (typically $\sim \frac{1}{2}h$) coming to rest, rather than as a stationary thin layer near the ground that then thickens vertically.

All the experiments just described were carried out by towing obstacles along tanks of finite length filled with stratified fluid. The columnar modes produced at the obstacle will reflect from the upstream end of the tank (McEwan & Baines 1974), but the observations were made before these returned to influence the observed field of flow significantly. Snyder et al. (1985) reported a series of observations of the density field upstream of two-dimensional obstacles (as well as other shapes), and they attributed upstream blocking to a "squashing" phenomenon. For their experiments, reflection from the upstream end (and in some cases, also the downstream end) was significant, so that the term "squashing" is applicable to their results. However, contrary to their suggestion, it is *not* applicable to the above-cited experiments that simulate a tank of infinite length (albeit for a finite time). Snyder et al. also pointed out that the most slowly moving upstream modes ($n, n-1, \ldots$) have significant amplitudes, so that the flow may take a long time to reach steady state at a fixed distance upstream. This is quite consistent with the flow-field observations of Baines (1979a,b), for example, who reported steady (or nearly steady) flow in the immediate vicinity of the obstacle near the end of the observing period.

A hydraulic model of the type described in the previous section, but one with 64 layers, has been developed to model flow over long obstacles with continuous stratification when H is not small (P. G. Baines & F. Guest, submitted for publication, 1986). Results are given in Figure 8, up to the point of blocking of the lowest layer, for $F_0 > 0.3$. In this parameter range, only modes $n = 1, 2$, and 3 may become subcritical (and hence propagate) upstream. A 64-layer model is a good approximation to continuous stratification when the upstream disturbances are small, but this is not necessarily the case when they are large, particularly near $F_0 = 1/n$; as slow-moving layers become thicker, their discreteness becomes significant. Some similarity between Figure 8 and the two- and three-layer calculations (Figures 5*b* and 6, respectively) is evident. The speeds of the upstream disturbances vary little with amplitude and are treated as rarefactions. These results have not yet been tested experimentally. This is partly because laboratory experiments with hydrostatic stratified flow are difficult because they require obstacles whose lengths are much greater than the fluid depth. Experiments described in Baines (1979a) for moderately short obstacles (length/depth ~ 1.5) give the curve for the onset of upstream disturbances shown (lightly) dashed in Figure 8; this implies that shorter, steeper obstacles may generate upstream disturbances for smaller h than longer obstacles. No results from fully numerical models for these finite-depth systems have yet been published.

We next consider the flow in a channel of width W past a two-dimensional transverse barrier with a small gap at one end of width w. This models a two-dimensional ridge with gaps of width $2w$ spaced periodically along the ridge at intervals of $2W$. If $w/W \ll 1$ we may expect the gap(s) to have negligible effect on the upstream motion. Experiments with this geometry have been reported for a particular obstacle (a short witch of Agnesi) by Baines (1979b) and Weil et al. (1981) for a range of gap sizes. If blocked

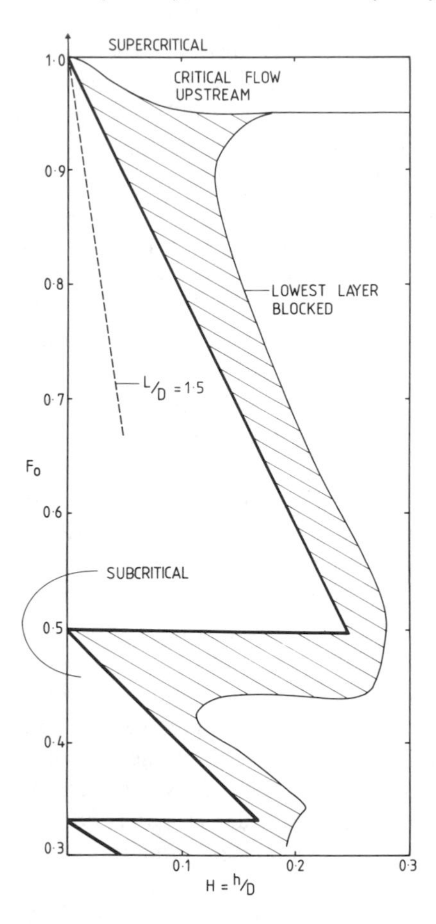

Figure 8 F_0-H diagram for the hydrostatic 64-layer model, approximating uniform stratification, up to the point of blocking for $F_0 > 0.3$. The "critical flow upstream" region $(0.95 < F_0 < 1)$ is an artifact of discrete layering. In the shaded region, upstream disturbances increase in amplitude with increasing H. The dashed line denotes the observed onset of steady upstream disturbances in uniform stratification for flow with obstacle-length/depth $\cong 1.5$ [i.e. nonhydrostatic flow (Baines 1979a, Figure 6*a*)] (from P. G. Baines & F. Guest, submitted for publication, 1986).

fluid is present upstream in the 2D case ($w = 0$), for $w/W \ll 1$ the "blocked" fluid will slowly converge on the gap and flow through it, but its upstream depth will only be affected slightly. If the gap is made wider, the depth of this nearly blocked fluid decreases as a result of increased leakage through the gap. The depth of the nearly blocked layer, z_s, a height that separates fluid flowing horizontally around the barrier from fluid above flowing over it, is quite sharply defined, as shown in Figure 9. For $w/W = 0.125$ this depth is given approximately in terms of Nh/U by

$$z_s/h = 1 - 2U/Nh.$$

If w/W is large enough there may be no permanent upstream disturbances at all. For long 3D obstacles (hydrostatic flow), in order to have upstream disturbances it is necessary for the flow to become critical at the minimum cross section, and the largest value of w/W for which this occurs will mark the change from flow that is 2D-like to 3D-like.

CONTINUOUSLY STRATIFIED FLOW—INFINITE DEPTH

This last case is the one of greatest relevance to the atmosphere. The upper radiation condition implies that there is no downward-propagating energy at the upper region of the fluid, so that (initially at least) a discrete spectrum of vertical modes does not exist; the spectrum of vertical wave numbers is continuous. Nevertheless, purely horizontally propagating linear internal waves are possible, provided that they have infinitely long horizontal wavelength. Furthermore, propagation of these waves in the upstream direction is possible for vertical wave numbers $n < N/U$, with wave speeds (both phase and group velocities) $c = N/n - U$ in the upstream direction (see, for example, Lighthill 1978, Section 4.12). The question is, Under what circumstances are they produced, given the absence of the resonance mechanism with discrete modes? For this system the important parameter is Nh/U, with the length and shape of the obstacle having only secondary significance.

Numerical studies of upstream effects in this system with N and U initially uniform have been reported by Pierrehumbert (1984) and Pierrehumbert & Wyman (1985), and laboratory studies have been described by Baines & Hoinka (1985). Earlier numerical studies of similar systems have concentrated on downstream effects, although some upstream disturbances are visible in the results of Peltier & Clark (1979). Pierrehumbert & Wyman employed a Boussinesq hydrostatic model with a terrain-following coordinate system and a sponge layer at the top to absorb wave energy. With obstacles of Gaussian shape and an

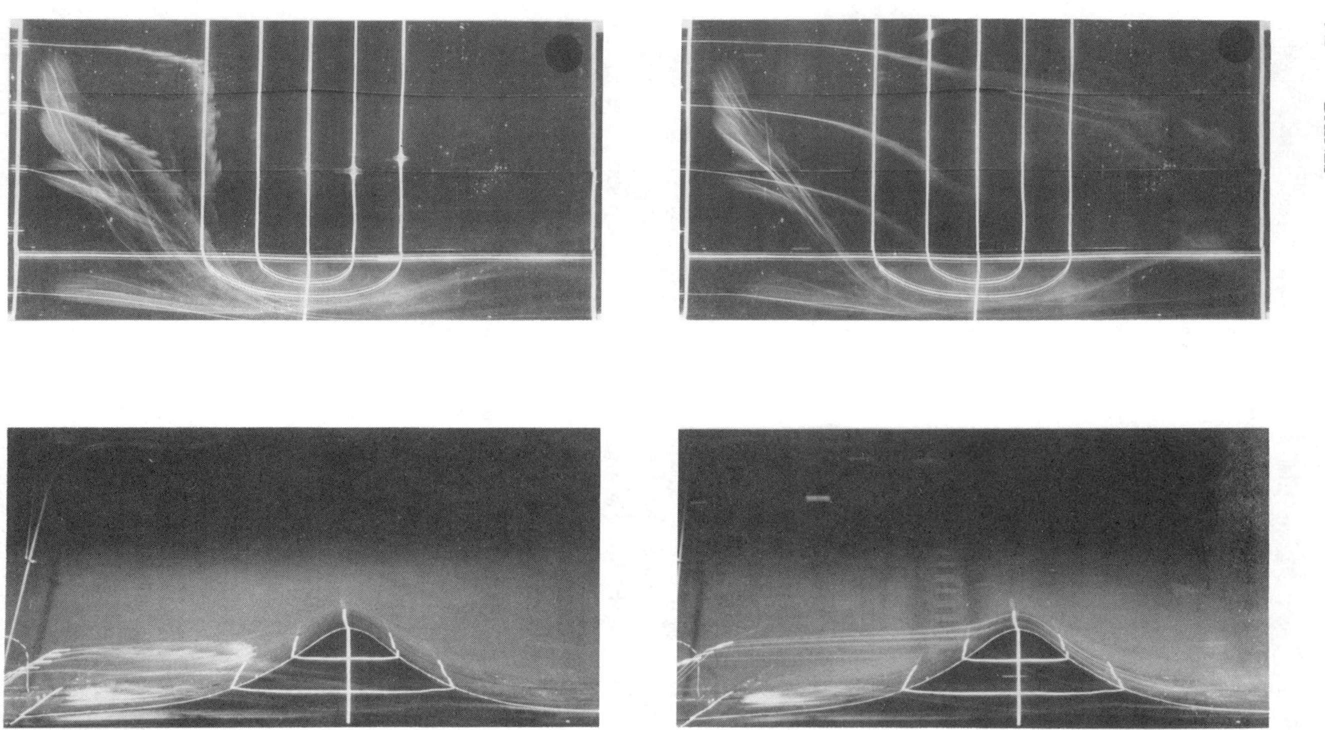

Figure 9 Plan (*upper*) and side (*lower*) views of uniformly stratified flow incident from the left on a barrier with a gap at one end with $Nh/U = 5.9$. The flow field is the same in both cases, but the dye is released from an upstream dye rake at a slightly higher level (9 mm higher, with a total obstacle height of 6.26 cm) in the right-hand frames. Flow in the left-hand frames passes around the obstacle, whereas flow in the right-hand frames passes over it, which demonstrates the abrupt change in flow character with height (from Baines 1979b).

impulsive start to the flow, they found that the steady-state flow was well described by the Long's model solution [a solution that extends steady-state linear theory to finite amplitude when N/U is constant (Long 1955, Lilly & Klemp 1979)] up to the point of overturning ($Nh/U < 0.75$). For $Nh/U > 0.75$, columnar upstream disturbances of *finite* amplitude were generated, and these increased in amplitude with Nh/U. Upstream blocking occurred near the obstacle for $Nh/U > 1.5$ (Gaussian shape) and $Nh/U > 1.75$ (witch of Agnesi shape), but upstream propagation of this blocked fluid was not observed until $Nh/U \gtrsim 2$. Figure 10 shows the time evolution of the flow field computed by Pierrehumbert & Wyman for $Nh/U = 2.0$.

Baines & Hoinka (1985) carried out towing experiments in a stratified tank similar to those described earlier, but with the difference that a radiation condition at the top of the working fluid was simulated with a novel geometrical arrangement. Experiments were carried out up to the point where the flow in the vicinity of the obstacle appeared to be steady and before this flow could be significantly affected by wave motion reflected from the upstream end of the tank. Five different obstacle shapes were used, and a broad range of Nh/U values were covered for each one. The obstacles were not long enough for the flow to be hydrostatic. Near-steady-state flow fields are shown in Figure 11 for the witch of Agnesi. The principal results were as follows. (*a*) For $0 < Nh/U < 0.5$ (± 0.2) no steady upstream effects were observed, and the steady-state flow was generally consistent with linear theory and Long's model solutions. (*b*) For $Nh/U > 0.5$ (± 0.2) steady upstream columnar disturbances were observed, with amplitude increasing from zero as Nh/U increased above 0.5. As the "error bars" indicate, this lower limit was only determined approximately because of the presence of upstream transients and the smallness of the signal. However, it seemed to be independent of obstacle shape and was not dependent on overturning in the lee-wave field, which was not observed until $Nh/U \gtrsim 1.5$. Upstream blocking was observed when Nh/U reached a value in the range 1.3 to 2.2, with the actual value depending on the obstacle shape, but for symmetric obstacles the value was approximately 2. As may be seen in Figure 11, reduced velocities and blocking at low levels upstream are accompanied by increased velocities above the level of the obstacle, and this velocity profile oscillates with decreasing amplitude as the height increases. The density gradient is very small in the slow-moving or blocked fluid, and it is correspondingly large in the overlying jet region; as Nh/U increases, this region becomes more like an interface that can support horizontally propagating waves, as shown on the lee side in the last two frames of Figure 11. For $Nh/U > 1.5$ a stagnant region (or "wave-induced critical level")

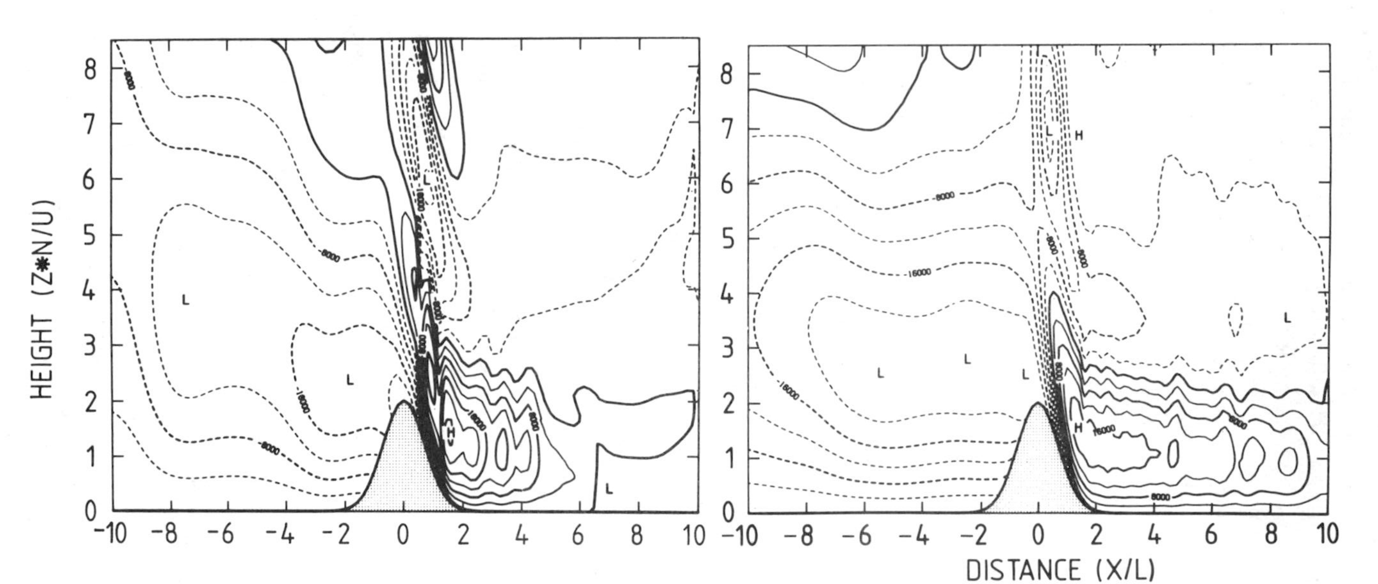

Figure 10 Time development of the computed stream-function perturbation (units of $m^2\ s^{-1}$) for hydrostatic flow with $Nh/U = 2.0$ (from Pierrehumbert & Wyman 1985). Left frame: $t = 7.2\ L/U$; right frame: $t = 14.4\ L/U$, where t denotes time from the impulsive commencement of motion.

exists above the jet over the lee side of the obstacle, and as Nh/U increases, the wave field at upper levels becomes less apparent. The flows then appear to be qualitatively similar to the finite-depth flows for the same Nh/U, provided $F_0 \ll 1$. The behavior shown in Figure 9 for 3D topography should also occur in the infinite-depth case.

There is, as yet, no mechanistic model that can explain and describe the upstream motions for this infinite-depth case. Unlike finite-depth systems, upstream effects are not observed unless Nh/U is sufficiently large, and the value at which this occurs is different in the numerical and laboratory

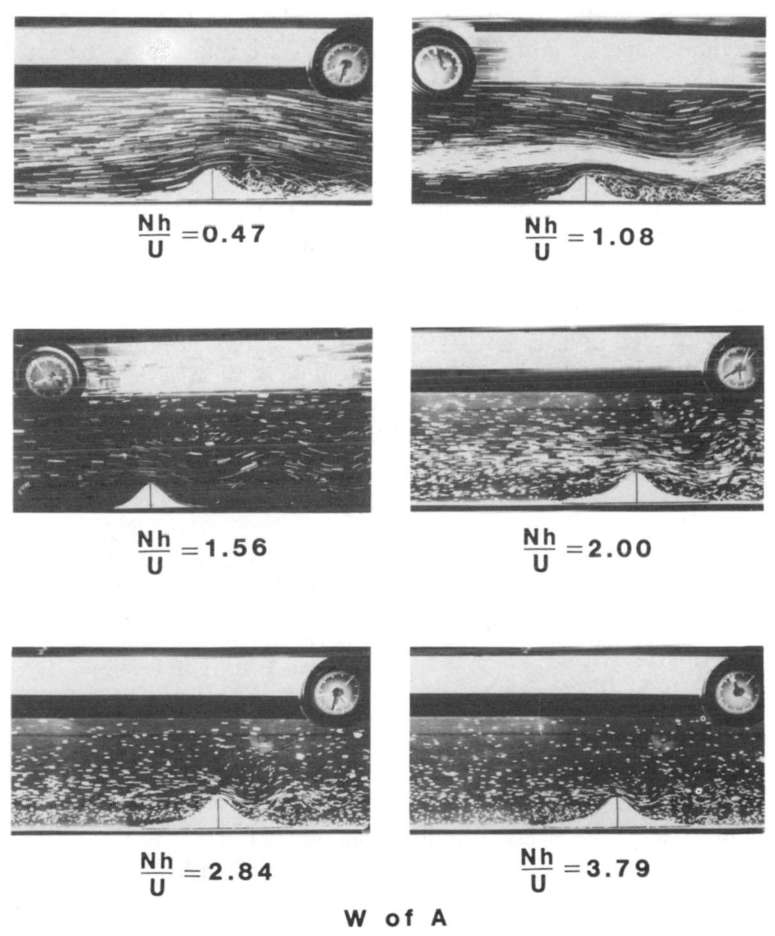

Figure 11 Near-steady-state streamlines for the witch of Agnesi for a range of Nh/U values. Flow is from left to right. Note the upstream blocking in the last three frames (from Baines & Hoinka 1985).

experiments. If, as the laboratory observations suggest, upstream motions may appear without lee-side overturning, then this result implies possible hysteresis in the system because the Long's model solutions are valid steady-state solutions up to the point of overturning. Recent non-hydrostatic computations by J. T. Bacmeister & R. T. Pierrehumbert (private communication) have investigated various start-up conditions, and the results suggest some steady upstream motion for $Nh/U > 0.5$ for a gradual commencement of motion, but the results are complicated by a slow approach to steady state.

Finally, two further aspects deserve mention, although space limitations preclude detailed discussion. Firstly, for application to the atmosphere, where time scales of more than a few hours are important, the Earth's rotation must be considered. This has been discussed for finite Nh/U by Pierrehumbert & Wyman (1985). Upstream effects are restricted to a distance of order Nh/f, where f is the Coriolis parameter. Secondly, the question of stagnant fluid versus sweeping out of periodic valleys in 2D stratified flow across the valleys has been studied by Bell & Thompson (1980) for finite-depth systems and by P. Manins & F. Kimura (private communication) for infinite-depth systems. (Both studies employed numerical and laboratory models.) Bell & Thompson found that blocking in the valleys occurred for $Nh/U \gtrsim 0.8$. Manins & Kimura observed that blocking in valleys was related to wave breaking and obtained a similar criterion, although the flow fields were different in many respects from those described by Bell & Thompson.

Literature Cited

Armi, L. 1986. The hydraulics of two flowing layers with different densities. *J. Fluid Mech.* 163: 27–58

Baines, P. G. 1977. Upstream influence and Long's model in stratified flows. *J. Fluid Mech.* 82: 147–59

Baines, P. G. 1979a. Observations of stratified flow over two-dimensional obstacles in fluid of finite depth. *Tellus* 31: 351–71

Baines, P. G. 1979b. Observations of stratified flow past three-dimensional barriers. *J. Geophys. Res.* 83: 7834–38

Baines, P. G. 1984. A unified description of two-layer flow over topography. *J. Fluid Mech.* 146: 127–67

Baines, P. G., Davies, P. A. 1980. Laboratory studies of topographic effects in rotating and/or stratified fluids. In *Orographic Effects in Planetary Flows, GARP Publ. No. 23*, pp. 233–99. Geneva: WMO/ICSU

Baines, P. G., Hoinka, K. P. 1985. Stratified flow over two-dimensional topography in fluid of infinite depth: a laboratory simulation. *J. Atmos. Sci.* 42: 1614–30

Bell, R. C., Thompson, R. O. R. Y. 1980. Valley ventilation by cross winds. *J. Fluid Mech.* 96: 757–67

Benjamin, T. B. 1970. Upstream influence. *J. Fluid Mech.* 40: 49–79

Benjamin, T. B., Lighthill, M. J. 1954. On cnoidal waves and bores. *Proc. R. Soc. London Ser. A* 224: 448–60

Benton, G. S. 1954. The occurrence of critical flow and hydraulic jumps in a multi-layered system. *J. Meteorol.* 11: 139–50

Chu, V. H., Baddour, R. E. 1977. Surges, waves and mixing in two-layer density stratified flow. *Proc. Congr. Int. Assoc. Hydraul. Res., 17th*, 1: 303–10

Grimshaw, R. H. J., Smyth, N. 1986. Resonant flow of a stratified fluid over topography. *J. Fluid Mech.* In press

Houghton, D. D., Isaacson, E. 1970. Mountain winds. *Stud. Numer. Anal.* 2: 21–52

Houghton, D. D., Kasahara, A. 1968. Nonlinear shallow fluid over an isolated ridge. *Commun. Pure Appl. Math.* 21: 1–23

Hunt, J. C. R., Snyder, W. H. 1980. Experiments on stably and neutrally stratified flow over a model three-dimensional hill. *J. Fluid Mech.* 96: 671–704

Keady, G. 1971. Upstream influence on a two-fluid system. *J. Fluid Mech.* 49: 373–84

Lawrence, G. A. 1985. The hydraulics and mixing of two-layer flow over an obstacle. *Rep. 85/02*, Hydraul. Eng. Lab., Univ. Calif., Berkeley

Lee, J. D., Su, C. H. 1977. A numerical method for stratified shear flows over a long obstacle. *J. Geophys. Res.* 82: 420–26

Lighthill, M. J. 1978. *Waves in Fluids*. Cambridge: Cambridge Univ. Press. 504 pp.

Lilly, D. K., Klemp, J. B. 1979. The effects of terrain shape on non-linear hydrostatic mountain waves. *J. Fluid Mech.* 95: 241–61

Long, R. R. 1954. Some aspects of the flow of stratified fluids. II. Experiments with a two-fluid system. *Tellus* 6: 97–115

Long, R. R. 1955. Some aspects of the flow of stratified fluids. III. Continuous density gradients. *Tellus* 7: 341–57

Long, R. R. 1970. Blocking effects in flow over obstacles. *Tellus* 22: 471–80

Long, R. R. 1972. Finite amplitude disturbances in the flow of inviscid rotating and stratified fluids over obstacles. *Ann. Rev. Fluid Mech.* 4: 69–92

Long, R. R. 1974. Some experimental observations of upstream disturbances in a two-fluid system. *Tellus* 26: 313–17

McEwan, A. D., Baines, P. G. 1974. Shear fronts and an experimental stratified shear flow. *J. Fluid Mech.* 63: 257–72

McIntyre, M. E. 1972. On Long's hypothesis of no upstream influence in uniformly stratified or rotating flow. *J. Fluid Mech.* 52: 209–43

Peltier, W. R., Clark, T. L. 1979. The evolution and stability of finite-amplitude mountain waves. Part II. Surface wave drag and severe downslope windstorms. *J. Atmos. Sci.* 36: 1498–1529

Pierrehumbert, R. T. 1984. Formation of shear layers upstream of the Alps. *Riv. Meteorol. Aeronaut.* 44: 237–48

Pierrehumbert, R. T., Wyman, B. 1985. Upstream effects of mesoscale mountains. *J. Atmos. Sci.* 42: 977–1003

Pratt, L. J. 1983. A note on nonlinear flow over obstacles. *Geophys. Astrophys. Fluid Dyn.* 24: 63–68

Pratt, L. J. 1984. On non-linear flow with multiple obstructions. *J. Atmos. Sci.* 41: 1214–25

Rouse, H., Ince, S. 1957. *History of Hydraulics*. Iowa City: Iowa Inst. Hydraul. Res. 269 pp.

Sheppard, P. A. 1956. Airflow over mountains. *Q. J. R. Meteorol. Soc.* 82: 528–29

Smith, R. B. 1976. The generation of lee waves by the Blue Ridge. *J. Atmos. Sci.* 33: 507–19

Snyder, W. H., Thompson, R. S., Eskridge, R. E., Lawson, R. E., Castro, I. P., et al. 1985. The structure of strongly stratified flow over hills: dividing streamline concept. *J. Fluid Mech.* 152: 249–88

Su, C. H. 1976. Hydraulic jumps in an incompressible stratified fluid. *J. Fluid Mech.* 73: 33–47

Wei, S. N., Kao, T. W., Pao, H.-P. 1975. Experimental study of upstream influence in the two-dimensional flow of a stratified fluid over an obstacle. *Geophys. Fluid Dyn.* 6: 315–36

Weil, J. C., Traugott, S. C., Wong, D. K. 1981. Stack plume interaction and flow characteristics for a notched ridge. *PPRP-61*, Environ. Cent., Martin Marietta Corp., Baltimore, Md

Wong, K. K., Kao, T. W. 1970. Stratified flow over extended obstacles and its application to topographical effect on vertical wind shear. *J. Atmos. Sci.* 27: 884–89

Wood, I. R., Simpson, J. E. 1984. Jumps in layered miscible fluids. *J. Fluid Mech.* 140: 329–42

Yih, C.-S., Guha, C. R. 1955. Hydraulic jump in a fluid system of two layers. *Tellus* 7: 358–66

Ann. Rev. Fluid Mech. 1987. 19 : 99–123

CAVITATION BUBBLES NEAR BOUNDARIES

J. R. Blake

Department of Mathematics, University of Wollongong, Wollongong, NSW 2500, Australia

D. C. Gibson

CSIRO Division of Energy Technology, Highett, Victoria 3190, Australia

> The impact of liquid jets, formed by involution of collapsing cavities is a primary factor in cavitation ... often outweighing in effect the better known implosion mechanism that Rayleigh demonstrated.
>
> One should always reason in terms of the Kelvin impulse, not in terms of the fluid momentum....
>
> An experimental fact, perhaps significant in this regard, is that cavitation damage induced by flow is often most severe in the neighbourhood of stagnation points downstream from the low pressure zone....
>
> Benjamin & Ellis (1966)

1. INTRODUCTION

Often cavitation damage may be responsible for the initiation of severe structural damage to ship propeller blades, turbomachinery, and hydraulic equipment. Graphic examples abound in the hydraulic-engineering literature (see, e.g., Knapp et al. 1970, Hammitt 1980, Arndt 1981a) of damage to impeller blades, valves, spillways, and propeller blades. An illustration of cavitation damage to a pump impeller blade may be found in Figure 1, and other examples are illustrated in a previous review in this series (Arndt 1981b). In turbomachinery, cavitation inception frequently occurs just downstream from the point of minimum pressure, prior to the separation point, so that cavitation bubbles are swept up over the "separation

0066–4189/87/0115–0099$02.00

bubble." Structural damage is then observed to occur near reattachment of the separated flow. An alternative explanation (Morch 1980) is that the cyclic detachment of the "separation bubble," consisting of a cluster of traveling cavities, is the principal source of cavitation damage. In this review we are not concerned with the circumstances leading to the formation or generation of cavitation bubbles; instead, we examine the physics of the growth and collapse of bubbles close to boundaries.

The first scientific study of cavitation is attributed to Reynolds (1894), who observed the growth and subsequent collapse of vapor cavities that were formed in water flowing through constricted tubes. Not many years later, severe structural damage was found on the propeller blades of fast steamships, which were often rendered useless after only a few hours of operation. This problem was such a major concern to the British Admiralty that a special commission was instituted to investigate the source of this mechanical damage. The commission reported that the damage was due primarily to the "hydraulic blows" that the blades suffered from collapsing cavitation bubbles. Extensive experimental studies by Sir Charles Parsons led to the implementation of multipropeller seacraft (see, e.g., Burrill 1951, Arndt 1981a). The first serious theoretical study of this problem was that of Lord Rayleigh (1917), who considered the collapse of a spherical bubble in an infinite fluid and thus developed the now well-known implosion mechanism of cavitation whereby extremely high

Figure 1 Example of severe cavitation damage to an agricultural pump impeller.

pressures are generated during the last moments of the collapse phase of the bubble.

Present-day interest in cavitation adjacent to boundaries can be traced back to pioneering works by Naude & Ellis (1961) and Benjamin & Ellis (1966) that showed beyond doubt that cavitation bubbles do not collapse spherically in the neighborhood of solid boundaries. This led to the controversial hypothesis that the liquid jet that threads the bubble and strikes the boundary in the last moments of collapse, or early in the rebound, is a prime cause of cavitation damage. Twenty years later, the controversy is still not completely resolved.

An additional dimension to the study of this problem appeared when Gibson (1968) showed that the jet formation, direction, and intensity are functions of bubble-boundary interaction that can be controlled by changing the character of the boundary impedance. Plesset & Chapman (1971) calculated the collapse of an initially spherical bubble in the neighborhood of a solid boundary and demonstrated that a very vigorous jet advances through the bubble toward the boundary relatively early in the collapse, long before compressibility effects could be of importance. Subsequent calculations and experiments have borne out this finding and have added to our understanding of bubble-boundary interactions (Kling & Hammitt 1972, Smith & Mesler 1972, Lauterborn & Bolle 1975, Chahine 1982).

Recent progress has been achieved through use of advanced theoretical and numerical techniques and increased computing capacity. Indeed, it is fair to say that whereas the 1970s saw experiment lead theory, the last five years have seen theory and computation draw ahead of experiment, particularly in relation to the behavior of bubbles near solid boundaries or a free surface. (Shima & Sato 1980, 1981, Shima et al. 1981, Guerri et al. 1982, Prosperetti 1982, Blake et al. 1986a,b). In this review, we restrict our discussion to attempts to understand the fundamental mechanisms that may be responsible for cavitation damage. Studies at this level have two principal aims: to determine the flow fields leading to potentially damaging phenomena, and to investigate methods to reduce or even eliminate damage. Our review is primarily concerned with experimental and theoretical studies of the growth and collapse of a single vapor bubble near boundaries in the presence of an ambient pressure gradient and a velocity field. The liquid is assumed to be "cold," with the contents of the bubble consisting of vapor and the fluid mechanics being dominated by inertia. Compressibility effects in both the liquid and vapor are not covered in this review. We use as a starting point the section in the review by Plesset & Prosperetti (1977) concerning the interaction of cavitation bubbbles with boundaries and concentrate on developments in this area since then.

2. EXPERIMENTAL TECHNIQUES

Cavitation in flowing liquids is a random, unsteady event. It first occurs when a gaseous impurity is swept through a region where the liquid pressure has been reduced below its vapor pressure by dynamic means. The uncertain origin, small size, and short life of the bubbles formed have provided a formidable challenge for experimenters studying the behavior of individual bubbles.

With the exception of Hammitt and co-workers (Kling & Hammitt 1982, Hammitt 1980), most experimenters have studied individual bubbles that expand and collapse in an otherwise stationary liquid. Bubble-generation techniques employed include a kinetic impulse (Benjamin & Ellis 1966), spark discharge (Gibson 1968, 1972b, Gibson & Blake 1980, 1982, Blake & Gibson 1981, Smith & Mesler 1972, Chahine 1977, 1982, Shima et al. 1981), and pulsed-laser discharge (Lauterborn & Bolle 1975, Lauterborn 1982, Lauterborn & Vogel 1984, Lauterborn & Hentschel 1985).

Although the kinetic-impulse technique most nearly models the natural inception process because the test liquid is momentarily put into tension, it suffers from the practical disadvantage that a small gaseous impurity or bubble must be accurately located in the test liquid prior to the impulse. While the problem of location and generation is overcome when a spark discharge is used, the high-voltage electrodes that intrude into the liquid invariably disrupt the bubble motion in the last critical moments of collapse when the bubble has become highly distorted. The pulsed-laser technique has all the advantages of spark discharge without the disadvantages of electrodes. Technically its only fault is the mode of generation, which, in common with spark discharge, involves intense local heating and vaporization of the liquid through application of a thermal impulse (Gibson 1972a).

High-speed photography has played an important part in the development of understanding of bubble-boundary interactions. A survey of the advances since the early 1950s is given in a recent paper by Lauterborn & Hentschel (1985). Over the years, rotating drum, mirror, and prism cameras have been used with exposure rates varying from 1000 to 1,000,000 frames s^{-1}, and more recently high-speed holography has been employed (Lauterborn & Vogel 1984). The greatest clarity is achieved when large bubbles are generated and illuminated with diffuse backlighting. Spectacular examples are given in the works of Benjamin & Ellis (1966), Gibson (1968), Chahine (1977, 1982), and Blake & Gibson (1981), who generated their bubbles in a liquid container held at a reduced pressure of order 0.1 atm. As a consequence they were able to generate bubbles with a maximum radius of order 10 mm and pulsation time of order 10 ms, which enabled careful examination of the evolution of the re-entrant jet that forms and threads the collapsing cavity. In these experiments the increased volume and lifetime of the bubble meant that buoyancy forces could play an important part in the developing motion (Gibson 1968, Blake et al. 1986a).

Buoyancy effects can be eliminated by conducting the experiments in a free-fall experimental apparatus. Figure 2 shows a free-fall apparatus developed originally by Benjamin & Ellis (1966) and subsequently refined by Gibson & Blake (1980) to remove the deleterious effects of buoyancy. The central part is a 370-mm-deep, 260-mm internal diameter plexiglass tank, filled with distilled water to a height of about 220 mm. The water is degassed by evacuating the tank and shaking vigorously in a vertical direction for about a half hour before each series of experiments. A high-voltage spark probe that can be traversed along the centerline extends into the tank from the floor. A 25-mm fine-wire square measurement scale is attached to the probe beneath the spark gap. The boundary to be examined is lowered down onto the water surface from the tank roof. The scale, probe, and boundary are viewed through a flat 25-mm-thick plexiglass window recessed into the sidewall of the tank.

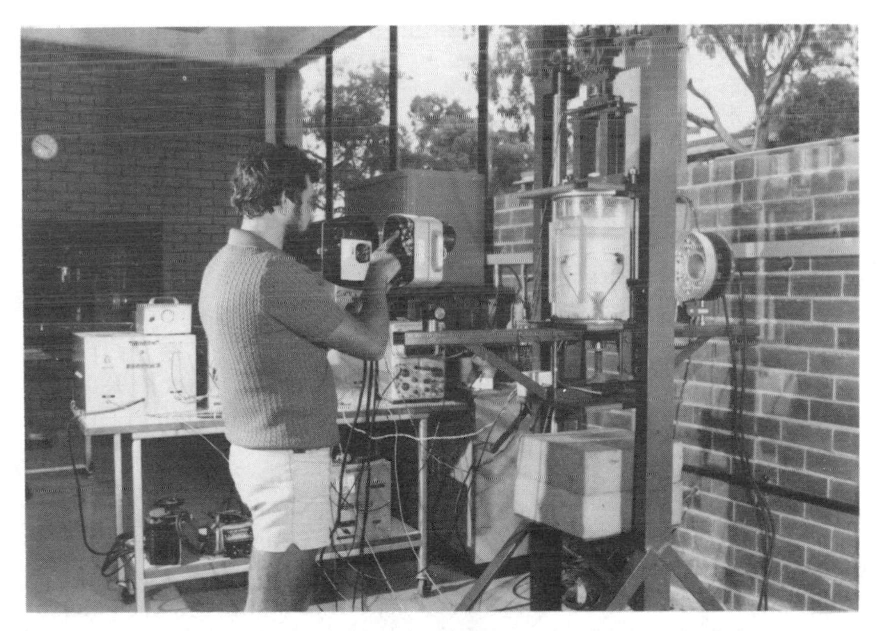

Figure 2 A free-fall cavitation-bubble rig. Bubble produced by spark discharge across electrodes. Resulting bubble shapes recorded by HYCAM cine camera.

The tank is mounted on a horizontal platform suspended between vertical guide rails by an electromagnet. A HYCAM high-speed cine camera fitted with a half-frame 16-mm prism and a PAL 2.4-kW continuous light are also mounted on the platform. The vapor bubble is generated by an electric spark discharged from a 0.25-μF condenser at about 8000 V. The spark probe is set at a prescribed depth beneath the boundary under test, the tank pressure is reduced to about 10 kPa, and the camera is set in motion. When the camera accelerates to the desired framing rate, it triggers the spark and releases the electromagnet. The platform falls from a state of rest while the bubble grows and collapses, with the camera recording the motion. All the bubble photographs presented in the following sections were recorded on this apparatus under free-fall conditions.

3. THEORETICAL DEVELOPMENTS

3.1 *Analytical*

In developing theoretical models of cavitation bubbles, it has been common practice to regard the fluid as incompressible, inviscid, and irrotational, thus neglecting compressibility and viscous effects, neither of which is likely to assume importance during the first pulsation. In spherically symmetric bubble growth and collapse, the radial velocity field may simply be obtained by integrating the continuity equation, which yields the expected source-sink-like flow (Besant 1859). This was exploited by Rayleigh (1917) to obtain the solution for cavitation bubbles in an infinite fluid. This theory showed how extremely large pressures may be generated during the collapse of a spherically symmetric bubble—now commonly known as the Rayleigh implosion mechanism. In addition, it is a relatively easy procedure to include surface tension, viscosity, and compressibility effects in the spherically symmetric bubble growth and collapse model. The Rayleigh solution and its variations are discussed in much greater depth in the previous Plesset & Prosperetti (1977) review and also in Hammitt's (1980) book.

When boundaries are introduced, analytical techniques have followed several paths. One approach has been to develop the solution in terms of spherical harmonics, with the coefficient of each term being a function of time and satisfying a nonlinear differential equation (see, e.g., Shima & Sato 1981). The first term in the expansion is the Rayleigh solution mentioned above, whereas higher-order terms introduce higher degrees of asymmetry into the bubble shape. Typically this approach will show the early stages of the formation of a liquid jet but cannot follow the development of the jet for any length of time because of the large number of terms that would be required in the series. An alternative approach is an asymptotic expansion in terms of a small parameter ε, which is the ratio of the maximum bubble radius to the distance from the boundary (see, e.g., Chahine & Bovis 1983). Again the first term in the expansion is the Rayleigh solution, with the higher-order terms introducing asymmetry into the bubble shape. Unfortunately, this approach is of little assistance in the region of prime interest around $\varepsilon = 1$. Because of the limitations of both of these approaches to large deformations of bubbles, a more robust approach via numerical techniques is needed.

3.2 *The Kelvin Impulse*

The Kelvin impulse is a particularly valuable dynamical concept in unsteady fluid mechanics. It has been used in a number of contexts in fluid mechanics (see, e.g., Voinov 1973, Yakimov 1973), but Benjamin & Ellis (1966) appear to be the first to have realized its value in cavitation bubble dynamics. The Kelvin impulse corresponds to the apparent inertia of the cavitation bubble and, like the linear momentum of a projectile, may be used to determine aspects of the gross bubble motion (such as the movement of the bubble centroid). It cannot be used to determine the deformation of the surface of the bubble except in the sense that the Kelvin impulse needs to be conserved after collapse, usually in the form of a vortex ring. Of course, situations may arise where a zero value of the Kelvin impulse is obtained, which in some circumstances may lead to two vortex rings of equal circulation but opposite sign being formed [e.g. the "hourglass" bubble (Gibson & Blake 1982, Lauterborn & Hentschel 1985)].

The Kelvin impulse $\mathbf{I}$ is defined as

$$\mathbf{I} = \rho \int_S \phi \mathbf{n} dS, \tag{1}$$

where ρ is the fluid density, ϕ is the velocity potential, S is the cavitation bubble surface, and $\mathbf{n}$ is the outward normal to the fluid (i.e. into the bubble). If we consider the conservation of linear momentum in a finite control volume enclosing the bubble and then let the exterior boundaries extend to infinity, we obtain the following relation for the Kelvin impulse:

$$\mathbf{I} = \int_0^t \mathbf{F}(t) dt. \tag{2a}$$

Here, we have

$$\mathbf{F}(t) = \rho g V \mathbf{e}_x + \rho \int_\Sigma \left\{ \frac{1}{2} (\nabla\phi)^2 \mathbf{n} - \frac{\partial \phi}{\partial n} \nabla\phi \right\} dS, \tag{2b}$$

where V is the volume of the bubble, $\mathbf{e}_x$ the unit vector in the x-direction, and Σ the surface of the boundary (Blake & Cerone 1982). The first term in (2b) is included to allow for buoyancy forces, which were certainly of importance for some experiments conducted under reduced pressures. Clearly this term would be of major significance in studies of underwater explosions and boiling. The second term represents the momentum flux at the boundary. It is apparent from (2a,b) that the Kelvin impulse is a function of time, starting from zero at inception and building up over the lifetime of the bubble. Contributions to the Kelvin impulse may come from the presence of nearby boundaries and the ambient velocity and pressure field. With this number of mechanisms contributing to its development, the Kelvin impulse may actually change sign during the lifetime of the bubble.

To illustrate the use of the Kelvin impulse, consider the location of the bubble to be at a relatively large distance h from the boundary (compared with the maximum bubble radius R_{m}; typically $h/R_{\mathrm{m}} = \gamma > 2$), where we may approximate the bubble for much of its lifetime by a time-varying source (a sink during the collapse phase) of strength $q(t)$. This may be substituted into (2b) to yield the following for the time rate of change of the Kelvin impulse for various boundaries or interfaces:

$$F_x(t) = \pm \rho g V + \frac{\rho q^2}{16\pi h^2}\mathbf{B},$$

$$\mathbf{B} = \begin{cases} -1: \text{ Rigid boundary}, \quad +1: \text{ Free surface}, \\ H(\alpha): \text{ Inertial boundary}, \quad A: \text{ Two-fluid interface}, \end{cases} \tag{3}$$

where $A = (\rho - \rho')/(\rho + \rho')$, with ρ' the density of the upper fluid not containing the bubble. We define $\alpha = \rho h/\sigma$, where σ is the mass per unit area of the boundary. In the above formulae, we have

$$H(\alpha) = 4\alpha - 1 - 8\alpha^2 e^{2\alpha} E_1(2\alpha),$$

where E_1 is an exponential integral and where $|H| < 1$, the limits imposed by the rigid boundary and free surface. Other examples exist for the Kelvin impulse, including the free surface with gravity and membrane boundaries, but these cases include convolution-type integrals for the source strength. In the absence of an imposed fluid motion, the second term in (3) may be regarded as the Bjerknes boundary-interaction term, i.e. bubbles in phase attract (rigid boundary), and those out of phase repel (free surface). This concept may be further exploited if we use the Rayleigh bubble expression for the source strength:

$$q(t) = \frac{dV}{dt} = \pm 4\pi R^2 \left[\frac{2}{3}\left(\frac{1}{R^3} - 1\right)\right]^{1/2}, \tag{4}$$

where $R(t)$ is the bubble radius (made dimensionless with respect to the maximum bubble radius R_{m}). If (3) and (4) are substituted into (2a), the integrals involving V and q may be integrated to yield the following for $I_x(t)$:

$$I_x(t) = \frac{\sqrt{6}\pi R_{\mathrm{m}}^5(\Delta p \cdot \rho)^{1/2}}{9h^2}\left[\pm \frac{2\rho g h^2}{\Delta p \cdot R_{\mathrm{m}}} C_1(t) + \mathbf{B} C_2(t)\right], \tag{5}$$

where

$$\left.\begin{aligned} C_1(t) &= B_a(11/6, 1/2) \\ C_2(t) &= B_a(7/6, 3/2) \end{aligned}\right\} \quad 0 \leq t \leq 0.915\ldots$$

$$\left.\begin{aligned} C_1(t) &= 2B(11/6, 1/2) - B_a(11/6, 1/2) \\ C_2(t) &= 2B(7/6, 3/2) - B_a(7/6, 3/2) \end{aligned}\right\} \quad 0.915\ldots < t \leq 1.83\ldots.$$

Here, Δp is the pressure difference between that at infinity and the saturated vapor pressure inside the bubble, $a(t) = R^3(t)$, and $B_a(z, w)$ and $B(z, w)$ are incomplete and complete Beta functions, respectively (see Abramowitz & Stegun 1965).

To illustrate the usefulness of this approach, let us consider the example of a buoyant bubble near a rigid boundary and calculate the Kelvin impulse at completion of the collapse phase (i.e. $T_{\mathrm{c}} = 1.83\ldots$). This yields

$$I_x(T_{\mathrm{c}}) = \frac{2\pi\sqrt{6}R_{\mathrm{m}}^5(\Delta p \cdot \rho)^{1/2}}{9h^2}[2\gamma^2\delta^2 B(11/6, 1/2) - B(7/6, 3/2)], \tag{6}$$

where $\gamma = h/R_{\mathrm{m}}$ and $\delta = (\rho g R_{\mathrm{m}}/\Delta p)^{1/2}$. The above arguments would suggest that the null Kelvin impulse line $I_x(T_{\mathrm{c}}) = 0$ will determine the regions in the γ-δ parameter space where bubbles will move either toward the rigid boundary or away from it (see later discussions on this aspect).

The results of Chahine & Bovis (1980) for bubbles near the interface between two fluids of different densities may also be interpreted in terms of the Kelvin impulse. The importance of the interface reaction will be reduced as the density of the upper fluid increases toward that of the lower fluid (i.e. $A \to 0$), and hence buoyancy forces will be more important. Thus, while a bubble located near the boundary will still be repelled, slightly farther away it will be seen to migrate toward the interface, as is so clearly evident in Chahine & Bovis' experiment.

The concept of the Kelvin impulse has only recently been fully exploited in unbounded multiphase flows (see van Wijngaarden 1976), and considerable development is still required, particularly in the areas of clouds and sheets of bubbles near boundaries. It is likely to provide valuable indicators as to the physical properties required of boundaries in order to reduce or eliminate cavitation damage.

In terms of theoretical developments, several phases of research activity can be identified, corresponding to the Rayleigh implosion mechanism, microjet phenomena, and more recently the interaction of clouds of bubbles and the dynamics of the flow field. Topics that require further extensive study include the influence of clouds of bubbles, the importance of ambient-fluid pressure gradients, and the role of toroidal bubbles in determining the intense pressure gradients near boundaries (Chahine & Genoux 1983). Already, preliminary studies by Chahine (1984) have shown the importance of clouds of bubbles in determining the pressure field and have suggested that the pressure increases by up to an order of magnitude for a cloud of bubbles over that calculated for a single bubble.

3.3 *Numerical*

In this section, we discuss only purely numerical solutions, even though some of the techniques involved in Section 3.1 ultimately depend on computation to obtain their results. Surprisingly, there have only been a relatively small number of attempts to develop numerical solutions of the "cavitation bubble" equations. Perhaps the highly nonlinear deformations of the bubble surface together with the nonlinear dynamic boundary conditions have previously dissuaded theoreticians from extensive numerical effort. In many respects, the free-surface problem is the axisymmetric analogue of the breaking-wave problem (see, e.g., Longuet-Higgins & Cokelet 1976, Longuet-Higgins 1983) with the development of jets, instantaneous stagnation points, and points of maximum pressure.

Most of the numerical solutions have so far been restricted to inviscid potential flow as a model, thus neglecting the effects of viscosity and compressibility. Viscous effects are typically very small because of the high Reynolds numbers and very thin boundary layers around the bubble, which may only be present in the late mobile stage of the lifetime of the bubble anyway. Compressibility effects do not appear to be important during the first pulsation, but shock waves are often observed during the first rebound. In most calculations it is commonplace to suppose that the pressure in the bubble is the saturated vapor pressure. Surface-tension forces also are usually neglected, even though they may be important in the early growth phase after inception and at the tip of the jet during collapse. Surface tension is also responsible for the Rayleigh instability that breaks up the free-surface "spike" into droplets. Nevertheless, even with all these assumptions, the numerical calculations appear to accurately predict many of the experimental observations over a wide range of physical conditions, which, after all, is an essential objective of any numerical study.

Most notable among the earlier attempts to solve the problem numerically were those by Mitchell & Hammitt (1973) and Plesset & Chapman (1971), who considered the collapse history of a cavitation bubble near a rigid boundary. Using a particle-in-cell technique, Plesset & Chapman were able to calculate the complete collapse history of an initially spherical bubble right through to the liquid jet striking the other side of the bubble. They predicted jet velocities in the range 130–180 m s^{-1}. In a later experimental study, Lauterborn & Bolle (1975) obtained results that compared favorably with these calculations (see also the previous review paper of Plesset & Prosperetti 1977). Plesset & Chapman's study has been the basis for comparison ever since (see, e.g., Lauterborn & Bolle 1975, Prosperetti 1982).

Over the last decade, the boundary-integral method has become the principal numerical technique used because of the ease with which it can follow the contortions of the bubble shape. Voinov & Voinov (1975), Lenoir (1976), Prosperetti (1982), Taib et al. (1984), and Blake et al. (1986a,b) have used variations of this technique to simulate the growth and/or collapse of cavitation bubbles near rigid boundaries, a free surface, or other bubbles. Considerable care needs to be taken with evaluating first-kind Fredholm integral equations, which have singular kernels (in axisymmetric flow the kernels involve complete elliptic integrals having a logarithmic singularity) that require special quadrature formulae (see Stroud & Secrest 1966). A Lagrangian description for a specified number of points on the bubble's surface is used to represent the movement of the bubble. Additionally, the time-stepping procedure needs to be controlled such that the change in potential $\Delta\phi$ is bounded at each time step. This leads to variations in the time step of several orders of magnitude over the computation, and in practice this restriction may be used to determine the actual time step. High fluid velocities require the time step to be very small for the initial and final stages of the bubble, whereas near maximum size the bubble is almost immobile and larger time steps may be employed. The updating procedure for the potential ϕ uses the Bernoulli pressure condition on the bubble to advantage, in that it allows for substitution of the partial time derivative into the total derivative. Thus, at each time step the new Lagrangian position and potential at the specified points may be obtained as follows:

$$\begin{aligned} \mathbf{x}(t+\Delta t) &= \mathbf{x}(t)+\mathbf{u}\Delta t+O(\Delta t^2), \\ \phi(t+\Delta t) &= \phi(t)+\left(\frac{\Delta p}{\rho}+\frac{1}{2}|\mathbf{u}|^2\right)\Delta t+O(\Delta t^2), \end{aligned} \tag{7}$$

where $\mathbf{u}$ is the velocity vector. Higher-order numerical schemes have been used by Lenoir (1976), Prosperetti (1982), and Taib (1985), but this simple Euler scheme appears to be the most robust for the special nature of the variation in time increments found in this problem. This allows us to trace the motion of a selected number of particles, examples of which are shown in the next section.

4. BUBBLES NEAR BOUNDARIES

We illustrate here some of the experimental and theoretical results by choosing selective examples to highlight fundamental aspects of cavitation bubble collapse near boundaries.

4.1 *Rigid Boundary*

After the pioneering studies of Naude & Ellis (1961) and Benjamin & Ellis (1966), a number of experimental studies have been undertaken on vapor bubble motion near a rigid boundary (Gibson 1968, Lauterborn & Bolle 1975, Chahine 1982). The bubble migrates toward the boundary, with the jet forming late in the collapse phase; depending on the distance from the boundary, the jet may strike the boundary after piercing the opposing side of the bubble. The close proximity of the rigid boundary lengthens the lifetime of the bubble.

There have been various estimates of the speed of the liquid jet that strikes a neighboring solid boundary during the collapse of an attached cavitation bubble. Working with collapse pressures of 1 atm, Naude & Ellis (1961) were unable to record the jet development directly. However, they measured peak surface pressures between 100 and 3000 psi (6.89×10^5–2.07×10^7 Pa). With reduced collapse pressures, large bubbles, and a prolonged collapse, both Benjamin & Ellis (1966) and Gibson (1968) were able to record the passage of the liquid jet through the collapsing bubble and measure its speed directly. Their estimates of the speed for a collapse pressure of 1 atm were 50 and 76 m s^{-1}, respectively. Experimenting at atmospheric pressure with very-high-speed photography, Lauterborn & Bolle (1975) and Shima et al. (1981) inferred jet speeds of 90–100 m s^{-1} by measuring the tip speed of the protrusion that develops on the opposite side of the bubble when the jet strikes it.

These speed estimates are less than the Plesset & Chapman (1971) calculated maximum jet speeds of 130–180 m s^{-1}. This discrepancy has recently been explained by Blake et al. (1986a), who calculated the growth and collapse of a bubble in the neighborhood of a rigid boundary. It transpires that the bubble-boundary interaction that takes place during bubble expansion is significant in determining the subsequent motion. These latest calculations agree well with early experiments. An example of an experimental record taken by Gibson & Blake (1980) is shown in Figure 3. In this sequence, the developing jet takes 10 successive frames of the film to traverse the interior of the bubble, and therefore its speeds can be estimated with some precision.

It is only when bubbles attach to the boundary during expansion that the jet strikes the boundary directly during collapse. At other times, the bubble becomes a toroidal cavity that moves toward the boundary while shrinking in volume. It is evident from the early work of Shutler & Mesler (1965) that the torus is unstable and breaks down into small elements in the last moments of the collapse. However, the damage potential of the collapsing torus has not yet been quantified. Extension of the recent theoretical study of Chahine & Genoux (1983) will help to resolve this issue.

Theoretical calculations yield far greater detail on the dynamics of cavitation bubble collapse than can be recorded through experiment. Some of the additional detail that can be obtained is illustrated in Figure 4 for

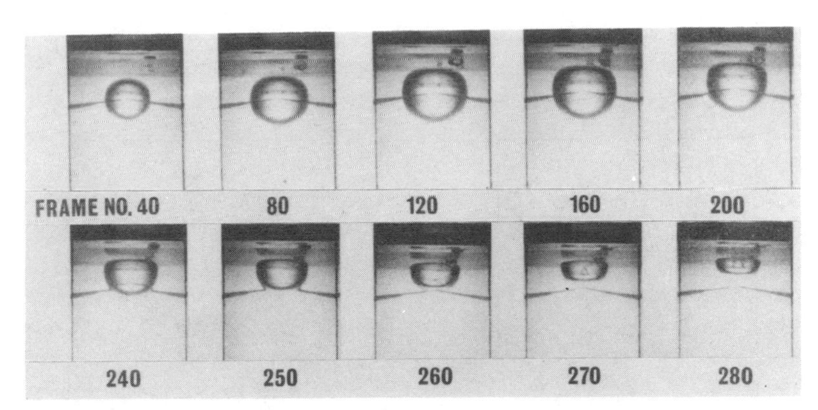

Figure 3 Photographic sequence for bubble shapes near a rigid boundary, which is located on the upper surface. The liquid jet can be clearly seen in frames 270 and 280. The frame numbers are shown on the diagram for a framing speed of 11,150 frames s^{-1}. Maximum bubble radius is 19 mm, with $\gamma = 0.95$.

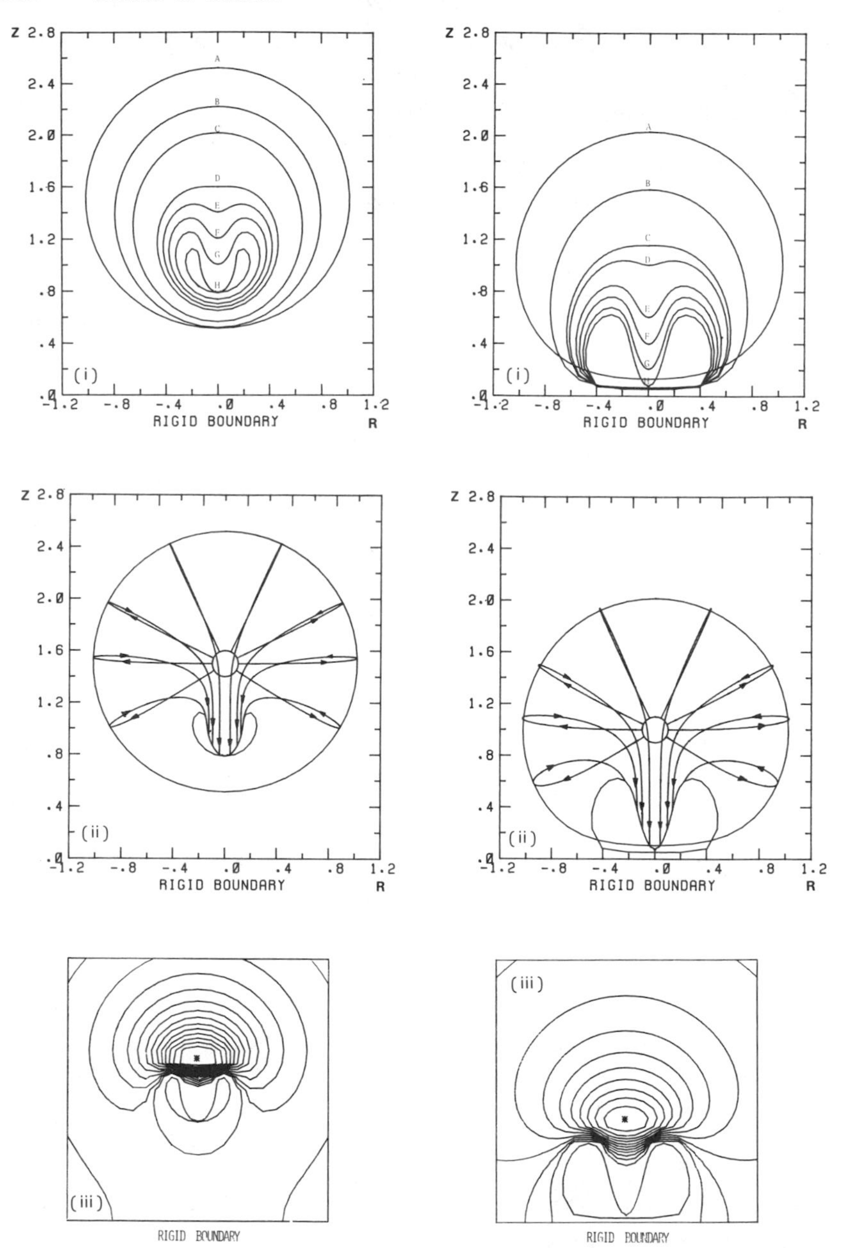

Figure 4 Theoretical calculations for the growth and collapse of vapor bubbles near a rigid boundary for (*left*) $\gamma = 1.5$ and (*right*) $\gamma = 1.0$. The diagrams illustrate (i) successive bubble shapes during collapse, (ii) particle paths through growth and collapse, and (iii) pressure contours late in the collapse phase.

the cases $\gamma = 1.0$ and 1.5. (γ is the ratio of the initial distance of the bubble centroid from the wall to the maximum bubble radius.) The high-speed jet is clearly evident during the collapse phase in both examples, although the maximum jet speed is slower in the $\gamma = 1.0$ case, having a value of $U_m = 8.6(\Delta p/\rho)^{1/2}$ against $11.0(\Delta p/\rho)^{1/2}$ in the $\gamma = 1.5$ example. These calculations compare favorably with the average value of $7.6(\Delta p/\rho)^{1/2}$ for bubbles very close to the boundary that was obtained by Gibson (1968) in his series of experiments. The higher velocities in the $\gamma = 1.5$ case may be attributed to the smaller size of the bubble just prior to its becoming multiply connected. Voinov & Voinov (1975) considered variations in the initial shape of the bubble and calculated substantially higher collapse speeds [up to $64(\Delta p/\rho)^{1/2}$] for only mild variations in the initial eccentricity of a spheroidal bubble. The particle pathlines are particularly interesting, showing the radial growth phase followed by a looping around near "half-life" into the parallel motion of the particles in the jet. Nearly all the particles on the initial bubble surface actually finish up in the jet.

The pressure contours show that the maximum pressure occurs on the axis above the bubble on the side opposite the rigid boundary. This maximum pressure has the effect of decelerating the fluid from infinity but also accelerating that small volume of fluid between it and the bubble surface and hence creating the high speed jet. The Kelvin impulse is directed toward the boundary as predicted in (3). It builds up steadily over the lifetime of the bubble to be finally conserved in a ring vortex around a toroidal bubble that forms after the jet has pierced the other side of the bubble.

Theory also allows the calculation of the motion of a bubble in an ambient velocity and pressure field. Various bubble shapes are shown in Figure 5 for the simplest examples of a buoyant bubble and a bubble in an axisymmetric stagnation-point flow. The dimensionless parameter $\delta = (\rho g R_m/\Delta p)^{1/2}$ determines the importance of buoyancy forces. Physically, it corresponds to the ratio of the time it takes an inviscid bubble of radius R_m to move one radius under the action of gravity to the half-life of a cavitation bubble. The dimensionless strength of the axisymmetric stagnation-point flow is given by the parameter $\alpha = cR_m(\rho/\Delta p)^{1/2}$, where c is the actual strength of the axisymmetric stagnation point flow. (The flow is a pure straining motion, characterized by a rate-of-strain tensor linearly dependent on the strength of the stagnation-point flow.) In the first two examples, buoyancy forces oppose the Bjerknes attraction force toward the rigid boundary: In Figure 5*a* the boundary attraction marginally dominates, yielding a column-shaped bubble, whereas in Figure 5*b* the buoyancy force is slightly stronger, yielding the classic "light bulb"-shaped bubble so familiar from chemical bubble reactors. Figure 5*c* shows an example of the case when both the Bjerknes force and the buoyancy force are acting in the same direction. In this case the jet develops much earlier in the collapse phase, leading to a premature collapse of the bubble where only slow jet speeds are predicted. As suggested by Benjamin & Ellis (1966) the premature collapse of a bubble will lead to a weak jet, which supports the view that only an extremely small percentage of cavitation bubbles are potentially damaging. In stagnation-point flow near a rigid boundary, a balance exists between the fluid motion, the pressure gradient acting away from the boundary, and the Bjerknes attraction. For an appropriate choice

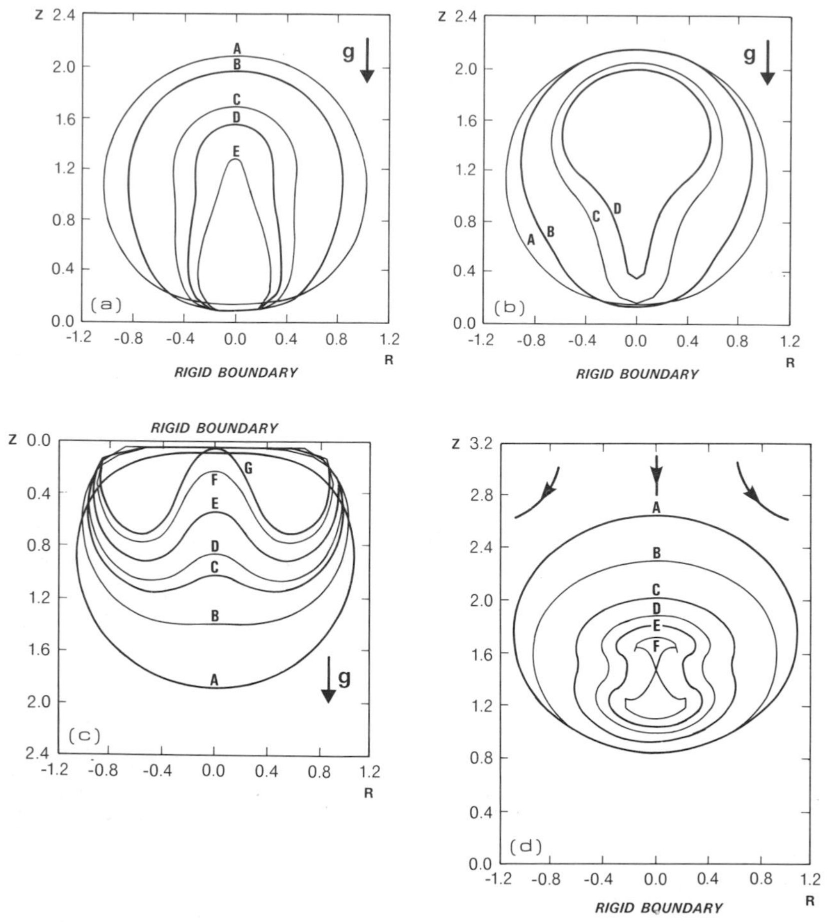

Figure 5 Further examples of the growth and collapse of a vapor bubble near a rigid boundary. In (*a*), (*b*), and (*c*) buoyancy forces are important, whereas in (*d*) the bubble is located in a forward stagnation-point flow. Dimensionless parameters are for (*a*) $\gamma = 1.0$, $\delta = 0.32$, (*b*) $\gamma = 1.0$, $\delta = 0.45$, (*c*) $\gamma = 1.0$, $\delta = -0.45$, and (*d*) $\gamma = 2.0$, $\alpha = 0.15$.

of parameters these contributions may be in balance, leading to unusual bubble shapes. An illustration of this may be found in Figure 5*d*. During the collapse phase the bubble is observed to neck off in the middle, which produces an "hourglass"-shaped bubble. This type of bubble shape has been reported in the literature on several occasions, for example by Ellis (1956) and later by Lauterborn & Hentschel (1985) and Chahine (1982) for a bubble midway between two parallel rigid boundaries and by Gibson & Blake (1982) for bubbles near flexible boundaries (see the discussion below). The maximum pressure in this example occurs as a ring around the central section (Taib 1985). Similar final bubble shapes are also obtained for the collapse of an initially oblate spheroid-shaped bubble (see, e.g., Voinov & Voinov 1975, Chapman & Plesset 1972, Shima & Sato 1981). Voinov & Voinov further analyzed the singular nature of the annular jet, speculating that the final state is two thin pointed jets "striking out" in opposite directions. Thus, a similar state can be achieved either by varying the initial kinematic state of the bubble or, alternatively, by varying the dynamic environment around the bubble.

By using the Kelvin impulse arguments of Section 3, we may postulate that the null-impulse line may divide the γ-δ parameter space into regions where the bubble is attracted toward the boundary and others where it is repelled. On applying this argument it is found that the dividing line is given by $\gamma\delta = 0.442\ldots$ In Figure 6 the direction of bubble migration is illustrated for a number of numerical experiments, together with the line obtained from the null impulse. The null-impulse line would appear to be an excellent discriminator in determining the direction of motion if one bears in mind the approximations that are used in obtaining this value.

4.2 *Free Surface*

The free surface provides a constant-pressure boundary in the near vicinity of a pulsating vapor bubble. The effect is both beautiful and complex. At large separations and in the absence of buoyancy, the bubble moves away from the free surface during collapse and the jet that forms during collapse threads the bubble in the direction of bubble migration. This phenomenon has been examined in detail by Voinov & Voinov (1975), Chahine (1977), Blake & Gibson (1981), and Blake et al. (1986b). An example of the interaction that occurs when a bubble is generated under these conditions, very close to the free surface, is shown in Figure 7.

The free surface develops a pronounced spike moving away from the bubble at the same time that the bubble becomes involuted from above. When buoyancy effects are present, the free-surface spike is greatly diminished in strength, and at a greater bubble separation the effects of buoyancy dominate the motion to such an extent that the bubble rises toward the

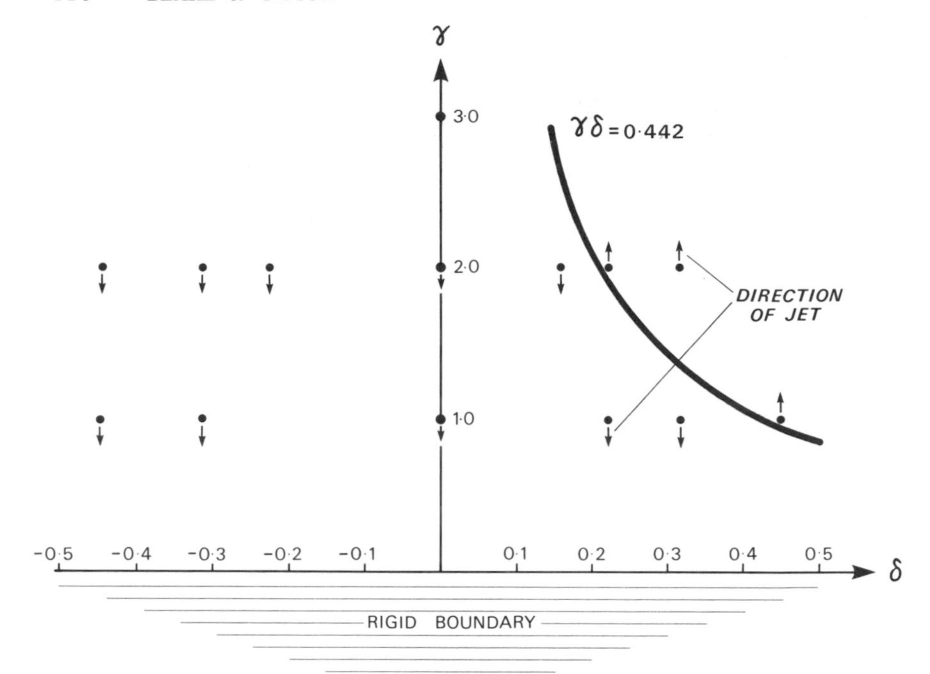

Figure 6 The γ-δ parameter space for buoyant vapor bubbles. The curve obtained from the null-Kelvin-impulse argument is indicated, together with a selection of parameters taken from numerical experiments. The arrows indicate the direction of bubble migration and jet formation.

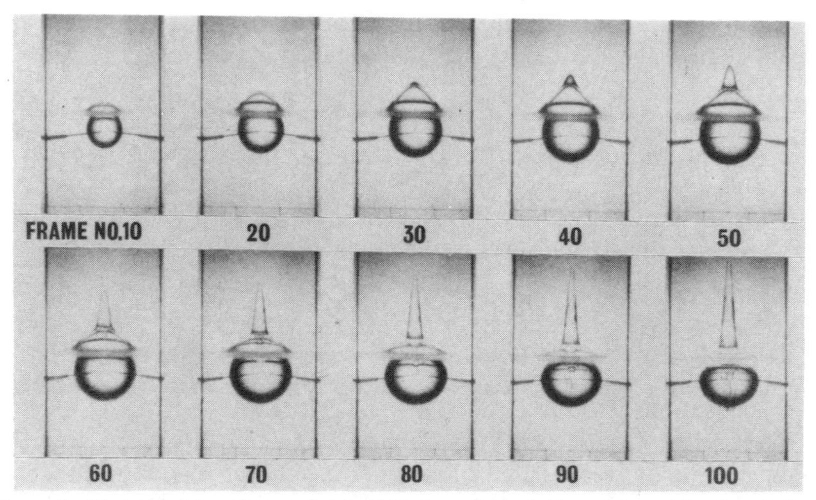

Figure 7 Photographic sequence for bubble shapes near a free surface. The free-surface "spike" and the liquid jet penetrating the bubble are clearly evident. The framing rate in this experiment is 12,640 frames s^{-1}, and the maximum bubble radius is 18 mm, with $\gamma = 0.56$.

surface during collapse. Further understanding of this phenomenon may be obtained from the diagrams in Figure 8, which show the theoretical calculations for bubble shape, particle pathlines, and pressures for the $\gamma = 1.5$ and 1.0 examples. There is a significantly different response between the $\gamma = 1.5$ and 1.0 examples, even though in both cases the Bjerknes effect is directed away from the free surface. In the $\gamma = 1.5$ example the theoretical predictions show that the collapsing bubble has a broad jet; this result compares favorably with the observations of Chahine & Bovis (1983). Furthermore, the free surface is seen to rise and fall with the growth and collapse of the bubble. However, in the $\gamma = 1.0$ example the liquid jet in the bubble is found to be very narrow (as observed in Figure 7), and additionally the free-surface spike continues to elongate during collapse (which is so clearly illustrated in the particle paths). The pressure contours reveal that the maximum pressure occurs between the bubble and free surface, while the instantaneous streamlines show an axially symmetric stagnation point flow (Cerone & Blake 1984).

4.3 *Interface Between Two Fluids of Differing Density*

Chahine & Bovis (1980) conducted a series of experiments on bubbles close to an interface between two fluids of differing density. The observed motion of the bubbles appeared to depend on the distance of the bubble from the interface. Bubbles close to the boundary were repelled by the boundary, whereas bubbles farther away were attracted toward the interface, with the final state appearing to be a vortex ring moving in the appropriate direction. It is suspected that buoyancy forces may play an important role in these observations in that while the light fluid repels the bubble when sufficiently close (as occurs for the free surface), farther away buoyancy forces dominate the motion and cause migration toward the interface. At very large distances from the interface, almost spherical bubbles are observed.

4.4 *Composite and Flexible Materials*

Fewer experimental studies have been conducted on flexible boundaries. One might expect a response between those of the two extremes of a rigid boundary and a free surface. Physically, the flexible boundary can extract energy from the fluid motion that may be dissipated within the boundary or returned to the fluid motion sometime later. The characteristic time scale of the flexible boundary is central to the response of the cavitation bubble. For light materials the response will be similar to that of the free surface; for heavier materials the response will tend toward that of a rigid boundary. An elastic restoring force can inject momentum back into the liquid.

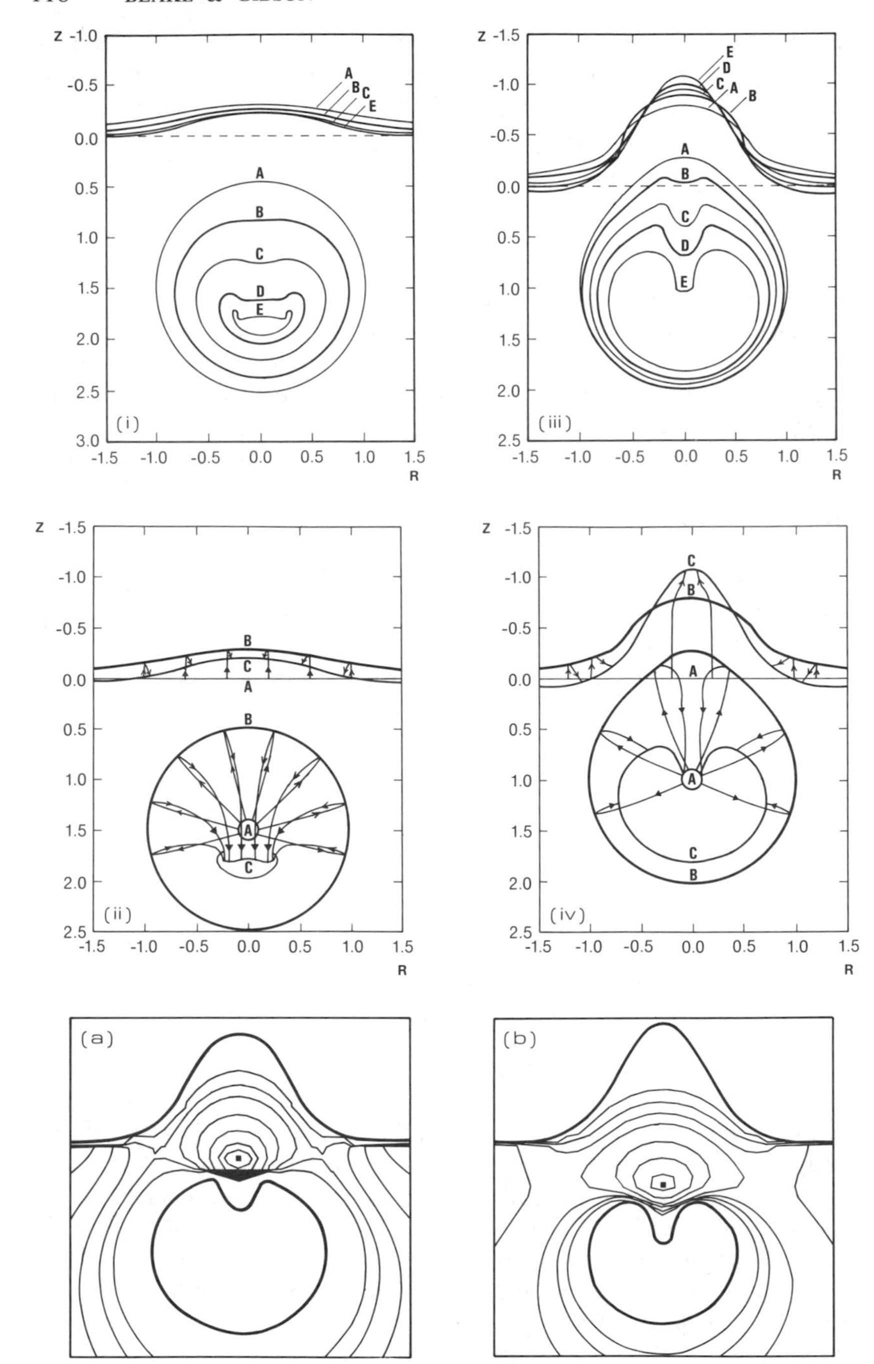

Z
R
A
B
C
D
E
(i)
(ii)
(iii)
(iv)
(a)
(b)

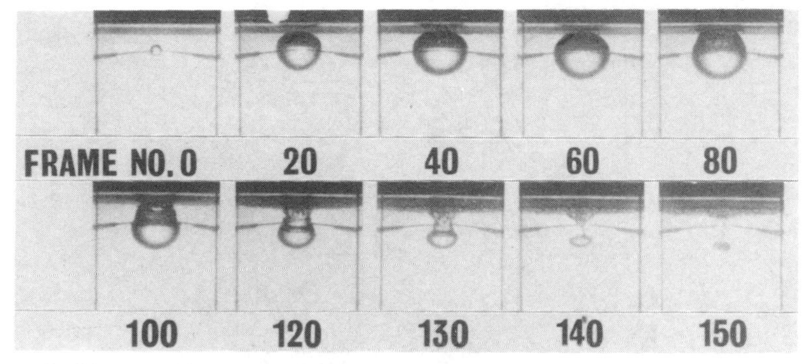

Figure 9 Photographic sequence for bubble shapes near a stretched vulcanized-rubber sheet with air backing. During the collapse phase, an "annular jet" yields an "hourglass"-shaped bubble. The framing rate is 13,540 frames s^{-1}, and the maximum bubble radius is $R_m = 16$ mm.

Gibson (1968) and Gibson & Blake (1980, 1982) have conducted a series of experiments to examine the growth and collapse of vapor bubbles in the vicinity of deformable surfaces. Their aim was to determine the type of surface coating that will repel a neighboring pulsating bubble in the same manner as a free surface. Calculations of the Kelvin impulse were helpful in the special cases of a rigid boundary, a free surface, the interface between two liquids, and an inertial boundary. Indeed, these calculations help to explain the strange behavior observed by Chahine & Bovis (1983) for collapse near the interface between two liquids. However, to date, theory has been inadequate to describe the detailed experimental results observed.

Figure 9 shows the behavior of a bubble grown and collapsed in close proximity to a stretched sheet of vulcanized rubber with an air gap on the other side. The expansion is very similar to the expansion adjacent to a rigid surface. The collapse, however, is very different. During collapse, the bubble contracts more rapidly from the sides toward the axis of symmetry and gradually develops an "hourglass" shape. This behavior is very similar to that calculated by Blake et al. (1986a) for a bubble collapsing near to a stagnation point (see Figure 5*d*).

Figure 10 summarizes the results reported by Gibson & Blake (1982) in an extensive study of bubble motion in the vicinity of deformable surfaces.

Figure 8 Theoretical calculations for the growth and collapse of a vapor bubble near a free surface. For $\gamma = 1.5$, successive bubble shapes are shown in (i) and particle pathlines in (ii). Corresponding features are shown in (iii) and (iv) for $\gamma = 1.0$. Pressure contours are illustrated for $\gamma = 1.0$ in (*a*) at dimensionless time $T = 1.186$ and in (*b*) at $T = 1.310$.

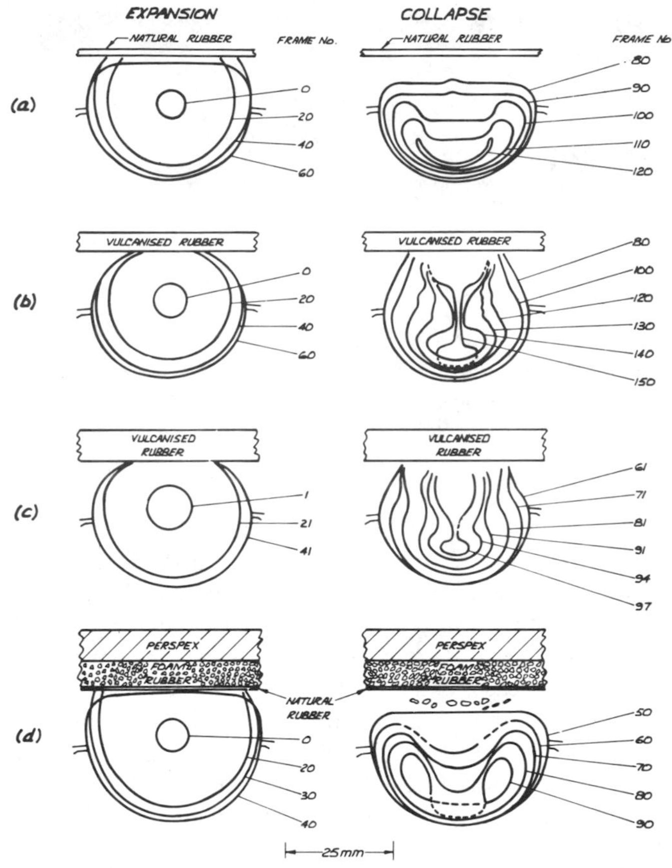

Figure 10 A range of responses for bubbles near different compliant boundaries. All bubbles were generated at frame zero. Camera speeds (frames ms^{-1}) were (*a*) 11.7 (*b*) 13.5, (*c*) 12.9, and (*d*) 12.3.

The figure shows that the bubble motion can be greatly modified by a wide range of deformable surfaces over a wide range of experimental conditions. The challenge now is to develop a robust numerical model that can cope with the mixed boundary conditions provided by a surface coating with inertia, damping, and stiffness.

5. RESEARCH TRENDS AND FUTURE DEVELOPMENTS

For many years experimental observation has led theoretical development in understanding the dynamics of cavitation bubble collapse. Even with the advent of very high-speed cameras, holography, and laser-generated bubbles, it now seems inevitable that computer simulation of bubble processes will in the immediate future lead to a much greater understanding of single-bubble-collapse mechanisms and their influence on boundaries and the flow domain. It is now possible to calculate velocities and pressures anywhere in the flow domain, not just in a few isolated locations. Most experimental observations to date have been restricted to a sequential picture of the bubble shape leading to estimates of particle velocities on the bubble surface. Few attempts have been made to measure pressures on the boundary. Computation, while in general agreement with bubble-shape observation, can indicate particle paths that, for example, show the development of a ring vortex, the parallel particle motion of the high-speed jet, and the development of a stagnation point in the free-surface example, as well as allowing for accurate prediction of particle velocities. In addition, the pressure field can be calculated.

Future developments in the experimental area will include more careful studies of cavitation bubbles near a compliant boundary, either in the form of a flexible or composite surface or of a gas bubble attached to the boundary. The effect of a cloud of bubbles on collapse characteristics will be central to further studies. Early results of Chahine (1984) suggest that pressures and jet velocities may be an order of magnitude larger in bubble clouds, primarily as a result of the much higher apparent far-field driving pressure created by the cloud of bubbles during the collapse phase. Perhaps also the more homogeneous structure of the cloud masks the presence of the boundary, allowing the bubble to maintain a spherical shape for a longer time and producing the much higher pressure that can occur through the Rayleigh implosion mechanism. Clearly a number of facets of bubble-cloud phenomena need to be studied more carefully before any definitive statements can be made about their impact on acknowledged damage mechanisms. The experimental study of cavitation bubbles in specially imposed flow fields may lead to further understanding of the kinematics and dynamic response of cavitation bubbles to different flow environments. The ability of the toroidal bubble/vortex ring system to damage boundaries needs to be studied thoroughly.

Numerical models based on the motion of an incompressible inviscid liquid are accurate when programmed to account for the correct nonlinear velocity and pressure boundary conditions imposed by adjacent surfaces. The effects of mean flow and imposed pressure gradients cannot be ignored when modeling cavitation. It is therefore important that future studies seek to introduce flow effects in a more realistic fashion, and that a systematic investigation be undertaken of the growth and collapse of bubbles in the vicinity of deformable surface coatings. This last work should be a computer study because the parameter space that needs to be covered is too great for an experimental search, at least in the initial stages.

ACKNOWLEDGMENTS

We wish to acknowledge the assistance we have received from Neil Hamilton with photographic aspects of this paper and from Bachok Taib and Greg Doherty with the numerical analysis. Thanks also go to our colleagues in this research area who provided us with details of their most recent research activities. Financial assistance from the Australian Research Grants Scheme in support of our research is gratefully acknowledged.

Literature Cited

Abramowitz, M., Stegun, I. A., eds. 1965. *Handbook of Mathematical Functions*. New York: Dover. 1046 pp.

Arndt, R. E. A. 1981a. Recent advances in cavitation research. *Adv. Hydrosci.* 12: 1–78

Arndt, R. E. A. 1981b. Cavitation in fluid machinery and hydraulic structures. *Ann. Rev. Fluid Mech.* 13: 273–328

Benjamin, T. B., Ellis, A. T. 1966. The collapse of cavitation bubbles and the pressures thereby produced against solid boundaries. *Philos. Trans. R. Soc. London Ser. A* 260: 221–40

Besant, W. H. 1859. *Hydrostatics and Hydrodynamics*. London: Cambridge Univ. Press

Blake, J. R., Cerone, P. 1982. A note on the impulse due to a vapour bubble near a boundary. *J. Aust. Math. Soc. B* 23: 383–93

Blake, J. R., Gibson, D. C. 1981. Growth and collapse of a vapour cavity near a free surface. *J. Fluid Mech.* 111: 123–40

Blake, J. R., Taib, B. B., Doherty, G. 1986a. Transient cavities near boundaries. Part I. Rigid boundary. *J. Fluid Mech.* 170: 479–97

Blake, J. R., Taib, B. B., Doherty, G. 1986b. Transient cavities near boundaries. Part II. Free surface. Submitted for publication

Burrill, L. C. 1951. Sir Charles Parsons and cavitation. *Trans. Inst. Mar. Eng.* 63(8): 149–67

Cerone, P., Blake, J. R. 1984. A note on the instantaneous streamlines, pathlines and pressure contours for a cavitation bubble near a boundary. *J. Aust. Math. Soc. B* 26: 31–44

Chahine, G. L. 1977. Interaction between an oscillating bubble and a free surface. *Trans. ASME, J. Fluids Eng.* 99: 709–15

Chahine, G. L. 1982. Experimental and asymptotic study of non-spherical bubble collapse. *Appl. Sci. Res.* 38: 187–98

Chahine, G. L. 1984. Pressure generated by a bubble cloud collapse. *Chem. Eng. Commun.* 28: 355–67

Chahine, G. L., Bovis, A. G. 1980. Oscillations and collapse of a cavitation bubble in the vicinity of a two-fluid interface. In *Cavitation and Inhomogeneities in Underwater Acoustics*, ed. W. Lauterborn. Berlin: Springer-Verlag. 319 pp.

Chahine, G. L., Bovis, A. G. 1983. Pressure field generated by non-spherical bubble collapse. *Trans. ASME, J. Fluids Eng.* 105: 356–63

Chahine, G. L., Genoux, P. F. 1983. Collapse of a cavitating vortex ring. *Trans. ASME, J. Fluids Eng.* 105: 400–5

Chapman, R. B., Plesset, M. S. 1972. Nonlinear effects in the collapse of a nearly spherical cavity in a liquid. *Trans. ASME, J. Basic Eng.* 94: 142–46

Ellis, A. T. 1956. Techniques for pressure pulse measurements and high speed photography in ultrasonic cavitation. In *Cavitation in Hydrodynamics, HMSO London Pap. 8*, pp. 1–32

Gibson, D. C. 1968. Cavitation adjacent to plane boundaries. *Proc. Aust. Conf. Hydraul. and Fluid Mech., 3rd*, pp. 210–14. Sydney: Inst. Eng.

Gibson, D. C. 1982a. The kinetic and thermal expansion of vapor bubbles. *Trans. ASME, J. Basic Eng.* 94: 89–96

Gibson, D. C. 1982b. The pulsation time of spark induced vapor bubbles. *Trans. ASME, J. Basic Eng.* 94: 248–49

Gibson, D. C., Blake, J. R. 1980. Growth and collapse of cavitation bubbles near flexible boundaries. *Proc. Aust. Conf.*

Hydraul. and Fluid Mech., 7th, pp. 283–86. Brisbane: Inst. Eng.

Gibson, D. C., Blake, J. R. 1982. The growth and collapse of bubbles near deformable surfaces. *Appl. Sci. Res.* 38: 215–24

Guerri, L., Lucca, G., Prosperetti, A. 1982. A numerical method for the dynamics of non-spherical cavitation bubbles. *Proc. Int. Colloq. Drops and Bubbles, 2nd*, pp. 175–81. Pasadena, Calif: JPL Publ.

Hammitt, F. G. 1980. *Cavitation and Multiphase Flow Phenomena*. New York: McGraw-Hill

Kling, C. L., Hammitt, F. G. 1972. A photographic study of spark-induced cavitation bubble. *Trans. ASME, J. Basic Eng.* 94: 825–33

Knapp, R. T., Daily, J. W., Hammitt, F. G. 1970. *Cavitation*. New York: McGraw-Hill. 578 pp.

Lauterborn, W. 1982. Cavitation bubble dynamics–new tools for an intricate problem. *Appl. Sci. Res.* 38: 165–78

Lauterborn, W., Bolle, H. 1975. Experimental investigations of cavitation bubble collapse in the neighbourhood of a solid boundary. *J. Fluid Mech.* 72: 391–99

Lauterborn, W., Hentschel, W. 1985. Cavitation bubble dynamics studied by high speed photography and holography. Part I. *Ultrasonics*. In press

Lauterborn, W., Vogel, A. 1984. Modern optical techniques in fluid mechanics. *Ann. Rev. Fluid Mech.* 16: 223–44

Lenoir, M. 1976. Calcul numérique de l'implosion d'une bulle de cavitation au voisinage d'une paroi ou d'une surface libre. *J. Méc.* 15: 725–51

Longuet-Higgins, M. S. 1983. Bubbles, breaking waves and hyperbolic jets at a free surface. *J. Fluid Mech.* 127: 103–22

Longuet-Higgins, M. S., Cokelet, E. D. 1976. The deformation of steep surface waves on water. 1. A numerical method of computation. *Proc. R. Soc. London Ser. A* 350: 1–26

Mitchell, T. M., Hammitt, F. H. 1973. Asymmetric cavitation bubble collapse. *Trans. ASME, J. Fluids Eng.* 95: 29–37

Morch, K. A. 1980. On the collapse of cavity clusters in flow cavitation. In *Cavitation and Inhomogeneities in Underwater Acoustics*, ed. W. Lauterborn, pp. 95–100. Berlin: Springer-Verlag

Naude, C. F., Ellis, A. T. 1961. On the mechanism of cavitation damage by nonhemispherical cavities in contact with a solid boundary. *Trans. ASME, J. Basic Eng.* 83: 648–56

Plesset, M. S., Chapman, R. B. 1971. Collapse of an initially spherical vapour cavity in the neighbourhood of a solid boundary. *J. Fluid Mech.* 47: 283–90

Plesset, M. S., Prosperetti, A. 1977. Bubble dynamics and cavitation. *Ann. Rev. Fluid Mech.* 9: 145–85

Prosperetti, A. 1982. Bubble dynamics: a review and some recent results. *Appl. Sci. Res.* 38: 145–64

Rayleigh, Lord. 1917. On the pressure developed in a liquid during the collapse of a spherical void. *Philos. Mag.* 34: 94–98

Reynolds, O. 1894. Experiments showing the boiling of water in an open tube at ordinary temperatures. *Br. Assoc. Adv. Sci. Rep. 564*. See also *Sci. Pap.* 2: 578, 1901

Shima, A., Sato, Y. 1980. The behaviour of a bubble between narrow parallel plates. *ZAMP* 31: 691–704

Shima, A., Sato, Y. 1981. The collapse of a spheroidal bubble near a solid wall. *J. Méc.* 20: 253–71

Shima, A., Takayama, Y., Tomita, Y., Miura, N. 1981. An experimental study on effects of a solid wall on the motion of bubbles and shock waves in bubble collapse. *Acustica* 48: 293–301

Shutler, N. D., Mesler, R. B. 1965. A photographic study of the dynamics and damage capabilities of bubbles collapsed near solid boundaries. *Trans. ASME, J. Basic Eng.* 87: 511–17

Smith, R. H., Mesler, R. B. 1972. A photographic study of the effect of an air bubble on the growth and collapse of a vapor bubble near a surface. *Trans. ASME, J. Basic Eng.* 94: 933–42

Stroud, A. H. Secrest, D. 1966. *Gaussian Quadrature Formulas*. Englewood Cliffs, NJ: Prentice-Hall

Taib, B. B. 1985. *Boundary integral methods applied to cavitation bubble dynamics*. PhD thesis. Univ. Wollongong, Aust.

Taib, B. B., Doherty, G., Blake, J. R. 1984. Boundary integral methods applied to cavitation bubble dynamics. *Proc. Cent. for Math. Anal. ANU Math. Program. and Numer. Anal. Workshop*, ed. S. A. Gustafson, R. S. Womersley, 6: 166–86. Canberra: ANU Press

Voinov, O. V. 1973. Force acting on a sphere in an inhomogeneous flow of an ideal incompressible fluid. *J. Appl. Mech. Tech. Phys.* 14: 592–94

Voinov, O. V., Voinov, V. V. 1975. Numerical method of calculating non-stationary motions of an ideal incompressible fluid with free surfaces. *Sov. Phys. Dokl.* 20: 179–80

van Wijngaarden, L. 1976. Hydrodynamic interaction between gas bubbles in liquid. *J. Fluid Mech.* 77: 27–44

Yakimov, Yu. L. 1983. Forces acting on a small body in a flowing incompressible liquid and equations of motion of a two-phase medium. *Fluid Dyn.* 8: 411–18

Ann. Rev. Fluid Mech. 1987. 19: 125–55

A DESCRIPTION OF EDDYING MOTIONS AND FLOW PATTERNS USING CRITICAL-POINT CONCEPTS

A. E. Perry and M. S. Chong

Department of Mechanical Engineering, University of Melbourne, Victoria, Australia

INTRODUCTION

There is a difficulty in conceptually visualizing three-dimensional steady and unsteady flow patterns. A description of flow patterns using critical-point concepts provides a framework and methodology for overcoming this difficulty. It also provides a language or vocabulary that enables complicated three-dimensional flow patterns to be described in an intelligible and unambiguous manner. Critical points are points in the flow field where the streamline slope is indeterminate and the velocity is zero relative to an appropriate observer. Asymptotically exact solutions of the Navier-Stokes and continuity equations can be derived close to the critical points, and these give a number of standard flow patterns.

A knowledge of critical-point theory is important for interpreting and understanding flow patterns whether they are obtained experimentally or computationally. For example, Figure 1 (*left*) shows the surface dye pattern on the downwind side of a missile-shaped body at an angle of attack. If we know the classification of critical points, the topologically correct pattern of limiting surface streamlines (i.e. lines of surface shear stress) can be immediately sketched as shown in Figure 1 (*right*). The pattern consists of an arrangement of nodes, saddles, and foci that will be elaborated on later.

Critical points are the salient features of a flow pattern; given a distribution of such points and their type, much of the remaining flow field

0066–4189/87/0115–0125$02.00

and its geometry and topology can be deduced, since there is only a limited number of ways that the streamlines can be joined. Even though a flow field may be unsteady, the instantaneous streamline patterns give an idea of the transport properties of an array of eddies in jets and wakes or in complicated three-dimensional separation patterns. Also, a knowledge of a flow field in one plane can often give a clue to the possible flows in other planes.

The motivation for developing this theory has grown considerably in recent years because of the large volume of three-dimensional steady and unsteady flow field data that can now be produced both experimentally and computationally with the advent of large computers and high-speed data-acquisition systems. These data need to be interpreted and understood. The correctness of a measurement or a computation can often be checked by noting whether the results conform with the known constraints given by critical-point theory.

The first systematic classification of critical points using local solutions of the Navier-Stokes and continuity equations was made by Oswatitsch (1958). Some important aspects of three-dimensional separation that incorporated critical-point descriptions were also studied by Maskell (1955, 1956) and many others, including Wang (1972, 1974a,b, 1975), Lighthill (1963), Legendre (1956, 1965), Werlé (1962, 1975), and Smith (1972). The Poincaré-Bendixson theorem was introduced to fluid-flow problems by Davey (1961) and extended by Hunt et al. (1978) for the study of surface patterns on obstacles attached to boundaries. This

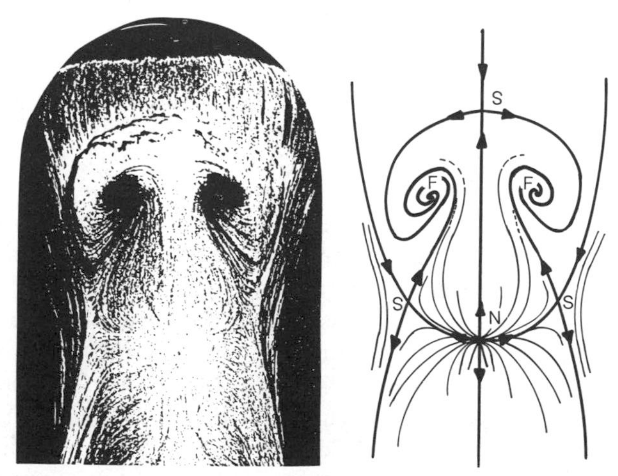

Figure 1 (*left*) Surface flow pattern on the downwind side of a missile-shaped body at an angle of attack. Courtesy of Dr. B. D. Fairlie. (*right*) Interpretation of flow pattern. Abbreviations as follows: F, focus; N, node; S, saddle.

theorem, when applied to bodies that are topologically equivalent to spheres, is often called the "hairy-sphere" theorem. The way the hair can be combed on a sphere is limited. The theorem simply states that the number of nodes less the number of saddles on a sphere must equal two. Foci count as nodes for the theorem. Other topological rules have been examined by Tobak & Peake (1982) (see later). All of the above works were concerned with critical points at no-slip boundaries. Perry & Fairlie (1974) applied phase-plane and phase-space techniques, which are used in the study of ordinary autonomous differential equations, to the study of critical points at no-slip boundaries and also to points within the fluid away from boundaries, such as those that occur in separation bubbles, jets, and wakes. Cantwell et al. (1978) applied critical-point theory to the description of the conditionally sampled velocity flow fields of turbulent spots. Cantwell (1978, 1981) and Cantwell & Allen (1984) applied critical-point theory to particle-path patterns normalized by similarity transformations in impulsively started jets, and found critical Reynolds numbers for changes in the jet-pattern topology. Recently, Kaynak et al. (1986) reported on work where experimental and computed transonic separated flows over low-aspect-ratio wings were interpreted with the aid of critical-point concepts.

APPLICATION OF CRITICAL-POINT CONCEPTS TO FLUID-FLOW PATTERNS

Critical-point theory (also known as the "phase-plane" or "phase-space" theory) has been used in examining solutions of autonomous ordinary differential equations (see, for example, Kaplan 1958, Pontryagin 1962, Andronov et al. 1966, Minorsky 1962). A brief summary of the technique as applied to fluid-flow patterns is given below.

Consider the Navier-Stokes and continuity equations in Cartesian tensor form given below for incompressible flow and uniform density:

$$\frac{\partial u_i}{\partial t} + u_j \frac{\partial u_i}{\partial x_j} = -\frac{1}{\rho}\frac{\partial p}{\partial x_i} + \nu \frac{\partial^2 u_i}{\partial x_j \partial x_j}, \tag{1}$$

$$\frac{\partial u_i}{\partial x_i} = 0. \tag{2}$$

Here ρ is fluid density, p is pressure, and ν is kinematic viscosity. It is assumed that the solutions of the uniform-density incompressible Navier-Stokes equations are regular at all points. Hence, an arbitrary point O in the flow field can be chosen and a Taylor series can be used to expand the velocity u_i in terms of the space coordinates x_j, with the origin for x_j located at O. This gives

$$u_i = \dot{x}_i = A_i + A_{ij}x_j + A_{ijk}x_jx_k + A_{ijkl}x_jx_kx_l + \ldots. \tag{3}$$

The coefficients A_i, A_{ij}, etc., are functions of time if the flow is unsteady, and they are symmetric tensors in all indices except the first. If O is located at a critical point, then, since the streamline slope is indeterminate,

$$\frac{u_i}{u_j} = \frac{0}{0} \quad \text{for} \quad i \neq j. \tag{4}$$

This means that the zeroth-order terms A_i are equal to zero. There are two types of critical points that we consider. The first type is the so-called no-slip critical point, such as a separation point on a no-slip boundary. The no-slip boundary condition means that $A_{ij} = 0$ and A_{ijk} is finite if $j = 3$ or $k = 3$ but zero for all other cases, given that x_3 is the space coordinate normal to the no-slip boundary. It is permissible to truncate (3) after the second-order terms and by substitution into (1) and (2) obtain relationships between the various coefficients A_{i3k} and A_{ij3}. Such a truncated expansion is an asymptotically exact solution to (1) and (2) about the point O.

The second type of critical point, the so-called free-slip critical point (Hornung & Perry 1984), occurs within the fluid away from no-slip boundaries—for example, in separation bubbles and in jet and wake patterns that are discussed later. This type of point was first considered by Perry & Fairlie (1974). The coefficient A_i is again zero but A_{ij} is finite. A local asymptotically exact solution to (1) and (2) can be obtained if terms up to third order are included. The higher-order terms, which introduce into the solution the effects of viscosity, have an influence on the relationships between the first-order coefficients A_{ij}. Perry & Fairlie (1974) discussed this type of critical point but truncated the series after the first-order terms. In general, this procedure is valid only for the Euler equations or for flows where vorticity is spatially uniform, and this includes irrotational flow (Perry 1984a).

In order to obtain an understanding of the geometry and topology of the flow immediately surrounding critical points, the three-dimensional streamline patterns, or "solution trajectories," are obtained by integrating the velocity field; only the lowest-order terms are required. For instance, the no-slip critical-point velocity field can be put in the form

$$\frac{u_i}{x_3} = B_{i\alpha}x_\alpha, \tag{5}$$

where $B_{i\alpha}$ is linearly related to A_{i3k} and A_{ij3}. For the limit $x_3 \to 0$, we obtain the "limiting streamlines" or lines of surface shear stress. This may be expressed as

$$\frac{dx_i}{d\tau} = B_{i\alpha}x_\alpha, \tag{6}$$

where the time variable $d\tau = x_3 dt$, and t is real time.

For the free-slip critical point, we have

$$u_i = \frac{dx_i}{dt} = A_{i\alpha}x_\alpha. \tag{7}$$

For unsteady flow the elements of $B_{i\alpha}$ and $A_{i\alpha}$ will be functions of time; since u_i is an instantaneous velocity, the solution trajectories are instantaneous streamlines.

The properties and classification of all the possible flow patterns that can occur are obtained by an examination of the elements of the 3×3 matrix represented by $A_{i\alpha}$ or $B_{i\alpha}$. The elements can be related to the local flow properties, such as pressure gradients, vorticity gradients, strain-rate coefficients, etc., at the critical points. Important properties of these matrices are the eigenvalues and the eigenvectors. If the eigenvalues and eigenvectors are real, then three planes can be defined by the eigenvectors, and these will be referred to as the eigenvector planes. These planes need not be mutually orthogonal, and in general they are the only planes that contain solution trajectories (i.e. some of the streamlines osculate to these planes close to the critical point). If the eigenvalues and eigenvectors are complex, then there will exist in general only one plane that contains solution trajectories. In these planes, simple phase-plane methods can be used to classify the critical-point patterns. The following equation is obtained by defining a new coordinate system in each plane in turn:

$$\begin{bmatrix} \dot{y}_1 \\ \dot{y}_2 \end{bmatrix} = \begin{bmatrix} a & b \\ c & d \end{bmatrix} \begin{bmatrix} y_1 \\ y_2 \end{bmatrix}, \tag{8}$$

or

$$\dot{\mathbf{y}} = \underset{\approx}{\mathrm{F}}\mathbf{y}.$$

Here the dot above the y denotes differentiation with respect to τ in the case of no-slip critical points, or to t in the case of free-slip critical points. The two quantities of importance are

$$\begin{aligned} p &= -(a+d) = -\text{trace } \underset{\approx}{\mathrm{F}}, \\ q &= (ad-bc) = \det \underset{\approx}{\mathrm{F}}. \end{aligned} \tag{9}$$

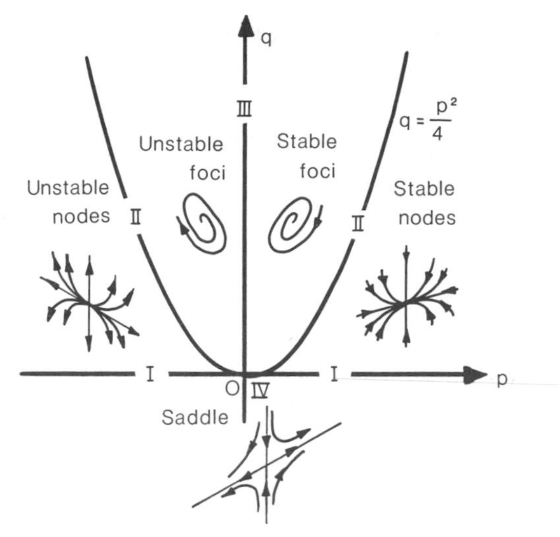

Figure 2 Classification of critical points on the *p-q* chart. For critical points on the boundaries I, II, and III and the origin IV, see Figure 5.

The types of critical points are classified on the *p-q* chart shown in Figure 2. Thus nodes, foci, or saddles can be obtained. The pattern depends on the region of location of a point defined by *p* and *q* on the *p-q* chart.

If all the eigenvalues are real, either nodes or saddles can be produced. These patterns in general will be in noncanonical form, i.e. the eigenvectors in the plane under consideration are nonorthogonal. Figure 3 (*left*) shows a node in noncanonical form, and Figure 3 (*right*) shows a node in canonical form. This is achieved by distorting the noncanonical pattern by an affine transformation, i.e. by a coordinate stretching (with a constant stretching factor) and differential rotation of the coordinates. If the eigen-

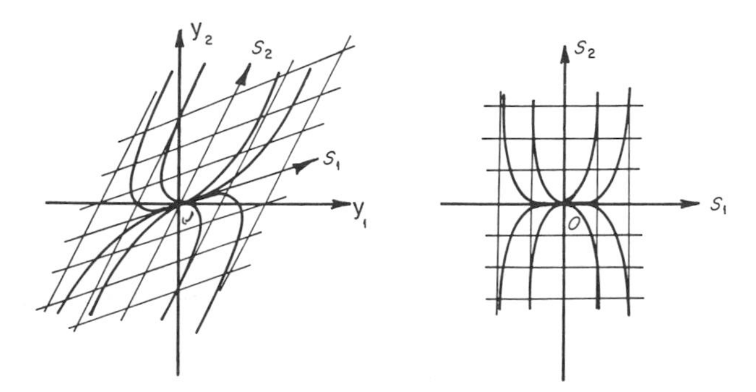

Figure 3 (*left*) Node in noncanonical form. (*right*) Node in canonical form. S_1 and S_2 are eigenvectors.

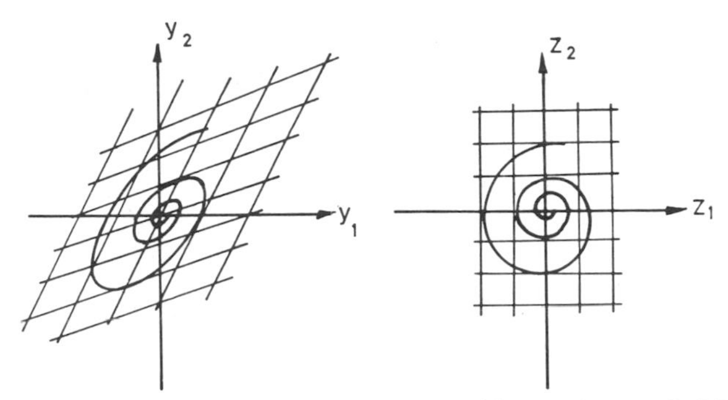

Figure 4 (*left*) Focus in noncanonical form. (*right*) Focus in canonical form.

values are complex and the y_1y_2 plane contains solution trajectories, a focus is obtained. Figure 4 (*left*) shows a noncanonical focus, and Figure 4 (*right*) shows a focus in canonical form. When in canonical form, nodes and saddles have solution trajectories that are simple power laws, i.e. $y_2 = Ky_1^m$, whereas foci reduce to simple logarithmic spirals. If the pattern occurs on the boundaries of the p-q chart (i.e. when $p^2 = 4q$, $p = 0$, or $q = 0$), then we have "borderline" cases. These are often referred to as "degenerate" critical points and are shown in Figure 5 (see Kaplan 1958).

Complicated three-dimensional critical-point patterns can therefore be understood by simply looking at the solution trajectories in each of the eigenvector planes. For example, Figure 6 shows irrotational free-slip critical points. For irrotational flow, all eigenvectors are mutually orthogonal, and so critical-point patterns in the eigenvector planes are canonical. This is not generally true with vorticity present. It should be noted that in axisymmetric and plane stagnation point flows, the patterns are degener-

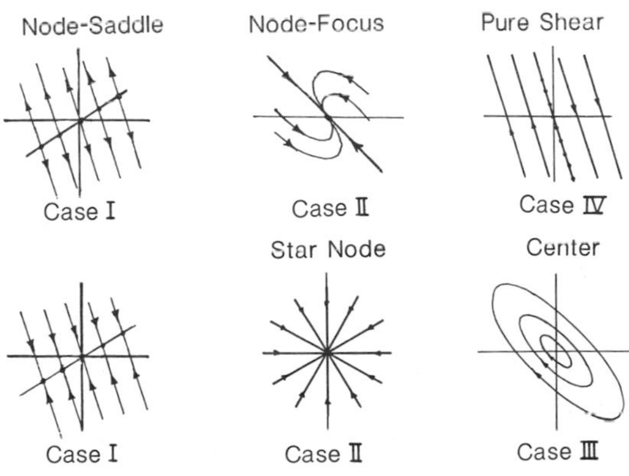

Figure 5 Degenerate critical points, or borderline cases.

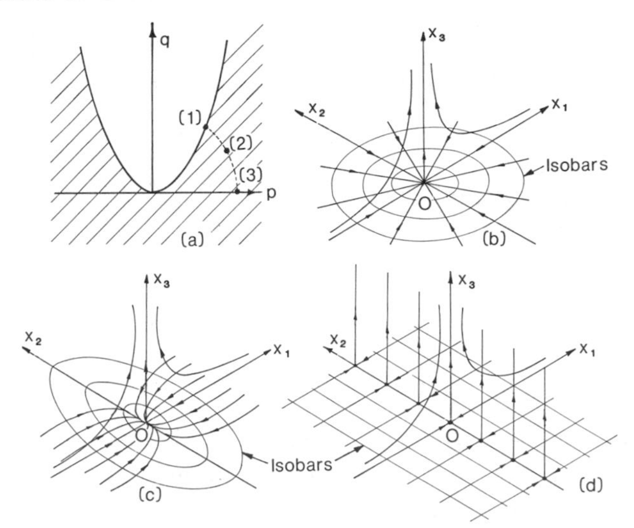

Figure 6 Critical points resulting from linear irrotational flow analysis. (*a*) Shaded zone shows allowable region for such points on the *p-q* chart for the x_1x_2 plane; (*b*) case (1); (*c*) case (2); (*d*) case (3).

ate. When all eigenvalues are real, two eigenvector planes will contain saddles and one plane will contain a node. This is true also for rotational flow and will be referred to as the "saddle-node combination" theorem. If the eigenvalues are complex, then one plane will contain a focus and solution trajectories will wrap around the one real eigenvector. Figure 7 shows such a pattern that often occurs in three-dimensional separation from a surface (see Lim et al. 1980).

A degenerate focus is a center, and this will occur only in flows that are locally two-dimensional. For two-dimensional flow, there are planes that

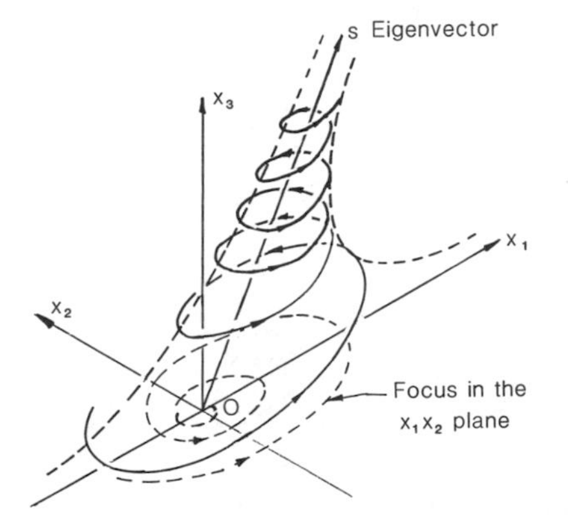

Figure 7 Critical point with complex eigenvalues.

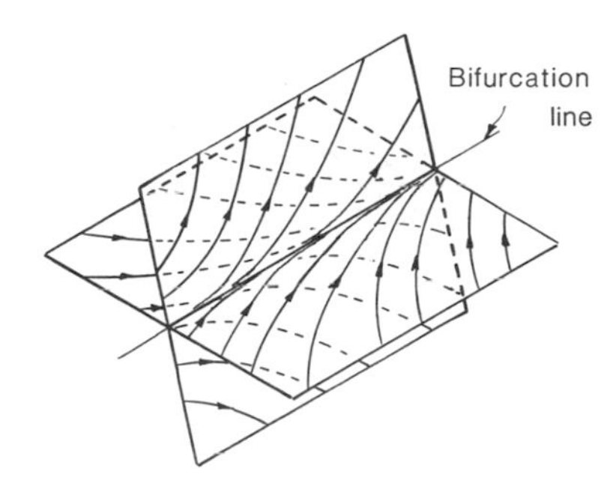

Figure 8 A bifurcation line.

contain streamlines where the divergence is zero, and the only critical points that can occur in these planes are centers and saddles.

Other flow-pattern features commonly occur and give insights into the physical processes. For instance, it is possible to have bifurcation lines, as discussed by Hornung & Perry (1984) and Perry & Hornung (1984) (see Figure 8). These are lines drawn in the flow toward which other trajectories are asymptotic. The bifurcation line locally is the intersection of two planes that contain solution trajectories. Locally, in some simple cases, the solution trajectories in one of the planes approach the bifurcation line exponentially, and in the other plane they diverge exponentially. A means of analytically defining the location of a bifurcation line in a given pattern has been discussed by Hornung & Perry (1984). A bifurcation line may "in the large" be curved and may be without any well-defined beginning or end. It may also form a closed curve to give a limit cycle, as shown in Figure 12. On a surface with three-dimensional separation, bifurcation lines are often called separation or reattachment lines, but such lines can also occur within the fluid away from boundaries. Most workers currently hold to the view that bifurcation lines are "asymptotes" rather than the Maskell (1956) or Buckmaster (1972) suggestion that they are "envelopes" to which other trajectories joint tangentially. In two-dimensional flow, in planes with zero divergence, bifurcation lines and limit cycles are not permitted.

A dislocated saddle (consisting of two half-saddles) is a feature that occurs on a vortex sheet, as shown in Figure 9 (*left*). This is an unsteady pattern and is an exact solution to the Euler equations. The two half-saddles approach each other exponentially in time, and the strength of the sheet (the velocity jump across it) diminishes exponentially in time. The sheet is being stretched across the vortex filaments (see Perry et al. 1980, Perry & Watmuff 1981). This can occur between two Kelvin-Helmholtz-like roll-ups, as shown in Figure 10. This is an irregular critical point, since it cannot be series expanded across the vortex sheet but it can be expanded

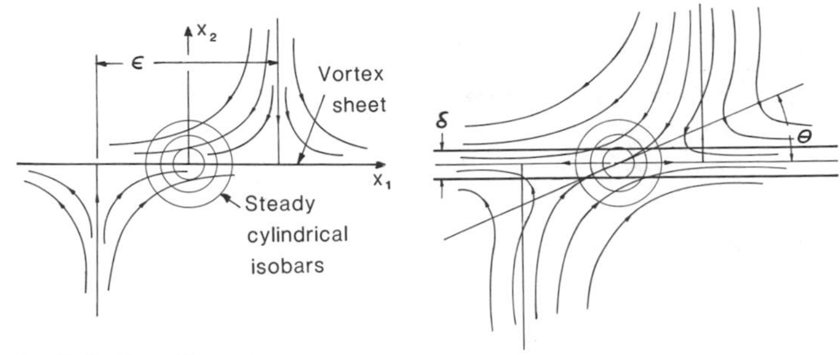

Figure 9 (*left*) Two-dimensional dislocated saddle on a vortex sheet, with ε decreasing exponentially with time. (*right*) Distorted saddle on a finite-thickness (δ) shear layer. Here θ is the angle between eigenvectors. As δ and $\theta \to 0$, we obtain the figure on the left.

on either side. However, if viscosity is present or if the pattern is an ensemble average of a large number of randomly positioned patterns, the discontinuity represented by the vortex sheet is "smeared out," and the pattern becomes a distorted regular saddle, as shown in Figure 9 (*right*). Such patterns are seen to occur in the computations of Patnaik et al. (1976) and Corcos & Sherman (1976) for the roll-ups of an initially plane shear layer.

Free-slip critical points require special attention here, particularly the focus. There are two important local solutions in the literature for vortex flows, namely the Rott (1958, 1959) solution and the Burgers (1948) solution. These solutions apply to a region in which the vorticity is unidirectional and aligned with the x_3 axis (say) and the vorticity magnitude is axisymmetrically distributed about the x_3 axis. Both give a similarity solution in vorticity magnitude that is a Gaussian function of radius from the x_3 axis (using cylindrical polar coordinates). In the Rott solution (which uses the Navier-Stokes equations), viscous diffusion causes the characteristic radius of the Gaussian distribution to grow with time [the radius being proportional to $(\nu t)^{1/2}$] and the vorticity magnitude to diminish in time. Thus, the flow is unsteady but the instantaneous streamlines in the x_1x_2 plane form a degenerate center. If the Euler equations are used, viscous diffusion is neglected and a steady pattern with a degenerate center is obtained.

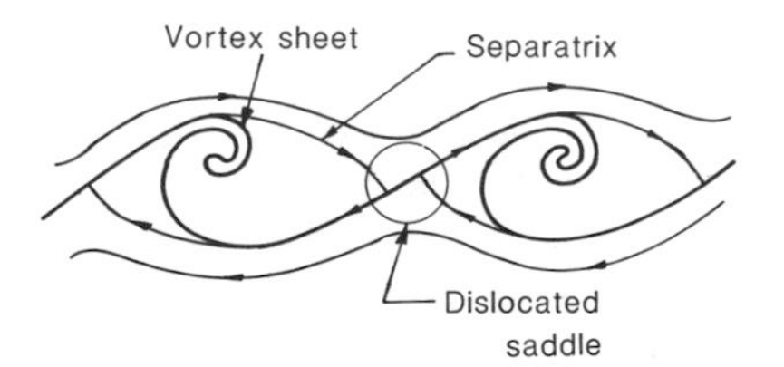

Figure 10 Dislocated saddle on a rolling-up vortex sheet.

If this region of vorticity is now subjected to a steady irrotational axisymmetrical strain rate (of the type shown in Figure 6*b* with the x_3 axis aligned with the vorticity), a similarity solution for vorticity is again obtained with a Gaussian form. On solving the Navier-Stokes and continuity equations, one finds that for very short times (e.g. see Perry & Chong 1982) the region of vorticity shrinks exponentially with time and the vorticity increases exponentially with time. The critical-point pattern in the x_1x_2 plane is a focus and is unsteady. However, after a sufficient time, the stretching effect and viscous diffusion effect "balance," and the distribution of vorticity becomes steady and a steady focus occurs. This is the Burgers vortex. If the Euler equations are solved, a focus is obtained but the pattern is never steady. The vorticity magnitude keeps increasing exponentially with time. It is possible to obtain a steady focus only if viscous diffusion is included in the formulation. This viscous diffusion of vorticity occurs only if we have a spatial variation of vorticity, even when solving the Navier-Stokes equations.

Another solution of interest that appears to be more applicable to flow in jets and wakes, which are considered later, is given by Perry & Chong (1982). When a vortex is being stretched with a constant extensional strain rate, a marked portion of vortex filament increases its length exponentially with time. In the flows cited above, a linear increase in length with time seems more applicable. This means that the strain rate is a function of time. For short times the region of vorticity shrinks exponentially, but after a sufficient time, the characteristic radius grows as $(\nu t)^{1/2}$.

Perry & Fairlie (1974) postulated that the steady complex eigenvalue free-slip critical points are located at peaks of vorticity. To the linearized approximation [using (7)] the vorticity is uniform, and when this is substituted into the Navier-Stokes and continuity equations, the only steady solution permitted is a degenerate center. For the flow to be steady with finite vorticity, local two dimensionality in planes orthogonal to the vorticity vectors is required, i.e. no stretching. Thus, Perry & Fairlie erroneously concluded that steady foci are not possible. However, the inclusion of higher-order terms in the series expansion introduces viscous terms that are of equal importance to the inertia terms. The mistake was to truncate the expansion to the first order. As mentioned earlier, for the more general form of free-slip critical points, we must have at least a third-order expansion.

If the vorticity is finite and uniform and the flow is steady, we are confined completely to the q axis of the p-q chart. All solutions are two dimensional and range from the irrotational saddle to solid-body rotation as q is increased, as shown in Figure 11. As we cross the p axis, we have a degenerate pure shear. Also shown in the figure is the solution for a region of uniform vorticity aligned with the x_2 axis and subjected to a nonaxisymmetrical irrotational strain rate of the type shown in Figure 6*c*, with all arrows reversed in Figure 6*c*. The pattern in the x_1x_3 plane becomes time dependent and follows the path labeled from 1 to 8. The pattern changes from a rotational saddle (eigenvectors are nonorthogonal) into a noncanonical node and then finally into a noncanonical focus. At points 3 and 5 we have degenerate patterns. This example serves to answer the question, "When does a region of vorticity constitute a 'vortex'?" The answer is, "When we have a focus, that is, when we are above the parabola on the *p*-*q* chart (i.e. when the eigenvalues are complex)." As the bound-

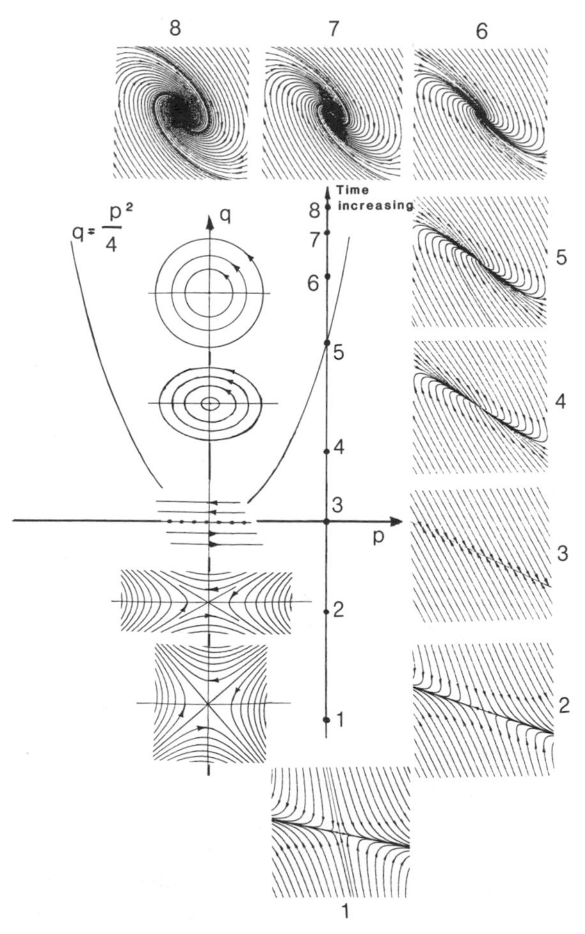

Figure 11 Possible patterns on the *q* axis. Also shown is a region of vorticity undergoing nonaxisymmetric stretching (labeled 1 to 8).

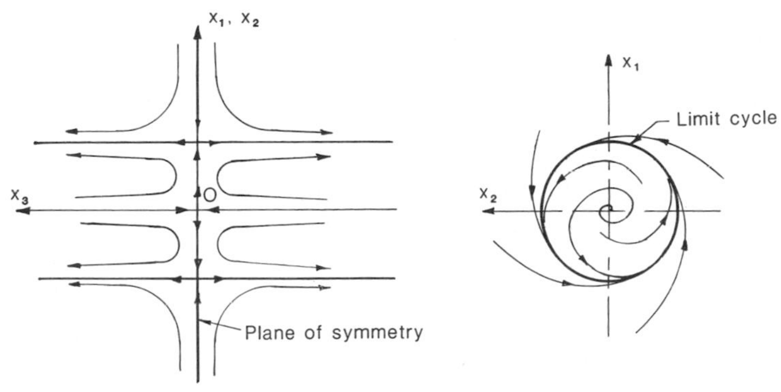

Figure 12 Streamline pattern for a Sullivan vortex. (*left*) Sectional streamlines. (*right*) Actual streamlines in the x_1x_2 plane.

aries of the *p-q* chart are crossed, the critical point is said to undergo a bifurcation (not to be confused with the bifurcation lines mentioned earlier). Dallmann (1983) has studied the bifurcation processes that can occur with an array of critical points.

Another important local solution is the Sullivan (1959) vortex. This is a two-celled vortex that is undergoing an axisymmetrical strain rate and that displays a limit-cycle pattern in the x_1x_2 plane, as shown in Figure 12 (*right*). The sectional streamlines in the x_1x_3 or x_2x_3 plane are shown in Figure 12 (*left*). Sectional streamlines are obtained by integrating in a sectioning plane the velocity components in that plane. It can be shown that such a vortex has the vortex filaments wrapped in a bihelical manner, as shown in Figure 13. This, therefore, might be a plausible interpretation of the limit cycles that appear in flow patterns recently observed in wakes by Perry & Steiner (1986) and Steiner & Perry (1986).

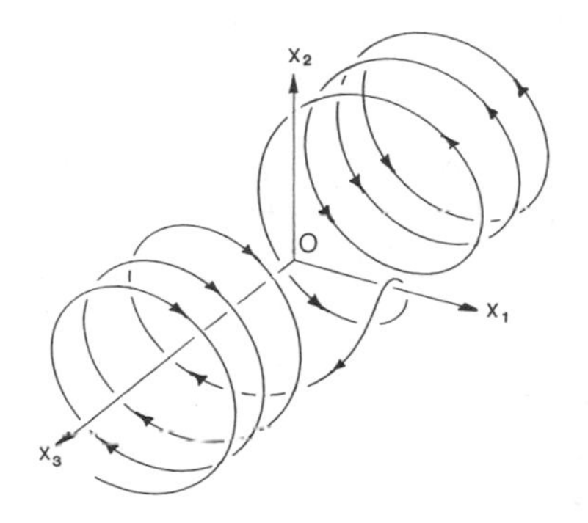

Figure 13 Typical vortex filament for the Sullivan vortex. Note the bihelical windings. Arrows indicate the direction of the vorticity vector using the right-hand screw rule.

Additional Physical Properties of Critical Points

Critical points have been defined as points where the streamline slope is indeterminate and the velocity is zero. The question arises as to whether such points have any other distinguishing physical feature or significance. Critical points on no-slip boundaries are points where the vorticity is zero (e.g. zero wall shear stress at a separation point), but the pressure gradient and gradient in vorticity is finite. This is true even in unsteady flow, provided the observer is at rest relative to the boundary. If the observer moves relative to the boundary, all critical points disappear from the boundary, since then the boundary would no longer have zero relative velocity.

In the case of the simple free-slip constant-vorticity critical points that are locally two dimensional (i.e. they are steady and lie on the q axis in Figure 11), the critical points correspond to extrema in pressure. Hence, vorticity and pressure gradients are zero, in contrast with the no-slip critical points.

For the more general class of free-slip critical points, the properties are more complex and still need to be explored, but the following general statements can be made. It is the presence of second-order terms alone that causes the extrema in vorticity and pressure to be displaced from the critical point. If, however, the viscosity is taken to zero, the vorticity extremum could remain displaced from the critical point, but the pressure extremum approaches the critical point. In the case of irrotational flow (which is an exact solution of the Navier-Stokes equations as well as the Euler equations), the pressure extremum (a maximum in this case) is always located at the critical point irrespective of whether or not second-order terms are present. Perry (1984b) erroneously concluded that the extremum was displaced from the critical point in irrotational flow with finite viscosity and finite second-order terms. We have since found that this is true only in rotational flow.

All free-slip critical points possess a certain type of rotational symmetry, provided we are in a region sufficiently close to the critical point where first-order terms dominate in the description of velocity magnitude and direction. Let $\mathbf{r}$ be a position vector relative to the critical point, and let the velocity at that position be $\mathbf{u}$. Then for the vector $-\mathbf{r}$, the velocity is $-\mathbf{u}$. The inclusion of odd-order terms (third, fifth, etc.) preserves this rotational symmetry, but even-order terms (second, fourth, etc.) destroy the rotational symmetry beyond a certain region. Thus a lack of rotational symmetry beyond a certain region indicates the possibility of significant second-order terms. The more significant these terms are, the smaller the region of rotational symmetry will be and the more the pressure and vorticity extrema will be displaced from the critical point.

In unsteady flow, if the free-slip critical point does not move relative to an observer who is moving in an inertial frame, all the above properties mentioned in this section should be true even though the pattern relative to this moving observer is unsteady, such as where vortex-stretching processes are occurring. Also, the unsteady "dislocated"-saddle solution mentioned earlier would be located at a pressure extremum (maximum). A collection of nonmoving critical points is sometimes observed in jets and wakes, provided the observer is "following the eddies" in an inertial frame. This point is discussed later.

In the case of free-slip critical points that occur in the cavity region of a vortex-shedding body, no simple statement can be made once such critical points start accelerating away from the body.

Structural Stability and Overidealization of Patterns

Two patterns are said to be topologically equivalent if they can be distorted into one another by a stretching process. Imagine that each pattern is drawn on a sheet of rubber that can be arbitrarily stretched without any tearing. If a critical point is infinitesimally close to the p axis of the p-q chart, then by an infinitesimal change in any of the relevant parameters, it could be changed from a saddle to node, or vice versa. A change in topology has occurred, and no amount of stretching can restore the original topology. The same is true for points infinitesimally close to the positive part of the q axis. Here a focus that is spiraling in could be changed to a focus that is spiraling out by an infinitesimal change in some parameter. Such critical points are said to be structurally unstable.

Trajectories that originate from saddle points are referred to as separatrices and divide the pattern into distinct regions that contain trajectories unique to that region (see Figures 16 and 18). A pattern having separatrices that join two saddle points is said to be structurally unstable for the same reasons as given above (see Tobak & Peake 1982, Perry & Hornung 1984). A spectacular change in the transport properties of a pattern can occur by a slight change of one of the relevant parameters.

Quite apart from these topological considerations, all degenerate critical points (i.e. points on any of the p-q chart boundaries) or saddle-to-saddle connections are extremely rare, so rare that they would never occur in practice except at an instant during a bifurcation process. Degenerate critical points and saddle-to-saddle connections are warning signals that the flow has been inadvertently idealized in some way, e.g. perfect two dimensionality or some form of symmetry has been assumed. Of course, in many situations this may not be important, since such assumptions may have been intended to simplify the analysis. The worker is then aware that the idealization has been made and that a more general result was not intended.

Jets and Wakes at Moderate Reynolds Numbers

Eddying motions have been observed in coflowing jets and wakes by Perry & Lim (1978) using flow visualization, and a classification of the simplest types of motions is shown in Figure 14. These sketches represent laser-sheet sections, down the plane of symmetry, using smoke issuing from a tube that is pointing downstream in a wind tunnel. If the Reynolds number based on tube diameter and velocity difference between the tunnel velocity and the tube exit velocity is of order 1000, it is found that these naturally occurring patterns, which modulate randomly in frequency and scale, can be "locked in" by a small vertical oscillation of the tube or by acoustic excitation. The patterns then become perfectly periodic in time. If we view the flow under stroboscopic light that is synchronized with the external excitation, the patterns appear frozen in time and therefore can be studied in great detail with strobed laser sheets. It should be noted that with wakes the eddies "point" downstream, and for jets the eddies point upstream. Also, if we use hot-wire anemometry and sample the data on the basis of the phase of the external oscillations, instantaneous-velocity vector fields can be mapped out.

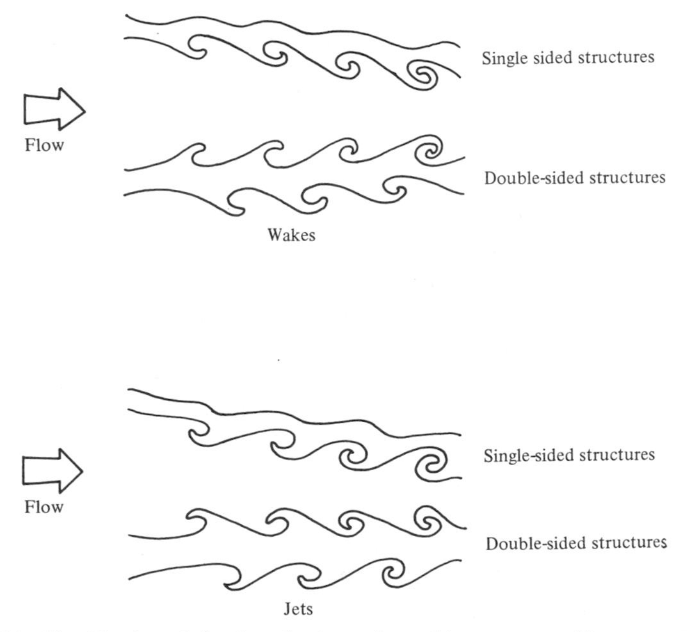

Figure 14 Classification of simple coflowing wake and jet structures. These are sections of smoke in the plane of symmetry.

It will be noticed from Figure 14 that vortex roll-ups occur only on one side for some structures (the single-sided structures) and on both sides for the other structures (the double-sided structures). The single sidedness is caused either by buoyancy effects (smoke being heavier or lighter than the surrounding air) or by a slight misalignment of the tube (an asymmetrical generation of vorticity). This was demonstrated by A. E. Perry & D. K. M. Tan (unpublished), who repeated some tests in a vertical wind tunnel with the tube pointing downward. The single-sided structures appear to consist of a pair of trailing vortices connected by a ladder-like network of cross strands of vorticity, and they bear a strong resemblance to cigarette smoke or chimney smoke in a cross stream (see Perry & Lim 1978, Van Dyke 1982). Such patterns have also been observed in the wakes behind three-dimensional bodies at an angle of attack [e.g. Calvert (1967), who used an inclined circular disk]. The double-sided structures can be produced by adjusting the flow rates so as to remove the buoyancy effect. These structures have been observed in wakes behind three-dimensional ellipsoidal bodies when the major axis has been set normal to the plane of the paper in Figure 14 (e.g. see Perry & Lim 1978). These patterns are three-dimensional versions of the classical two-dimensional Kármán vortex street as produced by Zdravkovich (1969).

Figure 15 shows the deduced shape of the smoke for the single-sided structures, as given in Perry et al. (1980); it consists of an array of three-dimensional Kelvin-Helmholtz-like roll-ups. Figure 16 (*top*) shows a typical smoke pattern passing a hot wire. Figure 16 (*bottom*) shows the instantaneous-velocity vector fields in the vertical plane of symmetry, as measured by Perry & Tan (1984). Nodes, saddles, and foci are apparent.

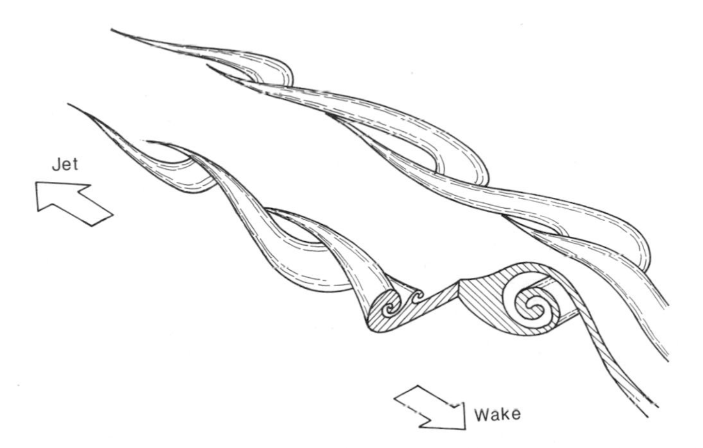

Figure 15 Oblique view of single-sided structures with sectionings of the smoke.

This pattern is as "seen" by an observer who is moving with the eddies. The correct horizontal convection velocity U_c for the eddies is difficult to define precisely, since the structures are developing with streamwise distance, and different parts of the pattern are being convected at slightly different velocities. Perry & Tan (1984) used a correlation technique that gave the velocity required for a moving observer to see a pattern with a minimum time variation. Cantwell & Coles (1983), in work to be discussed later, assumed U_c was given by the velocity of the centroids of vorticity

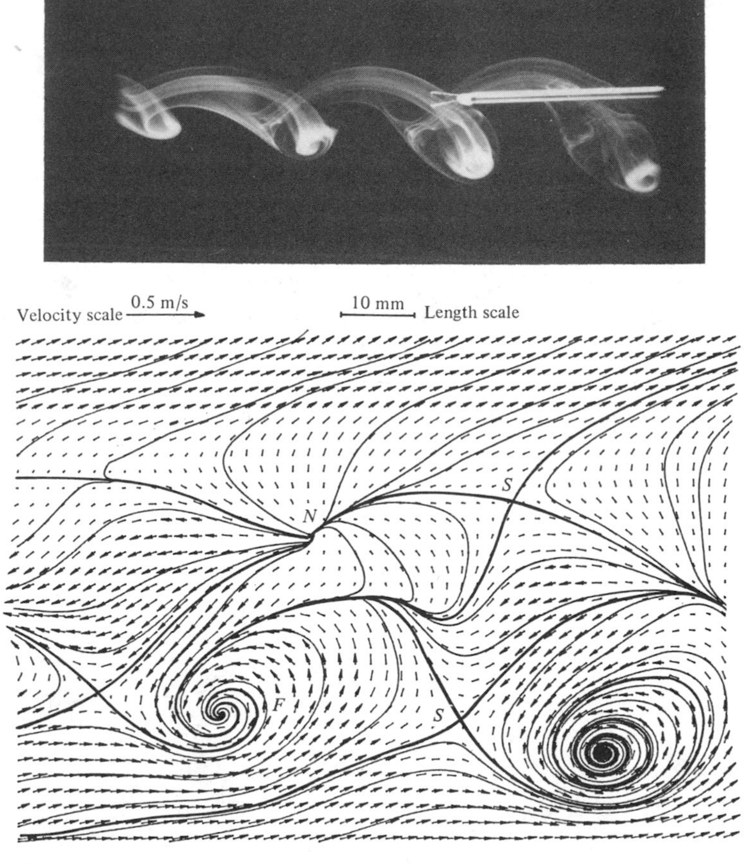

Figure 16 (*top*) Externally illuminated single-sided wake pattern passing-hot-wire probe. (*bottom*) Typical instantaneous (phase-averaged) velocity vector field for the smoke pattern above.

that are located in the various vortical regions of the pattern. Perry et al. (1980) assumed that the convection velocity is given by $U_c = \lambda f$, where λ is the measured wavelength of the structures and f is the frequency. The latter method proved slightly inaccurate. In the patterns shown, a slight vertical convection velocity was incorporated. This was estimated from the slope of the line passing through the centers of the eddies in the smoke photographs. It was found that the foci correspond closely with vorticity extrema, and that the nodes and saddles above the structures are approximately irrotational. Figure 17 shows the conjectured streamline patterns out of the plane of symmetry, as presented by Perry et al. (1980). This was sketched by using the saddle-node combination theorem as a guide.

Instantaneous streamlines in Figure 16 (*bottom*) were obtained by

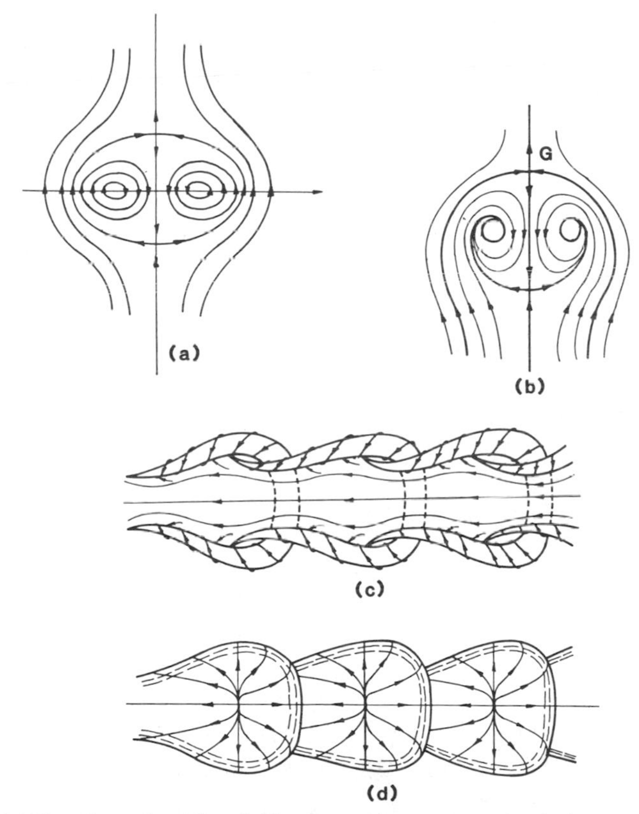

Figure 17 (*a*) Two-dimensional flow field generated by a vortex pair relative to an observer moving with the pair. (*b*) Conjectured sectional streamlines for pattern in Figure 16. Note the effect of divergence when compared with (*a*). (*c*) Upperside of conjectured streamlines near surface of smoke. (*d*) Underside of conjectured streamlines.

integrating the velocity vector field. This was achieved by fitting the velocity field to a fifth-order two-dimensional Taylor-series expansion over rectangular regions that overlapped. In subsequent work, splining techniques have been developed. The streamlines, or solution trajectories, were obtained by integrating in time the equations $u_1 = \dot{x}_1 = f_1(x_1, x_2)$ and $u_2 = \dot{x}_2 = f_2(x_1, x_2)$ using a predictor-corrector method. The functions f_1 and f_2 were obtained from fitting the velocity-field data. Unfortunately, the dislocated saddles between the foci reported by Perry et al. (1980) have been smeared out into regular saddles by this curve-fitting technique. Nevertheless, the method removes much of the subjectivity in sketching in the streamlines by hand ("eye-balling"), as was done in earlier work (see later).

One of the major disadvantages of using instantaneous streamline patterns is the sensitivity of such patterns to the choice of the velocity of the observer. For instance, a 10% change in convection velocity changes Figure 16 (*bottom*) to Figure 18, which looks more like the interpretation produced by Perry et al. (1980). The nodes and saddles above the structures have changed into a bifurcation line labeled G. The topology of the pattern has changed, but the overall transport properties are the same when describing the eddying motions. Both patterns are valid, but the pattern given in Figure 18 is more time dependent than that given in Figure 16

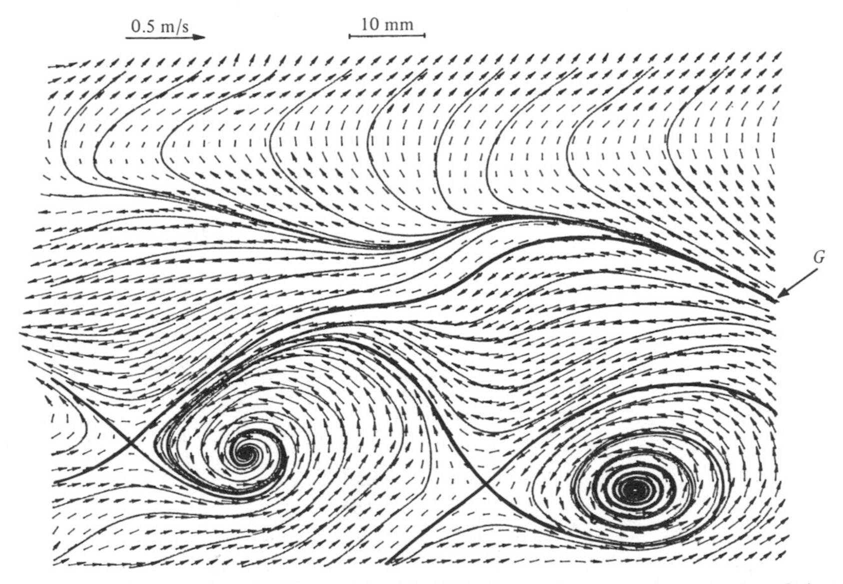

Figure 18 Pattern given in Figure 16 with 10% change in convection velocity. G is a bifurcation line.

(*bottom*). In many respects, it would be convenient if we could use a quantity that is invariant with the velocity of the observer. Such a quantity is vorticity. Unfortunately, vorticity is extremely difficult to measure directly, and so the vorticity field must be obtained by a differentiation of the velocity field. This technique is known to be inaccurate. Nevertheless, for simple flow cases, using the smoke photographs as a guide and computing vorticity by differentiation in the plane of symmetry, simple "vortex skeletons" of the flows can be constructed. These vortex skeletons are the "genetic code" of the flow field, since this requires very little specification and the Biot-Savart law can be used to generate the velocity field. Figure 19 (*top* and *middle*) shows the vortex skeleton conjectured by Perry & Tan (1984) for the single-sided structures. In this vortex-skeleton model, the solenoidal condition for vorticity is maintained. The velocity field obtained in the plane of symmetry using the Biot-Savart law is shown in Figure 19 (*bottom*). The vortex skeleton in effect gives two trailing vortices interconnected with cross strands of vorticity. Although the deduced velocity field is distorted from the experimental pattern as shown in Figure 16 (*bottom*), it has the same topology, i.e. a similar distribution and interconnection of critical points. The distortion is due to the approximations made in the construction of the vortex skeleton. Although the skeleton approximation is crude (sharp corners, etc.), it seems adequate for computing the velocity field, at an instant, in the plane of symmetry. Also, by assuming a Gaussian distribution of vorticity for the eddies, good qualitative agreement was found with the experimentally determined spectra and Reynolds stresses. This might be useful in turbulence models.

Figure 20 shows the deduced shape of the smoke pattern for double-sided structures, as given in Perry et al. (1980). Some of the detailed foldings in the smoke are conjectured. Figure 21 (*top*) shows the conjectured vortex skeleton, and Figure 21 (*bottom*) shows the velocity field using both the Biot-Savart law and the assumption of a Gaussian distribution of vorticity in the vortex rods. The surprising feature of this pattern is that the foci are spiraling out, whereas in the single-sided structures, the foci are spiraling in. This seems to agree with some of the data cited by Perry & Tan (1984), although recent studies by Perry & Steiner (1986) and Steiner & Perry (1986) have also revealed the existence of limit cycles of the type mentioned earlier. The fact that a spiraling out of trajectories occurs shows that vortex compression as well as vortex stretching is possible. In some of the early work in the interpretation of vector fields (e.g. Perry et al. 1980, Perry & Watmuff 1981), the streamlines were drawn in by hand with the preconceived idea that vorticity was always being stretched. Perry & Tan (1984) have reinterpreted some of these earlier results. All cases considered correspond to the simplest observed

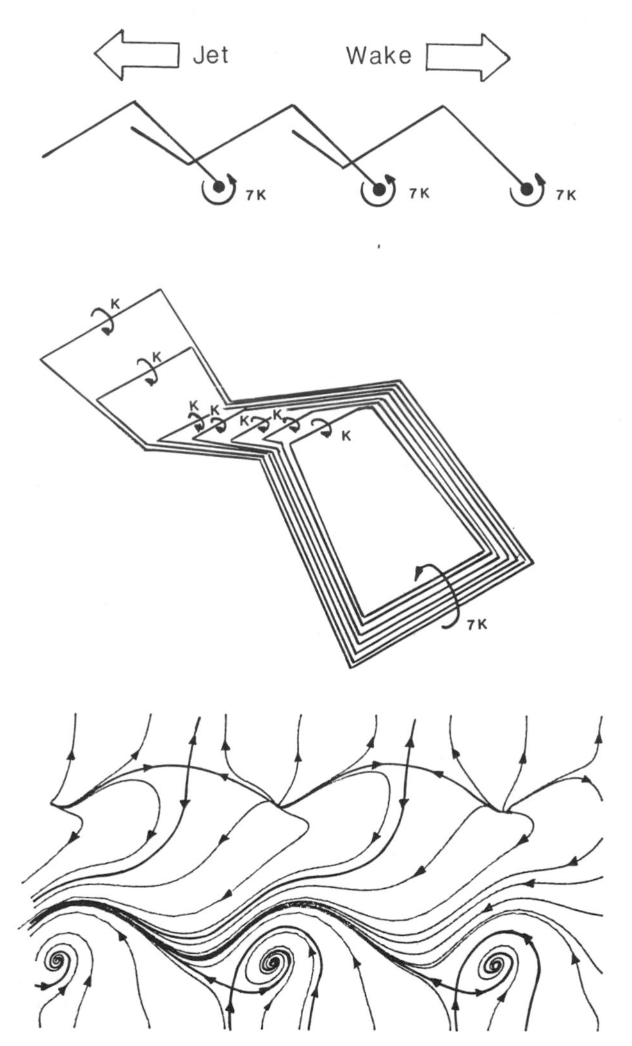

Figure 19 (*top*) Side view of vortex skeleton for single-sided structure. (*middle*) Oblique view of a typical cell. K denotes a unit of circulation. (*bottom*) Computed velocity field using the Biot-Savart law (as seen by an observer moving with the eddies).

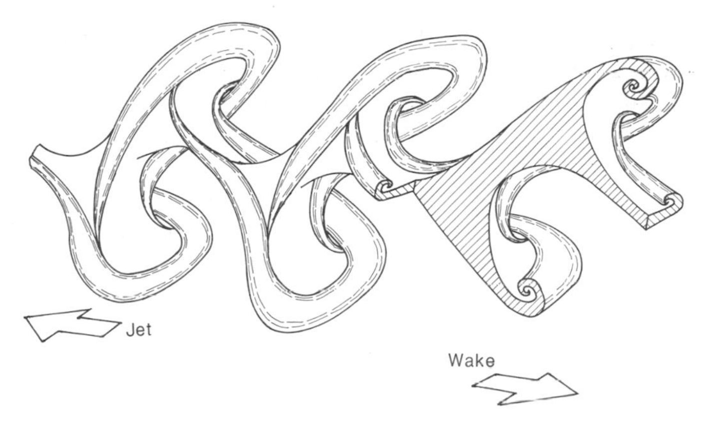

Figure 20 Conjectured smoke pattern for double-sided structures.

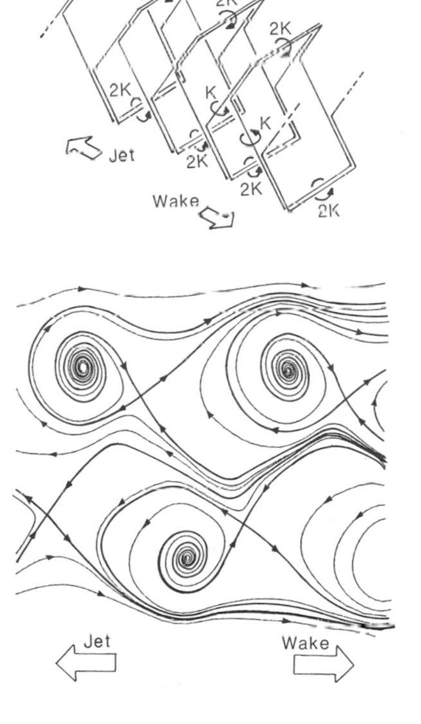

Figure 21 (*top*) Vortex skeleton for double-sided structures. K denotes a unit of circulation. (*bottom*) Computed velocity field using the Biot-Savart law (as seen by an observer moving with the eddies).

motions in jets and wakes. Vortex skeletons of far more complex flows have been produced in numerical computations that use Lagrangian vortex-element methods, as reviewed by Leonard (1985).

Wakes at High Reynolds Numbers

At high Reynolds numbers, where vortex shedding is occurring at the source and the flow is turbulent, it is possible to produce phase-averaged velocity vector fields based on the phase of the vortex shedding. Some sensing device is required near the vortex-shedding body to act as a phase detector. Due to the randomness of the flow, a large population of data at each position and phase is necessary for convergence. The only extensive and detailed velocity vector fields that have so far appeared in the literature are by Cantwell & Coles (1983) for flow behind a circular cylinder and by Perry & Watmuff (1981) for flow behind an ellipsoid. All of the flow-pattern features mentioned earlier also appear in these phase-averaged velocity vector fields. However, the laws governing these features will be different because of phase jitter and the superimposed fine-scale phase-averaged motions. Perry & Watmuff (1981) have investigated in some detail the effect of phase jitter on critical-point patterns. However, the effect of the superimposed fine-scale motions is unknown and introduces a vorticity-diffusion mechanism that has yet to be investigated. Temporal mean patterns also contain the same general flow-pattern features mentioned earlier, but vorticity diffusion, caused by both the large-scale and fine-scale motions, makes it impossible to carry out any analytical formulations with the equations of motion.

A MORE COMPLETE DESCRIPTION OF FLOW PATTERNS

So far, flow patterns have been described by Taylor series that are expanded about individual critical points, and this description is valid only in the immediate vicinity of each critical point. Dallmann (1983) has shown in some simple examples that if the series expansion is extended to higher-order terms, it is possible to describe in one formulation a "cluster" of critical points. With his analysis, he investigated some very interesting bifurcation processes where critical points merge as flow parameters are altered. Also, without having to solve completely the equations involved, he was able to deduce various types of separation bubbles that are topologically possible, using (among other constraints) the saddle-node combination theorem mentioned earlier. He also showed that a specification of the limiting streamline pattern does not uniquely define the pattern occurring above the surface. If complete solutions are required, so as to further investigate the properties of these separation patterns, we come across a major difficulty. This is the enormous amount of labor required to obtain three-dimensional solutions higher than third order. It might be possible to overcome this difficulty, since Perry (1984b) has recently developed an algorithm that in principle enables series-expansion solutions to the Navier-Stokes and continuity equations to be generated to arbitrary order. The series expansion given by (3) is substituted into (1) and (2), and by equating coefficients from equations generated by the continuity equation and by equating appropriate cross derivatives of pressure given by the Navier-Stokes equations (e.g. $\partial^2 p/\partial x_1 \partial x_2 = \partial^2 p/\partial x_2 x_1$, etc.), a series of equations for the relationships between the various coefficients in (3) can be found. By noting a pattern in the arrangement of the indices of the tensors, Perry (1984b) developed his algorithm, which enables computer software to be developed to do the algebra and generate the equations. If N is the order of the series expansion, N_c is the number of unknown coefficients, E_c is the number of equations generated by the continuity equation, and E_{ns} is the number of equations generated by the Navier-Stokes equations, Table 1 gives an idea of the complexity of the problem and the need for computer-generated algebra. It also shows that three-dimensional flow is an order of magnitude more complex than two-dimensional flow. It can be seen that the number of unknowns always exceeds the number of equations generated. Additional equations must be supplied from boundary conditions. The origin of the series expansion need not necessarily be a critical point, and so there can be zeroth-order terms in (3). The equations generated by the continuity equation are linear, and the equations generated by the Navier-Stokes equations contain both linear and quadratic terms. Some very interesting simple properties of these equations have been discovered. If the boundary conditions for u_i are specified (in the form of a series expansion) on the surfaces of a box that surrounds the origin of the expansion, an iterative procedure is required because of the nonlinearity mentioned above. If, however,

Table 1 Number of unknowns and equations

	Two dimensional				Three dimensional		
N	N_c	E_c	E_{ns}	N	N_c	E_c	E_{ns}
3	20	6	1	3	60	10	3
4	30	10	3	4	105	20	11
5	42	15	6	5	168	35	26
⋮	⋮	⋮	⋮	⋮	⋮	⋮	⋮
15	272	120	91	15	2448	680	1001

boundary conditions are specified on the orthogonal coordinate planes passing through the origin, a solution procedure can be found that causes at least one of the coefficients in each quadratic pair in the Navier-Stokes-generated equations to be known before these equations are required for solution. Thus, we effectively are solving a group of linear equations. Such "canonical" boundary conditions can be used to generate a whole variety of flow patterns analytically very rapidly on the computer. Once the basic equations are set up, the solution is found by successive substitutions without any iteration. These patterns are therefore known asymptotic solutions to the Navier-Stokes and continuity equations, and methods have been devised to determine the region where these truncated series solutions satisfy the full equations of motion to within a specified accuracy. For a given order of series expansion, the region of accuracy shrinks as the Reynolds number is increased. For a given Reynolds number, the region of accuracy grows as the order of the series expansion is increased.

Unsteady flow problems can also be solved using the algorithm. We obtain a set of ordinary first-order differential equations, and the boundary conditions are specified as functions of time. These become forcing functions, and predictor-corrector methods can be used for proceeding in time. The algorithm may be of use in computational schemes after further development, but so far it has been used for studying the geometry and topology of eddying motions, particularly for three-dimensional flow separation. Rather than using boundary conditions for generating flow patterns, we can apply a procedure for synthesizing flow patterns with various predetermined properties, as observed in experiments. This has been achieved by combining the algorithm with known properties of critical points. For instance, the no-slip critical points, using the formulation given by Equation (6), have been completely explored by Oswatitsch (1958) and Perry & Fairlie (1974). All the elements of $B_{i\alpha}$ are known and can be expressed in terms of both surface-vorticity gradients and the angle at which a streamline approaches or leaves a surface at a critical point.

Various three-dimensional flow-separation patterns that have been observed experimentally (using surface dye)—in particular, the patterns observed by Bippes & Turk (1983) and Fairlie (1980)—have been classified by Hornung & Perry (1984) and Perry & Hornung (1984). Vortex-skeleton models have been devised with an electromagnetic analogy using current-carrying wires and iron filings, and this method is described in detail by Perry & Hornung. One possible vortex skeleton for the pattern shown in Figure 1 is shown in Figure 22. This has been referred to as an owl-face pattern of the second kind. Such patterns and other types of separating patterns can now be synthesized analytically. For instance, by using a third-order expansion and the no-slip condition and by specifying (*a*) the

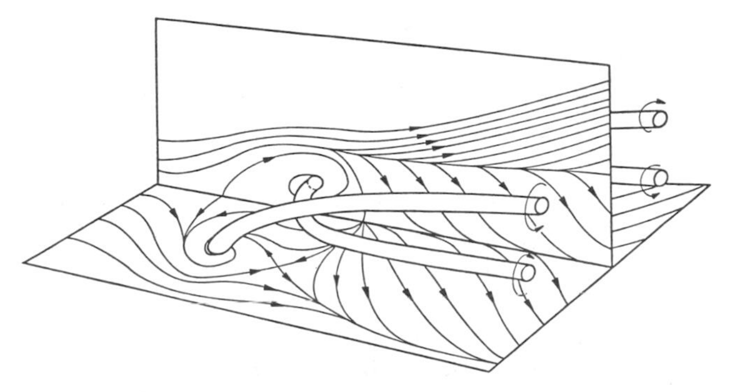

Figure 22 One possible vortex skeleton for an owl-face of the second kind. Image vortices under the horizontal surface are not shown.

surface-vorticity distribution, (*b*) the angle of the separating streamline and the reattaching streamline in the plane of symmetry, and (*c*) the location of a free-slip critical point on the plane of symmetry above the surface, we can obtain closure with the equations generated by the algorithm. This gives a simple U-shaped separation, as shown in Figure 23. Using a fourth- or (better still) fifth-order expansion and similar techniques to those described above, we can obtain an owl-face of the second

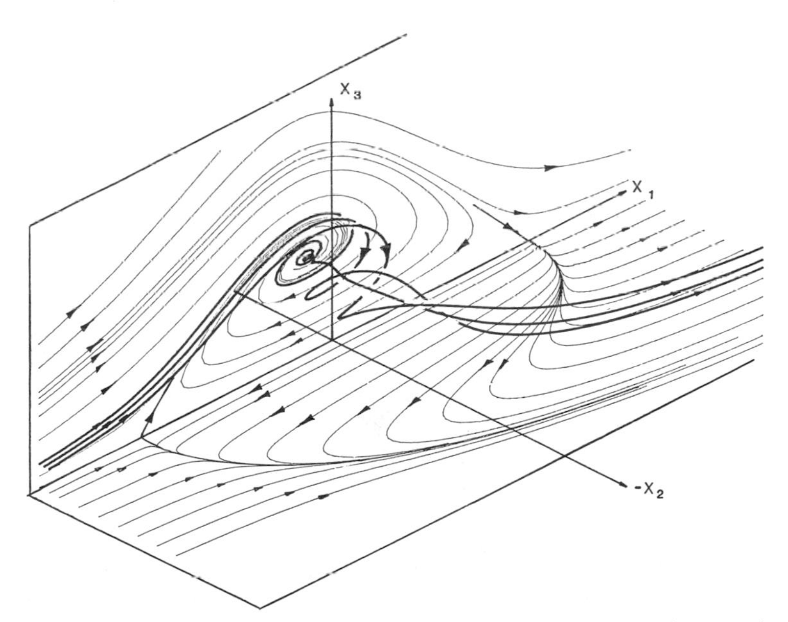

Figure 23 A computed simple U-shaped separation using a third-order series expansion.

kind, as shown in Figure 24. Some of this work has been reported by Perry et al. (1985), and further work should be directed toward extending this technique to higher orders of series expansion so that higher Reynolds numbers can be achieved. The owl-face pattern shown in Figure 24 has a Reynolds number of 45. As a check on the analysis, solutions can be used to generate boundary conditions for u_i on three orthogonal planes ($u_i = 0$ on the no-slip boundary). When combined with the algorithm, we must reproduce correctly the surface-vorticity distribution and the associated limiting streamline behavior that was initially specified. The vortex lines at the surface are orthogonal to the limiting streamlines (Lighthill 1963).

Unsymmetrical patterns can also be generated and the method used for exploring the topology of the vorticity field. At high Reynolds numbers the vortex-skeleton idea is probably satisfactory, but at low Reynolds

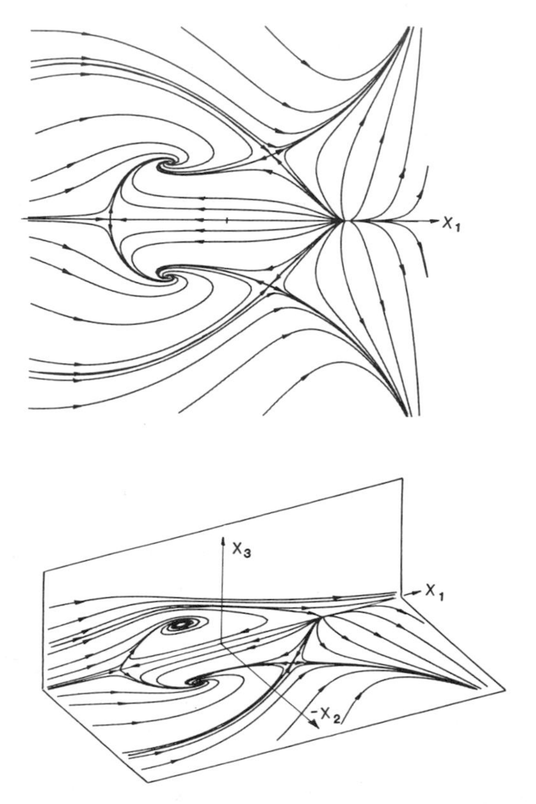

Figure 24 A computed owl-face of the second kind using a fourth-order series expansion. (*top*) Limiting streamlines on the surface, i.e. x_1x_2 plane. (*bottom*) Oblique view of the pattern.

numbers the vorticity field is not concentrated but is instead highly diffused. Vorticity fields have their own sets of critical points, and the properties of these have yet to be explored.

CONCLUDING REMARKS

We have attempted to show the usefulness of critical-point concepts in the understanding of flow patterns. A brief summary of the analytical method and some examples of its application to the study of flow patterns have been given. We feel that in future years some of the concepts developed will be directed toward the problem of turbulence modeling. This has already been attempted, to some limited extent, in the work of Cantwell & Coles (1983), Cantwell et al. (1978), Hussain (1984), and Coles (1984); these authors point out that in phase-averaged free-shear patterns, turbulence production is predominantly at the saddle points, and the produced turbulence is carried into the foci. Perry & Chong (1982) have made use of certain aspects of critical points in the understanding and modeling of wall turbulence.

ACKNOWLEDGMENTS

We wish to acknowledge the Australian Research Grant Scheme for supporting this project.

Literature Cited

Andronov, A. A., Vitt, A. A., Khaikin, S. E. 1966. *Theory of Oscillators*. Oxford: Pergamon. 818 pp.

Bippes, H., Turk, M. 1983. Oil flow patterns of separated flow on a hemisphere cylinder at incidence. *DFVLR Rep. IB 222-83 A07*, Göttingen, West Ger.

Buckmaster, J. 1972. Perturbation technique for the study of three-dimensional separation. *Phys. Fluids* 15: 2106–13

Burgers, J. M. 1948. A mathematical model illustrating the theory of turbulence. *Adv. Appl. Mech.* 1: 197–99

Calvert, J. R. 1967. Flow past an inclined disk. *J. Fluid Mech.* 29: 691–704

Cantwell, B. J. 1978. Similarity transformations for the two-dimensional unsteady stream function equations. *J. Fluid Mech.* 85: 257–71

Cantwell, B. J. 1981. Transition in the axisymmetric jet. *J. Fluid Mech.* 104: 369–86

Cantwell, B. J., Allen, G. A. 1984. Transition and mixing in impulsively started jets and vortex rings. In *Turbulence and Chaotic Phenomena in Fluids*, ed. T. Tatsumi, pp. 123–32. Amsterdam: North-Holland

Cantwell, B. J., Coles, D. E. 1983. An experimental study of entrainment and transport in the turbulent wake of a circular cylinder. *J. Fluid Mech.* 136: 321–74

Cantwell, B. J., Coles, D. E., Dimotakis, P. E. 1978. Structure and entrainment in the plane of symmetry of a turbulent spot. *J. Fluid Mech.* 87: 641–72

Coles, D. E. 1984. On one mechanism of turbulence production in coherent structures. In *Turbulence and Chaotic Phenomena in Fluids*, ed. T. Tatsumi, pp. 377–402. Amsterdam: North-Holland

Corcos, G. M., Sherman, F. S. 1976. Vorticity concentration and the dynamics of unstable free shear layers. *J. Fluid Mech.* 73: 241–64

Dallmann, U. 1983. Topological structures of three-dimensional flow separation. *DFVLR Rep. IB 221-82-A07*, Göttingen, West Ger.

Davey, A. 1961. Boundary layer flow at a

point of attachment. *J. Fluid Mech.* 10: 593–610 (see appendix)

Fairlie, B. D. 1980. Flow separation on bodies of revolution at incidence. *Proc. Aust. Hydraul. Fluid Mech. Conf., 7th, Brisbane*, pp. 338–41

Hornung, H. G., Perry, A. E. 1984. Some aspects of three-dimensional separation. Part I. Streamsurface bifurcations. *Z. Flugwiss. Weltraumforsch.* 8: 77–87

Hunt, J. C. R., Abell, C. J., Peterka, J. A., Woo, H. 1978. Kinematical studies of the flows around free or surface-mounted obstacles; applying topology to flow visualization. *J. Fluid Mech.* 86: 179–200

Hussain, A. K. M. F. 1984. Coherent structures and incoherent turbulence. In *Turbulence and Chaotic Phenomena in Fluids*, ed. T. Tatsumi, pp. 453–60. Amsterdam: North-Holland

Kaplan, W. 1958. *Ordinary Differential Equations*. Reading, Mass: Addison-Wesley

Kaynak, U., Cantwell, B. J., Holst, T. L., Sorenson, R. L. 1986. Numerical simulation of transonic separated flows over low aspect ratio wings. *AIAA Pap. 86-0508*

Legendre, R. 1956. Séparation de l'écoulement laminaire tridimensionel. *Rech. Aéronaut.* 54: 3–8

Legendre, R. 1965. Lignes de courant d'un écoulement continu. *Rech. Aérosp.* 105: 3–9

Leonard, A. 1985. Computing three-dimensional incompressible flows with vortex elements. *Ann. Rev. Fluid Mech.* 17: 523–59

Lighthill, M. J. 1963. Attachment and separation in three-dimensional flow. In *Laminar Boundary Layers*, ed. L. Rosenhead, pp. 72–82. Oxford: Oxford Univ. Press

Lim, T. T., Chong, M. S., Perry, A. E. 1980. The viscous tornado. *Proc. Aust. Hydraul. Fluid Mech. Conf., 7th, Brisbane*, pp. 250–53

Maskell, E. C. 1955. Flow separation in three dimensions. *RAE Rep. Aero 2565*

Maskell, E. C. 1956. The significance of flow separation in the calculation of a general fluid flow. *Congr. Int. Mech. Appl., 9th, Brussels*

Minorsky, N. 1962. *Nonlinear Oscillators*. Princeton, NJ: Van Nostrand. 714 pp.

Oswatitsch, K. 1958. Die Ablösungsbedingung von Grenzschichten. In *Grenzschichtforschung*, ed. H. Goertler, pp. 357–67. Berlin/Göttingen/Heidelberg: Springer-Verlag

Patnaik, P. C., Sherman, F. C., Corcos, G. M. 1976. A numerical simulation of Kelvin-Helmholtz waves of finite amplitude. *J. Fluid Mech.* 73: 215–40

Perry, A. E. 1984a. A study of degenerate and non-degenerate critical points in three-dimensional flow fields. *Forschungsber. DFVLR-FB 84-36*, Göttingen, West Ger.

Perry, A. E. 1984b. A series expansion study of the Navier-Stokes equations. *Forschungsber. DFVLR-FB 84-34*, Göttingen, West Ger.

Perry, A. E., Chong, M. S. 1982. On the mechanism of wall turbulence. *J. Fluid Mech.* 119: 173–217

Perry, A. E., Fairlie, B. D. 1974. Critical points in flow patterns. *Adv. Geophys.* 18B: 299–315

Perry, A. E., Hornung, H. G. 1984. Some aspects of three-dimensional separation. Part II. Vortex skeletons. *Z. Flugwiss. Weltraumforsch.* 8: 155–60

Perry, A. E., Lim, T. T. 1978. Coherent structures of coflowing jets and wakes. *J. Fluid Mech.* 88: 451–63

Perry, A. E., Steiner, T. R. 1986. Large-scale vortex structures in turbulent wakes behind bluff bodies. Part I. Vortex formation process. *J. Fluid Mech.* In press

Perry, A. E., Tan, D. K. M. 1984. Simple three-dimensional motions in coflowing jets and wakes. *J. Fluid Mech.* 141: 197–231

Perry, A. E., Watmuff, J. H. The phase-averaged large-scale structures in three-dimensional turbulent wakes. *J. Fluid Mech.* 103: 33–51

Perry, A. E., Lim, T. T., Chong, M. S. 1980. The instantaneous velocity fields of coherent structures in coflowing jets and wakes. *J. Fluid Mech.* 101: 243–56

Perry, A. E., Chong, M. S., Hornung, H. G. 1985. Local solutions of the Navier-Stokes equations for separated flows. *Symp. Numer. Phys. Aspects Aerodyn. Flows, 3rd, Long Beach, Calif*, pp. 8.25–8.32

Pontryagin, L. S. 1962. *Ordinary Differential Equations*. Reading, Mass: Addison-Wesley

Rott, N. 1958. On the viscous core of a line vortex. *ZAMP* 9b(5/6): 543–53

Rott, N. 1959. On the viscous core of a line vortex. II. *ZAMP* 10(1): 73–81

Smith, J. H. B. 1972. Remarks on the structure of conical flow. *Prog. Aeronaut. Sci.* 12: 241–72

Steiner, T. R., Perry, A. E. 1986. Large-scale vortex structures in turbulent wakes behind bluff bodies. Part II. Far wake structures. *J. Fluid Mech.* In press

Sullivan, R. D. 1959. A two-celled solution of the Navier-Stokes equations. *J. Aeronaut. Sci.* 26: 767–68

Tobak, M., Peake, D. J. 1982. Topology of three-dimensional separated flows. *Ann. Rev. Fluid Mech.* 14: 61–85

Van Dyke, M. 1982. *An Album of Fluid*

Motion, Stanford, Calif: Parabolic. 176 pp.
Wang, K. C. 1972. Separation patterns of boundary layers over an inclined body of revolution. *AIAA J.* 10: 1044–50
Wang, K. C. 1974a. Boundary layer over a blunt body at high incidence with an open type of separation. *Proc. R. Soc. London Ser. A* 340: 33–35
Wang, K. C. 1974b. Boundary layer over a blunt body at extremely high incidence. *Phys. Fluids* 17: 1381–85
Wang, K. C. 1975. Boundary layer over a blunt body at low incidence with circumferential reversed flow. *J. Fluid Mech.* 72: 49–65
Werlé, H. 1962. Separation on axisymmetrical bodies at low speed. *Rech. Aéronaut.* 90: 3–14
Werlé, H. 1975. Ecoulements decollés. Etude phénoménologique à partir de visualisations hydrodynamiques. *ONERA TP No. 1975-14*
Zdravkovich, M. M. 1969. Smoke observations of the formation of a Kármán vortex street. *J. Fluid Mech.* 37: 491–96

Ann. Rev. Fluid Mech. 1987. 19 : 157–82

VISCOELASTIC FLOWS THROUGH CONTRACTIONS

D. V. Boger

Department of Chemical Engineering, University of Melbourne, Parkville, Victoria 3052, Australia

1. INTRODUCTION

The basic elements for laminar flow through an abrupt circular contraction are illustrated in Figure 1. The flow progresses from being fully developed at a plane some distance upstream from the contraction to being fully developed in the downstream tube at a distance L_e from the contraction

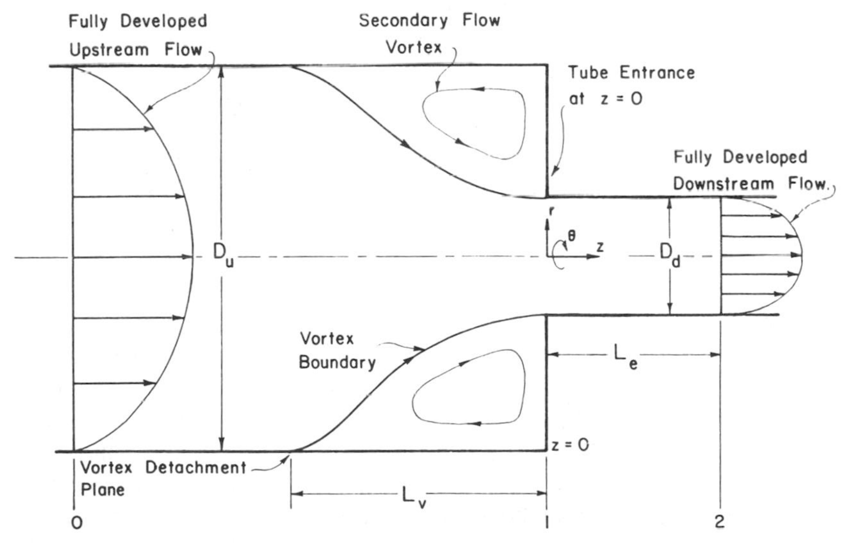

Figure 1 Basic elements of an entry flow for flow from a large tube through an abrupt entry into a smaller tube.

0066–4189/87/0115–0157$02.00

plane. The entry length L_e is the distance from the contraction plane ($z = 0$) required for the centerline velocity to become 98 or 99% of its fully developed value. Depending on the Reynolds number of the flow and the fluid type, a secondary-flow vortex may be present in the corner of the upstream tube. The characteristics of this vortex are of particular interest in the design of extrusion dies.

As with other confined flows, the major objective is to predict the detailed kinematics and pressure drop using the equations of motion and an appropriate constitutive equation to define the stress tensor. The ability to make such predictions is dependent on the entry-flow geometry itself, the Reynolds number of the flow, and the type of fluid flowing through the contraction.

Interest in the entry-flow problem extends back to the late 1800s, when early workers like Hagenback, Boussinesq, and Couette were interested in the pressure drop for circular entry flows. Their investigations were motivated by a need to develop the capillary rheometer for accurate viscosity measurement of Newtonian fluids [see Kestin et al. (1973) for a review on the use of a capillary rheometer for viscosity measurement of Newtonian fluids].

After the publication of boundary-layer theory by Prandtl in 1904, there followed a great number of papers considering the flow of a Newtonian fluid from a flat entry velocity profile at plane 1 in Figure 1 to a fully developed flow at plane 2. Flow upstream of the tube entry was not considered. This is the so-called *hydrodynamic entrance-region problem*, where the main concern is predicting the entry length L_e. There was little interest in establishing whether the entry velocity profile was indeed flat under any condition. Most authors were more concerned with developing approximate techniques (integral and numerical) for solution of the boundary-layer equations than in establishing the detailed kinematics of entry flows. A complete bibliography on the hydrodynamic entrance-region problem, published by Fan & Hwang (1966), listed 145 references on the subject.

Approximate boundary-layer techniques were first applied for the entry flow of inelastic non-Newtonian fluids (power-law fluids) by Bogue (1959) and Collins & Schowalter (1963). At about the same time, variational methods were first used to predict an upper bound on the pressure drop for creeping flow of a Newtonian fluid in the upstream tube of a circular contraction plane (Weissberg 1962). Duda & Vrentas (1972) extended the use of variational methods to power-law fluids. Such solutions for zero Reynolds number (creeping flow), which consider only the upstream flow, provide an upper bound on the pressure drop, but they do not provide detailed information on the kinematics.

Vrentas et al. (1966) were the first to present a numerical solution (finite difference) where the complete equations of motion were solved for a Newtonian fluid without the restriction imposed by the assumptions of either a flat entry velocity profile (boundary-layer theory) or a fully developed entry velocity profile (creeping flow). A comparison of the boundary-layer, creeping-flow, and earlier finite-difference solutions for the complete equations of motion with available experimental data for Newtonian and inelastic non-Newtonian fluids flowing through circular contractions is contained in a review published in 1982 (Boger 1982). Since these earlier and largely approximate solutions to the problem, there has been a revolution in digital computation to such an extent that the entry-flow problem, bounded by a region of fully developed flow upstream and a region of fully developed flow downstream, can now be solved for an inelastic fluid without any inherent assumptions in the governing differential equations. For Newtonian and shear-thinning (inelastic) non-Newtonian fluids, the entry-flow problem is now solved (see, for example, Kim-E et al. 1983).

The current challenge is to solve nonviscometric-flow problems for viscoelastic fluids, where many unique and unusual flow phenomena have been observed both in the entry flow and in other nonviscometric flows (Cochrane et al. 1981, Walters & Webster 1982). Understanding entry flows of viscoelastic fluids is of importance in fundamental flow-property measurement with a capillary rheometer (Meissner 1985) and in extrusion of polymer melts and solutions (White 1973, Petrie & Denn 1976). Of current major interest is the development of numerical methods in non-Newtonian fluid mechanics, where the circular entry flows is one of the problems being used to establish and test finite-element simulation techniques for viscoelastic fluids (Crochet & Walters 1983, Crochet et al. 1984).

This review briefly summarizes the solutions that are now available for Newtonian and inelastic shear-thinning fluids before concentrating on the progress that has been made in the solution of the entry-flow problem for viscoelastic fluids. Particular emphasis is placed on the interaction that now exists between experimental observation and numerical simulation for viscoelastic fluids in tubular entry flows.

2. NEWTONIAN FLUIDS

The entry of a Newtonian fluid through an abrupt circular contraction is now a solved problem. Sufficient experimental and theoretical work has been completed to define the detailed kinematics, the entry length L_e, and the entry presure drop ΔP_{EN}.[1] For contraction ratios ($\beta = D_u/D_d$) greater than or equal to two, the entry length was correctly predicted by the numerical calculations of Carter (1969) and Vrentas & Duda (1973):

$$\frac{L_e}{R} = 0.49 + 0.11\ N_{Re}, \tag{1}$$

where N_{Re} is the Reynolds number defined in terms of the downstream tube diameter D_d. For $\beta \geqq 4$ the entry pressure drop is defined by the prediction of Kestin et al. (1973) and Christiansen et al. (1972):

$$\frac{\Delta P_{EN}}{2\tau_w} = 0.0725\ N_{Re} + 0.69. \tag{2}$$

An almost identical result to Equation (2) was obtained from linear superposition of the pressure drop obtained from the approximate creeping-flow solution of Weissberg (1962) and the results of the various boundary-layer solutions (Boger 1982):

$$\frac{\Delta P_{EN}}{2\tau_w} = (C+1)\frac{N_{Re}}{32} + n_c = 0.0709\ N_{Re} + 0.589, \tag{3}$$

where C is the loss coefficient, n_c is the Couette correction, and τ_w is the wall shear stress in the downstream tube. Schmidt & Zelden (1969) tabulated the values of C computed by the various investigators; although there is little difference between the various computed values from boundary-layer theory, the Hornbeck (1961) value of $C = 1.269$ is the most appropriate, since the developing axial velocity profiles in the downstream tube predicted by Hornbeck were confirmed by experimental observation (Ramamurthy & Boger 1971). Theoretical and experimental values of n_c are compared in a recent review paper (Boger 1982). A complete finite-element simulation for the circular entry flow of a Newtonian fluid in a 4:1 contraction has confirmed Equation (3) for prediction of the excess pressure drop (Kim-E et al. 1983).

The detailed kinematics are also known from available experimental and theoretical results (Kim-E et al. 1983). The major characteristics are the upstream vortex present for small N_{Re} and the concavities in the entry-plane velocity profile for high N_{Re} when fluid inertia plays an important

[1] The entry pressure drop ΔP_{EN} is defined by $\Delta P_{EN} = p_0 - p_2 - \Delta P_{fd(0-1)} - \Delta P_{fd(1-2)}$, where $p_0 - p_2$ is the total pressure drop and ΔP_{fd} represents the pressure drop of the fully developed flow in the upstream and downstream tubes, respectively.

role in the problem. Figure 2 compares the streamlines and secondary-flow vortex observed by Nguyen (1978) at low N_{Re} with the finite-element creeping-flow predictions of Viriyayuthakorn & Caswell (1980). The excellent agreement between the predicted and observed streamlines is typical of the results that can now be obtained for inelastic fluids with careful use of commercially available finite-element simulation packages. For creeping flow ($N_{Re} \leqq 0.1$) the size of the upstream vortex, measured in terms of a dimensionless reattachment length, $X = L_v/D_u$ (see Figure 1), is independent of N_{Re} and β for $\beta \geqq 4$. The secondary flow has been observed (Boger et al. 1986) and identically predicted (Kim-E et al. 1983) to gradually disappear as fluid inertia becomes an important variable in the flow, while the concavities in the entry velocity profile observed by Burke & Burman (1969) at $N_{Re} = 108$ and $\beta = 4.65$ have also been successfully predicted by Kim-E et al. Clearly the laminar entry-flow problem is solved, and it is safe to assume that other entry-flow configurations can be examined by careful use of numerical (preferably finite-element) solution methods. As an example, existing finite-element packages should be useful in predicting the absence of the secondary-flow vortex, as is illustrated by the streak photography in Figure 3 for creeping flow of a Newtonian fluid in a 60° tapered entry.

For Newtonian fluids in circular entry flows the centerline velocity

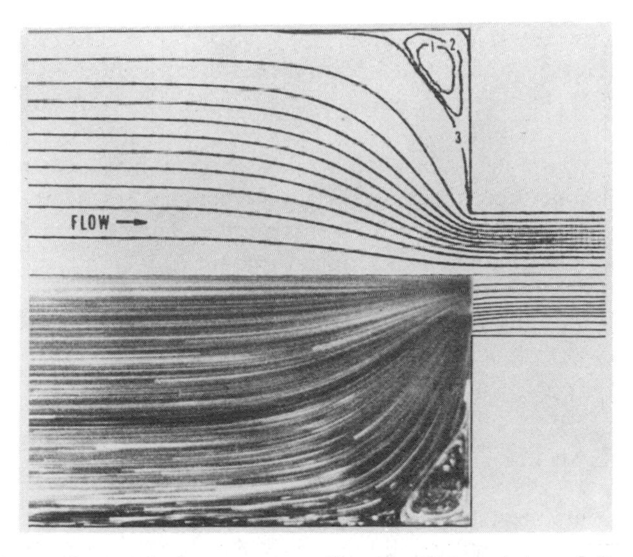

Figure 2 Comparison of the flow patterns predicted by Viriyayuthakorn & Caswell (1980) to the experimental observations of Nguyen (1978) for a Newtonian fluid in creeping flow in a 4:1 circular contraction.

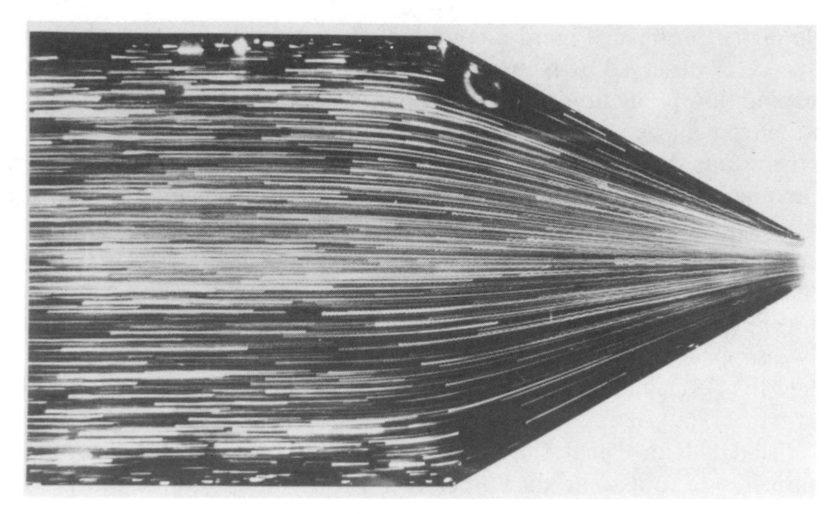

Figure 3 A streak photograph for creeping entry flow of a Newtonian fluid in a 60° tapered entry ($\dot{\gamma}_w = 100\ s^{-1}$, $N_{Re} = 0.007$) (Nguyen 1978).

development is essentially independent of Reynolds number for $N_{Re} \leqq 1$ and for $\beta \geqq 4$. At Reynolds numbers of this order and lower, the entry length is constant ($L_e = 0.49D_d/2$) and the entry-plane velocity profile is only slightly distorted from fully developed. The major characteristic of the flow at low N_{Re} is the secondary-flow vortex. Present in the corner of the upstream tube, the vortex has a reattachment length of $L_v/D_u = 0.17$ (see Figure 1) that is independent of the contraction ratio for $\beta \geqq 4$ and of the Reynolds number for $N_{Re} \leqq 1$. For $N_{Re} > 1$, fluid inertia becomes an important variable. The secondary-flow vortex is pushed into the corner of the upstream tube, reaching a reattachment length of about 0.05 at $N_{Re} = 100$. The entry length is Reynolds number dependent, and the entry-plane velocity profile makes a transition from *nearly* fully developed to *nearly* flat at $N_{Re} = 100$, where slight offcenter maxima are observed and predicted in the entry velocity profile. Creeping-flow assumptions are valid in the solution of this problem for $N_{Re} \leqq 1$, whereas fluid inertia must be considered for $N_{Re} > 1$.

3. INELASTIC NON-NEWTONIAN FLUIDS

Extending our interest to inelastic non-Newtonian fluids in entry flows introduces an additional variable—a viscosity that is a nonlinear function of the shear rate. Although many forms of this viscosity–shear-rate dependence have been observed, the simple shear-thinning fluid is of the greatest interest. Such materials are characterized by a constant viscosity at very low shear rates η_0, a viscosity that decreases with shear rate at intermediate shear rates, and a constant viscosity in the limit of a very high shear rate η_∞. A variety of constitutive equations are available to model shear-thinning behavior. Usually $\eta_\infty \lll \eta_0$, and an expression like the simplified Carreau equation is quite adequate to represent shear-thinning behavior (Bird et al. 1977):

$$\frac{\eta(\dot{\gamma})}{\eta_0} = [1 + (\lambda_s \dot{\gamma})^2]^{(n-1)/2}, \tag{4}$$

where $\dot{\gamma}$ is the strain rate, λ_s is a time constant, n is the dimensionless power-law index, and $n-1$ is the slope of the straight-line region of the log-log plot of η versus $\dot{\gamma}$. In many cases low-shear-rate viscosity data are not available, and shear-thinning fluids are characterized by the Ostwald and de Waale power-law model:

$$\eta(\dot{\gamma}) = K\dot{\gamma}^{(n-1)}, \tag{5}$$

where K is a constant of proportionality. The power-law viscosity has the obvious problem that $\eta \to \infty$ as $\dot{\gamma} \to 0$. Another viscosity equation used to represent shear-thinning behavior is the Ellis model (Bird et al. 1977):

$$\frac{\eta_0}{\eta(\dot{\gamma})} = 1 + \left(\frac{\tau}{\tau_{1/2}}\right)^{\alpha-1}, \tag{6}$$

where η_0 is the zero shear viscosity, $\tau_{1/2}$ is the value of the shear stress when $\eta = \eta_0/2$, and $\alpha - 1$ is the slope of $\eta_0/\eta - 1$ versus $\tau/\tau_{1/2}$ on a log-log plot. A time constant $\eta_0/\tau_{1/2}$ can be constructed for this model. Here α is equivalent to $1/n$.

Both Equations (4) and (6) can be written in dimensionless form for the purposes of numerical computation. With $\eta' = \eta/\eta_0$ and $\dot{\gamma}' = \dot{\gamma}/(V_d/R_d)$, we have

$$\eta' = [1 + (\mathrm{Cu}\ \dot{\gamma}')^2]^{(n-1)/2} \tag{7}$$

and

$$\left(\frac{1}{\eta'} - 1\right)(\eta')^{1-\alpha} = (\mathrm{El}\ \dot{\gamma}')^{\alpha-1}, \tag{8}$$

where $\mathrm{Cu} = \lambda_s V_d/R_d$ and $\mathrm{El} = \eta_0 V_d/\tau_{1/2} R_d$ are the Carreau and Ellis numbers, respectively. Both models describe the zero shear behavior and the power-law regions. For $\mathrm{Cu}\ \dot{\gamma}' \ll 1$ the Carreau equation predicts Newtonian behavior, and for $\mathrm{Cu}\ \dot{\gamma}' \gg 1$ power-law behavior is observed. Both models also predict essentially the same reduced viscosity behavior (η/η_0 versus $\dot{\gamma}'$), and thus the two are comparable when Cu = El (Kim-E et al. 1983).

Highly shear-thinning fluids that show no elastic characteristics are rather rare. Hence, experimental observations in tubular entry flows that unambiguously examine the interaction of fluid inertia and shear thinning in the absence of fluid elasticity hardly exist. The problem remains of interest, however, for establishing the role that shear thinning plays in determining the kinematics of tubular entry flows of viscoelastic fluids.

Following the treatment for Newtonian fluids, where the results obtained from boundary-layer theory were superposed with those obtained with creeping flow, the following expression was recommended by Boger (1982) as an upper bound for predicting the pressure drop for shear-thinning fluids in circular entry flow for $\beta \geq 4$:

$$\frac{\Delta P_{\mathrm{EN}}}{2\tau_{\mathrm{w}}} = (C'+1)\frac{N'_{\mathrm{Re}}}{32} + n'_{\mathrm{c}}, \tag{9}$$

where the generalized Reynolds number

$$N'_{\mathrm{Re}} = \frac{\rho D_{\mathrm{d}}^{n'} V_{\mathrm{d}}^{2-n'}}{8^{n'-1} K'} \tag{10}$$

is defined so that the Fanning function factor, $f = 16/N'_{\mathrm{Re}}$, is valid for any inelastic fluid. Here n' is the slope of a log-log plot of τ_{w} versus $8V/D$ obtained with a capillary rheometer, while K' is the intercept of this plot at $8V/D = 1\ \mathrm{s}^{-1}$. For a power-law fluid, we have $n = n'$ and $K = K'[4n/(3n+1)]^n$.

The entry-length predictions of Collins & Schowalter (1963) for a power-law fluid were verified experimentally for $\beta = 2$, $0.58 \leqq n \leqq 1$, and $190 \leqq N'_{\mathrm{Re}} \leqq 1940$ by Ramamurthy & Boger (1971). Thus, the values of the loss coefficient C', computed by Collins & Schowalter and listed in Table 1, are recommended for use in Equation (9). Values of the Couette correction $n_{\mathrm{c}} = \Delta P_{\mathrm{EN}}/2\tau_{\mathrm{w}}$ computed by Boger et al. (1978), M. J. Crochet (private communication, 1982), and Kim-E et al. (1983) are compared in Figure 4. The Boger et al. results, computed using the finite-element techniques developed by Tanner et al. (1975), are conservative on the high side because of the lack of grid mesh refinement in this early solution. Thus, the curve drawn through the combined and more recent results of Crochet and Kim-E et al. are recommended for describing the functional form of the Couette correction for power-law fluids.

Equation (9) is recommended as the upper bound for the excess pressure

Table 1 Values of the loss coefficient and Couette correction for use in Equation (9) to predict an upper bound on the excess pressure drop in tubular entry flows for inelastic shear-thinning fluids

Flow behavior index n	Loss coefficient C'	Couette correction n'_c
1.0	1.33	0.58
0.9	1.25	0.64
0.8	1.17	0.70
0.7	1.08	0.79
0.6	0.97	0.89
0.5	0.85	0.99
0.4	0.70	1.15
0.3	0.53	1.33

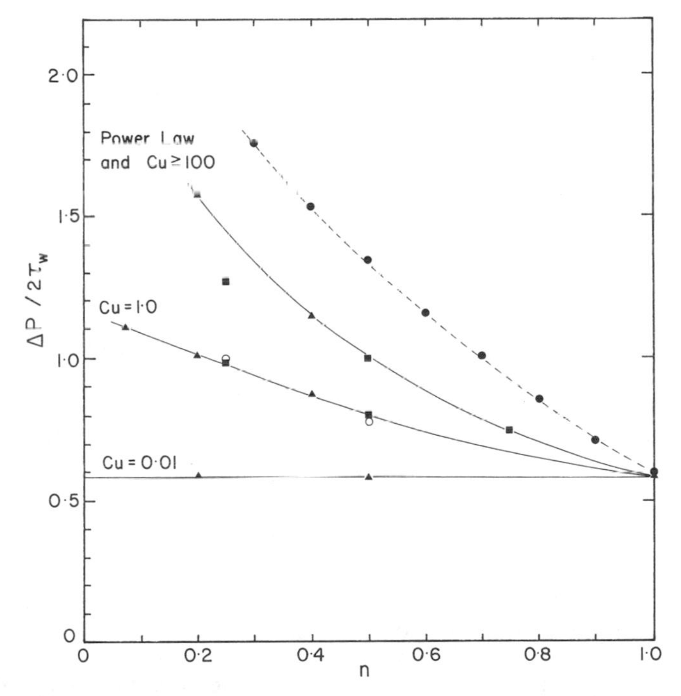

Figure 4 The Couette correction computed for power-law and Carreau fluids (● Boger et al. 1978, ■ M. J. Crochet, private communication, 1982, ▲ Kim-E et al. 1983, ○ Vrentas & Duda 1982).

loss for shear-thinning fluids in circular contractions. For high Reynolds number the upper bound on $\Delta P_{EN}/2\tau_w$ from Equation (9) should be quite accurate. However, for $N'_{Re} \leqq 30$ the excess pressure drop is also dependent on the detailed viscosity behavior of shear-thinning fluids at low shear rates, and an additional dimensionless group like the Carreau number is required to define the pressure drop (Kim-E et al. 1983, Vrentas et al. 1982, M. J. Crochet, private communication, 1982). Zero-Reynolds-number results from the various computational investigations are compared in Figure 4, where $\Delta P_{EN}/2\tau_w$ is shown as a function of n for Carreau numbers of 0.01 and 1.0. Kim-E et al., in their finite-element solution for a Carreau fluid, also considered the influence of fluid inertia by examining the flow in the 4:1 circular contraction at Reynolds numbers of 0.2, 2.0, and 10.[2] Crochet's finite-element computations, also for a Carreau fluid, were confined to zero Reynolds number, with the exception of one set of calculations for $N'_{Re} = 0.5$.[3] Vrentas et al. used finite-difference methods for an Ellis fluid and confined their attention to creeping flows. All the authors considered the case where $\beta = 4$.

Clearly from Figure 4, as the Carreau number increases at fixed n, the excess pressure drop increases from the Newtonian lower limit to the power-law upper limit (i.e. for a given shear-thinning fluid, $\Delta P_{EN}/2\tau_w$ increases with shear rate from 0.58 to a value dictated by the power-law index. Both Kim-E et al. and Crochet established that the power-law limit is reached for $\mathrm{Cu} \geqq 100$, whereas Vrentas et al. exceeded this limit in their computation for $\mathrm{El} \approx \mathrm{Cu} = 5000$. Kim-E et al. suggested that the difficulty was due to "numerical inertia." The three works do agree however for $\mathrm{Cu} \leqq 100$, as is illustrated in Figure 5, where $\Delta P_{EN}/2\tau_w$ is shown as a function of Cu for $n = 0.25$, 0.5, and 0.75. The detailed effect of shear thinning in the absence of fluid elasticity is now understood in creeping tubular entry flows. At particular values of n and Carreau number, fluid inertia increases the pressure drop, as is clearly demonstrated in Table 4 of the Kim-E et al. paper. Unfortunately, a direct comparison between the high-Reynolds-number results of Kim-E et al. and Equation (9) cannot easily be made because of the different definitions of Reynolds numbers employed. However, based on our understanding of Newtonian fluids, there is no reason to dispute Equation (9).

Although it was once thought that shear thinning increased the size of the secondary-flow vortex (Duda & Vrentas 1973, Boger 1982), it is now

[2] Kim-E et al. do not use the generalized Reynolds-number definition given by Equation (10) but instead define the Reynolds number in terms of the zero shear viscosity, $N_{Re} = \rho D_d V_d/\eta_0$.

[3] The results of Crochet presented here have not been published. They were generated in 1983 using POLYFLOW, a commercially available finite-element package.

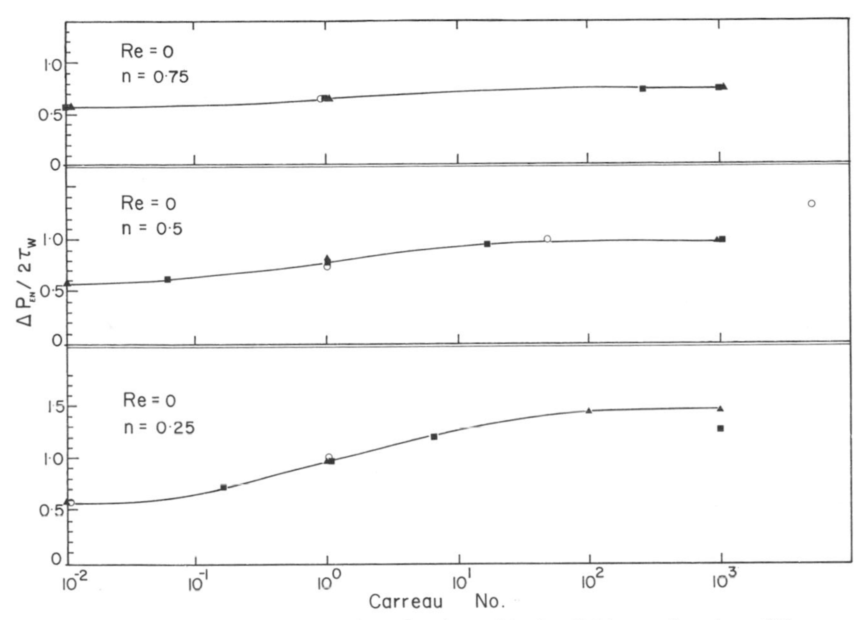

Figure 5 The excess entry pressure drop for shear-thinning fluids as a function of Carreau number (■ M. J. Crochet, private communication, 1982, ▲ Kim-E et al. 1983, ○ Vrentas & Duda 1982).

clear that the vortex decreases in size as the Carreau number is increased for a fixed n, and that the decay of the cell is enhanced as n decreases (Kim-E et al. 1983, M. J. Crochet, private communication, 1982). The creeping-flow corner vortex cannot be disassociated with $\mathrm{Cu} = \lambda V_d/R_d$. When Cu is small, Newtonian behavior prevails (see Figure 1), and when $\mathrm{Cu} \geqq 100$, when there is essentially no secondary-flow vortex, power-law behavior is observed. Kim-E et al. further demonstrated that shear thinning is not responsible for offcentered maxima in the velocity profile at the plane of the contraction, but rather that this is an inertial phenomenon, as observed and predicted at high Reynolds number for Newtonian fluids. Finally, these same authors did predict a small overshoot in the centerline velocity at the plane of the contraction, but the effect was very small compared with that found in the experimental observations of Kramer & Meissner (1980) for a viscoelastic (shear-thinning) low-density polyethylene.

The effects of shear thinning and the interaction of shear thinning and fluid inertia are now understood in tubular entry flow. In the absence of zero-shear-viscosity data and when the basic shear stress–shear-rate data can be fitted by a power-law, Equation (9) and the values of C' and n_c'

listed in Table 1 can be used to make a conservative prediction of the excess pressure drop $\Delta P_{EN}/2\tau_w$. If a more complete set of viscosity data is available (including values of η_0), then the Carreau equation [Equation (4)] should be fitted to the data, and the results in Figures 4 and 5 and/or in Table 4 of Kim-E et al. (1983) should be used for the pressure-drop determination. In terms of the now-known kinematics, shear-thinning effects cannot account for any of the flow phenomena observed for viscoelastic fluids.

The conclusions established here on the influence of shear thinning are based largely on the accuracy of the Kim-E et al. finite-element code. The extreme accuracy of this code has now been established by the excellent agreement that has been observed between laser-Doppler point-velocity measurements (radial and axial) and predictions using this code for Newtonian fluids flowing in a 4 : 1 contraction (Lawler et al. 1986).

4. VISCOELASTIC FLUIDS

Interest in viscoelastic-fluid flows is motivated largely by the importance of polymer-melt and solution flows in processing operations. Most viscoelastic fluids exhibit shear thinning but in addition also exhibit unequal normal stresses. They also show an extensional viscosity that is both strain and extensional strain-rate dependent; this is in contrast to a Newtonian fluid, where the extensional viscosity is constant and three times the steady shear value. The behavior of a viscoelastic fluid at any time is dependent on its recent deformation history. Common flow phenomena associated with fluid elasticity include the rod-climbing effect (Weissenberg effect), die swell, the open-channel siphon (Fano effect), stress-relaxation phenomena, and (in entry flows) vortex enhancement and flow instabilities that result in *melt fracture*. A discussion of viscoelastic-flow phenomena and the constitutive equations that have been developed in an attempt to describe their behavior is found in the book by Bird et al. (1977).

It is fair to say that our ability to predict observed and significant viscoelastic-flow phenomena in nonviscometric flows has been singularly unsuccessful until very recently. Experimental observations, often made with inadequately characterized materials, confuse shear-thinning, inertial, and elasticity effects; on the other hand, analytical treatments (usually numerical) have largely been confined to creeping flows of constant-viscosity elastic liquids, represented by constitutive equations such as the convected Maxwell and Oldroyd B models. However, with the discovery of a class of constant-viscosity elastic liquids, the gap between experimental observation and prediction has now narrowed considerably (Boger 1985).

Specification of the flow properties for adequate characterization of viscoelastic fluids is not simple. In principle, one should know the steady shear properties given by the viscosity $\eta(\dot{\gamma})$ and the first normal-stress difference $N_1(\dot{\gamma})$, the dynamic properties given by a storage modulus $G'(\omega)$ and loss modulus $G''(\omega)$, and some of the extensional-flow characteristics. In practice, extensional-flow properties are almost never known, and the storage and loss moduli are only now being measured on a more or less routine basis. It has thus become customary to characterize the elasticity of viscoelastic fluids on the basis of observed steady shear flow properties. Thus, in addition to a characteristic time λ_s and a power-law index n to describe shear thinning, one additional time constant, λ, is generally used to describe fluid elasticity. The definition of this time constant and the number of additional time constants required depend on the constitutive equation used. For a Maxwell fluid, we have

$$\lambda = \frac{N_1}{2\dot{\gamma}^2\eta}, \tag{11}$$

whereas for an Oldroyd B fluid (Oldroyd 1958), both a relaxation time λ_1 and a retardation time λ_2 arise. Thus we have

$$\lambda_1 = \frac{N_1}{2\dot{\gamma}^2\eta_p} \tag{12}$$

and

$$\lambda_2 = \frac{\lambda_1}{1+\alpha}, \tag{13}$$

where α is the retardation parametrer, $\alpha = \eta_p/\eta_s$, and η_p and η_s ($\eta = \eta_p+\eta_s$) are the polymer and solvent contributions to the viscosity, respectively.

Although Equations (11), (12), and (13) are useful in defining the constants from basic viscosity and first normal-stress difference data, neither the Maxwell nor the Oldroyd B constitutive equations allow for a shear-thinning viscosity. To allow for shear thinning, more complicated constitutive equations that introduce additional parameters are required. Such equations have rarely been used because of the complications they introduce in numerical simulations and because the additional experimental parameters for basic flow property measurements are generally not available. Thus, elastic effects in flows, such as the circular entry, are generally characterized by one dimensionless number, called the Weissenberg number by some authors and the Deborah number by others:

$$N_{We} = \lambda\dot{\gamma}, \tag{14}$$

where λ is defined by either Equation (11) or (12) and $\dot{\gamma}$ is a characteristic shear rate for the flow, usually the downstream wall shear rate in tubular entry flows. For a viscoelastic fluid with a constant viscosity, we have $\dot{\gamma} = 8V_d/D_d$.

In what follows, we first summarize the vast literature on observations of the behavior of polymer melts and solutions in tubular entry flows. This is followed by a discussion of the interaction between numerical prediction and experimental observation that is possible when both fluid inertia and shear thinning are eliminated in both numerical simulation and experimental observation. Inertia is generally not an important variable in polymer-processing operations, and thus inertial effects are eliminated entirely from our discussion. The experimental papers by Cable & Boger (1978a,b, 1979), where both point velocities and flow patterns are observed in stable and unstable flow of viscoelastic fluids in circular contractions, demonstrate how complex the entry flow can be when shear thinning, elasticity, and fluid inertia are all important variables.

Observations With Polymer Melts and Polymer Solutions

Flow-visualization studies, beginning with the work of Tordella (1957) and Bagley & Birks (1960), illustrated the existence of a large recirculating vortex in the corner of an abrupt circular entry flow at low Reynolds number. Extensive but largely qualitative studies of entry-flow patterns followed for a wide range of polymer melts and solutions. These works were reviewed by Tordella (1969), Dennison (1967), den Otter (1970), White (1973), Petrie & Denn (1976), White & Kondo (1977), and Boger (1982). The papers by J. L. White and coworkers are most informative. Many of the early works were for limited flow rates for uncharacterized materials and were more concerned with flow instabilities (melt fracture) than with stable flow patterns. However, it is clear that the large vortices are associated with excess entry pressure drops that are significantly higher than those expected and now predicted for Newtonian or inelastic shear-thinning fluids.[4] Also, it was observed that some viscoelastic molten polymers exhibit large vortices, whereas others do not (Ballenger & White 1971). Giesekus (1968) and Ramamurthy (1974) demonstrated for polymer solutions, and Ballenger & White (1971) for some polymer melts, that these vortices grow with increasing flow rate. In the latter paper, Ballenger & White also showed results for other molten polymers where the vortices

[4] Entry pressure-drop measurements for viscoelastic fluids in the absence of tube-flow exit losses are rarely available. An exception is the work from the laboratory of C. D. Han, where entry and exit losses have clearly been separated (Han & Charles 1971). Far more measurements of this type for well-characterized fluids are required.

are small and do not grow with flow rate. All workers suggested that fluid elasticity was responsible for the vortex growth, but the question as to why the vortex grows for some materials and is either not present or does not grow for others remained unresolved until the work of Cogswell (1972) and White & Kondo (1977). In both these papers, it was suggested that materials with an extensional viscosity that increases with extensional rate exhibit vortex enhancement (growth), whereas no (or only small) vortices are present for materials where the extensional viscosity decreases or remains constant with elongation rate. White & Baird (1986) confirmed this conclusion in a definitive paper where flow visualization was combined with good flow property measurement. Steady shear properties η and N_1, dynamic properties G' and G'', and extensional stress-growth data were measured for two different molten polymers. One polymer exhibited large vortices and vortex growth (low-density polyethylene), and the other (polystyrene) exhibited almost no secondary flows in a planar 5.9 : 1 contraction. At the conditions of observation both flows were characterized by about the same Weissenberg number [Equation (14)], yet the flow fields were quite different. The difference was in the extensional-flow properties of the polymers. Vortex growth was clearly associated with unbounded extensional stress growth with time at fixed extensional strain rates, whereas the absence of the vortex was associated with bounded extensional stress growth. This observation was consistent with the White & Kondo premise that large vortices and vortex growth are stress-relief mechanisms. Thus, all viscoelastic fluids do not exhibit large vortices and vortex growth in entry flows. The presence and characteristics of the vortex are strongly dependent on extensional-flow properties.

Constitutive equations that might be useful in characterizing shear thinning and elastic fluids, such as the molten polymers used in the experiments of White & Baird (1986), are at this stage normally too complex for use with the equations of motion for a numerical solution of the entry-flow problem. This presupposes, of course, that all material constants are available from the basic flow property measurement, which is hardly ever the case. Thus, the choice of a constitutive equation for examination of the entry flow of viscoelastic fluids has been a compromise between simplicity and practical reality. Although attempted solutions for creeping flow in a circular contraction for other constitutive equations are available, the most common choice of constitutive equations has been the convected Maxwell or the Oldroyd B model. Important papers that should be consulted are those of Mendelson et al. (1982) and Keunings & Crochet (1984), while numerical solution techniques and results for viscoelastic fluids in entry flows have been reviewed by Crochet & Walters (1983) and Crochet et al. (1984).

Influence of Elasticity in the Absence of a Shear-Thinning Viscosity

Both the convected Maxwell and Oldroyd B models represent constant-viscosity elastic liquids. Both models predict a steady-state extensional viscosity that grows with extensional rate, but both also suffer from singular points at fixed extensional rates above which the steady extensional viscosity is not defined. Given this inherent problem with both models, the hope still existed that these constitutive equations would be useful to predict (at least qualitatively) the vortex enhancement observed in tubular contractions for constant-viscosity elastic liquids—the so called Boger fluid (Walters 1979).

The series of streak photographs reproduced in Figure 6 is typical of the vortex-growth behavior observed in contracting flows for constant-viscosity elastic fluids (Nguyen & Boger 1979, Boger & Nguyen 1978, Cochrane et al 1981, Walters & Rawlinson 1982, Boger et al. 1986, Evans & Walters 1986). The fluid used for the observations shown in Figure 6 (a dilute solution of high-molecular-weight polyacrylamide dissolved in water and corn syrup) is characterized by a constant viscosity ($\eta = 97.5$ poise) and by $\lambda_0 = 0.308$ s and/or $\lambda_1 = 1.132$ s and $\alpha = 0.373$. Here λ_0 is the limiting or low-shear-rate Maxwell relaxation time, while the Oldroyd parameters, λ_1 and α, are adequate to describe the dynamic and steady shear properties of the fluid up to shear rates and frequencies of 3 s^{-1}. The polyacrylamide-in-corn-syrup fluids are now known to exhibit a steady-

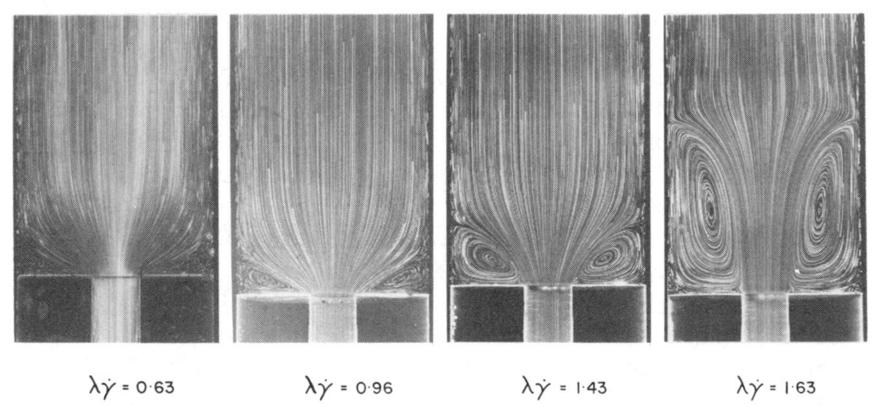

$\lambda\dot{\gamma} = 0{\cdot}63$ $\lambda\dot{\gamma} = 0{\cdot}96$ $\lambda\dot{\gamma} = 1{\cdot}43$ $\lambda\dot{\gamma} = 1{\cdot}63$

Figure 6 Streak-line photographs illustrating the changing vortex shape and growth as a function of $\lambda\dot{\gamma}$ for a constant-viscosity elastic liquid flowing in a 4.08 : 1 circular contraction. The fluid is characterized by $\eta = 97.5$ poise and by $\lambda_0 = 0.308$ s and/or $\lambda_1 = 1.132$ s and $\alpha = 0.373$. Complete steady and dynamic shear properties for the fluid are available (Boger et al. 1986).

state extensional viscosity that increases substantially with strain rate (Walters 1985). Thus, the observed vortex growth with increasing $\lambda\dot{\gamma}$ shown in Figure 6 is expected. As $\lambda\dot{\gamma}$ increases, the vortex grows from a Newtonian-like vortex with a reattachment length of 0.17 to a very large value at $\lambda\dot{\gamma} = 1.63$. For higher values of $\lambda\dot{\gamma}$, the secondary flow becomes asymmetric and ultimately rotates, which results in a three-dimensional flow (Nguyen & Boger 1979).

The observed reattachment length of the secondary-flow vortex as a function of $\lambda\dot{\gamma}$ is reproduced in Figure 7 for one shear-thinning (G) and six non-shear-thinning elastic polyacrylamide-in-corn-syrup solutions flowing in a 4:1 contraction (Boger 1985). The viscosity η and Maxwell (low-shear-rate) relaxation time λ_0 for the fluids ranged from $89 < \eta < 750$ poise and $0.1 < \lambda_0 < 0.9$ for the non-shear-thinning fluids, while the flow behavior index n for the shear-thinning elastic fluid was 0.64. Significant vortex growth is generated for the shear-thinning fluid only at very high values of $\lambda\dot{\gamma}$, where the influence of fluid elasticity is sufficient to overcome the influence of a shear-rate-dependent viscosity (see Section 3). Different functional forms of X as a function of $\lambda\dot{\gamma}$ are observed for the constant-viscosity elastic fluids with different characteristic times. Thus, more than one dimensionless group is required to correlate the vortex growth

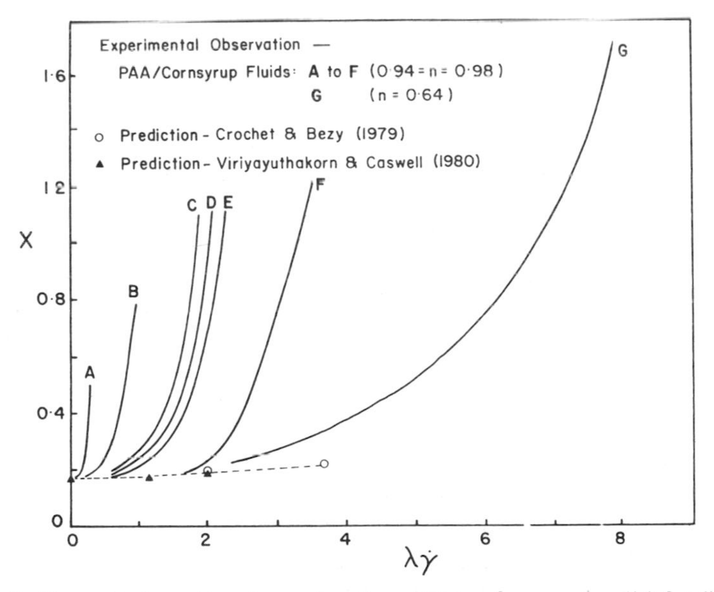

Figure 7 The reattachment length as a function of Weissenberg number ($\lambda\dot{\gamma}$) for different non-shear-thinning elastic fluids (polyacrylamide in corn syrup) flowing in a 4:1 contraction (Boger 1985).

observed for constant-viscosity elastic liquids, and a constitutive equation like the converted Maxwell model is not expected to be adequate to correlate the significant vortex growth observed except in a qualitative way. This conclusion is confirmed in Figure 7 by the very slight increase in cell size with $\lambda\dot{\gamma}$, which was predicted by Viriyayuthakorn & Caswell (1980) and by Crochet & Bezy (1979) using finite-element numerical simulation techniques for a Maxwell fluid in a circular 4:1 contraction. The experimental and numerical results are only in agreement at very low values of $\lambda\dot{\gamma}$ (for each fluid). This is to be expected, since the Maxwell model only describes the steady and dynamic flow properties of these fluids in the limit of low shear rates, whereas the Oldroyd B model is adequate to moderate shear rates (Gupta et al. 1983).

It was originally hoped that the qualitative failure of the Maxwell model was due to the limited domain of elasticity ($\lambda\dot{\gamma}$) where stable calculations could be completed, but the work of Keunings & Crochet (1984) for an Oldroyd B fluid allowed high values of $\lambda\dot{\gamma}$ to be reached, where unfortunately no significant vortex enhancement was predicted. Many other numerical workers have reached the same frustrating conclusion. With the exception of the Keunings & Crochet (1984) prediction, no significant vortex enhancement has been predicted for any viscoelastic fluid. Where then is the problem? Is it in the numerical calculations and mathematics, the constitutive equations themselves, and/or the experimental observations?

The first question is not addressed here, since it is far from the scope of this review and there are a number of very capable people examining the numerical mathematics associated with the tube-entrance flow problem (see, for example, Mendelson et al. 1982, Keunings 1986, Marchal & Crochet 1986). On the second question, the convected Maxwell constitutive equation has been unsuccessful (in terms of predicting significant and real effects) in many nonviscometric flows, and although the Oldroyd B constitutive equation could be dismissed on the basis of its poor performance in extension, there are two flow fields where some success has been obtained. In squeeze film flow and in fiber spinning, predictions of significant elastic effects agree at least qualitatively with experimental observation for constant-viscosity elastic fluids characterized by the Oldroyd B model (Phan-Thien et al. 1985, Sridhar et al. 1986). However, the Oldroyd B model only represents the steady and dynamic shear properties of the constant-viscosity elastic liquids up to moderate shear rates and fails at higher shear rates. Clearly more work is required on the development of *simple* but realistic constitutive equations, particularly with regard to predicting the behavior of constant-viscosity elastic liquids. In addition, techniques for extensional property measurement of mobile elastic liquids must be established.

Recent experimental observations on the kinematics of tubular entry flows do establish at least one reason why numerical simulation of this flow field has been unsuccessful. A further examination of the streak photographs shown in Figure 6 shows that at low values of $\lambda\dot{\gamma}$, the vortex is virtually identical in shape and size to that observed and predicted for an inelastic Newtonian fluid (see Figure 2). Here the vortex boundary is concave, with its center of rotation near the corner of the upstream tube. As $\lambda\dot{\gamma}$ increases, the center of rotation of the cell shifts toward the tube entrance and the cell boundary straightens until at higher $\lambda\dot{\gamma}$ the cell boundary is convex, with the center of rotation of the vortex positioned close to the tube entrance, or *lip*. This lip vortex now grows, and the reattachment length of the cell increases for $\lambda\dot{\gamma} \geqq 1$. Such behavior has been consistently observed for the constant-viscosity polyacrylamide-in-corn-syrup solutions for $\beta \geqq 4$, where the vortex growth is not a significant function of β. The change in shape of the vortex boundary without a change in the reattachment length of the cell has not been predicted by any numerical simulation. A lip vortex such as that illustrated in Figure 6 was first observed in entry flows for viscoelastic fluids by Giesekus (1968), more recently by Dembek (1982), and consistently by Walters and co-workers in circular entry flows and in other entry-flow geometries (Cochrane et al. 1981, Walters & Rawlinson 1982, Walters 1984, Evans & Walters 1986).

The transition in flow from a corner to a lip vortex and a mechanism for the change in shape of the vortex boundary are not apparent from the streak photographs shown in Figure 6, nor were these apparent from experiments conducted with the polyacrylamide-in-corn-syrup solutions conducted in contraction ratios up to 15. However, the transition and a mechanism become apparent from the streak photographs shown in Figure 8 for another constant-viscosity elastic liquid with steady and dynamic flow properties similar to those of the fluid used for the observations shown in Figure 6 for flow in a 4.08 : 1 circular contraction. Here the reattachment length and shape of the cell remain similar to the Newtonian cell with increasing $\lambda\dot{\gamma}$ to a value of about 1.6. At this stage the reattachment length of the secondary-flow vortex *decreases* as $\lambda\dot{\gamma}$ increases, and it effectively reaches a zero value at $\lambda\dot{\gamma} = 2.37$ (see Boger et al. 1986). The decrease in the size of the corner vortex is not an inertial effect, as is expected for Newtonian fluids when $N_{\mathrm{Re}} > 1$, since the downstream N_{Re} never exceeds 0.0192 for the streak photographs shown in Figure 8. The decrease in the reattachment length X with increasing $\lambda\dot{\gamma}$ occurs at the same time as the formation of an independent vortex (at $\lambda\dot{\gamma} = 2$) that emanates from the tube-entrance lip. The lip vortex ultimately destroys the corner vortex, and vortex growth occurs for $\lambda\dot{\gamma} > 2.40$. Thus, two distinctly different flow-

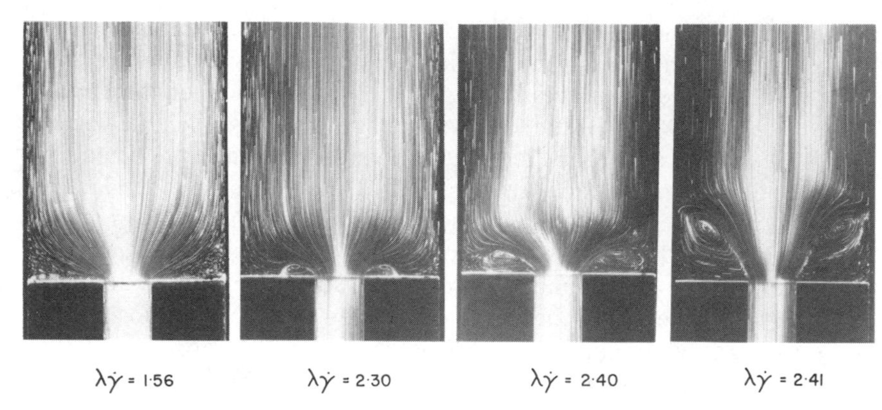

Figure 8 Streak-line photographs illustrating the changing vortex shape and growth as a function of $\lambda\dot{\gamma}$ for a constant-viscosity elastic liquid flowing in a 4.08 : 1 circular contraction. The fluid (polyisobutylene-polybutene) is characterized by $\eta = 251$ poise and by $\lambda_0 = 0.149$ s and/or $\lambda_1 = 1.19$ s and $\alpha = 0.143$. Complete steady and dynamic shear properties for the fluid are available (Boger et al. 1986).

pattern developments for two constant-viscosity elastic fluids with essentially the same time constants are observed for creeping flow in the same circular 4.08 : 1 contraction. Clearly the two materials are not similar and differ in some unmeasured flow property—most likely the extensional viscosity.

The influence of extensional strain and extensional strain rate can be qualitatively assessed by close examination of the observed flow patterns as a function of contraction ratio. Such observations have been made and are compared (qualitatively) for both materials in Figures 9 and 10. The entry-flow pattern for both materials develops in the same way if the contraction ratio is high enough, i.e. $\beta \geqq 4$ for the polyacrylamide–corn syrup fluid and $\beta \geqq 16$ for the polyisobutylene-polybutene fluid. A gradual change in the shape of the vortex boundary is observed with increasing $\lambda\dot{\gamma}$ from concave curvature to convex curvature while the vortex size remains essentially constant. At higher values of $\lambda\dot{\gamma}$ the convex curvature becomes more pronounced, and vortex growth is observed. We believe that interaction of the two vortices occurs in all contractions, but the effect is less obvious at high contraction ratios, where high levels of extensional strain are experienced by the fluid. At lower contraction ratios, corresponding to lower levels of extensional strain (i.e. $\beta \leqq 2$ for the polyacrylamide–corn syrup fluid and $\beta \leqq 4$ for the polyisobutylene-polybutene fluid), the interaction of the two vortices is very apparent. In fact, in the lower contraction ratios for both fluids, when the lip vortex is dominant and the corner is essentially absent, a very pronounced rotation of the lip vortex is observed. The rotating lip vortex grows in size as the shear rate is increased, as does the frequency of rotation until at a critical value of $\lambda\dot{\gamma}$ the flow again becomes two dimensional. Beyond this value of $\lambda\dot{\gamma}$ the vortex of the two-dimensional steady flow grows significantly until the next level of transient three-dimensional flow is observed. The latter three-dimensional transient flow has been observed on many occasions, whereas the three-dimensional transient at low values of $\lambda\dot{\gamma}$ is only apparent from very recent work (Boger et al. 1986, Lawler et al. 1986). Lawler et al. observed precisely the same flow phenomena with a constant-viscosity elastic polyisobutylene-polybutene fluid. Using an automated two-color laser-Doppler velocimeter, they were able to make very accurate axial- and radial-velocity measurements in a circular 4 : 1 contraction for both Newtonian and elastic liquids. The experimental techniques were demonstrated by comparing the two-component velocity measurements for a Newtonian fluid to the results obtained using the finite-element code developed by Kim-E et al. (1983). The agreement between the velocity observations and numerical prediction is very impressive. For the ideal elastic fluid the authors established that for a lower critical value of Weissenberg number (or Deborah number), the flow changes from a steady two-dimensional motion, nearly identical to the Newtonian kinematics, to a time-periodic flow with a tangential velocity component that fluctuates about zero. Multiple time-periodic (three-dimensional) motions were observed for an increasing range of Weissenberg number until at a second critical Weissenberg number the flow suddenly reverted back to two-

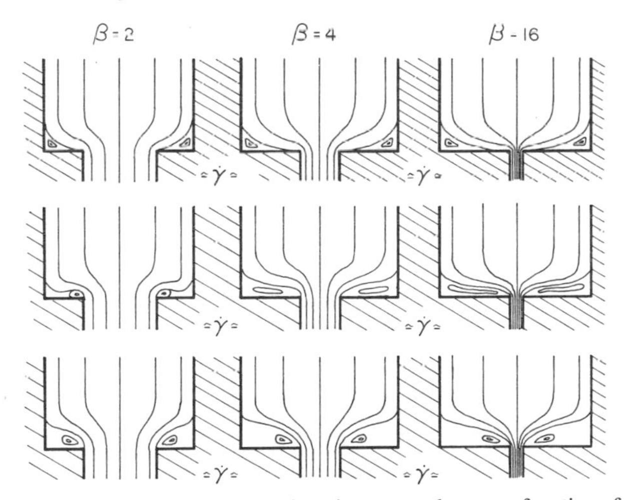

Figure 9 A qualitative comparison of the changing vortex shape as a function of contraction ratio for the polyisobutylene-polybutene constant-viscosity elastic liquid used for the observations shown in Figure 8.

dimensional, time-independent motion. The observations of Lawler et al. (1986) and our own (Boger et al. 1986) are indeed identical. Both show that an elastic liquid in tubular entry flow can experience two transitions to a time-dependent three-dimensional flow—one transition at low Weissenberg number and another transition at significantly higher values of Weissenberg number. In between the two critical Weissenberg-number regions, vortex growth in a steady two-dimensional flow is observed. All attempts at numerical simulation fail at approximately the lower critical conditions observed for the onset of the first unsteady three-dimensional flow. Although far more experimentation is required to characterize the multiple unsteady flows observed at the lower critical Weissenberg number, it is clear that any numerical simulation aimed at predicting the observed stable vortex-growth flow must be generalized to three dimensions and time dependence, even at low Weissenberg number.

5. CONCLUSION

The tubular entry-flow problem has a history of about 120 years. It is now solved for Newtonian and inelastic shear-thinning fluids, both with and without inertia in the flow, and great strides have been made in gaining an understanding of the complexity of tubular entry flows of viscoelastic fluids. This understanding has led to further challenges in mathematics, in numerical methods, in the development of simple but effective constitutive equations, and in the definition of the precise experimentation required. It

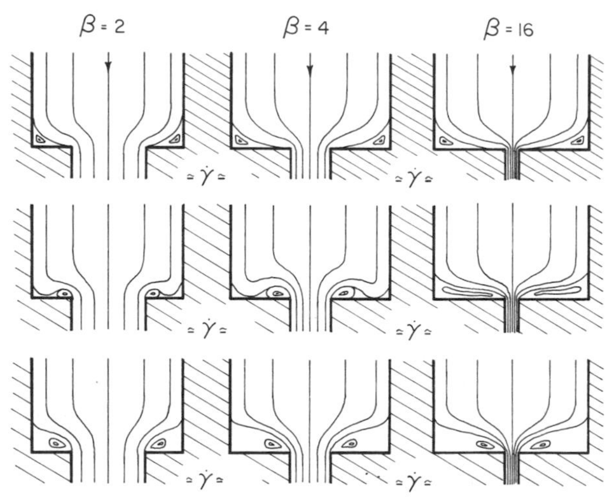

Figure 10 A qualitative comparison of the changing vortex shape as a function of contraction ratio for the constant-viscosity elastic liquid used for the observations shown in Figure 6.

is clear that entry flows of viscoelastic liquids are dominated by extensional effects, and that we must come to grips with the measurement of extensional-flow properties. Also, our conception of nonslip boundary conditions at surfaces in regions of high stress may be inadequate. Ramamurthy (1986) deals with the nonslip boundary-condition question in entry flows of molten polymers and clearly demonstrates slip flows at conditions

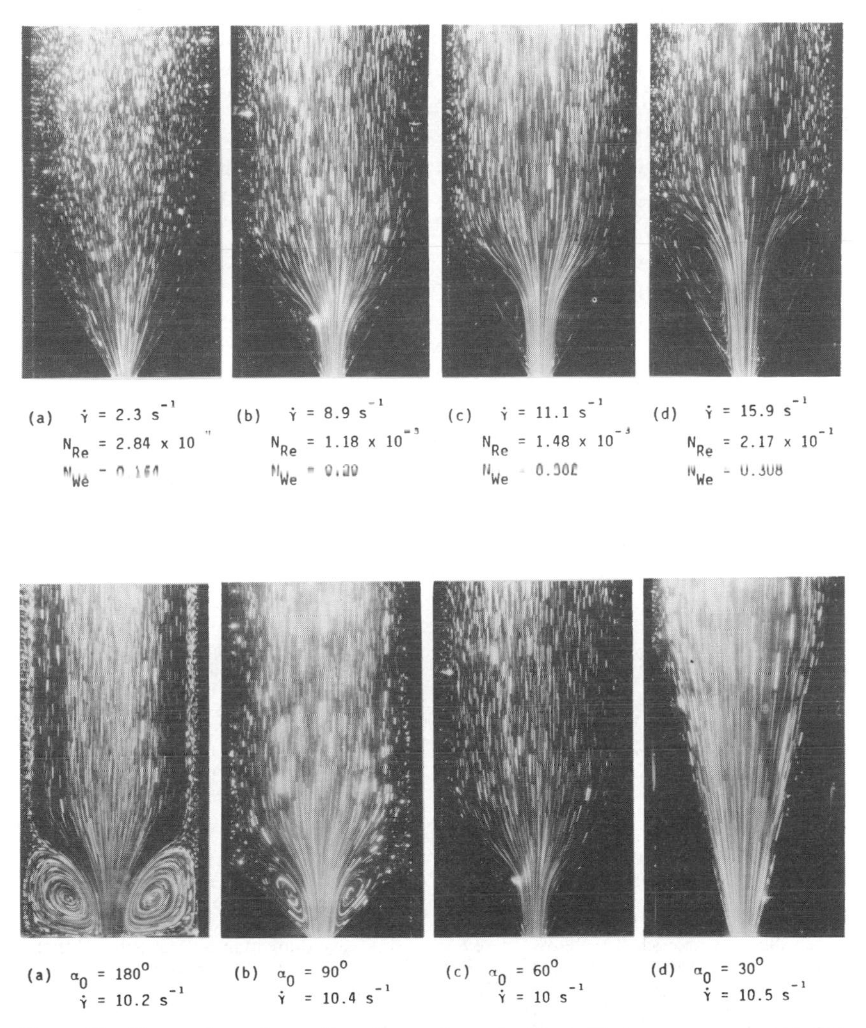

Figure 11 Flow-pattern observations in angular entry flows. Upper sequence: vortex formation and growth in a 60° tapered entry. Lower sequence: comparison of stable entry flow with different included entry angles at fixed shear rate.

of unstable flow. After scores of other publications, Ramamurthy presents the first documented and logical explanation for melt-flow instability.

Viscoelastic-fluid entry flows are not only of interest as an appropriate test problem for developing the fluid mechanics of viscoelastic fluids, but they are also of interest and of great importance in polymer processing. Here the ultimate aim is to predict the influence of the entry-flow geometry on the kinematics and pressure drop in order to both minimize the latter and optimize the former by eliminating secondary flows and regions of high stress. For such predictions to be possible, good basic flow property measurements are required for use with an adequate constitutive equation in the equations of motion in order to predict flow phenomena such as those illustrated by the photographs shown in Figure 11. The upper sequence of photographs illustrates the influence of Weissenberg number on the secondary-flow vortex observed for a constant-viscosity elastic liquid in a 60° circular entry flow, while the lower sequence examines the influence of entry angle on the secondary-flow vortex at fixed downstream shear rate (Nguyen 1978). Prediction of flow phenomena such as these will be possible in the next few years.

ACKNOWLEDGMENTS

Research in non-Newtonian fluid mechanics in Chemical Engineering at the University of Melbourne is supported by the Australian Research Grants Scheme. The help of Rod Binnington in the organization of this review is gratefully acknowledged.

Literature Cited

Bagley, E. B., Birks, A. M. 1960. Flow of polyethylene into a capillary. *J. Appl. Phys.* 31(3): 556–61

Ballenger, T. F., White, J. L. 1971. The development of the velocity field in polymer melts in a reservoir approaching a capillary die. *J. Appl. Polym. Sci.* 15(8): 1949–62

Bird, R. B., Armstrong, R. C., Hassager, O. 1977. *Dynamics of Polymeric Liquids*, Vol. 1, *Fluid Mechanics*. New York: Wiley. 470 pp.

Boger, D. V. 1982. Circular entry flows of inelastic and viscoelastic fluids. In *Advances in Transport Processes*, ed. A. S. Mujumdar, R. A. Mashelkar, 2:43–98. New Delhi: Wiley Eastern

Boger, D. V. 1985. Model polymer fluid systems. *Pure Appl. Chem.* 57(7): 921–30

Boger, D. V., Gupta, R., Tanner, R. I. 1978. The end correction for power-law fluids in the capillary rheometer. *J. Non-Newtonian Fluid Mech.* 4: 239–48

Boger, D. V., Hur, D. U., Binnington, R. J. 1986. Further observations of elastic effects in tubular entry flows. *J. Non-Newtonian Fluid Mech.* 20: 31–49

Boger, D. V., Nguyen, H. 1978. A model viscoelastic fluid. *Polym. Eng. Sci.* 18(13): 1037–43

Bogue, D. C. 1959. Entrance effects and prediction of turbulence in non-Newtonian flow. *Ind. Eng. Chem.* 51(7): 874–78

Burke, J. P., Burman, N. S. 1969. Entrance flow development in circular tubes at small axial distances. *ASME Pap. No. 69-WA/FE-13*

Cable, P. J., Boger, D. V. 1978a. A comprehensive experimental investigation of tubular entry flow of viscoelastic fluids: Part 1. Vortex characterization in stable flow. *AIChE J.* 24(5): 869–79

Cable, P. J., Boger, D. V. 1978b. A comprehensive experimental investigation of tubular entry flow of viscoelastic fluids: Part 2. The velocity field in stable flow. *AIChE J.* 24(6): 992–99

Cable, P. J., Boger, D. V. 1979. A comprehensive experimental investigation of tubular entry flow of viscoelastic fluids: Part 3. Unstable flow. *AIChE J.* 25(1): 152–59

Carter, T. R. 1969. *Laminar flow from a reservoir up to and through a tube entrance region.* PhD thesis. Univ. Utah, Salt Lake City. 319 pp.

Christiansen, E. A., Kelsey, S. J., Carter, T. R. 1972. Laminar tube flow through an abrupt contraction. *AIChE J.* 18(2): 372 80

Cochrane, T., Walters, K., Webster, M. F. 1981. On Newtonian and non-Newtonian flow in complex geometries. *Philos. Trans. R. Soc. London Ser. A* 301: 163–81

Cogswell, F. N. 1972. Converging flows of polymer melts in extrusion dies. *Polym. Eng. Sci.* 12: 64–73

Collins, M., Schowalter, W. R. 1963. Behaviour of non-Newtonian fluids in the inlet region of a pipe. *AIChE J.* 9(1): 804 9

Crochet, M. J., Bezy, M. 1979. Numerical solution for the flow of viscoelastic fluids. *J. Non-Newtonian Fluid Mech.* 5: 201–19

Crochet, M. J., Davies, A. R., Walters, K. 1984. *Numerical Simulation of Non-Newtonian Flow, Rheology Ser.*, Vol. 1. Amsterdam: Elsevier. 352 pp.

Crochet, M. J., Walters, K. 1983. Numerical methods in non-Newtonian fluild mechanics. *Ann. Rev. Fluid Mech.* 15: 241–60

Dembek, G. 1982. Structural changes of polyisobutylene solutions induced by orifice flow. *Rheol. Acta* 21(4/5): 553–55

Dennison, M. T. 1967. Flow instability in polymer melts: a review. *Plast. Inst. Trans. J.* 35: 803–8

den Otter, J. L. 1970. Mechanisms of melt fraction. *Plast. Polym.* 38: 155–68

Duda, J. L., Vrentas, J. S. 1972. Pressure losses in non-Newtonian entrance flows. *Can. J. Chem. Eng.* 50: 671–74

Duda, J. L., Vrentas, J. S. 1973. Entrance flows of non-Newtonian fluids. *Trans. Soc. Rheol.* 17: 89–108

Evans, R. E., Walters, K. 1986. Flow characteristics associated with abrupt changes in geometry in the case of highly elastic liquids. *J. Non-Newtonian Fluid Mech.* 20: 11–29

Fan, L. T., Hwang, C. L. 1966. Bibliography of hydrodynamic entrance region flow. Special Report 67. *Kansas State Univ. Bull.* 50(3): 1–17

Giesekus, H. 1968. Non-linear effects in the flow of viscoelastic fluids through slits and circular apertures. *Rheol. Acta* 7(27): 127–38

Gupta, R. K., Prilutski, G., Sridhar, T., Rejon, M. E. 1983. Model viscoelastic fluids. *J. Non-Newtonian Fluid Mech.* 12: 233–41

Han, C. D., Charles, M. 1971. Entrance and exit-correction in capillary flow of molten polymers. *Trans. Soc. Rheol.* 15: 371–84

Hornbeck, R. W. 1961. Laminar flow in the entrance region of a pipe. *Appl. Sci. Res. Sect. A* 13: 224 32

Kestin, J., Sokolov, M., Wakeham, W. 1973. Theory of capillary viscometers. *Appl. Sci. Res.* 27: 241–64

Keunings, R. 1986. On the high Weissenberg number problem. *J. Non-Newtonian Fluid Mech.* 20: 209–26

Keunings, R., Crochet, M. J. 1984. Numerical simulation of the flow of a viscoelastic fluid through an abrupt contraction. *J. Non-Newtonian Fluid Mech.* 14: 279–99

Kim-E, M. E., Brown, R. A., Armstrong, R. C. 1983. The roles of inertia and shear-thinning in flow of an inelastic liquid through an axisymmetric sudden contraction. *J. Non-Newtonian Fluid Mech.* 13: 341–63

Kramer, H., Meissner, J. 1980. Applications of laser Doppler velocimetry to polymer melt flow studies. In *Rheology*, ed. G. Astarita, G. Marrucci, L. Nicolais, 2: 463 68. New York: Plenum. 677 pp.

Lawler, J. V., Muller, S. J., Brown, R. A., Armstrong, R. C. 1986. Laser doppler velocimetry measurements of velocity field and transition in viscoelastic fluids. *J. Non-Newtonian Fluid Mech.* 20: 51–92

Marchal, J. M., Crochet, M. J. 1986. Hermitian finite elements for calculating viscoelastic flow. *J. Non-Newtonian Fluid Mech.* 20: 187–207

Meissner, J. 1985. Rheometry of polymer melts. *Ann. Rev. Fluid Mech.* 17: 45–64

Mendelson, M. A., Yeh, P. W., Brown, R. A., Armstrong, R. C. 1982. Approximation error in finite element calculation of viscoelastic fluid flows. *J. Non-Newtonian Fluid Mech.* 10: 31–54

Nguyen, T. H. 1978. *The influence of elasticity on die entry flow.* PhD thesis. Monash Univ., Clayton, Victoria, Aust. 272 pp.

Nguyen, H., Boger, D. V. 1979. The kinematics and stability of die entry flows. *J. Non-Newtonian Fluid Mech.* 5: 353–68

Oldroyd, J. G. 1958. Non-Newtonian effects in steady motion of some idealised elastico-viscous liquids. *Proc. R. Soc. London Ser. A* 245: 278–97

Petrie, C. J. S., Denn, M. M. 1976. Instabilities in polymer processing. *AIChE J.* 22(2): 209–36

Phan-Thien, N., Dudek, J., Boger, D. V., Tirtaatmadja, V. 1985. Squeeze film flow of ideal elastic liquids. *J. Non-Newtonian Fluid Mech.* 18: 227–54

Ramamurthy, A. V. 1974. Flow instabilities in a capillary rheometer for an elastic polymer solution. *Trans. Soc. Rheol.* 18(3): 431–52

Ramamurthy, A. V. 1986. Toward a coherent theory for polymer melt fracture. *J. Rheol.* 30(2): 337–57

Ramamurthy, A. V., Boger, D. V. 1971. Developing velocity profiles on the downstream side of a contraction for inelastic polymer solutions. *Trans. Soc. Rheol.* 15(4): 709–29

Schmidt, F. W., Zelden, B. 1969. Laminar flows in inlet sections of tubes and ducts. *AIChE J.* 15(4): 612–14

Sridhar, T., Gupta, R. K., Boger, D. V., Binnington, R. J. 1986. Steady spinning of the Oldroyd fluid B. II: experimental results. *J. Non-Newtonian Fluid Mech.* 21: 115–26

Tanner, R. I., Nickell, R. E., Bilger, R. W. 1975. Finite element methods for the solution of some incompressible non-Newtonian fluid mechanics problems with free surfaces. *Comput. Meth. Appl. Mech. Eng.* 6: 155–74

Tordella, J. P. 1957. Capillary flow of molten polyethylene—a photographic study of melt fracture. *Trans. Soc. Rheol.* 1: 203–12

Tordella, J. P. 1969. Unstable flow of molten polymers. In *Rheology*, ed. F. R. Eirich, 5: 57–92. New York: Academic

Viriyayuthakorn, M., Caswell, B. 1980. Finite element simulation of viscoelastic flow. *J. Non-Newtonian Fluid Mech.* 6: 245–67

Vrentas, J. S., Duda, J. L., Hong, S.-A. 1982. Excess pressure drops in entrance flows. *J. Rheol.* 26(4): 349–57

Vrentas, J. S., Duda, J. L., Bargeron, K. G. 1966. Effect of axial diffusion of vorticity on flow development in circular conducts. *AIChE J.* 12(5): 837–44

Vrentas, J. S., Duda, J. L. 1973. Flow of a Newtonian fluid through a sudden contraction. *Appl. Sci. Res.* 28: 241–59

Walters, K. 1979. Developments in non-Newtonian fluid mechanics—a personal view. *J. Non-Newtonian Fluid Mech.* 5: 113–24

Walters, K. 1984. Some modern developments in non-Newtonian fluid mechanics. In *Advances in Rheology*, ed. B. Mena, A. Garcia-Rejon, C. Rangel-Nafaile, 1: 31–38. Mexico City: Univ. Nac. Auton. Mex. 730 pp.

Walters, K. 1985. Overview of macroscopic viscoelastic flow. In *Viscoelasticity and Rheology*, ed. A. S. Lodge, M. Renardy, J. A. Nohel, pp. 47–79. New York: Academic

Walters, K., Rawlinson, D. M. 1982. On some contracting flows for Boger fluids. *Rheol. Acta* 21: 547–52

Walters, K., Webster, M. F. 1982. On dominating elastico-viscous response in some complex flows. *Philos. Trans. R. Soc. London Ser. A* 308: 199–218

Weissberg, H. L. 1962. End correction for slow viscous flow through long tubes. *Phys. Fluids* 5(9): 1033–36

White, J. L. 1973. Critique on flow patterns in polymer fluids at the entrance of a die and instabilities leading to extrudate distortion. *Appl. Polym. Symp.* 20: 155–74

White, J. L., Kondo, A. 1977. Flow patterns in polyethylene and polystyrene melts during extrusion through a die entry region: measurement and interpretation. *J. Non-Newtonian Fluid Mech.* 3: 41–64

White, S. A., Baird, D. G. 1986. The importance of extensional flow properties on planar entry flow patterns of polymer melts. *J. Non-Newtonian Fluid Mech.* 20: 93–102

Ann. Rev. Fluid Mech. 1987. 19 : 183–215

THEORY OF SOLUTE TRANSPORT BY GROUNDWATER

Gedeon Dagan

Department of Fluid Mechanics and Heat Transfer,
Faculty of Engineering, Tel Aviv University, Ramat Aviv 69979, Israel

1. INTRODUCTION

The transport of solutes in groundwater flow has been studied with increasing intensity in the last two decades as a result of growing concern about water quality and pollution. The intensification of groundwater exploitation, on the one hand, and the increase in solute concentration in aquifers due to saltwater intrusion, leaking repositories, use of fertilizers, etc., on the other, have made this a subject of immediate and wide interest.

The phenomenon of solute transport is quite complex, as it depends on several factors, such as the complicated geohydrological structures of aquifers, the nonuniformity and unsteadiness of flow, the physico-chemical interactions between solutes and matrix, and the mechanism of solute spreading.

The field study of solute transport also faces serious difficulties. First, measurements must be carried out by drilling numerous observation wells and by monitoring the concentration, which is quite costly and time consuming. Second, the spreading of the solutes is a very slow process, and an experiment may last many years if one wishes to investigate the long-range transport process.

Under these circumstances, the theory plays an important role, being instrumental in interpreting field tests and in predicting the fate of solutes under new conditions. The aims of the theory are to identify the main factors that influence transport and to provide the mathematical tools that permit one to compute the spatial distribution and the time evolution of the solute concentration, given the flow conditions.

0066–4189/87/0115–0183$02.00

Following a traditional path, the experimental support for the development of the theory at its beginning was provided by laboratory experiments. In a typical experiment, a uniform flow is created in a laboratory column, and solute at constant concentration is introduced at the inlet. By measuring subsequently the concentration at the outlet, a breakthrough curve is obtained. The effect of the porous structure upon transport is to enhance the mixing process, resulting in an increased effective diffusion, which has been termed hydrodynamic dispersion (for a review of the early development of the subject, see Fried & Combarnous 1971). In essence, it was found that the concentration C (defined as mass of solute per volume of solution) satisfies the transport equation

$$\frac{\partial C}{\partial t} + V_j \frac{\partial C}{\partial x_j} = \frac{\partial}{\partial x_j}\left(D_{jl} \frac{\partial C}{\partial x_l}\right), \qquad (j, l = 1, 2, 3), \tag{1.1}$$

where $\mathbf{V} = \mathbf{q}/n$ is the fluid filtration velocity, $\mathbf{q}$ is the specific discharge, n is the porosity, and D_{jl} are the components of the dispersion tensor. Here and in what follows, the summation convention for repeated indices is adopted.

In an isotropic medium, D_{jl} reduces to two components, the longitudinal D_{L} and the transverse D_{T}. In turn, these can be written as $D_{\mathrm{L}} = D_{\mathrm{d}} + U\alpha_{\mathrm{L}}$, $D_{\mathrm{T}} = D_{\mathrm{d}} + U\alpha_{\mathrm{T}}$, where D_{d} is the effective molecular diffusion coefficient and α_{L}, α_{T} are known as dispersivities. The magnitudes of the latter quantities have been found from experiments with uniform granular materials to be of the order of the pore size, $\alpha_{\mathrm{L}}/\alpha_{\mathrm{T}}$ being much larger than unity (say, by 20 or more). Recent experiments with samples of natural, nonuniform porous media (Klotz et al. 1980) gave higher values, of the order of centimeters for α_{L}.

Equation (1.1) has served, and it is still used, to solve problems at the field scale. Toward this aim, involved codes that allow for nonuniform velocity fields $\mathbf{V}$ and for complex flow-domain boundaries have been developed.

Field experiments (see Section 2) have shown in a consistent manner, however, that apparent effective dispersion coefficients are larger by orders of magnitude than those determined with the aid of laboratory samples. (This enhanced spreading has been termed "megadispersion.") Furthermore, it has also been found that the apparent dispersivity may grow with the travel time of the solute body. These findings have cast doubts on the applicability of (1.1) to large natural formations, and it has become quite apparent that the spread of solute is dominated by large-scale heterogeneity, rather than by pore-scale heterogeneity. Hence, new concepts and equations had to be developed to account for these large-scale heterogeneities. Although some early models were suggested (e.g. Mercado 1967, Buyevich et al. 1969), a systematic and concerted effort has become possible only in the last decade, in the framework of the emerging new field of stochastic modeling of groundwater flow in heterogeneous formations (for a recent overview, see Dagan 1986). The main aim of the present review is to present these latest developments.

Two fundamental heterogeneity scales of porous formations have been previously suggested (Dagan 1984, 1986): the local and the regional scales. The local scale refers to the spatial variation of the hydraulic conductivity K in a domain whose size is of the order of the aquifer depth in the vertical direction and of the same order in the horizontal plane. If the point value of permeability is viewed as that of a core extracted at a point, the local heterogeneity scale is defined as the distance over which the values of K are correlated. This distance has been found in a few available field studies to be of the order of meters. In contrast, the regional scale refers to the entire aquifer, whose horizontal extent (of the order of tens of kilometers) is much larger than the depth. At this scale, flow variables are averaged over depth, as in the shallow-water approximation, and the flow is viewed as two-dimensional in the horizontal plane. The pertinent property, defined as a point variable, is now the transmissivity, determined as a rule by pumping tests. A recent extensive survey of various aquifers (Hoeksema & Kitanidis 1985) has arrived at transmissivity heterogeneity scales of the order of kilometers.

Solute transport has been studied so far, both experimentally and theoretically, mostly at the local scale, and it is this scale that is the focus of the present review. The important problem of transport at the regional scale, where the uncertainty of concentration prediction may become quite large, is briefly discussed in Section 8 along the lines of the article by Dagan (1984).

This review focuses mainly on the theoretical development of the transport theory, with the field data only briefly recalled to illustrate salient points. Furthermore, the discussion is limited to the transport of inert solutes at low concentrations, which do not decay or interact with the solid matrix, nor do they influence fluid properties. Because of space limitations, neither numerical approaches (e.g. Smith & Schwartz 1980) nor the many outstanding problems are reviewed here.

2. A FEW FIELD FINDINGS

A typical field test consists of injecting into an aquifer a volume of solution having concentration different from the ambient one. Under a natural gradient flow the solute is carried by groundwater, and the concentration has to be monitored with the aid of downstream observation wells. By assuming that the concentration obeys the dispersion equation (1.1) and by using a best-fit procedure, one may determine from such measurements both the average velocity and the apparent dispersion coefficient. However, the procedure is error prone, and the values obtained for dispersivity should be viewed as approximate unless they are corroborated by a large number of observations. In an injection-withdrawal test, the solute body is pumped out either by the same well or by a neighboring one. Such a test is simpler and quicker, but it suffers from two limitations: It reflects the short-time behavior of the transport process only [mainly the influence of pore-scale dispersion (see Section 6)], and the results are dependent on the radial flow conditions created by the wells. The identification of the transport coefficients is even more difficult and error prone in such a test.

With these reservations in mind, we present in Figure 1 a graph reproduced from Gelhar (1986) displaying an extensive compilation of longi-

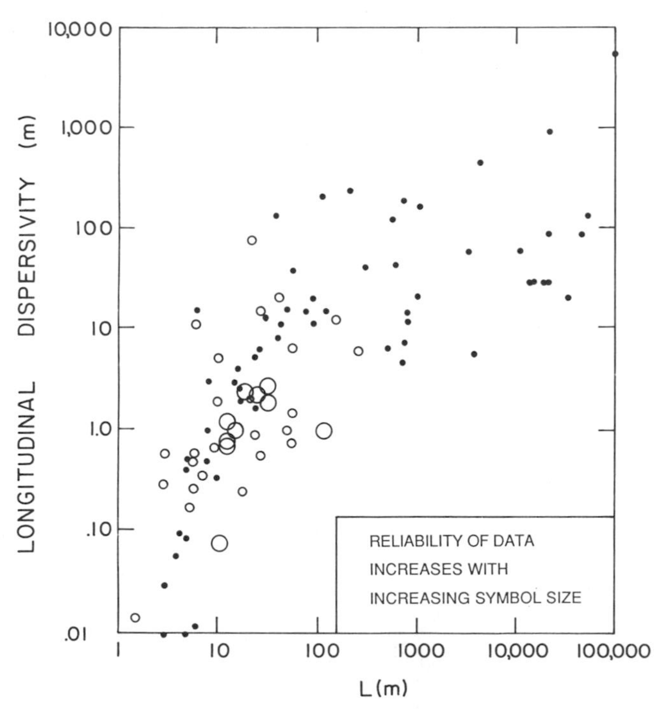

Figure 1 A compilation of longitudinal dispersivity values as a function of the distance L traveled by the solute body, inferred from various field tests (reproduced, with permission, from Gelhar 1986).

tudinal dispersivity data from many field sites around the world. The striking features, already mentioned in Section 1, are (*a*) the large values of field dispersivities as compared with laboratory results, and (*b*) the increase of the dispersivity with the distance traveled by the solute body. [A similar graph can be found in Lallemand-Barrés & Peaudecerf (1978).]

Since the main aim of the theory is to predict transport on the basis of the heterogeneous aquifer properties, a field test in which the latter are not mapped in detail is of limited usefulness for validating theoretical models. Such elaborate tests, in which aquifer properties and concentrations are measured separately, have been undertaken only recently. The one employed here for purposes of illustration was carried out by Freyberg and Sudicky at the Borden tracer site (Sudicky 1985, 1986, Freyberg 1986). The results of these tests are discussed in Section 7 and are represented in Figures 3 and 4. We mention here the findings concerning the formation heterogeneity; these results were manifested in the spatial disribution of the hydraulic conductivity K, which has been mapped extensively in this experiment.

Thus, it has been found that $Y = \ln K$, regarded as a random space function, is approximately stationary and normal. Its covariance $C_Y(\mathbf{x})$ can be represented by

$$C_Y(x, y, 0) = \sigma_Y^2 \exp\left[-(x^2/I_1^2 + y^2/I_2^2)^{1/2}\right];$$
$$C_Y(0, 0, z) = \sigma_Y^2 \exp(-|z|/I_3), \tag{2.1}$$

where $\mathbf{x}(x, y, z)$ is the separation vector between two points, σ_Y^2 is the variance, I_1 and I_2 are correlation scales in the horizontal (x, y) plane, and I_3 is the vertical correlation scale. In the case of the Borden site (E. A. Sudicky, personal communication), the data are $I_1 = I_2 = I_{\mathrm{h}} = 2.8$ m, $I_3 = I_{\mathrm{v}} = 0.10$ m, and $\sigma_Y^2 = 0.38$. While the log-normality of K has been found to be prevalent in many formations (Freeze 1975), the field determination of C_Y at the local scale has seldom been carried out.

The disparity between the vertical correlation scale and the horizontal one is indicative of stratification. It is important to recognize that the results indicate, in addition to isotropy in the horizontal plane, that stratification is not continuous, in the sense that Y ceases to be correlated at a few meters separation in the x, y-directions. [Less detailed measurements, but similar conclusions, have been reported for another site by Moltyaner (1985).] The relevance of these findings for solute transport will become apparent in what follows. More such laborious field experiments are needed, however, before these findings can be regarded as being of a general nature.

3. BRIEF REVIEW OF STOCHASTIC THEORY OF TRANSPORT BY CONTINUOUS MOTIONS

General

We present now the general theoretical approach to transport in formations of heterogeneous structures. The basic idea is to regard the formation properties as space-stationary random functions, as mentioned already for a particular case in Section 2 (for a more detailed discussion, see Dagan 1986). Since the water-filtration velocity **V** depends on the hydraulic conductivity, **V** is also a space random function. In Section 5 we analyze the relationships between the statistical moments of the velocity field and those of the formation properties. For the sake of completeness, a general framework based on Taylor's theory of diffusion by continuous motions (Taylor 1921) is briefly reviewed first.

We consider a porous formation and a solute body of concentration C_0 introduced at time $t = t_0$ in a volume V_0 (Figure 2*a*); the ambient concentration is $C = 0$. We aim to determine the concentration field at $t > t_0$. A solute particle is viewed as an indivisible infinitesimal body of mass $dM = n_0 C_0 d\mathbf{a}$ that moves along a trajectory of equation $\mathbf{x} = \mathbf{X}(t; \mathbf{a}, t_0)$, such that $\mathbf{X}(t_0; \mathbf{a}, t_0) = \mathbf{a}$, where $\mathbf{a}$ is the initial coordinate vector of the particle (Figure 2*a*). The concentration distribution associated with the particle may be written as follows:

$$\Delta C(\mathbf{x}, t; \mathbf{a}, t_0) = \frac{dM}{n}\,\delta(\mathbf{x}-\mathbf{X}), \quad \text{i.e.} \quad \Delta C = \frac{n_0}{n}\,C_0\delta(\mathbf{x}-\mathbf{X})d\mathbf{a}, \tag{3.1}$$

where δ is the Dirac distribution and n is the effective porosity. Since n changes very little as compared with K, we assume that it is constant, i.e. $n = n_0$. To simplify matters further, C_0 is also assumed to be constant.

In general we do not seek C, but rather the average concentration $\bar{C}$ over

Figure 2 (*a*) Schematical representation of the initial solute body and of the total displacement of a particle. (*b*) The decomposition of the total displacement into convection by the mean velocity, convection by the velocity fluctuation, and a "Brownian motion"–type component.

a volume V whose centroid is at $\mathbf{x}$. Under these quite general conditions, the concentration field is expressed as follows:

$$\bar{C}(\mathbf{x}, t; t_0) = \frac{1}{V}\int_V C(\mathbf{x}', t) d\mathbf{x}' = \frac{1}{V}\int_V\int_{V_0} \Delta C(\mathbf{x}', t; \mathbf{a}, t_0)\, d\mathbf{a}\, d\mathbf{x}'$$

$$= \frac{C_0}{V}\int_V\int_{V_0} \delta[\mathbf{x}' - \mathbf{X}(t; \mathbf{a}, t_0)]\, d\mathbf{a}\, d\mathbf{x}'. \tag{3.2}$$

A similar expression can be written for a plume (i.e. for continuous injection of the solute) by introducing $dC_0 = \dot{m}\, dt_0$ in (3.1) and integrating over t_0. Without loss of generality, we refer here to (3.2) only.

We evaluate now the expected value (cnsemble average) $\langle \bar{C} \rangle$ of the space random function $\bar{C}$ (3.2). Since integration and ensemble averaging are commutative, it is sufficient to derive $\langle \Delta C \rangle$ (3.1). Let $f(\mathbf{X}; t, t_0, \mathbf{a})$ be the probability density function (pdf) of $\mathbf{X}$, i.e. $f\, d\mathbf{X}$ is the probability of a particle originating at $\mathbf{x} = \mathbf{a}$ at time t_0 to be within $d\mathbf{X}$ at time t. By the definition of the expected value, we obtain from (3.1) that

$$\langle \Delta C(\mathbf{x}, t; \mathbf{a}, t_0) \rangle = \frac{dM}{n} f(\mathbf{x}; t, t_0, \mathbf{a}); \tag{3.3}$$

this fundamental result can be described as follows: The concentration expected value is given by the pdf of the particle's trajectory, which is regarded as a function of $\mathbf{x}$ and t. It is readily seen that $\langle C \rangle$ and $\langle \bar{C} \rangle$ are obtained from (3.2) by integration of (3.3) with respect to $\mathbf{a}$ and $\mathbf{x}$. Equation (3.2) may serve as the starting point for deriving the higher statistical moments of $\bar{C}$ as well, but this point is deferred to Section 6. It is also seen that by (3.3) the expected value and covariances of $\mathbf{X}$ are proportional to the first and second spatial moments of $\langle \Delta C \rangle$.

A particular, but important, pdf of $\mathbf{X}$ is the stationary multivariate normal one, i.e.

$$f(\mathbf{X}; t, t_0, \mathbf{a}) = \frac{1}{(2\pi)^m |X_{jl}|} \exp\left[-\frac{1}{2}\sum_{j=1}^{m}\sum_{l=1}^{m} (X_j - \langle X_j \rangle)(X_l - \langle X_l \rangle) X_{jl}^{-1}\right], \tag{3.4}$$

where $\mathbf{X}' = \mathbf{X} - \langle \mathbf{X} \rangle$ is the residual (fluctuation). Furthermore, in the stationary case the covariance $X_{jl}(t; \mathbf{a}, t_0) = \langle X'_j(t; \mathbf{a}, t_0) X'_l(t; \mathbf{a}, t_0) \rangle$ does depend only on the time lag $t - t_0$, and we can take $t_0 = 0$ and $X_{jl} = X_{jl}(t)$. Here and in what follows, $m = 2, 3$ is the number of dimensions of the space of the flow domain, $|X_{jl}|$ is the determinant of X_{jl}, and X_{jl}^{-1} is the inverse of the matrix $\|X_{jl}\|$. It is easy to ascertain (e.g. by using the characteristic function) that for the Gaussian (3.4), $\langle \Delta C \rangle$ (3.3) satisfies the convection-diffusion type of equation

$$\frac{\partial \langle \Delta C \rangle}{\partial t} + \left\langle \frac{dX_j}{dt} \right\rangle \frac{\partial \langle \Delta C \rangle}{\partial x_j} = D_{jl} \frac{\partial^2 \langle \Delta C \rangle}{\partial x_j \partial x_l}, \qquad (j, l = 1, \ldots, m), \tag{3.5}$$

$$D_{jl}(t) = \frac{1}{2} \frac{dX_{jl}}{dt}. \tag{3.6}$$

Taylor's theory is concerned with the relationship between $f(\mathbf{X})$ and the fluid velocity field. The basic approach is to relate the particle displacements $\mathbf{X}$ to the Lagrangian velocity field. This line of reasoning has been pursued in the groundwater context by Dieulin et al. (1981a,b), who applied the results to interpret a few field and laboratory measurements.

In groundwater applications, the velocity field that is accessible either by direct measurements or by computations is the Eulerian one. We concentrate here, therefore, on the relationship between transport and the Eulerian velocity field.

The Eulerian Framework

THE DIRECT APPROACH Let $\mathbf{V}(\mathbf{x}, t)$ be the Eulerian velocity field, with $\mathbf{V} = \mathbf{U} + \mathbf{u}(\mathbf{x}, t)$, $\mathbf{U} = \langle \mathbf{V} \rangle$, and $\mathbf{u}$ a random space function, deterministic in t. At this point we separate the particle total displacement $\mathbf{X}_t$ into two components: $\mathbf{X}_t = \mathbf{X} + \mathbf{X}_d$ (Figure 2*b*). Here $\mathbf{X}_d$ is associated with a "Brownian motion" type of transport, such that $X_{d,jl} = 2D_{d,jl}t$, whereas $\mathbf{X}$ is the displacement originating from convection by the fluid. The displacement $\mathbf{X}$ is related to $\mathbf{u}$ by

$$\frac{d\mathbf{X}_t}{dt} = \mathbf{U} + \mathbf{u}(\mathbf{X}_t, t) + \frac{d\mathbf{X}_d}{dt}, \quad \text{i.e.} \quad \frac{d\mathbf{X}}{dt} = \mathbf{U} + \mathbf{u}(\mathbf{X}_t, t);$$

$$\frac{d\mathbf{X}'}{dt} = \mathbf{u}(\mathbf{X}_t, t), \tag{3.7}$$

which, for given $\mathbf{u}$, constitutes an integral equation for $\mathbf{X}_t$ (Figure 2*b*). Hence, generally the covariance X_{jl} cannot be determined directly from the statistical moments of $\mathbf{u}$ by using (3.7). Various approximate schemes have been suggested in order to overcome this fundamental difficulty, but before recalling them here, we first recast the problem in Fourier-transform space, i.e. by using the spectral method. This method has been widely applied to turbulent-diffusion theory, and we review here briefly the procedure by following Lundgren & Pointin (1975). A general and rigorous derivation in an unbounded domain requires using Stieltjes-Fourier integrals, since a random function does not generally have an ordinary Fourier integral transform. The same final results are obtained, however, by employing ordinary Fourier series in a finite cube whose dimensions are allowed to expand to infinity or by generalizing Fourier transforms to include the Dirac distribution; we adopt the latter approach (see Dagan 1985). Hence, by assuming an unbounded flow domain, we find that $\hat{u}$, the Fourier transform of u, is defined by

$$\hat{\mathbf{u}}(\mathbf{k}, t) = \frac{1}{(2\pi)^{m/2}} \int \mathbf{u}(\mathbf{x}, t) e^{i\mathbf{k}\cdot\mathbf{x}}\, d\mathbf{x};$$

$$\mathbf{u}(\mathbf{x}, t) = \frac{1}{(2\pi)^{m/2}} \int \hat{\mathbf{u}}(\mathbf{k}, t) e^{-i\mathbf{k}\cdot\mathbf{x}}\, d\mathbf{k}, \qquad (m = 1, 2, \text{ or } 3), \tag{3.8}$$

and since $\mathbf{u}$ is real, it follows that $\hat{\mathbf{u}} \exp(-i\mathbf{k}\cdot\mathbf{x}) = \hat{\mathbf{u}}^* \exp(i\mathbf{k}\cdot\mathbf{x})$, with $\hat{u}^*$ the complex conjugate of $\hat{u}$. Here $\mathbf{k}$ is the wave-number vector of components (k_1, k_2, k_3) or (k_x, k_y, k_z), integration is from $-\infty$ to $+\infty$, and m is the space dimensionality.

By using (3.8), the last relationship in (3.7) becomes

$$\frac{d\mathbf{X}'}{dt} = \frac{1}{(2\pi)^{m/2}} \int \hat{\mathbf{u}}(\mathbf{k}, t) \exp(-i\mathbf{k}\cdot\mathbf{X}_t)\, d\mathbf{k}. \tag{3.9}$$

Furthermore, limiting the scope to statistical Lagrangian stationarity, we obtain from (3.7) and (3.9)

$$\frac{d^2 X_{jl}(t)}{dt^2} = 2\langle u_j(0, 0) u_l(\mathbf{X}_t, t)\rangle$$

$$= \frac{2}{(2\pi)^m} \iint \langle \hat{u}_j(\mathbf{k}', 0) \hat{u}_l^*(\mathbf{k}'', t) \exp(i\mathbf{k}''\cdot\mathbf{X}_t)\rangle\, d\mathbf{k}'\, d\mathbf{k}'',$$

$$(j, l = 1, \ldots, m;\ m = 1, 2, \text{ or } 3). \tag{3.10}$$

This basic equation relates the particle-displacement covariance to the Fourier components of the Eulerian velocity field. However, due to the statistical dependence of $\mathbf{X}$ upon $\mathbf{u}$, X_{jl} (3.10) cannot be generally expressed in terms of the spectrum of $\mathbf{u}$, and we face the same difficulty as before. A few approximate schemes have been devised in the literature on turbulent diffusion in order to overcome it, e.g. Corrsin's conjecture (Lundgren & Pointin 1975). That scheme has not yet been applied to transport in groundwater, and we limit here the discussion to two other approximations investigated in the past. But first we somewhat simplify (3.10) by substituting $\mathbf{X}_t = \mathbf{X} + \mathbf{X}_d$ and by observing that the "Brownian motion" component $\mathbf{X}_d$, of zero mean and of covariance $X_{d,jl} = 2D_{d,jl}t$, is independent of $\mathbf{u}$. Furthermore, since $\mathbf{X}_d$ is a Gaussian process, we obtain by a well-known relationship (e.g. Lundgren & Pointin 1975) from (3.10)

$$\frac{d^2X_{jl}}{dt^2} = \frac{2}{(2\pi)^m} \iint \langle \hat{u}_j(\mathbf{k}', 0)\hat{u}_l^*(\mathbf{k}'', t)e^{i\mathbf{k}''\cdot\mathbf{X}'}\rangle$$

$$\times \exp\left(i\mathbf{k}''\cdot\langle\mathbf{X}\rangle - D_{d,pq}k_p''k_q''t\right)\, d\mathbf{k}'\, d\mathbf{k}'',$$

$$(j, l = 1, \ldots, m;\ m = 2 \text{ or } 3) \quad (3.11)$$

The general Equation (3.11) is the starting point for the discussion of the two following approximations.

The stratified aquifer One of the earliest models of transport at formation scale is that of a stratified aquifer (Mercado 1967, Marle et al. 1967), and the study of such models is still pursued (Matheron & de Marsily 1980, Pickens & Grisak 1981, Gelhar et al. 1979, Güven et al. 1984). In this model the hydraulic conductivity is a function of z only, and except for Matheron & de Marsily (1980), whose work is discussed in Section 5, most workers assume the flow to be in the x-direction. Furthermore, most studies are concerned with the Taylor-Aris limit (Taylor 1953, Aris 1956), which may be attained in an aquifer of finite thickness B. With $L = Ut$ the distance traveled by the solute body, where U is the average horizontal velocity, this asymptotic regime is attained for $L/B \gg B/\alpha_T$, i.e. after a travel time that is large compared with the one required to ensure complete lateral mixing by pore-scale dispersion across the entire thickness. With $\bar{C}(x,t) = (1/B)\int_0 C(x,z)\, dz$ and with $X_{11}(t)$ defined here as the spatial second moment of $\bar{C}$, the Taylor-Aris main result is that X_{11} increases linearly with time and $\bar{C}$ obeys a convection-dispersion equation. In the latter the velocity is U, and the constant effective dispersion coefficient $D_{11} = (1/2)\, dX_{11}/dt$ is proportional to B^2U/α_T.

Although this mechanism offers, in principle, a simple explanation for the large magnitude of dispersion coefficients observed in field tests, it is doubtful that the asymptotic regime is ever attained in natural formations. Indeed, because of the smallness of α_T, the limiting travel distance $L \gg B^2/\alpha_T$ is very large, and it is improbable that continuous, parallel layers of constant properties persist in nature over such large distances. In fact, when a spatial, three-dimensional correlation analysis is carried out, field findings indicate that $K(\mathbf{x})$ has the anisotropic structure depicted by (2.1), with finite, rather than infinite, horizontal correlation scales. Hence, I believe that the realistic models of transport are those of two- or three-dimensional structures, and thus I do not dwell here upon the results derived for the above regime in layered aquifers.

In contrast, it is of both theoretical and some practical interest to examine transport in a stratified formation at early times. Two asymptotic regimes are of interest now: The first is the Taylor early stage, for $L/I \ll I/\alpha_T$, where I is the vertical correlation scale of K. In this stage both the effects of lateral pore-scale dispersion and of aquifer finite thickness are immaterial. The second regime of interest is that for $I/\alpha_T \ll L/I \ll B/\alpha_T$, a regime that is possible if $I \ll B$, which is generally the case. This regime occurs for a time that is large enough to ensure lateral spreading across the layers, but that is sufficiently small to render the influence of the boundaries negligible. This regime has been studied by Matheron & de Marsily (1980), who considered an infinite thickness from the outset. The early-time stages are examined now with the aid of the spectral method.

The flow is assumed to take place in the x, z-plane, and the Eulerian field is given by $V_1 = U + u(z)$ and $V_2 = V$, where U, V are constant and $u(z)$ is a stationary space-random function of zero mean and covariance $u_{11}(z)$. Furthermore, the "Brownian motion" displacement component $\mathbf{X}_d$ has covariances $X_{d,11} = 2D_L t$, $X_{d,22} = 2D_T t$, whereas the convective displacement residual has only a horizontal component $X'_1 = \int_0^t u(Vt + X_{d,2})\, dt$. Under these conditions we have to substitute in (3.10)

$$\hat{u}_1(\mathbf{k}) = (2\pi)^{1/2} \hat{u}_1(k_2)\delta(k_1);$$

$$\langle \hat{u}_1(k'_2)\hat{u}_1^*(k''_2)\rangle = (2\pi)^{1/2}\hat{u}_{11}(k''_2)\delta(k'_2 - k''_2),$$

where $\hat{u}_{11}(k)$, the spectrum, is the Fourier transform of $u_{11}(z)$. With $m = 2$, Equation (3.10) yields, after integrating over the Dirac distributions of the preceding expression for $\langle \hat{u}_1 \hat{u}_1^* \rangle$,

$$\frac{d^2 X_{11}}{dt^2} = \left(\frac{2}{\pi}\right)^{1/2} \int_{-\infty}^{\infty} \hat{u}_{11}(k) \exp\,(ikVt - D_T k^2 t)\, dk, \tag{3.12}$$

which is essentially the result obtained by Matheron & de Marsily (1980). We discuss the application of (3.12) to stratified aquifers in Section 5.

First-order perturbation or iterative solution In the more realistic cases of two- or three-dimensional heterogeneous structures, it is generally not possible to separate the covariance $\hat{u}_{jl}$ from the exponential term in (3.10). Various approximate schemes have been suggested in the literature in order to overcome this fundamental difficulty, and we employ here the iterative one investigated systematically by Phythian (1975). The first approximation to $\mathbf{X}$ is its ensemble average, $\mathbf{X}_t^{(0)}(t; \mathbf{a}, t_0) = \int_{t_0}^t \mathbf{U}(\mathbf{X}_t^{(0)}, t')\, dt' + \mathbf{a}$. Subsequently, the iterative scheme resulting from (3.7) is $\mathbf{X}'^{(n)}_t(t) = \int_{t_0}^t \mathbf{u}(\mathbf{X}_t^{(n-1)}, t')\, dt' + \mathbf{X}_d$. The first approximation, the only one to be used here, is therefore given by

$$\mathbf{X}^{(0)} = \int_{t_0}^{t} \mathbf{U}\, dt' + \mathbf{a}; \qquad \mathbf{X}'^{(1)} = \int_{t_0}^{t} \mathbf{u}(\mathbf{X}^{(0)}, t')\, dt';$$

$$\mathbf{X}_t^{(1)} = \mathbf{X}^{(0)} + \mathbf{X}'^{(1)} + \mathbf{X}_d. \tag{3.13}$$

Hence, in this first approximation we have to replace $\mathbf{X}_t$ in the exponential term of (3.10) by $\langle \mathbf{X}_t \rangle = \mathbf{X}^{(0)}$. We can arrive, however, at a more general result by replacing it by $\langle \mathbf{X} \rangle + \mathbf{X}_d$, i.e. by leaving the "Brownian motion" component as part of the zeroth-order term in (3.13). The result in (3.10) is

$$\frac{d^2 X_{jl}}{dt^2} = \frac{2}{(2\pi)^m} \iint \langle \hat{u}_j(\mathbf{k}', 0) \hat{u}_l^*(\mathbf{k}'', t) \rangle \exp\,[i\mathbf{k}'' \cdot \langle \mathbf{X} \rangle - D_{d,pq} k_p'' k_q'' t]\, d\mathbf{k}'\, d\mathbf{k}'',$$

$$(j, l = 1, \ldots, m;\ m = 2 \text{ or } 3). \tag{3.14}$$

For a spatially homogeneous Eulerian velocity field, we also have

$$\langle \hat{u}_j(\mathbf{k}', 0) \hat{u}_l^*(\mathbf{k}'', t) \rangle = (2\pi)^{m/2} \hat{u}_{jl}(\mathbf{k}'', t) \delta(\mathbf{k}' - \mathbf{k}'') \tag{3.15}$$

with $\hat{u}_{jl}(\mathbf{k}, t)$, the spectrum, being the Fourier transform of the Eulerian covariance $u_{jl}(\mathbf{x}, t)$. Hence, the final result in (3.15) is

$$\frac{d^2 X_{jl}}{dt^2} = \frac{2}{(2\pi)^{m/2}} \int \hat{u}_{jl}(\mathbf{k}, t) \exp\,[i\mathbf{k} \cdot \langle \mathbf{X} \rangle - D_{d,pq} k_p k_q t]\, d\mathbf{k},$$

$$(j, l = 1, \ldots, m;\ m = 2 \text{ or } 3) \tag{3.16}$$

rendering the displacement covariance in terms of the Eulerian velocity covariance. Equation (3.16) is quite general, being valid for nonuniform average as well as time-dependent flows. Phythian (1975) has compared the results of numerical simulations with the iterative solution and has found that the latter converges for smooth velocity spectra.

The same approximation could be jusified by a perturbation expansion scheme. Indeed, let us assume that the velocity field is characterized by a few parameters: U, the mean velocity, σ_u^2, its variance, I_u, a velocity correlation scale, D_d, the "Brownian motion" diffusivity, etc. On dimensional grounds, we can write in a general manner $\langle C \rangle / C_0 = \text{funct}(U\mathbf{x}/I_u, \sigma_u^2/U^2, D_d/UI_u^2, \ldots)$. If σ_u^2/U^2 is assumed to be small, we expand $\langle C \rangle / C_0$ in an asymptotic power series in σ_u^2/U^2, i.e. $\langle C \rangle = \langle C^{(0)} \rangle + \sigma_u^2/U^2 \langle C^{(1)} \rangle + \ldots$, where $\langle C^{(0)} \rangle, \langle C^{(1)} \rangle, \ldots$ are independent of σ_u^2/U^2. Approximation (3.16) is precisely the one leading to the first-order term $\langle C^{(0)} \rangle + \sigma_u^2/U^2 \langle C^{(1)} \rangle$, since retaining $\mathbf{X}'$ in the exponential of (3.10) would result in higher-order terms in σ_u^2/U^2 for X_{jl}. In fact, in groundwater applications D_d/UI_u is generally much smaller than σ_u^2/U^2, and there is no a priori justification in expanding $\langle C\rangle$ in the latter expression only. (This point is considered again in Section 5.)

Equation (3.16) can be further simplified if the flow is steady and the average velocity is constant, which leads to

$$\frac{d^2X_{jl}}{dt^2}=\frac{2}{(2\pi)^{m/2}}\int \hat{u}_{jl}(\mathbf{k})\exp(i\mathbf{k}\cdot\mathbf{U}t-D_{\mathrm{d},pq}k_pk_qt)\,d\mathbf{k},$$

$$(j,l=1,\ldots,m;\,m=2\text{ or }3),\quad(3.17)$$

which can be integrated for various $\hat{u}_{jl}$. Furthermore, by inverting $\hat{u}_{jl}$ in (3.17), we obtain the expression for X_{jl} in terms of u_{jl} as follows:

$$\frac{d^2X_{jl}}{dt^2}=\frac{2}{(2\pi)^{m}}\iint u_{jl}(\mathbf{x}')\exp[-i\mathbf{k}\cdot(\mathbf{x}'-\mathbf{U}t)-D_{\mathrm{d},pq}k_pk_qt]\,d\mathbf{k}\,d\mathbf{x}'.\quad(3.18)$$

Integration over $\mathbf{k}$ in (3.18) leads to the alternative formulation

$$\frac{d^2X_{jl}}{dt^2}=2^{1-m}\pi^{-m/2}|D_{\mathrm{d},pq}|^{-1/2}\int \mathrm{u}_{jl}(\mathbf{x}')$$

$$\times\exp[-(x'_p-U_pt)(x'_q-U_qt)D^{-1}_{\mathrm{d},pq}]\,d\mathbf{x}',\quad(3.19)$$

where $|D_{\mathrm{d},pq}|=D_{\mathrm{L}}D_{\mathrm{T}}^2$ is the determinant and $D^{-1}_{\mathrm{d},pq}$ is the inverse of the matrix $D_{\mathrm{d},pq}$.

If we neglect $D_{\mathrm{d},pq}$ in (3.17) or (3.18), we immediately have

$$X_{jl}(t)=\int_{t_0}^{t}\int_{t_0}^{t}u_{jl}[\mathbf{U}(t'-t'')]\,dt'\,dt''=2\int_0^t(t-t')u_{jl}(\mathbf{U}t')\,dt',$$

$$X_{\mathrm{t},jl}=X_{jl}+2D_{\mathrm{d},jl}t,\quad(3.20)$$

which is the starting point in Dagan (1982b, 1984).

An important point to be mentioned here is that the first approximation $\mathbf{X}'^{(1)}$ (3.13) is Gaussian if the Eulerian field is, since $\mathbf{X}'^{(1)}$ results from a linear operation applied to $\mathbf{u}$. Furthermore, in the general case $\mathbf{X}$ tends to normality for a travel distance L much larger than the correlation scale of the velocity field $\mathbf{u}$. Indeed, in such a case $\mathbf{X}$ can be regarded as a sum of many independent steps, and normality can be invoked by arguments relying on the central limit theorem. In both cases $\langle C\rangle$ satisfies the convection-dispersion equation (3.5).

4. THE RELATIONSHIP BETWEEN THE EULERIAN VELOCITY FIELD AND THE HETEROGENEOUS STRUCTURE

In the preceding section we discussed in general terms the relationship between the Eulerian velocity field and the concentration. As stated already in Section 1, the main aim of the transport theory is to relate the concentration field to the heterogeneous structure, which can be achieved in turn by deriving the relationship between the velocity field and the formation properties. We present here a brief outline of this topic; the details can be found in the literature on groundwater flow through heterogeneous formations (see, e.g., Dagan 1986).

We consider a porous formation within a domain Ω bounded by a surface Σ. With neglect of elastic storage and recharge, the flow equations are

$$\mathbf{q} = -K\nabla\Phi; \qquad \nabla\cdot\mathbf{q} = 0, \tag{4.1}$$

where $\Phi = p/\gamma + z$ is the water head, p is the pressure, γ is the fluid specific weight, z is the elevation, and $\mathbf{q}$ is the specific discharge. At the local scale K is the hydraulic conductivity, generally a function of x, y, and z. The same equations apply at the regional scale, provided that Φ and $\mathbf{q}$ are averaged over depth, but K now becomes the ratio between transmissivity and depth, and flow is two-dimensional in the x, y-plane.

In a typical flow problem, Φ or its normal derivative is given on the boundary Σ, and Φ has to be determined in Ω by solving (4.1) after eliminating $\mathbf{q}$. Subsequently, the Eulerian velocity at each point can be calculated from its definition $\mathbf{V} = \mathbf{q}/n = -K\nabla\Phi/n$.

In a heterogeneous formation in which K is viewed as a random space function, the solution of (4.1) is difficult and can be generally achieved only by numerical methods. Our purpose here is to discuss approximate solutions, which lend themselves to simple formulations of the transport problem and which allow one to grasp the phenomena of interest. Toward this aim, and for the sake of simplicity, we assume that a uniform head gradient $-\mathbf{J}$ is applied on the boundary. Defining the head fluctuation ϕ by $\phi = \Phi + \mathbf{J}\cdot\mathbf{x}$, elimination of $\mathbf{q}$ from (4.1) yields

$$\nabla^2\phi + \nabla Y\cdot\nabla\phi = \mathbf{J}\cdot\nabla Y, \quad (\mathbf{x}\in\Omega); \qquad \phi = 0, \quad (\mathbf{x}\in\Sigma), \tag{4.2}$$

where $Y = \ln K$. Obviously, in the case of a homogeneous formation, it follows that $\nabla Y = 0$ and the flow is uniform. The above approximation is, therefore, justified if the solute body moves throughout a region of essentially uniform average flow. Next, Y is assumed to be a stationary random function of given expected value $\langle Y\rangle$ and of two-point covariance $C_Y(\mathbf{x}'', \mathbf{x}''')$.

A general solution of (4.2) is still difficult; a notable exception is the case of a stratified formation, which is discussed first. Selecting $x_1 = z$ normal to the layers, and $\mathbf{J}$ in the x, z-plane, we can represent the flow as a superposition of a horizontal flow of velocity $V_1 = U+u$ caused by J_x and a vertical uniform flow $V_2 = V$ caused by J_z, with

$$U = \frac{K_A}{n} J_x; \qquad V = \frac{K_H J_z}{n}; \qquad u = \frac{K' J_x}{n}, \tag{4.3}$$

where $K_A = \langle K \rangle$ is the arithmetic mean and $K_H = \langle K^{-1} \rangle^{-1}$ is the harmonic mean of K, whereas $K' = K - \langle K \rangle$. These relationships can be obtained directly from (4.1). Although we can easily handle the spatial variability of the porosity n, we assume that n is constant in view of its small changes in comparison with those of K. Thus, the first two moments of the velocity are given by

$$\langle V_1 \rangle = U; \qquad \langle V_2 \rangle = V; \qquad u_{11}(z) = \frac{J_x^2}{n^2} C_K(z) = U^2 \frac{C_K(z)}{K_A^2};$$

$$u_{12} = u_{22} = 0, \tag{4.4}$$

where $C_K(z)$ is the two-point covariance of the stationary random function $K(z)$.

These results, which express the statistics of the velocity field in terms of K, are employed in Section 5, together with the results of Section 3, in order to analyze transport.

No such simple results can be achieved in the more general and realistic case of genuine two- or three-dimensional heterogeneous formations. The solution of the stochastic differential equation (3.2), rendering the moments of the random function ϕ in terms of those of Y, is one of the central problems of the theory of heterogeneous materials (see, e.g., Beran 1968). In particular, the covariances

$$C_{Y\phi}(\mathbf{x}'', \mathbf{x}'''; \sigma_Y^2) = \langle Y'(\mathbf{x}'')\phi(\mathbf{x}''') \rangle; \; C_{\phi}(\mathbf{x}'', \mathbf{x}'''; \sigma_Y^2) = \langle \phi(\mathbf{x}'')\phi(\mathbf{x}''') \rangle \tag{4.5}$$

depend in a nonlinear fashion upon σ_Y^2, the log-conductivity variance. A considerable simplification of the computations is achieved if a first-order term of an expansion of (4.5) in σ_Y^2 is adopted as an approximation of ϕ. Furthermore, a recent study (Dagan 1985) has shown that such approximations may be quite accurate for σ_Y^2 as large as unity, which makes the results applicable to many actual aquifers. Under these conditions, the moments (4.5) can be obtained explicitly from the following linearized version of (4.2):

$$\nabla^2 \phi = \mathbf{J} \cdot \nabla Y, \quad (\mathbf{x} \in \Omega); \qquad \phi = 0, \quad (\mathbf{x} \in \Sigma). \tag{4.6}$$

The solution of (4.6) can be expressed in terms of the Green function

$G(\mathbf{x}'', \mathbf{x}''')$ for the Laplace equation, along the lines of Dagan (1982a, 1984), as follows:

$$\phi(\mathbf{x}) = -\mathbf{J}\cdot\int_{\Omega} \nabla Y'(\mathbf{x}')G(\mathbf{x},\mathbf{x}')\, d\mathbf{x}' = \mathbf{J}\cdot\int_{\Omega} Y'(\mathbf{x}')\nabla G(\mathbf{x},\mathbf{x}')\, d\mathbf{x}', \tag{4.7}$$

or, for an infinite domain Ω, by taking the Fourier transform of (4.6). This last line of attack has been pursued in the pioneering work of Buyevich et al. (1969), who viewed the randomness of K as following from the randomness of n, and who also considered a more general linearized momentum equation than Darcy's law (4.1). The application of the first-order approximation to groundwater flow has been pursued vigorously in a series of works by Gelhar, Gutjahr, and their coworkers (e.g. Bakr et al. 1978, Gutjahr et al. 1978, and mainly Gelhar & Axness 1983).

By the same linearized approximation of Darcy's law (4.1), the velocity $\mathbf{V}$ is related to Y and ϕ by

$$\mathbf{V} = \frac{1}{n}\, e^{Y}(\mathbf{J}-\nabla\phi) = \frac{1}{n}\, e^{\langle Y\rangle}(1+Y'+\ldots)(\mathbf{J}-\nabla\phi), \tag{4.8}$$

which leads to the final linearized expressions

$$\langle\mathbf{V}\rangle = \mathbf{U} = \frac{1}{n}\, e^{\langle Y\rangle}\mathbf{J}; \qquad \mathbf{u} = \mathbf{V}-\mathbf{U} = \frac{1}{n}\, e^{\langle Y\rangle}(Y'\mathbf{J}-\nabla\phi). \tag{4.9}$$

Equations (4.9) and (4.7) lead to expressions of the velocity covariance u_{jl} as functions of the covariances C_Y, $C_{Y\phi}$, and C_{ϕ}; these in turn can be expressed in terms of C_Y only, achieving the goal set at the beginning of this section. The alternative (Fourier-transform formulation) yields for an infinite domain and for stationary Y, by (4.6) and (4.9),

$$\hat{u}_{jl}(\mathbf{k}) = U_p U_q\left(\delta_{pj} - \frac{k_j k_p}{k^2}\right)\left(\delta_{ql} - \frac{k_l k_q}{k^2}\right)\hat{C}_Y(\mathbf{k}), \tag{4.10}$$

which again achieves the same goal.

5. THE DEPENDENCE OF THE CONCENTRATION EXPECTED VALUE UPON THE FORMATION HETEROGENEOUS STRUCTURE

By combining the results of the last two sections, we are now in a position to derive the relationship between the expected value of the concentration field and the spatial statistical moments of the permeability or transmissivity.

The Stratified Formation (Matheron & de Marsily 1980)

We return to this exact solution because of its fundamental interest, the limitations of its applicability to actual formations notwithstanding. We consider first flow parallel to the layers, i.e. $J_z = 0$ in (4.3). Substitution of u_{11} (4.4) into (3.12) and integrating once over t yields

$$\frac{dX_{11}}{dt} = \left(\frac{2}{\pi}\right)^{1/2} \frac{1}{D_T} \int_{-\infty}^{\infty} \hat{u}_{11}(k) \frac{1-\exp(-k^2 D_T t)}{k^2}\, dk;$$

$$\frac{dX_{t,11}}{dt} = \frac{dX_{11}}{dt} + D_L. \tag{5.1}$$

Integration in (5.1) can be carried out in a closed form for simple $\hat{u}_{11}(k)$, but we prefer to concentrate here on some general, asymptotic results. With I the hydraulic conductivity integral scale, i.e. $I = (1/\sigma_K^2)\int_0^\infty C_K(z)\, dz$, the small travel-time $D_T t/I^2 \ll 1$ limit of (5.1) is obtained by expanding the exponential. The leading-order term is

$$\frac{dX_{11}}{dt} \to \frac{2J_x^2 \sigma_K^2 t}{n^2} = 2\frac{\sigma_K^2}{K_A^2} U^2 t. \tag{5.2}$$

This is the well-known Taylor short-time limit in which the apparent dispersion coefficient grows linearly with time, and its application to transport has been discussed in Section 3.

The asymptotic, long-time limit of (5.1), valid for $D_T t/I^2 \gg 1$, is obtained by letting $k \to 0$, i.e.

$$\frac{dX_{11}}{dt} \to \left(\frac{2}{\pi}\right)^{1/2} \frac{\hat{u}_{11}(0)}{D_T} \int_{-\infty}^{\infty} \frac{1-\exp(-k^2 D_T t)}{k^2}\, dk = \frac{4}{\pi^{1/2}} \frac{\sigma_K^2}{K_A^2} U^2 \left(\frac{tI^2}{D_T}\right)^{1/2}. \tag{5.3}$$

This is the main result of the analysis of Matheron & de Marsily (1980). They have assumed that the formation is of large extent in the z-direction and C is averaged across layers, such that by ergodic arguments (see Section 6) and for large t, $D_{11} = (1/2)\, dX_{11}/dt$ represents the dispersion coefficient in Equation (4.5). Their main conclusion was that for such a formation the transport is not Fickian, since D_{11} (5.3) does not tend to a constant limit but grows like $t^{1/2}$. Thus, although lateral pore-scale dispersion across the layers diminishes the dispersive effect of stratification as expressed by (5.2), it does not render a constant D_{11}. We recall that a layered formation does not qualify for the asymptotic conditions of Lagrangian transport of Section 3, since velocities are correlated for an indefinite distance in the x-direction. This is also the main limitation of the model for actual porous formations, as mentioned before. It is emphasized that (5.3) could not be obtained by seeking a "steady-state" solution for $t = \infty$ in (5.1). If this is done unwarrantedly, a finite result is obtained only if $\hat{C}_K$ vanishes for $k \to 0$, i.e. C_K is restricted to a "hole" function with a zero-integral scale.

Next, following Matheron & de Marsily (1980), we assume that the flow is tilted with respect to the x-axis, i.e. $V \neq 0$ in (4.4). Substituting the latter in (3.12) yields now

$$\frac{dX_{11}}{dt} = \left(\frac{2}{\pi}\right)^{1/2} \frac{U^2}{K_A^2} \int_{-\infty}^{\infty} \hat{C}_K(k) \frac{1-\exp(ikVt - D_T k^2 t)}{k(-iV + D_T k)} dk. \tag{5.4}$$

Various asymptotic results can be obtained from (5.4), depending on the transverse Peclet number VI/D_T. If we assume that the latter is $O(1)$, then the short-time limit of (5.4) is the same as in (5.2), i.e. the transverse flow does not affect transport for $Vt/I \ll 1$. In contrast, the long-time limit of (5.4) is completely different from (5.3). Indeed, for $Vt/I \gg 1$ we get in (5.4) by letting $k \to 0$

$$\frac{dX_{11}}{dt} \to i\left(\frac{2}{\pi}\right)^{1/2} \frac{J_x^2 \hat{C}_K(0)}{n^2 V} \int_{-\infty}^{\infty} \frac{1-\exp(ikVt)}{k} dk = 2\frac{\sigma_K^2}{K_A^2} \frac{U^2 I}{V}. \tag{5.5}$$

Hence, the important theoretical result obtained by Matheron & de Marsily (1980) is that tilting of the flow relative to bedding leads to a Fickian regime, with a longitudinal dispersion coefficient D_{11} half that of (5.5). This result can be interpreted using the Langrangian theory by following two particles that move across the layers with velocity V: For $Vt/I \gg 1$, their trajectories become uncorrelated and (5.5) is expected. Still, this regime may be reached only after a considerable time if the transversal flow is slow.

Formations of Two- and Three-Dimensional Structures (Dagan 1982b, 1984, Gelhar & Axness 1983)

These are the cases of paramount interest for application to groundwater transport. They have been investigated by Dagan (1982b, 1984) and Gelhar & Axness (1983) by adopting the same first-order approximations of the flow field (Section 4) and concentration (Section 3). Otherwise, the two procedures differ in a few respects: Dagan followed the Lagrangian approach, employed the direct method (Equation 3.20), and used Green functions (Equation 4.7), whereas Gelhar & Axness started from the Eulerian formulation (1.1), assumed an asymptotic infinite-time regime from the outset, and calculated the effective dispersion coefficient (macrodispersivity) by spectral methods. We show in the sequel that these apparent disparate procedures can be presented as particular cases of the unified approach discussed in Sections 3 and 4. Toward this aim we make use of the first-order approximation of the displacement covariance (3.17), expressed in terms of the spectrum $\hat{u}_{jl}$, and the first-order approximation of the latter (4.10), to obtain

$$\frac{d^2 X_{jl}}{dt^2} = \frac{2}{(2\pi)^{m/2}} U_p U_q \int \left(\delta_{pj} - \frac{k_j k_p}{k^2}\right)\left(\delta_{ql} - \frac{k_l k_q}{k^2}\right)$$

$$\times \hat{C}_Y(\mathbf{k}) \exp(i\mathbf{k}\cdot\mathbf{U}t - D_{\mathrm{d},rs} k_r k_s t)\, d\mathbf{k},$$

$$(j, l = 1, \ldots, m;\ m = 2 \text{ or } 3), \quad (5.6)$$

which renders $X_{jl}(t)$ as functions of the log-conductivity spectrum $\hat{C}_Y$, the average velocity $\mathbf{U}$, and the "Brownian motion" component. The various approximations underlying (5.6) are described in Sections 3 and 4.

DAGAN (1982b, 1984) A further simplification was achieved by assuming that the log-conductivity correlation scale is much larger than the dispersivities associated with $D_{\mathrm{d},jl}$. The latter represent the effect of pore-scale dispersion when transport at local scale is modeled by (5.6), or the effect of the local heterogeneous structure when (5.6) is related to regional transport and Y stands for log-transmissivity. In fact, representing $D_{\mathrm{d},jl}$ as a "Brownian motion" type of transport is underlain by the assumption of this disparity of scales, which is supported by the field findings mentioned in Section 2. Under these conditions, the second term in the exponential of (5.6) has been neglected as compared with the first one. It is easy to ascertain, by using the definition of the Fourier transform, that this leads to (3.20), which was originally the starting point of Dagan's derivations.

Dagan (1984) derived closed-form expressions for the displacement covariances for the particular, but important, case of exponential covariances, i.e.

$$C_Y(\mathbf{x}) = \sigma_Y^2 \exp(-|\mathbf{x}|/I);$$

$$\hat{C}_Y(\mathbf{k}) = 2^{(3m/2-1)} \pi^{1-m/2} I^m (1 + |\mathbf{k}|^2)^{-(m+1)/2}, \qquad (m = 2 \text{ or } 3). \qquad (5.7)$$

For the isotropic C_Y (5.7), the only three-dimensional components of X_{jl} are the longitudinal X_{11} and the lateral $X_{22} = X_{33}$ ones, with $\mathbf{U}(U, 0, 0)$. Their expressions are given by Dagan (1984, Equation 4.9), where they are also represented graphically. The small-time limits for $tU/I \ll 1$ are

$$X_{\mathrm{t},11} \to \frac{8}{15} \sigma_Y^2 U^2 t^2 + 2D_{\mathrm{L}} t; \qquad X_{\mathrm{t},22} \to \frac{1}{15} \sigma_Y^2 U^2 t^2 + 2D_{\mathrm{T}} t, \qquad (5.8)$$

which have the general structure of Taylor's short-time limit and display pronounced anisotropy, with $X_{22}/X_{11} = 8$. The more interesting long-time limits are

$$X_{\mathrm{t},11} \to 2\sigma_Y^2 U I t + 2D_{\mathrm{L}} t + O(I/Ut);$$

$$X_{\mathrm{t},22} \to \frac{2}{3} \sigma_Y^2 I^2 + 2D_{\mathrm{T}} t + O(I/Ut), \qquad (tU/I \gg 1). \qquad (5.9)$$

It is seen that the longitudinal covariance has the form of the Taylor long-time limit, with $D_{11} = \sigma_Y^2 UI$. In contrast, the lateral covariance X_{22} tends to a constant limit, and the further lateral growth of $\langle C \rangle$ is controlled by the slow pore-scale mechanism. This result might be related to the nature of the first-order approximation, in which the lateral covariance $u_{22}(x, 0, 0)$ is of zero-integral scale. Still, if linear terms in time are present, they should at least be of order σ_Y^4 or $\sigma_Y^2 D_d$, i.e. quite small.

The corresponding two-dimensional results are (Dagan 1982a, 1984)

$$X_{11} \to \frac{3}{8}\sigma_Y^2 U^2 t^2; \qquad X_{22} \to \frac{1}{8}\sigma_Y^2 U^2 t^2, \qquad (tU/I \ll 1),$$

$$X_{11} \to 2\sigma_Y^2 IUt\left[1 - \frac{3}{2}\frac{\ln(tU/I)}{(tU/I)} + O(I/Ut)\right];$$

$$X_{22} \to \sigma_Y^2 I^2 [\ln(tU/I) + 0.933 + O(I/Ut)], \qquad (tU/I \gg 1). \tag{5.10}$$

Again, the small-time limit has the expected form, and the same is true for the longitudinal covariance X_{11} for $tU/I \gg 1$. At the latter limit, X_{22} grows logarithmically with time, and although the $\langle C \rangle$ distribution expands laterally, D_{22} tends to zero like $1/t$. The constant $D_{11} = \sigma_Y^2 IU$ [long-time limit in (5.10)] is attained only after the solute body has traveled a considerable distance, say tens of conductivity integral scales.

It is seen that the longitudinal dispersion coefficient tends to the same limit, for large t, in both two- (5.9) and three-dimensional (5.10) flows. In fact, this limit originates from the term $\delta_{pj}\delta_{ql}$ in (5.6), which leads to the result $X_{11} \to 2\sigma_Y^2 UI$ for any covariance function of finite integral scale. Furthermore, inspection of X_{11} for two- and three-dimensional flows (see Figure 1 in Dagan 1984) reveals that they are quite close for any time, and this result suggests that longitudinal dispersion is quite insensitive to anisotropy also. In contrast, the lateral covariances X_{22} differ markedly in two- and three-dimensional flows.

Although of fundamental interest, the three-dimensional solution is of limited applicability in the case of unconsolidated formations that display strong anisotropy. In contrast, the two-dimensional solution may be applied to analyze field tests in two important cases:

1. It may be used for transport at local scale in which C_Y is isotropic in the horizontal plane but has a much smaller correlation scale I_v in the vertical direction. If, furthermore, C is defined as C averaged in the vertical direction over a distance of the order of the vertical integral scale, then $\langle C \rangle$ has (5.10) as displacement covariances, with $I = I_h$ in the horizontal

plane. Indeed, the presence of layers of low conductivity prevents convection in the vertical direction, and consequently the solute "particle," defined by the above average over the vertical, moves effectively in the horizontal plane and is subjected to random motions due to the randomness of Y in the same plane. Furthermore, in this process the uncorrelated random variations of Y, which are manifest as a "nugget" in C_Y, are also wiped out and σ_Y^2, which is used in the theoretical results, is somewhat reduced. Finally, longitudinal pore-scale dispersion is somewhat enhanced by layering and by defining a particle as above. The two-dimensional solution (Dagan 1984, Equations 4.5, 4.6) is applied to a field test fulfilling these conditions in Section 7.

2. The two-dimensional solution is also used for regional flow and for the large-scale heterogeneity related to the transmissivity variations. In this case, a "particle" is defined by averaging over the vertical across the entire solute body, and local heterogeneity manifests itself in a "Brownian motion" effect. This type of transport is discussed in Section 8.

GELHAR & AXNESS (1983) Their starting point was Equation (1.1), in which $\mathbf{V}$ has been regarded as the random Eulerian velocity field related to the heterogeneous structure, whereas the D_{jl} are pore-scale dispersion coefficients. By adopting a first-order perturbation approximation for the concentration C and for the velocity field, and by assuming an infinite-time "steady state" in which $\langle C \rangle$ satisfies a Fickian transport equation with constant coefficients, Gelhar & Axness arrived at expressions of the effective dispersion coefficients related to heterogeneity. In spite of the difference in approach, we show now that their results are a particular case (long-time limit) of the Lagrangian formulation (5.6). Indeed, integrating (5.6) once, we have

$$\frac{dX_{jl}}{dt} = \frac{2}{(2\pi)^{m/2}} U_p U_q \int \frac{(\delta_{pj} - k_j k_p k^{-2})(\delta_{ql} - k_p k_q k^{-2})}{i\mathbf{k} \cdot \mathbf{U} - D_{\mathrm{d},rs} k_r k_s} \times \hat{C}_Y(\mathbf{k}) [\exp(i\mathbf{k} \cdot \mathbf{U}t - D_{\mathrm{d},rs} k_r k_s t) - 1] \, d\mathbf{k},$$

$$(j, l = 1, \ldots, m; m = 2 \text{ or } 3) \quad (5.11)$$

at any time t. Gelhar & Axness' (1983) central formula is obtained by letting $t \to \infty$ in (5.11) and, therefore, by neglecting the exponential term. With the effective dispersion coefficients equal to $(1/2)(dX_{jl}/dt)$ [their additional factor $\gamma = 1 + O(\sigma_Y^2)$ is taken, for consistency, to be equal to unity here], their results are valid for a travel time satisfying either $t \gg I/U$ or $t \gg I^2/D_\mathrm{d}$. It is emphasized that the limit procedure just described is a delicate one, since k passes through zero in (5.11), and no matter how large is t, the argument of the exponential also goes through zero. In fact,

$\hat{C}_Y(\mathbf{k})$ has to be finite for $\mathbf{k} \to 0$ in two-dimensional flow to ensure the uniformity of the limit, and this is generally the case. The procedure may fail, however, if it is applied to X_{jl} [which is obtained by an additional integration in Equation (5.11)], unless $\hat{C}_Y$ vanishes for $\mathbf{k} \to 0$.

Gelhar & Axness (1983) have used the above limit of (5.11) in order to investigate a large number of cases by considering the anisotropic log-conductivity covariances and also by allowing for an angle between the average velocity $\mathbf{U}$ and the principal axes of anisotropy. Here, we show a few of their analytical results, which may be compared with the time-dependent solutions (5.8)–(5.10).

For the exponential and isotropic log-conductivity covariance (5.7) and for small $\varepsilon = \alpha_L/I$, Gelhar & Axness (1983) obtained

$$D_{11} = \frac{1}{2}\frac{dX_{11}}{dt} = \sigma_Y^2 IU;$$

$$D_{22} = D_{33} = \frac{1}{2}\frac{dX_{22}}{dt} = \frac{1}{15}\sigma_Y^2\alpha_L\left(1+4\frac{\alpha_T}{\alpha_L}\right)U \tag{5.12}$$

for three-dimensional flow, and

$$D_{11} = \sigma_Y^2 IU; \qquad D_{22} = \frac{1}{8}\sigma_Y^2\alpha_L\left(1+3\frac{\alpha_T}{\alpha_L}\right)U \tag{5.13}$$

for two-dimensional flow, after neglecting terms $O(\varepsilon)$ compared with $O(1)$. Comparison of (5.12) and (5.13) with (5.9) and (5.10), respectively, shows that results for the longitudinal component D_{11} coincide at the common limit of large t and small ε. In contrast, the constants D_{22} in (5.12) and (5.13) do not appear as linear terms in t in (5.9) and (5.10), since they are of order $\sigma_Y^2 D_d$ and were neglected in the analysis leading to the latter equations. Conversely, the constant and logarithmic terms present in X_{22} for $t \to \infty$ in (5.9) and (5.10), respectively, are lost in Gelhar & Axness' analysis, since the derivatives of these terms tend to zero for $t \to \infty$. It is instructive to evaluate the time for which the lateral-displacement covariance would be approximated accurately by (5.12) and (5.13), rather than by (5.9) and (5.10). The order of magnitude of t is given by

$$2D_{22}t \gg \frac{2}{3}\sigma_Y^2 I^2, \quad \text{i.e.} \quad tU/I \gg 5I/\alpha_L,$$

$$2D_{22}t \gg \sigma_Y^2 I^2 \ln(tU/I), \quad \text{i.e.} \quad \frac{tU/I}{\ln(tU/I)} \gg 4I/\alpha_L, \tag{5.14}$$

where the lower and upper equations in (5.14) are for two- and three-dimensional flow, respectively. These limiting times are exceedingly large in most conceivable field applications.

Additional results of interest were obtained in the case of the anisotropic covariance of type (2.1) for uniform flow parallel to one of the principal directions, say x_1, and for $I_3 = I_v$ much smaller than $I_1 = I_2 = I_h$. One of the central results is that the "macrodispersion" coefficient D_{11} is equal to those of (5.12) and (5.13), provided that mixing across I_v by lateral pore-scale dispersion is quicker than convection over I_h.

Gelhar & Axness (1983) have analyzed extensively various cases of flow in which the average velocity $\mathbf{U}$ is tilted with respect to the principal axes of anisotropy of C_Y. The result of principle is similar to that found by Matheron & de Marsily (1980) for longitudinal dispersion in layered media, namely, that the lateral "macrodispersivities" are convection-dominated and are much larger than in (5.12) and (5.13), being different from zero even if D_d is neglected. The lateral coefficients were found to be as large as 8% of the longitudinal coefficients for a tilt of, say, 30° (Gelhar & Axness 1983, Figure 7).

It is worthwhile to mention here that the basic first-order expression that relates the asymptotic, infinite-time expression of D_{jl} to the velocity field has also been obtained by Winter et al. (1984) using a different methodology.

6. CONCENTRATION VARIANCE AND ERGODICITY

In the preceding sections we have analyzed the relationship between the expected value of the concentration field $\langle C(\mathbf{x}, t)\rangle$ and the random heterogeneous structure. This is sufficient to characterize $C(\mathbf{x}, t)$ if the ergodic hypothesis is obeyed, i.e. if it can be assumed that the space average $\bar{C}$ (3.2) and the ensemble mean $\langle C\rangle$ are interchangeable, which is a generally accepted tenet in diffusion theories. The concentration C in the given realization can then be taken equal to the ensemble average, the statistical model being rather a mathematical vehicle for computing an essentially deterministic quantity. Skipping here a detailed and rigorous analysis, we assume that the ergodic hypothesis is fulfilled if the concentration variance $\sigma_{\bar{C}}^2$ tends to zero. (This is exact in the case of a normal pdf of $\bar{C}$.) The ergodic hypothesis might not be obeyed in some important applications of groundwater transport, and it is of paramount importance to evaluate $\sigma_{\bar{C}}^2$ in such cases. (We do not discuss here higher-order statistical moments.) The brief analysis of this topic that is presented here follows Dagan (1984).

To derive an explicit expression for the concentration variance, we start from the definition of the random function $\bar{C}$ (3.2) and compute its variance by the procedure that led to $\langle \Delta C \rangle$ (3.3). The immediate, and fundamental, result is

$$\sigma_{\bar{C}}^2(\mathbf{x}, t) = \frac{C_0^2}{V^2} \int_{V_0} \int_{V_0} \int_V \int_V [f_t(\mathbf{x}'', \mathbf{x}'''; \mathbf{a}, \mathbf{b}, t) \\ - f_t(\mathbf{x}''; \mathbf{a}, t) f_t(\mathbf{x}'''; \mathbf{b}, t)] \, d\mathbf{x}'' \, d\mathbf{x}''' \, d\mathbf{a} \, d\mathbf{b}, \tag{6.1}$$

where the only new quantity is $f_t(\mathbf{X}'', \mathbf{X}'''; \mathbf{a}, \mathbf{b}, t)$, which is defined as the joint pdf of the total displacements of two particles; here $\mathbf{X}''(\mathbf{a}, t)$ and $\mathbf{X}'''(\mathbf{b}, t)$ originate at time $t_0 = 0$ from $\mathbf{x} = \mathbf{a}$ and $\mathbf{x} = \mathbf{b}$, respectively. To second order, f_t $(\mathbf{X}'', \mathbf{X}'''; \mathbf{a}, \mathbf{b}, t)$ is characterized by the covariances $X_{t,jl} = \langle X_j' X_l' \rangle + 2D_{d,jl}t$ for $\mathbf{a} = \mathbf{b}$, and $Z_{t,jl} = \langle X_j' X_l' \rangle$ for $\mathbf{a} \neq \mathbf{b}$. In particular, f_t is completely determined by these covariances if $\mathbf{X}'$ is Gaussian, in which case it has the expression (3.4), with a covariance matrix $\| X_{t,jl}, Z_{t,jl} \|$ that has 36 and 16 elements in three- and two-dimensional flow, respectively. We derive next the equations rendering Z_{jl} in terms of C_Y by the procedure of Section 5.

To grasp some important features, we consider first the case of a stratified aquifer, for which Z_{jl} can be derived exactly. In the simple case of motion parallel to layers, i.e. $V = 0$ in (4.3), and by reasoning similar to that which led to (5.1), we have for $Z_{11} = \langle X_1'(t; a, 0) X_1'(t; b, 0) \rangle$

$$\frac{dZ_{11}}{dt} = \left(\frac{2}{\pi}\right)^{1/2} \frac{1}{D_T} \int_{-\infty}^{\infty} \hat{u}_{11}(k) \frac{1 - \exp(-k^2 D_T t)}{k^2} e^{ik|b-a|} \, dk. \tag{6.2}$$

It is easy to derive the small- and large-time limits of (6.2), precisely as was done in Section 5. Thus, for the early time $tD_T/I \ll 1$, we obtain

$$\frac{dZ_{11}}{dt} = \left(\frac{2}{\pi}\right)^{1/2} t \int_{-\infty}^{\infty} \hat{u}_{11}(k) \, e^{ik(b-a)} \, dk = \frac{2U^2 C_K(a-b)t}{K_A^2} \\ = \frac{dX_{11}}{dt} C_K(b-a), \qquad \left(\frac{tD_T}{I^2} \ll 1\right) \tag{6.3}$$

which obviously degenerates into (5.2) for $a = b$. On the basis of (6.3), and for a given pdf of K, a few interesting cases can be singled out in (6.1). We now present a qualitative discussion, with some possible implications for field tests.

In a first, extreme case, let us assume that solute is inserted over a volume V_0, which extends across the layers over a distance L_0 much larger than the integral scale I. Furthermore, $\bar{C}$ is defined by space averaging over a volume V of a similar magnitude. In this case, it can be shown that $\sigma_{\bar{C}}^2/\langle\bar{C}\rangle^2 \rightarrow 0$ like $O(I/L_0)$ and that it does not depend on z. Hence, ergodicity may be assumed to prevail, i.e. $\bar{C} \cong \langle C\rangle$. In a field test, this would correspond to an injection well and a neighboring observation or pumping well, both of length $L_0 \gg I$. However, if the same well serves for both injection and withdrawal of the solute, none of these effects will be observed, since transport is reversible in the early stage. The only dispersive effect will be manifest in pore-scale dispersion, and the effect of heterogeneity cannot be observed in such a test.

In the second, extreme case we consider a test in which the solute is inserted across layers over a distance L_0 that is small compared with the integral scale I, and the same is true for the observation well. Assuming, furthermore, for illustration that K is Gaussian, we can easily obtain a closed-form expression for $\sigma_{\bar{C}}^2(z, t; a)$. Indeed, under the stipulated conditions, we have $Z_{11} \cong X_{11}$ for $a \neq b$ ($|a-b| \ll I$) and $Z_{11} = X_{11}+2D_{\mathrm{T}}t$ for $a = b$. Substituting these values in the bivariate normal distribution f_t and integrating in (6.1) for finite $|a-b| \ll I$ yields

$$\frac{\sigma_{\bar{C}}^2}{\langle\bar{C}\rangle^2} \cong \frac{1}{2}\frac{\sigma_K}{K_{\mathrm{A}}}\left(\frac{Ut}{D_{\mathrm{L}}}\right)^{1/2} \exp\left[\frac{(z-a)^2 D_{\mathrm{T}}t}{X_{11}(X_{11}+D_{\mathrm{L}}t)}\right]. \tag{6.4}$$

Hence, the coefficient of variation of the concentration may be quite large for $z = a$, and it increases even quicker for $|z-a| > 0$, for which case the ergodicity is obviously not ensured. In a field test this would correspond to injecting solute over a very short screen at $z = a$ and sampling over a small volume around z. From the nature of the early-stage transport it is obvious that no solute will be detected if the sampling point is not close to the layer, whereas for $z = a$ one would measure in the given realization a concentration distribution corresponding to the one-dimensional transport at velocity $u(a)$ and pore-scale dispersion D_{L}. These measurements, of course, may differ considerably from $\langle\bar{C}\rangle$, a fact that is reflected in the large values in (6.4).

Various intermediate cases between the two extreme ones described above are possible. Thus, in the case of injection over $L_0 \gg I$, but sampling over a small volume, the concentration variance will be given by (6.4) for $z = a$, and ergodicity is generally not warranted. In a field test one would measure again the local one-dimensional dispersion, with no detection of the large-scale heterogeneity effect (for a similar discussion, see Pickens & Grisak 1981).

The situation is quite different for the large-time limit. Indeed, in this case we have

$$\frac{dZ_{11}}{dt} \cong \frac{dX_{11}}{dt} \{\exp(-\lambda^2) - \pi^{1/2}\lambda[1 - \mathrm{erf}(\lambda)]\};$$

$$\lambda^2 = \frac{(b-a)^2}{4D_T t}, \quad \left(\frac{tD_T}{I^2} \gg 1\right), \tag{6.5}$$

and, for instance, in the case of injection over $L_0 \gg (D_T t)^{1/2}$, the point sampling will be of small variance, which can be understood in view of the effect of the pore-scale dispersion on spreading across the layers. In this stage, transport is no longer reversible, and pumping of the solute by the injecting well will reveal the effect of heterogeneity, again provided that layering is continuous over a sufficiently large distance.

A similar analysis can be carried out for the case in which transverse flow of velocity V is present. Again, similar to the result of Section 5, V has a negligible influence at the early stage. In the long-time limit we get the simple result of Equation (6.3), with dX_{11}/dt given by (5.5). Hence, ergodicity is warranted under the same conditions as for (6.3) above.

From the analysis of this simple type of flow, it is seen that the extents of the initial solute body V_0 and of the averaging volume V (as compared with the conductivity correlation scale on the one hand, and the dimensionless time tU/I on the other) play an essential role in determining the magnitude of σ_C^2.

To extend this analysis to two- and three-dimensional structures, we must adopt some simplifying assumptions. Under the conditions of Section 5, it is easy to write down the expressions of Z_{jl}. Thus, equations similar to (3.16), (3.17), and (3.18) are obtained, except that the integrands are multiplied by $\exp[i\mathbf{k}\cdot(\mathbf{a}-\mathbf{b})]$, whereas in (3.20) the argument of u_{jl} is now $\mathbf{U}t' + \mathbf{a} - \mathbf{b}$.

Dagan (1984) has computed Z_{jl} for two-dimensional flow and the exponential covariance (5.7); the results are given in his Figure 3. The main result is that Z_{jl} tends to zero as time increases if initially the two particles are at a distance of a few integral scales normal to the direction of the mean flow, whereas the trajectories remain correlated for a considerable time if they are aligned with the mean flow. Conclusions and regimes similar to those prevailing for layered aquifers apply to two- and three-dimensional flows, and ergodicity is ensured if V_0 and V are of a large transversal extent compared with I. Again, if V_0 and V are small compared with I, then $\sigma_C^2/\langle C\rangle^2$ becomes large and the solute concentration is subjected to a large degree of uncertainty.

A partial differential equation can be written down for the integrand in (6.1), which we denote by $\Delta\sigma^2(\mathbf{x}'', \mathbf{x}''', t) = f_t(\mathbf{x}'', \mathbf{x}'''; \mathbf{a}, \mathbf{b}, t) - f_t(\mathbf{x}''; \mathbf{a}, t) \times$

$f_t(\mathbf{x}''';\mathbf{b},t)$ in the case of a multivariate normal pdf of $\mathbf{X}$. Indeed, by using the characteristic function, it can be shown that $\Delta\sigma^2$ satisfies

$$\frac{\partial(\Delta\sigma^2)}{\partial t}+\mathbf{U}\cdot\nabla(\Delta\sigma^2)-\frac{1}{2}\nabla\cdot\left[\frac{d\underline{\underline{Z}}}{dt}\cdot\nabla(\Delta\sigma^2)\right]=\frac{dZ_{jl}}{dt}\frac{\partial\langle C\rangle}{\partial x''_j}\frac{\partial\langle C\rangle}{\partial x'''_l}, \tag{6.6}$$

where the gradient operator applies to the generalized space coordinate $\mathbf{x}'', \mathbf{x}'''$, and where $\underline{\underline{Z}}$ on the left-hand side is for both $\mathbf{a}=\mathbf{b}$ $(Z_{jl}=X_{jl})$ and $\mathbf{a}\neq\mathbf{b}$. Hence, $\Delta\sigma^2$ satisfies a dispersion equation in six dimensions, with a source term. Since $\Delta\sigma^2(\mathbf{x}'',\mathbf{x}''';0)=0$, a nontrivial solution is obtained only if dZ_{jl}/dt is different from zero on the right-hand side of (6.6).

The salient question is whether in typical cases of transport by groundwater, the ergodic hypothesis is obeyed. Although the preceding discussion shows that the answers may differ, depending on the specific conditions, ergodicity is probably ensured at the local scale, since injection wells, observation wells, and repositories are generally of dimensions larger than the local heterogeneity correlation scale. Still, the concentration variance may not be negligible in some cases, and its computation may be required in order to assess the interval of confidence of the concentration estimate.

The situation is completely different for transport at the regional scale, where concentration variances may be quite large. This topic is discussed briefly in Section 8.

7. COMPARISON BETWEEN A RECENT FIELD TEST AND THEORY

In Section 2, we mentioned a very elaborate field test of solute transport that was recently conducted at the Borden site in Canada (for details, see Sudicky 1985, 1986, Freyberg 1986). This test is particularly suitable for comparison with theory, since both hydraulic conductivity and concentration have been mapped extensively in space and time. The main findings about $Y=\ln K$ have been described in Section 2, namely the disparity between the vertical ($I_v=0.10$ m) and horizontal ($I_h=2.8$ m) integral scales on the one hand, and the existence of an isotropic covariance in the horizontal plane on the other.

The motion of the solute body, of initial volume $V_0=12$ m^3, has been analyzed by Freyberg (1986). The average horizontal velocity was $U=0.09$ m day^{-1}, and the spatial second-order moments of the solute body with respect to its center of mass were determined at various times. One of the main findings was that vertical spreading is negligible, with dispersion occurring only in the horizontal plane. A possible interpretation of this effect is that thin, horizontal layers of very low conductivity prevented vertical velocity fluctuations. As was suggested in Section 5, the solute transport can be modeled in such a case as being two dimensional, provided that a fluid "particle" is defined by space averaging the motion in the vertical direction (say, over an integral scale I). The relationship between the covariance of $\bar{Y} = (1/I_v)\int_0^{I_v} Y(x, y, z)\, dz$ and that of Y is as follows:

$$C_{\bar{Y}}(x, y) = \frac{2}{I_v^2}\int_0^{I_v} (I_v - z)C_Y(x, y, z)\, dz. \tag{7.1}$$

To simplify the calculations, the following relationship for C_Y, which is compatible with measurements, is assumed:

$$C_Y(x, y, z) = \sigma_Y^2 \rho_{Yh}(x, y)\rho_{Yv}(z);$$

$$\rho_{Yh} = \exp[-(x^2+y^2)^{1/2}/I_h], \quad \rho_{Yv} = \exp(-|z|/I_v), \tag{7.2}$$

which, after substitution in (7.1), leads to $C_{\bar{Y}}(x, y) = 0.74\,\sigma_Y^2\rho_{Yh}$ In other words, in this approximate scheme, we may regard the transport as two dimensional and characterized by the isotropic exponential covariance $\sigma_{\bar{Y}}^2\rho_{Yh}$, provided that the log-conductivity variance is reduced to $\sigma_{\bar{Y}}^2 = 0.74\,\sigma_Y^2$. An additional effect of this process is an increase of the "Brownian motion" longitudinal component beyond the pore-scale effect, but this is neglected in view of the larger contribution of heterogeneity in the horizontal plane.

The solute-body initial dimensions were much larger than I_v and also somewhat larger than I_h. It can be assumed, therefore, that the concentration field is not far from satisfying the ergodic requirements, once the concentration is defined by a suitable space average (Section 6). In Figures 3 and 4 we have reproduced the experimental results from Freyberg (1986), in which the spatial moments S_{jl} of the concentration with respect to the center of mass of the solute body in the horizontal principal directions (i.e. for $x = x_1$ parallel to U and $y = x_2$ normal to it) are represented as functions of travel time. Since these moments in the horizontal plane imply that the concentration is integrated along the vertical, they can definitely be compared with the theoretical results for the expected value $\langle C \rangle$. With the latter satisfying Equation (3.5), the S_{jl} are related to the displacement covariances X_{jl} by $S_{jl}(t) = S_{jl}(0) + X_{jl}(t)$. In Figures 3 and 4 the theoretical solution of Dagan (1982b, 1984), also given in (5.10) here, has been plotted by using the value $U = 0.091$ m day^{-1} and by searching for the values of $\sigma_{\bar{Y}}^2 = 0.74\,\sigma_Y^2$ and I_h that led to a best fit between theory and experiment. These values were found to be equal to $\sigma_Y^2 = 0.24$ and $I_h = 2.7$ m. It is seen that the agreement for the integral scale is very good, whereas the required value of the log-conductivity variance is somewhat smaller than that determined from cores. This discrepancy may be due to some uncorrelated

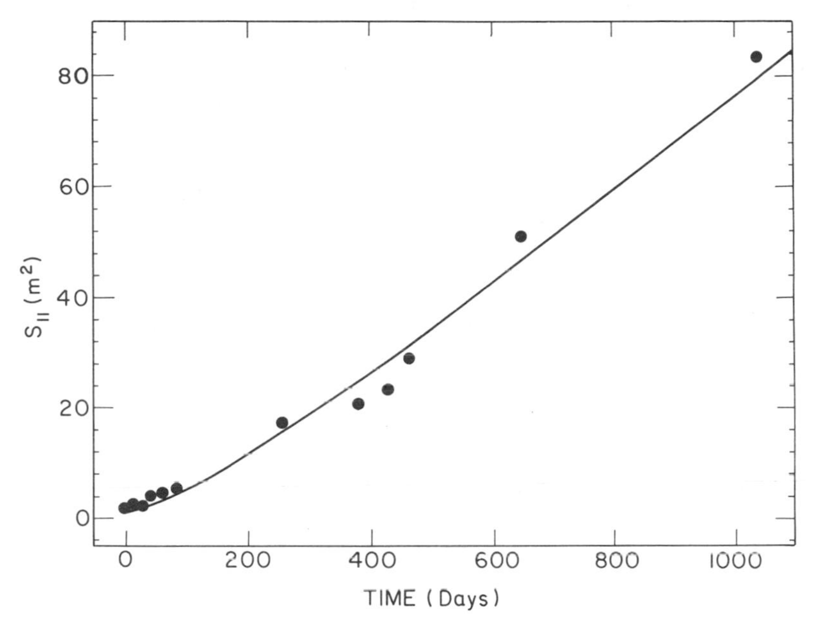

Figure 3 Comparison between measured longitudinal spatial moments S_{11} (●) for the Borden site experiment (Freyberg 1986) and computed X_{11} (curve) from Dagan (1982a, 1984) for best-fitted values $\sigma_{Y'}^2 = 0.74\,\sigma_Y^2$, $\sigma_Y^2 = 0.24$, and $I_h = 2.7$ m (reproduced, with permission, from Freyberg 1986).

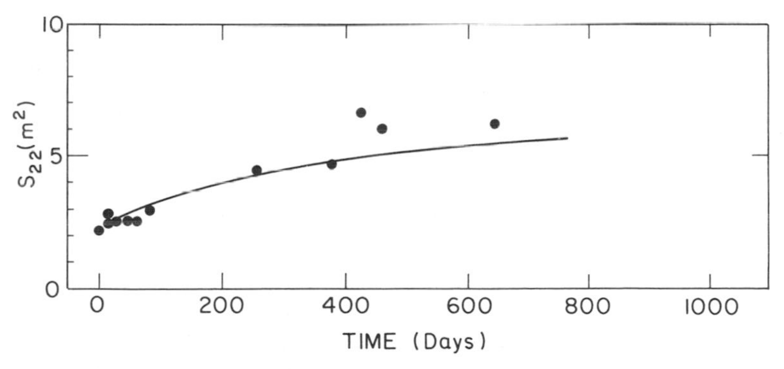

Figure 4 Same as in Figure 3 for the lateral spatial moment S_{22} and X_{22}, respectively (reproduced, with permission, from Freyberg 1986).

noise in the Y-measurements that does not contribute to random motions, or to the approximate manner in which $\sigma_{Y'}^2$ has been computed above in order to make the two-dimensional results applicable to this test.

The agreement between the theory and the somewhat scattered measurements is satisfactory in view of the various approximations involved in the process. A calculation of the apparent instantaneous dispersivity $X_{jl}/2Ut$, as perceived by an observer who assumes that at each t the concentration is represented by the solution of Equation (3.5) with constant coefficients, leads to the conclusion that the asymptotic regime for which $X_{11}/2Ut \rightarrow D_{11} =$ constant (5.10) has been reached toward the end of the period of Figure 3. In contrast, the lateral apparent dispersivity follows the trend displayed in Dagan (1984, Figure 1*a*), i.e. it eventually decreases with time, as a result of the logarithmic behavior of X_{22} (5.10).

Although these results should be regarded as preliminary, since the experiment was still going on at the time of writing, they strengthen our confidence in the ability of the theory to represent the transport phenomena. More such elaborate experiments, in formations of larger log-conductivity variances and correlation scales and for extended periods of time, are needed in order to arrive at more definitive conclusions.

8. TRANSPORT AT REGIONAL SCALE AND MODELING BY CONDITIONAL PROBABILITY

So far, most field and theoretical transport studies have dealt with the effect of local-scale heterogeneity upon solute spreading, and this was the main topic of the preceding sections. As was already mentioned in the Introduction, a different and much larger heterogeneity scale prevails at the regional scale, the spatially varying property now being the transmissivity $T(x, y)$. The survey by Hoeksema & Kitanidis (1985) indicates that $Y = \ln T$ is normal and that C_Y can be approximated by an exponential, isotropic function (5.7), with I (the integral scale) of the order of kilometers.

In view of these findings, and from a theoretical standpoint, the transport of an inert solute may be modeled along the lines of Section 5, with the concentration expected value based on the solution of two-dimensional flow (5.10). Furthermore, C should be regarded as the spatial average over the solute body in the vertical direction, and the effect of the local-scale heterogeneity may be represented as a "Brownian motion" type of transport, in view of the disparity between the local and the regional scales.

As was shown in Section 6, however, the expected value $\langle \bar{C} \rangle$ is close to the space average $\bar{C}$ in a given realization and for realistic travel time only if the solute input zone V_0, as well as its space-averaging volume V, have transverse dimensions large compared with I. While this may happen for solute sources distributed over large areas, it is definitely not true in the more common case of injecting wells or repositories, whose dimensions are much smaller than I and which can be regarded as point sources. As we have shown in Section 6, the concentration coefficient of variation may be very large in such a case, and since only one realization is available, the prediction of the concentration expected value is quite useless. This may be easily understood if we consider a continuous source, which creates a plume in a flowing aquifer. This plume experiences lateral and longitudinal dispersion by the local heterogeneity effect and is slowly winding in the plane as a result of the large-scale regional heterogeneity. The theoretical expected value $\langle C \rangle$, an ensemble average for all possible plumes, will result instead in a huge, laterally dispersing plume of low concentration.

An obvious approach to overcoming this fundamental difficulty is to regard the large-scale effect as deterministic and to incorporate it into the convective term of Equation (1.1). This requires, however, a detailed and deterministic mapping of Y to compute the velocity field by solving Equation (4.2). Such detailed information is seldom available in practice, and Y is subject to uncertainty.

This dilemma, common to many problems involving point sources in large heterogeneous systems, can be resolved by using the concept of conditional probability (or the so-called geostatistical approach). Only a few points of principle are discussed here, and further details can be found in Dagan (1984).

In the theory of diffusion by continuous motions, and in the material discussed so far here, it is tacitly assumed that heterogeneity is modeled by random space functions that are defined in terms of their unconditional probability distribution functions. Thus, Y has been regarded as a multivariate normal function, completely characterized by its expected value and its two-point covariance, which in turn depends on a few parameters. In practice, these parameters are determined by statistical inference by using measurements of the transmissivity T. At the measurement points, however, T (i.e. Y) is fixed and known deterministically except for measurement errors, and this knowledge is not accounted for by the unconditional pdf, which regards Y as subject to the same degree of uncertainty at any point. This additional information can be incorporated into the statistical model in a systematic manner by using the concept of conditional probability (or the associated, and somewhat more restricted, geostatistical technique of kriging). The basic idea is to replace $f(Y)$ by the conditional $f^{(c)}(Y|Y_1, Y_2, \ldots, Y_M)$, where $Y_1, \ldots, Y_M$ are the measured values of Y at the points $\mathbf{x}_1, \mathbf{x}_2, \ldots, \mathbf{x}_M$, respectively. The well-known and fundamental formula rendering $f^{(c)}$ is

$$f^{(c)}(Y|Y_1, Y_2, \ldots, Y_M) = \frac{f(Y, Y_1, Y_2, \ldots, Y_M)}{f(Y_1, Y_2, \ldots, Y_M)}. \tag{8.1}$$

Subsequently, $f^{(c)}$ is used as the input pdf to compute the velocity field, which in turn leads to the concentration expected value and variance, conditioned on the measurements of Y.

Dagan (1984) has examined the effect of measurements both of Y and of the water head Φ at a few points upon the concentration field, and mainly upon the reduction of its variance. A few fundamental results are the following: (*a*) the conditional pdf is no longer statistically stationary (homogeneous). Both the expected value and the variance depend on $\mathbf{x}$, even if the unconditional pdf is stationary. More precisely, the statistical moments depend on the relative position of the point $\mathbf{x}$ with respect to the measurement points. (*b*) In particular, the variance of Y is equal to zero at conditioning points, and the expected value is equal to the measured one. In contrast, at sufficiently distant points their values tend to the unconditional ones. (*c*) In view of the last remark, the conditional probability achieves a continuous transition from the unconditional case to a nearly deterministic representation when measurements are carried out on a sufficiently dense grid. Some of these properties apply to the concentration field, which depends on Y and Φ through the velocity field. These topics are examined in a quantitative manner by Dagan (1984), and this new subject, which is not covered by the conventional literature on diffusion, deserves further investigation.

ACKNOWLEDGMENTS

This review was written during the author's sabbatical leave at the Department of Mechanical and Environmental Engineering at the University of California at Santa Barbara, whose hospitality is acknowledged with gratitude.

The author is thankful to Dr. E. A. Sudicky (University of Waterloo) and Dr. D. Freyberg (Stanford University) for putting at his disposal the data on the Borden site tracer experiment.

Partial support from EWA Inc. is also acknowledged with gratitude.

Literature Cited

Aris, R. 1956. On the dispersion of a solute in a fluid flowing in a tube. *Proc. R. Soc. London Ser. A* 235: 67–77

Bakr, A. A., Gelhar, L. W., Gutjahr, A. L., MacMillan, J. R. 1978. Stochastic analysis of spatial variability in subsurface flow. 1. Comparison of one- and three-dimensional flows. *Water Resour. Res.* 14: 263–71

Beran, M. J. 1968. *Statistical Continuum Theory*. New York: Interscience

Buyevich, Yu. A., Leonov, A. J., Safrai, V. M. 1969. Variations in filtration velocity due to random large-scale fluctuations of porosity. *J. Fluid Mech.* 37: 371–81

Dagan, G. 1982a. Stochastic modeling of groundwater flow by unconditional and conditional probabilities. 1. Conditional simulation and the direct problem. *Water Resour. Res.* 18: 813–33

Dagan, G. 1982b. Stochastic modeling of groundwater flow by unconditional and conditional probabilities. 2. The solute transport. *Water Resour. Res.* 18: 835–48

Dagan, G. 1984. Solute transport in heterogeneous porous formations. *J. Fluid Mech.* 145: 151–77

Dagan, G. 1985. A note on the higher-order corrections of the head covariances in steady aquifer flow. *Water Resour. Res.* 21: 573–78

Dagan, G. 1986. Statistical theory of groundwater flow and transport: pore- to laboratory-, laboratory- to formation- and formation- to regional-scale. *Water Resour. Res.* In press

Dieulin, A., Matheron, G., de Marsily, G., Beaudoin, B. 1981a. Time dependence of an "equivalent dispersion coefficient" for transport in porous media. In *Flow and Transport in Porous Media*, pp. 199–202. Rotterdam: A. A. Balkema

Dieulin, A., Matheron, G., de Marsily, G. 1981b. Growth of the dispersion coefficient with the mean travelled distance in porous media. *Sci. Total Environ.* 21: 319–28

Freeze, R. A. 1975. A stochastic-conceptual analysis of one-dimensional groundwater flow in nonuniform homogeneous media. *Water Resour. Res.* 11: 725–41

Freyberg, D. 1986. A natural gradient experiment on solute transport in a sand aquifer. 2. Spatial moments and advection and dispersion of nonreactive tracers. *Water Resour. Res.* In press

Fried, J. J., Combarnous, M. A. 1971. Dispersion in porous media. *Adv. Hydrosci.* 7: 169–282

Gelhar, L. W. 1986. Stochastic subsurface hydrology from theory to applications. *Water Resour. Res.* In press

Gelhar, L. W., Axness, C. L. 1983. Three-dimensional stochastic analysis of macrodispersion in aquifers. *Water Resour. Res.* 19: 161–80

Gelhar, L. W., Gutjahr, A. L., Naff, R. L. 1979. Stochastic analysis of macrodispersion in a stratified aquifer. *Water Resour. Res.* 15: 1387–97

Gutjahr, A. L., Gelhar, L. W., Bakr, A. A., MacMillan, J. R. 1978. Stochastic analysis of spatial variability in subsurface flow. 2. Evaluation and application. *Water Resour. Res.* 14: 953–59

Güven, O., Molz, F. J., Melville, J. G. 1984. An analysis of dispersion in a stratified aquifer. *Water Resour. Res.* 20: 1337–54

Hoeksema, R. J., Kitanidis, P. K. 1985. Analysis of spatial structure of properties of selected aquifers. *Water Resour. Res.* 21: 563–72

Klotz, D., Seiler, K. P., Moser, H., Neumaier, F. 1980. Dispersivity and velocity relationship from laboratory and field experiments. *J. Hydrol.* 45: 169–84

Lallemand-Barrés, P., Peaudecerf, P. 1978. Recherche des relations entre la valeur de la dispersivité macroscopique d'un milieu aquifère, ses autres caractéristiques et les conditions de mesure. Étude bibliographique. *Bull. Bur. Rech. Geol. Minières Sect.* 3 4: 277–84

Lundgren, T. S., Pointin, Y. B. 1975. Turbulent self-diffusion. *Phys. Fluids* 19: 355–61

Marle, C., Simandoux, P., Pacsirsky, J., Gaulier, C. 1967. Étude du déplacement de fluides miscibles en milieu poreux stratifié. *Rev. Inst. Fr. Pét.* 22: 272–94

Matheron, G., de Marsily, G. 1980. Is transport in porous media always diffusive? A counterexample. *Water Resour. Res.* 16: 901–17

Mercado, A. 1967. The spreading pattern of injected waters in a permeability stratified aquifer. *IAHS/AISH Publ. No. 72*, pp. 23–36

Moltyaner, G. L. 1985. *Stochastic versus deterministic: a case study*. Presented at Symp. Stochastic Approach to Subsurf. Flow, Int. Assoc. Hydrol. Res., Montvillargenne, Fr.

Phythian, R. 1975. Dispersion by random velocity fields. *J. Fluid Mech.* 67: 145–53

Pickens, J. F., Grisak, G. E. 1981. Scale-dependent dispersion in a stratified granular aquifer. *Water Resour. Res.* 17: 1191–1211

Smith, L., Schwartz, F. W. 1980. Mass transport. 1. Stochastic analysis of macrodispersion. *Water Resour. Res.* 16: 303–13

Sudicky, E. A. 1985. *Spatial variability of hydraulic conductivity at the Borden tracer site*. Presented at Symp. Stochastic Approach to Subsurf. Flow, Int. Assoc. Hydrol. Res., Montvillargenne, Fr.

Sudicky, E. A. 1986. A natural-gradient experiment on solute transport in a sand aquifer: spatial variability of hydraulic conductivity and its role in the dispersion process. *Water Resour. Res.* In press

Taylor, G. I. 1921. Diffusion by continuous movements. *Proc. London Math. Soc.* 2(20): 196–212

Taylor, G. I. 1953. Dispersion of soluble matter in solvent flowing slowly through a tube. *Proc. R. Soc. London Ser. A* 219: 186–203

Winter, C. L., Newman, C. M., Neuman, S. P. 1984. A perturbation expansion for diffusion in a random velocity field. *SIAM J. Appl. Math.* 44(2): 411–24

Ann. Rev. Fluid Mech. 1987. 19 : 217–36

TSUNAMIS

S. S. Voit

P. P. Shirshov Institute of Oceanology, Academy of Sciences of the USSR, Krasikova ul. 23, 117218 Moscow, USSR

INTRODUCTION

The giant destructive waves caused by underwater earthquakes and sometimes by volcanic activity or underwater landslides are traditionally called "tsunamis," a Japanese word that means an unusually big wave in seaports and harbors.

The main cause of tsunamis is the abrupt vertical shift of some areas of the ocean bottom or the displacement in a horizontal direction of steep and long underwater slopes. Tsunamis may also result from a seismic pulse or elastic oscillations of the ocean bottom that are not followed by residual alteration of the bottom shape, from volcanic explosions, or even from a coastal landslide.

Three stages of tsunami development from the moment of its appearance are usually distinguished: formation of the wave due to the initial cause and its propagation near to the source; free propagation of the wave in the open ocean at large depths; and propagation of the wave in the region of continental shallows and shelf, where as a result of shallowness a strong deformation of the wave profile takes place right up to its breaking and run-up onto the beach. Corresponding to these three stages, one differentiates the physical and mathematical methods of studying tsunamis. The second stage has been thoroughly investigated from the theoretical point of view, the linear wave theory being quite suitable for this stage. Observations have been made mainly for the third stage. There have been cases of artificial tsunamis created by underwater and onwater atomic explosions (Van Dorn 1961). The process of tsunami excitation by real earthquakes (the first stage) has not yet been sufficiently studied, although it has been ascertained that the characteristics of a tsunami depend

0066–4189/87/0115–0217$02.00

mainly on the details of its origin, on its geometry, on the magnitude and duration of the bottom deformation, and on the strength of the earthquake.

In the spectrum of marine gravitational waves, tsunamis occupy an intermediate position between tides and ripples. The period of a tsunami varies between 2 and 200 min (more often between 2 and 40 min), and its wavelength λ varies from tens up to hundreds of kilometers; the velocity is near $c = (gh)^{1/2}$, where h is the ocean depth. Most investigators consider that the initial elevation of the free surface above the source of the tsunami does not exceed some meters. When tsunamis come into shallow water, their velocity decreases. The amplitude increases simultaneously, approaching its maximum value near the shoreline. The greatest increase of the waves is observed in narrow bays, river outflows, and near capes jutting out into the sea (Murty 1977, Voit 1978).

FUNDAMENTAL EQUATIONS FOR TSUNAMIS

In mathematical descriptions of tsunamis one uses the equations of motion for an ideal incompressible fluid bounded by a free surface $z = \zeta(x, y, t)$ and a solid bottom[1] $z = -h(x, y)$. Let us assume the coordinate system to be oriented so that the (x, y)-plane coincides with the undisturbed free surface of the liquid and the z-axis is directed vertically upward. The fundamental equations of motion are then given as follows:

$$\frac{\partial u}{\partial x} + \frac{\partial v}{\partial y} + \frac{\partial w}{\partial z} = 0, \tag{1}$$

$$\frac{d\mathbf{u}}{dt} + \nabla p = 0. \tag{2}$$

The boundary conditions are

$$w = \frac{d\zeta}{dt}, \qquad p = g\zeta \quad \text{when} \quad z = \zeta, \tag{3}$$

$$w + u\frac{\partial h}{\partial x} + v\frac{\partial h}{\partial y} = 0 \quad \text{when} \quad z = -h, \tag{4}$$

where $\mathbf{u}(u, v, w)$ is the velocity of the liquid and $p = p_n/\rho + gz$ (p_n is the total pressure).

Let us introduce the following nondimensional variables

$$x = \lambda x', \qquad y = ly', \qquad z = Hz', \qquad t = \lambda t'/c, \qquad \zeta = a\zeta',$$

[1] Later, in some problems $h(x, y, t)$ will also occur.

$$h = Hh', \qquad u = \frac{ca}{H} u', \qquad v = \frac{ca\lambda}{Hl} v', \qquad w = \frac{ca}{\lambda} w', \qquad p = agp',$$

where a is the amplitude, λ the wavelength, c the phase velocity, H the average ocean depth, and l a characteristic length along the y-axis.

Omitting the primes, we can rewrite (1)–(4) as

$$\frac{\partial u}{\partial x} + \gamma \frac{\partial v}{\partial y} + \frac{\partial w}{\partial z} = 0,$$

$$\frac{\partial u}{\partial t} + c_0^2 \frac{\partial p}{\partial x} + \alpha \left(u \frac{\partial u}{\partial x} + \gamma v \frac{\partial u}{\partial y} + w \frac{\partial u}{\partial z} \right) = 0,$$

$$\frac{\partial v}{\partial t} + c_0^2 \frac{\partial p}{\partial y} + \alpha \left(u \frac{\partial v}{\partial x} + \gamma v \frac{\partial v}{\partial y} + w \frac{\partial v}{\partial z} \right) = 0,$$

$$\beta \frac{\partial w}{\partial t} + c_0^2 \frac{\partial p}{\partial z} + \alpha\beta \left(u \frac{\partial w}{\partial x} + \gamma v \frac{\partial w}{\partial y} + w \frac{\partial w}{\partial z} \right) = 0, \tag{5}$$

with boundary conditions

$$w = \frac{\partial \zeta}{\partial t} + \alpha \left(u \frac{\partial \zeta}{\partial x} + \gamma v \frac{\partial \zeta}{\partial y} \right), \qquad p = \zeta \quad \text{when} \quad z = \alpha\zeta,$$

$$w + u \frac{\partial h}{\partial x} + \gamma v \frac{\partial h}{\partial y} = 0 \quad \text{when} \quad z = -h; \tag{6}$$

here

$$\alpha = \frac{a}{H}, \qquad \beta = \frac{H^2}{\lambda^2}, \qquad c_0^2 = \frac{gH}{c^2}, \qquad \gamma = \frac{\lambda^2}{l^2}.$$

For tsunamis in the open ocean, we have $a \ll \mathrm{H}$ and $H \ll \lambda$. Assuming that $\alpha = 0$ in (5)–(6), corresponding to the neglect of nonlinear terms, we can obtain the equations of linear theory; assuming further that $\beta = 0$ (the hydrostatic approximation) and in addition by considering only flows for which $\partial u/\partial z = \partial v/\partial z = 0$, we obtain the shallow-water equations. We can deduce from (5)–(6) the equations describing long, gently sloping waves in an ocean with smoothly changing depth if we neglect terms of order β^2 and $\alpha\beta$ but retain the terms of lowest order; this means that the relation of the depth of the ocean to the wavelength is rather small but not negligible (Peregrine 1967). In addition, we neglect the terms $\beta(\partial h/\partial x)$ and $\beta(\partial h/\partial y)$, which correspond to a smooth variation of the ocean depth.

Thus we obtain

$$\frac{\partial \zeta}{\partial t}+\frac{\partial}{\partial x}[(\alpha\zeta+h)\tilde{u}]+\gamma\frac{\partial}{\partial y}[(\alpha\zeta+h)\tilde{v}]=0,$$

$$\frac{\partial \tilde{u}}{\partial t}+c_0^2\frac{\partial \zeta}{\partial x}+\alpha\left(\tilde{u}\frac{\partial \tilde{u}}{\partial x}+\gamma\tilde{v}\frac{\partial \tilde{u}}{\partial y}\right)+\frac{1}{3}\beta h\frac{\partial^3\zeta}{\partial t^2\partial x}=0,$$

$$\frac{\partial \tilde{v}}{\partial t}+c_0^2\frac{\partial \zeta}{\partial y}+\alpha\left(\tilde{u}\frac{\partial \tilde{v}}{\partial x}+\gamma\tilde{v}\frac{\partial \tilde{v}}{\partial y}\right)+\frac{1}{3}\beta h\frac{\partial^3\zeta}{\partial t^2\partial y}=0, \tag{7}$$

where

$$\tilde{u}=\frac{1}{h}\int_{-h}^{0}u\,dz,\qquad \tilde{v}=\frac{1}{h}\int_{-h}^{0}v\,dz.$$

The system (7) is called the Boussinesq system (Whitham 1974). An equivalent system was obtained by Miles & Salmon (1985), who used the least-action principle. Plane waves corresponding to (7) can propagate in either a forward or a reverse direction. Further simplifications are connected with the transformation to equations describing almost-plane waves ($\gamma \ll 1$) traveling in only one direction. From (7) it is evident that if $h = 1$, there are solutions for which

$$\frac{\partial \zeta}{\partial \tau}+\frac{\partial \zeta}{\partial x}=O(\alpha,\beta,\gamma),\qquad \tau=c_0 t; \tag{8}$$

that is, in this approximation no wave propagates in the negative x-direction. Neglecting second-order terms in the small parameters α, β and γ, we obtain from (7) the two-dimensional equation of Korteweg–de Vries (KdV) (Kadomtsev & Petviashvili 1970, Miles 1981):

$$\frac{\partial}{\partial x}\left(\frac{\partial \zeta}{\partial \tau}+\frac{\partial \zeta}{\partial x}+\frac{3}{2}\alpha\zeta\frac{\partial \zeta}{\partial x}+\frac{1}{6}\beta\frac{\partial^3\zeta}{\partial x^3}\right)+\frac{\gamma}{2}\frac{\partial^2\zeta}{\partial y^2}=0. \tag{9a}$$

In the one-dimensional case ($\gamma = 0$), this equation transforms into the KdV equation

$$\frac{\partial \zeta}{\partial \tau}+\frac{\partial \zeta}{\partial x}+\frac{3}{2}\alpha\zeta\frac{\partial \zeta}{\partial x}+\frac{1}{6}\beta\frac{\partial^3\zeta}{\partial x^3}=0. \tag{9b}$$

In cylindrical coordinates the equation has the following form (Johnson 1980):

$$\frac{\partial}{\partial r}\left(\frac{\partial \zeta}{\partial \tau}+\frac{\partial \zeta}{\partial r}+\frac{1}{2r}\zeta+\frac{3}{2}\alpha\zeta\frac{\partial \zeta}{\partial r}+\frac{1}{6}\beta\frac{\partial^3\zeta}{\partial r^3}\right)+\frac{\gamma}{2r^2}\frac{\partial^2\zeta}{\partial \theta^2}=0. \tag{10}$$

It may be noted that in the last two terms of (9), which contain the parameters α and β, x-derivatives may be replaced by τ-derivatives without reducing the accuracy that results from (8).

Equations (9a) and (9b) have been thoroughly studied, and sets of exact solutions have been found (Novikov et al. 1984). In particular, the KdV equation has cnoidal and solitary waves as steady solutions.

Equation (9) has solutions in the form of algebraic solitary waves [$\zeta \sim (x^2+y^2)^{-1}$ when $(x^2+y^2) \to \infty$ (Novikov et al. 1984)]. The initial-value problem for the KdV equation may be solved by the inverse scattering method. The relation between (9) and (10) and their modifications was considered by Johnson (1980), who also discussed the region of their validity.

The approximate self-similar solution of (10) in the form

$$\zeta = r^{-2/3}\mathscr{F}\left(\frac{r-\tau}{\sqrt[3]{r}}\right)+O(\beta^2) \tag{11}$$

was found by Miles (1978), who assumed that $\alpha \sim \beta \sim r^{-2/3}$.

When it is necessary to take into account the variable depth of the ocean, the general equation for plane waves with smoothly varying depth is used. It is deduced analogously from (7) (Ostrovskiy & Pelinovskiy 1970, Kakutani 1971):

$$\frac{\partial \zeta}{\partial \tau}+\sqrt{h}\left(1+\frac{3\alpha\zeta}{2h}\right)\frac{\partial \zeta}{\partial x}+\frac{\beta}{6}h^2\sqrt{h}\frac{\partial^3\zeta}{\partial x^3}+\frac{\sqrt{h}}{4h}\zeta\frac{\partial h}{\partial x}=0. \tag{12}$$

An example of numerical computations performed on the basis of this equation and their comparison with laboratory experiments (Svendsen & Hansen 1978) is presented in Figure 1.

The influence of nonlinearity and dispersion on tsunami propagation in the open ocean has been the subject of numerous investigations. Practically all the computations show that nonlinearity and dispersion may be neglected in offshore conditions (Alexeev et al. 1978). Analytical estimates (Nagashima 1977, Kajiura 1979, Hammack & Segur 1978) are very different and do not allow drawing simple conclusions. In any case the main results concerning the propagation of tsunamis in the open ocean have been obtained by the methods of linear theory. All questions open to investigation for this problem should not be considered as settled. The

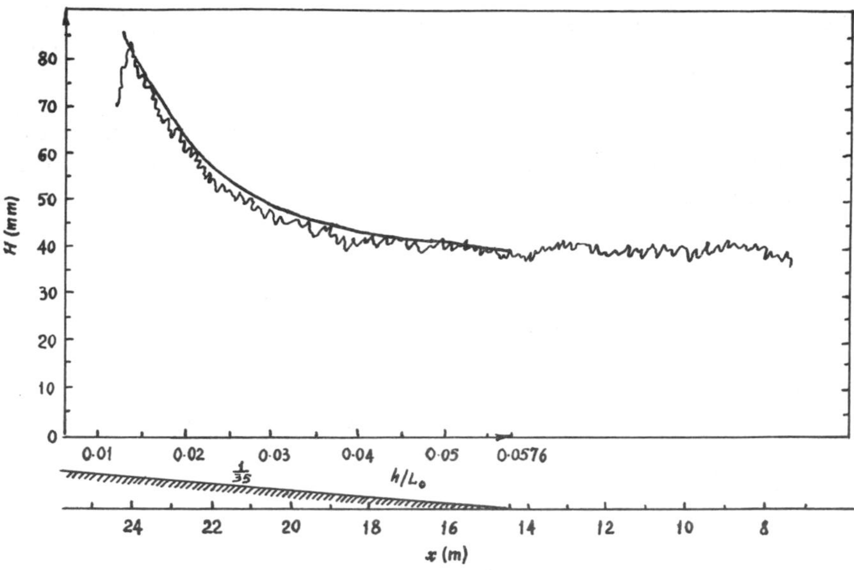

Figure 1 The wave-height variation over a slope of 1/35. Irregular curve, measured values; regular curve, theoretical values (Svendsen & Hansen 1978).

large amount of research in this field during recent years is evidence of the truth of this statement.

INVESTIGATION OF TSUNAMIS ON THE BASIS OF LINEAR THEORY

Assuming that the motion starts from a state of rest, one can introduce a velocity potential $\varphi(x, y, z, t)$. Then the boundary-value problem (1)–(4) in the linear formulation will take the following form:

$$\Delta\varphi = 0,$$

$$\frac{\partial\varphi}{\partial z} = \frac{\partial\zeta}{\partial t}, \qquad \frac{\partial\varphi}{\partial t} = -g\zeta - p/\rho, \qquad (z = 0)$$

$$\frac{\partial\varphi}{\partial z} = w_{\mathrm{B}}(x, y, t), \qquad (z = -h) \tag{13}$$

where $p(x, y, t)$ is the pressure applied to the free surface, and $w_{\mathrm{B}}(x, y, t)$ is the prescribed vertical component of the velocity at points of the bottom in some region. The disturbances that cause the motion of the fluid are given in the form of an initial deviation from the equilibrium state or in the form of impulsive or continuous-in-time pressures applied on the free surface or the bottom of the ocean. The boundary-value problem under consideration is well known in the literature as the Cauchy-Poisson problem, and its solution may be easily found with the aid of Fourier transforms in space and Laplace transforms in time. For tsunamis, this problem has been solved in numerous scientific papers (Gazaryan 1955, Hwang & Divoky 1970, Braddock et al. 1973). An unusual approach to this problem was made by Kajiura (1963), who used the method of Green's functions depending on time (Stoker 1957). The peculiarity of Kajiura's solution is that in his final formulas the effects connected with the character of the source of disturbance are distinguished from those connected with the dispersion properties of the medium. This approach is similar to that suggested by Voit & Sebekin (1968). They propose introducing transfer functions $w(t)$ that characterize the response of the ocean to a definite standard action in the form of an "instantaneous" impulse given as a Dirac delta-function. Then the reaction of the ocean to any disturbance is obtained as a convolution of the transfer function $w(t)$ with the function $f(t)$ describing the development of this disturbance in time. Knowing the transfer function, one may calculate the response of a given hydro dynamical system to any disturbance. Methods for such calculations have been thoroughly developed in the theory of automatic control. The works cited above and many others have allowed the theoretical investigation of the features of tsunami propagation in an open ocean of constant depth. They have also permitted the study of the leading wave, the laws of its attenuation with time and distance from the source, and the consideration of different geometric forms of the region of initial disturbance. The direction of radiation, connected with geometrical form, is determined in terms of the different dimensions of the source in two mutually perpendicular directions. This directive effect takes place only within limited distances from the source.

Another way of studying tsunami propagation in the open ocean is based upon the linear equations for shallow water. Suppose in (5) and (6) that $\alpha = \beta = 0$. Writing the equations in the dimensional form, we obtain

$$\frac{\partial u}{\partial t} = -g\frac{\partial \zeta}{\partial x}, \qquad \frac{\partial v}{\partial t} = -g\frac{\partial \zeta}{\partial y},$$

$$\frac{\partial \zeta}{\partial t} = w_B(x, y, t) - h\left(\frac{\partial u}{\partial x} + \frac{\partial v}{\partial y}\right). \tag{14}$$

The last equation here is supplemented with the term $w_B(x, y, t)$ representing the vertical velocity of the point (x, y) of the basin bottom. This system can be reduced to the equation

$$\frac{\partial^2 \zeta}{\partial t^2} = \frac{\partial}{\partial x}\left(gh\frac{\partial \zeta}{\partial x}\right) + \frac{\partial}{\partial y}\left(gh\frac{\partial \zeta}{\partial y}\right) + \frac{\partial w_B}{\partial t}, \tag{15}$$

where $c^2 = gh(x, y)$, an equation used in many papers dealing with tsunamis.

The appropriate calculations for areas of initial disturbance of rectangular, circular, and elliptical form and for homogeneous displacement were made by Kajiura (1970), including numerical illustration. He also compared his results with analogous ones obtained on the basis of a linear-potential model. The comparison showed that the long-wave approximation repeats in general the common tendency of dispersing wave groups (Figure 2).

However, the frequency dispersion of gravity waves, not accounted for in the linear shallow-water theory, even in the deep ocean leads to a gradual accumulation of errors and cannot give reasonable results at large distances from the source of disturbance. At the same time the simplicity and clarity of the linear wave model continue to attract investigators' attention, in particular for numerical studies of particular tsunamis that have happened. The long-wave linear model also calls forth other critical remarks. For example, Ichiye (1983) notes that if the horizontal dimensions of a tsunami source are comparable with the depth of the basin or if the motion of the bottom is too fast or too slow in comparison with the time of passage of the long waves above the disturbance region, then the long-wave model leads to large errors in both the near and far zones. As a confirmation, he compares the data of the long-wave model with the solution of the corresponding one-dimensional Cauchy-Poisson problem.

TRAPPED WAVES

The propagation of tsunamis is strongly affected by the topography of the bottom. Underwater ridges in the open ocean and the shelf zone, bounded by the beach on one side and the zone of sharply increasing depth on the other, serve as waveguides, along which the wave energy is concentrated and which form the primary direction of wave propagation. Munk & Arthur (1952) had already considered this phenomenon, and the theory (without using a hydrostatic approximation) has been further developed in Garipov's (1965) work. It was shown that the influence of an underwater ridge on the propagation of waves does not reduce only to a simple increase of amplitude, but that it essentially determines the process of wave propagation itself, altering the character of its attenuation along the ridge.

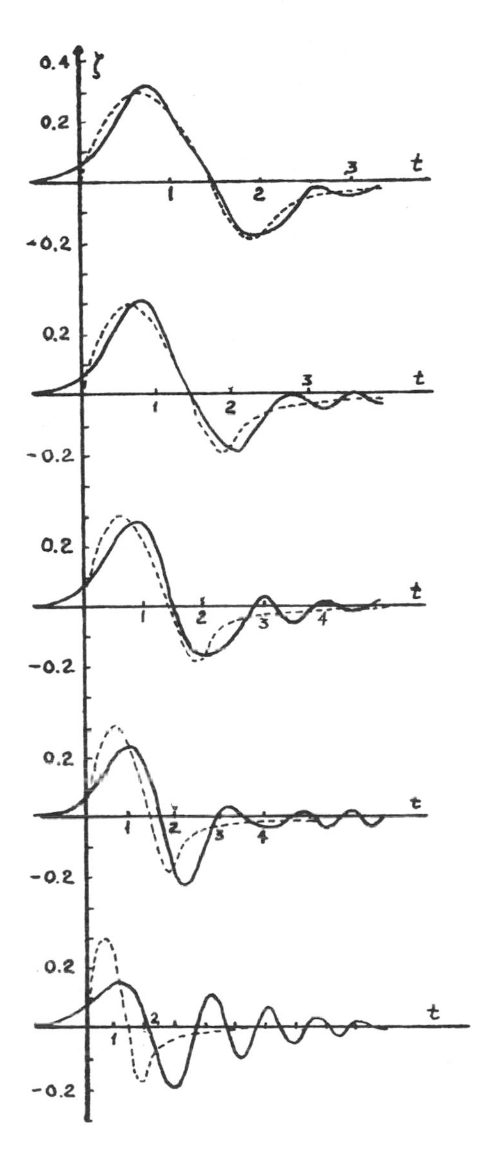

Figure 2 Comparison of scaled waveforms. Potential model, ----; shallow-water model, ——; t, nondimensional time (measured from the first arrival of a nondispersive wave at the observation point); ζ, scaled elevation (Kajiura 1970).

Kajiura (1973) studied the effect of directivity of the energy of a tsunami generated near the continental shelf; the shelf becomes a waveguide and captures the tsunami energy, thus leading to the formation of gravity edge waves.

In the framework of linear shallow-water theory, Kajiura studied in detail the formation of edge waves for a steplike shelf. However, the qualitative conclusions made in this work may be expanded to the case of more arbitrary shelf profiles. For the waves trapped by an underwater ridge, there is a discrete set of frequencies, and the propagation of the waves is followed by their dispersion, which results in the long waves lagging behind the shorter waves. One more problem of interest deals with the trapping of waves in the case of a source located in a shelf zone. In the work of King & LeBlond (1982) a shelf of constant depth falls down abruptly to a deep region also of constant depth. It is easy in this case to construct wave rays, which have the form of segments in this simple model. If a wave gauge is placed above the shelf or on the beach, then in addition to the direct ray between the source and the gauge, there will be another ray passing through the deep water and having the form of a broken line. Thus if the distance along the beach from the source to the gauge is large enough, the side wave may reach the gauge earlier than the direct one. Murty & Loomis (1983), using these concepts, attempted to interpret the results of numerical computations of a similar problem in a basin with a piecewise-linear bottom. In this case, besides the direct ray, the disturbance comes to the observation point along some broken line consisting of rectilinear and cycloidal segments. The disturbance propagating along this complex path may arrive at the observation point earlier than one along the direct path. This outdistancing may in principle serve as forewarnings of approaching tsunamis.

The urgency of all problems dealing with the task of wave trapping is due to the fact that the majority of all catastrophic tsunamis in the Pacific Ocean arise in the neighborhood of the shelf and continental slope along the Pacific-basin seismic region. Ishii & Abe (1980) believe that the action of the Kamchatka tsunami near the Japan coastline in 1952 was caused just by trapped waves.

RAY METHOD, TWO-DIMENSIONAL NUMERICAL MODELS

The classical methods of analytical calculation of tsunamis are applicable to a constant-depth ocean. One may carry out some modifications of these methods that allow one, for example, to take into account rectilinear large-scale inhomogeneities of the bottom relief or to consider axisymmetrical problems. However, such calculations may present only qualitative pictures.

The investigation of tsunamis as they move into the shallow-water coastal zone is especially difficult. Here a number of factors comparable, as a rule, with each other in their influence on the wave propagation must be considered. Primary among these factors are nonlinearity, dispersion, inhomogeneity connected with the bottom topography, and dissipation (Ostrovskiy & Pelinovskiy 1976).

The ray method is widely used in practice. This method allows one to calculate wave parameters along large ocean routes with only a small expenditure of computer time. With the system (7) as a basis, let us substitute the variable $s = \tau(x, y) - t$ for t. For a simple definition of the variables, physical premises are necessary. In addition, let us consider that the radius of curvature of the wave front is large ($\gamma \gg 1$) (quasiplanar two-dimensional wave). Then it seems natural to suppose that the solution depends mainly on only the variable s. (In the absence of inhomogeneity for a plane wave, it is the ordinary variable $t - x/c$.) The inhomogeneity of the bottom leads to a weak dependence of the solution on x and y. As a result, one may obtain two independent equations. The first is the well-known eikonal equation that enables us to calculate the ray along which the wave moves:

$$\nabla\tau^2 = (gh)^{-1}.$$

It is essential that the eikonal equation include only the local depth of the ocean, so that the field of rays in this approximation be the same as in the linear problem (Ostrovskiy & Pelinovskiy 1975). The second equation allows one to calculate the variation of the amplitude of the wave along the ray tube. In dimensional variables, it may be written as follows:

$$\frac{\partial\zeta}{\partial t} + \sqrt{gh}\left(1 + \frac{3\zeta}{2h}\right)\frac{\partial\zeta}{\partial l} + \frac{h^2\sqrt{gh}}{6}\frac{\partial^3\zeta}{\partial l^3} + \frac{\sqrt{gh}}{4h\Delta^2}\frac{d(h\Delta^2)}{dl} = 0, \tag{16}$$

where l is the length of the ray, Δ is the distance between rays (the width of a ray tube) determined by the eikonal equation in the usual manner (Babich & Buldyrev 1972). In the case of a plane wave in an ocean of constant depth, this equation may be reduced to the KdV equation. Hence it is called the generalized KdV equation (Shen & Keller 1973, Shuto 1974). If this equation is used for shallow water, one may add a term characterizing energy losses caused by molecular and turbulent viscosity. There are a number of methods for finding this term that are based on semiempirical turbulence models. The applicability of the ray method is restricted to some extent because of the fact that in certain kinds of bottom relief the construction of rays leads to an excessive expansion of the ray tubes, when use of a one-dimensional equation along a tube becomes problematic. Besides, during the propagation of a tsunami over a considerable distance the interaction between the tsunami and islands or coastline protrusions becomes inevitable, leading to inapplicability of the ray method. This is why in developing numerical methods, we use finite-difference analogues of the two-dimensional (in the horizontal coordinates) equations of shallow-water theory. For the study of wave behavior in the ocean with a large-scale change of depth, the Boussinesq system (7) has been applied successfully. It has been studied in detail for the axisymmetric case

$$r\frac{\partial\zeta}{\partial r}+\frac{\partial}{\partial r}[(1+\alpha\zeta)ru]=0,$$

$$\frac{\partial u}{\partial t}+\alpha u\frac{\partial u}{\partial r}+c_0^2\frac{\partial\zeta}{\partial r}-\frac{\beta}{3}\frac{\partial^2}{\partial r\partial t}\left[\frac{1}{r}\frac{\partial}{\partial r}(ru)\right]=0. \tag{17}$$

Calculations made by Chwang & Wu (1977) illustrate the run-up of cylindrical waves onto an island circular in plan, with depth decreasing linearly. Figure 3 shows the slow deformation of a solitary wave. Chwang & Power (1983) considered the focusing and reflection of the cylindrical solitary wave in a coastal zone of constant depth. The analytical solutions obtained correspond to the numerical results of Chwang & Wu (1977). The Boussinesq equations are quite suitable for numerical methods right up to shallow water as long as the wave does not become very steep. This is why numerical algorithms are being developed for these equations, with

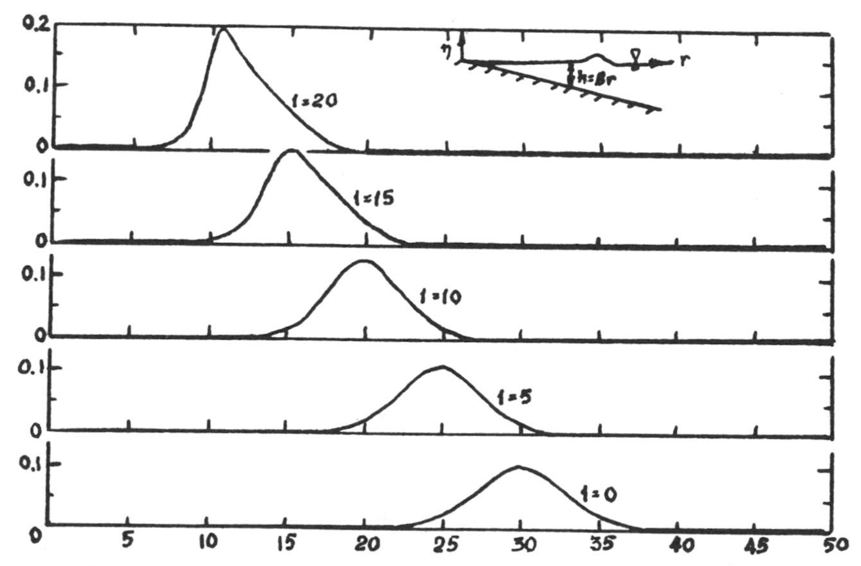

Figure 3 A cylindrical solitary wave converging on a submerged conical island. η, nondimensional amplitude; r, nondimensional distance from the coast (see the scheme at top) (Chwang & Wu 1977).

some of the algorithms containing dissipative effects (Marchuk et al. 1983). The Boussinesq system describing gently sloping long waves may be written in various forms. The nonlinear dispersive system obtained by Peregrine (1967) seems to be very promising, as it gives the best agreement with the exact dispersive equation of linear waves.

The model proposed by Wu (1979) is also rather general and presents a systematic derivation of Boussinesq-type equations, taking into account bottom oscillations and atmospheric pressure changes. In some particular cases this system provides long-wave models: a linear dispersive model, a nonlinear nondispersive model, and a linear nondispersive model. The applicability of any of these systems is connected with the Ursell number $\mathrm{Ur} = \alpha/\beta = a\lambda^2/H^3$, which represents the ratio of the amplitude dispersion α to the phase dispersion β. If $\mathrm{Ur} \ll 1$, the linear theory is applicable. In the other limiting case ($\mathrm{Ur} \gg 1$), the nonlinear nondispersive theory (Airy model) should be used. If $\mathrm{Ur} \sim 1$, so that nonlinear and phase dispersion are of the same order, one should use the complete Boussinesq system. Each of these cases (including the special case $\mathrm{Ur} \sim 4$) was investigated in detail by Pelinovskii (1982). He considered the degree of applicability of different solutions (depending on the value Ur) to the tsunami problem and compared them with observations.

THE INTERACTION OF A TSUNAMI WITH A SLOPING BEACH

Investigation of the last stages of the propagation of a tsunami—its transformation in the coastal zone and its run-up—is associated with great difficulties in principle.

The analysis of observational materials on run-up of tsunamis reveals that approximately 80% (Mazova et al. 1982) of tsunamis undermine the coast without breaking. Run-up of tsunamis in the form of a "wall" take place relatively rarely and as a rule only for waves of amplitude more than 10 m. This is why it seems reasonable to study nonbreaking waves. For comparatively open seashore regions, it is possible to work within the framework of one-dimensional theory.

The first nonlinear problem on the run-up of long waves on a slope was solved by Carrier & Greenspan (1958). Taking as a basis the shallow-water equations in the following form

$$\frac{\partial u}{\partial t} + u\frac{\partial u}{\partial x} + g\frac{\partial \zeta}{\partial x} = 0,$$

$$\frac{\partial \zeta}{\partial t} + \frac{\partial}{\partial x}[(h+\zeta)u] = 0, \tag{18}$$

and using the Legendre transformation in the case of the plane slope $hx = -\alpha x$, they reduced this system to a linear wave equation and found particular solutions in the form of nonlinear standing waves on the sloping beach. Later, Spielvogel (1976) also considered a particular case of this solution. The Legendre transformation was also used in the works of Mazova et al. (1982) and Kaistrenko et al. (1985), who showed that the problem of determining tsunami run-up without breaking may be effectively solved with the help of this transformation. These authors also showed that the maximum height of the water on the coastline from the linear approximation is equal to the maximum height of the vertical run-up of the wave from nonlinear theory. This also suggests a method of determining the vertical run-up of waves by solving the linear problem.

It is worth mentioning that the condition of single-valuedness of the Legendre transformation demands that the Jacobian of the transformation be nonzero. This imposes a limitation on the steepness of slopes (that is, the waves do not break and remain single valued when the slopes are only moderately steep). By Fourier superposition of particular solutions, one may construct the run-up of solitary tsunami waves.

For numerical modeling of tsunami behavior in the coastal zone, the main problem is the choice of the correct boundary condition at the wave front of the tsunami running up on the beach. Matsutomi (1983) proposed postulating the condition by assuming that the head of the tsunami is similar to the wave from a broken dam. Goto & Shuto (1983) constructed numerical models of tsunamis running up a breach with breaking both present and absent. If the wave at the boundary front does not break, the Lagrange approach is applied, making it easier to describe the boundary condition. However, numerical solutions obtained by them show a lack of agreement with solutions obtained by others using an Eulerian formulation. These questions require additional discussion. Here it is appropriate to mention some important problems that have not yet been solved, in particular the following: the calculation of the velocity characteristics in the run-up of unsteady waves, and the solution of two-dimensional problems of run-up.

Many papers have been written on wave run-up with breaking, including numerical studies as well as theoretical investigations. Those interested in these problems may refer to the detailed review by Peregrine (1983). Further improvement in the study of tsunami run-up may be carried out by taking into account the dissipation of waves. There are two mechanisms of tsunami energy absorption in the coastal zone. One of them is dissipation in the boundary layer, and the other is connected with the breaking of the wave front. It is necessary to introduce empirical laws for the study of such processes. In this connection, numerical methods for solving problems of tsunami run-up have been extended by the application of widely developed methods in hydraulics. These works are beyond the scope of the present review.

INVESTIGATION OF TSUNAMIS INCLUDING ELASTIC PROPERTIES OF THE OCEAN BOTTOM

So far we have considered only models of tsunamis that are based upon a study of the origin and propagation of gravity waves in the ocean generated by given disturbances of its free surface and by bottom movements, without consideration of tectonic processes occurring in the Earth's crust.

An attempt to establish a mutual relation between a seismic center in the elastic bottom of the basin and the movement of the ocean waters caused by it was made by Pod''yapol'skii (1970, 1978). He showed that the conjecture that there is a close connection between tsunami excitation and substantial deformation of the bottom or landslides and other non-elastic phenomena accompanying earthquakes is not complete. It follows from the work of Pod''yapol'skii that the source of the type of motion along the discontinuity in a purely elastic medium provides intensive enough tsunami waves and even explains some peculiarities of these waves. The mechanism of tsunami excitation can be examined with the help of the following model: a layer of compressible heavy liquid of depth H overlies an elastic half-space (rocks of the Earth's crust). The field of dynamic displacement of particles is considered in relation to the initial static equilibrium in a gravitational field. In linear approximation the equations of motion and the boundary conditions are written for a water layer and an elastic half-space, taking gravity into account.

The initial conditions for the dynamic part of the field are assumed to be zero. The field is excited by a point source that begins acting at the instant $t = 0$ and is situated in the solid medium at an arbitrary depth l beneath the ocean bottom. Pod''yapol'skii used the Nabarro source as the source that simulated a seismic center. This source is used in seismology for approximating real epicenters and takes gravity into account. It simulates the motion along a plane of rupture and is characterized by its orientation apart from its time dependence. This orientation is determined by giving the plane of rupture and the direction of motion in that plane. For such a source, waves of two major types, space and resonance waves, may be distinguished at sufficiently long distances with the help of methods known from the theory of elasticity. The intensity of resonance waves as they move away from the source decreases only proportional to $r^{-1/2}$ because of cylindrical divergence. Mathematically, resonance waves are connected with the real roots of the dispersion equation $\Delta(w, \theta) = 0$ (poles of the integrand), which gives the dependence of the phase velocity θ^{-1} of the wave on its frequency. This dispersion equation determines an infinite sequence of roots, which remain when neglecting the gravitational terms. They determine the main and the highest modes of the Rayleigh wave. Apart from the Rayleigh wave, the dispersion equation has two real roots essentially connected with gravitational terms and having no analogues in the usual theory of elasticity. One of them is retained under the passage to the limiting case of an incompressible liquid and corresponds to the classical dispersion relation for gravity waves. This root corresponds to a tsunami. The second root corresponds to a gravity wave arising at the interface of the liquid layer and the underlying half-space.

Asymptotic analysis of the solution allows one to draw a number of important conclusions. In particular, it turns out that there is a certain "optimum" depth of the source (approximately 30–50 km) at which the amplitude of displacement of the gravitational wave has its maximum. The amplitudes of the gravitational waves greatly depend on characteristics of the source orientation—that is, the position of the rupture plane and the direction of motion.

Of principal importance is the estimation (although it is rather approximate) of the portion of the energy of the seismic center that is spent on the generation of a tsunami. The result obtained (about 1%) agrees with an empirical estimation given by Iida (1963).

For an incompressible liquid the major problem may, with a sufficient degree of accuracy, be divided into two independent problems. The first one deals with the determination of the field of displacement in a solid medium for a given source with the assumptions of a free boundary at the ocean bottom. This problem is solved with the quasi-static approximation. The second problem concerns the determination of the displacement field in a water layer with a given displacement of the bottom. The connection between these two problems may be reduced to the fact that the solution of the first problem is taken as the given displacement for the second. This division was greatly modified and used for the numerical solution of a number of quite concrete problems (Alexeev et al. 1978, Gusyakov 1978). The potentialities of such a method are great, for each problem may be solved in a formulation complicated enough to approch the concrete data of seismology, and may furthermore be solved by applying the most improved numerical methods of hydrodynamics.

We also mention some critical remarks concerning this model. The difficulty is that the bottom of the ocean is covered with inelastic sedimentary materials whose influence on the mechanical interaction of the ocean waters with the bottom demands additional explanation. We have not yet obtained exhaustive information on the energy spectrum of elastic waves at the frequencies corresponding to tsunamis.

An interesting mechanism for the excitation of long waves by a group of short waves in an uneven elastic bottom has been proposed by Mei & Benmoussa (1984). If the change of depth of the ocean occurs in only one direction and is localized in a finite region, the induced long waves have another direction of displacement than short waves on the ocean bottom. It may happen that the short waves will pass through an underwater ridge, whereas the resulting long waves may be trapped by the ridge and propagate along it. We note one additional point testifying to the fact that the study of tsunamis as a joint problem of hydrodynamics and the theory of elasticity has undoubted prospects. This is that the study of the theoretical spectra of Rayleigh waves reveals their great dependence on the depth of the seismic source and the orientation of the motion in it. This gives in principle the possibility of using Rayleigh waves to determine these earthquake parameters, with the aim of enhancing tsunami forecasting reliability (see Gusyakov 1983).

THE ROLE OF HORIZONTAL IMPULSE IN THE DIRECTION OF WAVE RADIATION

Let us consider one more approach to the tsunami generation problem, one that allows us to find the ocean response to tectonic processes occurring in the Earth's crust. The essence of this approach is in the following: the reaction of the ocean to a tectonic disturbance consists in the deflection of its level from a state of relative equilibrium and in the acceleration of water particles. The distribution of these disturbances depends, of course, on a great number of unknown factors. However, as has already been mentioned, it is known from current estimates that approximately 1% of the energy of an earthquake passes from the bottom to the ocean. This is why tectonic and hydrodynamic processes may be interconnected with the help of such a universal quantity as energy. This concept was used to explain the direction of tsunami radiation from the source of its generation. The proposed mechanism of tsunami generation is connected with the horizontal faults on the continental slope in the event that the earthquake embraces regions of land adjacent to the sea. A typical example of an earthquake that led to a directional tsunami was the one that took place in Chile in May, 1960. The generated wave practically did not attenuate in the direction perpendicular to the South American coastline, but it attenuated very quickly in other directions. According to data from Solov'ev & Go (1974), the elevation of the water in the region of directional radiation was of a similar order for the coasts of both Chile and the

Hawaiian Islands and even for the coast of Japan. Such directivity of a tsunami is explained in works of the present author and coworkers through the transfer of momentum in the horizontal direction to the water mass. The acceleration obtained as a result of this impulse is simulated by the application of mass forces to the water particles. It was shown in the work of Voit et al. (1982) and Lebedev & Sebekin (1982) that within the limits of the linear theory of long waves, the disturbing mass force normal to the rectilinear coastline generates a nondecaying wave. Voit et al. (1985) proposed methods to calculate the density of the time-integrated energy flux in its dependence on the azimuth of the point of observation. This enables one in turn to calculate the amplitude of the nondecaying wave in the direction perpendicular to the coast.

CONCLUSION

The tsunami problem is a many-sided one and is closely connected with many other scientific disciplines, such as geophysics and geology. Here I have taken up only those aspects of the problem that concern fluid mechanics. The main aim of investigators in the end is to predict the probable characteristics of tsunamis for different points of a coastline and to learn how to forecast tsunamis by using data from the tectonic processes causing them. The hydrodynamic aspect of this problem may be successfully solved by numerical methods, but analytical methods must still be developed. Even if the problem of the travel time of a tsunami from its source to any point of the coastline is successfully solved, the calculation of tsunami amplitude characteristics is an enormously more difficult problem, still far from solution, because of uncertainty of the wave shape at the source and insufficient development of the methods of describing the wave run-up on the beach.

Literature Cited

Alexeev, A. S., Gusyakov, V. K., Chubarov, L. B., Shokin, Yu. I. 1978. Numerical investigation of tsunami generation and propagation in the ocean with real bathymetry. Linear model. In *A Study of Tsunami Waves in the Open Ocean*, pp. 5–20. Moscow: Nauka

Babich, V. M., Buldyrev, V. S. 1972. *Asymptotic Methods in Problems of Diffraction of Short Waves*. Moscow: Nauka. 456 pp.

Braddock, R. D., van den Driessche, P., Peady, G. W. 1973. Tsunami generation. *J. Fluid. Mech.* 59: 817–28

Carrier, G. F., Greenspan, H. P. 1958. Water waves of finite amplitude on a sloping beach. *J. Fluid Mech.* 4: 97–109

Chwang, A. T., Wu, T. Y. 1977. Cylindrical solitary waves. *Lect. Notes Phys.* 64: 80–90

Chwang, A. T., Power, H. 1983. Focusing and reflection of a cylindrical solitary wave. See Iida & Iwasaki 1983, pp. 251–63

Garipov, R. M. 1965. Unsteady waves above an underwater ridge. *Dokl. Akad. Nauk SSSR* 161: 547–50

Gazaryan, J. L. 1955. Surface waves in the ocean excited by underwater earthquakes. *Akust. Zh.* 1: 203–17

Goto, G., Shuto, N. 1983. Numerical simulation of tsunami propagations and run-up. See Iida & Iwasaki 1983, pp. 439–51

Gusyakov, V. K. 1978. Survey of works on the excitation of tsunamis. In *Methods of Calculation of Tsunami Origin and Propagation*, pp. 18–29. Moscow: Nauka. 143 pp.

Gusyakov, V. K. 1983. Investigation of Rayleigh wave spectra for a set of tsunamigenic and nontsunamigenic earthquakes. See Iida & Iwasaki 1983, pp. 25–36

Hammack, J. L., Segur, H. 1978. Modelling criteria for long water waves. *J. Fluid Mech.* 84: 359–73

Hwang, L. S., Divoky, D. 1970. Tsunami generation. *J. Geophys. Res.* 75: 6802–17

Ichiye, T. 1983. Tsunami generation as finite depth Cauchy-Poisson problem or long wave problem. See Iida & Iwasaki 1983, pp. 265–74

Iida, K. 1963. Magnitude, energy, and generation mechanisms of tsunamis and a catalogue of earthquakes associated with tsunamis. *Proc. Tsunami Meet. Assoc. Pac. Sci. Congr., 10th, Honolulu, 1961*, pp. 7–18. Paris: Inst. Géograph. Nat.

Iida, K., Iwasaki, T., eds. 1983. *Tsunamis: Their Science and Engineering, Proc. Int. Symp., Sendai-Ofunato-Kamaishi, Jpn., 1981*. Tokyo: Terra, Dordrecht: Reidel. xiv + 563 pp.

Ishii, H., Abe, K. 1980. Propagation of tsunami on a linear slope between two flat regions. Part I. Eigenwave. *J. Phys. Earth* 28: 531–41

Johnson, R. S. 1980. Water waves and Korteweg–de Vries equation. *J. Fluid Mech.* 97: 701–9

Kadomtsev, V. B., Petviashvili, V. I. 1970. Concerning the spectrum of acoustic turbulence. *Dokl. Akad. Nauk SSSR* 208: 794–96

Kaistrenko, V. M., Pelinovskiy, E. N., Simonov, K. V. 1985. The wash and the transformation of tsunami waves on shallow water. *Meteorol. Gidrol.* 1985(10): 68–75

Kajiura, K. 1963. The leading wave of a tsunami. *Bull. Earthquake Res. Inst. Tokyo Univ.* 41: 535–71

Kajiura, K. 1970. Tsunami source, energy and directivity of wave radiation. *Bull. Earthquake Res. Inst. Tokyo Univ.* 48: 835–69

Kajiura, K. 1973. The directivity of energy radiation of the tsunami generated in the vicinity of a continental shelf. *Trans. Tsunami Symp. Gen. Assem. MGGM, 15th, Moscow, 1971*, pp. 5–26

Kajiura, K. 1979. Tsunami generation. In *Tsunamis, Proc. Natl. Sci. Found. Workshop*, ed. L. S. Hwang, Y. K. Lee, pp. 15–35. Pasadena, Calif: TetraTech

Kakutani, T. 1971. Effect of an uneven bottom on gravity waves. *J. Phys. Soc. Jpn.* 30: 272–76

King, D. R., LeBlond, P. H. 1982. The lateral wave at a depth discontinuity in the ocean and its relevance to tsunami propagation. *J. Fluid Mech.* 117: 269–82

Lebedev, A. N., Sebekin, B. I. 1982. Generation of an offshore directed tsunami wave. *Izv. Acad. Sci. USSR Atmos. Oceanic Phys.* 18: 305–10

Marchuk, An. G., Chubarov, L. B., Shokin, Yu. I. 1983. *The Numerical Simulation of Tsunami Waves*. Novosibirsk: Nauka. 175 pp.

Matsutomi, H. 1983. Numerical analysis of the run-up of tsunamis on dry bed. See Iida & Iwasaki 1983, pp. 479–93

Mazova, R. Kh., Pelinovskii, E. N., Shavratskii, S. H. 1982. One-dimensional theory of run-up of nonbreaking tsunami waves. In *Processes of Tsunami Generation and Propagation*, pp. 98–103. Moscow: Acad. Sci. USSR Inst. Oceanol.

Mei, C. C., Benmoussa, C. 1984. Long waves induced by short-wave groups over an uneven bottom. *J. Fluid Mech.* 139: 219–35

Miles, J. W. 1978. An axisymmetric Boussinesq wave. *J. Fluid Mech.* 84: 181–91

Miles, J. W. 1981. The Korteweg–de Vries equation: a historical essay. *J. Fluid Mech.* 106: 131–47

Miles, J., Salmon, R. 1985. Weakly dispersive nonlinear gravity waves. *J. Fluid Mech.* 157: 519–31

Munk, W. H., Arthur, R. S. 1952. Wave intensity along a refracted ray. In *Gravity Waves*, pp. 95–108. *US Natl. Bur. Stand. Circ. No. 521*

Murty, T. S., 1977. *Seismic Sea Waves—Tsunamis*. Ottawa: Fish. Mar. Serv. x + 337 pp.

Murty, T. S., Loomis, H. G. 1983. Diffracted long waves along continental shelf edges. *Proc. Tsunami Symp. Hamburg*, pp. 211–27. Seattle: NOAA, Pac. Mar. Environ. Lab.

Nagashima, H. 1977. Deformation of nonlinear shallow water waves. *Sci. Pap. Inst. Phys. Chem. Res.* 71: 13–14

Novikov, S. P., Manakov, L. P., Pitaevskii, L. P., Zakharov, V. E. 1984. *Theory of Solitons*. New York: Plenum. 288 pp.

Ostrovskiy, L. A., Pelinovskiy, E. N. 1970. Wave transformation on the surface of a fluid of variable depth. *Izv. Acad. Sci. USSR Atmos. Oceanic Phys.* 6: 552–55

Ostrovskiy, L. A., Pelinovskiy, E. N. 1975. Refraction of nonlinear ocean waves in a beach zone. *Izv. Acad. Sci. USSR Atmos. Oceanic Phys.* 11: 36–41

Ostrovskiy, L. A., Pelinovskiy, E. N. 1976.

Nonlinear evolution of tsunami waves. *Proc. Tsunami Res. Symp., 1974*, pp. 203–11. Paris: UNESCO

Pelinovskii, E. N. 1982. *The Nonlinear Dynamics of Tsunami Waves*. Gorkii: Akad. Nauk SSSR Inst. Prikl. Fiz. 226 pp.

Peregrine, D. H. 1967. Long waves on a beach. *J. Fluid Mech.* 27: 815–27

Peregrine, D. H. 1983. Breaking waves on beaches. *Ann. Rev. Fluid Mech.* 15: 149–78

Podyapolsky, G. S. 1970. Generation of the tsunami wave by the earthquake. In *Tsunamis in the Pacific Ocean*, ed. W. M. Adams, pp. 19–32. Honolulu: East-West Cent.

Pod"yapol'skii, G. S. 1978. Excitation of tsunamis by an earthquake. In *Methods of Calculation of Tsunami Origin and Propagation*, pp. 30–87. Moscow: Nauka. 143 pp.

Shen, M. C., Keller, J. B. 1973. Ray method for nonlinear wave propagation in a rotating fluid of variable depth. *Phys. Fluids* 16: 1565–72

Shuto, N. 1974. Nonlinear long waves in a channel of variable section. *Coastal Eng. Jpn.* 17: 1–12

Solov'ev, S. L., Go, Ch. N. 1974. *A Catalog of Tsunamis on the East Coast of the Pacific*. Moscow: Nauka. 204 pp.

Spielvogel, L. Q. 1976. Single wave run-up on a sloping beach. *J. Fluid Mech.* 74: 685–94

Stoker, J. J. 1957. *Water Waves*. New York: Interscience. 567 pp.

Svendsen, I. A., Hansen, J. B. 1978. On the deformation of periodic long waves over a gently sloping bottom. *J. Fluid Mech.* 87: 433–48

Van Dorn, W. G. 1961. Some characteristics of surface gravity waves in the sea produced by nuclear explosions. *J. Geophys. Res.* 66(11): 3845–62

Voit, S. S. 1978. Tsunami waves. In *Physics of the Ocean*, Pt. 2. *Hydrodynamics of the Ocean*, pp. 229–54. Moscow: Nauka

Voit, S. S., Sebekin, B. I. 1968. Some hydrodynamical models of unsteady waves of tsunami type. *Morsk. Gidrofiz. Issled.* 1: 137–45

Voit, S. S., Lebedev, A. N., Sebekin, B. I. 1982. Tsunamis with directivity and its generation by horizontal shift. In *Processes of Tsunami Generation and Propagation*, pp. 18–23. Moscow: Acad. Sci. USSR Inst. Oceanol.

Voit, S. S., Lebedev, A. N., Sebekin, B. I. 1985. The energy of forced waves in rotating stratified media. *Morsk. Gidrofiz. Zh.* 1: 26–33

Whitham, G. B. 1974. *Linear and Nonlinear Waves*. New York: Wiley. xvii + 636 pp.

Wu, T. Y. 1979. On tsunami propagation—evaluation of existing models. In *Tsunamis, Proc. Natl. Sci. Found. Workshop*, ed. L. S. Hwang, Y. K. Lee, pp. 110–43. Pasadena, Calif: TetraTech

Ann. Rev. Fluid Mech. 1987. 19 : 237–70

TURBULENT PREMIXED FLAMES

S. B. Pope

Sibley School of Mechanical and Aerospace Engineering,
Cornell University, Ithaca, New York 14853

1. INTRODUCTION

Turbulent premixed flames exhibit phenomena not found in other turbulent flows. In some circumstances a thin flame sheet (thinner than the Kolmogorov scale) forms a connected but highly wrinkled surface that separates the reactants from the products. This flame surface is convected, bent, and strained by the turbulence and propagates (relative to the fluid) at a speed that can depend on the local conditions (surface curvature, strain rate, etc.). Typically, the specific volume of the products is seven times that of the reactants, the flame surface being a volume source. Because of this volume source there is a pressure field associated with the flame surface that affects the velocity field and hence indirectly affects the evolution of the surface itself. For the simplest case of a plane laminar flame, this feedback mechanism tends to make the flame unstable.

As well as looking at the detailed structure of a turbulent premixed flame, we can examine mean quantities. Here too, in comparison to other turbulent flows, there are some unusual observations, the most striking being countergradient diffusion. Within the flame there is a mean flux of reactants due to the fluctuating component of the velocity field. Contrary to normal expectations and observations in other flows, it is found that this flux transports reactants up the mean-reactants gradient, away from the products (hence countergradient diffusion). A second notable phenomenon is the large production of turbulent energy within the flame: Behind the flame the velocity variance can be 20 times its upstream value (Moss 1980). Both these phenomena result from the large density difference

0066–4189/87/0115–0237$02.00

between reactants and products and from the pressure field due to volume expansion.

There is a wide variety of theories and models for premixed turbulent flames. Some take as their prime objective the determination (or correlation) of the turbulent-flame speed u_T as a function of the relevant parameters (Abdel-Gayed & Bradley 1981, Tabaczynski et al. 1980, Andrews et al. 1975). More ambitious are the probabilistic field theories that attempt to calculate statistical properties of the flame as functions of position and time. As an example, in the pdf approach the statistic calculated is the one-point joint probability density function (pdf) of the velocities and compositions.

Here we review, first, our knowledge of the structure of turbulent premixed flames and the fundamental processes involved. Second, we review the application of probabilistic field theories (primarily the pdf approach) to turbulent premixed flames to reveal their achievements, shortcomings, and issues yet to be resolved.

By definition, in a premixed flame the gaseous fuel and oxidant are homogeneously mixed prior to combustion. In applications, because of the explosion hazard, premixing is generally avoided. Nevertheless, there are several important applications of turbulent premixed combustion; the principal one is the (homogeneously charged) spark-ignition engine. Other examples are reheat systems in jet engines, industrial tunnel burners, and gaseous explosions in a turbulent atmosphere (Bray 1980).

There have been several experiments in which the flame within a spark-ignition engine has been studied [see Tabaczynski (1976), Keck (1982), and Abraham et al. (1985) for references]. But most quantitative information about premixed flames comes from experiments in a variety of simpler configurations. Most closely related to engine flames are statistically spherical flames, ignited by a spark and propagating outward into turbulent reactants (Mickelsen & Ernstein 1956, Bolz & Burlage 1960, Palm-Leis & Strehlow 1969, Hainsworth 1985). A variant is the double-kernel technique, in which two spark-ignited flame balls propagate into turbulent reactants and eventually collide (see, e.g., Abdel-Gayed et al. 1984, Groff 1986). Like flames in spark-ignition engines, both the single- and double-kernel flames are not statistically stationary and do not depend upon stabilization.

Jet flames have been extensively studied since the early work of Damköhler (1940). The reactants flow up a cylindrical burner tube, sometimes through a turbulence-generating grid. The approximately conical flame is then stabilized on the rim of the burner. (An annular hydrogen flame may also be used for stabilization.) The early work on these flames has been reviewed by Stambuleanu (1976). Recent investigations have been performed by Moss (1980), Yoshida & Tsuji (1982), Shepherd & Moss (1982), Gunther (1983), Suzuki & Hirano (1984), and Cheng & Shepherd (1986), among others.

Other configurations to which we give less consideration in what follows are ducted flames stabilized on a cylinder, say, held perpendicular to a high-speed stream (see, e.g., Wright & Zukoski 1962), unconfined flames stabilized in low-speed streams (see, e.g., Dandekar & Gouldin 1982, Cheng 1984, Gulati & Driscoll 1986a,b), and flames stabilized in stagnation flow (Cho et al. 1986).

Recent reviews on turbulent combustion in general have been provided by Williams (1985a), by Libby & Williams (1976, 1980, 1981), and by Jones & Whitelaw (1982, 1984). For turbulent premixed flames, Bray (1980) provides an excellent review and exposition of the fundamental issues. Useful material can also be found in the books of Kuo (1986), Williams (1985b), Strehlow (1968, 1985), Stambuleanu (1976), and Lewis & von Elbe (1961).

In the next section the fundamentals of turbulent premixed combustion are outlined. This includes a brief consideration of the governing equations, laminar premixed flames, and of the characterization of turbulence. Section 3 starts with a consideration of the important dimensionless parameters that are used to identify different regimes of turbulent premixed combustion. Then, for the important regimes, we examine the detailed structure and fundamental propagation processes and how probabilistic field theories succeed or fail in representing them. The effect of combustion on the turbulence is examined in Section 4. The primary effect is through the pressure field induced by the volume expansion. In the discussion, the major areas of uncertainty are identified and possible research approaches indicated.

2. FUNDAMENTALS

2.1 *Governing Equations*

A typical premixed flame may contain scores of species (mostly intermediates) that take part in hundreds of elementary reactions. Taking full account of this complexity, Warnatz has made useful and impressive computations of simple laminar flames (e.g. Warnatz 1984). But needless to say, sweeping simplifying assumptions are essential to progress in the analysis of laminar flames or in any approach to turbulent flames. The assumptions usually made fall into four categories:

1. General assumptions
2. Reaction scheme

3. Transport properties
4. Detailed assumptions.

The general assumptions, common to virtually all approaches, are single-phase (gaseous) flow, low Mach number, and negligible radiative heat transfer. The low-Mach-number assumption is particularly important: It implies that spatial differences in pressure Δp are much smaller than the absolute value of the pressure p_0. Hence the pressure enters the thermochemistry only through p_0, whereas only pressure differences affect the velocity field.

In most theories of laminar or turbulent premixed flames, the complex chemical reactions are modeled by a one-step overall reaction (for exceptions, see Williams 1985b). Such a sweeping assumption clearly has a limited range of validity. Clavin (1985) suggests that the assumption is generally satisfactory, but it is inadequate to describe chemical-kinetic extinction, pollutant formation, and sensitization or inhibition of the reaction by additives. In addition, a one-step mechanism is inadequate to describe ignition.

Equally sweeping assumptions are made concerning the molecular transport processes. In many laminar-flame studies, the diffusion coefficient of each species is assumed to be the same, Soret and Dufour effects are neglected, and the Lewis number is assumed constant. (The Lewis number Le is the ratio of the thermal to mass diffusivities.)

Finally, detailed assumptions are needed concerning the specific form of the reaction rate and the dependence of the density and diffusivity on temperature and species concentrations.

With all these assumptions, the thermochemistry of a premixed flame can be described by just two transport equations. The two dependent variables can be chosen to be a reaction progress variable $c(\mathbf{x}, t)$ and the enthalpy $h(\mathbf{x}, t)$. By definition, c is zero in the reactants and unity in the products. If the reactants were to burn homogeneously, then the enthalpy would remain constant: In particular, the enthalpy of the products is the same as that of the reactants. In a flame, the only mechanism by which the enthalpy changes is the differential diffusion of heat and mass. Consequently, in laminar-flame studies the Lewis number is of prime importance (see, e.g., Clavin 1985, Buckmaster & Ludford 1982).

For turbulent flames, it has generally been assumed (Bray & Moss 1974, Pope & Anand 1984) that the Lewis number is unity. Then the enthalpy is constant and uniform, and the thermochemistry is described by the single variable $c(\mathbf{x}, t)$. The assumption of unit Lewis number is clearly useful in reducing the number of dependent variables, but it must be borne in mind that it excludes phenomena that may be important—in particular, it excludes the diffusive-thermal instability (see, e.g., Williams 1985b).

With the above assumptions, the transport equation for the reaction progress variable $c(\mathbf{x}, t)$ is

$$\rho \frac{Dc}{Dt} = \rho\left(\frac{\partial}{\partial t} + \mathbf{U}\cdot\nabla\right)c = \nabla\cdot(\rho D \nabla c) + \rho S, \tag{1}$$

where $\mathbf{U}(\mathbf{x}, t)$ is the fluid velocity. The density ρ, the diffusivity D, and the reaction rate S are all given functions of c. Typically ρ may be given by

$$1/\rho(c) = 1/\rho_r + c(1/\rho_p - 1/\rho_r), \tag{2}$$

where ρ_r and ρ_p are the densities in the reactants and products, respectively. The ratio $R \equiv \rho_r/\rho_p$ is typically in the range 5–10. A typical reaction rate is (Pope & Anand 1984)

$$S(c) = S^*(c)/\tau_R, \tag{3}$$

where τ_R is the reaction time scale and

$$S^*(c) = 6.11 \times 10^7\ c(1-c)\exp\{-30{,}000/(300+1800c)\}. \tag{4}$$

This expression, which is plotted in Figure 1, corresponds to an activation temperature of 30,000 K and reactant and product temperatures of 300 K and 2100 K, respectively. The numerical constant is chosen so that the maximum of $S^*(c)$ is unity.

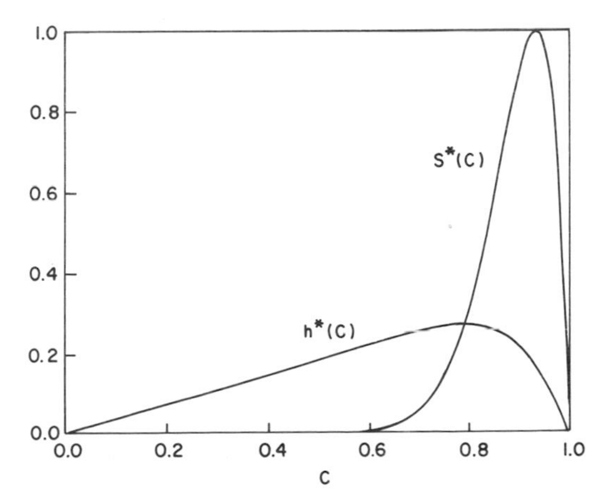

Figure 1 Normalized reaction rate $S^*(C)$ [Equation (4)] and laminar-flame function $h^*(C)$ [Equation (27)].

2.2 *Laminar Premixed Flames*

In some instances of premixed turbulent combustion, the flame surface behaves locally like a laminar flame. Studies of laminar flames have a bearing, therefore, on the turbulent case, especially those studies concerned with the effects of straining and curvature on the flame.

The simplest case is that of a plane laminar flame propagating steadily into quiescent reactants. The propagation speed (measured relative to the reactants) is the laminar-flame speed u_L, which is uniquely determined by the thermochemical state of the reactants. There are many experimental methods for determining u_L (Rallis & Garforth 1980), and there are abundant data in the literature (e.g. Metghalchi & Keck 1982). With appropriate boundary conditions, Equations (1–4) can be solved for a plane laminar flame, and u_L emerges as an eigenvalue. But, neglecting the variation of ρ and D with c, dimensional analysis suffices to yield the well-known result that u_L scales as $(D/\tau_R)^{1/2}$. Alternatively, this relation can be inverted to define a laminar-flame time scale

$$\tau_L \equiv D_r/u_L^2, \tag{5}$$

where D_r is the thermal diffusivity of the reactants. (Recall that with the unity-Lewis-number assumption, the thermal and mass diffusivities are equal.) Equation (5) is useful because, for a given fuel/oxidant mixture, D_r and u_L are usually known, whereas τ_R (even if well defined) is not generally known.

Figure 2 (adapted from Abraham et al. 1985) shows the temperature and heat-release profiles through a stoichiometric propane-air flame at

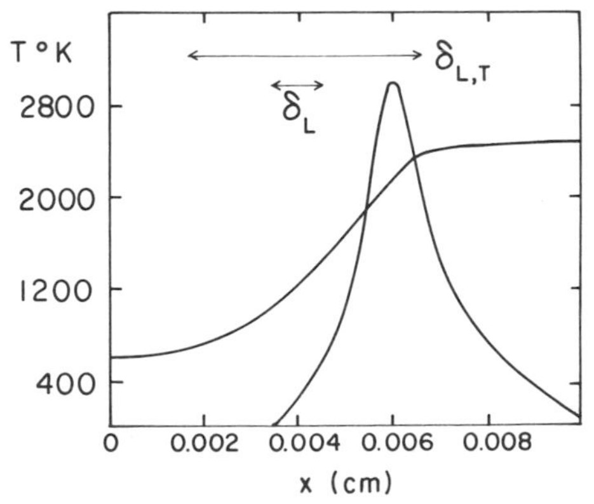

Figure 2 Temperature and heat-release (arbitrary scale) profiles for a propane-air laminar premixed flame, showing the thicknesses δ_L and $\delta_{L,T}$. Equivalence ratio = 1.0, reactant temperature = 600 K, pressure = 5 atm (after Abraham et al. 1985).

atmospheric pressure and an initial temperature of 600 K. In the preheat zone ($x < 0.004$ cm, say) the reaction rate is negligible, and so convection and diffusion are in balance. In the reaction zone (where heat release is significant), reaction and diffusion are the dominant processes.

The laminar-flame thickness can be defined in many ways. Abraham et al. (1985) compared seven definitions, of which we consider two. A natural definition is $\delta_{L,T}$—the distance between the positions of 5% and 95% temperature rise. For the flame considered, $\delta_{L,T}$ is 0.0046 cm. But the determination of $\delta_{L,T}$ requires that the temperature profile be known. From Equations (1–4), again on dimensional grounds, it follows that the flame thickness (however defined) scales with $(D\tau_R)^{1/2}$ or, equivalently, with $(D_r\tau_L)^{1/2} = D_r/u_L$. Hence, we use the definition

$$\delta_L \equiv D_r/u_L. \tag{6}$$

For the flame considered, δ_L is 0.0011 cm or $\delta_L \approx \frac{1}{4}\delta_{L,T}$.

Above all it should be realized that premixed laminar flames are very thin: In the flame considered, $\delta_{L,T}$ is about 1/20 mm. Away from stoichiometric conditions, or with lower initial temperatures, the thickness increases. But, on the other hand, the thickness is inversely proportional to the pressure and hence can be yet smaller in spark-ignition engines.

Compared with the rudimentary description given here, Peters (1986) provides a more detailed account of the internal structure of premixed laminar flames.

2.3 *Characterization of Turbulence*

It is natural to suppose that a premixed flame is strongly influenced by the turbulence into which it is propagating. Hence, we need to characterize the turbulence field in the reactants ahead of the flame. This we do below, and we use the results subsequently. But it should be borne in mind that there is a two-way interaction between the flame and the turbulence; and the turbulence within the flame may be substantially different from that ahead of it. Indeed it may be possible for a turbulent flame to propagate into nonturbulent reactants (Wright & Zukoski 1962, Sivashinsky 1979).

In the reactants ahead of the flame, the density ρ_r and kinematic viscosity ν are uniform. At any location, the principal characteristics of the turbulence are the turbulence intensity u' and the dissipation rate ε. Let $\mathbf{u}(\mathbf{x}, t)$ be the fluctuating component of velocity, and let angled brackets denote means. Then we have

$$u' \equiv (\langle u_i u_i \rangle /3)^{1/2} \tag{7}$$

and

$$\varepsilon \equiv \nu\left\langle \frac{\partial u_i}{\partial x_j}\frac{\partial u_i}{\partial x_j}\right\rangle. \tag{8}$$

In terms of these quantities we can further define length and time macroscales

$$l \equiv u'^3/\varepsilon \quad \text{and} \quad \tau \equiv u'^2/\varepsilon = l/u', \tag{9}$$

the Taylor (length) microscale

$$\lambda \equiv (15\nu u'^2/\varepsilon)^{1/2}, \tag{10}$$

and the Kolmogorov length and time microscales

$$\eta \equiv (\nu^3/\varepsilon)^{1/4} \quad \text{and} \quad \tau_k \equiv (\nu/\varepsilon)^{1/2}. \tag{11}$$

While the approximate significance of these scales is well known, some care is needed in providing more precise interpretations of them. In moderate-Reynolds-number grid turbulence, the longitudinal integral scale L is simply proportional to l. [$L = 1.2l$ can be deduced from the data of Comte-Bellot & Corrsin (1971).] Since the energy-containing scales are not universal, the constant of proportionality depends on the way the turbulence is produced. The Taylor scale has no clear physical significance (Tennekes & Lumley 1972), although it has been ascribed a significance in some theories of turbulence (Tennekes 1968) and turbulent combustion (Chomiak 1976, Tabaczynski et al. 1980).

According to the Kolmogorov hypotheses (see Monin & Yaglom 1975) the smallest turbulent motions are of size of order η. In order to be more precise, we need to define precisely a length scale l_s that characterizes the size of the smallest motions. A reasonable definition is that l_s is the wavelength corresponding to the centroid of the dissipation spectrum. Then, using a standard model of the energy spectrum in high-Reynolds-number turbulence (Tennekes & Lumley 1972, Equation 8.4.6), we obtain

$$l_s \approx 13\eta. \tag{12}$$

Thus, accepting l_s as a measure of the smallest scales, we see that η underestimates by an order of magnitude the size of the smallest motions.

The inverse of the Kolmogorov time scale τ_k is the root-mean-square (rms) velocity gradient: From Equations (8) and (11) we have

$$\tau_k^{-1} = \left\langle \frac{\partial u_i}{\partial x_j}\frac{\partial u_i}{\partial x_j}\right\rangle^{1/2}. \tag{13}$$

Both the symmetric part of $\partial u_i/\partial x_j$ (the rate of strain) and the antisymmetric part (the vorticity) contribute to the right-hand side of Equation (13). While there are velocity gradients on all scales, the dominant contributions are from the smallest scales. Below, we use τ_k^{-1} as a measure of the rate of strain. But it should be borne in mind that this is the rms: At high Reynolds numbers, in view of internal intermittency, much higher strain rates can occur.

While the Taylor microscale has no clear physical significance, the inverse time scale u'/λ is a measure of the rms strain rate. Indeed, from Equations (10–11) we have

$$u'/\lambda = \tau_k^{-1}/(15)^{1/2}. \tag{14}$$

3. TURBULENT PREMIXED-FLAME STRUCTURE

3.1 *Regimes of Combustion*

Different conditions can give rise to qualitatively different regimes of combustion in which different physical processes occur. Some understanding of these regimes is provided by a consideration of the most important dimensionless groups.

Two dimensional quantities are needed to give a basic description of the thermochemistry. We choose the laminar-flame time scale τ_L and the diffusivity D_r. Since the Prandtl number $Pr \equiv \nu/D_r$ is generally close to unity and has little effect on the combustion, we can replace D_r with ν. For the turbulence in the reactants, a basic characterization is provided by u', l, and ν. From the four dimensional quantities τ_L, ν, u', and l, two dimensionless groups can be formed, though their choice is not unique. Bray (1980) chose u'/u_L and the Reynolds number $R_l \equiv u'l/\nu$; Williams (1985b) chose u'/u_L and η/δ_L. (Note that u_L, δ_L, and η can be expressed in terms of the four dimensional quantities.) Here we follow McNutt (1981) and Abraham et al. (1985) in choosing the Reynolds number R_l and the Damköhler number

$$Da \equiv \tau/\tau_L. \tag{15}$$

This Damköhler number is the ratio of the (large-scale) turbulent time scale to the laminar-flame time scale.

Before examining the significance of different values of R_l and Da, we note that other quantities may also be important. Among these are the density ratio ρ_r/ρ_p, the Lewis number Le, and of course the geometry of the flame. In addition, a given flame may behave differently at different locations and times. Some of these additional effects are illustrated in Section 3.2.

Figure 3 shows the Reynolds-number/Damköhler-number plane. Given R_l and Da, any other dimensionless group can be determined. The loci on the plane where various ratios are unity are shown on the figure: Each of these ratios increases with Da (at fixed R_l). The open symbols correspond to conditions in particular spark-ignition engine experiments, and the dashed rectangle encloses all engine operating conditions (Abraham et al. 1985). We restrict our attention to moderate and high Reynolds numbers.

In the flame-sheet regime, the smallest turbulent motions are larger than the laminar-flame thickness ($\eta > \delta_L$), and the time scale of turbulent straining is large compared with the laminar-flame time scale ($\tau_k > \tau_L$). This suggests that combustion can, indeed, occur in thin ($\sim\delta_L$) flame sheets. If, on the other hand, η were significantly less than δ_L, then turbulent motions within the reaction sheet could disrupt the convective-diffusive balance in the preheat zone. Or if τ_k were less than τ_L, the straining might extinguish the flame, as originally suggested by Karlovitz et al. (1953). Such considerations led Kovasznay (1956), Klimov (1963), and Williams (1976) to suggest $\tau_k/\tau_L = 1$ or equivalently $\eta/\delta_L = 1$ as the boundary of the flame-sheet regime. In fact, as Abraham et al. (1985) observed, $\eta/\delta_L > 1$ is certainly a sufficient condition for flame-sheet combustion, but it may not be necessary: This is discussed further in Section 3.2.

Williams (1985b) and Abraham et al. (1985) refer to this as the "reaction-sheet" regime, but the term "flame-sheet" seems preferable, since the preheat zone as well as the reaction zone is contained in the sheet. These

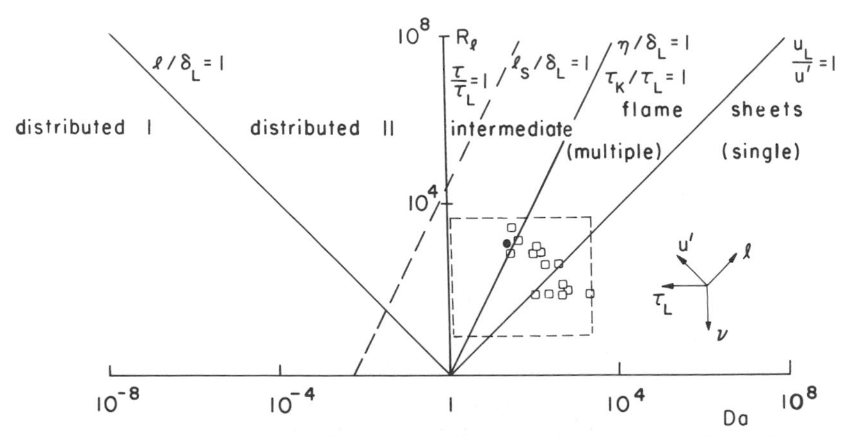

Figure 3 Reynolds-number/Damköhler-number plane, showing regimes of turbulent premixed combustion. Arrows indicate effect of changing one variable while holding the other three fixed. Hainsworth's (1985) experiment ●; engine experiments □ (from Abraham et al. 1985; rectangle encloses engine operating conditions).

authors also suggest a subdivision of the regime into single sheets ($u_L/u' > 1$) and multiple sheets ($u_L/u' < 1$).

At the other end of the Damköhler-number range there is distributed combustion. The criterion $l/\delta_L < 1$ proposed by Damköhler (1940) defines a region of distributed combustion, denoted "distributed I" on Figure 3. (This region is somewhat unnatural, in that for a turbulent flame to exist, the flow field must be much larger than the integral length scale and must endure for many time scales.) In this regime the structure of the flame is similar to that of the laminar flame, but with a turbulent viscosity ν_T (of order $u'l$) replacing ν. Hence the turbulent-flame speed u_T is of order $(\nu_T/\tau_L)^{1/2} \sim u'\mathrm{Da}^{1/2} \ll u'$; and the turbulent-flame thickness δ_T is of order $(\nu_T\tau_L)^{1/2} \sim l\mathrm{Da}^{-1/2} \gg l$. Note that the time scale δ_T/u_T is simply τ_L, which is greater than τ by a factor of $\mathrm{Da}^{-1} \gg 1$. As a consequence, fluctuations in thermochemical quantities (e.g. c) are very small: Their dissipation rate ($\sim\tau^{-1}$) is much greater than their production rate ($u'l/\delta_T^2$).

The above physical arguments used to justify the existence of distributed combustion rely only on the criteria $u_T \ll u'$ and $\delta_T \gg l$. These criteria are satisfied provided that $\mathrm{Da} \ll 1$. Hence, the region denoted "distributed II" on Figure 3 also corresponds to distributed combustion, and there is no transition across the line $l/\delta_L = 1$. McNutt (1981) made calculations, based on a modeled transport equation for the pdf of c, that support this extended region of distributed combustion: For all Damköhler numbers less than 0.1 the calculated turbulent-flame speed and thickness agree with Damköhler's theory, and the fluctuations in c are less than 1%.

About the remaining intermediate regime (defined by $0.1 < \mathrm{Da} < \mathrm{R}_l^{1/2}$), little is known with certainty. And as Williams (1985b) observes, there may be more than one regime of combustion within the region. One possibility is that (for $1 \ll \mathrm{Da} \ll \mathrm{R}_l^{1/2}$) there is a region of distributed preheating but localized reaction. This possibility is discussed further in Section 3.4.

3.2 *Flame-Sheet Regime*

There is little doubt that spark-ignition engines and most laboratory experiments operate in the flame-sheet regime. In order to study the fundamental processes in this regime and the types of theory that are applicable, we examine one flame in some detail.

Hainsworth (1985) performed an experiment on a statistically spherical methane-air flame propagating into (nominally) homogeneous isotropic turbulence. The experimental conditions are given in Table 1, and the corresponding (R_l, Da) point is plotted on Figure 3.

It is observed (by Schlieren photography) that the initial flame kernel is a smooth sphere of radius 1.5 mm. The subsequent evolution of the flame-

Table 1 Conditions at the time of ignition in Hainsworth's (1985) experiment[a]

$u' = 1.93\ \mathrm{m\ s^{-1}}$	$\tau = 4.23$ ms
$u_L = 0.29\ \mathrm{m\ s^{-1}}$	$\tau_k = 0.13$ ms
$l = 8.16$ mm	$\tau_L = 0.18$ ms
$\eta = 0.044$ mm	$R = 5.3$
$\delta_L = 0.052$ mm	$R_l = 1070$
	Da $= 24$

[a] Methane-air mixture, equivalence ratio 0.8, atmospheric temperature and pressure.

ball radius $R_f(t)$ is shown in Figure 4. For the first millisecond (about 7 Kolmogorov time scales) the rate of change of radius $\dot{R}_f$ is just the same as for a laminar flame ball, strongly suggesting that the flame sheet propagates at the laminar-flame speed and remains approximately spherical. While the flame ball as a whole is convected by the large-scale velocity fluctuations (of order $u' \approx 7u_L$), the structure of the flame is affected only by the turbulent motions of size $R_f(t)$ or less. It is these less energetic, smaller-scale motions that initially wrinkle the flame.

It is possible that the initial wrinkling of the surface is not due to turbulence, but instead is due to the thermal-diffusive instability

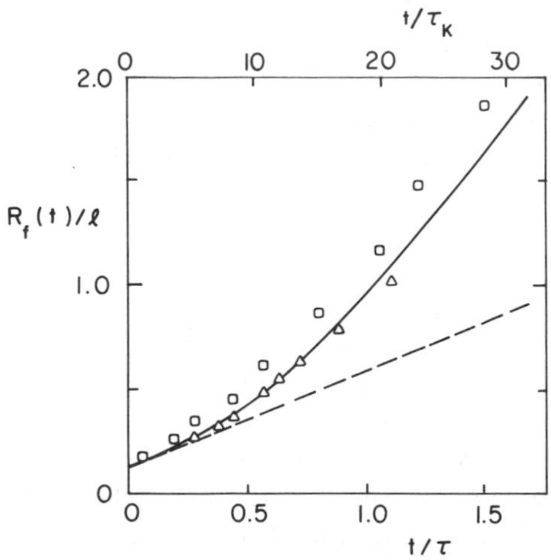

Figure 4 Flame-ball radius $R_f(t)$ versus time t. Solid line, pdf calculation (Pope & Cheng 1986); dashed line, laminar-flame-ball radius; symbols, two experimental realizations (Hainsworth 1985). Normalizing variables (l, τ, τ_k) are evaluated at $t = 0$.

(Strehlow 1968, Sivashinsky 1983). However, for most fuel-air mixtures, thermal-diffusive effects ($Le \neq 1$) are likely to be stabilizing (Abraham et al. 1985).

For some time the surface, although it is convected, bent, and strained by the turbulence, remains regular and singly connected. (This observation illustrates the limitations of representing regimes of combustion in terms of R_l and Da alone.) Several theoretical approaches have been based on the representation of the flame as a propagating surface. Most of these are *global* in that they attempt to describe the evolution of the surface as a whole (e.g. Matalon & Matkowsky 1982, Clavin & Williams 1979, 1982, Clavin 1985). Pope (1986) has developed a *local* description of propagating surfaces in which the evolution of surface-element properties is studied. These properties are the principal curvatures, principal directions, and fractional area increase of the surface element. In global theories, with a complete description of the surface, the effect of surface shape, etc., on the propagation speed w can be accounted for more accurately. But the simpler local description leads to a tractable probabilistic approach (Pope 1986). Neither approach has been used to make detailed calculations of turbulent flames. (The propagation speed w, which may vary over the surface, is defined relative to the reactants just ahead of the flame.)

In Hainsworth's (1985) experiment the reaction sheet propagates (initially) with the laminar-flame speed (i.e. $w = u_L$), even though the turbulent strain rates τ_k^{-1} are comparable to τ_L^{-1}. In other experiments (notably Fox & Weinberg 1962) it is found that the propagation speed is significantly different from u_L. Clearly, for theories based explicitly on reaction sheets it is essential to know the dependence of propagation speed on straining, curvature, and other relevant parameters.

In the reactants just ahead of a point on the flame, let γ be the rate of strain in the plane of the flame. Then $\gamma\tau_L$ is a nondimensional measure of the straining. Let H be the mean curvature of the surface, with H positive if the surface is concave to the reactants. Then $H\delta_L$ is a nondimensional measure of curvature. The fractional rate of area increase of the surface is called the flame stretch K (see Karlovitz et al. 1953, Matalon 1983, Chung & Law 1984). In terms of γ, H, and the propagation speed w, the flame stretch is (Pope 1986)

$$K = \gamma - 2Hw, \tag{16}$$

and a nondimensional measure of flame stretch is

$$\kappa \equiv K\tau_L. \tag{17}$$

The many analytical studies of laminar flames with strain and curvature have been reviewed by Clavin (1985) and Williams (1985b). A general conclusion of these studies is that when both strain ($\gamma\tau_L$) and curvature ($H\delta_L$) are small (compared with unity), then the propagation speed is

$$w/u_L = 1 + A\kappa + O(\kappa^2), \tag{18}$$

where A (which is of order unity) depends on the thermochemistry. Experiments on strained laminar flames (Mendes-Lopes & Daneshyar 1985, Wu & Law 1984, Law et al. 1986) support this linear, order-one dependence on strain rate.

In the flame-sheet regime, since τ_k^{-1} is smaller than τ_L^{-1}, straining is generally weak, and it certainly is so in the regime of single sheets: It is generally assumed that the curvature is small also. In particular, Klimov (1975) assumes that the mean curvature H is of order η^{-1}, which is small compared with δ_L^{-1} (in this regime). If strain and curvature are small, then so also is κ, and to a good approximation the propagation speed is simply

$$w = u_L(1 + A\kappa). \tag{19}$$

This expression emerges from global theories of flame surfaces and can readily be incorporated in local theories.

The support for the assumption that the curvature ($H\delta_L$) is small is not clear. Even for the simpler case of a material surface, it has not been proved or demonstrated that the curvature is no greater than of order η^{-1}. The most that can be stated with confidence is that the curvature of a material surface cannot increase more than exponentially with time (Pope 1986). The principal curvature k_1 of a propagating surface increases at a rate wk_1^2 as a result of propagation. Hence, a singularity ($k_1 = \infty$), such as a cusp, can form in finite time. Schlieren photographs of some turbulent flames clearly show cusplike regions (Fox & Weinberg 1962, Keck 1982).

Returning to the statistically spherical flame, as time progresses the wrinkling increases, cusps are possibly formed, and eventually the flame sheet collides with itself. If the flow were two dimensional, then as the flame sheet collided with itself, an island of reactants would form. In three dimensions, contrary to some authors' assertions, islands do not form once a collision has taken place. Rather, just after the first collision, the reactants, the products, and the surface remain connected, but they are each doubly connected. Disconnected regions of reactants can form subsequently, but, in the absence of extinction, the products remain connected.

Given the geometric complexity of multiply connected flame sheets, it would be difficult to construct a quantitative theory based on the explicit representation of the flame surface.

In Hainsworth's (1985) flame, the initial turbulent-flame speed u_T is equal to the laminar-flame speed u_L and subsequently increases. But at large times, does the turbulent-flame speed depend on u_L? The observable times in the experiment are too short to answer this question. In a different (double-kernel) experiment, at the highest Reynolds numbers measured ($R_l = 3000$–4000), Abdel-Gayed & Bradley (1981) observed $u_T \approx 2u'$ independent of u_L for u_L/u' as small as 1/35. Since in this regime combustion depends on the propagation of the flame sheet, can such a result (i.e. $u_T \approx 2u'$) be valid in the limit $(u_L/u') \to 0$? This important question was addressed by Klimov (1975), who obtained an expression of the form

$$u_T/u' \sim (u_L/u')^{0.3}, \tag{20}$$

which suggests that u_T/u' tends to zero in the limit.

Klimov's (1975) analysis pertains to notional turbulence of a single length and time scale. The above result stems from the claim that the time required for combustion to take place (τ_c) tends to infinity as u_L tends to zero. (We can define τ_c as the inverse of the mean reaction rate.) In fact, using the same physical arguments as Klimov but taking account of the different scales of turbulence, we now show that the appropriately normalized combustion time τ_c/τ is of order unity as u_L/u' tends to zero. First, we observe from the relations $u_L/u' = \mathrm{Da}/R_l$ and (in the flame-sheet regime) $R_l^{1/2} < \mathrm{Da} < R_l$ that the limit $(u_L/u') \to 0$ corresponds to $R_l \to \infty$.

For simplicity we consider a constant-density, statistically homogeneous flow. Initially ($t = 0$) the surface area of the flame sheet per unit volume, Σ_0, is of order l^{-1}. The effect of straining on the sheet is assumed to be the same as on a material surface—that is, the area increases exponentially with time on the Kolmogorov time scale (see Monin & Yaglom 1975). Hence the surface-to-volume ratio at time t is

$$\Sigma(t) = \Sigma_0 \exp(at/\tau_k), \tag{21}$$

where a is of order unity. Assuming that the propagation speed is $w = u_L$ and for the moment neglecting collisions of the surface, we find that the fraction of the volume burnt in time t is

$$B(t) = \int_0^t u_L \Sigma(t')\, dt' = \frac{u_L \Sigma_0 \tau_k}{a}\,[\exp(at/\tau_k) - 1]. \tag{22}$$

A characteristic combustion time τ_c' is obtained by solving the equation $B(\tau_c') = 1$:

$$\tau_c' = \frac{\tau_k}{a} \ln\left(1 + \frac{a}{\Sigma_0 \tau_k u_L}\right). \tag{23}$$

(This equation is the same as Klimov's Equation (3), with τ_k/a being his time scale.) When nondimensionalized, Equation (23) becomes

$$\tau_c'/\tau = \frac{R_l^{-1/2}}{a} \ln(1 + aR_l \mathrm{Da}^{-1/2}). \tag{24}$$

Given the inequality $R_l^{1/2} < \mathrm{Da} < R_l$ in the flame-sheet regime, it may be seen that according to Equation (24), τ_c'/τ tends to zero as R_l tends to infinity.

This analysis is an oversimplification because of the neglect of collisions, which is equivalent to the assumption that the surface fills the space uniformly. While the rate of area increase scales with τ_k^{-1}, the rate of dispersion of the surface throughout the volume scales with τ^{-1} [note that $\tau^{-1} \ll (\tau_c')^{-1} \ll \tau_k^{-1}$]. This, then, is the rate-controlling process, and τ—not τ_c'—is the appropriate estimate of the mean reaction time τ_c.

The observation that τ_c/τ is of order unity is far from a proof that u_T/u' is independent of u_L/u' as R_l tends to infinity. It does, however, invalidate Klimov's (1975) claim that u_T must depend on u_L for $u_L/u' \ll 1$. Experiments have not provided a clear answer to the question. Klimov cites Russian experiments in support of Equation (20), while other experiments (e.g. Abdel-Gayed & Bradley 1981) suggest $u_T/u' \approx 2$, independent of u_L/u'. The more recent data of Abdel-Gayed et al. (1984) show that u_T/u' is constant for moderate values of u'/u_L but that it decreases for large values. However, this decrease is associated with the experimental conditions approaching the intermediate regime (τ_L/τ_k approaching unity).

We now examine in more detail the Klimov-Williams criterion $\delta_L/\eta < 1$ (or equivalently $\tau_L < \tau_k$) for flame-sheet combustion. As the boundary $\delta_L/\eta = 1$ is approached from the flame-sheet regime, the flame stretch κ becomes of order unity and Equation (19) is no longer valid. At this boundary, or perhaps further into the intermediate regime, two qualitative changes could also occur. First, the thermochemical fields in the flame sheets could cease to be essentially one dimensional (i.e. no longer varying appreciably only normal to the sheet) because of velocity variations (on a scale δ_L or less) within the flame. Second, because of large strain rates $\gamma\tau_L \gtrsim 1$, the flame sheet could be extinguished locally.

In Section 2.3 it was suggested that rather than η, the length scale $l_s \approx 13\eta$ more precisely measures the size of the small-scale motions. Thus $\delta_L/l_s = 1$ may be a better criterion for the breakdown of essentially one-dimensional flame sheets. (The line $l_s/\delta_L = 1$ is shown on Figure 3.) Several factors cloud the picture, however: Because of strain, the flame-sheet thickness may be less than δ_L; since the kinematic viscosity in the flame may be 10 times that of the reactants, even l_s may underestimate the size of the small-scale motions; and because of the intermittent nature of the small scales, some motions may be much smaller than l_s. At moderate

Reynolds number, the third point may not be important: The first two suggest that the breakdown of flame sheets occurs to the left of the line $l_s/\delta_L = 1$ on Figure 3.

It has long been speculated (Karlovitz et al. 1953) that extinction occurs when a flame sheet is strained sufficiently rapidly ($\gamma\tau_L > 1$). The evidence has to be examined carefully. One flow that has been extensively analyzed is an infinite, plane, strained laminar flame between semi-infinite bodies of reactants and products, the products being at the adiabatic flame temperature. The analyses—recently reviewed by Williams (1985b)—indicate that the flame cannot be extinguished by straining except if the Lewis number is unusually large. But these analyses are almost all based on one-step kinetics that may be inadequate to study extinction. Numerical calculations by Warnatz & Peters (1984) incorporating detailed kinetics show that a rich hydrogen-air flame ($Le \approx 3$) can be extinguished, and the calculations of Rogg (reported by Peters 1986) based on a four-step scheme show that a stoichiometric methane-air flame can also be extinguished by straining.

A second relevant flow is a pair of infinite, plane laminar flames between two semi-infinite, counterflowing reactant streams. As the flow rate (and hence the strain rate) increases, the two flames move closer together. If the Lewis number is greater than unity, extinction can occur before the flames merge on the plane of symmetry: For $Le < 1$, extinction occurs as the flames merge. Both analyses and experiments support this picture [see Williams (1985b) for references].

These observations suggest that straining can cause extinction locally in a turbulent flame sheet. There is also direct experimental evidence (Abdel-Gayed et al. 1984) that turbulent straining ($\gamma\tau_L \approx 1$) can cause global extinction of the flame. As theory suggests, flames with large Lewis numbers are most susceptible to extinction (Abdel-Gayed & Bradley 1985).

3.3 *Calculations of Flame-Sheet Combustion*

Standard turbulence models—mean-flow or second-order closures—experience severe difficulties when applied to premixed flames (except in the least important case $Da \ll 1$). A major problem is that the mean reaction rate $\langle S(c) \rangle$ (recall Equations 1–4) cannot be approximated in terms of a few moments of c, because $S(c)$ is highly nonlinear (Figure 1).

Two approaches have proved more successful. The first is the Bray-Moss-Libby model, which is a second-order closure with special closure approximations appropriate to flame-sheet combustion. Libby (1985) provides a recent review of the model and calculations based on it. The second approach is the pdf method, in which a modeled transport equation is solved for the joint probability density function (pdf) of the velocities and the reaction progress variable. Pope (1985) provides a comprehensive review of the theory and modeling involved in pdf methods.

The structure of thin, multiply connected flame sheets presents a challenge to any probabilistic field theory. In this section we describe how this challenge is met in pdf methods, first in the limit $R_l \to \infty$ (Pope & Anand 1984). Then the application of the pdf method to Hainsworth's (1985) flame is described (Pope & Cheng 1986). (Consideration of the effect of combustion on the turbulence is postponed to Section 4.)

Pope & Anand (1984) considered the idealized case of a statistically one-dimensional and stationary constant-density flame in nondecaying homogeneous turbulence. The appropriate joint pdf is $f(\mathbf{V}, C; x)$—the joint probability density of $\mathbf{u}(\mathbf{x}, t) = \mathbf{V}$, $c(\mathbf{x}, t) = C$ at $x_1 = x$, where $\mathbf{u}$ is the velocity fluctuation. The derivation, modeling, and solution of pdf transport equations are fully described by Pope (1985). Here we consider just the modeling concerned with flame sheets.

Pope & Anand (1984) considered "flamelet combustion" defined by $1 \ll R_l^{1/2} \ll \mathrm{Da} \ll R_l$, which is essentially the case discussed at the end of the previous subsection (i.e. $R_l \to \infty$, $u_L/u' \to 0$). Since the flame sheets are thin ($\delta_L/\eta \ll 1$) and the straining is weak ($\tau_k^{-1}/\tau_L^{-1} \ll 1$), it is assumed that locally (i.e. on a scale δ_L) the flame-sheet structure is the same as that of a plane, unstrained laminar flame. [Implicitly, it is assumed that regions of high curvature ($H\delta_L \gtrsim 1$) and regions of flame-sheet collision account for a negligible fraction of the total sheet area.]

In the pdf method, the relevant term that has to be modeled is the conditional expectation of the right-hand side of Equation (1):

$$\hat{h}(C, \mathbf{x}, t) \equiv \langle \nabla \cdot (\rho D \nabla c) + \rho S | c(\mathbf{x}, t) = C \rangle. \tag{25}$$

As is now shown, the assumption made about the flamelet structure is sufficient to determine $\hat{h}$.

In a plane, unstrained laminar flame, the scalar quantity $[\nabla \cdot (\rho D \nabla c) + \rho S]$ is uniquely related to c (since c increases monotonically through the flame). That is, there is a function h (that can be determined from the laminar-flame solution) such that

$$[\nabla \cdot (\rho D \nabla c) + \rho S]_{c=C} = h(C). \tag{26}$$

The nondimensional function

$$h^*(C) \equiv \tau_L h(C), \tag{27}$$

obtained from the solution of Equations (1–4) (with constant ρ and D), is shown on Figure 1. With the assumption that the turbulent flamelets have the laminar structure, we obtain simply

$$\hat{h}(C, \mathbf{x}, t) = h(C), \tag{28}$$

a known quantity. The striking conclusion is that with the flamelet assumption, the reaction and diffusion terms in the pdf equation are closed without ad hoc or empirical modeling assumptions.

Unfortunately, this closure is flawed because, for somewhat subtle reasons, the flamelet assumption is too strong. An examination of the resulting modeled pdf equation shows that if a fluid element is initially specified to be pure reactants ($c = 0$), it will never burn, irrespective of the state of the surrounding fluid. This problem arises because the flamelet assumption breaks down in the far preheat zone. At a distance $\Delta \gg \delta_L$ from the flamelet (on the reactants side), according to the laminar-flame assumption, the reaction progress variable and its gradient are

$$c = \exp(-\Delta/\delta_L) \tag{29}$$

and

$$|\nabla c| = \exp(-\Delta/\delta_L)/\delta_L. \tag{30}$$

From Equation (29) we see that the specification $c = 0$ implies that the fluid element in question is infinity far from a flamelet ($\Delta = \infty$) and hence will not burn in finite time. In fact, the fate of any fluid element is predetermined by its initial condition through the ordinary differential equation

$$\frac{dc}{dt} = h(c) \tag{31}$$

(see Pope 1985). Strictly, c is greater than zero at all finite distances from the flame, and hence the initial condition $c = 0$ is incorrect: But a model that requires initial and boundary conditions to be specified to such precision is not useful.

In the regime considered, δ_L is much smaller than η. Consequently, from Equation (30) it may be seen that at a distance η from a flamelet the gradient is small compared with η^{-1}. At such distances it is no longer reasonable to assume, therefore, that the progress-variable field is uniquely determined by the laminar-flame structure independent of the turbulence. To account for the additional effect of turbulent mixing remote from flamelets, Pope & Anand (1984) added a standard mixing model to the modeled pdf equation. This causes pure reactants ($c = 0$) to be preheated to some extent ($c > 0$) at a rate proportional to τ^{-1}.

Subsequent reaction takes place rapidly. Equation (31) can be rewritten

$$\tau \frac{dc}{dt} = \text{Da}\, h^*(c), \tag{32}$$

where $h^*(c)$ (Figure 1) is of order unity. Once mixing has increased c just slightly ($c \gtrsim \text{Da}^{-1} \ll 1$), reaction takes place ($c \to 1$) in a short time of order $\tau\, \text{Da}^{-1}$. Consequently, there is only a small probability (of order Da^{-1}) of c adopting intermediate values ($\text{Da}^{-1} < c < 1 - \text{Da}^{-1}$). That is, to a good approximation, the pdf of c adopts a double-delta-function distribution, as assumed in the Bray-Moss-Libby model and as an inevitable consequence of the assumption that the flame sheets occupy a small fraction of the volume.

In summary, the virtues of the rigorous closure Equation (28) are eclipsed by the necessity to add a mixing model, which is rate controlling. The details of the function $h^*(c)$ are unimportant, since they only affect intermediate values of c that have negligible probability. Pope & Anand's (1984) result that the turbulent-flame speed scales with u' (specifically $u_T = 2.1u'$) is a direct consequence of the assumption that the mixing rate is proportional to τ^{-1} independent of u_L/u'.

Even though the rate-controlling combustion process is not modeled in a fundamental way, nevertheless the modeled joint pdf equation appears to yield solutions in accord with observations. For example, Pope & Cheng (1986) applied the method (with some refinements) to Hainsworth's (1985) statistically spherical flame. The flame radius R_f as a function of time is calculated quite accurately (see Figure 4). This is not an easy flow for a model to deal with: the initial flame radius (1.5 mm) is small compared with the turbulence scale ($l \approx 8$ mm), and initially the flame is convected a significant distance compared with its radius.

In Pope & Anand's analysis the laminar-flame speed is assumed to be small ($u_L/u' \ll 1$). Even though in Hainsworth's flame this ratio is quite small ($u_L/u' \approx 1/7$), nevertheless at early times laminar propagation is the dominant process. Pope & Cheng (1986) accounted for this in an ad hoc way. It may be possible to construct a better probabilistic model, valid for all u_L/u', based on local flame-surface properties (Pope 1986).

3.4 *Intermediate Regime*

For spark-ignition engines, it may be seen from Figure 3 that higher speeds (increased u') and leaner mixtures (increased τ_L) drive the combustion from the flame-sheet regime into the intermediate regime. Similarly, while a rod-stabilized V-flame may be in the flame-sheet regime, the higher velocities used in ducted stabilized flames may result in combustion in the intermediate regime.

There is no consensus on the nature of combustion in this regime, and indeed there may be several regimes. Here we discuss the idealized limiting case $1 \ll \mathrm{Da} \ll \mathrm{R}_l^{1/2}$.

The limit $1 \ll \mathrm{Da} \ll \mathrm{R}_l^{1/2}$ corresponds to high Reynolds numbers and conditions remote from both the distributed-reaction and flame-sheet regimes. We have the strong inequalities $u_{\mathrm{L}}/u' \ll 1$, $\delta_{\mathrm{L}}/\eta \gg 1$, and most importantly $\tau_\eta^{-1} \gg \tau_{\mathrm{L}}^{-1} \gg \tau^{-1}$. That τ_{L}^{-1} is much greater than τ^{-1} suggests that reaction is not the rate-limiting process: that τ_η^{-1} is much greater than τ_{L}^{-1} suggests that on the small scales, turbulent straining—not reaction—causes the steepest gradients of the reaction progress variable c.

In this regime it is highly likely that the mean time for combustion τ_{c} scales with τ. One reason is that τ is the longest relevant time scale; another is that plausible models of distributed combustion (McNutt 1981) and of flame-sheet combustion (Pope & Anand 1984) yield $\tau_{\mathrm{c}} \sim \tau$ as the intermediate regime is approached from each side.

It follows immediately from the scaling $\tau_{\mathrm{c}} \sim \tau$ that reaction cannot be distributed but must be localized in space or time—or both. We can define a space-time point in the flame as "reactive" or not depending on whether the reaction rate is greater than one tenth (say) of the maximum reaction rate. Then the fraction F of space-time that is "reactive" is

$$F = \mathrm{Prob}\,\{S(c) > 0.1/\tau_{\mathrm{R}}\}, \tag{33}$$

where $S(c)$ is the reaction rate with maximum value τ_{R}^{-1} (see Figure 1). The mean reaction rate τ_{c}^{-1} then scales as F/τ_{R}. (This assumes that the dominant contribution to the mean reaction comes from the "reactive" regions. An order-of-magnitude analysis of the pdf equation for c confirms the validity of this assumption.) The two scaling relations for τ_{c} combine to yield

$$F \sim 1/\mathrm{Da} \ll 1. \tag{34}$$

Reaction is not the rate-limiting process, since in reactive regions the reaction rate is larger than τ^{-1} by order Da. The rate-limiting process is the mixing process by which reactants ($c = 0$) are preheated to such an extent ($c \geq c_{\mathrm{r}}$) that reaction becomes rapid [$S(c_{\mathrm{r}}) = 10/\tau$, say].

The structure of combustion in this regime and the precise nature of the rate-limiting mixing process are unknown. It is likely that a satisfactory description must take account of the intermittent nature of the smaller turbulent scales (Klimov 1975). Possibly an important process is the turbulent mixing that occurs when a region of high dissipation rate intersects the boundary between regions of reactants and products. Since the local mixing rate (τ_η^{-1}) is large compared with the reaction rate (τ_{R}^{-1}), well-mixed reactive regions (of unknown shape and size) can form and subsequently burn.

The idea that turbulent mixing is the rate-controlling process is central to the eddy breakup model of Spalding (1971). It is also an inevitable consequence of Pope & Anand's (1984) modeling of the joint pdf equation for this regime (which they, perhaps inappropriately, referred to as the distributed-combustion regime). In the pdf method, as in the flame-sheet regime, the relevant term to be modeled is the conditional expectation $\langle \nabla \cdot (\rho D \nabla c) + \rho S | c(\mathbf{x}, t) = C \rangle$. The part involving the reaction rate is in closed form [i.e. $\rho(C)S(C)$], while Pope & Anand argue that a standard mixing model that ignores the presence of reaction is appropriate to the first term, since (locally) mixing is rapid compared with reaction ($\tau_\eta^{-1} \gg \tau_R^{-1}$).

Pope & Anand's (1984) pdf calculations in this regime are for an idealized, one-dimensional, constant-density flame. Figure 5 shows $f_c(C)$, the calculated pdf of c at the location where $\langle c \rangle = 1/2$, for Da $= 10^4$. There are spikes (with probabilities 0.10 and 0.42) at zero and unity corresponding to pure reactants and products, respectively. Where the reaction rate $S(c)$ is large compared with τ^{-1} ($c_r = 0.55 < c < 0.99$) there is negligible probability. (In fact, here the pdf of c is of order $[\mathrm{Da}\ S^*(C)]^{-1}$.) But where the reaction rate is relatively small ($c < 0.45$) there is significant probability of partially preheated reactants. Thus, unlike in the flame-sheet regime, the pdf of the reaction progress variable is not a double-delta function.

The value of c at which reaction becomes rapid, c_r, decreases weakly with Damköhler number. Consequently, as Da increases, the amount of preheating needed before reaction takes place decreases. For this reason, although turbulent mixing is the rate-controlling process, Pope & Anand (1984) found a weak dependence of the turbulent-flame speed on Da. Specifically, over the range studied ($1 \leq \mathrm{Da} \leq 10^4$) they obtained

$$u_T/u' = 0.25 + 1.25 \log_{10} \mathrm{Da}. \tag{35}$$

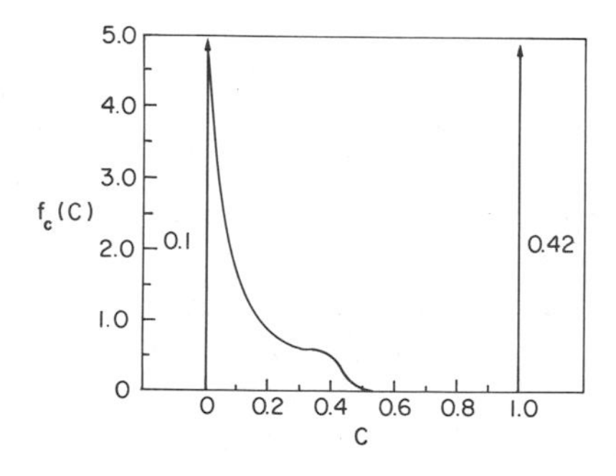

Figure 5 Progress variable pdf in the center of the flame ($\langle c \rangle = 1/2$). From joint pdf calculations (Pope & Anand 1984) in the intermediate regime for Da $= 10^4$.

The conclusions that there is significant probability of partially preheated reactants and that u_T/u' depends (albeit weakly) on Da rest on the details of the mixing model. They are, therefore, subject to confirmation.

4. EFFECTS OF COMBUSTION ON TURBULENCE

Combustion affects the turbulent velocity field through the large increases in specific volume and viscosity resulting (mainly) from the large temperature rise. Typically the ratios ρ_r/ρ_p and ν_p/ν_r are 7 and 10, respectively.

The main effects of the increase in viscosity with temperature are on the small scales of turbulence. At the moderate (cold) Reynolds numbers often encountered in turbulent premixed flames, these effects could be significant. With few exceptions (e.g. Wu et al. 1985) these effects have not been studied.

To an extent, the effects of density variations have been successfully accounted for both in the Bray-Moss-Libby model and in the pdf approach. As is described in the following subsection, the models yield countergradient diffusion and large turbulence-energy production in accord with experimental observations. But the neglect of the fluctuating pressure field (associated with density variations) is a weakness in the modeling. We return to this point in Section 4.2

4.1 *Countergradient Diffusion and Energy Production*

The original Bray-Moss model (Bray & Moss 1974, 1977) assumed gradient diffusion. But an improved version was developed by Libby & Bray (1981) and applied by Bray et al. (1981). The improved version, reviewed by Libby (1985), is a second-order closure that avoids gradient-diffusion assumptions, not only for the second moments, but for the third moments as well. An extension of the model from one-dimensional flames to the general case is presented by Bray et al. (1985).

The Bray-Moss-Libby model calculations of Bray et al. (1981) and Libby (1985) and the pdf calculations of Anand & Pope (1986) pertain to a statistically stationary and one-dimensional flame in the flame-sheet regime with $u_L/u' \ll 1$. The reactants and products have densities ρ_r and ρ_p, and their ratio $R \equiv \rho_r/\rho_p$ is the dominant parameter in the problem. In this regime the pdf of the reaction progress variable adopts a double-delta-function distribution—there is negligible probability of partial reactedness.

From the Euler equations

$$\frac{D\mathbf{U}}{Dt} = -\frac{1}{\rho}\nabla p, \tag{36}$$

it is readily seen that a given pressure gradient accelerates the light products more than the heavier reactants. This mechanism is responsible both for countergradient diffusion and for turbulent energy production. For a model to represent these processes accurately it must, therefore, take proper account of the effects of density variations on convection and on the pressure field.

In the velocity-composition joint pdf equation, the convective term is in closed form even in variable-density flows (Pope 1985). In second-order closures—such as the Bray-Moss-Libby model—this is not the case. But the closure problem is greatly alleviated by the use of density-weighted (or Favre) averaging (see, e.g., Libby & Williams 1980). For the reaction progress variable $c(\mathbf{x}, t)$, the Favre mean and fluctuation are

$$\tilde{c} \equiv \langle \rho c \rangle / \langle \rho \rangle \quad \text{and} \quad c'' \equiv c - \tilde{c}. \tag{37}$$

Libby (1985) neglects the effects of pressure fluctuations completely, while Anand & Pope (1986) retain a model appropriate to constant-density flow. (This modeled term is found to have little effect on the calculations.) Thus in both models it is the mean pressure gradient that is responsible for the differential acceleration of reactants and products. For a statistically stationary one-dimensional flame, the reactants flow into the flame at the turbulent-flame speed u_T. In view of mass conservation the products leave at speed $u_T \rho_r / \rho_p = R u_T$; and momentum conservation shows that there is a pressure drop of magnitude $\rho_r u_T^2 (R-1)$.

Figure 6 shows Anand & Pope's calculations of the turbulent flux of products $\widetilde{u''c''}$ plotted against $\tilde{c}$ (which of course increases monotonically with x—the distance through the flame). For the constant-density case ($R = 1$) it may be seen that this flux is negative everywhere, which indicates gradient diffusion (i.e. $\widetilde{u''c''}\, d\tilde{c}/dx < 0$). But for density ratios of 4 and above, the favorable pressure gradient preferentially accelerates the lighter products in the flow direction, thus yielding a positive flux of products nearly everywhere. At the cold boundary there is, of necessity, a region of gradient diffusion (Libby 1985).

According to the calculations for $R = 10$, the variance of the axial velocity $\widetilde{u''^2}$ increases by a factor of 17 through the flame, while the variances of the other two components increase by just 50%. Figure 7 shows the budget of $\langle \rho \rangle \widetilde{u''^2}$ through the flame. The large source, it may be seen, is due to the mean-pressure-gradient term $-2\langle u'' \rangle d\langle p \rangle / dx$. Since $\langle u'' \rangle$ is proportional to $\widetilde{u''c''}$ (Libby 1985), countergradient diffusion and energy production go hand in hand. At small density ratios ($R < 4$) there is gradient diffusion and the pressure-gradient term is a sink. For all density ratios the dilatation term $-2\langle \rho \rangle \widetilde{u''^2}\, d\tilde{U}/dx$ is a sink.

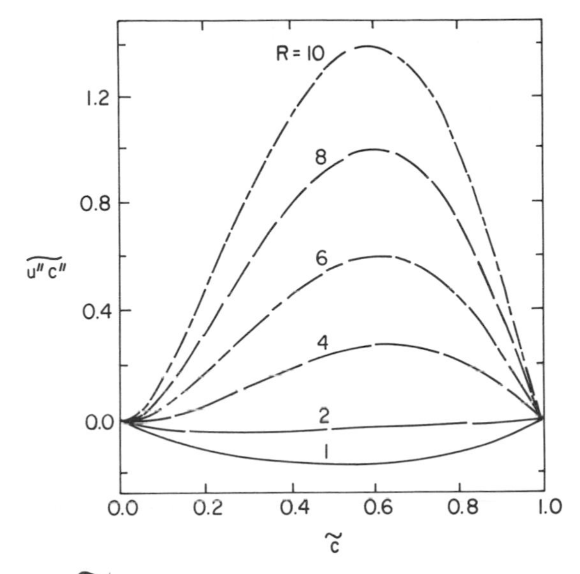

Figure 6 Scalar flux $\widetilde{u''c''}$ versus $\tilde{c}$ as a function of density ratio R. From joint pdf calculations (Anand & Pope 1986) in the flame-sheet regime. (Here u'' is normalized with the upstream turbulence intensity.)

Compared with the Bray-Moss-Libby model, the pdf method has the advantages that fewer processes have to be modeled and more information can be extracted from the solution. But for this flame, the two methods give similar results.

A statistically stationary and one-dimensional flame has not been realized experimentally, and so the calculations cannot be compared directly with data. Although it involves some uncertainty, Bray et al. (1981) and Libby (1985) compared their calculations with the data of Moss (1980) obtained in a conical flame. In general there is good agreement. But in order to provide an unambiguous, quantitative test of the modeling, more accurate data are needed, and the calculations should correspond more closely to the experimental configuration.

4.2 *Pressure Field in Flame-Sheet Combustion*

In turbulent combustion in general, our knowledge of the statistics of the pressure field and of their effect on the turbulence is slight. The experimental problems are severe: The pressure has to be measured on the smallest length and time scales of turbulence; and, rather than the pressure itself, its gradients and their correlation with the velocity are the prime quantities of importance. Though some progress has been made

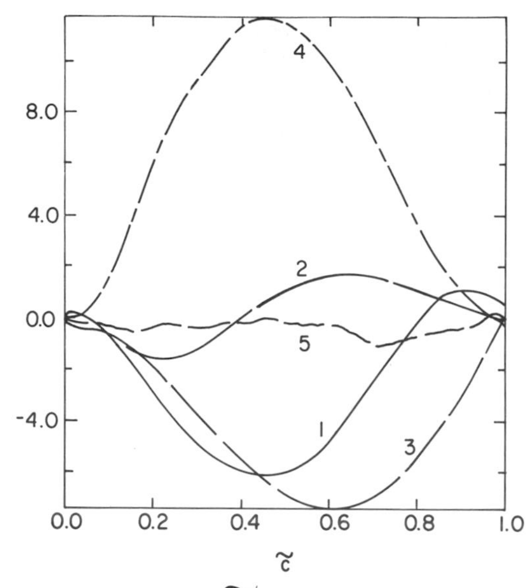

Figure 7 Budget of axial component $\langle\rho\rangle\widetilde{u''^2}$ of turbulence energy versus $\tilde{c}$. From joint pdf calculations (Anand & Pope 1986) in the flame-sheet regime for $R = 10$. (1) Convection: $-\langle\rho\rangle\tilde{U}\,d\widetilde{u''^2}/dx$; (2) diffusion: $-d(\langle\rho\rangle\widetilde{u''^3})/dx$; (3) dilatation: $-2\langle\rho\rangle\widetilde{u''^2}d\tilde{U}/dx$; (4) mean pressure gradient: $-2\langle u''\rangle d\langle p\rangle/dx$; (5) remainder: modeled redistribution term and statistical error. (All quantities are normalized with the upstream density, intensity, and length scale.)

(Komerath & Strahle 1983), accurate measurements of the velocity–pressure-gradient correlation are not in sight.

In constant-density flows, the usual theoretical approach is to relate one-point pressure statistics to two-point velocity statistics through the Poisson equation for pressure. But for variable-density reactive flows, the Poisson equation contains additional source terms that, in general, make this approach intractable. A different approach has been explored by Strahle (1982).

First, we present an argument that suggests that the neglect of pressure fluctuations in the model calculations is a serious omission. Then we show that because of the special structure of the density field in flame-sheet combustion, useful information can be obtained from the Poisson equation. Specifically, one-point pressure statistics can be related to two-point velocity-velocity and velocity–flame-front statistics.

A direct implication of the neglect of pressure fluctuations is that the pressure field accelerates an element of products more by a factor of $R = \rho_r/\rho_p$ than it accelerates an element of reactants (at the same position and time but in a different realization). Simply, the instantaneous Euler equation becomes

$$D\mathbf{U}/Dt = -\rho^{-1}\nabla\langle p\rangle. \quad (38)$$

The differential acceleration results in countergradient diffusion and energy production in accord with experimental observations. But the factor of R is too large, since it ignores acceleration reaction[1] (see, e.g., Batchelor 1967). That is, an acceleration of the light products is accompanied by a proportionate acceleration of the displaced heavier reactants. Part of the work done by the applied force goes to accelerate the reactants, and hence the acceleration of the products is less than that implied by Equation (38).

The magnitude of the overestimate of the differential acceleration could be large. By analogy, consider Equation (38) applied to an initially stationary, spherical air bubble randomly located within a quiescent body of water. According to Equation (38) the bubble will accelerate vertically at the rate $g(R'-1)$, where g is the gravitational acceleration and $R' \approx 1000$ is the density ratio (water to air). But taking due account of acceleration reaction, we find that the true acceleration of the bubble is just $2g(R'-1)/(R'+2) \approx 2g$ (see Batchelor 1967).

Although the errors resulting from the neglect of pressure fluctuations may be large, the calculations reported in Section 4.1 appear to be in agreement with the data of Moss (1980). This apparent conflict emphasizes the need for more direct comparisons between model calculations and experimental data.

In the development of models for constant-density flows, the fluctuating pressure is eliminated by use of the Poisson equation $\nabla^2 p = -\rho(\partial U_i/\partial x_j)(\partial U_j/\partial x_i)$. This approach, introduced by Chou (1945), has been used to deduce both the form of pressure correlations and some exact results (see, e.g., Rotta 1951, Launder et al. 1975, Pope 1981). For a general variable-density turbulent reacting flow, the Poisson equation contains additional source terms that, to date, have nullified the usefulness of this approach. But for turbulent premixed flames in the flame-sheet regime, the density field has a special structure that allows the Poisson equation to be expressed in a useful form. This is demonstrated for a simple case.

Figure 8 shows a sketch of the flame sheet at an instant in an unbounded, statistically spherical turbulent flame, such as that of Hainsworth (1985). Provided that the flame-sheet thickness ($\sim\delta_L$) is much smaller than other relevant length scales, the flame sheet can be regarded as a mathematical surface separating constant-density regions of reactants and products (see, e.g., Markstein 1964, Matalon & Matkowsky 1982, Pope 1986). The flame

[1] I am indebted to Dr. J. C. R. Hunt for this observation.

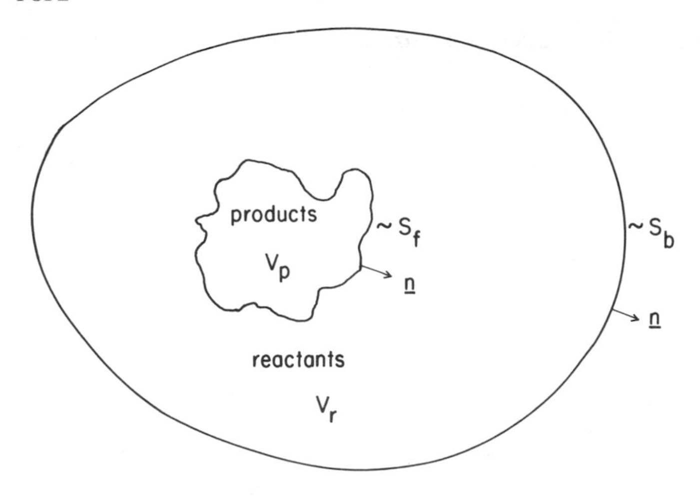

Figure 8 Sketch of closed flame sheet S_f.

surface S_f is assumed to be regular and has local normal **n** (pointing into the reactants) and propagates at the local speed w (in the direction of **n**, relative to the reactants just ahead). Enclosed by the surface S_f is the volume V_p of constant-density (ρ_p) products. For mathematical convenience we define a closed surface S_b remote from S_f. Between S_f and S_b is the volume V_r of constant-density (ρ_r) reactants.

With this construction it follows from Green's third identity (Kellogg 1967, pp. 219–21) that the pressure at any point **x** in V_r or V_p can be decomposed into five contributions,

$$p(\mathbf{x}) = p_r(\mathbf{x}) + p_p(\mathbf{x}) + p_{fw}(\mathbf{x}) + p_{fa}(\mathbf{x}) + p_b(\mathbf{x}), \tag{39}$$

corresponding, respectively, to integrals over V_r, V_p, S_f, S_f, and S_b.

The first two integrals are

$$4\pi p_{r,p}(\mathbf{x}) = -\iiint\limits_{V_{r,p}} r^{-1}\nabla^2 p \, dV = \rho_{r,p} \iiint\limits_{V_{r,p}} r^{-1} \frac{\partial U_i}{\partial x_j}\frac{\partial U_j}{\partial x_i} \, dV, \tag{40}$$

where integration is over all points **y** in V_r or V_p, r is the distance $|\mathbf{y}-\mathbf{x}|$, and the integrands are evaluated at **y**. The right-hand expression in Equation (40) follows from the momentum equation with the assumption of uniform density and Newtonian viscosity. These terms are familiar from constant-density studies.

The first integral over the flame surface is

$$4\pi p_{\mathrm{fw}}(\mathbf{x}) = \iint_{S_{\mathrm{f}}} [p] \frac{\partial r^{-1}}{\partial n} \, dS = \rho_{\mathrm{r}}(R-1) \iint_{S_{\mathrm{f}}} w^2 \frac{\partial r^{-1}}{\partial n} \, dS, \tag{41}$$

where integration is over all points $\mathbf{y}$ in S_{f}, square brackets denote the jump across the surface (reactant-side value minus product-side value), n is the coordinate in the direction of $\mathbf{n}$, and the integrands are evaluated at $\mathbf{y}$.

The right-hand expression in Equation (41) follows from the known jump conditions across the surface (Markstein 1964) with the neglect of viscous terms. Thus the contribution p_{fw} arises from the motion of the surface relative to the fluid. [For a material surface ($w = 0$), p_{fw} is zero.]

It is interesting to note that in the products, if w is constant ($w = u_{\mathrm{L}}$, say) then p_{fw} is a constant and hence causes no acceleration. Let $d\Omega$ be the solid angle subtended at $\mathbf{x}$ by the surface element dS at $\mathbf{y}$ and having the sign of $\mathbf{n} \cdot (\mathbf{y}-\mathbf{x})$. Then, from Equation (41) we obtain

$$p_{\mathrm{fw}}(\mathbf{x}) = -\frac{\rho_{\mathrm{r}}(R-1)}{4\pi} \iint_{S_{\mathrm{f}}} w^2 \, d\Omega = -\rho_{\mathrm{r}}(R-1) \, u_{\mathrm{L}}^2, \tag{42}$$

the integral expression holding in general (Kellogg 1967, p. 67), the right-hand expression for points in V_{p} for the case $w = u_{\mathrm{L}}$.

The second contribution from S_{f} is

$$4\pi p_{\mathrm{fa}}(\mathbf{x}) = -\iint_{S_{\mathrm{f}}} \left[\frac{\partial p}{\partial n}\right] r^{-1} \, dS = \rho_{\mathrm{r}}(1-R^{-1}) \iint_{S_{\mathrm{f}}} r^{-1} a_n \, dS, \tag{43}$$

where again integration is over all points $\mathbf{y}$ in S_{f} and the integrands are evaluated at $\mathbf{y}$. The jump in $\partial p / \partial n$ is related to the acceleration of the surface (again neglecting viscous terms): a_n is the component of the acceleration of a surface point in the direction of $\mathbf{n}$.

The final contribution,

$$4\pi p_{\mathrm{b}}(\mathbf{x}) = \iint_{S_b} r^{-1} \frac{\partial p}{\partial n} - p \frac{\partial r^{-1}}{\partial n} \, dS, \tag{44}$$

need not be considered in detail, at least for unbounded flows. The surface S_{b} can be chosen to be remote from $\mathbf{x}$ (i.e. many integral scales away), and then p_{b} makes a negligible contribution to quantities of interest such as $\langle u_i'(\mathbf{x}) p(\mathbf{x}) \rangle$. (This is because $\langle u_i'(\mathbf{x}) p(\mathbf{x}+\mathbf{r}) \rangle$ and its derivatives are negligible at large $|\mathbf{r}|$.)

It may be seen then that for the case considered, a useful formal solution for the pressure $p(\mathbf{x})$ is obtained. One-point statistics of p and its derivatives can be expressed in terms of two-point statistics of the velocity field and of the flame surface. Beyond the case considered, the solution is valid for multiple, open or closed, nonintersecting surfaces (Kellogg 1967).

5. DISCUSSION AND CONCLUSION

Over the past 10 years significant progress has been made in the development of probabilistic field theories for turbulent premixed flames. Both the Bray-Moss-Libby model and the pdf approach are able to account for countergradient diffusion and energy production. As far as can be deduced from the imperfect comparison with experimental data, both approaches yield quantitatively plausible results. Nevertheless, there remain two major areas of uncertainty: the nature of combustion in the intermediate regime, and the effect of combustion-induced pressure fluctuations.

In spite of the advances over the past decade, our predictive abilities for turbulent premixed flames are modest and uncertain compared with our abilities for turbulent-diffusion flames. In 1975 models were capable of calculating, reasonably accurately, the basic features of simple jet diffusion flames (Lockwood & Naguib 1975, Kent & Bilger 1976). Now, for these simple flames, more refined calculations with multistep kinetics have been performed (e.g. Correa et al. 1984, Pope & Correa 1986, Jones & Kollmann 1986). And the basic model has been applied to the three-dimensional flow in a gas-turbine combustion chamber (Coupland & Priddin 1986).

From a theoretical viewpoint, turbulent diffusion flames are inherently simpler than premixed flames. But four other reasons can be identified for the relatively rapid progress for diffusion flames:

1. There is a simple canonical flow—a fuel jet in a coflowing airstream—for which there is a good data base (Bilger 1980).
2. Standard turbulence models (incorporating gradient diffusion) can be applied.
3. The boundary-layer approximations can be applied to mean equations.
4. Laser diagnostics have been extensively applied to the flows (e.g. Correa et al. 1984).

In contrast, for turbulent premixed flames, the canonical one-dimensional stationary flame considered in theories has not been realized in practice. Laser diagnostics are now being applied to premixed flames (e.g. Dandekar & Gouldin 1982, Cheng 1984, Gulati & Driscoll 1986a,b), but the configurations being studied have drawbacks. In particular, for stationary, unconfined, stabilized flames the boundary-layer approximations are not valid because of the rapid volume expansion. Thus the computationally more demanding elliptic equations are appropriate. Nevertheless, because direct quantititative comparison of calculations with experiments is vital to the development of theories, it appears that model calculations of these (elliptic) flows is an unavoidable step toward significant progress. This conclusion is reinforced by the numerous recent experiments on these flames. To the same end, further experiments on the computationally simpler statistically spherical flames would be valuable.

The theories described here are in some ways quite limited: They apply only to homogeneously premixed reactants; different modeling is required in the flame-sheet and intermediate regimes; and without ad hoc modifications, they are valid only for $u_L/u' \ll 1$. Consequently, further developments are needed before the models can be used in some important applications—stratified-charge spark-ignition engines, for example.

As in the past, it can be expected that laminar-flame analyses and experiments will contribute to our understanding of flame-sheet turbulence. It can also be expected that Full Turbulence Simulations (Rogallo & Moin 1984) will play an expanding role in elucidating the structure and mechanisms of combustion in both the flame-sheet and intermediate regimes.

ACKNOWLEDGMENTS

I am grateful to Professors P. A. Libby, N. Peters, and F. A. Williams for their comments on the draft of this review.

This work was supported in part by the US Army Research Office (grant number DAAG29-84-K-0020), by the US Air Force Office of Scientific Research (grant number AFOSR-85-0083), and by the Department of Energy (contract number AC 02-83ER1303A).

Literature Cited

Abdel-Gayed, R. G., Bradley, D. 1981. A two-eddy theory of premixed turbulent flame propagation. *Philos. Trans. R. Soc. London Ser. A* 301: 1–25

Abdel-Gayed, R. G., Bradley, D. 1985. Criteria for turbulent propagation limits of premixed flames. *Combust. Flame* 62: 61–68

Abdel-Gayed, R. G., Bradley, D., Hamid, M. N., Lawes, M. 1984. Lewis number effects on turbulent burning velocity. *Symp. (Int.) Combust., 20th*, pp. 505–12. Pittsburgh: Combust. Inst.

Abraham, J., Williams, F. A., Bracco, F. V. 1985. A discussion of turbulent flame structure in premixed charges. *SAE Pap. 850345*

Anand, M. S., Pope, S. B. 1986. Calculations of premixed turbulent flames by pdf methods. *Combust. Flame*. In press

Andrews, G. E., Bradley, D., Lwakabamba, S. B. 1975. Turbulence and turbulent flame propagation—a critical appraisal. *Combust. Flame* 24: 285–304

Batchelor, G. K. 1967. *An Introduction to Fluid Dynamics*. Cambridge: Cambridge Univ. Press. 615 pp.

Bilger, R. W. 1980. Turbulent flows with nonpremixed reactants. In *Turbulent Reacting Flows*, ed. P. A. Libby, F. A. Williams, pp. 65–113. Berlin: Springer-Verlag

Bolz, R. E., Burlage, H. Jr. 1960. Propagation of free flames in laminar and turbulent flow fields. *NASA TND-551*

Bray, K. N. C. 1980. Turbulent flows with

premixed reactants. In *Turbulent Reacting Flows*, ed. P. A. Libby, F. A. Williams, pp. 115–83. Berlin: Springer-Verlag

Bray, K. N. C., Moss, J. B. 1974. A unified statistical model of the premixed turbulent flame. *Rep. AASU 335*, Univ. Southampton, Engl.

Bray, K. N. C., Moss, J. B. 1977. A unified statistical model of premixed turbulent flame. *Acta Astronaut.* 4: 291–319

Bray, K. N. C., Libby, P. A., Masuya, G., Moss, J. B. 1981. Turbulence production in premixed turbulent flames. *Combust. Sci. Technol.* 25: 127–40

Bray, K. N. C., Libby, P. A., Moss, J. B. 1985. Unified modeling approach for premixed turbulent combustion—Part I: general formulation. *Combust. Flame* 61: 87–102

Buckmaster, J. D., Ludford, G. S. S. 1982. *Theory of Laminar Flames*. Cambridge: Cambridge Univ. Press. 266 pp.

Cheng, R. K. 1984. Conditional sampling of turbulence intensities and Reynolds stresses in premixed turbulent flames. *Combust. Sci. Technol.* 41: 109–42

Cheng, R. K., Shepherd, I. G. 1986. Intermittency and conditional velocities in premixed conical turbulent flames. *Rep.*, Univ. Calif., Berkeley

Cho, P., Law, C. K., Hertzberg, J. R., Cheng, R. K. 1986. Structure and propagation of turbulent premixed flames stabilized in a stagnation flow. *Symp. (Int.) Combust., 21st*. Pittsburgh: Combust. Inst. In press

Chomiak, J. 1976. Dissipation fluctuations and the structure and propagation of turbulent flames in premixed gases at high Reynolds number. *Symp. (Int.) Combust., 16th*, pp. 1665–72. Pittsburgh: Combust. Inst.

Chou, P. Y. 1945. On velocity correlations and the solution of the equations of turbulent fluctuation. *Q. Appl. Math.* 3: 38–54

Chung, S. H., Law, C. K. 1984. An invariant derivation of flame stretch. *Combust. Flame* 55: 123–25

Clavin, P. 1985. Dynamic behavior of premixed flame fronts in laminar and turbulent flows. *Prog. Energy Combust. Sci.* 11: 1–59

Clavin, P., Williams, F. A. 1979. Theory of premixed flame propagation in large scale turbulence. *J. Fluid Mech.* 90: 589–604

Clavin, P., Williams, F. A. 1982. Effects of molecular diffusion and of thermal expansion on the structure and dynamics of premixed flames in turbulent flows of large scale and low intensity. *J. Fluid Mech.* 116: 251–82

Comte-Bellot, G., Corrsin, S. 1971. Simple Eulerian time correlations of full- and narrow-band velocity signals in grid-generated "isotropic" turbulence. *J. Fluid Mech.* 42: 273–337

Correa, S. M., Drake, M. C., Pitz, R. W., Shyy, W. 1984. Prediction and measurement of a non-equilibrium turbulent diffusion flame. *Symp. (Int.) Combust., 20th*, pp. 337–43. Pittsburgh: Combust. Inst.

Coupland, J., Priddin, C. H. 1985. Modelling the flow and combustion in a production gas turbine combustor. *Symp. Turbul. Shear Flows, 5th*, pp. 10.1–11. Ithaca, NY: Cornell Univ.

Damköhler, G. 1940. Der Einfluss der Turbulenz auf die Flammengeschwindigkeit in Gasgemishen. *Z. Elektrochem.* 46: 601–26. Transl., 1947, as *NACA TM 1112*

Dandekar, K. V., Gouldin, F. C. 1982. Temperature and velocity measurements in premixed turbulent flames. *AIAA J.* 20: 652–59

Fox, M. D., Weinberg, F. J. 1962. An experimental study of burner-stabilized turbulent flames in premixed reactants. *Proc. R. Soc. London Ser. A* 268: 222–39

Groff, E. G. 1986. An experimental evaluation of an entrainment flame-propagation model. *Rep.*, General Motors Res. Lab., Warren, Mich.

Gulati, A., Driscoll, J. F. 1986a. Velocity-density correlations and Favre averages measured in a premixed turbulent flame. *Combust. Sci. Technol.* In press

Gulati, A., Driscoll, J. F. 1986b. Flame-generated turbulence and mass fluxes: effect of varying heat release. *Symp. (Int.) Combust., 21st*. Pittsburgh: Combust. Inst. In press

Gunther, R. 1983. Turbulence properties of flames and their measurements. *Prog. Energy Combust. Sci.* 9: 105–54

Hainsworth, E. 1985. *Study of free turbulent premixed flames*. MS thesis. Mass. Inst. Technol., Cambridge

Jones, W. P., Kollmann, W. 1986. Multiscalar pdf transport equations for turbulent diffusion flames. In *Turbulent Shear Flows 5*. Berlin: Springer-Verlag. In press

Jones, W. P., Whitelaw, J. H. 1982. Calculation methods for turbulent reactive flows: a review. *Combust. Flame* 48: 1–26

Jones, W. P., Whitelaw, J. H. 1984. Modelling and measurements in turbulent combustion. *Symp. (Int.) Combust., 20th*, pp. 233–49. Pittsburgh: Combust. Inst.

Karlovitz, B., Denniston, D. W. Jr., Knappschaefer, D. H., Wells, F. E. 1953. Studies in turbulent flames. *Symp. (Int.) Combust., 4th*, pp. 613–20. Baltimore: Williams & Wilkins

Keck, J. C. 1982. Turbulent flame structure and speed in spark-ignition engines. *Symp. (Int.) Combust., 19th*, pp. 1451–66. Pittsburgh: Combust. Inst.

Kellogg, O. D. 1967. *Foundations of Potential Theory*. Berlin: Springer-Verlag. 384 pp.

Kent, J. H., Bilger, R. W. 1976. The prediction of turbulent diffusion flame fields and nitric oxide formation. *Symp. (Int.) Combust., 16th*, pp. 1643–56. Pittsburgh: Combust. Inst.

Klimov, A. M. 1963. Laminar flame in a turbulent flow. *Zh. Prikl. Mekh. Tekh. Fiz.* 3: 49–58

Klimov, A. M. 1975. Premixed turbulent flames—interplay of hydrodynamic and chemical phenomena. *Dokl. Akad. Nauk SSSR* 221: 56–59. Transl., 1975, in *Sov. Phys. Dokl.* 20: 168

Komerath, N. M., Strahle, W. C. 1983. Measurement of the pressure-velocity correlation in turbulent reactive flows. *AIAA Pap. No. 83-0400*

Kovasznay, L. S. G. 1956. A comment on turbulent combustion. *Jet Propul.* 26: 485

Kuo, K. K. 1986. *Principles of Combustion*. New York: Wiley. 810 pp.

Launder, B. E., Reece, G. J., Rodi, W. 1975. Progress in the development of a Reynolds-stress turbulence closure. *J. Fluid Mech.* 68: 537–66

Law, C. K., Zhu, D. L., Yu, G. 1986. Propagation and extinction of stretched premixed flames. *Symp. (Int.) Combust., 21st*. Pittsburgh: Combust. Inst. In press

Lewis, B., von Elbe, G. 1961. *Combustion, Flames, and Explosions of Gases*. New York: Academic. 731 pp. 2nd ed.

Libby, P. A. 1985. Theory of normal premixed flames revisited. *Prog. Energy Combust. Sci.* 11: 83–96

Libby, P. A., Bray, K. N. C. 1981. Countergradient diffusion in premixed turbulent flames. *AIAA J.* 19: 205–13

Libby, P. A., Williams, F. A. 1976. Turbulent flows involving chemical reactions. *Ann. Rev. Fluid Mech.* 8: 351–76

Libby, P. A., Williams, F. A., eds. 1980. *Turbulent Reacting Flows*. Berlin: Springer-Verlag. 243 pp.

Libby, P. A., Williams, F. A. 1981. Some implications of recent theoretical studies in turbulent combustion. *AIAA J.* 19: 261–74

Lockwood, F. C., Naguib, A. S. 1975. The prediction of the fluctuations in the properties of free, round jet, turbulent, diffusion flames. *Combust. Flame* 24: 109–24

Markstein, G. M. 1964. *Non-Steady Flame Propagation*. New York: Macmillan. 328 pp.

Matalon, M. 1983. On flame stretch. *Combust. Sci. Technol.* 31: 169–81

Matalon, M., Matkowsky, B. J. 1982. Flames as gasdynamic discontinuities. *J. Fluid Mech.* 124: 239–59

McNutt, D. G. 1981. *A study of premixed turbulent flames*. MS thesis. Mass. Inst. Technol., Cambridge

Mendes-Lopes, J. M. C., Daneshyar, H. 1985. Influence of strain fields on flame propagation. *Combust. Flame* 60: 29–48

Metghalchi, M., Keck, J. C. 1982. Burning velocities of mixtures of air with methanol, isooctane, and indolene at high pressure and temperature. *Combust. Flame* 48: 191–210

Mickelsen, W. R., Ernstein, N. E. 1956. Growth rates of turbulent free flames. *Symp. (Int.) Combust., 6th*, pp. 325–33. New York: Reinhold

Monin, A. S., Yaglom, A. M. 1975. *Statistical Fluid Mechanics*, Vol. 2. Cambridge, Mass: MIT Press. 874 pp.

Moss, J. B. 1980. Simultaneous measurements of concentration and velocity in an open premixed flame. *Combust. Sci. Technol.* 22: 119–29

Namazian, M., Hansen, S., Lyford-Pike, E., Sanchez-Barsse, J., Heywood, J., Rife, J. 1980. Schlieren visualization of the flow and density fields in the cylinder of a spark-ignition engine. *SAE Trans.* 89: 276–303

Palm-Leis, A., Strehlow, R. A. 1969. On the propagation of turbulent flames. *Combust. Flame* 13: 111–29

Peters, N. 1986. Laminar flamelet concepts in turbulent combustion. *Symp. (Int.) Combust., 21st*. Pittsburgh: Combust. Inst. In press

Pope, S. B. 1981. Transport equation for the joint probability density function of velocity and scalars in turbulent flow. *Phys. Fluids* 24: 588–96

Pope, S. B. 1985. PDF methods for turbulent reactive flows. *Prog. Energy Combust. Sci.* 11: 119–92

Pope, S. B. 1986. The evolution of surfaces in turbulence. *Rep. FDA-86-05*, Cornell Univ., Ithaca, N.Y.

Pope, S. B., Anand, M. S. 1984. Flamelet and distributed combustion in premixed turbulent flames. *Symp. (Int.) Combust., 20th*, pp. 403–10. Pittsburgh: Combust. Inst.

Pope, S. B., Cheng, W. K. 1986. Statistical calculations of spherical turbulent flames. *Symp. (Int.) Combust., 21st*. Pittsburgh: Combust. Inst. In press

Pope, S. B., Correa, S. M. 1986. Joint pdf calculations of a non-equilibrium turbulent diffusion flame. *Symp. (Int.) Combust., 21st*. Pittsburgh: Combust. Inst. In press

Rallis, C. J., Garforth, A. M. 1980. The determination of laminar burning velocity. *Prog. Energy Combust. Sci.* 6: 303–29

Rogallo, R. S., Moin, P. 1984. Numerical simulation of turbulent flows. *Ann. Rev. Fluid Mech.* 16: 99–137

Rotta, J. C. 1951. Statistische Theorie nichthomogener Turbulenz. *Z. Phys.* 129: 541

Shepherd, I. G., Moss, J. B. 1982. Measurements of conditioned velocities in a turbulent premixed flame. *AIAA J.* 20: 566–69

Sivashinsky, G. I. 1979. On self-turbulization of a laminar flame. *Acta Astronaut.* 6: 569–91

Sivashinsky, G. I. 1983. Instabilities, pattern formation, and turbulence in flames. *Ann. Rev. Fluid Mech.* 15: 179–99

Spalding, D. B. 1971. Mixing and chemical reaction in steady confined turbulent flames. *Symp. (Int.) Combust., 13th*, pp. 649–57. Pittsburgh: Combust. Inst.

Stambuleanu, A. 1976. *Flame Combustion Processes in Industry*. Tunbridge Wells, Engl: Abacus. 567 pp.

Strahle, W. C. 1982. Estimation of some correlations in a premixed reactive turbulent flow. *Combust. Sci. Technol.* 29: 243–60

Strehlow, R. A. 1968. *Fundamentals of Combustion*. Scranton, Pa: Int. Textbook. 465 pp.

Strehlow, R. A. 1985. *Combustion Fundamentals*. New York: McGraw-Hill. 554 pp.

Suzuki, T., Hirano, T. 1984. Dynamic characteristics of flame fronts in turbulent premixed flame zone. *Symp. (Int.) Combust., 20th*, pp. 437–44. Pittsburgh: Combust. Inst.

Tabaczynski, R. J. 1976. Turbulence and turbulent combustion in spark-ignition engines. *Prog. Energy Combust. Sci.* 2: 143–65

Tabaczynski, R. J., Trinker, F. H., Shannon, A. S. 1980. Further refinement and validation of a turbulent flame propagation model for spark-ignition engines. *Combust. Flame* 39: 111–21

Tennekes, H. 1968. Simple model for the small-scale structure of turbulence. *Phys. Fluids* 11: 669–70

Tennekes, H., Lumley, J. L. 1972. *A First Course in Turbulence*. Cambridge, Mass: MIT Press. 300 pp.

Warnatz, J. 1984. Chemistry of high temperature combustion of alkanes up to octane. *Symp. (Int.) Combust., 20th*, pp. 845–56. Pittsburgh: Combust. Inst.

Warnatz, J., Peters, N. 1984. Stretch effects in planar premixed hydrogen-air flames. *Prog. Astronaut. Aeronaut.* 95: 61–74

Williams, F. A. 1976. Criteria for existence of wrinkled laminar flame structure of turbulent premixed flames. *Combust. Flame* 26: 269–70

Williams, F. A. 1985a. Turbulent combustion. In *The Mathematics of Combustion*, ed. J. D. Buckmaster, pp. 97–131. Philadelphia: SIAM

Williams, F. A. 1985b. *Combustion Theory*. Menlo Park, Calif: Benjamin-Cummings. 680 pp.

Wright, F. M., Zukoski, E. E. 1962. Flame spreading from bluff-body flame holders. *Symp. (Int.) Combust., 8th*, pp. 933–43. Pittsburgh: Combust. Inst.

Wu, C. K., Law, C. K. 1984. On the determination of laminar flame speeds from stretched flames. *Symp. (Int.) Combust., 20th*, pp. 1941–49. Pittsburgh: Combust. Inst.

Wu, C.-T., Ferziger, J. H., Chapman, D. R. 1985. Simulation and modelling of homogeneous, compressed turbulence. *Symp. Turbul. Shear Flows*, pp. 17.13–19. Ithaca, NY: Cornell Univ.

Yoshida, A., Tsuji, H. 1982. Characteristic scale of wrinkles in turbulent premixed flames. *Symp. (Int.) Combust., 19th*, pp. 403–12. Pittsburgh: Combust. Inst.

Ann. Rev. Fluid Mech. 1987. 19 : 271–311

VISCOUS FINGERING IN POROUS MEDIA

G. M. Homsy

Department of Chemical Engineering, Stanford University, Stanford, California 94305

INTRODUCTION

Scope

"Viscous fingering" generally refers to the onset and evolution of instabilities that occur in the displacement of fluids in porous materials. In most but not all cases, the mechanism of the instability is intimately linked to viscosity variations between phases or within a single phase containing a solute, hence the term "viscous fingering." Shown in Figure 1 are three examples of the complex and intriguing patterns that evolve as a result of this instability. Figure 1*a* shows the fingering pattern that occurs when a more viscous material is displaced by a less viscous one fully miscible with it by injecting from one corner and withdrawing from the diagonal corner of a horizontal square Hele-Shaw cell; in this case the fluid consists of water injected into glycerine. Although the mixture is fully miscible, it is obvious that less viscous material tends to penetrate and finger through the more viscous material. In this particular example, the patterns are driven by the difference in viscosity and influenced by the diffusive mixing between the fluids. In Figure 1*b*, we show an example of the extreme patterns that are formed when a less dense, less viscous fluid penetrates a more dense, more viscous fluid *immiscible* with it when a large Hele-Shaw cell is tipped into a vertical position. In this case both gravity and viscosity are important forces in driving the instability. Finally, in Figure 1*c* we show the pattern that results when a low-viscosity Newtonian fluid injected from a source penetrates a Hele-Shaw cell filled with a miscible but strongly non-Newtonian fluid. We discuss these examples in more detail below, but even though the experiments were done in the relatively simple geometry of Hele-Shaw cells, the detailed dynamics leading to the observed patterns

0066–4189/87/0115–0271$02.00

are not fully understood. However, one feature in common in these examples is the fact that the physical conditions of the experiments allow a wide spectrum of length scales to occur. Below we provide a detailed discussion of the mechanisms that govern these flows, which we refer to as *shielding*, *spreading*, and *splitting*, that will enable us to at least qualitatively understand these fascinating patterns.

Such phenomena are important in a wide variety of applications, including secondary and tertiary oil recovery, fixed bed regeneration in chemical processing, hydrology, and filtration. Indeed, the phenomena are expected to occur in many of the myriad of fields of science and technology in which fluids flow through porous materials, and thus the literature is a diverse one. Many combinations of configurations, important fluid-mechanical forces, and boundary conditions have been studied. Thus, we cannot provide an exhaustive review, and many important areas of research are omitted from our discussion. In addition, recent activity in the field has been explosive to such an extent that this review is destined to be out of date, perhaps seriously so, by the time it has appeared. Luckily, the editors of this series have provided me with both a time deadline and a page limit,

Figure 1 Examples of viscous fingering in Hele-Shaw cells: (*a*) miscible flow in a five-spot geometry (E. L. Claridge, personal communication, 1986); (*b*) immiscible flow in gravity-driven fingering (Maxworthy 1986); (*c*) miscible flow of a non-Newtonian fluid in radial source flow (Daccord et al. 1986). With permission.

behind which I take refuge against inevitable criticisms of the timeliness and scope of this review.

The areas to be covered are as follows: We are interested in viscous fingering in homogeneous porous materials and thus do not discuss the important and emerging area of the interaction between viscous fingering

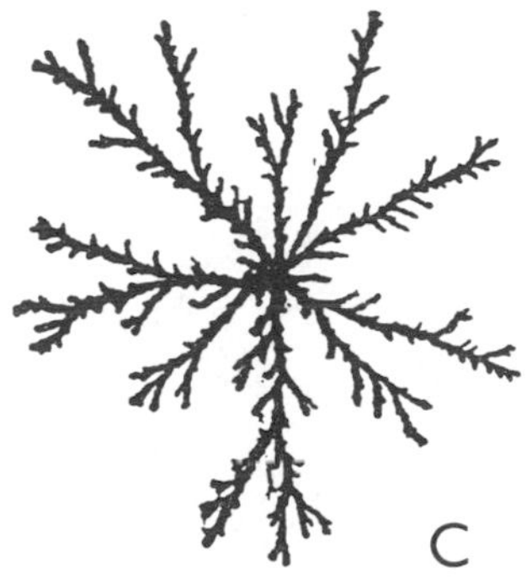

and permeability heterogeneities. Furthermore, we cannot give a treatment of all possible geometries and boundary conditions. Thus, we deal exclusively with the three simple geometries of rectilinear displacement, radial source flow, and the so-called five-spot pattern, which are sketched in Figure 2. In each case there is a characteristic macroscopic length L, a characteristic velocity U, a permeability K, and a characteristic viscosity μ. We are primarily interested in two-dimensional flows, as little has been done on three-dimensional problems, especially nonlinear ones. We also confine ourselves to cases in which the orientation of gravity, should it be important, is colinear with the displacement direction. In limiting the review in this manner, we hope to emphasize mechanisms and current

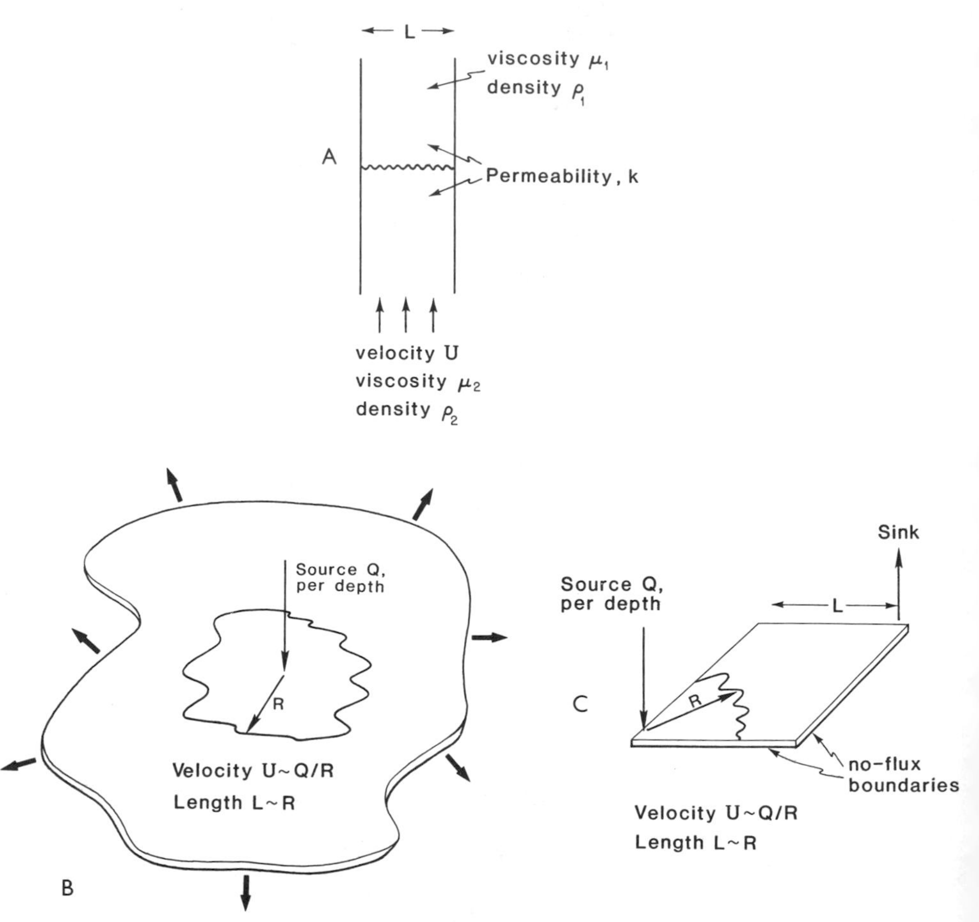

Figure 2 Common geometries and their defining parameters: (*a*) rectilinear flow; (*b*) radial source flow; (*c*) "five-spot" pattern.

research for this restricted class of problems. We deemphasize almost entirely the case in which instabilities are driven solely by buoyancy, although this is a field with an equally rich range of phenomena.

A subject of current focus is the extent to which viscous fingering may be analogous to other free-boundary problems that exhibit similar pattern formation, notably the growth of crystals from melt or solution. Unfortunately, this is a question that, while intriguing, is outside the scope of the present review, although many of the more recent references given here discuss this issue. Also, in the discussion of immiscible flow in porous materials, we touch only briefly on those aspects of displacement that are dominated by capillarity in the microscopic pore space, and that rightly form a subset of percolation phenomena, on which there exists an extensive literature. Some aspects of viscous fingering have been well reviewed elsewhere, and we simply refer the reader to those sources as appropriate. These include the review by Wooding & Morel-Seytoux (1976) on the more general subject of multiphase flow in porous media; that by Aref (1986) on modern approaches to simulations of such flows, and features these flows have in common with other well-studied nonlinear problems in fluid mechanics; the monograph edited by Ewing (1983), in which some of the numerical-analysis aspects of the simulation of displacement processes are discussed; and the reviews by Bensimon et al. (1986) and Saffman (1986), which deal exclusively with fingering in Hele-Shaw flows.

Mechanisms of Viscous Fingering

Consider a displacement in a homogeneous porous medium, characterized by a constant permeability K. The flow will typically involve the displacement of a fluid of viscosity μ_1 and density ρ_1 by a second of viscosity μ_2 and density ρ_2. These differences in physical properties may result from using two different, immiscible phases, or from injection of a solvent fully miscible with fluid 1. It is the *variation* of these properties across some front that is important. As noted above, we limit ourselves to the forces of gravity, viscosity, and (if the fluids are immiscible) surface tension. In the case of miscible systems in which differences in viscous forces are due to differences in solute concentration, we must also consider the molecular diffusion and mechanical dispersion of the solute.

The following simple argument may be made in order to understand the basic mechanism of the instability. Under suitable continuum assumptions, the flow may be taken to satisfy Darcy's law, which for a one-dimensional steady flow may be written

$$\frac{dp}{dx} = -\mu U/K + \rho g. \tag{1}$$

Now consider a sharp interface or zone where density, viscosity, and solute concentration all change rapidly, e.g. a zone such as that shown in Figure 2*a*. Then the pressure force (p_2-p_1) on the displaced fluid as a result of a virtual displacement δx of the interface from its simple convected location is

$$\delta p = (p_2-p_1) = [(\mu_1-\mu_2)U/K+(\rho_2-\rho_1)g]\delta x. \tag{2}$$

If the net pressure force is positive, then any small displacement will amplify, leading to an instability. Thus we see that a combination of unfavorable density and/or viscosity ratios and flow *direction* can conspire to render the displacement unstable. For example, for downward vertical displacement of a dense, viscous fluid by a lighter, less viscous one, we have $(\mu_1-\mu_2) > 0$, $(\rho_2-\rho_1) < 0$, and $U > 0$. Thus, gravity is a stabilizing force, while viscosity is destabilizing, leading to a critical velocity U_c above which there is instability:

$$U_c = (\rho_1-\rho_2)gK/(\mu_1-\mu_2). \tag{3}$$

There are three other obvious cases depending upon the signs of $\Delta\rho$, U, and $\Delta\mu$: one in which gravity drives the instability and viscosity stabilizes it, and the two cases when both basic forces are either stabilizing or destabilizing.

A simpler statement may be made when the gravity force is absent, e.g. in a horizontal displacement. In this case, instability *always* results when a more viscous fluid is displaced by a less viscous one, since the less viscous fluid has the greater mobility. Thus we see that the two basic forces responsible for the instability are gravity and viscosity. More refined analyses, discussed below, will show that surface tension and/or dispersion can modify but not stabilize a flow characterized as unstable by this simple criterion.

As in many areas of fluid mechanics, interest lies in the behavior of the flow for conditions that exceed the critical limits, and perhaps the most fascinating behaviors are those that occur for highly supercritical conditions. In this respect, viscous fingering is no exception.

A Historical Note

It is interesting to trace the literature in order to establish priority for the discovery and understanding of viscous fingering in terms of the fluid mechanics involved. Despite the fact that this instability is discussed in many fluid-mechanics textbooks and literature papers as the "Saffman-Taylor Instability" and is attributed to Saffman & Taylor (1958), the phenomenon had been noted and recorded in many earlier works, although not always with a clear understanding of the mechanics and the basic mechanism. The first scientific study of viscous fingering can reasonably be attributed to Hill (1952), who not only published the simple "one-dimensional" stability analysis given above, but who also conducted a series of careful and quantitative experiments for both gravity-stabilized viscous fingering in vertical downflow and the curious counter-case of viscous stabilization of a gravitationally unstable configuration in the case of vertical downward displacement of a light, less viscous fluid by a heavy, more viscous one, which can be stabilized for velocities above U_c. In all cases, the critical displacement velocity measured via flow visualizations was in *quantitative* agreement with Equation (3). Those interested in either the history of viscous fingering or an example of concise scientific writing should consult this beautiful paper. The next significant development occurred in the late 1950s, when Chouke et al. (1959) and Saffman & Taylor (1958) published their now-classical papers. Both these papers, submitted within six months of one another, contain essentially identical linear-instability analyses of one-dimensional displacement, leading to Equation (7) below. Significantly, Saffman & Taylor state that "the result is not essentially new, and that mining engineers and geologists have long been aware of it," and they further attribute the inclusion of surface tension in the analysis to "Dr. Chouke." They then go on to study experimentally the evolution and shape of the now-famous single dominant finger and to discuss its theoretical description. The paper of Chouke et al., on the other hand, refers only to a presentation of the linear-instability analysis by Chouke at a technical meeting in 1958, with no mention of Saffman & Taylor, although they do reference Hill's work. Given the delays involved in publishing research from an industrial laboratory, it seems plausible that credit for the first rigorous stability *analysis* of viscous fingering be given to Chouke, but that the phenomenon under discussion should almost certainly be called the "Hill Instability." There is unfortunately little chance of this designation gaining general acceptance.

HELE-SHAW FLOWS

The Simplification of Hele-Shaw Flows

Since most porous materials are opaque, a convenient analogue to study is that of the Hele-Shaw model, the geometry of which is shown in the definition sketch in Figure 3. Flow takes place in a small gap of thickness b, and there is a second, macroscopic dimension L. Hele-Shaw theory and experiments seek to describe the two-dimensional features of the flow, *depth-averaged* over the thin gap. It is well known that in *single-phase* flow, these two-dimensional equations are

$$\nabla \cdot \bar{\mathbf{u}} = 0, \tag{4a}$$

$$\nabla \bar{p} = -12\mu\bar{\mathbf{u}}/b^2 + \rho\mathbf{g}, \tag{4b}$$

which hold in the limit of low-Reynolds-number flow, and $e \equiv b/L \to 0$. We see that the flow satisfies Darcy's law, in which the equivalent permeability of the medium is simply $b^2/12$. Thus, *single-phase* Hele-Shaw flow is analogous to two-dimensional incompressible flow in porous media. It might be supposed that the same analogy would hold for viscous fingering in porous materials. However, as we shall see, the analogy is imperfect. In the case of miscible fluids with concentration gradients, Taylor dispersion will occur due to the velocity profile in the thin dimension, making the mixing or dispersion characteristics in Hele-Shaw flow highly anisotropic in ways that porous materials may not be. In particular, if mechanical dispersion is important, it usually affects both the longitudinal and transverse dispersion coefficients in porous media, whereas in Hele-Shaw flow, only the longitudinal component is affected.

The analogy fails completely in the case of flow of immiscible fluids, since flow in porous materials in this case is truly *multiphase*, as opposed to two phase, and the forces associated with the propagation of menisci through the pore space comprising the medium cannot be neglected and are not modeled in the Hele-Shaw geometry. Furthermore, the geometry of the solid matrix can also influence the fingering patterns observed. Nevertheless, the case of Hele-Shaw flow is of fundamental interest in its own right and permits us to establish some useful concepts. Furthermore,

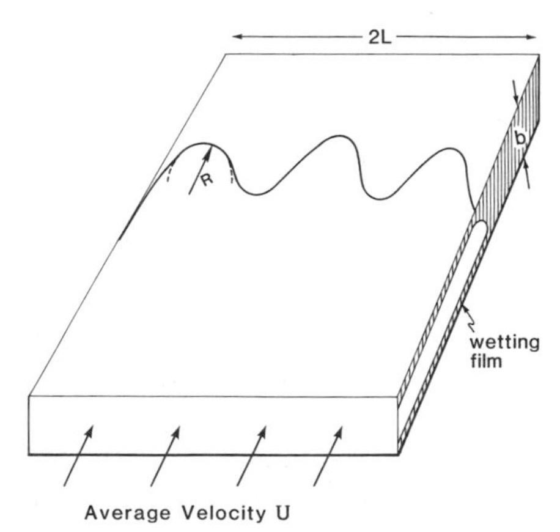

Figure 3 Detailed schematic of Hele-Shaw flow in the case where the displaced fluid wets the wall. The interface moves with normal velocity U, surface tension γ acts on the interface, and R is the radius of curvature in the plane.

it allows us to discuss the processes of *shielding*, *spreading*, and *splitting* alluded to in the Introduction.

Immiscible Displacements in Hele-Shaw Cells

In the Appendix of their paper, Saffman & Taylor (1958) showed that equations of the form of (4) also govern the flow of two *immiscible* phases in Hele-Shaw cells, provided that the film of displaced fluid left on the plates is of constant thickness, but that the effective viscosities and densities appearing therein depend upon the property ratios as well as the thickness ratio. Thus we adopt them as the field equations in each phase in regions away from the interface between the phases. It remains to give boundary conditions that apply at the location of the interface, taken as a *two-dimensional depth-averaged* surface. The only analyses of the details of the flow in the thin dimension, which must then be depth averaged in order to give jump conditions for the fields $\bar{\mathbf{u}}$ and $\bar{p}$, are those of Park & Homsy (1984) and Reinelt (1986). For the case in which the displaced fluid wets the wall, these conditions depend upon the local capillary number of the flow, $\mathrm{Ca} = \mu U/\gamma$, as well as on the magnitude of the surface tension γ. In terms of the quantities defined in Figure 3, these conditions are, for small Ca,

$$[|\bar{p}|] = \frac{2\gamma}{b}(1 + 3.8\,\mathrm{Ca}^{2/3} + \quad) + \frac{\gamma}{R}[\pi/4 + O(\mathrm{Ca}^{2/3})], \tag{5a}$$

$$[|n\cdot\bar{\mathbf{u}}|] = O(\mathrm{Ca}^{2/3}). \tag{5b}$$

Most analyses of Hele-Shaw flows pertain to the limit $\mathrm{Ca} \equiv 0$. In this case, the equations simplify to

$$[|\bar{p}|] = 2\gamma/b + \gamma\pi/(4R), \tag{6a}$$

$$[|n\cdot\bar{\mathbf{u}}|] = 0. \tag{6b}$$

Much of the literature on the subject uses these boundary conditions, with the constant 1.0 appearing in place of $\pi/4$, but a simple rescaling of surface tension allows these existing results to be carried over without change. This constant may be simply computed as a function of the contact angle in the thin dimension (Park 1985) and is set to unity in what follows. Furthermore, the leading constant in (6a) is often set to zero without loss of generality. Equations (4), together with these simpler jump conditions [Equations (6)], have been referred to as the *Hele-Shaw equations*, and they hold asymptotically in the limit of small capillary number and small ratios of the gap thickness to any macroscopic dimension. These equations have been the object of much study, and many features of their solutions have been discussed elsewhere (Aref 1986, Bensimon et al. 1986, Saffman 1986).

The Hele-Shaw equations may be scaled in a straightforward way by using a macroscopic length as the characteristic length and L/U as the characteristic time to show that solutions may be expected to depend upon three basic parameters:

$$\mathrm{Ca}' = \frac{12\mu_1 U}{\gamma}(L/b)^2 \quad \text{modified capillary number,}$$

$$A = \frac{(\mu_1 - \mu_2)}{(\mu_1 + \mu_2)} \quad \text{viscosity contrast,}$$

$$G = \frac{(\rho_1 - \rho_2)gb^2}{12(\mu_1 + \mu_2)U} \quad \text{modified Darcy-Rayleigh number.}$$

The modified capillary number measures the viscous forces relative to surface tension. Other equivalent definitions for this parameter are also in use, involving other numerical factors depending upon the choice of length and velocity scales, and the inverse of Ca′ is sometimes used, as it occurs naturally in the boundary conditions of the problem. Unfortunately no universal convention for this controlling parameter has been adopted in the literature, and care must be taken in comparing results of different investigators. The group G similarly measures the relative importance of buoyant to viscous forces. The property ratio A is self-explanatory.

The linear-stability analysis of one-dimensional rectilinear displacements with the action of surface tension is due to Chouke et al. (1959). In the usual fashion, if disturbances are taken to be of the form of normal modes proportional to $\exp(\sigma t + iky)$, one finds the following dispersion relation for the growth constant of the instability:

$$(\mu_1 + \mu_2)\sigma = \left[U(\mu_1 - \mu_2) + \frac{(\rho_1 - \rho_2)gb^2}{12}\right]k - \frac{\gamma b^2 k^3}{12}, \tag{7a}$$

or in dimensionless form,

$$\sigma = (A + G)k - \frac{(A+1)k^3}{2\,\mathrm{Ca}'}. \tag{7b}$$

It may be seen that the critical speed for which $\sigma > 0$ is given accurately by the simple analysis of Hill (1952), since it holds for long waves, but that for unstable situations, the inclusion of surface tension leads to a wave number of maximum growth rate and a cutoff wave number. For the simple case of both gravity and viscosity driving the instability, the maximum growth rate occurs for a wave number

$$k_{\mathrm{m}} = \left[\frac{2(A+G)\,\mathrm{Ca}'}{3(A+1)}\right]^{1/2}. \tag{8}$$

The simple physical interpretation of these results is that surface tension will damp short waves, whereas the basic mechanism favors them, leading to competing effects and the occurrence of a preferred mode in a manner similar to many other stability problems. The ratio of unstable length scales to the macroscopic scale decreases as Ca′ increases, which implies that many scales of motion are allowed in the limit of very large capillary number (small surface tension). Thus we expect the dynamics of fingering to become complex in this limit.

It has proven difficult to quantitatively verify the dispersion relation [Equation (7)], but the available body of information is in general agreement with it. Many of the experiments pertain to relatively simple measurements of the apparent wavelength of fingers in their early stages of growth, as typified by those of White et al. (1976). These measurements are found to compare favorably with those given by Equation (8). Park et al. (1984) have provided the only available measurements of growth rates and summarize previous investigations on the initial development of the instability. Recently, Schwartz (1986) has reanalyzed the linear-stability problem using Equation (5a) rather than (6a), which results in an improved agreement between theory and the experiments of Park et al., but the data are not sufficiently accurate to provide a critical test of the theory leading to (6a).

The linear instability of radial source flow was first treated by Wilson (1975), and thereafter by Paterson (1981). The velocity field in radial source flow is given by

$$u = Q/r, \tag{9}$$

where Q is the two-dimensional source strength. The characteristic scalings change somewhat for this case, but a modified capillary number still determines the cutoff scales. Since both the velocity and characteristic length change with time, the analysis of the linear instability is based upon a quasi-static analysis that treats the velocity U at a radius R as locally constant. The results, in dimensionless form, are

$$\sigma = Am - 1 - \mathrm{Ca}'m(m^2-1)(A+1)/2, \tag{10}$$

where A and Ca′ are defined above using an instantaneous velocity and radius, and m is a discrete azimuthal wave number. Analysis of Equation

(10) is simple only when $A = 1$, in which case all waves with wave numbers satisfying

$$m \leq 1/2[1+(1+4\text{Ca}')^{1/2}] \tag{11}$$

are unstable, with a wave number of maximum growth rate, as before.

Experiments by Paterson (1981), shown in Figure 4*a*, show the evolution of fingering patterns in source flow for displacement of glycerine with air. For short times, there is reasonable agreement with the expectations of the linear theory. Radial source flow is also a good short-time approximation to the flow in a five-spot pattern, and the above theoretical results should hold for that case as well.

The classical experimental study of Saffman & Taylor focused on the nonlinear evolution of fingers in the limit $A = 1$, $G = 0$. They found that after the initial instability, a single finger appeared to dominate the flow. The reason for this can be understood simply in terms of *shielding*. Since the tendency is for fingers of mobile fluid to grow in the direction of the pressure gradient in the more viscous fluid, a finger slightly ahead of its neighbors quickly outruns them and shields them from further growth. (An equivalent argument can be made using the fact that the pressure in the less mobile phase is harmonic and the interface is nearly isopotential, leading to a larger flux of fluid near the tip of any finger that is ahead of neighboring ones.)

Steady solutions describing finite-amplitude solutions in rectilinear flow have been much discussed in the recent literature, which we briefly summarize here. Saffman & Taylor (1958) sought solutions of the Hele-Shaw equations in the limit $A = 1$, $\text{Ca}' = \infty$, $G = 0$. They found that there was a continuous family of solutions, all of which were linearly unstable (Taylor & Saffman 1958). The traditional parameter characterizing these solutions is λ, the ratio of the finger width to the characteristic macroscopic length. Thus λ is not uniquely determined for $\text{Ca}' = \infty$. McLean & Saffman (1981) and, more recently Vanden Broeck (1983) solved the steady free-boundary problem numerically for finite Ca' and examined the limit $\text{Ca}' \gg 1$. There exists an apparently infinite family of discrete solutions, $\lambda_n(\text{Ca}')$, all of which tend toward the same shape with $\lambda_n = 1/2$ as $\text{Ca}' \to \infty$.

Large-scale numerical simulations of the evolution of such finger shapes have been provided by Tryggvason & Aref (1985), DeGregoria & Schwartz (1986), and Liang (1986). Tryggvason & Aref (1983, 1985) applied vortex-in-cell methods to study a number of problems in viscous fingering, and their simulations show the typical growth of the dominant finger by shielding. The results they obtain for the steady shape, for the cases in which they have good numerical accuracy ($\text{Ca}' \leq 50$), are in agreement with the

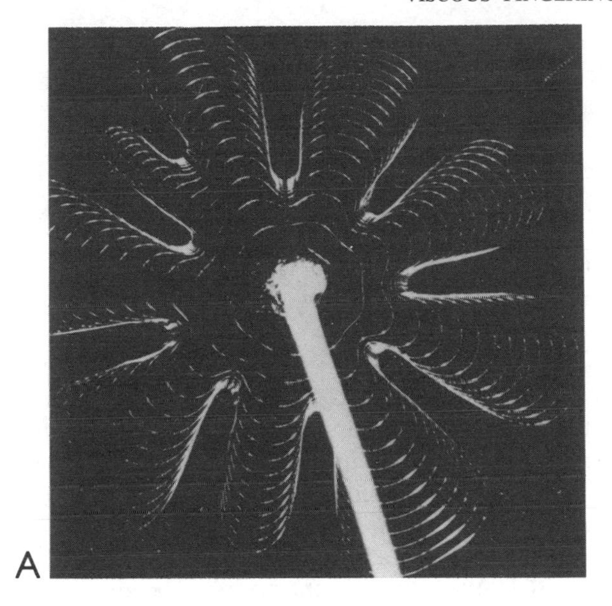

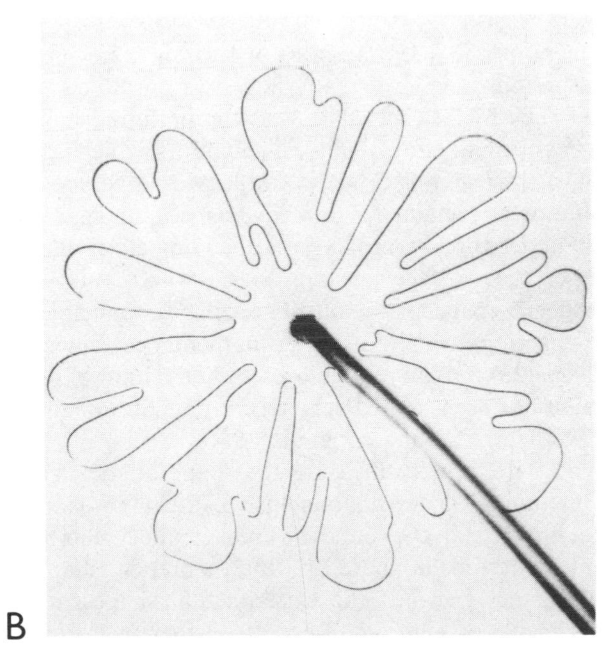

Figure 4 Fingering in immiscible radial source flow into a Hele Shaw cell. (*a*) Multiple exposure showing growth at early time; (*b*) tip splitting at later time (Paterson 1981). With permission.

branch calculated by McLean & Saffman (1981). DeGregoria & Schwartz (1986) developed a solution technique based on the fact that from Equations (4), the pressure satisfies Laplace's equation if the viscosity is constant almost everywhere. Thus, boundary integral techniques are appropriate. They too reproduce the McLean & Saffman branch, extending the range of numerical solutions to $\mathrm{Ca}' \cong O(10^3)$. Finally, Liang (1986) has recently discussed a numerical method based upon random-walk simulations of the solution of Laplace's equation, which is claimed to be superior to conventional techniques for very large capillary numbers and which goes over to solutions of diffusion-limited aggregation (DLA), discussed below, when $\mathrm{Ca}' = \infty$. The physical significance of certain key features of this method are obscure, however. His simulations show shielding, of course, and evolve to the McLean & Saffman branch of solutions. Thus, at present the other branches of steady shapes have not been realized in any initial-value calculation, which has led to speculation that these branches are unstable; calculations in support of this speculation are given by Kessler & Levine (1986c). Most intriguing in the simulations of DeGregoria & Schwartz and of Liang is the fact that as Ca′ becomes large, the steady finger shapes are not always stable. We discuss these observations in more detail below.

We also mention the recent analysis of Kessler & Levine (1986b), who show that a periodic array of steady fingers is unstable to any spanwise modulation. They further speculate that the preferred mode is one corresponding to an annihilation of nearest neighbors, which leads to a pairing process that, if the resulting finger is stable, would persist until only one finger from the array remained. The mechanism of this spanwise instability can be easily understood in terms of the shielding effect discussed above.

Thus we see that surface tension, *however weak*, acts to spread the dominant finger to a particular width. This will be important in interpreting patterns of fingering at high Ca′. The mathematical description of how this *selection mechanism* can remain present in the limit of extremely small surface tension has been treated very recently by Shraiman (1986), Hong & Langer (1986), and Combescot et al. (1986).

Experimental attempts to measure the shape of the steady dominant finger as a function of Ca′ are not conclusive. Pitts (1980) reports a curious experimental finding that the shapes at finite capillary number form a self-similar family, a result not shared by the theoretical shapes. There also exists a discrepancy between the experimental shapes and any of the theoretical branches of solutions. Recently, Tabeling & Libchaber (1986) have suggested that this may be due to the fact that experiments have not been done in the regime of applicability of the Hele-Shaw equations. With reference to Equations (5) and (6), the Hele-Shaw equations pertain to the double limit $e \to 0$, $\mathrm{Ca} \to 0$, $\mathrm{Ca}' = e^{-2}$ Ca finite. Uniform neglect of the additional pressure jump due to the existence of a wetting film, the $O(\mathrm{Ca}^{2/3})$ term in Equation (5a), requires $e\ \mathrm{Ca}^{-2/3} \gg 1$. It is difficult to satisfy all of these restrictions in experiments. Generally, the experiments of Saffman & Taylor and of Pitts were carried out for *finite* Ca, with e moderately small. The experiments of Tabeling et al. (1986), in which e was varied systematically, were again generally for small e but for values of Ca as large as 1.0. The experiments of Park & Homsy (1985), which utilized a very small aspect-ratio cell, were done for small Ca and $e\ \mathrm{Ca}^{-2/3} \approx O(0.5)$. Thus we see that many of the widely quoted experimental studies have not been conducted in the region of validity of the Hele-Shaw equations, and comparison between experiment and theory must be approached with some caution. Numerical solutions of the equations that include the parameters e and Ca separately instead of the combination $\mathrm{Ca}' = e^{-2}$ Ca are not available, but the ability of Tabeling & Libchaber (1986) to collapse available data on finger width using a modified scaling parameter based upon Equation (5a), is encouraging. Quantitative investigation of the effect of the wetting layer must await solution to a set of equations that take explicit account of its existence.

There are also unanswered questions regarding the stability of the single dominant finger. Taylor & Saffman (1958) and McLean & Saffman (1981) provided stability analyses of the steady nonlinear shapes that indicated that such shapes should be *unstable* for infinite and finite Ca′, respectively. However, the experiments of Saffman & Taylor and of Pitts indicated stability. Recently, a number of workers have shown experimentally that the single finger is subject to a tip-splitting instability (Nittmann et al. 1985, Park & Homsy 1985, Tabeling et al. 1986). Partial rationalization of this paradox between the observation of stable fingers on the one hand and prediction of instability on the other has been provided by Kessler & Levine (1986a,c), who suggest that previous analyses of the stability of the single finger are inaccurate. [The stability calculations of McLean & Saffman are apparently incorrect (Saffman 1986).] Kessler & Levine analyze the full spectrum of the linear operator for instability modes that grow in space, and they solve the resulting eigenvalue problem by numerical means. The range of accurate numerical results is limited to $\mathrm{Ca}' \leq O(10^3)$, for which steady fingers are stable. Unfortunately, the numerical solution of the linear-stability problem becomes ill conditioned at high Ca′, in a manner similar to the Orr-Sommerfeld equation at high Reynolds number. Thus reliable results for very large Ca′ are not available, but Kessler & Levine speculate that stability persists to $\mathrm{Ca}' = \infty$. The numerical results of DeGregoria & Schwartz (1986) show that the McLean & Saffman branch of steady solutions is apparently stable for $\mathrm{Ca}' < 10^3$

but unstable to a symmetrical tip-splitting mode for $Ca' > 10^3$. They also provide numerical evidence that the onset of the tip-splitting instability depends upon the amplitude of the imposed perturbation, indicating that the instability is a *subcritical bifurcation*. They further argue that the bifurcation, in addition to being subcritical, is from infinity, i.e. fingers are stable to infinitesimal perturbations for all but infinite Ca', a point of view that is only partially supported by their calculations. DeGregoria & Schwartz (1986), using an improved version of their algorithm that allows asymmetrical solutions, have observed tip splitting above $Ca' = 1.25 \times 10^3$. Liang (1986), using random-walk simulations, has also observed both steady fingers and tip splitting. However, as noted above, several aspects of the algorithm used are not clear. As an apparent result of insufficient spatial resolution, he observes splitting for very low Ca', for which the dominant finger should be linearly stable. An attempt to analyze the nature of the bifurcation has been made by Bensimon (1986). His analysis clearly indicates the subcritical nature of the instability, but because of the qualitative nature of the model, in which he considers perturbations that are not dynamically admissible, further work is needed to settle the issue of the nature of the bifurcation.

Careful experimental studies of the tip-splitting instability are only now beginning. Park & Homsy (1985) were the first to attempt to measure the critical Ca' for onset of splitting, which they report to be $Ca' = 600$. Since the instability is subcritical, this value may not stand the test of time; indeed, Tabeling et al. (1986) report delay in the onset of splitting to $Ca' = 2\text{–}3 \times 10^3$ by reducing the noise level in the experiment. Much more experimental work needs to be done in order to characterize the critical conditions for onset of tip splitting, as well as the regime of flow in the postinstability regime. Figure 5 shows a comparison between the experiments of Park (1985) and the calculations of DeGregoria & Schwartz (1985) and of Liang (1986). The qualitative similarity between the patterns observed is encouraging in the sense that the experimental behavior appears to be contained within the Hele-Shaw equations.

The mechanism by which such patterns are formed can be understood in the following way. As surface tension becomes weak, the front of the steady finger is susceptible to a viscous-fingering instability by the basic mechanism associated with a less viscous fluid displacing a more viscous one. After a split, each of the new lobes of the finger is stable as a result of their being thinner than the finger from which they split. As a result of *shielding*, one of these lobes will eventually outgrow the other and, owing to surface tension, will then *spread* to occupy the appropriate width of the cell. In the process, it reaches a width that is again unstable to *splitting*, and the pattern repeats. Thus, surface tension plays a subtle but essential dual role; it must be weak enough for the tip front to be unstable, but it is also the physical force causing the spreading and ensuing repeated branching. In this sense, it is analogous to the dual role of viscosity in the instability of parallel shear flows. We thus arrive at a scenario in which *shielding*, *spreading*, and *splitting* are all important processes in determining the dynamics of viscous fingering.

Once the existence of the tip-splitting instability is known and the mechanism understood at least qualitatively, more complicated patterns such as those observed by Maxworthy (1986), shown in Figure 1*b*, and those of Paterson (1981), in Figure 4, can be understood. Maxworthy's experiments (which incidentally are far outside the region of validity of the Hele-Shaw equations) were done at extremely large Ca′ ($\mathrm{Ca}' = 5 \times 10^4$), values for which tip splitting takes place over a range of length scales, leading to a cascade of splittings, including secondary splittings of the side branches. Such primary and secondary events may be clearly identified in the sequence of photographs leading to Figure 1*a* (Maxworthy 1986; see also the photographs published by Reed 1985). This cascade continues in time, leading to geometrical shapes that are not simple and that may have statistical features describable by a fractal geometry [see Maxworthy (1986) and Nittmann et al. (1985) for a further exposition of this idea].

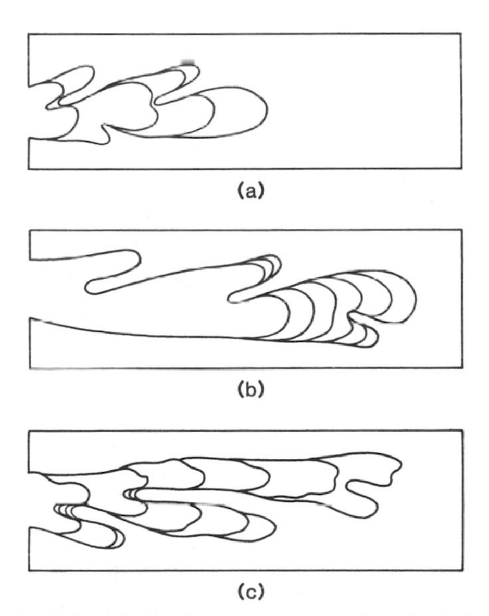

Figure 5 Comparison of tip splitting in experiments and in simulations of the Hele-Shaw equations: (*a*) tracings from experiments for $\mathrm{Ca}' = 1.3 \times 10^3$ (Park 1985); (*b*) simulations for $\mathrm{Ca}' = 1.3 \times 10^3$ (DeGregoria & Schwartz 1985); (*c*) simulations for $\mathrm{Ca}' = 600$ (Liang 1986). With permission.

The observations of Paterson (1981), shown in Figure 4*b*, as well as similar ones by Ben-Jacob et al. (1985) in radial source flow may be similarly understood. For a given Ca′, there will be a preferred azimuthal wave number from Equation (10) when the critical radius is reached. Following growth of the preferred mode, the lobes of the pattern again spread laterally as a result of surface tension until they exceed a local stability limit and undergo another fingering instability at the tip. Although this will presumably continue indefinitely, little experimental work past the first one or two generations of splittings has been reported, and there is no evidence one way or the other for a regime of fractal splitting.

Few studies have been made to characterize the effect of the other two dynamic groups, G and A, on fingering. Maher (1985) reports observations in small Hele-Shaw cells. By using two fluids near a thermodynamic critical point, he is able to achieve small A while maintaining immiscible conditions. Because the mobilities of both phases are nearly equal, the strong shielding present when one phase essentially controls the pressure drop is reduced or eliminated, with the result that all fingers grow, at least for short times, and the pattern is not characterized by the emergence of the single dominant finger. It is not clear if this regime persists for long times and large displacements, but at short times it is clearly qualitatively different from the $A = 1$ case. Furthermore, the experimental observations are in good qualitative agreement with the calculations of Tryggvason & Aref (1983) for small A. Without the growth of the dominant finger and its subsequent splitting instability, the patterns of fingering are also correspondingly different.

Although the effect of the gravity group G is superficially similar to that of an imposed flow in that density differences impose a vertical pressure gradient far from the boundary of the finger, the dynamic effects of gravity and viscous forces are not interchangeable except when the interface is nearly flat and perpendicular to the motion, as indicated in Equation (7b) above. Thus, much work remains to be done to fully characterize the parametric dependence of nonlinear viscous fingering upon the remaining groups G and A.

Miscible Displacements in Hele-Shaw Cells

An important consequence of immiscibility in the case of Hele-Shaw flow is that the viscosity is constant in each phase, with a jump at the interface. This is not necessarily the case for miscible displacements. In miscible displacements, there is no interface, and as a result a *single-phase* Darcy's law [Equation (4)] holds throughout the domain. Viscous fingering may still be driven, however, by variations in viscosity that result from variations in the concentration of a chemical component in the fluid. If we let the concentration be denoted by c, scaled between zero and one, then the equations governing the system are

$$\nabla \cdot \mathbf{u} = 0, \tag{12a}$$

$$\nabla p = -\mu(c)\mathbf{u}/K + \rho g, \tag{12b}$$

$$\frac{Dc}{Dt} = \nabla \cdot (\mathbf{D} \cdot \nabla c), \tag{12c}$$

$$\mu = \mu(c). \tag{12d}$$

(Note that although we are discussing Hele-Shaw flow, we have written the permeability as K rather than $b^2/12$, as the theory for homogeneous porous media, treated in the next section, is *identical* to Hele-Shaw theory in the special case of isotropic dispersion.) The concentration is taken to obey a convection-diffusion equation, perhaps with an anisotropic dispersion tensor, and the relation between concentration and viscosity must also be given. If we select a macroscopic length L, a viscosity μ_1, and a velocity U as scaling parameters, then the solution to these equations depends upon the following dimensionless parameters:

$\mathrm{Pe} = UL/D_0$	Peclet number,
$A = \dfrac{\mu_1 - \mu_2}{\mu_1 + \mu_2}$	Viscosity contrast,
$G = \dfrac{\Delta\rho g K}{(\mu_1 + \mu_2)U}$	Gravity group,
$\mathbf{D}^*(u)$	Dimensionless dispersion function,
$\mu^*(c);\ M^{-1} < \mu^* < 1$	Dimensionless viscosity function, where M is the viscosity, or mobility, ratio μ_1/μ_2.

In these definitions, μ_1, μ_2 are the viscosities without solvent and with the maximum solvent concentration, respectively; D_0 is a reference value of the dispersion coefficient, usually taken as the zero-velocity diffusion limit or the longitudinal dispersion at the reference velocity; and $\mathbf{D}^*$, μ^* are dimensionless functions describing the material behavior of the medium and the solute/solvent mixture, respectively.

Discussion of the linear stability theory for this case is complicated by the fact that, unlike the case of miscible displacements, there is no simple *steady* solution to the relevant equations, since dispersion will always act to render the concentration profile and hence the viscosity profile time dependent. Thus we must deal with the stability of time-dependent base states, a subject of some difficulty. However, consider for the moment the case of infinite Peclet number (zero dispersion). For rectilinear flow, the solvent concentration will be independent of time in a convected coordinate system, and the above stability results, specialized for the case of zero surface tension, apply. We see that

$$\sigma = (A+G)k, \tag{13}$$

a nonphysical result that indicates that smaller and smaller wavelengths are increasingly unstable. Clearly, dispersion or some other physical effect will act on such short-wavelength disturbances, leading to a cutoff. The first analysis of the effect of dispersion was given by Chouke almost 30 years ago, and reported recently in the Appendix to Gardner & Ypma (1982). Expressed in the present variables, for the case of a jump in viscosity (i.e. a base-state profile corresponding to zero axial dispersion), but allowing both axial and transverse dispersion to act on disturbances, Chouke's result reads

$$\sigma = 1/2[Ak - \mathrm{Pe}^{-2}k^3 - k(\mathrm{Pe}^{-2}k^2 + 2A\mathrm{Pe}^{-1}k)^{1/2}] \tag{14}$$

with a cutoff wave number

$$k_c = \mathrm{Pe}A/4, \tag{15}$$

while the growth rate is a maximum at

$$k_m = \mathrm{Pe}(2\sqrt{5}-4)A/(4) \approx 0.12\mathrm{Pe}A. \tag{16}$$

The effect of gravity on these results may be simply included through the parameter G, as in the above discussion. As in the case of immiscible displacements, there is a physical parameter that leads to a cutoff length scale, in this case the Peclet number. Correspondingly, we again expect complex behavior when the Peclet number becomes very large. Thus we see that transverse dispersion is responsible for controlling the length scales of fingers, even when axial dispersion has not yet distorted the concentration profile.

There have been many subsequent attempts to analyze the stability characteristics of miscible displacements, including the dispersive widening of the zone of viscosity variation. Heller (1966) approximated the profiles by straight-line segments and invoked a quasi-static approximation in neglecting the time dependence of the base state relative to the growth of disturbances. Schowalter (1965) dealt with fingering driven by both density and viscosity variations and used a constant-thickness diffusive zone to describe the mobility profile. He assumed a *steady* base state (which is not allowed by the equations), special boundary conditions, and a competition between viscosity and density stratifications. Both these works give dispersion relations with cutoff wave numbers, but the questionable assumptions make their general validity suspect. Wooding (1962) made one of the few attempts to treat the stability of a time-dependent base state. He considered buoyancy-driven fingering and expressed the disturbance quantities as a Hermite expansion. By analyzing a one-term truncation, he concluded that the dispersion relation has the general characteristics of Equation (14) above, but that the cutoff length shifts with time to larger values, corresponding to the dispersive widening of the profile, and that ultimately all disturbances must decay if dispersion is given an infinite time to act. Recently, Tan & Homsy (1986) have discussed this problem in the limit $G = 0$, $\mu^*(c) = \exp(-c \ln M)$ for fingering in rectilinear flow in unbounded domains. They treat both isotropic and highly anisotropic media and solve for the stability of the time-dependent mobility profiles by utilizing both a quasi-static assumption and a numerical solution of the initial-value problem for the growth of small-amplitude fingers. They find that for the isotropic case, the analysis of Chouke is essentially correct in predicting the magnitudes of the growth rates and preferred wave numbers, but that as time proceeds dispersion acts to shift the scales to larger wave lengths and to stabilize the entire flow somewhat. Figure 6 shows typical results for the quasi-static growth rates for fingering in infinite domains, which indicate these general trends. Tan & Homsy also utilize Chouke's technique to determine how anisotropic dispersion characteristics influence the linear-stability characteristics. Not surprisingly, small transverse dis-

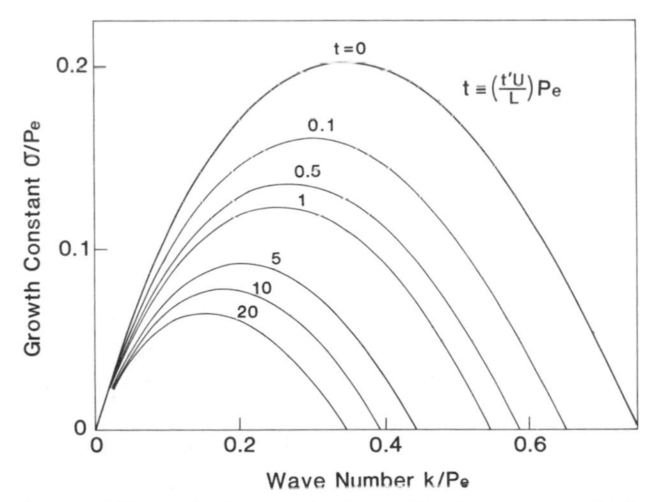

Figure 6 Linear-stability results for fingering in miscible displacements with isotropic dispersion $M = 20$ (Tan & Homsy 1986). With permission.

persion will result in a shift to smaller length scales. Recently, Hickernell & Yortsos (1986) have analyzed the stability of rectilinear displacements for the case when there is a zone of mobility variation of finite thickness but zero dispersion. This corresponds to the situation in which the solvent is injected in varying amounts over time, resulting in a spatially varying mobility profile. In this case, the finite thickness of the zone of viscosity stratification provides a cutoff scale, similar to the "regularization" of other ill-posed problems [see Aref (1986) for examples].

Curiously, there are no experimental studies of fingering in rectilinear flows in Hele-Shaw cells with which the theory can be *quantitatively* compared, although there have been a large number of experiments, characterized by those of Benham & Olson (1963), which record results of engineering importance, such as the time history of the production of solute. Experiments in porous media, which are closely related, are discussed below. Buoyancy-driven fingering, while outside the primary scope of this review, deserves limited mention here. Wooding (1969) has reported observations of such instabilities for $A \ll 1$, $G \neq 0$ in Hele-Shaw cells and has discussed the evolution of nonlinear fingers. Some of his photographs are reproduced in Figure 7. Here dispersion plays a dual and subtle role similar to that of surface tension in the immiscible case. Transverse dispersion sets the initial length scale of fingering, but it also leads to the lateral *spreading* of fingers in the nonlinear regime. As spreading occurs, the tips of fingers may become unstable as their characteristic breadth exceeds the cutoff scale. Thus *tip splitting* may occur in *miscible* as well as in *immiscible* flows, and it is the primary mechanism of pattern formation. The similarity lies in the fact that physical phenomena—dispersion and surface tension, respectively—are responsible both for determining the allowable scales of *splitting* and for causing the tip *spreading*. An important difference, at least as far as these experiments are concerned, is that substantial *shielding* does not take place, as it does when $A = 1$. In Figure 7*a*, the Peclet number is apparently below a critical value, *which is presently unknown*, and tip splitting does not occur, whereas Figure 7*b* shows the tip splitting that occurs at higher Peclet number. We mention here the simulations of Tryggvason & Aref (1983) in the limit of small A and large Ca$'$, which bear a superficial resemblance to the fingering pattern of Wooding. However, as they discuss, these simulations cannot describe spreading and the continuous change in the length scales caused by dispersion. Numerical methods capable of accurately describing the long-time effects of weak dispersion at very high Peclet numbers are not presently available. In fact, the pioneering attempt at such a simulation by Peaceman & Rachford (1962) failed because of the dominance of numerical errors and has led to a misconception in the petroleum-engineering literature that

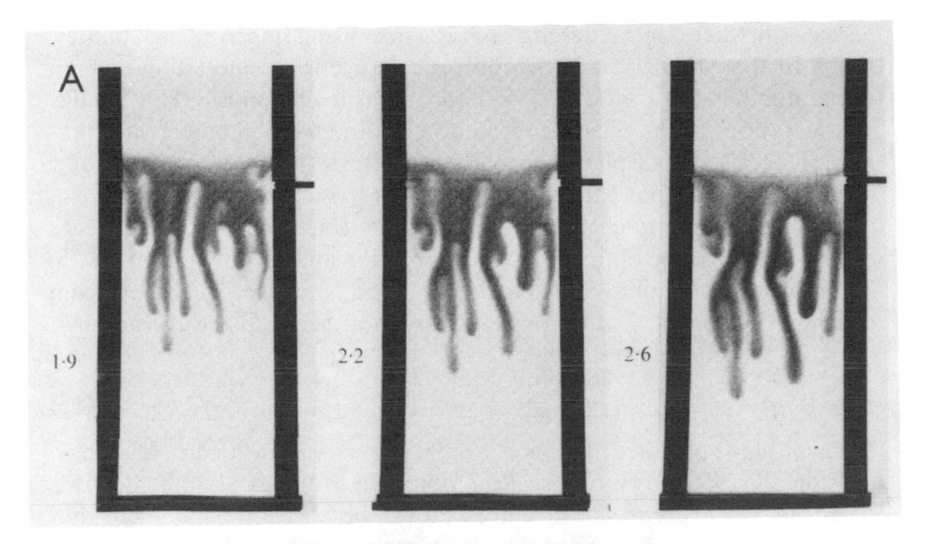

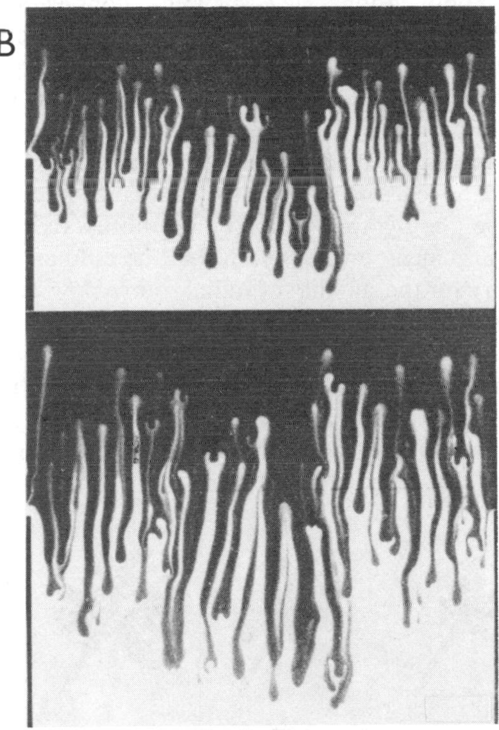

Figure 7 Fingering in gravity-driven miscible displacements showing (*a*) nonlinear fingering with transverse spreading at moderate Peclet number, and (*b*) tip splitting at high Peclet number (Wooding 1969). The different panels represent a time sequence. With permission.

permeability heterogeneities are *essential* in causing fingering instabilities. For a further discussion of the difficulty in accurate simulation at high Peclet number, see Claridge (1972) and Christie & Bond (1986). Similar considerations of the accurate resolution of dispersion in numerical simulations of miscible displacements also pertain to other geometries; for a discussion of five-spot patterns, see Ewing (1983).

The linear-stability characteristics of radial source flow in the absence of dispersion can be obtained directly from the analysis of Wilson (1975) and have been summarized by Paterson (1985). Not surprisingly, we again obtain a dispersion relation that has no cutoff azimuthal wave number, i.e. from Equation (10) we obtain

$$\sigma = Am - 1. \tag{17}$$

Paterson has suggested that as a result, fingers will occur in Hele-Shaw cells on all scales down to a scale comparable to the gap width. If this is true, then in experiments, a cutoff occurs because of phenomena not included in the two-dimensional Hele-Shaw equations leading to Equation (17) above. He gives a heuristic argument based upon energy dissipation in the gap, by which this cutoff, which scales with the gap width b, may be computed. Shown in Figure 8 is a set of photographs from his paper that display fingering in miscible displacement of glycerine by water. The process of areal spreading followed by tip splitting is a familiar one, being reminiscent of that occurring in immiscible radial source flow. The small tick marks on the photographs indicate the cutoff wavelength estimated by the heuristic argument, and it is seen that the estimate is qualitatively correct. No theory for the stability of radial source flow that incorporates dispersion is currently available by which one could distinguish cutoff by dispersion vs. cutoff by three-dimensional effects in these experiments, but it seems clear that the latter predominate, at least in this range of Peclet numbers.

Let us return to Figure 1*a*, which shows the fingering observed in

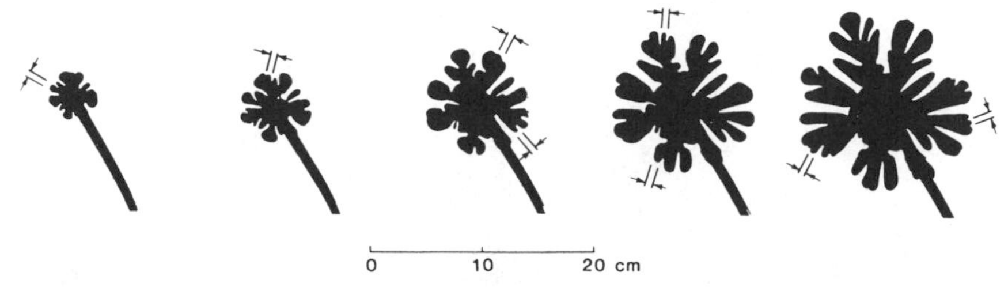

Figure 8 Fingering in miscible radial Hele-Shaw source flow (Paterson 1985). The tick marks indicate the scale of the gap width. With permission.

extremely thin-gap Hele-Shaw cells (E. L. Claridge, personal communication, 1986). [Numerous tracings of similar experiments, which unfortunately do not capture the finer details of these patterns, are to be found in Stoneberger & Claridge (1985).] Examination of a time sequence of such patterns reveals that fingers grow, spread by transverse dispersion, and split at the tips, in the same manner as described by Wooding (1969) for buoyancy-driven fingering. In this case, fingering does not occur down to the scale of the gap width, this length being exceedingly thin in these experiments (0.01 cm). This would seem to indicate that the heuristic argument of Paterson (1985) that fingering occurs at all scales down to the Hele-Shaw gap width is not generally applicable. Unfortunately, too few experiments that specifically study the effect of the Peclet number are available, and in many cases insufficient data exist to allow a calculation of Peclet numbers for the conditions of a given experiment. This is a curious situation, given the key role dispersion plays in fingering in miscible systems, and it is hoped that careful experiments at high Peclet numbers might be reported in the near future. In summary, we have seen that the phenomena of *shielding*, *spreading*, and *splitting* are all important in determining the dynamics of fingering in miscible displacements, that the parameter analogous to the modified capillary number is the Peclet number, and that interesting and complex behavior occurs at high Peclet number.

We close this subsection by discussing the intriguing work of Nittmann et al. (1985) and Daccord et al. (1986). They conducted experiments in both rectilinear flow and radial source flow of water into a Hele-Shaw cell initially containing a non-Newtonian, viscoelastic fluid. The fingering patterns observed are qualitatively different from those observed with conventional fluids—compare Figure 1*c* with Figures 1*b* and 8. In these experiments the pattern grows by extremely localized tip splitting, and the shielding is much stronger than in the case of Newtonian fluids. They present data suggesting that the geometry of the patterns has a fractal structure, and the measured fractal dimension is close but not equal to that obtained from simulations of diffusion-limited aggregation (DLA), discussed below. Visual comparison with DLA clusters reveals a lesser degree of side branching in the experiments. They argue that such a choice of fluid pairs is necessary in order to obtain such structures, since Newtonian fluids do not have sufficiently high viscosity contrast. Such a claim does not stand up under close scrutiny, since the viscosity ratio enters the problem as the parameter A, not the absolute ratio, and for large viscosity ratios, the dominant length scales become sensibly independent of this ratio. Thus, systems of any reasonable viscosity contrast, such as those for the systems used by Paterson and by Claridge, have $A = 1$, but do not exhibit such ramified patterns. It is quite probable that these results, while fascinating, are linked to the viscoelastic nature of the displaced fluid used in their experiments. It is tempting to speculate on the mechanism that leads to dramatic structures such as those shown in Figure 1*c*, but until more work is done, these structures will remain unexplained and in sharp contrast to the results available for Newtonian fluids.

POROUS MEDIA FLOWS

There is a distinction to be drawn, where appropriate, between viscous fingering in model systems like Hele-Shaw cells and "real" porous materials, since some of the phenomena are distinctly different and some of the mathematical models used are less fundamentally based. Many of the mechanisms discussed above still pertain; however, in many cases it becomes necessary to recognize the particulate nature of the medium. In this section we review the available understanding of fingering in real media, with the anticipation that this knowledge will be incomplete in significant ways. The description of fingering will in turn depend strongly on the length scales at which we choose to view the flow. Thus we face at the outset a fundamental issue of the level of detail of description of pore-level fluid mechanics that is desirable or even possible. There is at present no general agreement on this issue, and therefore we briefly discuss both continuum approaches, which seek to describe the fingering on scales large compared with typical pore dimensions, and discrete approaches, which are essentially discontinuous descriptions. As we shall see, both approaches involve large elements of modeling and simulation, as opposed to fundamental prediction obtained as a result of solving an agreed-upon set of equations.

Miscible Displacements in Porous Media

If we begin by adopting a continuum description of miscible displacements in which we average over scales comparable to the scales of pores or grains of the medium, then the single-phase Darcy equations with variable viscosity hold [Equations (12)]. Furthermore, we recognize that the effective dispersion tensor of the medium arises as a result of mixing and mechanical dispersion on the pore scale. Thus we have in effect accomplished whatever averaging is necessary to obtain the continuum description, which will certainly be valid in describing any instability that is smooth on the continuum length scale. The theoretical studies of fingering in miscible systems reviewed above pertain *equally* to Hele-Shaw cells and to porous media as long as the dispersion tensor is evaluated appropriately. The dynamic dimensionless groups are as we have indicated above, with the Peclet number the main determinant of the scale of fingering. We expect, therefore, to find a close similarity between Hele-Shaw results and porous-media results in the limit of isotropic dispersion.

As we have seen, both transverse and longitudinal dispersion lead to preferred wavelengths for fingering, and that in the strict absence of dispersion, fingering will occur on all scales, with growth rates that increase with decreasing scale. Scales will ultimately be reached for which the continuum hypothesis no longer holds, and a pore-level description is then appropriate. Thus the nature of fingering at very high Peclet number may be more conveniently described in noncontinuum terms. As noted above, there are very few continuum analyses available of either the linear-stability characteristics or the nonlinear miscible fingering. Those that exist indicate that as dispersion acts to spread the mobility profile, there is a shift to longer wavelength fingers. No analyses to date have been able to describe tip splitting in the nonlinear regime.

There have been many experimental studies of fingering in porous materials, most of which have appeared in the petroleum-engineering literature. In these studies, great care has typically been taken to ensure the homogeneity of the media by constructing them from sand or other unconsolidated materials. Since porous materials are generally opaque, very few visualizations of fingers exist in thick media. We take special note of some early X-ray studies by Slobod & Thomas (1963) and Perkins et al. (1965). As an illustrative example, we show in Figure 9 two of the X-ray pictures of Slobod & Thomas taken during the same experiment in quasi-two-dimensional horizontal displacement. Viscous fingers whose characteristic length is many times the characteristic pore size are clearly evident, as is the trend toward longer scales as time progresses. Note the similarity between these patterns and those of Wooding (1969) in Hele-Shaw cells, shown in Figure 7*a*. Unfortunately, insufficient time-resolved information is available to determine the mode of finger growth in these experiments, i.e. whether it occurs by tip splitting or simply by spreading due to transverse dispersion. Since the transverse Peclet number is comparatively low, the latter seems the more probable. Tan & Homsy (1986) have compared the apparent wavelengths in visualizations like Figure 9*a* to the linear-stability calculations shown in Figure 6, with reasonable agreement between theory and experiment.

Although the X-ray visualizations apparently did not show tip splitting, there are other experiments that do, most notably the celebrated five-spot experiments of Habermann (1960), done in relatively thin media and shown in Figure 10*a*. Although the Peclet number is not known for these experiments and no time sequences are shown, it is obvious that significant growth of the pattern has occurred at the tips due to shielding, and that repeated tip splitting has also occurred, resulting in the characteristic pattern shown. Note the similarity with the Hele-Shaw experiments of Claridge (Figure 1*a*), also done at high Peclet numbers. It is also interesting to note that the fractional recovery of the displaced fluid as a function of mobility ratio M is nearly equal in the porous-media experiments of Habermann and the Hele-Shaw experiments of Stoneberger & Claridge (1985), which indicates the similarity of the mechanics in the two cases. There are many empirical correlations of the fractional recovery that indicate that this quantity is relatively insensitive to the Peclet number for sufficiently large Pe, which implies that the large-scale *geometric* properties of the pattern become independent of Pe as $\mathrm{Pe} \to \infty$. This is a concept familiar to researchers who study the large-scale features of turbulent flows at high Reynolds numbers.

In spite of the geometrical complexity of Figures 1*a* and 10*a*, in both cases the dominant length scales are much larger than any microscopic pore scale. This is not true when one studies fingering at ever-increasing Peclet numbers at ever-decreasing scales. Paterson et al. (1982) have studied miscible fingering in packed beds that shows patterns similar to those occurring in DLA. Obviously, for a given pore scale there is a crossover between the continuum and discrete descriptions, but very little quantitative information is available to indicate when this crossover occurs.

A challenging and interesting question pertains to the structure of nonlinear fingering in the limit in which phenomena leading to a cutoff scale are entirely missing, i.e. the case of zero dispersion or molecular diffusion in the examples above. Because of the properties of the linear-stability theory in this limit, the initial-value problem is ill posed, but we may seek solutions that have discontinuities or other singularities. Neglecting dispersion preserves sharp jumps in viscosity profiles for all times, with the result that the pressure obeys Laplace's equation, with prescribed boundary conditions at the moving boundary. Various types of singularities can occur in the solution as a result, and very little work exists on

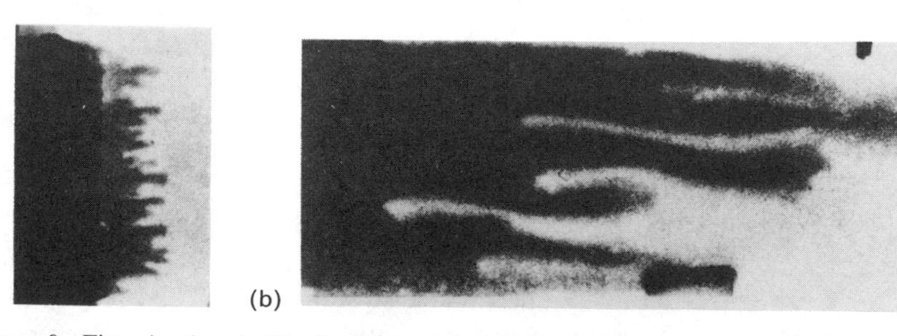

Figure 9 Fingering in miscible displacement in opaque media, at two different times, as visualized by X-ray absorption (Slobod & Thomas 1963). Compare Figure 9*b* with Figure 7*a*. © 1963 by SPE-AIME: with permission.

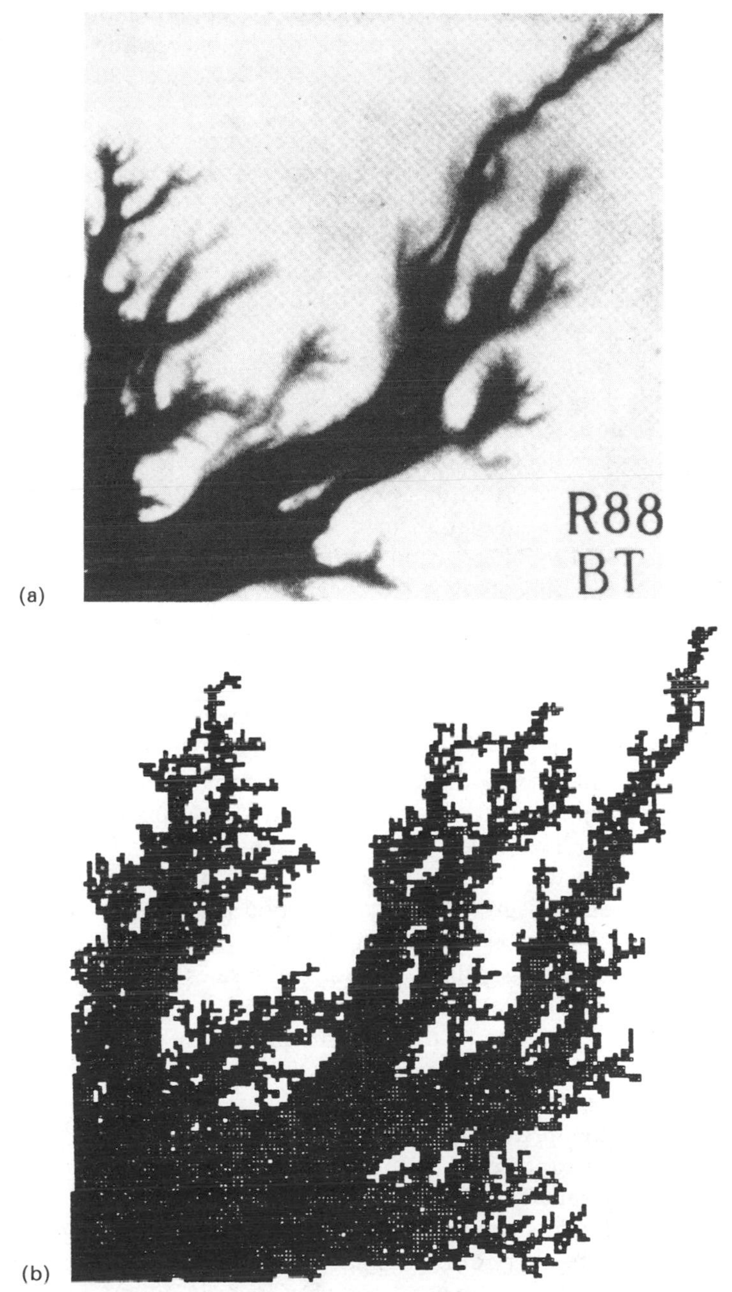

Figure 10 Fingering in five-spot patterns: (*a*) two-dimensional porous-media experiments of Habermann (1960) © 1960 by SPE-AIME; (*b*) lattice simulations of Sherwood (1986b). Compare these two with Figure 1*a*. With permission.

the subject. However, we can anticipate that different nonuniformities will appear in different boundary-value problems. In this regard, we mention the work of Shraiman & Bensimon (1984) and Howison (1986a,b) on cusp formation in fingering problems.

Also of interest are the large number of recent papers that attempt to construct discontinuous solutions by random-walk calculations of the type used in DLA. The first observation of the relevance of DLA to fingering in porous media is due to Paterson (1984), who commented on the formal equivalence between DLA and fingering in miscible systems at infinite Peclet numbers and $A = 1$, and who reported simulations in which he claimed to describe fingering in five-spot patterns. Similar approaches, but modified to account for different geometries, boundary conditions, and finite mobility ratios, have been discussed by Sherwood & Nittmann (1986), DeGregoria (1985), and Sherwood (1986a,b). For $A = 1$, these approaches capture the effect of shielding quite accurately, and growth of fingers is seen to occur primarily at the tips. Figure 10*b* shows the results of a two-dimensional simulation in a five-spot pattern at the point of breakthrough for $A = 1$ (Sherwood 1986a). When this simulation is compared with the experiments of E. L. Claridge (personal communication, 1986) and of Habermann (1960), a striking geometrical similarity is observed. The resulting pattern, since it can access all scales above the scale of the lattice on which the simulation is performed, has the geometrical properties of a fractal object. The analogy with DLA is not so clear in the case of finite mobility, as the rules for sticking the random walkers to the aggregates must be modified to account for the finite mobility of the "aggregate" phase. However, Sherwood & Nittmann (1986) and DeGregoria (1985) have suggested rules for simulations at finite mobility ratio that advance the interface at a point with a probability proportional to the local pressure gradient. Sherwood & Nittmann argue that such rules model physical dispersion, but the physical significance of some aspects of these algorithms, including the rules for trapping of the displaced phase, is not clear.

While these approaches to discrete simulations of viscous fingering seem to hold much promise, there are at least two drawbacks to their quantitative accuracy. First, at their present state of development, they do not account for the *spreading* of fingers due to weak dispersion, and as a result they cannot capture all of the physics of tip splitting that are so important in determining much of the fascinating structure of fingering patterns. Second, many parameters of scientific and engineering interest, e.g. the fractional recovery of the displaced fluid, depend upon the lattice size, sometimes sensitively so (see, e.g., DeGregoria 1985). Thus, these simulations involve an artificial cutoff related to a lattice size, which at present cannot be related to any quantity of physical significance and which must be used as a fitting parameter in order to obtain results other than those represented solely by the *geometry* of the fingered pattern, i.e. other than the fractal dimension. DeGregoria (1985) has argued that once this size is fit to a given physical system, the effect of varying the mobility ratio may be *quantitatively* predicted, and he gives a calculation of the fractional recovery as a function of mobility ratio that agrees well with experiments. Further research needs to be done to connect these discrete simulations to smooth but steep solutions of the continuum equations.

Immiscible Displacements in Porous Media

The problem of viscous fingering in immiscible systems is arguably one of the most difficult problems pertaining to porous media flow. This is so because of the wide variety of phenomena occurring that involve a myriad of pore-scale phenomena, including the details of wetting behavior and wetting films, the movement of contact lines and dynamic contact angles, the static stability of capillary bridges, the dynamic instability of blobs of immiscible, nonwetting fluid, the propagation of these blobs, the phase transitions during flow, the heat and mass transfer across interfaces, and many others. Some of these areas have been touched on in the review by Wooding & Morel-Seytoux (1976), and some of them are sufficiently rich that they have been the subject of specialized reviews in this series [e.g. the review on "ganglia" mechanics by Payatakes (1982), that on contact lines and contact angles by Dussan V. (1979), and that on capillary instabilities by Michael (1981). Further compounding the complexity of the subject is the fact that the distributions of menisci within the material, together with the configuration of phases within the medium, are under some circumstances strongly coupled to the topology of the underlying solid matrix, the description of which is a subject of much current research in the field of disordered media. It is clear that we cannot give a detailed account of all these issues, so our discussion is necessarily limited to those situations in which *viscosity* plays a major role. As we have commented above, this focus excludes a large and interesting emerging literature on the slow, capillary-dominated propagation of menisci on lattices and within disordered materials, which finds a natural setting within the framework of percolation phenomena.

Once again, we face the major problem that "real" media are opaque, and as a result very few visualizations are available, certainly not enough for us to understand the detailed dependence of viscous fingering upon the parameters of the process. As in the case of miscible fingering, instructive observations are again ones that were done some years ago, in this instance by van Meurs (1957). Some of these visualizations, accomplished by working with displaced fluids of the same index of refraction as the medium, have already been reproduced in the article by Wooding & Morel-Seytoux (1976), and the interested reader is urged to consult that reference. This beautiful time sequence shows that the processes of *shielding*, *spreading*, and *splitting* are present in real porous materials and determine the patterns of fingering in essential ways. Shown in Figure 11 are similar observations of fingering from Chouke et al. (1959), which show the variation of the length scales of fingering with increasing velocity and viscosity contrast, i.e. with increasing capillary number. Fingering apparently takes place on many scales, including a macroscopic one, and thus one might assign a characteristic macroscopic length scale (if not a wavelength) to what is seen. It is obvious from these and other visualizations that the scale of fingering changes, becoming smaller when either the velocity or the viscosity of the displaced fluid is increased. From these and other observations on both three- and quasi-two-dimensional media (see, e.g., Peters & Flock 1981, Paterson et al. 1984a,b, White et al 1976, Måløy et al. 1986, and the references quoted therein), the following may be observed:

1. Wetting properties are important: There is a qualitative difference in fingering when the invading fluid does or does not wet the medium. In the former case fingering is characterized by some macroscopic continuum scale, whereas in the latter fingering is more likely to be confined to the pore scale, with shielding dominating over spreading, as might be expected.
2. Characteristic macroscopic scales, if present, decrease with increasing capillary number $\mathrm{Ca} = \mu_1 U/\gamma$.
3. When the invading fluid is nonwetting, the pattern is a probe of the topology of the microstructure and is characteristic of percolation behavior, with a backbone that may have a fractal character.
4. There is simultaneous flow of both phases in a zone behind a displacement front, assuming such a front may be identified.

The above observations are oversimplifications of complex behavior, but they will help to fix our ideas.

The first attempt to provide a theoretical analysis of the onset of fingering was by Chouke et al. (1959). They assumed that there was *complete* displacement of one fluid by the other and ignored the zone of partial saturation or volume concentration of the displacing fluid behind the front. This reduces the equations in each phase to the single-phase Darcy equations [Equations (4)]. In order to avoid the short-wave catastrophe that, as we have seen, occurs, Chouke took the bold step of applying the jump condition [Equation (6a)] at the front, but with a constant γ^* that is different from the molecular surface tension. He provides a heuristic justification based upon energy arguments, which is incorrect. This has been referred to as "the Chouke boundary condition," and the resulting predictions will be referred to as Hele-Shaw-Chouke theory. This boundary condition has been used in similar contexts by White et al. (1976), Peters & Flock (1981), Paterson et al. (1984a), and many others. It is by no means clear how surface tension, acting as it must on menisci at the

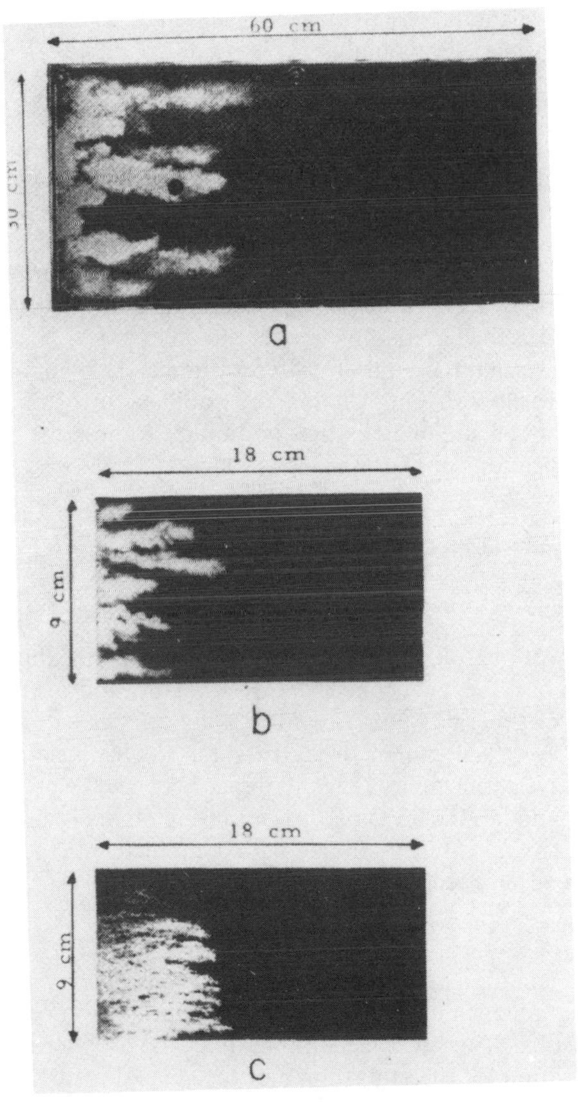

Figure 11 Fingering in immiscible displacement in porous media (Chouke et al. 1959): (*a*) Ca = 2×10^{-5}; (*b*) Ca = 1.2×10^{-4}; (*c*) Ca = 3×10^{-3}. © 1959 by SPE-AIME: with permission.

pore level, can provide a restoring force proportional to macroscopic curvature, as implied by Equation (6a). However, it is found that estimates of the characteristic macroscopic length from observations such as those shown in Figure 11 compare well with those given by Equation (8) if the empirical "effective surface tension" γ^*, which is found to depend upon wetting conditions, is fit at one value of the capillary number. This is essentially equivalent to the statement that the length scale of fingers scales with the modified capillary number $\mathrm{Ca}' = (\mu_1 UL^2/\gamma^* K)$ as

$$k \sim O(\mathrm{Ca}'^{1/2}). \tag{18}$$

Alternative theoretical descriptions of fingering attempt to model the fact that variations in the saturation behind the front can reduce the mobilities of both phases there, and that because of these saturation gradients, there will be a tendency, depending upon the wetting conditions, for one fluid to displace the other by the action of capillary imbibition. This leads to a process of *spreading*, which appears in the equations [Equations (22) below] as a diffusive effect.

While it is clear that the continuum description of such pore-scale events of simultaneous flow of both phases and capillary invasion is not well understood, the set of equations that is thought to describe these events involves modifications of Darcy's law as follows:

$$\nabla p_i = -\lambda_i^{-1}\mathbf{u}_i, \qquad i = 1, 2, \tag{19}$$

where

$$\lambda_i = K_i(c)/\mu_i \tag{20}$$

is the mobility of phase i, and the quantity K_i is the permeability of phase i, dependent upon the local concentration or saturation of the phases, c. Without loss of generality, we take c to be the relative saturation of the invading phase. It is common to factor K_i as the intrinsic permeability times the relative permeability. The difference between the pressures in the phases, which models the invasion by the wetting phase, is also dependent upon saturation and is taken to scale with a characteristic pore scale and the molecular surface tension, i.e.

$$p_1 - p_2 \equiv \gamma P_c(c)/K^{1/2}, \tag{21}$$

which is the defining equation for P_c, the dimensionless capillary pressure. These equations must be augmented by continuity equations for the individual phases. With suitable manipulations, the problem reduces to an evolution equation for the saturation c of the general form

$$\frac{\partial c}{\partial t} + \nabla \cdot \left(\frac{\mathbf{u}_0}{1 + M^{-1}} \right) = \mathrm{Ca}^{*-1} \nabla \cdot [G(c) \nabla c] \tag{22a}$$

with

$$M(c) = \lambda_2(c)/\lambda_1(c) \tag{22b}$$

and

$$G(c) = -\left(\frac{dP_c}{dc}\right)\left(\frac{1}{1+M^{-1}}\right). \tag{22c}$$

Here $\mathbf{u}_0$ is the volume average velocity of the mixture, and the modified capillary number Ca* is defined as

$$\text{Ca}^* \equiv \mu_1 UL/\gamma K^{1/2}, \tag{23}$$

which differs from that appearing in Hele-Shaw theory by the factor involving the ratio of macroscopic to microscopic length scales. As noted, these equations have some physical intuition embodied in them and are routinely used in the petroleum-engineering literature, but they cannot be rigorously justified and have never been derived from a pore-level description, even for simple geometries. For a further discussion of these equations, see Wooding & Morel-Seytoux (1976) and the references therein.

Solutions of these equations will depend upon the modified capillary number, the mobility function $M(c)$, and the capillary pressure function $P_c(c)$. It will be important later to note that models from percolation theory and data suggest that $M(c)$ vanishes at a finite value of c, related to the percolation threshold. Analyses of these equations in the limit of $\text{Ca}^* = \infty$ go under the general designation of Buckley-Leverett theory. Since Equation (22a) with $\text{Ca}^* = \infty$ is a nonlinear hyperbolic system, it is not surprising that the solutions develop shocks at a finite time, with a rarefaction wave of saturation that disperses as time proceeds. For specific choices of the function $M(c)$, simple one-dimensional solutions are well known. For large but finite Ca*, the problem is a singularly perturbed one, and Wooding (1975) has analyzed the continuous structure of the front region using matched expansions. He finds that this region is $O(\text{Ca}^{*-1})$ in thickness, as opposed to the more probable $O(\text{Ca}^{*-1/2})$ for nonlinear diffusion problems.

Stability analyses of these one-dimensional solutions have only just recently begun (Yortsos & Huang 1984, Huang et al. 1984, Jerauld et al. 1984a,b). There are several conceptual as well as practical difficulties in performing such analyses. The first is that the profiles are inherently time dependent, as in the case of miscible displacements, and thus one must discuss the stability of time-dependent flows. This has been circumvented in some cases by applying inflow boundary conditions corresponding to injection of both phases at saturations that allow steady solutions to the

Buckley-Leverett equations. Even then, the predictions are specific to the choices of the material functions $M(c)$ and $P_c(c)$, so only a few features of a general nature are known. The first is that, unlike Hele-Shaw-Chouke theory, it is incorrect to infer stability or instability from evaluation of the *viscosity* ratio of the fluids; rather, it is the *mobility* contrast at the front, together with the saturation profile behind the front, that determines stability. In this sense the stability characteristics share much in common with graded mobility processes analyzed by Gorell & Homsy (1983) for immiscible processes, and by Hickernell & Yortsos (1986) for miscible displacements at infinite Peclet number. Thus, depending upon the mobility function, a displacement that has an unfavorable viscosity ratio may still be linearly stable, even at infinite Ca*. Yortsos & Huang (1984) have analyzed the linear stability by using a quasi-static approximation for the saturation profiles, and straight-line approximations to the mobility profiles, in the spirit of Heller (1966). They find that without capillary pressure, there is no cutoff scale, which is a familiar property of sharp fronts with mobility jumps but with no dissipative mechanism. With capillary pressure, they find a cutoff wave number

$$k_c \sim O(\mathrm{Ca}^*), \tag{24}$$

as opposed to that given by Hele-Shaw-Chouke theory [Equation (18)]. Huang et al. (1984) have analyzed more general spatially varying profiles. They find certain profiles that are neutrally stable even in the absence of capillary pressure. They also solve the eigenvalue problem by numerical means, a process that becomes difficult at high Ca*, in a manner similar to the ill conditioning that occurs in Hele-Shaw theory at high Ca′, in miscible displacements at high Peclet number, and in Orr-Sommerfeld theory at high Reynolds numbers. They find that the critical viscosity ratio below which displacement is stabilized by the graded mobility profile may be simply estimated from the inlet saturations and the mobility jump from Buckley-Leverett theory. Chikhliwala & Yortsos (1985) have provided long-wave expansions of the eigenvalue problem that they claim capture most of the qualitative features of the numerical calculations. Jerauld et al. (1984a,b) have analyzed a class of steady solutions of the Buckley-Leverett equations in a manner similar to that of Yortsos & Huang (1984) and Chikhliwala & Yortsos (1985), including long-wave expansions and numerical solutions to the related variable-coefficient eigenvalue problem, and find similar results of stabilization due to spatially varying mobility profiles, and stability at viscosity ratios greater than one. They also comment on the differences in predicted scales given in Equations (18) and (24). The linear-stability calculations done to date must be considered as providing only preliminary understanding of the solution properties of a set of *model* equations, which must be further tested and validated by comparison with more extensive experimentation.

As can be seen, there is a formal similarity between immiscible displacements, described by the fractional flow equations (22) above, and miscible displacements, described by Equations (12). Differences appear in the material behavior of the function $M(c)$ or $\mu^*(c)$, in the fact that $M(c)$ has a zero, and in the fact that the dispersive term in each continuity equation has different functionalities, being isotropic and dependent upon saturation on the one hand, and anisotropic and dependent upon velocity on the other. However, it is probable that these differences become only ones of detail in the limits of $\mathrm{Pe} \to \infty$ and $\mathrm{Ca}^* \to \infty$, respectively.

Insufficient experimental evidence exists to adequately test any of these predictions. Both approaches provide scalings, given in Equations (18) and (24), which capture the qualitative trends shown, e.g., in Figure 11 of decreased scales of fingers with increasing velocity and displaced fluid viscosity, but careful experiments that vary all the parameters over wide ranges and that distinguish the differences between Equations (18) and (24) have yet to be reported. However, the observations by Chouke et al. (1959), White et al. (1976), Peters & Flock (1981) and Paterson et al. (1984a) provide compelling evidence in favor of the Hele-Shaw-Chouke theory, which in this author's opinion is presently unexplained.

Qualitative experimental studies on rectilinear displacements in two-dimensional etched networks by Lenormand & Zarcone (1985) have suggested that continuum model equations have a restricted range of validity. The authors conducted displacement experiments for both favorable and unfavorable mobility ratios as a function of capillary number $\mathrm{Ca} = \mu_1 U/\gamma$, where U is the average interstitial velocity. For favorable mobility ratios, they found that below $\mathrm{Ca} \sim 10^{-4}$–10^{-3}, displacement was by invasion percolation, for which a continuum description is not possible. For $\mathrm{Ca} > 10^{-3}$, displacement proceeds in a fashion appropriate to Buckley-Leverett theory. For unfavorable mobility ratios for which viscous fingering is possible, they observed a crossover at $\mathrm{Ca} = 10^{-8}$–10^{-7}, from invasion percolation behavior to that characterized by DLA, without observing a continuum regime. Lenormand (1985) has proposed a "phase diagram" of these different regimes that involves mobility ratio and capillary number, which can at present be considered to be only qualitative and specific to the apparatus used, and most certainly dependent upon wetting conditions.

Fingering resulting from invasion of nonwetting fluids into porous media remains a poorly understood and complicated subject. Recently, Måløy et al. (1986) have reported fingering in radial source flow at high capillary number, defined using the superficial velocity, in which the pattern is claimed to have a fractal geometry with dimension appropriate to DLA. Since this is an immiscible displacement of nonwetting fluid, it is not clear why DLA should apply at all, and since their apparatus consisted of only a monolayer of particles between solid plates, the results cannot be considered to have applicability to porous materials.

Recently, King (1985) and King & Scher (1986) have presented some probabilistic lattice simulations for both rectilinear and five-spot flow, similar in spirit to those of Sherwood and of DeGregoria discussed above, but for the Buckley-Leverett equations [Equations (22) in the singular limit of $Ca^* \to \infty$]. As stability analyses show, this is in general an ill-posed problem unless the mobility function and inlet saturations are specially chosen, and a cutoff due to the finite lattice spacing exists. The patterns produced are as intriguing and suggestive as those shown in Figure 10*b*, but as the authors point out, fractional recovery is sensitively dependent upon the lattice spacing. Again, this spacing must be connected to some parameter of physical significance before this approach can be generally useful. Finally, we note briefly the comments by Chen & Wilkinson (1985), who point out that the probabilistic rules used by DeGregoria, Sherwood, and King can be considered as models for the natural variation of pore-scale features such as pore radius, since patterns in deterministic network models with variable geometrical features are directly comparable to those observed in probabilistic simulations.

ACKNOWLEDGMENTS

Our work on viscous fingering and the preparation of this article were sponsored by the US Department of Energy, Office of Basic Energy Sciences, whose support is gratefully acknowledged. I am also pleased to acknowledge helpful discussions of viscous fingering with H. Aref, M. King, T. Maxworthy, P. G. Saffman, L. Schwartz, J. Sherwood, and Y. Yortsos.

Literature Cited

Aref, H. 1986. Finger, bubble, tendril, spike. *Pol. Acad. Sci. Fluid Dyn. Trans.* 13: In press

Benham, A. L., Olson, R. W. 1963. A model study of viscous fingering. *Soc. Pet. Eng. J.* 3: 138–44

Ben-Jacob, E., Godbey, R., Goldenfeld, N. D., Koplik, J., Levine, H., et al. 1985. Experimental demonstration of the role of anisotropy in interfacial pattern formation. *Phys. Rev. Lett.* 55: 1315–18

Bensimon, D. 1986. On the stability of viscous fingering. *Phys. Rev. A* 33: 1302–8

Bensimon, D., Kadanoff, L. P., Liang, S., Shraiman, B., Tang, C. 1986. Viscous flows in two dimensions. *Rev. Mod. Phys.* In press

Chen, J.-C., Wilkinson, D. 1985. Pore-scale viscous fingering in porous media. *Phys. Rev. Lett.* 55: 1892–95

Chikhliwala, E. D., Yortsos, Y. C. 1985. Theoretical investigations on finger

growth by linear and weakly nonlinear stability analysis. *SPE 14367*, Soc. Pet. Eng., Dallas, Tex.
Chouke, R. L., van Meurs, P., van der Poel, C. 1959. The instability of slow, immiscible, viscous liquid-liquid displacements in permeable media. *Trans. AIME* 216: 188–94
Christie, M. A., Bond, D. J. 1986. Detailed simulation of unstable processes in miscible flooding. *SPE/DOE 14896*, Soc. Pet. Eng., Dallas, Tex.
Claridge, E. L. 1972. Discussion of the use of capillary tube networks in reservoir performance studies. *Soc. Pet. Eng. J.* 12: 352–61
Combescot, R., Dombre, T., Hakim, V., Pomeau, Y. 1986. Shape selection of Saffman-Taylor fingers. *Phys. Rev. Lett.* 56: 2036–39
Daccord, G., Nittmann, J., Stanley, H. E. 1986. Radial viscous fingers and diffusion-limited aggregation: fractal dimension and growth sites. *Phys. Rev. Lett.* 56: 336–39
DeGregoria, A. J. 1985. A predictive Monte Carlo simulation of two-fluid flow through porous media at finite mobility ratio. *Phys. Fluids* 28: 2933–35
DeGregoria, A. J., Schwartz, L. W. 1985. Finger break-up in Hele-Shaw cells. *Phys. Fluids* 28: 2313–14
DeGregoria, A. J., Schwartz, L. W. 1986. A boundary-integral method for two-phase displacement in Hele-Shaw cells. *J. Fluid Mech.* 164: 383–400
Dussan V., E. B. 1979. On the spreading of liquids on solid surfaces: static and dynamic contact lines. *Ann. Rev. Fluid Mech.* 11: 371–400
Ewing, R. E., ed. 1983. *The Mathematics of Reservoir Simulation.* Philadelphia: SIAM. 186 pp.
Gardner, J. W., Ypma, J. G. J. 1982. An investigation of phase behavior–macroscopic bypassing interaction in CO_2 flooding. *SPE 10686*, Soc. Pet. Eng., Dallas, Tex.
Gorell, S., Homsy, G. M. 1983. A theory of the optimal policy of oil recovery by secondary displacement processes. *SIAM J. Appl. Math.* 43: 79–98
Habermann, B. 1960. The efficiency of miscible displacement as a function of mobility ratio. *Trans. AIME* 219: 264–72
Heller, J. P. 1966. Onset of instability patterns between miscible fluids in porous media. *J. Appl. Phys.* 37: 1566–79
Hickernell, F. J., Yortsos, Y. C. 1986. Linear stability of miscible displacement processes in porous media in the absence of dispersion. *Stud. Appl. Math.* 74: 93–115
Hill, S. 1952. Channelling in packed columns. *Chem. Eng. Sci.* 1: 247–53
Hong, D. C., Langer, J. S. 1986. Analytic theory of the selection mechanism in the Saffman-Taylor problem. *Phys. Rev. Lett.* 56: 2032–35
Howison, S. D. 1986a. Cusp development in Hele-Shaw flow with a free surface. *SIAM J. Appl. Math.* 46: 20–26
Howison, S. D. 1986b. Fingering in Hele-Shaw cells. *J. Fluid Mech.* In press
Huang, A. B., Chikhliwala, E. D., Yortsos, Y. C. 1984. Linear stability analysis of immiscible displacement including continuously changing mobility and capillary effects: Part II—general basic flow profiles. *SPE 13163*, Soc. Pet. Eng., Dallas, Tex.
Jerauld, G. R., Davis, H. T., Scriven, L. E. 1984a. Stability fronts of permanent form in immiscible displacement. *SPE 13164*, Soc. Pet. Eng., Dallas, Tex.
Jerauld, G. R., Nitsche, L. C., Teletzke, G. F., Davis, H. T., Scriven, L. E. 1984b. Frontal structure and stability in immiscible displacement. *SPE 12691*, Soc. Pet. Eng., Dallas, Tex.
Kessler, D. A., Levine, H. 1986a. Stability of finger patterns in Hele-Shaw cells. *Phys. Rev. A* 32: 1930–33
Kessler, D. A., Levine, H. 1986b. Coalescence of Saffman-Taylor fingers: a new global instability. *Phys. Rev. A* 33: 3625–27
Kessler, D. A., Levine, H. 1986c. Theory of the Saffman-Taylor finger. I & II. *Phys. Rev. A* 33: 2621–33, 2634–39
King, M. J. 1985. Probabilistic stability analysis of multiphase flow in porous media. *SPE 14366*, Soc. Pet. Eng., Dallas, Tex.
King, M., Scher, H. 1986. A probability approach to multi-phase and multi-component fluid flow in porous media. *Phys. Rev. A.* In press
Lenormand, R. 1985. Différents mécanismes de désplacements visqueux et capillaires en milieu poreaux: diagramme de phase. *C. R. Acad. Sci. Paris Ser. II* 310: 247–50
Lenormand, R., Zarcone, C. 1985. Two-phase flow experiments in a two-dimensional permeable medium. *PhysicoChem. Hydrodyn.* 6: 497–506
Liang, S. 1986. Random walk simulations of flow in Hele-Shaw cells. *Phys. Rev. A* 33: 2663–74
Maher, J. V. 1985. Development of viscous fingering patterns. *Phys. Rev. Lett.* 54: 1498–1501
Måløy, K. J., Feder, J., Jøssang, T. 1986. Viscous fingering fractals in porous media. *Phys. Rev. Lett.* 55: 2688–91
Maxworthy, T. 1986. The non-linear growth

of a gravitationally unstable interface in a Hele-Shaw cell. *J. Fluid Mech.* In press
McLean, J., Saffman, P. G. 1981. The effect of surface tension on the shape of fingers in a Hele-Shaw cell. *J. Fluid Mech.* 102: 455–69
Michael, D. H. 1981. Meniscus stability. *Ann. Rev. Fluid Mech.* 13: 189–215
Nittmann, J., Daccord, G., Stanley, H. E. 1985. Fractal growth of viscous fingers: quantitative characterization of a fluid instability phenomenon. *Nature* 314: 141–44
Park, C.-W. 1985. *Theory and experiment for instabilities in Hele-Shaw cells*. PhD thesis. Stanford Univ., Stanford, Calif. 123 pp.
Park, C.-W., Homsy, G. M. 1984. Two-phase displacement in Hele-Shaw cells: theory. *J. Fluid Mech.* 139: 291–308
Park, C.-W., Homsy, G. M. 1985. The instability of long fingers in Hele-Shaw flows. *Phys. Fluids* 28: 1583–85
Park, C.-W., Gorell, S., Homsy, G. M. 1984. Two-phase displacement in Hele-Shaw cells: experiments on viscously driven instabilities. *J. Fluid Mech.* 141: 257–87. Corrigendum: 144: 468–69
Paterson, L. 1981. Radial fingering in a Hele-Shaw cell. *J. Fluid Mech.* 113: 513–29
Paterson, L. 1984. Diffusion-limited aggregation and two-fluid displacements in porous media. *Phys. Rev. Lett.* 52: 1621–24
Paterson, L. 1985. Fingering with miscible fluids in a Hele-Shaw cell. *Phys. Fluids* 28: 26–30
Paterson, L., Hornof, V., Neale, G. 1982. A consolidated porous medium for the visualization of unstable displacements. *Powder Technol.* 33: 265–68
Paterson, L., Hornof, V. Neale, G. 1984a. Water fingering into an oil-wet porous medium saturated with oil or connate water saturation. *Rev. Inst. Fr. Pet.* 39: 517–21
Paterson, L., Hornof, V., Neale, G. 1984b. Visualization of a surfactant flood of an oil saturated porous medium. *Soc. Pet. Eng. J.* 24: 325–27
Payatakes, A. C. 1982. Dynamics of oil ganglia during immiscible displacement in water-wet porous media. *Ann. Rev. Fluid Mech.* 14: 365–93
Peaceman, D. W., Rachford, H. H. Jr. 1962. Numerical calculation of multidimensional miscible displacement. *Soc. Pet. Eng. J.* 2: 327–39
Perkins, T. K., Johnston, O. C., Hoffman, R. N. 1965. Mechanics of viscous fingering in miscible systems. *Soc. Pet. Eng. J.* 5: 301–17
Peters, E. J., Flock, D. L. 1981. The onset of instability during two-phase immiscible displacement in porous media. *Soc. Pet. Eng. J.* 21: 249–58
Pitts, E. 1980. Penetration of fluid into a Hele-Shaw cell. *J. Fluid Mech.* 97: 53–64
Reed, H. 1985. Gallery of fluid motion. *Phys. Fluids* 28: 2631–40
Reinelt, D. A. 1986. Interface conditions for two-phase displacement in Hele-Shaw cells. Submitted for publication
Saffman, P. G. 1986. Viscous fingering in Hele-Shaw cells. *J. Fluid Mech.* In press
Saffman, P. G., Taylor, G. I. 1958. The penetration of a fluid into a porous medium or Hele-Shaw cell containing a more viscous liquid. *Proc. R. Soc. London Ser. A* 245: 312–29
Schowalter, W. R. 1965. Stability criteria for miscible displacement of fluids from a porous medium. *AIChE. J.* 11: 99–105
Schwartz, L. 1986. Stability of Hele-Shaw flows: the wetting layer effect. *Phys. Fluids*. In press
Sherwood, J. D. 1986a. Island size distribution in stochastic simulations of the Saffman-Taylor instability. *J. Phys. A* 19: L195–200
Sherwood, J. D. 1986b. Unstable fronts in a porous medium. *J. Comput. Phys.* In press
Sherwood, J. D., Nittmann, J. 1986. Gradient-governed growth: the effect of viscosity ratio on stochastic simulations of the Saffman-Taylor instability. *J. Phys.* 47: 15–22
Shraiman, B. I. 1986. Velocity selection and the Saffman-Taylor problem. *Phys. Rev. Lett.* 56: 2028–31
Shraiman, B., Bensimon, D. 1984. Singularities in nonlocal interface dynamics. *Phys. Rev. A* 30: 2840–42
Slobod, R. L., Thomas, R. A. 1963. Effect of transverse diffusion on fingering in miscible-phase displacement. *Soc. Pet. Eng. J.* 3: 9–13
Stoneberger, M., Claridge, E. L. 1985. Graded viscosity bank design with pseudoplastic fluids. *SPE 14230*, Soc. Pet. Eng., Dallas, Tex.
Tabeling, P., Libchaber, A. 1986. Film draining and the Saffman-Taylor problem. *Phys. Rev. A* 33: 794–96
Tabeling, P., Zocchi, G., Libchaber, A. 1986. An experimental study of the Saffman-Taylor instability. Submitted for publication
Tan, C.-T., Homsy, G. M. 1986. Stability of miscible displacements in porous media: rectilinear flow. *Phys. Fluids*. In press
Taylor, G. I., Saffman, P. G. 1958. Cavity flows of viscous fluids in narrow spaces. *Symp. Nav. Hydrodyn., 2nd*, pp. 277–91
Tryggvason, G., Aref, H. 1983. Numerical experiments on Hele-Shaw flow with a sharp interface. *J. Fluid Mech.* 136: 1–30

Tryggvason, G., Aref, H. 1985. Finger-interaction mechanisms in stratified Hele-Shaw flow. *J. Fluid Mech.* 154: 287–301

Vanden Broeck, J.-M. 1983. Fingers in a Hele-Shaw cell with surface tension. *Phys. Fluids* 26: 2033–34

van Meurs, P. 1957. The use of transparent three-dimensional models for studying the mechanism of flow processes in oil reservoirs. *Trans. AIME* 210: 295–301

White, I., Colombera, P. M., Philip, J. R. 1976. Experimental study of wetting front instability induced by gradual change of pressure gradient and by heterogeneous porous media. *Soil Sci. Soc. Am. J.* 41: 483–89

Wilson, S. D. R. 1975. A note on the measurement of dynamic contact angles. *J. Colloid Interface Sci.* 51: 532–34

Wooding, R. A. 1962. The stability of an interface between miscible fluids in a porous medium. *ZAMP* 13: 255–66

Wooding, R. A. 1969. Growth of fingers at an unstable diffusing interface in a porous medium or Hele-Shaw cell. *J. Fluid Mech.* 39: 477–95

Wooding, R. A. 1975. Unsaturated seepage flow from a horizontal boundary. *Q. Appl. Math.* 33: 143–60

Wooding, R. A., Morel-Seytoux, H. J. 1976. Multiphase fluid flow through porous media. *Ann. Rev. Fluid Mech.* 8: 233–74

Yortsos, Y. C., Huang, A. B. 1984. Linear stability analysis of immiscible displacement including continuously changing mobility and capillary effects: Part I—simple basic flow profiles. *SPE 12692*, Soc. Pet. Eng., Dallas, Tex.

Ann. Rev. Fluid Mech. 1987. 19 : 313–37

COMPUTATION OF FLOWS WITH SHOCKS

Gino Moretti

G.M.A.F., Inc., PO Box 184, Freeport, New York 11520

1. *Introduction*

Before the advent of electronic computers, numerical analysis of inviscid, compressible flows with shocks was a forbidding task. Emmons' pioneering work of 1944–1948 by the relaxation method (Emmons 1948) should be acknowledged as courageous and successful but not repeatable. Today, the literature on the subject has grown well beyond reading affordability, reflecting a vast amount of theoretical work and numerical experimentation. Fortunately, well-differentiated standpoints were taken and were so conspicuously and, at times, hotly debated that we can attempt a rough classification and sketch a brief history of trends and techniques. The current state of the art proves that we have a much clearer view of the numerical treatment of flows with shocks, and that we are founding a measure of common understanding. It also shows that it is possible to choose between various techniques. We attempt to assess the importance, validity, and efficiency of the work accomplished so far. We are, however, well aware of the fact that our current techniques may have to be revised in view of recent advances in computer hardware and software.

2. *Early Philosophies*

The equations of conservation in integral form are valid for all flows, including the ones with finite jumps. By applying Green's formulas, we can recast the same equations into a partial differential form (known as the "divergence form"); but the partial differential equations have a more restricted range of validity. Indeed, partial differential equations cease to be valid wherever the functions cannot be differentiated; therefore, they can describe regions of continuous flows but not shock waves. Nevertheless, if the concept of derivative is generalized in the spirit of distribution theory, it has been shown (Lax 1954) that the equations in divergence form

0066–4189/87/0115–0313$02.00

admit "weak" solutions (that is, solutions containing jumps), and that such jumps appear in the right places. Consequently, if they have to move, they move at the right speed. The same cannot be said if the equations of motion are recast in any other form.

2.1 TWO BASIC CHOICES At this stage, one can proceed along two different lines of thought. The first, which is known as "shock capturing," relies on the proven mathematical legitimacy of weak solutions. All types of flows, including flows with shocks, should be computed by using the same discretization of the equations in divergence form at all nodes. The second, known as "shock fitting," prefers to use one scheme to integrate the partial differential equations wherever they are legitimate in a traditional sense and to treat the shocks (and any other discontinuity, for that matter) as real discontinuities, governed by their own algebraic equations. In this case, there is no need for maintaining the equations of motion in divergence form, and any other, algebraically equivalent form can be used.

The shock-capturing school is by far the more popular, for a number of reasons. First, one would like to have a code that can describe any flow, no matter how complicated, by repeating the same set of operations at all nodes without having to set up special logics to detect and track down discontinuities. (The emphasis here is on conceptual simplicity.) In addition, the mathematical properties of such a code can be discussed by a local analysis, and general conclusions can be drawn about order of accuracy, stability, growth factors, phase shifts, and convergence. (The emphasis here is on reliability.) In shock fitting, the interaction between discretized codes and original algebraic codes is much harder to analyze formally. We can say that the purpose of research in shock capturing is mathematical foundations of general-purpose codes, whereas the purpose in shock fitting is physical justifications of codes to make shockless calculations interact with jump analyses.

2.2 EARLY ATTEMPTS AT SHOCK CAPTURING In the 1960s, high hopes for simplicity and reliability were shattered. In a mesh interval containing a jump, the "derivatives" (which, as was mentioned above, must be interpreted in the spirit of distribution theory to produce a weak solution) are approximated by differences of values between opposite sides of the jump and are, in principle, wrong. Truly, if a first-order discretizing scheme is used, all discontinuities are smeared out, replaced by monotonic transitions over several mesh intervals. This happens because the truncation error acts as an artificial viscosity with extremely low Reynolds number. A classical analysis (von Mises 1950) shows that a shock can be considered as a sharp, but smooth, continuous transition having a certain thickness if the flow inside the shock is viscous and heat conducting. Unfortunately, first-order schemes produce shocks that are too thick, with some unpleasant consequences:

1. Extremely fine meshes are needed for a realistic approximation of inviscid flows; this results in large memory requirements and in very long computations.
2. Otherwise, the distortion of the flow pattern may spread far from the jump region, and the results become hopelessly garbled if more than one shock is present, as in the case of impinging or coalescing shocks.
3. Finally, the presence of an artificial viscosity of a very low Reynolds number casts legitimate doubts on the possibility of extending first-order schemes to evaluate real viscous effects with high Reynolds numbers.

In the early 1960s, such defects were compounded by the relative slowness and limited memory capacity of the computers. Attention was thus focused on second-order schemes that do not have second-order viscosity produced by truncation errors. Good methods of this type [such as the widely used Lax-Wendroff (1960) and MacCormack (1969) schemes] converge to the solution "in the mean." That is, they have the same mode of convergence as a Fourier series or any series of orthogonal functions. In the absence of discontinuities, such series converge uniformly as well, but the uniform convergence is lost when the function to be expanded has a jump. On either side of the jump, oscillations appear that do not invalidate the convergence in the mean (the Gibbs phenomenon). *Mutatis mutandis*, similar oscillations appear in any problem with shocks. The result is not only hard to interpret at times, but also the addition of nonlinear effects can turn a local nuisance into a major disaster.

To avoid abandoning the attempt, one must search for a code that automatically introduces monotonicity where shock-induced oscillations tend to appear. The idea is extremely simple and appears in what is perhaps the first important paper on the subject (von Neumann & Richtmyer 1950). From the standpoint of viscous flows, a shock has a thickness that, for fairly high Reynolds numbers, is well below the length of a mesh interval. In practice, we can settle for shocks having a thickness equal to the length of a mesh interval, or at most two or three times as big. We may hope to achieve the goal by adding artificial-viscosity terms to the equations of motion, defined from a numerical evaluation of second derivatives and properly weighted to minimize their influence at points not in the proximity of jumps. We return to this idea later on.

2.3 EARLY CALCULATIONS BY SHOCK FITTING Curiously, whereas most of the systematic analysis and experimentation of shock capturing was confined to one-dimensional problems (Gary 1964), efficient and reliable codes using shock fitting had been produced for some two- and three-dimensional, time-dependent problems by the present author and his collaborators. The problems in question were the blunt-body problem (Moretti & Abbett 1966, Moretti & Bleich 1967) and the three-dimensional, steady, supersonic flow around simple three-dimensional bodies (Moretti 1963) or pointed cones at an angle of attack (Moretti 1967). In all these cases, the problem is made simple by the fact that there is a single shock, and that it can be considered as a boundary of the computational region. Therefore, some second-order accurate scheme can be used at all interior mesh points, and a direct treatment of the algebraic conditions across the shock is confined to boundary points. Two major advantages of shock fitting over shock capturing were made evident by our early calculations, and they appeared to dispel a prevailing belief that accuracy could be achieved only on very fine meshes. Indeed, the second-order scheme acts only on a region of smooth flow; therefore, a coarse mesh can be used. In addition, the jump conditions can be satisfied exactly without introducing errors into the computational region. The most impressive example was the calculation of a blunt-body flow at $M = 4$, on a mesh of 2×4 intervals, with errors not larger than 1%.

How were the shock points computed to produce such good results? On the low-pressure side, conditions at infinity in the supersonic flow are known. Two jumps in any two thermodynamical parameters and one (vector) jump in the velocity must be determined, together with the local shock velocity. The Rankine-Hugoniot conditions are one scalar equation short to perform the task. One additional piece of information has to be obtained from the interior of the high-pressure region. Success was provided by using a compatibility equation along a generator of the characteristic conoid reaching the shock. Variations on the characteristic theme, aiming to make the shock calculation more similar to the calculation of interior points by replacing the algebraic Rankine-Hugoniot conditions with their time derivatives, were proposed by Kentzer (1970) and later by de Neef & Moretti (1980). The technique was accepted and adopted for the treatment of shocks as boundaries. In many cases, shock fitting for a boundary shock (for example, a bow shock) was coupled with a shock-capturing technique for the calculation of interior points, where embedded shocks could appear.

Indeed, for a large number of years, shock fitting for embedded shocks proved to be a hard bone to chew. The reason was essentially topological. Shock fitting had been proven to perform well with the shock as a boundary. When one embedded shock is present, it can be turned into an internal boundary, but the latter must be extended all the way to an external boundary and two meshes must be defined, each with its own normalization. If more than two embedded shocks are present, the number of computational regions may grow beyond reasonable limits; the bookkeeping is cumbersome. A courageous and successful attempt in that direction was taken by Marconi & Salas (1973). They computed a three-dimensional, steady, supersonic flow about an aircraft. Cross sections of the flow field at successive stations are shown in Figure 1. In addition to the bow shock, there are up to three embedded shocks, as shown in the

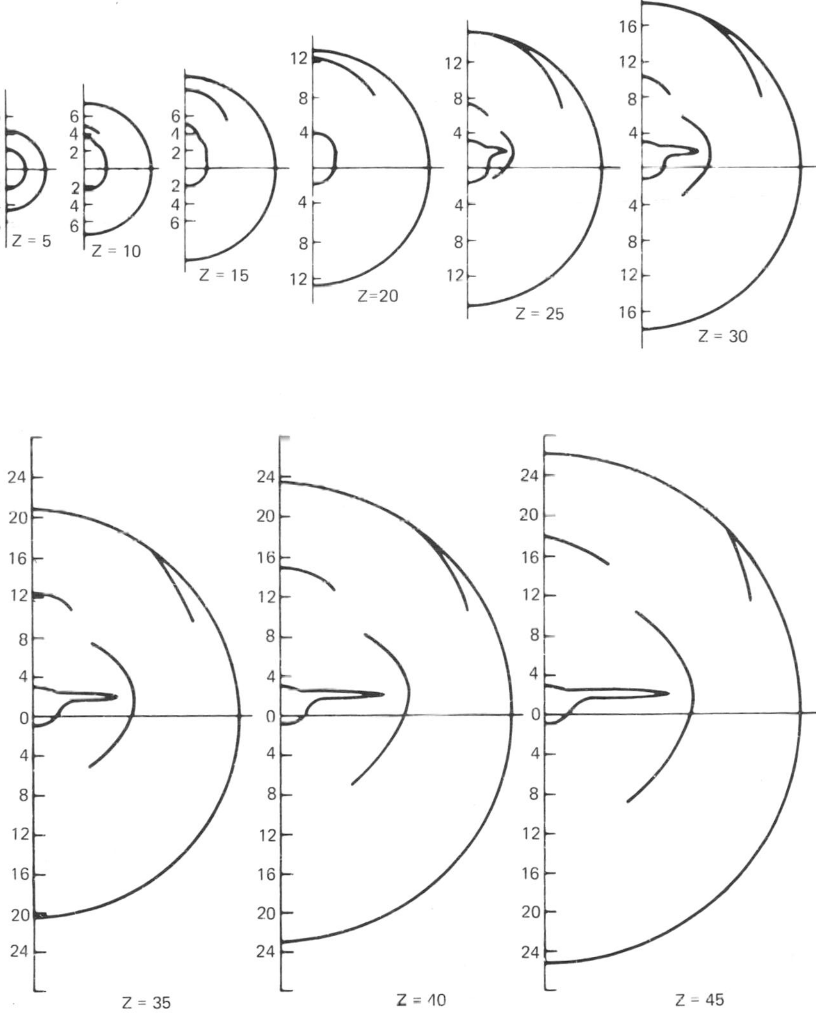

Figure 1 Cross-sectional views of a supersonic aircraft and shock patterns.

figure. On the negative side, we see that all such embedded shocks are of the same type—that is, wrapped around the body. There are no embedded shocks stemming out of the body (such as a wing recompression shock). Therefore, we are still far from reaching a critical topology. The Marconi & Salas paper remains an almost unique example, much as Emmons' paper had been when "computer" denoted a person, not a machine.

3. *Recent Trends—One-Dimensional Studies*

After more than a decade of preliminary empiricism, the possibility of replacing costly, uncertain, difficult, or altogether impossible experiments by computing was accepted. First-rate analysts undertook to legitimize numerical procedures and to provide practical codes. The first goal was pursued essentially on one-dimensional problems, where some mathematical proofs could be set forth. Two-dimensional and, less frequently, three-dimensional exercises were also performed to find (mostly in an empirical way) whether one-dimensional techniques, proven to be correct, would remain valid in more than one space dimension.

We discuss the one-dimensional and two-dimensional activities in that order. Obviously, there is no way to discuss shock treatment per se, and we are bound to dwell on general integration techniques, which is a complicated subject in its own right. For brevity's sake, we borrow definitions and notations from a recent paper by Roe (1986). In particular, we write the equations of one-dimensional motion in divergence form as

$$\mathbf{w}_t + \mathbf{F}_x = 0 \tag{1}$$

(where $\mathbf{w}$ and $\mathbf{F}$ are vectors representing unknowns and fluxes, respectively), and in nonconservation form as

$$\mathbf{w}_t + A\mathbf{w}_x = 0, \tag{2}$$

where

$$A = \partial\mathbf{F}/\partial\mathbf{w}. \tag{3}$$

The characteristic slopes $dx/dt = \lambda_k$, $k = 1, 2, 3$, are the eigenvalues of the matrix A, and the differences in fluxes between two adjacent nodes are

$$\Delta\mathbf{F} = \sum \alpha_k \lambda_k \mathbf{e}_k, \tag{4}$$

where the $\mathbf{e}_k$ are the right eigenvectors of A, and α_k is the strength of the kth wave.

3.1 CHARACTERISTIC-BASED SCHEMES In recent years, the attention of numerical analysts has been focused on a variety of schemes that all can be categorized as belonging to the same family, because all are based on the concept of characteristics. The concept itself, of course, is far from being new, and an application to finite-difference techniques had been proposed long ago by Courant et al. (1952). The related scheme is known as the CIR scheme (an acronym on the authors' names). Characteristic-based techniques blossomed in the late 1970s and can be classified into two main groups.

3.2 APPROXIMATE RIEMANN SOLVERS The first group, to which the "flux-vector splitting" and "flux-difference splitting" techniques belong, aims to capture shocks by considering balances of fluxes and weak solutions. The concept of characteristics intervenes insofar as each flux is considered as the resultant of separate contributions, conveyed along different characteristics. Three operations must be performed:

1. Finding the slopes of the characteristics at each point.
2. Splitting the x-derivative of each flux into three contributions, carried by a different characteristic.
3. Adding all such contributions to define the time derivatives of the three unknowns and then integrating in time.

Different techniques are obtained by using different concepts in any of the three steps above. Of course, all these techniques aim to capture shocks; therefore, they must work well in nonlinear environments. To this effect, the Riemann problem is brought into the picture. The Riemann problem consists of finding the flow at time $t+\Delta t$ if a flow consisting of two semi-infinite states is given at time t. If the Riemann problem is solved exactly in each mesh interval, assuming that the two initial states are defined by the values at its end points, the slopes of all characteristics, contact discontinuities, and shock waves are defined, and all fluxes across the boundaries are computable. In this case, all three steps above are replaced by an algebraic solution of the Riemann problem, and the solution is exact under any circumstances. The idea of solving a Riemann problem at every mesh interval and at every step is due to Godunov (1959) and is obviously related to the fact that sound waves, both compressive and expansive, can be considered as shocks of vanishing strength. Therefore, the presence of a shock can be assumed at every interval; if an interval contains a shock of finite strength, the code should be able to capture it. Unfortunately, the exact solution of the Riemann problem is cumbersome to compute, and the idea is used (in an approximate form) in the first two steps only. Godunov himself, in his original paper, demoted it to a first-order scheme.

In Roe's simplification (Roe 1981) A is replaced by another matrix $\tilde{A}$ that satisfies (4) as well as

$$\Delta \mathbf{w} = \sum \alpha_k \mathbf{e}_k. \tag{5}$$

Then the linearized problem is solved. The interesting point is that if only one wave is present and it is a shock wave, the technique yields the exact solution. Unfortunately, the method, in its simpler form, may produce expansion shocks. Remedies have been found (Harten 1983, Roe 1985), but these we do not discuss, since they are not directly related to the subject of shock calculations.

Osher (Osher & Solomon 1982), generalizing an earlier work by Engquist & Osher (1981), follows a different path. He simplifies the Riemann problem by assuming that both expansion and compression waves (including shock waves) are isentropic. The entropy jump is confined to the contact discontinuity. Not being encumbered by complicated Rankine-Hugoniot conditions, the physical parameters in each of the regions into which the original Riemann-problem jump is split can be determined in a simple way. The next step consists of determining which of such regions belongs to the node in question at time $t+\Delta t$. This operation requires many possible configurations to be examined, with particular care paid to sonic transitions. More details, including a graphic representation of the Riemann problem and its numerical counterpart in the phase space, can be found in Pandolfi (1984).

3.2.1 *Monotonicity and second-order schemes* Once the flux derivatives have been split and the third step has to be taken, different schemes can be used to integrate (1). A discussion of possible choices falls outside the limits of the present review. Here we confine ourselves to a few remarks.

First-order schemes, as was stated above, have very little accuracy in the vicinity of shocks. Second-order schemes produce oscillations; this has been proved by Godunov (1959) (see Roe 1986). To find a remedy, one must again resort to reducing the order of accuracy in the vicinity of a shock. Consequently, we must find where the shock is, in one way or another. Generality—one of the major advantages of shock capturing—is, although not lost, definitively weakened. The proposed devices to maintain monotonicity have a common background. Some of the flux differences, which are responsible for the production of oscillations, are weighted by a multiplier, which is called a "limiter function" and depends on a ratio of the differences of the unknowns on two adjacent intervals. Various forms of the limiter function have been given (van Leer 1974, Roe & Baines 1982, 1984, Roe 1985, Sweby 1984); they have somewhat different effects on the results in the vicinity of a shock, but they do not seem to invalidate an overall second-order accuracy. Of course, the definition of the limiter function at every node and the manipulation of the flux differences that it affects add further computational burdens to codes that are already rather complex.

3.2.2 *The MUSCL and PPM schemes* Considering again the sequence of operations mentioned at the beginning of Section 3.2, we single out two very interesting techniques; in both of these schemes, the authors prefer to complicate the first two phases in order to maintain simplicity in the third. Additionally, in both cases, on either side of the jump used in the Riemann problem the simple semi-infinite constant states are replaced by states that vary with x. In the MUSCL code (van Leer 1979), not only the values of the unknowns but also their slopes are considered at every step. A rather complicated algorithm is used to describe the collapse of the jump in the presence of a linearly variable environment. From values so obtained of the unknowns at half-intervals, the corresponding fluxes are calculated, and then Equation (1) is integrated by a simple formula. The new slopes are obtained by simply differencing the new unknowns between half-steps. The procedure gives monotonic results at shocks.

Colella & Woodward (1984) introduced the PPM technique. Here the original values of the components of $\mathbf{w}$ are fitted by parabolas from the middle of one interval to the middle of the next. In turn, the coefficients of the parabolas are obtained from a higher-order interpolation on four successive nodes. At every middle point, an equivalent Riemann problem is defined, using values that are averaged from the parabolas over suitable intervals defined by the slopes of the characteristics. Obviously, different formulas have to be written according to the local pattern of characteristics defined by the Riemann problem. Also, fine-coding details are needed to correct parabolic fits that are not monotonic. Finally, some additional dissipation has to be added, in the spirit of "fine tuning." After this elaborate preparation of the data, the integration of (1) is straightforward, and the results are remarkably good.

3.3 THE λ-SCHEME The second group of characteristic-based techniques, which is a direct offspring of the CIR idea and was named the λ-scheme by Moretti (1979), is a way of writing the classical compatibility equations along characteristics in a finite-difference form (which is easily made second-order accurate). The scheme basically consists of diagonalizing A

$$D = LAL^{-1}, \tag{6}$$

letting $\mathbf{R} = L\mathbf{w}$, and solving the equation

$$\mathbf{R}_t + D\mathbf{R}_x = 0, \tag{7}$$

with x-derivatives approximated by upwind differencing. The upwind directions are defined by the eigenvalues of D, which are the same λ_k as mentioned above. The scheme emphasizes the propagation of signals (defined by the vector **R**) along characteristics. It does not directly impose balances of fluxes, and therefore it is unable to capture shocks correctly. The complete Euler solver must contain a local application of the λ-scheme and some shock-fitting device. We now briefly analyze the shock to aid in understanding how the two ideas can be coupled efficiently.

3.3.1 *Analysis of a normal shock* A shock is controlled by its environment, regardless of the number of space dimensions involved. Most of the basic facts can be made clear in a one-dimensional context. The Rankine-Hugoniot conditions can be cast in a form that describes the jumps across a shock as functions of the Mach number of the shock alone. Let u, a, γ, R, and W be the gas velocity, the speed of sound, the ratio of specific heats, the gas constant, and the shock velocity, respectively. Let also $\delta = (\gamma - 1)/2$, S be the entropy, (divided by γR), A and B denote the low-pressure and high-pressure side of the shock, respectively, and $\Delta f = f_{\mathrm{B}} - f_{\mathrm{A}}$. Then, with the Mach number of the shock defined as

$$M = (u_{\mathrm{A}} - W)/a_{\mathrm{A}}, \tag{8}$$

a set of independent Rankine-Hugoniot conditions is

$$\Delta a/a_{\mathrm{A}} = [(\gamma M^2 - \delta)(1 + \delta M^2)]^{1/2}/[(1+\delta)|M|] - 1, \tag{9a}$$

$$\Delta u/a_{\mathrm{A}} = (1 - M^2)/[(1+\delta)M], \tag{9b}$$

$$\Delta S = \{\ln[(\gamma M^2 - \delta)/(1+\delta)] - \gamma \ln[(1+\delta)M^2/(1+\delta M^2)]\}/(2\gamma\delta). \tag{9c}$$

Clearly, if a_{A} and M are known, all jumps can be determined. Conversely, if at least one of the jumps or a combination of them were known, together with a_{A}, the Mach number of the shock could be calculated.

In the context of numerical analysis, how could we exploit the environment to define the shock (in particular, its Mach number and velocity)? Clearly, we must choose the jump of a quantity that can be determined with great accuracy on both sides of the shock despite the presence of the discontinuity, and also a jump that is large enough to provide M without uncertainties due to truncation and/or round-off errors. To make the first point more precise, we must be able to determine the quantity in question correctly by using information that proceeds, at A, from upstream only and, at B, from downstream only. Here the concept of characteristics is very useful. Physically, each point in space and time depends on information conveyed along characteristics. In one-dimensional problems, three pieces of information are carried to a point. One is the entropy, carried along a particle path, and the other two are certain combinations of velocity and pressure, carried along lines having the slopes $dx/dt = u \pm a$. Such combinations can be expressed by their increments

$$a\Delta \ln p \pm \gamma \Delta u, \tag{10}$$

or, in an equivalent and more convenient form,

$$\Delta a/\delta \pm \Delta u - a\Delta S. \tag{11}$$

For numerical purposes, the form (11) is better than the form (10), since the nonlinear term $a\Delta S$ is generally less relevant than the nonlinear term $a\Delta \ln p$, and it vanishes altogether in isentropic regions. To simplify, the variables $R_1 = a/\delta + u$ and $R_2 = a/\delta - u$ depend on information conveyed along the lines $dx/dt = \lambda_1 = u + a$ and $dx/dt = \lambda_2 = u - a$, respectively. [Note that R_1 and R_2 are two components of the vector **R** in Equation (7).] If, for argument's sake, we consider the case $u > 0$, we see that neither R_1 nor its x-derivative changes appreciably across a shock, whereas R_2 has a well-defined jump. Moreover, R_2 is defined by right-running information where $u > a$, and by left-running information where $u < a$. Let us consider now Figure 2, where the flow Mach numbers, $M_A = (u/a)_A$ and $M_B = (u/a)_B$, are used as abscissas and ordinates, respectively. The AB-line is defined by $M_A = 1$, and the BN-line and HM-line by $M_B = \pm 1$. In the region to the right of and below ABN, we have $u > a$ at A and $u < a$ at B. All shocks defined by points in this region can be computed with ease. First, R_2 is determined at A and B from opposite sides of the jump; then we evaluate

$$\Sigma = \Delta R_2 / a_A \tag{12}$$

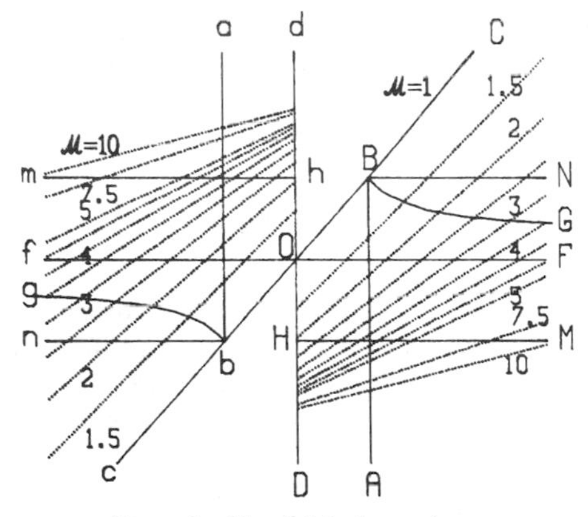

Figure 2 Shock Mach numbers.

and obtain M from a combination of (9a) and (9b), noting that

$$\Sigma = (\Delta a/\delta - \Delta u)/a_{\mathrm{A}}. \tag{13}$$

From (9), all individual jumps of a, u, and S are computed, and (8) provides the shock velocity. The shock location can be updated accordingly. Note that the BG-line, describing a stationary shock, is contained in this region.

If the shock is represented by a point within DOBA (this is a shock moving to the left, in a subsonic environment), then we have $u < a$ at both A and B. The value of R_2 at B is still determined correctly, but not at A. In this case, we can use

$$\Sigma_1 = (R_{2\mathrm{B}} + R_{1\mathrm{A}})/a_{\mathrm{A}} \tag{14}$$

instead of Σ and proceed in a similar way. An example of this type of problem is the motion produced by an originally sinusoidal wave between two rigid walls. Baum & Levine (1985) presented an interesting comparison of various shock-capturing techniques, concluding that the best results are obtained using a correction of fluxes in the vicinity of the shock, in the spirit of Harten's artificial-compression method (Harten 1977, Yee et al. 1985).

If the point representing the shock falls in the region delimited by CBN, then we have $u > a$ at both A and B. The value of R_2 at A is determined correctly, but not at B. In this case, one should force $R_{2\mathrm{B}}$ to be computed using downstream information only; the gambit is justified in a frame relative to the shock, and the case is, anyway, of no practical interest. The analysis that we have outlined is intended to show that determining a normal shock from its environment is a very simple task, which can be accomplished to a high degree of accuracy and requires negligible computation time.

3.4 CONCLUSIONS The problem of one-dimensional flows with shocks and other discontinuities is now well understood from a physical point of view, and excellent results have been obtained with various numerical techniques. The mathematics of shock capturing is still in an evolutionary stage, but relevant progress has been made and neat results have been obtained with sharp discontinuities and no spurious oscillations. From the viewpoint of efficiency, however, I believe that simple upwind schemes derived from the equations of motion in nonconservative form plus discontinuity fitting are far superior to discontinuity-capturing schemes. In fact, a λ-scheme is, at points of continuous flow, more accurate than any scheme derived from the conservation equations (Favini & Zannetti 1986). The number of computational nodes necessary to obtain the same overall accuracy is a couple of orders of magnitude lower. Compare, for example, typical calculations of Moretti & DiPiano (1983) using a total of about 50 points with an (otherwise remarkable) example shown by Woodward & Colella (1984) using a total of about 3600 points. Other advantages of the λ-cum-shock-fitting technique are the total absence of Gibbs phenomena (a natural consequence of not trying to approximate derivatives generalized in the sense of distribution theory) and (for related reasons) no requirements of "fine tuning." Finally, a shock-capturing technique must contain some explicit special treatment in the vicinity of a shock (which belies the spirit of a shock-capturing code), or it must analyze a Riemann problem at all mesh intervals when the analysis is necessary only at intervals containing a shock (which is a waste of work).

In this connection, it is proper to mention some attempts to hybridize the λ-scheme with a local analysis of the Riemann problem, more or less in the spirit of flux-difference splitting (Dadone & Magi 1985, Pandolfi 1985). By so doing, the good qualities of the λ-scheme are preserved, and tracking of the shock in its motion through the nodes is not needed; on the other hand, however, the shock is smeared over two intervals instead of one, and some local accuracy may be lost.

4. *Two-Dimensional Problems*

In moving from one-dimensional problems to two-dimensional problems, new difficulties arise. Some are related to the integration schemes, whether or not the flow is smooth, and some are connected with the shocks themselves.

4.1 BASIC DIFFICULTIES OF TWO-DIMENSIONAL CALCULATIONS The problems with integration schemes can be summarized as follows. Is a wave, propagating in a direction oblique with respect to the computational grid, properly described by the discretized scheme? Does the scheme itself retain the properties that have been proven in a one-dimensional context? In particular, is its accuracy impaired by the metric of the grid? The answers to these questions are far from simple or straightforward. The grid may or may not be orthogonal, but orthogonal grids are the best choice and should be adopted whenever possible. Nevertheless, even with orthogonal grids, changes in the metric between adjacent cells can be so large that truncation errors of different terms may conflict with each other. Other possible causes of errors to be explored stem from strong stretching of coordinates.

As mentioned above, all techniques carefully analyzed in one-dimensional cases (not just the ones mentioned in this brief review) have been extended somewhat empirically to two-dimensional calculations, but no serious analysis has been attempted yet. There is a drastic difference between one-dimensional problems and two-dimensional problems. In one-dimensional problems, the matrix A of (3) can be diagonalized. In mathematical terms, one can split flux differences into contributions oriented as the eigenvectors of A. In physical terms, one can think of signals conveyed along different characteristics. In two-dimensional problems, the basic equation is

$$\mathbf{w}_t + \mathbf{F}_x + \mathbf{G}_y = 0, \tag{15}$$

and $\mathbf{F}$ and $\mathbf{G}$ cannot be diagonalized by the same operator. The splitting of flux differences in two directions becomes arbitrary. In physical terms again, the characteristics are replaced by a characteristic conoid, and once more the choice of conoid generators in the discretization process becomes arbitrary. In the same vein, there is no clear understanding for what a two-dimensional Riemann problem could be. Most of the current procedures seem to accept the possibility of computing $\mathbf{w}_t$ as the sum of contributions from each of the other terms in (15), which are then split separately according to one-dimensional criteria. The possibility of missing nonlinear effects and the close dependence of the splitting of (15) on an arbitrary choice of a frame of reference, unrelated to the physical local situation, are still open questions.

The λ-scheme is not free from such criticisms, despite some justification of the choice of generators given by Zannetti & Colasurdo (1981) in the case of isentropic flows. With some more elaborate preliminary work, but no major additional burden on the coding, the technique can be extended to nonisentropic flows, and the splitting of the waves can be made more consistent with the geometry of rigid boundaries (Moretti 1986). The advantages of directness and coding simplicity of the one-dimensional scheme are retained.

Even in the absence of good mathematical guidelines for two-dimensional coding, however, approximate Riemann solvers, the λ-scheme, earlier second-order techniques based on central differences, and finite-volume techniques all seem to produce good results for shockless flows. It should be noted, however, that many calculations have been conducted on Cartesian grids, evenly spaced in both directions. More work is certainly necessary to justify the use of approximate Riemann solvers on general grids.

4.2 TWO-DIMENSIONAL FLOWS WITH SHOCKS When shocks are present, shock-capturing techniques have, until now, played a preeminent role. As was mentioned in Section 2.3, shock fitting in two dimensions may create topological and logical difficulties if one tries to define every shock as a boundary. Such difficulties disappear if the shocks are allowed to float over the computational grid, and coding becomes particularly simple when the λ-scheme is used for ordinary points (Moretti 1986). Briefly, once the normal to the shock at one of its points is known, the analysis of Section 3.3.1 holds, provided that the normal component of the velocity is used. (The tangential component, of course, remains unchanged across the shock.) In practice, it is not even necessary to reformulate the equations; Σ can be defined using the velocity component along one of the grid lines, and the slope of the normal appears only in the calculation of M from Σ. It remains only to define the normal at any shock point P. Best values are obtained using the coordinates of two shock points, M and N, bracketing P, and then applying centered differences to define the slope of the shock wave. Since the shock information is stored in one-dimensional arrays, and shock points may be generated erratically in a complicated flow, one has to find where M and N are located in the arrays once P is given. A simple logical device helps overcome this difficulty, using only a half-dozen lines of coding. Therefore, the program can handle flows having shocks in any part of the field and in any number.

4.3 SOME REPRESENTATIVE PROBLEMS We consider here some of the most representative problems and briefly review the most recent work, including some yet unpublished results obtained by shock-fitting methods. We choose the following problems as particularly noteworthy:

1. Transonic flow in a channel with a bump, as considered at a GAMM Workshop in Stockholm (Rizzi & Viviand 1980).
2. Supercritical flow about a circle.
3. Supercritical flow about airfoils.
4. Transonic flows in ducts and (less frequently) in and about air intakes; the flows are generally supersonic upstream and often are supersonic downstream as well.
5. Transonic flow about a forward-facing step (supersonic at upstream infinity).
6. Complex Mach reflection.
7. Flow about a circle in a supersonic stream.

Problems 1, 3, and 5 are essentially oriented toward steady-state solutions. Problem 7 may also be considered as a steady-flow case, although a closer inspection may prove that it is not. Problem 2 is, according to the most recent evidence, a case of unsteady flow. Problem 4 may have steady solutions, but the most interesting features appear in unsteady cases. Finally, problem 6 is unsteady but the solution is self-similar, and therefore it can be considered as steady in an (x/t, y/t, $\ln t$) frame (Jones et al. 1951).

From the viewpoint of shock-pattern complexity, problems 1, 2, 3, and 5 are the simplest. They contain a single shock or two shocks that are fairly separated and do not interfere with each other. In particular, in the first three problems the shocks are generally almost oriented along a coordinate line. Next in order of complexity are problems 4 and 6. The number of shocks may be very large; some shocks are oblique, and others are normal or almost normal to the flow. Regular reflections and Mach reflections may occur and turn into one another as the flow evolves. Shocks also appear and disappear. Similar complications are present in problem 7, compounded by the fact that some shocks are almost radial and others are partially wrapped around the circle (for example, the bow shock).

4.3.1 *Airfoil problems* Substantial progress has been made since 1979 in simple cases such as problems 1 and 3. The 1979 Stockholm GAMM Workshop showed a remarkable discrepancy between results obtained under the assumption of potential flow and results obtained by numerical Euler solvers. A critical analysis of the potential approach (Salas et al. 1983) eventually led to its replacement by Euler solvers and a renewed effort to improve them. The bump in a channel and the NACA 0012 airfoil, with or without incidence, have been popular testing grounds for shock-capturing and shock-fitting techniques as well.

There is little difference among results obtained by various shock-capturing techniques. Two remarks, however, are in order.

1. Even slight variations in shock location may affect the drag coefficient; therefore, the airfoil designer wants a stronger assurance of reliability.
2. Again, from a designer's standpoint, computational speed is a basic requirement. It seems that none of the sophisticated codes created as extensions of the characteristic-based techniques mentioned above has had a winning chance. Indeed, they require too many computations and are very slow. The palm for speed seems to go to Jameson's codes (Schmidt & Jameson 1985), which do not at all belong to the categories mentioned above, but instead rely on a finite-volume approach without splitting, an efficient updating scheme, the use of local time increments, and a multigrid technique. The shock is captured by adding artificial-viscosity terms of second and fourth order. In brief, the techniques are a replay of earlier ideas with more efficient coding and a fine mesh. These codes are very satisfying as regards speed, but they are still open to some questions about the proper location of the shock and its strength (Roe 1986). In addition, the artificial-viscosity terms are affected by a number of coefficients that, at times, must be finely tuned to assure convergence and increase speed of convergence.

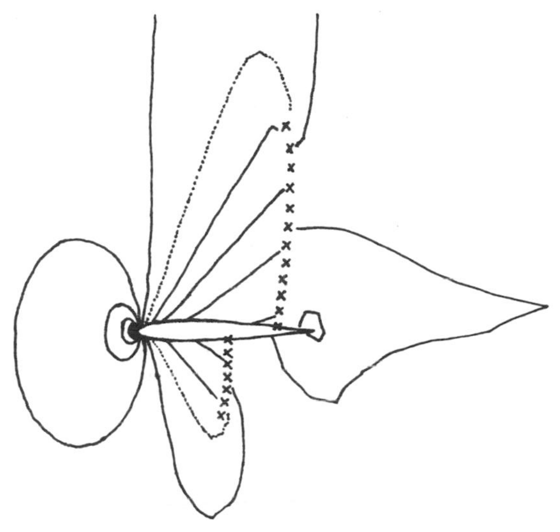

Figure 3 Isomachs around an NACA 0012 airfoil ($M_\infty = 0.85$, $\alpha = 1°$).

The λ-scheme, either with shock fitting or with the hybridization mentioned in Section 3.4, performs very well both for airfoil calculations and on the bump-in-a-channel problem. For brevity, we show only a plot of isobars (Figure 3) and the C_p distribution (Figure 4) for a NACA 0012 airfoil at $M = 0.85$, $\alpha = 1°$, obtained with shock fitting using a C-grid in the domain shown in Figure 3, a 32×128 mesh, and local time increments. Preliminary tests on a CRAY 1 computer show convergence within engineering standards in a computational time of the order of 30 s. (No multigrid technique is used.)

4.3.2 *Supercritical flow about a circle* From a computational viewpoint, this problem is similar to problems 1 and 3, and solutions for it exist by

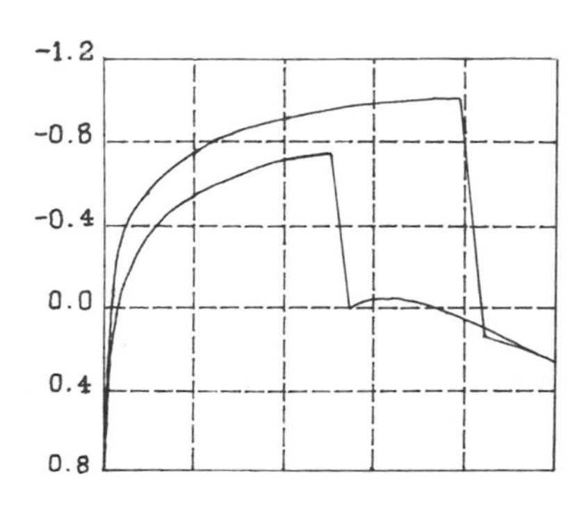

Figure 4 C_p distribution on the airfoil of Figure 3.

all techniques mentioned above. The special interest of the problem lies in the strong vorticity produced by the shock and the consequent inviscid generation of a recirculation zone (Salas 1982). Careful studies contributed to the abandonment of the popular belief that separation is produced by viscous effects. Indeed, separation does not occur if the computational code is free of artificial viscosity and the flow is subcritical. Once separation occurs, the flow does not reach a steady configuration. If the flow is computed over a half-plane only, the recirculation bubble vibrates when $M_\infty = 0.5$. At $M_\infty = 0.6$, the vibrations become more violent, and vortical bubbles are shed periodically. If the flow over the entire plane is computed and the code is made as symmetrical as possible, the flow again tends to reach a symmetrical pattern. This pattern is, however, unstable; at a length, any unperceivable disturbance destabilizes the flow, generating a periodic pattern of vortices, alternately shed from above and from below and forming a Kármán street with the usual Strouhal number. M. Pandolfi gave a detailed analysis of all the cases above using a hybridized λ-scheme; his results will appear in a future volume of *Notes on Numerical Fluid Mechanics* (Vieweg) containing the Proceedings of a GAMM Workshop at Rocquencourt in 1986.

4.3.3 *Ducts and inlets* A simple case of unsteady flows in ducts, which can be found in the literature, is provided by two straight, semi-infinite ducts of different cross-sectional areas connected by a straight transition. Montagné (1985) presents a set of results obtained with a characteristic-based code, using some correction à la Harten. The transition has a 20% slope, and the upstream Mach number is 1.6. Two grids have been used, one with 75×25 intervals, the other with 150×50 intervals. The two calculations are consistent with each other, but a sharp definition of the shocks is obtained only with the finer mesh. A. Kumar (personal communication) has computed the same case using a MacCormack scheme and artificial viscosity (Kumar 1981) on a 150×50 mesh. As in the airfoil problem, it seems that a judicious use of old-fashioned methods yields results that are comparable with those of more sophisticated techniques, obviously in a shorter computational time; Kumar's results (Figure 5) indeed are not dissimilar from Montagné's.

The λ-scheme also provides similar results, but with a much coarser mesh. Figure 6 shows results obtained with a 50×20 mesh. The locations of shocks are denoted by small crosses. The entire calculation takes a little more than 8 s; the shock manipulations require about 1 s (on a CRAY 1). If a finer mesh were used, the ratio of shock-to-ordinary points running time would decrease substantially.

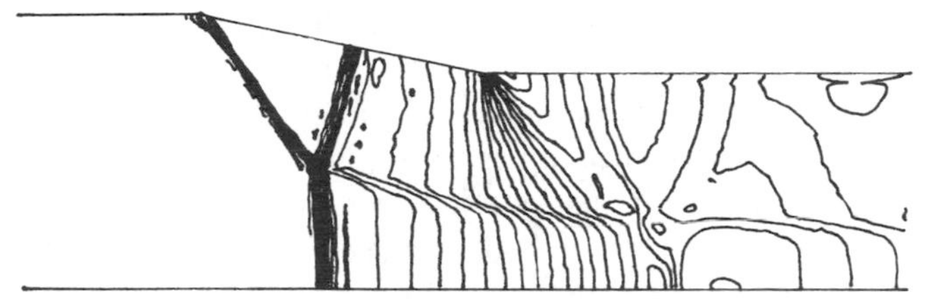

Figure 5 Isomachs in a duct computed by a MacCormack scheme and shock capturing on a 150 × 50 mesh (A. Kumar).

4.3.4 *Complex Mach reflection* The diffraction of a planar shock impinging on a wedge has been a subject of great interest in the last decade. Minor changes in the shock Mach number and the wedge angle provide a large variety of cases, ranging from a regular reflection to various types of complex Mach reflections (Glass 1986). An analytical solution (with one, rather innocuous, empirical relation) has been given and discussed by Ben-Dor (1980). Accurate and detailed experimental results are available (Glass 1986). Many calculations have already been performed on a variety of cases from this family. A great appeal of this problem, from a numerical viewpoint, is that it can be worked out in a Cartesian mesh. This is also a

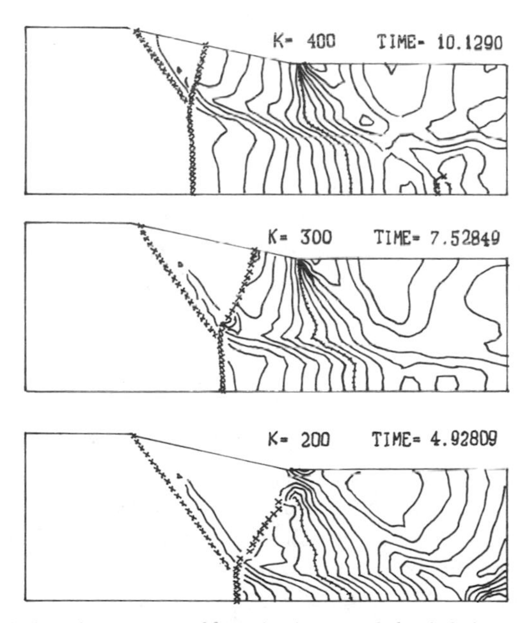

Figure 6 Isomachs in a duct computed by a λ-scheme and shock fitting on a 50 × 20 mesh.

great shortcoming, since it is not a test of the validity of codes in non-Cartesian meshes. As a proving ground for numerical techniques, the problem has a large number of interesting features, and it can be used for comparison of different techniques. A rapidly moving impinging shock coexists with stationary or quasi-stationary shocks, with a variety of orientations with respect to the grid. Triple points, kinks, contact discontinuities, and a possible formation of vortices make this problem an analyst's delightful nightmare. Woodward & Colella (1984) experimented with a variety of techniques, all belonging to the shock-capturing family. The reader is referred to their paper for more bibliographical information. The paper is a commendable piece of work, covering all the most interesting techniques in detail and providing an abundant graphical display of results in a form that makes comparisons easy. Other cases of the same type, using a Godunov-like technique of the second order with downgrading to the first order in the vicinity of shocks, have been presented by Glaz et al. (1985). A representative time for the calculation of one case on a CRAY 1 is given as 20–30 min; it must be noted that some cases deal with real gas effects, so that the calculation may be, at times, more elaborate than for a polytropic gas. The number of grid points is not mentioned, but a visual inspection suggests a mesh in the order of 100×300 intervals. Woodward & Colella (who work with a polytropic gas) mention their meshes but do not quote computational times.

Shock fitting is a newcomer on this type of problem, but it is not without good perspectives. An interesting step in this direction is found in a paper by Yamamoto et al. (1984), which unfortunately is so poor in details as to be of very little practical use. The authors chose to work in self-similar coordinates; the problem, thus, is oversimplified because all shocks in the solution are stationary and a good initial guess is easy. Within such limitations, the paper has two features worth mentioning. One is the introduction of a fitting for a contact discontinuity, which is allowed to float among grid lines. Despite the obscurity of the presentation, it is clear that the technique has a good potential. Its implementation is made easy by the use of an integration scheme of the λ-family (Chakravarthy et al. 1980) and the explicit use of entropy as a basic unknown. A tracking device, similar to the one mentioned in Section 4.2 for shocks, could allow more than one contact discontinuity to be sharply outlined. The other interesting feature is the coarseness of the mesh used for the calculation. The mesh has 10 intervals between the wall and the reflected shock, and 20 intervals along the wall. Woodward & Colella's grids, in the finest case, average about 50×300 points to cover the same area. The computed results, despite the coarseness of the mesh, show certain details very well

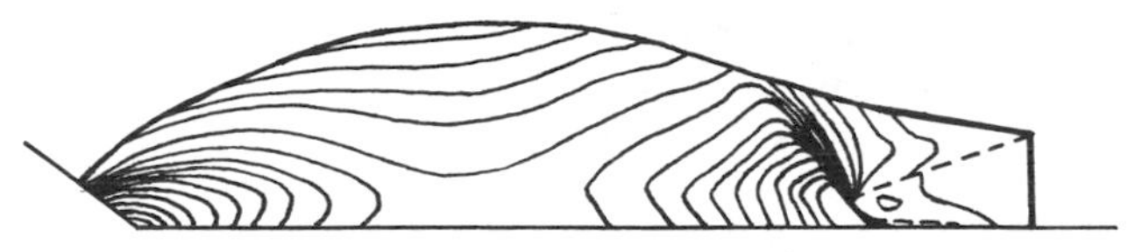

Figure 7 Complex Mach reflection computed by Yamamoto et al. (1984) using shock and contact-discontinuity fittings.

(see, for example, Figure 7, which is reproduced from Figure 17 of Yamamoto et al. 1984). A direct comparison of these results with results obtained by shock-capturing techniques is not possible because of different shock Mach numbers and wedge angles.

Figure 8 is a plot of isomachs for the problem considered by Woodward & Colella (1984), obtained by the present author using a λ-scheme and shock fitting.

4.3.5 *Circle in a supersonic stream* The calculation of the diffraction of a shock wave by a circular cylinder, presented by Yang et al. (1986), is a state-of-the-art product that gives a perfect idea of what can be obtained with a combination of flux-vector splitting (Steger & Warming 1981, van Leer 1982), flux-difference splitting, and Harten's artificial compression (Harten 1983). The program is indeed complicated, but the result is an impressive set of pictures showing the entire evolution of the flow, from a few instants prior to the impact of the shock wave on the cylinder to a time when the impinging shock is way behind the body, a bow shock has formed, and a complicated pattern of secondary shocks and contact discontinuities in the wake is seen. We cannot describe these results in detail here as a result of space limitations, and thus we recommend that the reader examine the plots in the original paper, soon to appear in the *AIAA Journal*. The ability of current shock-capturing techniques to capture shocks (on a 300 × 100 mesh) is evident. Contact discontinuities are not sharp. So far, there is no calculation of this problem by shock-fitting techniques.

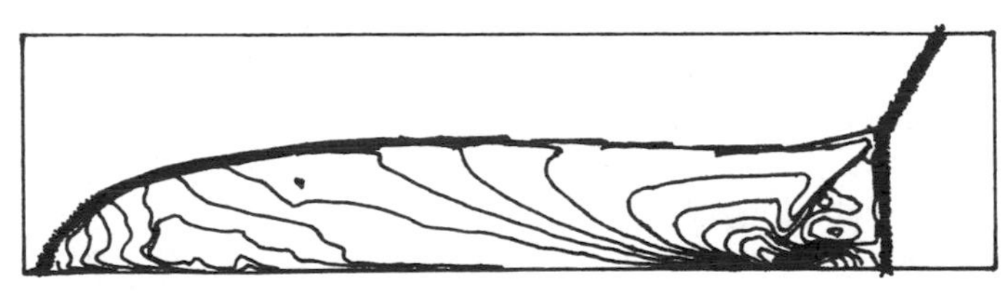

Figure 8 Complex Mach reflection computed using a λ-scheme and shock fitting on a 240 × 60 mesh.

4.4 CONCLUSIONS The current matters of concern in the calculation of two-dimensional flows with shocks are the following:

1. Robustness.
2. Proper physical description of continuous flows.
3. Shock sharpness.
4. Computational speed.
5. Easiness of usage.

The current techniques can be classified into three major divisions:

1. Old-fashioned, noncharacteristic-based schemes with artificial viscosity and, possibly, artificial compression.
2. Schemes based on Riemann solvers, particularly of the PPM type.
3. λ-schemes with shock fitting or local hybridization at the shocks.

Comparisons are hard to make and mostly premature. In principle, one should run all the codes on the same computer for a large number of cases. Relying on the scant information that I could collect, I dare advance a few tentative assessments.

Old-fashioned techniques are robust and fast. They may not be the easiest codes to work with because of fine-tuning impedimenta. They do not seem to behave badly in regions of continuous flow. Shocks are captured at the proper place and with the proper speed, but the resolution is fair to poor. Such techniques should not be discarded, in view of recent progress in computational memory and speed. For example, if a grid with twice the number of points in both directions were used, the calculation would take a time eight times as long. This time, however, would still be practically the same as the time used by a PPM technique on the original grid, and the results could be close. Computers of the near future may allow a further doubling of the grid, in which case I do not see any advantage in using PPM rather than an old-fashioned technique.

Among the Riemann-solver techniques, my own choice would be, so far, one of the PPM codes; they seem to be robust and correct. I continue to have some doubt about the possibility of generalizing PPM to non-Cartesian grids. These techniques are definitively slow and cumbersome, and I have no way of judging how much fine tuning they may need.

Despite technological advances, I still believe that extremely fine grids are a waste of space and time. We must keep in mind that the final goal of computational gasdynamics is three dimensional and that the gas may be viscous and nonpolytropic. Added complications will require more memory and computation time, and some initial restraint seems to be in order.

My personal bias is in favor of shock fitting and of the λ-scheme that makes it simple. The λ-scheme is correct, fast, and robust, and it does not require fine tuning. The computation for ordinary points is fully vectorizable, and the computation of the shock takes a minimal additional amount of time. The advantage is particularly felt when fine meshes are used (because the number of shock points grows linearly, and the number of grid points quadratically). The most important advantage of shock fitting (and, if necessary, contact-discontinuity fitting) resides in the possibility of using coarse meshes. Most of the techniques of shock fitting are still in a phase of development, including the hybridizations. It is premature to say anything about robustness; probably, the hybridized schemes are by their own nature more robust than the sharp shock fitting as a result of some damping produced at the shock by the Riemann solver; the Rankine-Hugoniot conditions, as they have been coded, offer no damping, and convergence to a steady state may be made difficult to understand because of local, almost invisible, vibrations of the shocks.

Literature Cited

Baum, J. D., Levine, J. N. 1985. A critical study of numerical methods for the solution of nonlinear hyperbolic equations for resonance systems. *J. Comput. Phys.* 58: 1–28

Ben-Dor, G. 1980. Analytical solution of double-Mach reflection. *AIAA J.* 18: 1036–43

Chakravarthy, S. R., Anderson, D. A., Salas, M. D. 1980. The split coefficient matrix method for hyperbolic systems of gasdynamic equations. *AIAA Pap. No. 80-0268*

Colella, P., Woodward, P. 1984. The piecewise parabolic method (PPM) for gasdynamical simulations. *J. Comput. Phys.* 54: 174–201

Courant, R., Isaacson, E., Rees, M. 1952. On the solution of nonlinear hyperbolic differential equations by finite differences. *Commun. Pure Appl. Math.* 5: 243–49

Dadone, A., Magi, V. 1985. A quasi-conservative lambda formulation. *AIAA Pap. No. 85-0088*

de Neef, T., Moretti, G. 1980. Shock fitting for everybody. *Comput. Fluids* 8: 327–34

Emmons, H. W. 1948. Flow of a compressible fluid past a symmetrical airfoil in a wind tunnel and in free air. *NACA TN 1746*

Engquist, B., Osher, S. 1981. One-sided difference approximations for non-linear conservation laws. *Math. Comput.* 36: 321–51

Favini, B., Zannetti, L. 1986. On conservative properties and nonconservative forms of Euler solvers. *Int. Conf. Numer. Gasdynam., 10th, Beijing*. In press

Gary, J. 1964. On certain finite difference schemes for hyperbolic systems. *Math. Comput.* 18: 1–18

Glass, I. I. 1986. Some aspects of shock-wave research. *AIAA J.* In press (*AIAA Pap. No. 86-0306*)

Glaz, H. M., Colella, P., Glass, I. I., Deschambault, R. L. 1985. A numerical study of oblique-shock wave reflections with experimental comparisons. *Proc. R. Soc. London Ser. A* 398: 117–40

Godunov, S. K. 1959. A finite difference method for the numerical computation of discontinuous solutions of the equations of fluid dynamics. *Mat. Sb.* 47: 271–306

Harten, A. 1977. The artificial compression method for computation of shocks and contact discontinuities. I. Single conservation laws. *Commun. Pure Appl. Math.* 30: 611–38

Harten, A. 1983. High resolution schemes for hyperbolic conservation laws. *J. Comput. Phys.* 49: 357–93

Jones, D. M., Martin, P. M., Thornhill, C. K. 1951. A note on the pseudo-stationary flow behind a strong shock diffracted or

reflected at a corner. *Proc. R. Soc. London Ser. A* 209: 238–48
Kentzer, C. P. 1970. Discretization of boundary conditions on moving discontinuities. *Lect. Notes Phys.* 8: 108–13
Kumar, A. 1981. Numerical analysis of the scramjet-inlet flow field by using two-dimensional Navier-Stokes equations. *NASA TP 1940*
Lax, P. D. 1954. Weak solutions of nonlinear hyperbolic equations and their numerical computation. *Commun. Pure Appl. Math.* 7: 159–93
Lax, P. D., Wendroff, B. 1960. Systems of conservation laws. *Commun. Pure Appl. Math.* 13: 217–37
MacCormack, R. W. 1969. The effect of viscosity in hypervelocity impact cratering. *AIAA Pap. No. 69-354*
Marconi, F., Salas, M. 1973. Computation of three dimensional flows about aircraft configurations. *Comput. Fluids* 1: 185–95
Montagné, J. L. 1985. Second order accurate flux splitting scheme in two-dimensional gasdynamics. *Lect. Notes Phys.* 218: 406–11
Moretti, G. 1963. Three-dimensional supersonic flow computations. *AIAA J.* 1: 2192–93
Moretti, G. 1967. Inviscid flowfield about a pointed cone at an angle of attack. *AIAA J.* 5: 789–91
Moretti, G. 1979. The λ-scheme. *Comput. Fluids* 7: 191–205
Moretti, G. 1986. A technique for integrating two-dimensional Euler equations. *Comput. Fluids.* In press
Moretti, G., Abbett, M. 1966. A time-dependent computational method for blunt body flows. *AIAA J.* 4: 2136–41
Moretti, G., Bleich, G. 1967. Three-dimensional flow around blunt bodies. *AIAA J.* 5: 1558–62
Moretti, G., DiPiano, M. T. 1983. An improved lambda-scheme for one-dimensional flows. *NASA Contract Rep. No. 3712*
Osher, S., Solomon, F. 1982. Upwind difference schemes for hyperbolic systems of conservation laws. *Math. Comput.* 38: 339–74
Pandolfi, M. 1984. A contribution to the numerical prediction of unsteady flows. *AIAA J.* 22: 602–10
Pandolfi, M. 1985. The merging of two different ideas: a shock fitting performed by a shock capturing. *Int. Symp. Comput. Fluid Dyn., Jpn. Soc. Comput. Fluid Dyn., Tokyo*, pp. 484–95
Rizzi, A., Viviand, H., eds. 1980. *Notes on Numerical Fluid Mechanics*, Vol. 3. Braunschweig: Vieweg
Roe, P. L. 1981. Approximate Riemann solvers, parameter vectors, and difference schemes. *J. Comput. Phys.* 43: 357–72
Roe, P. L. 1985. Some contributions to the modelling of discontinuous flows. *Proc. Semin. Appl. Math., 15th*, pp. 163–93. Providence, RI: Am. Math. Soc.
Roe, P. L. 1986. Characteristic-based schemes for the Euler equations. *Ann. Rev. Fluid Mech.* 18: 337–65
Roe, P. L., Baines, M. J. 1982. Algorithms for advection and shock problems. In *Notes on Numerical Fluid Mechanics*, ed. H. Viviand, 5: 281–90. Braunschweig: Vieweg
Roe, P. L., Baines, M. J. 1984. Asymptotic behaviour of some nonlinear schemes for linear advection. In *Notes on Numerical Fluid Mechanics*, ed. M. Pandolfi, R. Piva, 7: 283–90. Braunschweig: Vieweg
Salas, M. D. 1982. Recent developments in transonic Euler flow over a circular cylinder. *NASA TM 83282*
Salas, M. D., Jameson, A., Melnik, R. E. 1983. A comparative study of the nonuniqueness problem of the potential equation. *Proc. AIAA Comput. Fluid Dyn. Conf., 6th*, pp. 48–60 (*AIAA Pap. No. 83-1888*)
Schmidt, W., Jameson, A. 1985. Euler solvers as an analysis tool for aircraft aerodynamics. In *Advances in Computational Transonics*, ed. W. G. Habashi, pp. 371–404. Swansea, Wales: Pineridge
Steger, J. L., Warming, R. F. 1981. Flux vector splitting of the inviscid gasdynamic equations with application to finite difference methods. *J. Comput. Phys.* 40: 263–93
Sweby, P. K. 1984. High resolution schemes using flux limiters for hyperbolic conservation laws. *SIAM J. Numer. Anal.* 21: 995–1011
van Leer, B. 1974. Towards the ultimate conservative difference scheme, II. Monotonicity and conservation combined in a second-order scheme. *J. Comput. Phys.* 14: 361–76
van Leer, B. 1979. Towards the ultimate conservative difference scheme, V. A second-order sequel to Godunov's method. *J. Comput. Phys.* 32: 101–36
van Leer, B. 1982. Flux-vector splitting for the Euler equations. *Lect. Notes Phys.* 170: 507–12
von Mises, R. 1950. On the thickness of a steady shock wave. *J. Aeronaut. Sci.* 17: 551–55
von Neumann, J., Richtmyer, R. D. 1950. A method for the numerical calculation of hydrodynamic shocks. *J. Appl. Phys.* 21: 232–57

Woodward, P., Colella, P. 1984. The numerical simulation of two-dimensional fluid flow with strong shocks. *J. Comput. Phys.* 54: 115–73

Yamamoto, O., Anderson, D. A., Salas, M. D. 1984. Numerical calculations of complex Mach reflection. *AIAA Pap. No. 84-1679*

Yang, J.-L., Liu, Y., Lomax, H. 1986. A numerical study of shock wave diffraction by a circular cylinder. *AIAA Pap. No. 86-0272*

Yee, H. C., Warming, R. F., Harten, A. 1985. Implicit total variation diminishing (TVD) schemes for steady-state calculations. *J. Comput. Phys.* 57: 327–60

Zannetti, L., Colasurdo, G. 1981. Unsteady compressible flow: a computational method consistent with the physical phenomena. *AIAA J.* 19: 852–56

Ann. Rev. Fluid Mech. 1987. 19: 339–67

SPECTRAL METHODS IN FLUID DYNAMICS[1]

M. Y. Hussaini

Institute for Computer Applications in Science and Engineering, NASA Langley Research Center, Hampton, Virginia 23665-5225

T. A. Zang

NASA Langley Research Center, Hampton, Virginia 23665-5225

INTRODUCTION

In certain areas of computational fluid dynamics, spectral methods have become the prevailing numerical tool for large-scale calculations. This is certainly the case for such three-dimensional applications as direct simulation of homogeneous turbulence, computation of transition in shear flows, and global weather modeling. For many other applications, such as heat transfer, boundary layers, reacting flows, compressible flows, and magnetohydrodynamics, spectral methods have proven to be a viable alternative to the traditional finite-difference and finite-element techniques.

Spectral methods are characterized by the expansion of the solution in terms of global and, usually, orthogonal polynomials. Since the mid-nineteenth century this has been a standard analytical tool for linear, separable differential equations. Nonlinearities present considerable algebraic difficulties, even on a modern computer. These difficulties were surmounted effectively in the early 1970s, and only then did spectral methods become competitive with alternative algorithms. By the present time, however, spectral methods have been refined and extended to the point where many problems in fluid mechanics are only tractable by these techniques.

Numerical spectral methods for partial differential equations were originally developed by meteorologists. Though this approach was proposed by Blinova (1944) and Haurwitz & Craig (1952), the first numerical computations were conducted by Silberman (1954). The expense of computing nonlinear terms remained a severe drawback until Orszag (1969) and Eliasen et al. (1970) developed the transform methods that still form the backbone of many large-scale spectral computations.

These methods and others used in fluid mechanics prior to 1970 are now termed spectral Galerkin methods: The fundamental unknowns are the expansion coefficients, and the equations for these are derived by the techniques used in classical analysis. The advent of computers made feasible an alternative discretization, termed the spectral collocation technique, in which the fundamental unknowns are the solution values at selected collocation points, and the series expansion is used solely for the purpose of approximating derivatives. This approach was proposed by Kreiss & Oliger (1971) and Orszag (1972).

Many useful versions of spectral methods have been developed since 1971 and especially during the 1980s. This review discusses many of the recent innovations and focuses on the collocation technique, since it is the version most readily applicable to nonlinear problems. We survey applications to both compressible and incompressible flows, to viscous as well as inviscid flows, and also to chemically reacting flows. In the interests of brevity we do not cover the applications to meteorology, magnetohydrodynamics, astrophysics, and other related fields. Moreover, we restrict ourselves to the three-dimensional applications of well-established algorithms while discussing some two- and even one-dimensional applications of more novel spectral methods.

We mention here some other articles for those interested in additional historical references, applications in other fields, and theoretical developments in the numerical analysis of spectral methods. The monograph by Gottlieb & Orszag (1977) describes the theory and applications developed prior to 1977. It will be referenced hereafter as GO. The following five years are covered in the proceedings edited by Voigt et al. (1984). Fluid-dynamical applications, especially multigrid techniques, are discussed by Zang & Hussaini (1985c). Compressible-flow applications are covered by Hussaini et al. (1985a). The role of spectral methods in meteorology is explained by Jarraud & Baede (1985). The book by Canuto et al. (1987) contains a detailed description of many spectral algorithms and presents an exhaustive discussion of the theoretical aspects of these numerical methods. It is referenced hereafter as CHQZ.

FUNDAMENTALS

The motivation for the use of spectral methods in numerical calculations stems from the attractive approximation properties of orthogonal polynomial expansions. Suppose, for example, that a function $u(x)$ is expanded in a truncated Chebyshev series on $[-1, 1]$:

$$u_N(x) = \sum_{n=0}^{N} a_n T_n(x), \tag{1}$$

where $T_n(x) = \cos(n \arccos x)$. The classical form of the expansion coefficients (or spectra) is

$$a_n = (2/c_n) \int_{-1}^{1} u(x) T_n(x) (1-x^2)^{-1/2} \, dx, \tag{2}$$

where $c_0 = 2$, and $c_n = 1$ for $n \geq 1$. The substitution $x = \cos\theta$ converts this into a Fourier cosine series. A simple integration-by-parts argument (GO, Ch. 3) reveals that

$$n^p a_n \to 0 \quad \text{as } n \to \infty, \quad \text{for all } p > 0, \tag{3}$$

provided that u is infinitely differentiable. Consequently, the approximation error decreases faster than algebraically. This rapid convergence is referred to as infinite-order accuracy, exponential convergence, or spectral accuracy. Our primary concern in this review is on numerical methods for partial differential equations that exhibit spectral accuracy for infinitely differentiable solutions.

The approximation just described is typical of spectral Galerkin methods. An alternative approximation, termed spectral collocation, is one of interpolation. It retains the expansion (1) but replaces condition (2) for the expansion coefficients with the condition

$$u_N(x_j) = u(x_j), \tag{4}$$

where x_j are special, so-called collocation points in $[-1, 1]$. For most problems, the optimal choice of these collocation points is

$$x_j = \cos(\pi j/N), \qquad j = 0, 1, \ldots, N. \tag{5}$$

This choice of collocation points yields an extremely accurate approximation (CHQZ, Ch. 2) to the integral appearing in Equation (2):

$$a_n = (2/N\bar{c}_n) \sum_{j=0}^{N} \bar{c}_j^{-1} u(x_j) T_n(x_j), \tag{6}$$

where $\bar{c}_0 = \bar{c}_N = 2$, and $\bar{c}_n = 1$ otherwise. No matter whether (2) or (6) is used for the expansion coefficients, the expansion (1) is differentiated analytically to form the approximations to whatever derivatives are required for the problem at hand.

A graphical distinction between traditional approximations and spectral ones is provided in Figure 1 for the simple task of estimating the derivative of the function $1+\sin(2\pi x+\pi/4)$ on $[-1, 1]$ from the values of the function at a finite number of grid points. A finite-difference or finite-element method uses local information to estimate derivatives, whereas a spectral method uses global information. In this figure a second-order (central)

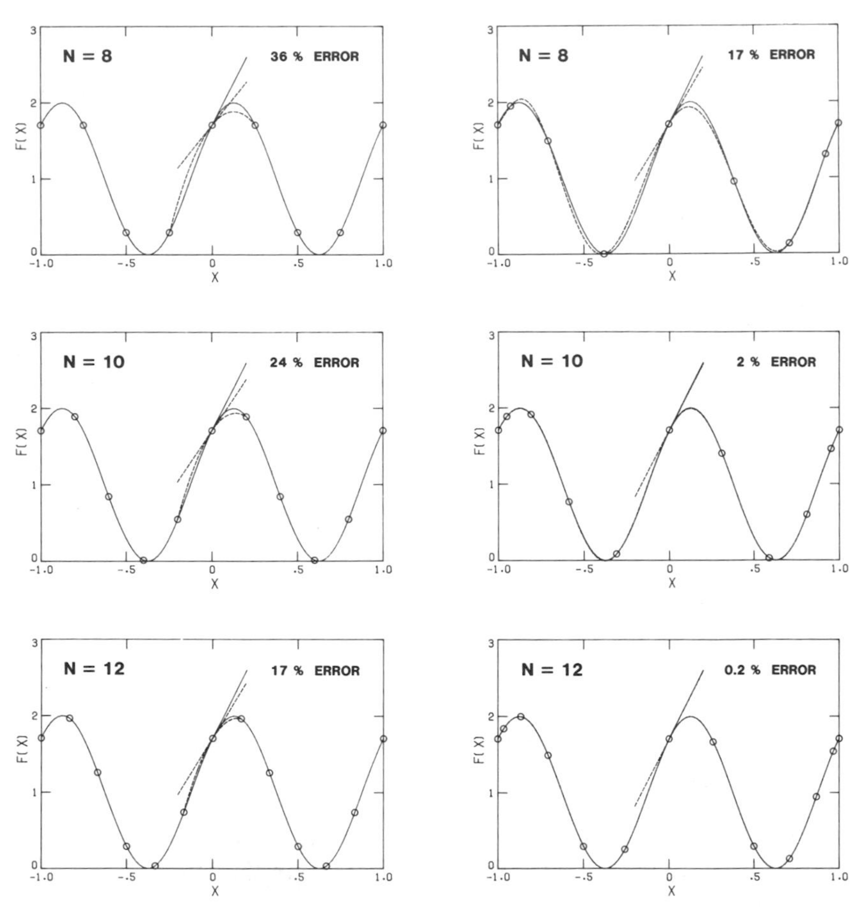

Figure 1 Comparison of finite-difference (*left*) and Chebyshev spectral (*right*) differentiation. The solid curves represent the exact function, and the dashed curves are their numerical approximations. The solid lines are the exact tangents at $x = 0$, and the dashed lines the approximate tangents. The error in slope is noted, as is the number of intervals N.

finite-difference method is compared with a Chebyshev spectral collocation method. The finite-difference approximation estimates the derivative at, say, $x = 0$ from the parabola that interpolates the function at $x = 0$ and the two adjacent grid points. A separate parabola is used at each grid point. The spectral approximation, on the other hand, uses all the available information about the function. If there are $N+1$ grid points, then the interpolating polynomial from which the derivative is extracted has degree N, and the same polynomial is used for all the grid points. Note that the local method produces a second-order accurate derivative, with the error decreasing as $1/N^2$, whereas the error from the global method decreases exponentially.

An essential aspect of any spectral method is the choice of expansion functions. Consider first the case of a bounded, Cartesian domain. Fourier series are the most familiar expansion functions, but they are only appropriate for problems with periodic boundary conditions. The appropriate collocation points on $[0, 2\pi]$ are

$$x_j = 2\pi j/N, \qquad j = 0, 1, \ldots, N-1. \tag{7}$$

In the general, nonperiodic case, normalized to $[-1, 1]$, the appropriate class of functions is the Jacobi polynomials. The proper collocation points are generally -1, $+1$, and the extrema of the last polynomial retained (CHQZ, Ch. 2). The most commonly used Jacobi polynomials are the Chebyshev and Legendre ones.

On an unbounded domain, the obvious choices of Laguerre or Hermite polynomials are rarely advisable. Not only are fast transforms unavailable, but these expansion functions have relatively poor resolution properties (GO, Ch. 3). A better approach is to combine a mapping with a Fourier or Chebyshev series in the mapped variable. Boyd (1986) has shown that spectral accuracy can be achieved for $u(x)$ on $(-\infty, \infty)$ with the mapping $x = x_* \cot \xi$ and a full Fourier series in ξ, provided that $u(x)$ exhibits at least algebraic decay at ∞. Moreover, if $u(x)$ has exponential decay, then a Fourier cosine series will suffice. The latter case is equivalent to a Chebyshev series in η with $x = x_*\eta/(1-\eta^2)^{1/2}$. Spalart (1984) noted that the odd (or even) Chebyshev polynomials work well on $[0, \infty)$ when combined with an exponential mapping, provided that $u(x)$ decays faster than exponentially.

The process of numerical differentiation is particularly simple when the expansion functions are trigonometric polynomials. Starting from u_j, the values of u at x_j, one computes

$$a_k = (1/N) \sum_{j=0}^{N-1} u_j \exp(-ikx_j), \qquad k = -\frac{N}{2}, -\frac{N}{2}+1, \ldots, \frac{N}{2}-1 \tag{8}$$

and then uses

$$\sum_{k=-N/2}^{N/2-1} ika_k \exp(ikx_j) \tag{9}$$

to approximate du/dx at x_j. The Fast Fourier Transform (FFT) can be used to evaluate both of the sums given above. The total cost of computing the derivative in this manner is $5N \log_2 N + N$ real operations. [All operation counts given in this review presume, for simplicity, that N is a power of 2 and that the complex FFT is used; however, FFTs that allow prime factors of 3 and 5 are just as efficient and are widely available, and real to half-complex FFTs offer a 20% savings (Temperton 1983).] The FFT can also be used to differentiate functions that are expanded in Chebyshev series, since expansions in these special Jacobi polynomials reduce to cosine series. Moreover, in terms of the Chebyshev coefficients, derivatives are obtained by simple recursion relations (CHQZ, Ch. 2). For Chebyshev series the total operation count for differentiation is $5N \log_2 N + 16N$.

For the classical expansion functions, the matrix that represents differentiation, i.e. $d^q u/dx^q = D^q u$, is known in closed form (Gottlieb et al. 1984). Unlike the differentiation matrices for alternative, local discretizations, these matrices are full. Hence, the matrix-vector multiplication that produces the derivative at the collocation points costs $2N^2$ operations. These operation counts suggest that for $N \geq 16$, transform methods are faster for differentiation than matrix-vector multiplication. On modern scalar and vector computers the transform methods become faster than the matrix-vector multiply methods for N between 16 and 32 (CHQZ, Ch. 2).

An important issue in many applications of Chebyshev spectral methods is the manner in which the boundary conditions are enforced. Dirichlet boundary conditions are generally straightforward. Neumann boundary conditions may be enforced by altering the boundary values to ensure the desired normal derivative or by building the boundary condition into the differential operator (Streett et al. 1985). For hyperbolic systems, characteristic boundary conditions are a virtual necessity (Gottlieb et al. 1981, Hussaini et al. 1985a). Canuto & Quarteroni (1986) discuss how to implement characteristic boundary conditions for implicit time discretizations. Chebyshev spectral methods have the advantages (over standard finite-difference schemes) that they require the same number and type of boundary conditions as the analytical formulations of the problem, and that no special difference formulas are required at the boundary.

The spectra of the discrete differentiation operators D^q are an important characteristic of numerical methods. For Fourier approximations to periodic problems, these are obvious: purely imaginary and growing as $N/2$ for D^1, negative real and growing as $N^2/4$ for D^2. Indeed, for periodic problems such as $u_t + u_x = 0$, the Fourier eigenvalues are exactly equal to their analytic counterparts. This means that Fourier spectral methods propagate the numerical solution with zero phase error. This is illustrated in Figure 2 for the problem whose solution is $u(x, t) = \sin[\pi \cos(x-t)]$. The lagging phase of the finite-difference solution is apparent, whereas the Fourier solution is indistinguishable from the true one. Of course, in realistic problems, variable coefficients or nonlinear terms will introduce nonzero (but still relatively small) phase errors.

Figure 3 displays the eigenvalues of a Chebyshev approximation to d/dx on $[-1, 1]$ with homogeneous Dirichlet boundary conditions at $x = +1$. The eigenvalues are predominantly imaginary but do have negative real parts. The absolute value of the largest eigenvalue grows as N^2. These eigenvalues may be surprising at first sight. After all, the periodic discrete problem has purely imaginary eigenvalues, whereas the nonperiodic continuous problem has no discrete eigenvalues. Nevertheless, Figure 3 does convey the nature of the eigenvalues of the discrete problem, and these are crucial for both time-differencing methods and iterative schemes. The eigenvalues of Chebyshev approximations to d^2/dx^2 with homogeneous Dirichlet boundary conditions at $x = -1$ and $x = +1$ are real and negative, and the largest eigenvalue grows as N^4 (Gottlieb & Lustman 1983).

In practice, when one is solving an evolution problem such as $u_t = Lu$, where the operator L contains all the spatial derivatives, one combines a spectral discretization of L with a standard finite-difference technique for the time derivative. The Leap Frog, Adams-Bashforth, Crank-Nicolson,

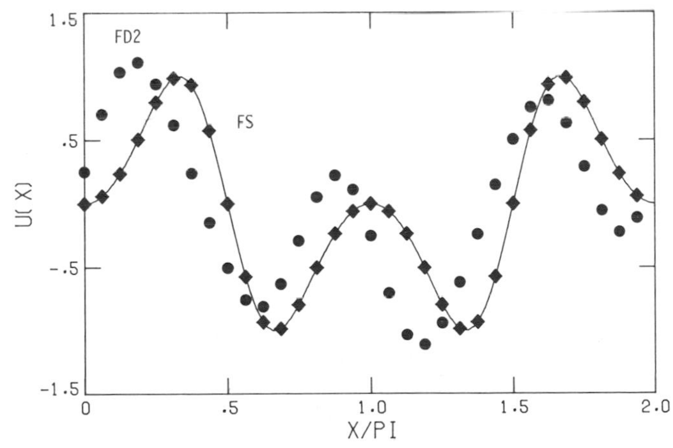

Figure 2 Finite-difference (circles) and Fourier spectral (diamonds) approximations after one period to a simple wave equation whose exact solution is represented by the curve.

and Runge-Kutta schemes are the ones most commonly used (CHQZ, Ch. 4). The stability regions of these schemes depend upon the spatial operators. The stability properties of Fourier methods are qualitatively the same as those for second-order central-difference spatial operators. However, the precise stability limit is typically a factor of $(1/\pi)^n$ smaller for Fourier approximations, where n is the order of the highest spatial derivative that appears in L.

The stability properties of Chebyshev methods are more subtle. For example, Leap Frog is unconditionally unstable for advection problems, such as $u_t + u_x = 0$, since the discrete eigenvalues of the spatial operator have negative real parts. On the other hand, second-order Adams-Bashforth and Runge-Kutta methods are strictly stable (and not weakly unstable like their Fourier counterparts) for the same reason.

The typical time-step limitations on Chebyshev methods are $1/N^2$ for first-derivative operators and $1/N^4$ for second-derivative ones. These are far more stringent than the analogous restrictions for uniform-grid finite-difference approximations. They arise from the crowding of the collocation points near the boundaries (see Figure 1). Although this crowding necessitates small time steps, it is required for the high spatial resolution of the method and is quite advantageous for problems with boundary layers.

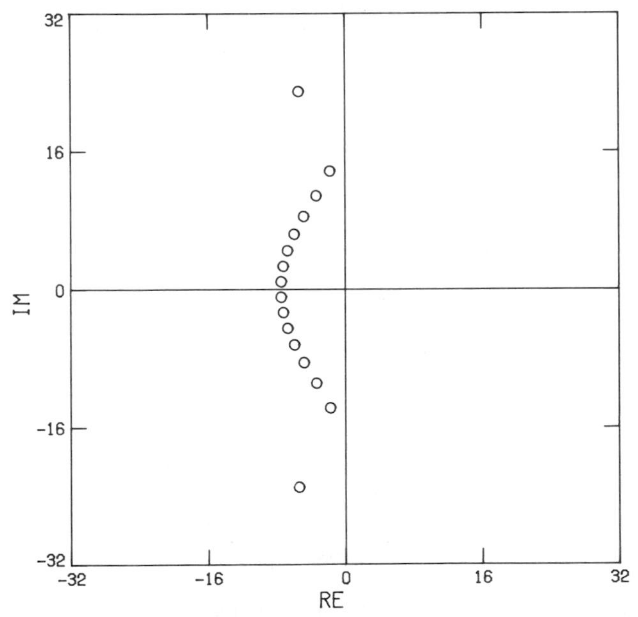

Figure 3 Eigenvalues of the Chebyshev first-derivative operator for $N = 16$.

This is, however, a substantial disadvantage for problems with very little structure near the boundaries. It can be alleviated to some degree by mapping, but a mapping to a uniform grid is counterproductive because it destroys the spatial accuracy.

This Chebyshev time-step limitation disappears when implicit time discretizations are employed. The principal difficulty is obtaining efficient solutions of the resulting implicit equations, since the matrices that represent the differentiation operators are full. In some special cases fast, direct solution methods are available. These typically require low-order polynomial coefficients and, in multidimensional problems, at most one nonperiodic direction (GO, Chs. 9 and 10, Moser et al. 1983).

The use of implicit techniques in more general situations requires iterative methods. This has been one of the major developments of the current decade (CHQZ, Ch. 5). Let us denote a typical linear, implicit system arising from a spectral discretization as $L_{sp}u = f$. The simplest iterative scheme—Richardson's method—is just

$$u \leftarrow u + \omega(f - L_{sp}u), \tag{10}$$

where ω is an acceleration parameter. The matrix L_{sp} will be full and will have eigenvalues that grow rapidly as the number of grid points increases. The fullness of the matrix does not preclude iterative methods, since transform techniques for differentiation permit the matrix-vector product $L_{sp}u$ to be computed in $O(N \log_2 N)$ operations rather than $O(N^2)$. The slow convergence that results from the large eigenvalues of L_{sp} can be ameliorated by preconditioning. In this case the basic iterative scheme is

$$u \leftarrow u + \omega H^{-1}(f - L_{sp}u), \tag{11}$$

where H is a preconditioning matrix. This will accelerate convergence if H is a good approximation to L_{sp}, and it will be relatively inexpensive if H is readily inverted. The former condition is met by low-order finite-difference (Orszag 1980) and finite-element (Deville & Mund 1985) approximations to L_{sp}. Although the latter condition certainly holds for one-dimensional problems, these particular preconditionings become increasingly expensive to invert as the dimensionality of the problem increases.

The most attractive approach to very large problems is to combine a less accurate but more readily inverted preconditioning with multigrid techniques. Spectral multigrid methods take advantage of the fact that most iterative methods are highly effective in reducing the error components corresponding to the upper half of the eigenvalue spectrum, but are very inefficient on the remaining, low-frequency components. Thus, in a multigrid method one combines iterations on the desired grid with (much cheaper) iterations on successively coarser grids. The details of this method are admittedly subtle, but they have been carefully described in a series of papers (Zang et al. 1982, 1984, Streett et al 1985, Phillips et al. 1986). Brandt et al. (1985) have demonstrated that many periodic problems can be successfully solved in this manner without the need for any preconditioning.

Another recent innovation has been the development of spectral multidomain techniques. These allow spectral methods to be applied to geometries for which a single, global expansion is either impossible or else inadvisable because of resolution requirements that vary widely over the domain. In a multidomain technique the full domain is partitioned into (not necessarily disjoint) subdomains. These may be patched together at interfaces or else they may overlap. The crucial part of the patched multidomain methods are the interface conditions. These may be expressed explicitly as continuity conditions (Orszag 1980, Kopriva 1986), may arise from a variational principle (Patera 1984), may consist of integral constraints (Macaraeg & Streett 1986), or may be enforced by a penalty method (Delves & Hall 1979). The spectral-element method of Patera is to date the most highly developed of these. Many techniques, such as isoparametric elements (Korczak & Patera 1986), have been borrowed from conventional finite-element methodology. Indeed, there are many similarities in this approach to the p-version of the finite-element method (Babuska & Dorr 1981). Figure 4 illustrates a spectral-element grid as well

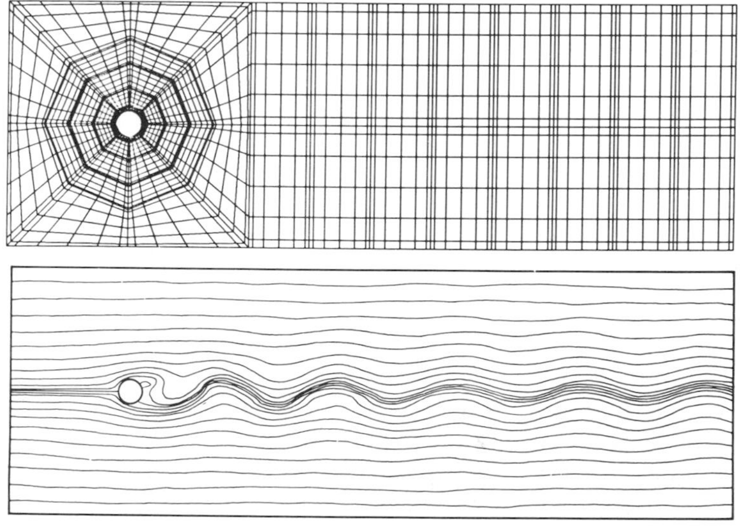

Figure 4 A spectral-element grid (*top*) and the corresponding numerical solution (*bottom*) for flow past a circular cylinder (courtesy of G. E. Karniadakis and A. T. Patera).

as the computed solution for flow past a cylinder (Karniadakis et al. 1986). In all cases, convergence is achieved with a fixed number of subdomains while the number of grid points on each subdomain increases. The spectral overlapping-subdomain methods were devised by Morchoisne (1983) and are currently being investigated extensively in Europe.

INVISCID FLOW

Perhaps the simplest fluid-dynamical problems are those that are steady, inviscid, incompressible, and irrotational. In terms of the velocity potential ϕ, these are described by the Laplace equation

$$\nabla^2\phi = 0, \tag{12}$$

with Neumann conditions on the boundaries. Spectral methods can be quite effective on such elliptic problems and also on the slightly more general class of problems described by

$$\nabla^2\phi - \lambda\phi = f \tag{13}$$

with Dirichlet, Neumann, or mixed boundary conditions. These more general methods could easily be applied to the idealized flow problem described above.

Spectral methods have been developed for such Poisson/Helmholtz problems in a variety of geometries. Direct methods are straightforward when at most one of the directions requires nonperiodic boundary conditions and hence a Chebyshev polynomial representation. Constant-coefficient equations become diagonal in the periodic directions. In a Cartesian nonperiodic direction, the equation can be reduced to a quasi-tridiagonal form (GO, Ch. 10) if the domain is finite and to a pentadiagonal form if it is infinite and the cotangent mapping is used (Cain et al. 1984). Otherwise, a matrix-diagonalization technique can be employed (Murdock 1977, Haidvogel & Zang 1979). Direct methods for problems with two or more nonperiodic directions have been discussed by Haidvogel & Zang (1979), Haldenwang et al. (1984), and LeQuere & de Roquefort (1985). Some extensions to three nonperiodic directions are described by Haldenwang et al. (1984) and Tan (1985). Iterative methods allow efficient treatment of more general geometries, especially for exterior problems [see Canuto et al (1985) and Deville & Mund (1985) for some standard techniques]. Especially for very large problems of this type, spectral multigrid methods appear to be the most efficient (Zang et al. 1982, 1984).

Compressible potential flow is described by a similar, but nonlinear, equation:

$$\nabla \cdot (\rho \nabla \phi) = 0, \tag{14}$$

where the density ρ is a quadratic function of $\nabla\phi$. For subsonic flow this problem is elliptic. Streett et al. (1985) have demonstrated the great efficiency that spectral multigrid methods achieve for this case. They have applied these techniques to the two-dimensional flow past a circular cylinder. Using a mere 2000 grid points, they have obtained an estimate for the free-stream Mach number at which the flow first becomes sonic. It agrees to six digits with the results of Van Dyke & Guttmann (1983) based on a Rayleigh-Jensen expansion.

For transonic flow, the potential equation is of mixed type, with a supersonic pocket embedded in a subsonic flow. There will be a sonic line and usually a shock that terminates the supersonic region. The challenging numerical task is to obtain a converged solution to the discrete, nonlinear potential equation. Spectral multigrid methods have proven competitive with finite-difference methods and have achieved substantial economies in storage (Streett et al. 1985).

Still within the confines of inviscid flow, one can obtain the effects of vorticity by resorting to the Euler equations

$$\frac{\partial \rho}{\partial t} + \nabla \cdot (\rho \mathbf{q}) = 0,$$

$$\frac{\partial \mathbf{q}}{\partial t} + \mathbf{q} \cdot \nabla \mathbf{q} = -\frac{1}{\rho} \nabla p,$$

$$\frac{\partial S}{\partial t} + \mathbf{q} \cdot \nabla S = 0, \tag{15}$$

where $\mathbf{q}$ is the velocity, p is the pressure, S is the entropy, and $p = \rho^{\gamma} e^{S/S_0}$. As is the case for all numerical methods, the real delicacy is the treatment of sonic lines and shock waves. The discontinuities arising from shocks are especially troublesome for spectral methods. The global nature of these approximations induces oscillations in the solution that are essentially of a Gibbs-phenomenon type. The high-frequency component of the solution decays very slowly. This part of the spectrum must be filtered to produce a presentable approximation. A detailed mathematical analysis of filtering techniques in Fourier spectral methods for linear, hyperbolic problems with discontinuous solutions has been presented by Majda et al. (1978). A postprocessing procedure that involves matching the computed solution with simple discontinuities has been discussed by Abarbanel et al. (1985).

The first applications of spectral methods to compressible flows focused on the treatment of shock waves in one-dimensional problems (Gottlieb et al. 1981, Zang & Hussaini 1981, Taylor et al. 1981). As is the case with finite-difference methods, spectral methods for problems involving shocks require some type of explicit or implicit numerical dissipation. In solutions to partial differential equations, the explicit dissipation may take the form of a linear, spectral filter, or it may consist of an artificial-viscosity term that is added to the Euler equations. This artificial viscosity may be nonlinear. Approximations based on Chebyshev polynomials may be stable without any explicit dissipation, since the Chebyshev derivative operator contains implicit dissipation (Gottlieb et al. 1981).

Most investigations have confined themselves to problems whose solutions (even in two dimensions; Sakell 1984) were either piecewise constant or else piecewise linear. No one has yet exhibited a spectral solution to a problem with both shock waves and complex flow structure in which spectral accuracy was attained (Hussaini et al. 1985b).

The difficulties that shock-capturing spectral methods exhibit are not due to any intrinsic difficulty in resolving transonic and supersonic flows. Kopriva et al. (1984) solved the Ringleb flow problem, which is a smooth two-dimensional transonic flow with a closed-form solution, by a Chebyshev spectral method. They were able to exhibit the usual spectral accuracy on this class of problems.

The shock-fitting approach popularized by Moretti (1968) for finite-difference schemes was adapted to spectral discretizations by Salas et al. (1982). This technique avoids the Gibbs phenomenon by treating the shock as a boundary rather than as an interior region of the flow. It is applicable to flows that contain a few, geometrically simple shocks. Some problems for which high-resolution results have been obtained by this method are the shock-vortex interaction (Salas et al. 1982), the shock-turbulence interaction (Zang et al. 1983), and the blunt-body problem (Hussaini et al. 1985c).

BOUNDARY LAYER

In many aerodynamic applications the boundary-layer equations are an economical and useful model of viscous effects, especially when coupled interactively with an inviscid model for the outer flow (AGARD 1981). In similarity variables, the two-dimensional boundary layer is described by

$$\frac{\partial}{\partial\eta}\left(\nu\frac{\partial f}{\partial\eta}\right)-v\frac{\partial f}{\partial\eta}-\beta(f^2-1)-2\xi f\frac{\partial f}{\partial\xi}=0,\quad \frac{\partial v}{\partial\eta}+f+2\xi\frac{\partial f}{\partial\xi}=0, \tag{16}$$

where f is the nondimensional streamwise velocity, v is the normal velocity, η is the normal coordinate, ξ is the streamwise coordinate, ν is the kinematic viscosity, and β is the pressure-gradient vector. The boundary conditions are $f = v = 0$ at $\eta = 0$ and $f \to 1$ as $\eta \to \infty$. An inflow condition is required at some $\xi = \xi_0$.

Chebyshev spectral approximations to a similar version of this system ($\partial/\partial\xi = 0$) are fairly straightforward to obtain by simple, preconditioned iterative schemes (Streett et al. 1984). This work demonstrated that a combination of domain truncation (typically at $\eta = 15$) and grid stretching (to pack grid points near the solid boundary) is quite effective. A mere 20 collocation points will usually yield values for the wall shear and displacement thickness that have three-digit accuracy, while 30 points produce five-digit accuracy.

The full nonsimilar equations are more challenging, since there is a Chebyshev approximation in two directions. Streett et al. (1984) used an alternating-direction type of preconditioning to obtain a solution. For nonsimilar flow, roughly 20 polynomials in ξ (coupled with 25 in η) are required for three-digit accuracy. Streett et al. found that the Chebyshev approximation in ξ produced a substantial improvement over a simpler, mixed scheme that used finite differences in ξ together with Chebyshev

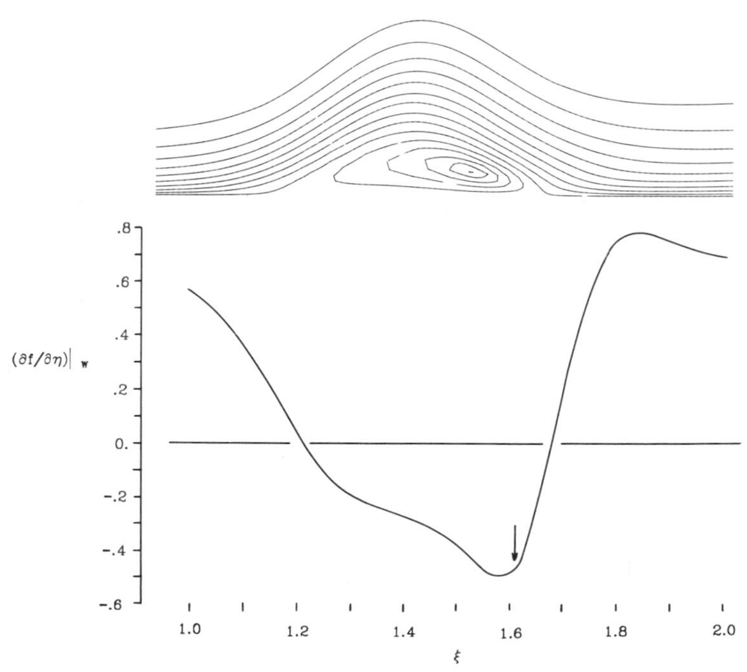

Figure 5 Streamlines (*top*) and skin friction (*bottom*) from a Chebyshev spectral solution of the boundary-layer equations (courtesy of C. Streett).

collocation in η. The global nature of the streamwise approximation is especially useful for handling separated flow. In this case, marching techniques in ξ are ineffective for finite-difference approximations. Figure 5 displays the streamlines and skin friction from a fully spectral solution of separated boundary-layer flow. The arrow marks the region of the flow that is most sensitive to the numerical resolution. To obtain four-digit accuracy in the skin friction here requires 40 collocation points in the normal direction and 26 in the streamwise direction. The corresponding requirements for a standard second-order finite-difference method are 240 and 200 points, respectively. Moreover, the spectral solution requires only 10% of the CPU time taken by the finite-difference method.

NAVIER-STOKES FLOW

Much of the current enthusiasm for spectral methods is attributable to their success on simple, yet computationally intensive, problems in viscous, time-dependent incompressible flow. The pioneering simulations of three-dimensional homogeneous, isotropic turbulence by Orszag & Patterson (1972) were particularly influential. Subsequent calculations of three-dimensional transition and turbulence in simple wall-bounded flows have also been persuasive. Algorithms for these problems are substantially more difficult and time consuming than those for homogeneous flows. The presence of nonperiodic boundary conditions makes purely Fourier methods inappropriate, and detailed simulations of transition problems typically require an order of magnitude more time steps than do turbulence problems. The simplest class of such problems consists of flows that are assumed to be periodic in two directions, e.g. Poiseuille flow and Taylor-Couette flow for cylinders of infinite length. In these cases, one needs a Chebyshev discretization in only one direction. Several types of multi-domain spectral techniques are currently being explored to extend further the class of viscous problems that are amenable to spectral methods.

In many applications the preferred version of the Navier-Stokes equations is

$$\frac{\partial \mathbf{q}}{\partial t} + \omega \times \mathbf{q} = -\nabla P + \nu \nabla^2 \mathbf{q},$$

$$\nabla \cdot \mathbf{q} = 0, \tag{17}$$

where $\mathbf{q}$ is the velocity, p is the pressure, $\omega = \nabla \times \mathbf{u}$ is the vorticity, $P = p + (1/2)|\mathbf{q}|^2$ is the pressure head, and ν is the kinematic viscosity. This so-called rotation form is favored because, as noted by Orszag (1972), the use of the rotation form guarantees that Fourier collocation methods conserve kinetic energy. One can easily show that momentum is conserved as well. The conservation of kinetic energy is especially important for numerical reasons. In practice, it means that if the time-differencing scheme is operated at time steps below its stability limit, then nonlinear instabilities will not occur.

Homogeneous Turbulence

Homogeneous, isotropic turbulence is perhaps the one fluid-dynamical problem for which strictly periodic boundary conditions in all spatial directions are justifiable. Hence, Fourier spectral methods are ideally suited for this class of problems. Moreover, since the nonlinearities of the Navier-Stokes equations are at worst quadratic, Fourier Galerkin methods are the most natural and efficient spectral techniques for this problem (Orszag & Patterson 1972). Rogallo (1981) developed a linear coordinate transformation that permits simulation of flows with constant strain, shear, and rotation within the confines of periodic boundary conditions. Rogallo (1981) and Basdevant (1983) have discussed techniques for minimizing the storage, CPU time, and I/O costs of such algorithms.

The original simulations of Orszag & Patterson were on 32^3 grids. By the early 1980s, 64^3 simulations were fairly routine. Rogallo (1981), Kerr (1985), and Lee & Reynolds (1985) have performed numerous 128^3 simulations. By fully exploiting the special symmetries of the Taylor-Green vortex, Brachet et al. (1983) achieved a simulation of this flow at a Reynolds number of 3000 with an effective resolution of 256^3.

Fourier collocation approximations to this problem are also possible. For these approximations, use of the rotation form of the Navier-Stokes equations is crucial. (Galerkin approximations to the inviscid form of these equations will automatically conserve momentum and kinetic energy in the absence of time-differencing errors).

The review by Rogallo & Moin (1984) discusses many applications of these techniques to problems in homogeneous turbulence. Here, we need mention only the most recent applications. A primary goal of most of the simulations of isotropic turbulence has been to establish numerically the existence of an inertial range. The inertial range has, of course, been established experimentally, but only for Reynolds numbers exceeding 10,000. Even though the high-resolution calculations of Brachet et al. (1983) were performed at a Reynolds number of 3000, which is uncomfortably low by experimental standards, they did achieve the first plausible inertial range in a numerical simulation of turbulence. Bardina et al. (1985) and Dang & Roy (1985) have simulated the evolution of turbulence intensity in rotating flow. Wu et al. (1985) have performed calculations of compressed turbulence. Lee & Reynolds (1985) have analyzed the structure of turbulence in axisymmetrically contracting and expanding flow. Moin et al. (1985) have used numerical simulations to extract the large-scale vortical structures of some turbulent shear flows. Kerr (1985) has examined high-order correlations and small-scale structure in isotropic turbulence involving passive scalars.

A few applications, all using the collocation technique, have been made to compressible, homogeneous turbulence. Feiereisen et al. (1981) simulated subsonic turbulent flows with uniform shear. They used a collocation method, in part because a Galerkin method is much more cumbersome and costly for problems with more than quadratic nonlinearities. Compressible, two-dimensional turbulence has been investigated by Leorat et al. (1985) and by Delorme (1984), the former with a fairly standard scheme and the latter with an implicit time-differencing method based on the ideas of Lerat et al. (1982).

Linear Stability

Most investigations of stability and transition in wall-bounded flows rely, at least in part, upon the results of linear-stability theory. The Orr-Sommerfeld equation has been the basis for many investigations of the stability of incompressible parallel flows (Drazin & Reid 1981). This eigenvalue problem is described by a fourth-order ordinary differential equation. The Chebyshev approximation developed by Orszag (1971) for the temporal-stability problem has been adopted and extended by many investigators. [A separate development of Chebyshev methods for ordinary differential eigenvalue problems has been conducted by Ortiz. For further details, the reader should consult Chaves & Ortiz (1968) and, more recently, Ortiz & Samara (1983).] Leonard & Wray (1982) developed a Galerkin method for pipe flow that uses special Jacobi polynomials. Spalart (1984) demonstrated that for exterior flows (such as the parallel boundary layer), the use of only half the usual Chebyshev basis was advisable. Boyd (1985) has developed methods in the complex plane that are useful for flows in which the critical layer is well separated from the wall. Von Kerczek (1982) has used Chebyshev polynomials for assessing the stability of oscillatory plane Poiseuille flow. Mac Giolla Mhuiris (1986) has used the Galerkin technique to examine the linear stability of some axisymmetric flows that are relevant to the vortex-breakdown problem.

The spatial-stability versions of these problems are more difficult because the eigenvalue enters nonlinearly. Chebyshev methods for time-independent but spatially growing perturbations of Poiseuille flow are discussed by Bramley & Dennis (1982). Bridges & Morris (1984a,b) used a spectral method to solve the more difficult, general spatial-stability problem of self-similar boundary layers.

These methods have been extended, in the manner of Floquet theory, to include weakly nonlinear effects. In addition to a Chebyshev discretization in the direction normal to the wall, one includes several Fourier harmonics in the streamwise direction. Orszag & Patera (1983) and Herbert (1983a) have used this approach to determine the neutral-stability surface of finite-amplitude, two-dimensional Tollmien-Schlichting waves in channel flow. In turn, the linear stability of these neutral finite-amplitude waves can be examined. Thus, the linear stability of some special, temporally and spatially varying flows can be investigated. Orszag & Patera (1983) have used this technique to study the interaction of two-dimensional and three-dimensional Tollmien-Schlichting waves in channel flow. Herbert (1983a,b, 1984) has performed a detailed study of channel and boundary-layer flows. He has unraveled the details of fundamental and subharmonic instabilities in parallel flows.

Transition

Transition to turbulence is highly nonlinear, and a full simulation of the Navier-Stokes equations is required for its investigation. The primary difficulty of algorithms for incompressible flows is the simultaneous enforcement of the incompressibility constraint and the no-slip boundary condition. This constraint is most easily, but least rigorously, satisfied in splitting methods, of which the Orszag & Kells (1980) algorithm is the prototype. The splitting errors of this method are $O(1)$ near the boundary for the normal pressure gradient and diffusion terms (Deville 1985). They appear to cause no serious errors in the channel-flow problem. [See the summary provided by Gottlieb et al. (1984) of the yet unpublished work by S. A. Orszag, M. Deville & M. Israeli.] However, Marcus (1984a) decisively demonstrated that the boundary errors produce serious inaccuracies in Taylor-Couette flow—as both the spatial and temporal discretizations are refined, the algorithm appears to converge to answers that disagree with experiments in the third digit. Marcus (1984a) and Kleiser & Schumann (1984) devised an influence-matrix technique that completely eliminates the splitting errors at a modest extra cost. Marcus found that the results of this algorithm agreed with the experimental results to the full four digits that were available. He ascribed the sensitivity of the rotating-cylinder problem to the fact that its dynamics are driven by the motion of the boundary rather than by a mean pressure gradient.

A procedure for reducing, although not entirely eliminating, the splitting errors at the boundary was devised by Fortin et al. (1971) for finite-element methods (rediscovered later by Kim & Moin 1985) and applied to spectral algorithms by Zang & Hussaini (1986). It consists of modifying the boundary conditions for the intermediate steps of the algorithm so that both the no-slip and divergence-free conditions are satisfied at the end of the full time step to a higher order in the size of the time step.

The big advantage of these splitting techniques is that they require the solution of only Poisson equations (for the pressure) or Helmholtz equations (from a Crank-Nicolson discretization of the viscous term). These positive-definite, scalar equations are much easier to solve numerically than the indefinite, coupled equations that arise in unsplit methods. The Orszag-Kells, Marcus, and Kleiser-Schumann algorithms resort to direct solution methods of the type discussed in the section on inviscid flow. The Zang-Hussaini algorithm employs iterative techniques so that it is applicable to a wider class of problems. The most sophisticated and powerful of the iterative techniques is the spectral multigrid method. It makes the cost of a single time step of order $N^3 \log_2 N$, even for problems with variable geometric terms and transport coefficients. In contrast, a parallel-flow problem, even with uniform transport coefficients, requires order N^4 operations per step by direct methods.

One way to avoid the splitting errors is to integrate the incompressible Navier-Stokes equations in a single step that couples the divergence-free constraint with the momentum equations. The numerical difficulty of this approach is that one must invert a larger set of equations (it involves the pressure as well as the three velocity components), which is indefinite. In a few special cases, direct techniques are viable (Moin & Kim 1980). The preconditioned iterative scheme of Malik et al. (1985) has been applied to channel flow (Zang & Hussaini 1985a) and to the heated boundary layer (Zang & Hussaini 1985b), a problem that involves variable transport coefficients, and has also been used in a verification of weakly nonlinear stability theory for stagnation point flow (Hall & Malik 1986).

Many of the numerical problems caused by the incompressibility constraint can be avoided by an expansion in functions that are divergence free (Ladyzhenskaya 1969, Temam 1977). Leonard & Wray (1982) first applied this idea to spectral methods. They devised a set of basis functions for pipe flow that are both divergence free and satisfy no-slip boundary conditions. Similar basis functions have been developed for straight and curved channels (Moser et al. 1983) and for the parallel boundary layer (Spalart 1984). This class of methods can be quite economical of storage, since only two variables per grid point are required to specify the flow field. (However, in actual implementations it may be more efficient in terms of CPU time to store several additional quantities per grid point.) The efficiency of these methods depends upon the bandwidth of the matrices that arise from the implicit treatment of the viscous terms. In the examples cited above, the bandwidth is quite small, roughly of order 10.

This requirement has dictated the use of special Jacobi polynomials, rather than Chebyshev ones, in pipe and boundary-layer flow. As a consequence, transform methods are not applicable in the nonperiodic direction. Hence, the cost of evaluating the nonlinear terms increases as N^4 rather than as $N^3 \log_2 N$. Moreover, in even slightly more general cases, the matrices can be completely full.

Orszag & Patera (1983) performed a parametric study of the secondary instability in channels and pipes, demonstrating that subcritical instabilities exist at Reynolds numbers as low as 1100. Kleiser & Schumann (1984) replicated many of the features of the Nishioka et al. (1980) experiments on channel-flow transition. Both groups also obtained good quantitative agreement with the predictions of weakly nonlinear theory. Rozhdestvensky & Simakin (1984) have exhibited a variety of secondary flows in plane channels. The subharmonic instabilities that were predicted by Herbert's (1983b, 1984) weakly nonlinear analysis [and that are also in evidence in boundary-layer experiments (Saric et al. 1984)] were reproduced by Spalart (1985) and Laurien (1986) for the boundary layer and by Zang & Hussaini (1985a) and Singer et al. (1986) for channel flow. The existence of a similar nonlinear instability of center modes in channel flow was uncovered by Zang & Hussaini (1985a). A detailed comparison of nonlinear effects on the laminar-flow control techniques of pressure gradient, suction, and heating in boundary-layer flow was made by Zang & Hussaini (1985b). Krist & Zang (1986) have performed a detailed study of the resolution requirements for simulation of the later stages of transition to turbulence in channel flow. The spanwise direction places the greatest demands on the resolution because of the very sharp spanwise gradients that occur near the tip of the characteristic hairpin vortex. Figure 6, which is extracted from that work, illustrates the structure.

Marcus (1984a,b) has performed a careful numerical study of nonaxisymmetric instabilities in classical Taylor-Couette flow. He has produced four-digit agreement with the wave speeds measured by King et al. (1984) for both the one wavy-vortex and the two wavy-vortex states. Marcus & Tuckerman (1986a,b) have simulated axisymmetric spherical Couette flow. Unlike previous workers, they did not assume equatorial symmetry. This was a crucial factor in their success in reproducing the transitions between 0, 1, and 2 vortex states observed by Wimmer (1976).

Inhomogeneous Turbulence

In several cases these algorithms have been used to simulate turbulence in wall-bounded flows. Orszag & Patera (1983) performed a 64^3 simulation of turbulent channel flow that reproduced the turbulent velocity profile, including the law-of-the-wall behavior. Moser & Moin (1984) computed turbulent flow in a curved channel on a $128^2 \times 64$ grid. They reproduced some of the data on low-order turbulence statistics and exhibited some of the effects of curvature. Spalart & Leonard (1985) have done some analyses of pressure-gradient effects in turbulent boundary layers.

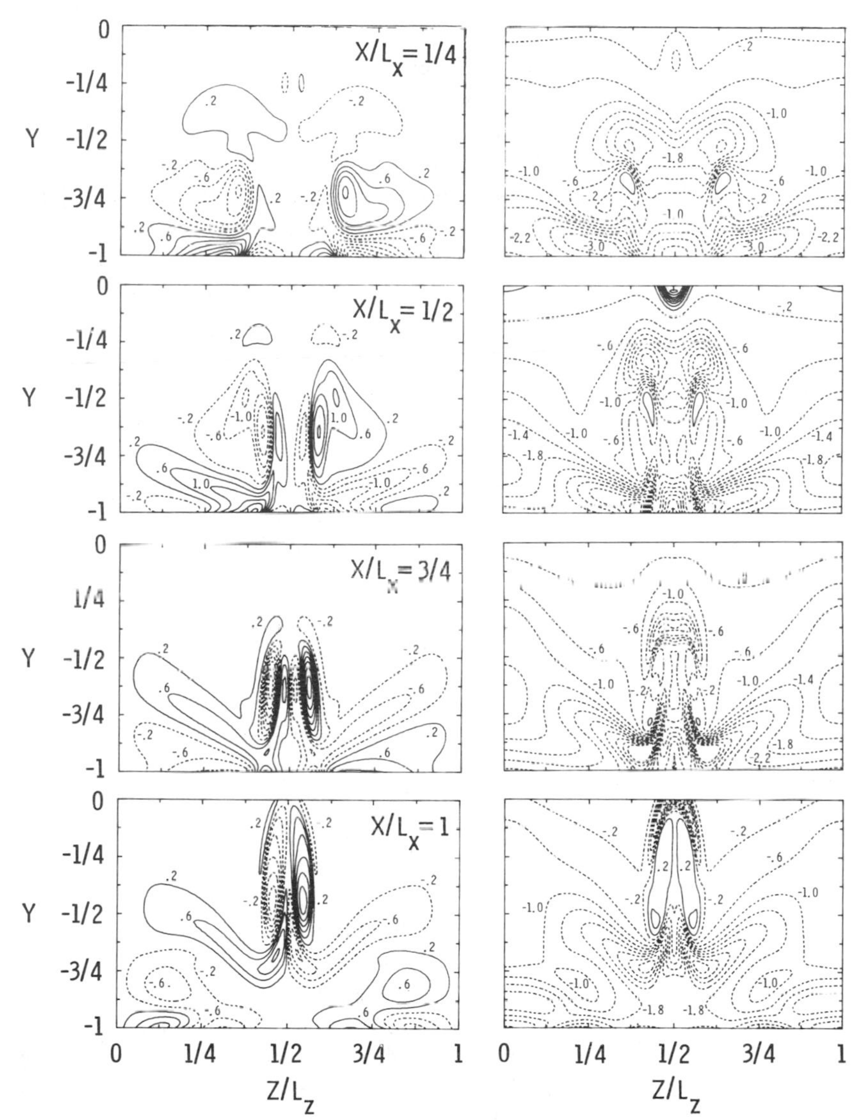

Figure 6 Streamwise (*left*) and spanwise (*right*) vorticity at four streamwise locations for a hairpin vortex in low-Reynolds-number channel-flow transition. Only the lower half of the channel is shown.

More Realistic Geometries

As noted above, there is a substantial increase in cost when there is more than one inhomogeneous direction in the problem. The Kleiser-Schumann influence-matrix technique has been extended to two nonperiodic directions by LeQuere & de Roquefort (1985), who used it to study thermal convection in a square cavity. Streett & Hussaini (1987) similarly extended the split algorithm of Zang & Hussaini (1986) and used it to study the effect of finite-length cylinders in Taylor-Couette flow. Ku et al. (1987) have developed an algorithm for three nonperiodic directions. This method presently treats only the pressure term implicitly. Thus, there can be a severe time-step limitation arising from the viscous terms. Morchoisne (1984) has developed a number of methods for problems with more than one nonperiodic direction. In general, iterative techniques are used for solving the resulting implicit equations. There has not yet been any systematic comparison of these methods. Leonard (1984) has derived a set of divergence-free basis functions for two nonperiodic directions, but an efficient solution technique for the implicit equations has not yet been devised.

Several of the multidomain spectral methods have been applied to viscous problems. Morchoisne (1984) has performed some sample calculations of channel flow. The spectral element has been used to calculate heat transfer in a two-dimensional, grooved channel (Ghaddar et al. 1984) and to investigate stability and resonance phenomena in embedded cavities in channel flows (Ghaddar et al. 1986a,b). Other applications include two-dimensional flow past a cylinder and flow past three-dimensional roughness elements (Karniadakis et al. 1986).

Spectral/Finite-Difference and Quasi-Spectral Methods

Heretofore, this review has been confined to numerical fluid-dynamical work that employed spectral discretizations in all coordinate directions. There have, of course, been numerous computations that used mixed spectral/finite-difference methods, i.e. algorithms with spectral discretizations in some directions and finite differences in the others. The parallel boundary-layer transition calculations of Wray & Hussaini (1984) fall into this category. They used a Fourier spectral method in two periodic directions and second-order finite differences in the normal direction. They demonstrated that despite the neglect of nonparallel effects, these simulations could reproduce features observed experimentally by Kovasznay et al. (1962), up to the so-called two-spike stage of transition. A slightly different spectral/finite-difference method was used by Moin & Kim (1982) in their large-eddy simulations of turbulent channel flow and by Biringen

(1985) in a study of active control in channel flows. More recently, Eidson et al. (1986) have used a similar algorithm in a high-resolution direct simulation of a turbulent Rayleigh-Bénard flow.

Another alternative to true spectral methods is what might be termed quasi-spectral methods. Such algorithms employ Fourier expansions in all directions, but infinite-order accuracy is not attained as a result of nonperiodic physical boundary conditions in at least one direction. The simulations by Riley & Metcalfe (1980) of a time-developing mixing layer fall into this category. In this idealized flow the mean velocity is solely a function of the transverse coordinate y. Although the flow extends to $y = \pm\infty$, Riley & Metcalfe computed on a finite domain in y and used sine or cosine expansions to enforce free-slip boundary conditions in y. Quasi-spectral methods have also been used by Curry et al. (1984) to study Bénard convection.

True spectral methods have been developed for the time-developing mixing layer. Cain et al. (1984) used a cotangent transformation in y combined with a Fourier method. Metcalfe et al. (1986) applied hyperbolic tangent or algebraic transformations combined with a Chebyshev method.

Riley & Metcalfe (1980) have found that large-amplitude, two-dimensional disturbances have a pronounced effect upon the evolution of a turbulent mixing layer. Metcalfe et al. (1986) have observed that the mixing layer exhibits three-dimensional secondary instabilities similar to those that occur in wall-bounded flows. These instabilities appear to account for the mushroom-shaped features that are observed experimentally. Cain et al. (1981) have performed large-eddy simulations of this problem.

REACTING FLOWS

An emerging application field for spectral methods is reacting flows. These flows are especially challenging because they contain sharp gradients in both space and time and because most real flows involve dozens or even hundreds of species. Flame fronts and shock waves are an additional complication. Some of the important features are mixing rates, ignition, and flame holding.

There are a number of simplifying assumptions that lead to more tractable, but less realistic, models of reacting flows. The most drastic of these is that the reactions proceed without heat release and that the Mach number is so low that the flow may be treated as incompressible. Riley et al. (1986) have performed some three-dimensional simulations of a two-species, time-developing mixing layer. They used a quasi-spectral method and obtained good agreement with both similarity theory and experimental data.

McMurtry et al. (1986) employed a low-Mach-number approximation that includes some mild heat-release effects but neglects the acoustic modes. They performed some two-dimensional calculations that indicate that the first-order effect of heat release is to reduce the rate of mixing.

Drummond et al. (1986) applied a Chebyshev spectral method to a supersonic quasi-one-dimensional diverging nozzle flow with a simple but quite stiff two-species hydrogen-air reaction. The spectral method proved to be quite economical compared with a benchmark finite-difference result. The Chebyshev grid-point distribution was quite well adapted to the sharp gradients at the nozzle inflow but less well suited to the fairly uniform outflow region.

PERSPECTIVE

A decade ago spectral methods appeared to be well suited only to problems governed by ordinary differential equations or by partial differential equations with periodic boundary conditions. And, of course, the solution itself needed to be smooth. Some of the obstacles to wider application of spectral methods were (*a*) sensitivity to boundary conditions, (*b*) treatment of discontinuous solutions, (*c*) resolution and time-step limitations imposed by the standard spectral grids, and (*d*) drastic geometric constraints.

Substantial progress has been made on the implementation of Neumann boundary conditions, on characteristic boundary conditions for hyperbolic systems, and on the use of pressure and intermediate boundary conditions in incompressible flow. There have been some theoretical advances on filtering techniques for discontinuous solutions to linear problems. Moreover, the development of shock-fitting techniques has opened a new field of applications to compressible flows with shock waves. Some efficient direct solution techniques have been devised that enable severe viscous time-step limitations to be overcome in certain special geometries. The development of preconditioned iterative methods and, in particular, spectral multigrid techniques have radically expanded the class of problems that can be handled efficiently by spectral methods. Moreover, they lend much greater flexibility (combined with mapping techniques) to the grid-point distribution. Finally, various multidomain techniques have expanded the range of spectral methods to many problems of real, practical interest.

Literature Cited

Abarbanel, S., Gottlieb, D., Tadmor, E. 1985. Spectral methods for discontinuous problems. *NASA CR 117974*, NASA Langley Res. Cent., Hampton, Va.

AGARD. 1981. *Computation of Viscous-Inviscid Interactions, Conference Proceedings No. 291*

Babuska, I., Dorr, M. R. 1981. Error estimates for the combined *h* and *p* versions of the finite element method. *Numer. Math.* 37 : 257–77

Bardina, J., Ferziger, J. H., Rogallo, R. S. 1985. Effect of rotation on isotropic turbulence: computation and modelling. *J. Fluid Mech.* 154 : 321–36

Basdevant, C. 1983. Technical improvements for direct numerical simulation of homogeneous three-dimensional turbulence. *J. Comput. Phys.* 50 : 209–14

Biringen, S. 1985. Active control of transition by periodic suction-blowing. *Phys. Fluids* 27 : 1345–47

Blinova, E. N. 1944. Hydrodynamic theory of pressure and temperature waves and center of action of the atmosphere. *Transl. No. 113*, Reg. Control Off., Second Weather Reg., Patterson Field, Dayton, Ohio

Boyd, J. P. 1985. Complex coordinate methods for hydrodynamic instabilities and Sturm-Liouville eigenproblems with an interior singularity. *J. Comput. Phys.* 57 : 453–71

Boyd, J. P. 1986. Spectral methods using rational basis functions on an infinite interval. *J. Comput. Phys.* In press

Brachet, M. E., Meiron, D. I., Orszag, S. A., Nickel, B. G., Morf, R. H., Frisch, U. 1983. Small-scale structure of the Taylor-Green vortex. *J. Fluid Mech.* 130 : 411–52

Bramley, J. S., Dennis, S. C. R. 1982. The calculation of eigenvalues for the stationary perturbations of Poiseuille flow. *J. Comput. Phys.* 47 : 179–98

Brandt, A., Fulton, S. R., Taylor, G. D. 1985. Improved spectral multigrid methods for periodic elliptic problems. *J. Comput. Phys.* 58 : 96–112

Bridges, T. J., Morris, P. J. 1984a. Differential eigenvalue problems in which the parameter appears nonlinearly. *J. Comput. Phys.* 55 : 437–60

Bridges, T. J., Morris, P. J. 1984b. Spectral calculations of the spatial stability of non-parallel boundary layers. *AIAA Pap. No. 84-0437*

Cain, A. B., Reynolds, W. C., Ferziger, J. H. 1981. A three-dimensional simulation of transition and early turbulence in a time-developing mixing layer. *Rep. No. TF-14*, Dep. Mech. Eng., Stanford Univ., Calif.

Cain, A. B., Ferziger, J. H., Reynolds, W. C. 1984. Discrete orthogonal function expansions for non-uniform grids using the fast Fourier transform. *J. Comput. Phys.* 56 : 272–86

Canuto, C., Quarteroni, A. 1986. On the boundary treatment in spectral methods for hyperbolic systems. *NASA CR 178055*, NASA Langley Res. Cent., Hampton, Va.

Canuto, C., Hariharan, S. I., Lustman, L. 1985. Spectral methods for exterior elliptic problems. *Numer. Math.* 46 : 505–20

Canuto, C., Hussaini, M. Y., Quarteroni, A., Zang, T. A. 1987. *Spectral Methods in Fluid Dynamics*. Berlin: Springer-Verlag. In press

Chaves, T., Ortiz, E. L. 1968. On the numerical solution of two-point boundary value problems for linear differential equations. *ZAMM* 48 : 415–18

Curry, J. H., Herring, J. R., Loncaric, J., Orszag, S. A. 1984. Order and disorder in two- and three-dimensional Bénard convection. *J. Fluid Mech.* 147 : 1–38

Dang, K., Roy, P. 1985. Direct and large-eddy simulation of homogeneous turbulence submitted to solid body rotation. *Proc. Symp. Turbul. Shear Flows, 5th, Ithaca, N.Y.*, pp. 17.1–6

Delorme, P. 1984. Numerical simulation of homogeneous, isotropic, two-dimensional turbulence in compressible flow. *Rech. Aérosp.* 1984(1) : 1–13

Delves, L. M., Hall, C. A. 1979. An implicit matching procedure for global element calculations. *J. Inst. Math. Its Appl.* 23 : 223–34

Deville, M., Mund, E. 1985. Chebyshev pseudospectral solution of second-order elliptic equations with finite element preconditioning. *J. Comput. Phys.* 60 : 517–33

Drazin, P. G., Reid, W. M. 1981. *Hydrodynamic Stability*. Cambridge: Cambridge Univ. Press. 527 pp.

Drummond, J. P., Hussaini, M. Y., Zang, T. A. 1986. Spectral methods for modelling supersonic chemically reacting flow fields. *AIAA J.* 24 : 1461–67

Eidson, T. M., Hussaini, M. Y., Zang, T. A. 1986. Simulation of the turbulent Rayleigh-Bénard problem using a spectral/finite difference technique. *NASA CR 178027*, NASA Langley Res. Cent., Hampton, Va.

Eliasen, E., Machenauer, E., Rasmussen, E. 1970. On a numerical method for integration of the hydrodynamical equations with a spectral representation of the horizontal fields. *Rep. No. 2*, Dep. Meteorol., Copenhagen Univ., Den.

Feiereisen, W. J., Reynolds, W. C., Ferziger, J. H. 1981. Numerical simulation of compressible, homogeneous turbulent shear flow. *Rep. No. TF-13*, Dep. Mech. Eng., Stanford Univ., Calif.

Fortin, M., Peyret, R., Temam, R. 1971. Résolution numérique des équations de Navier-Stokes pour un fluide incompressible. *J. Mec.* 10 : 357–90

Ghaddar, N., Patera, A. T., Mikic, B. 1984. Heat transfer enhancement in oscillatory flow in a grooved channel. *AIAA Pap. No. 84-0495*

Ghaddar, N. K., Korczak, K. Z., Mikic, B. B., Patera, A. T. 1986a. Numerical investigation of incompressible flow in grooved channels, Part 1. Stability and self-sustained oscillations. *J. Fluid Mech.* 163: 99–127

Ghaddar, N. K., Korczak, K. Z., Mikic, B. B., Patera, A. T. 1986b. Numerical investigation of incompressible flow in grooved channels, Part 2. Resonance and oscillatory heat transfer. *J. Fluid Mech.* 168: 541–67

Gottlieb, D., Lustman, L. 1983. The spectrum of the Chebyshev collocation operator for the heat equation. *SIAM J. Numer. Anal.* 20: 909–21

Gottlieb, D., Orszag, S. A. 1977. *Numerical Analysis of Spectral Methods: Theory and Applications, CBMS-NSF Reg. Conf. Ser. in Appl. Math.*, Vol. 26. Philadelphia: SIAM. 170 pp.

Gottlieb, D., Lustman, L., Orszag, S. A. 1981. Spectral calculations of one-dimensional inviscid compressible flows. *SIAM J. Sci. Stat. Comput.* 2: 296–310

Gottlieb, D., Hussaini, M. Y., Orszag, S. A. 1984. Theory and applications of spectral methods. See Voigt et al. 1984, pp. 1–54

Haidvogel, D. B., Zang, T. A. 1979. The accurate solution of Poisson's equation by expansion in Chebyshev polynomials. *J. Comput. Phys.* 30: 167–80

Haldenwang, P., Labrosse, G., Abboudi, S., Deville, M. 1984. Chebyshev 3-D spectral and 2-D pseudospectral solvers for the Helmholtz equation. *J. Comput. Phys.* 55: 115–28

Hall, P., Malik, M. R. 1986. On the instability of a three-dimensional attachment line boundary layer: weakly nonlinear theory and a numerical approach. *J. Fluid Mech.* 163: 257–82

Haurwitz, B., Craig, R. A. 1952. Atmospheric flow patterns and their representation by spherical surface harmonics. *AFCRL Geophys. Res. Pap. No. 14*, Hanscom AFB, Mass.

Herbert, T. 1983a. Stability of plane Poiseuille flow: theory and experiment. *Fluid Dyn. Trans.* 11: 77–126

Herbert, T. 1983b. Secondary instability of plane channel flow to subharmonic three-dimensional disturbances. *Phys. Fluids* 26: 871–74

Herbert, T. 1984. Analysis of the subharmonic route to transition in boundary layers. *AIAA Pap. No. 84-0009*

Hussaini, M. Y., Salas, M. D., Zang, T. A. 1985a. Spectral methods for inviscid, compressible flows. In *Advances in Computational Transonics*, ed. W. G. Habashi, pp. 875–912. Swansea, Wales: Pineridge

Hussaini, M. Y., Kopriva, D. A., Salas, M. D., Zang, T. A. 1985b. Spectral methods for the Euler equations: Part 1. Fourier methods and shock-capturing. *AIAA J.* 23: 64–70

Hussaini, M. Y., Kopriva, D. A., Salas, M., Zang, T. A. 1985c. Spectral methods for the Euler equations: Part 2. Chebyshev methods and shock-fitting. *AIAA J.* 23: 234–40

Jarraud, M., Baede, A. P. M. 1985. The use of spectral techniques in numerical weather prediction. *Lect. Appl. Math.* 22: 1–41

Karniadakis, G. E., Bullister, E. T., Patera, A. T. 1986. A spectral element method for solution of the two- and three-dimensional time-dependent incompressible Navier-Stokes equations. *Proc. Eur.-US Conf. Finite Element Methods for Nonlinear Probl., Norway*. Berlin: Springer-Verlag

Kerr, R. M. 1985. Higher-order derivative correlations and the alignment of small-scale structures in isotropic numerical turbulence. *J. Fluid Mech.* 153: 31–58

Kim, J., Moin, P. 1985. Application of a fractional-step method to incompressible Navier-Stokes equations. *J. Comput. Phys.* 59: 308–23

King, G. P., Li, Y., Lee, W., Swinney, H. L., Marcus, P. S. 1984. Wave speeds in wavy Taylor-vortex flow. *J. Fluid Mech.* 141: 365–90

Kleiser, L., Schumann, U. 1984. Spectral simulation of the laminar-turbulent transition process in plane Poiseuille flow. See Voigt et al. 1984, pp. 141–63

Kopriva, D. A. 1986. A spectral multidomain method for the solution of hyperbolic systems. *Appl. Numer. Math.* In press

Kopriva, D. A., Zang, T. A., Salas, M. D., Hussaini, M. Y. 1984. Pseudospectral solution of two-dimensional gas-dynamics problems. *Proc. GAMM Conf. Numer. Methods in Fluid Mech., 5th*, ed. M. Pandolfi, R. Piva, pp. 185–92. Braunschweig/Wiesbaden: Friedr Vieweg

Korczak, K. Z., Patera, A. T. 1986. Isoparametric spectral element method for solution of the Navier-Stokes equations in complex geometry. *J. Comput. Phys.* 62: 361–82

Kovasznay, L. S., Komoda, H., Vasudeva, B. R. 1962. Detailed flow field in transition. *Proc. Heat Transfer and Fluid Mech. Inst.*, pp. 1–26. Stanford, Calif: Stanford Univ. Press

Kreiss, H.-O., Oliger, J. 1971. Comparison of accurate methods for the integration of hyperbolic equations. *Rep. No. 36*, Dep. Comput. Sci., Uppsala Univ., Swed.

Krist, S., Zang, T. A. 1986. Numerical

simulation of channel flow transition—the structure of the hairpin vortex. *NASA TP*, NASA Langley Res. Cent., Hampton, Va.

Ku, H. C., Taylor, T. D., Hirsch, R. S. 1987. Pseudospectral methods for solution of the incompressible Navier-Stokes equations. *Comput. Fluids*. In press

Ladyzhenskaya, O. A. 1969. *The Mathematical Theory of Viscous Incompressible Flow*. New York: Gordon & Breach. 224 pp.

Laurien, E. 1986. Numerische Simulation zur aktiven Beeinflussung des laminar-turbulenten Übergangs in der Plattengrenzschichtströmung. *DFVLR-FB 86-05*

Lee, M. J., Reynolds, W. C. 1985. Numerical experiments on the structure of homogeneous turbulence. *Proc. Symp. Turbul. Shear Flows, Ithaca, N.Y.*, pp. 17.7–12

Leonard, A. 1984. Numerical simulation of turbulent fluid flows. *NASA TM 84320*

Leonard, A., Wray, A. 1982. A new numerical method for the simulation of three-dimensional flow in a pipe. *Proc. Int. Conf. Numer. Methods in Fluid Dyn., 8th*, ed. E. Krause, pp. 335–41. Berlin: Springer-Verlag

Leorat, J., Pouquet, A., Poyet, J. P., Passot, T. 1985. Spectral simulations of 2D compressible flows. *Proc. Int. Conf. Numer. Methods in Fluid Dyn., 9th*, ed. Soubbarameyer, J. P. Boujet, pp. 369–74. Berlin: Springer-Verlag

LeQuere, P., de Roquefort, T. A. 1985. Computation of natural convection in two-dimensional cavities with Chebyshev polynomials. *J. Comput. Phys.* 57: 210–28

Lerat, A., Sides, J., Daru, V. 1982. An implicit finite volume method for solving the Euler equations. *Proc. Int. Conf. Numer. Methods in Fluid Dyn., 8th*, ed. E. Krause, pp. 343–49. Berlin: Springer-Verlag

Macaraeg, M., Streett, C. L. 1986. Improvements in spectral collocation through a multiple domain technique. *Appl. Numer. Math.* In press

Mac Giolla Mhuiris, N. 1986. Calculations of the stability of some axisymmetric flows proposed as a model of vortex breakdown. *Appl. Numer. Math.* In press

Majda, A., McDonough, J., Osher, S. 1978. The Fourier method for nonsmooth initial data. *Math. Comput.* 32: 1041–81

Malik, M. R., Zang, T. A., Hussaini, M. Y. 1985. A spectral collocation method for the Navier-Stokes equations. *J. Comput. Phys.* 61: 64–88

Marcus, P. S. 1984a. Simulation of Taylor-Couette flow, Part 1. Numerical methods and comparison with experiment. *J. Fluid Mech.* 146: 45–64

Marcus, P. S. 1984b. Simulation of Taylor-Couette flow, Part 2. Numerical results for wavy-vortex flow with one traveling wave. *J. Fluid Mech.* 146: 65–113

Marcus, P. S., Tuckerman, L. S. 1986a. Simulation of flow between concentric rotating spheres, Part I. Steady states. *J. Fluid Mech.* In press

Marcus, P. S., Tuckerman, L. S. 1986b. Simulation of flow between concentric rotating spheres, Part II. Transitions. *J. Fluid Mech.* In press

McMurtry, P. A., Jou, W.-H., Riley, J. J., Metcalfe, R. W. 1986. Direct numerical simulations of a reacting mixing layer with chemical heat release. *AIAA J.* 24: 962–70

Metcalfe, R. W., Orszag, S. A., Brachet, M. E., Menon, S., Riley, J. J. 1986. Secondary instability of a temporally growing mixing layer. *J. Fluid Mech.* In press

Moin, P., Kim, J. 1980. On the numerical solution of time-dependent viscous incompressible fluid flows involving solid boundaries. *J. Comput. Phys.* 35: 381–92

Moin, P., Kim, J. 1982. Numerical investigation of turbulent channel flow. *J. Fluid Mech.* 118: 341–77

Moin, P., Rogers, M. M., Moser, R. D. 1985. Structure of turbulence in the presence of uniform shear. *Proc. Symp. Turbul. Shear Flows, 5th, Ithaca, N.Y.*, pp. 17.21–26

Morchoisne, Y. 1983. Resolution des equations de Navier-Stokes par une methode spectrale de sous-domaines. *Proc. Int. Conf. Numer. Methods in Sci. and Eng., 3rd*

Morchoisne, Y. 1984. Inhomogeneous flow calculations by spectral methods: mono-domain and multi-domain techniques. See Voigt et al. 1984, pp. 181–208

Moretti, G. 1968. Inviscid blunt body shock layers. *PIBAL Rep. No. 68-15*, Polytech. Inst. Brooklyn, New York, N.Y.

Moser, R. D., Moin, P. 1984. Direct numerical simulation of curved, turbulent channel flow. *Rep. No. TF-20*, Dep. Mech. Eng., Stanford Univ., Calif.

Moser, R. D., Moin, P., Leonard, A. 1983. A spectral numerical method for the Navier-Stokes equations with applications to Taylor-Couette flow. *J. Comput. Phys.* 52: 524–44

Murdock, J. W. 1977. A numerical study of nonlinear effects on boundary-layer stability. *AIAA J.* 15: 1167–73

Nishioka, M., Asai, M., Iida, S. 1980. An experimental investigation of the secondary instability in laminar-turbulent transition. In *Laminar-Turbulent Transition*, ed. R. Eppler, H. Fasel, pp. 37–46. Berlin: Springer-Verlag

Orszag, S. A. 1969. Numerical methods for

the simulation of turbulence. *Phys. Fluids* 12: 250–57 (Suppl. II)

Orszag, S. A. 1971. Accurate solution of the Orr-Sommerfeld equation. *J. Fluid Mech.* 50: 689–703

Orszag, S. A. 1972. Numerical simulation of incompressible flows within simple boundaries: I. Galerkin (spectral) representations. *Stud. Appl. Math.* 50: 293–327

Orszag, S. A. 1980. Spectral methods for problems in complex geometries. *J. Comput. Phys.* 37: 70–92

Orszag, S. A., Kells, L. C. 1980. Transition to turbulence in plane Poiseuille and plane Couette flows. *J. Fluid Mech.* 96: 159–205

Orszag, S. A., Patera, A. T. 1983. Secondary instability of wall-bounded shear flows. *J. Fluid Mech.* 128: 347–85

Orszag, S. A., Patterson, G. S. 1972. Numerical simulation of three-dimensional homogeneous isotropic turbulence. *Phys. Rev. Lett.* 28: 76–79

Ortiz, E. L., Samara, H. 1983. Numerical solution of differential eigenvalue problems with an operational approach to the tau method. *Computing* 31: 95–103

Patera, A. T. 1984. A spectral element method for fluid dynamics: laminar flow in a channel expansion. *J. Comput. Phys.* 54: 468–88

Phillips, T. N., Zang, T. A., Hussaini, M. Y. 1986. Preconditioners for the spectral multigrid method. *IMA J. Numer. Anal.* In press

Riley, J. J., Metcalfe, R. W. 1980. Direct numerical simulation of a perturbed, turbulent mixing layer. *AIAA Pap. No. 80-0274*

Riley, J. J., Metcalfe, R. W., Orszag, S. A. 1986. Direct numerical simulations of chemically reacting turbulent mixing layers. *Phys. Fluids*. In press

Rogallo, R. S. 1981. Numerical experiments in homogeneous turbulence. *NASA TM-81315*

Rogallo, R. S., Moin, P. 1984. Numerical simulation of turbulent flows. *Ann. Rev. Fluid Mech.* 16: 99–137

Rozhdestvensky, B. L., Simakin, I. N. 1984. Secondary flows in a plane channel: their relationship and comparison with turbulent flows. *J. Fluid Mech.* 147: 261–89

Sakell, L. 1984. Pseudospectral solutions of one- and two-dimensional inviscid flows with shock waves. *AIAA J.* 22: 929–34

Salas, M. D., Zang, T. A., Hussaini, M. Y. 1982. Shock-fitted Euler solutions to shock-vortex interactions. *Proc. Int. Conf. Numer. Methods in Fluid Dyn., 8th*, ed. E. Krause, pp. 461–67. Berlin: Springer-Verlag

Saric, W. S., Kozlov, V. V., Levchenko, V. Y. 1984. Forced and unforced subharmonic resonance in boundary-layer transition. *AIAA Pap. No. 84-0007*

Silberman, I. 1954. Planetary waves in the atmosphere. *J. Meteorol.* 11: 27–34

Singer, B., Reed, H. L., Ferziger, J. H. 1986. Investigation of the effects of initial disturbances on plane channel transition. *AIAA Pap. No. 86-0433*

Spalart, P. R. 1984. A spectral method for external viscous flows. *Contemp. Math.* 28: 315–35

Spalart, P. R. 1985. Numerical simulation of boundary-layer transition. *Proc. Int. Conf. Numer. Methods in Fluid Dyn., 9th*, ed. Soubbarameyer, J. P. Boujot, pp. 531–35. Berlin: Springer-Verlag

Spalart, P. R., Leonard, A. 1985. Direct numerical simulation of equilibrium turbulent boundary layers. *Proc. Symp. Turbul. Shear Flows, 5th, Ithaca, N.Y.*, pp. 9.35–40

Streett, C. L., Hussaini, M. Y. 1987. Finite length effects on Taylor-Couette flow. In *Stability of Time-Dependent and Spatially Varying Flows*, ed. D. L. Dwoyer, M. Y. Hussaini. New York: Springer-Verlag. In press

Streett, C. L., Zang, T. A., Hussaini, M. Y. 1984. Spectral methods for solution of the boundary-layer equations. *AIAA Pap. No. 84-0170*

Streett, C. L., Zang, T. A., Hussaini, M. Y. 1985. Spectral multigrid methods with applications to transonic potential flow. *J. Comput. Phys.* 57: 43–76

Tan, C. S. 1985. Accurate solution of three-dimensional Poisson's equation in cylindrical coordinates by expansion in Chebyshev polynomials. *J. Comput. Phys.* 59: 81–95

Taylor, T. D., Myers, R. B., Albert, J. H. 1981. Pseudospectral calculations of shock waves, rarefaction waves and contact surfaces. *Comput. Fluids* 9: 469–73

Temam, R. 1977. *Navier-Stokes Equations*. Amsterdam: North-Holland. 519 pp.

Temperton, C. 1983. Self-sorting and mixed-radix fast Fourier transforms. *J. Comput. Phys.* 52: 1–23

Van Dyke, M. D., Guttmann, A. J. 1983. Subsonic potential flow past a circle and the transonic controversy. *J. Aust. Math. Soc. Ser. B* 24: 243–61

Voigt, R. G., Gottlieb, D., Hussaini, M. Y., eds. 1984. *Spectral Methods for Partial Differential Equations*. Philadelphia: SIAM-CBMS

von Kerczek, C. H. 1982. The instability of oscillatory plane Poiseuille flow. *J. Fluid Mech.* 116: 91–114

Wimmer, M. 1976. Experiments on a viscous fluid flow between concentric rotating spheres. *J. Fluid Mech.* 78: 317–35

Wray, A., Hussaini, M. Y. 1984. Numerical experiments in boundary-layer stability. *Proc. R. Soc. London Ser. A* 392: 373–89

Wu, C.-T., Ferziger, J. H., Chapman, D. R. 1985. Simulation and modeling of homogeneous compressed turbulence. *Proc. Symp. Turbul. Shear Flows, Ithaca, N.Y.*, pp. 17.13–19

Zang, T. A., Hussaini, M. Y. 1981. Mixed spectral-finite difference approximations for slightly viscous flows. *Proc. Int. Conf. Numer. Methods in Fluid Dyn., 7th*, ed. W. C. Reynolds, R. W. MacCormack, pp. 461–66. Berlin: Springer-Verlag

Zang, T. A., Hussaini, M. Y. 1985a. Numerical experiments on subcritical transition mechanisms. *AIAA Pap. No. 85-0296*

Zang, T. A., Hussaini, M. Y. 1985b. Numerical experiments on the stability of controlled shear flows. *AIAA Pap. No. 85-1698*

Zang, T. A., Hussaini, M. Y. 1985c. Recent applications of spectral methods in fluid dynamics. *Lect. Appl. Math.* 22: 379–409

Zang, T. A., Hussaini, M. Y. 1986. On spectral multigrid methods for the time-dependent Navier-Stokes equations. *Appl. Math. Comput.* 19: 359–72

Zang, T. A., Wong, Y.-S., Hussaini, M. Y. 1982. Spectral multigrid methods for elliptic equations. *J. Comput. Phys.* 48: 485–501

Zang, T. A., Kopriva, D. A., Hussaini, M. Y. 1983. Pseudospectral calculation of shock turbulence interactions. *Proc. Int. Conf. Numer. Methods in Laminar and Turbul. Flow, 3rd*, ed. C. Taylor, pp. 210–20. Swansea, Wales: Pineridge

Zang, T. A., Wong, Y.-S., Hussaini, M. Y. 1984. Spectral multigrid methods for elliptic equations II. *J. Comput. Phys.* 54: 489–507

Ann. Rev. Fluid Mech. 1987. 19: 369–402

DYNAMICS OF TORNADIC THUNDERSTORMS

Joseph B. Klemp

National Center for Atmospheric Research, Boulder, Colorado 80307

1. INTRODUCTION

Tornadic thunderstorms are the most intense and most damaging type of convective storm. Whereas ordinary convective cells grow, produce rain, and then decay over a period of 40 min to an hour, certain thunderstorms may develop into a nearly steady-state structure that persists for several hours, producing heavy rain, large hail, damaging surface winds, and tornadoes. Although tornadoes may arise in a variety of storm conditions, these long-lived storms produce tornadoes most frequently and generate virtually all of the most damaging ones. Prominent features of these tornadic thunderstorms, which are particularly common in the Great Plains and midwestern regions of the United States, are illustrated in an idealized schematic in Figure 1.

Our understanding of tornadic storms has evolved gradually over the years as technological advances have been made in observing systems and computer models. The first indications that certain storms exhibit a special behavior came from early studies of data from upper-air soundings and surface stations. While most thunderstorms moved with the mean winds over the lower and middle troposphere (Byers & Braham 1949), certain large storms were found to propagate consistently to the right of the mean winds (Byers 1942, Newton & Katz 1958). Observing the shift in the wind direction as storms passed by surface stations, Byers (1942) and later Brooks (1949) surmised that these intense storms have a *cyclonic* circulation (counterclockwise rotation about a vertical axis, viewed from above). Using data from scanning radar and ground observations, case studies of a severe thunderstorm near Wokingham, England (Browning & Ludlam 1962), and another near Geary, Oklahoma (Browning & Donaldson 1963), revealed numerous distinctive features that were strikingly

0066–4189/87/0115–0369$02.00

similar in the two storms. Browning & Donaldson proposed that "these two storms may be representative of an important class of local storms; namely those developing within a strongly sheared environment which remain persistently intense and tend toward a steady state circulation."

Browning (1964) presented a conceptual model to explain the structure of these severe right-moving storms, which he named *supercells*. These storms develop when there is strong vertical shear of the environmental wind, as illustrated in Figure 2*a*, which plots the west-east (U) component of the wind field on the x-axis versus the south-north (V) component on the y-axis. Here, low-level winds (L) from the south turn gradually with height, becoming westerly at higher levels (H) in the upper troposphere; the supercell storm moves to the right of the mean winds, in an easterly direction. Notice that in a coordinate framework moving with the storm, low-level air approaches the storm from the southeast, midlevel air from the south, and upper-level air from the west. Browning's airflow model, shown schematically in Figure 2*b*, depicts a three-dimensional, nearly steady-state circulation (relative to the storm motion) in which warm, moist low-level air feeds continuously into a single large updraft, which is driven by buoyancy derived from the latent heating of condensing water vapor. Evaporative cooling within the region of heaviest precipitation just north of the updraft drives the main downdraft, which ingests air passing around in front of the eastward-moving storm. As the colder and drier

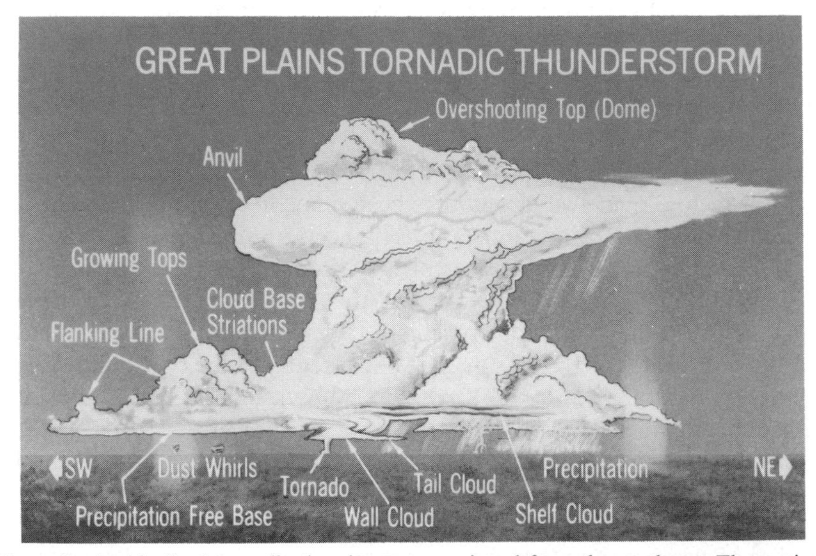

Figure 1 An idealized tornadic thunderstorm as viewed from the southeast. The vertical scale is exaggerated by about a factor of two. (From Joe Golden, NWS-NOAA, personal communication.)

downdraft air spreads out beneath the storm, it collides with the warm moist inflow along a line called the *gust front*; this convergence promotes the lifting of potentially buoyant air into the updraft and sustains the convection.

The important feature of this storm structure is the physical separation of the updraft and downdraft circulations such that each branch supports, rather than disrupts, the other. In weakly sheared environments, precipitation forms within the updraft and produces negative buoyancy that destroys the convection. Subsequent research has largely substantiated the basic features of Browning's model.

Occasionally, severe storms are observed that move faster than, and to the left of, the mean winds (cf. Hitschfeld 1960, Newton & Fankhauser

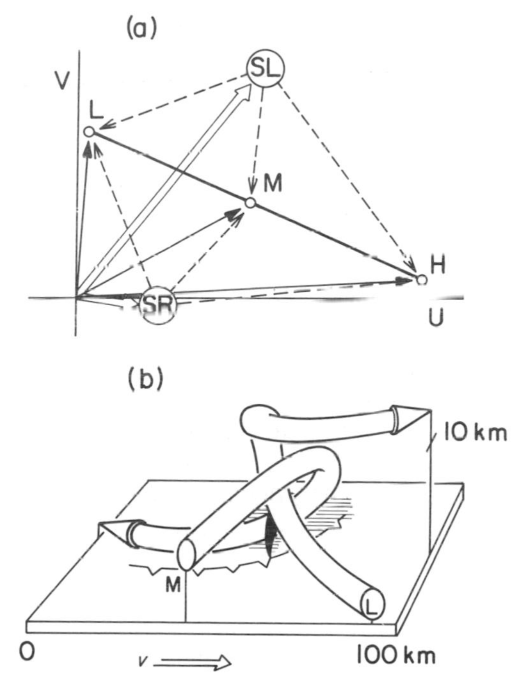

Figure 2 Browning's conceptual model for a right-moving supercell thunderstorm (SR). (*a*) Wind plot illustrating low-(L), middle-(M), and high-(H) level winds relative to the ground (solid arrows) and relative to the storm (dashed arrows). Motion of the SR storm is shown with an open arrow. The motion and relative wind vectors for a possible left-moving supercell (SL) are also shown. (*b*) Three-dimensional airflow within the SR storm viewed from the south-southeast. Updraft and downdraft circulations are shown relative to the storm motion, depicted with a five-fold exaggeration in the vertical scale. Also depicted are the approximate extent of precipitation at the ground (hatched area), and positions of the gust front (barbed line) and tornado (when present). (Adapted from Browning 1964.)

1964, Hammond 1967). Browning (1968) noted that the environmental winds shown in Figure 2*a* could also promote a left-moving severe storm (SL). This SL storm experiences relative winds that are a mirror image of the winds in the SR storm about the line LMH. Thus, Browning speculated that the SL storm would similarly have a structure in mirror image to the SR storm shown in Figure 2*b* and that a single thunderstorm might split into two parts, one an SR storm and the other an SL storm, which would move along diverging paths. Hammond (1967) and later others have confirmed this type of SL storm structure, while numerous investigators have documented the splitting and subsequent divergence of severe storms (cf. Hitschfeld 1960, Fujita & Grandoso 1968, Achtemeier 1969, Charba & Sasaki 1971, Fankhauser 1971).

By the mid-1970s, technological advances in radar and computers made it possible to investigate the internal structure of supercell thunderstorms in far greater detail. Scanning storms simultaneously with two Doppler radars permitted calculation of the three-dimensional wind field within storms (Ray et al. 1975, Miller 1975), while three-dimensional numerical models began to simulate storms in highly idealized environments (Wilhelmson 1974, Miller & Pearce 1974, Schlesinger 1975). More recently, simulations of specific supercell thunderstorms have reproduced many of the important features of the observed storms (Klemp et al. 1981, Wilhelmson & Klemp 1981). With more comprehensive data, new theories arose to explain the mechanisms that govern the important physical processes in these severe thunderstorms, such as the development of rotation, storm splitting, storm propagation, the preferential enhancement of right-moving cyclonically rotating storms, and the intensification of low-level rotation as a storm enters its tornadic phase.

In the following sections, I discuss the essential fluid processes promoting these special storm features that characterize the tornadic supercell thunderstorm. In doing so, I draw heavily from the results of three-dimensional numerical-modeling studies, in which storms can be generated under controlled conditions and which provide complete kinematic and thermodynamic data both in and around a storm. However, although these models have demonstrated good qualitative agreement with observed storms, some of the mechanisms derived from the detailed analyses of simulated storms must still be tested against future data that will be obtained from the increasingly sophisticated storm-observing systems.

Over the past several years, a number of excellent reviews have comprehensively documented the progress in convective-storm research (cf. Lilly 1979, Houze & Hobbs 1982, Kessler 1985). In this more limited review, I focus on selected aspects of tornadic thunderstorm dynamics, without attempting to represent all points of view, in order to avoid undue distraction for the reader unfamiliar with this topic.

2. EARLY DEVELOPMENT OF ROTATION

During its early development, an isolated cumulus grows as a buoyant thermal (or perhaps as a cluster of thermals). At this stage, it has long been known (Byers & Braham 1949) that wind shear inhibits the convection, tending to tear apart the rising thermals. However, when a vigorous updraft does develop in a sheared environment, it invariably exhibits significant rotation about a vertical axis, typically in the form of a pair of counterrotating vortices (Wilhelmson 1974, Schlesinger 1975, Kropfli & Miller 1976, Wilhelmson & Klemp 1978). This vortex-pair circulation arises as a result of the tilting into the vertical of horizontal vortex lines embedded initially in the environmental shear. Rotunno (1981) pointed out that the manner in which tilting produces vertical vorticity in the early convection may differ significantly from that in the mature thunderstorm propagating transverse to the mean winds.

To illustrate the evolution of rotation, we consider the basic governing equations that describe the relevant storm processes. For a compressible atmosphere, the appropriate analogue to the incompressible Boussinesq equations may be written in the following form (Ogura & Phillips 1962, Lilly 1979):

$$\frac{\partial \mathbf{v}}{\partial t} + \mathbf{v} \cdot \nabla \mathbf{v} + \nabla \pi = B\hat{\mathbf{k}} + \mathbf{F}, \tag{1}$$

$$\nabla \cdot \bar{\rho} \mathbf{v} = 0, \tag{2}$$

where $\mathbf{v} = (u, v, w)$ is the three-dimensional velocity vector in Cartesian coordinates (x, y, z), $\mathbf{F}$ represents the turbulent mixing, $\hat{\mathbf{k}}$ is the unit vector in the vertical direction, and B is the total perturbation buoyancy, including the influences of temperature and water vapor as well as drag caused by liquid water. Here, $\pi = p'/\bar{\rho}$, where p' is the pressure perturbation about a mean state corresponding to an adiabatic atmosphere (i.e. the mean pressure $\bar{p}$ and mean density $\bar{\rho}$ are related by $\bar{p} \sim \bar{\rho}^{c_p/c_v}$), and henceforth we refer to π as the pressure. The Coriolis force has been omitted from (1), since it does not play a fundamental role in the storm dynamics to be discussed (Klemp & Wilhelmson 1978).

Taking the curl of (1) then yields expressions for the vertical (ζ) and horizontal ($\boldsymbol{\omega}_\mathrm{h}$) components of vorticity $\boldsymbol{\omega} = \nabla \times \mathbf{v}$:

$$\frac{d\zeta}{dt} = \underbrace{\boldsymbol{\omega}_\mathrm{h} \cdot \nabla_\mathrm{h} w}_{\text{tilting}} + \underbrace{\zeta \frac{\partial w}{\partial z}}_{\text{stretching}} + \underbrace{F'_\zeta}_{\text{mixing}}, \tag{3}$$

$$\frac{d\boldsymbol{\omega}_\mathrm{h}}{dt} = \underbrace{\boldsymbol{\omega} \cdot \nabla \mathbf{v}_\mathrm{h}}_{\substack{\text{tilting and}\\ \text{stretching}}} + \underbrace{\nabla \times (B\hat{\mathbf{k}})}_{\substack{\text{baroclinic}\\ \text{generation}}} + \underbrace{\mathbf{F}'_\mathrm{h}}_{\text{mixing}}, \tag{4}$$

where $\mathbf{v}_h = (u, v)$ and $F'_\zeta, \mathbf{F}'_h$ are the respective mixing terms with $\mathbf{F}' = \nabla \times \mathbf{F}$, and d/dt is the Lagrangian time derivative.

Focusing on the generation of vertical vorticity, we see that the first term on the right-hand side of (3) contributes to ζ by tilting horizontal vortex lines into the vertical, while the second term alters ζ through the vertical stretching of vortex tubes. Clearly, if convection begins in an environment containing no vertical vorticity, then the initial production of ζ must arise through the tilting of horizontal vorticity contained in the ambient wind shear. The important parameter characterizing the vertical wind shear is the environmental wind-shear vector $\mathbf{S} = d\mathbf{V}/dz$, where $\mathbf{V} = (U, V)$. $\mathbf{S}$ has two components: speed shear represented by $|\mathbf{S}|$, and directional shear caused by turning of the shear vector. Notice that the wind field shown in Figure 2*a* has only speed shear, with $\mathbf{S}$ being oriented parallel to the line LMH. Numerical storm simulations have shown that a unidirectional wind shear is sufficient to produce supercell-like storms with characteristic structure similar to that shown in Figure 2*b* (Klemp & Wilhelmson 1978). However, directional shear also exerts important influences on storm evolution, as is discussed in Section 4.

For simplicity, consider first an isolated cumulus growing in unidirectional wind shear in which the westerly velocity U increases with height and $V = 0$. In its early convective growth, the cloud moves roughly with the mean westerly flow (averaged over the depth of the cloud). Thus, in a storm-relative framework, low-level inflow approaches the cloud from the east while upper-level outflow returns toward the east, as illustrated in Figure 3*a*. A vortex-pair circulation develops as south-north-oriented vortex lines are swept into the updraft and tilted into the vertical.

This mechanism is contained in the linearized vertical vorticity equation given by (ignoring the mixing term)

$$\frac{d\zeta}{dt} = \frac{dU}{dz}\frac{\partial w}{\partial y}. \tag{5}$$

Positive (cyclonic) vertical vorticity is generated along the southern flank of the updraft ($\partial w/\partial y > 0$), while negative (anticyclonic) vorticity is produced on the northern flank ($\partial w/\partial y < 0$). As the updraft intensifies, vorticity that has been tilted into the vertical can be amplified substantially by the stretching of vortex tubes, which is a nonlinear effect.

3. STORM SPLITTING

Although the cloud in Figure 3*a* is still a rather ordinary cumulus, forcing influences are already promoting its transition to a supercell storm. As precipitation accumulates within the updraft, increasing negative buoy-

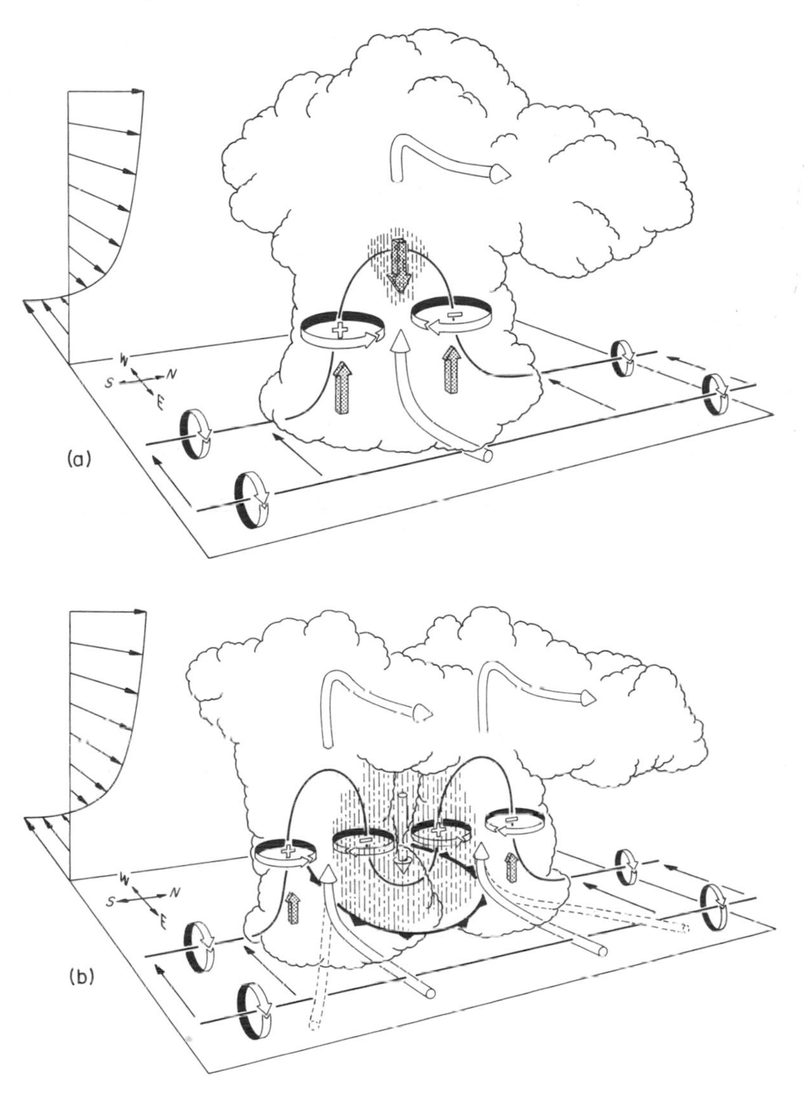

Figure 3 Schematic depicting how a typical vortex tube contained within (westerly) environmental shear is deformed as it interacts with a convective cell (viewed from the southeast). Cylindrical arrows show the direction of cloud-relative airflow, and heavy solid lines represent vortex lines with the sense of rotation indicated by circular arrows. Shaded arrows represent the forcing influences that promote new updraft and downdraft growth. Vertical dashed lines denote regions of precipitation. (*a*) Initial stage: Vortex tube loops into the vertical as it is swept into the updraft. (*b*) Splitting stage: Downdraft forming between the splitting updraft cells tilts vortex tubes downward, producing two vortex pairs. The barbed line at the surface marks the boundary of the cold air spreading out beneath the storm. (Adapted from Rotunno 1981.)

ancy produces a downdraft within the cloud. Enhanced by evaporative cooling, the downdraft outflow at the surface, being colder than the surrounding environment, spreads out beneath the storm. In weak wind shear, this outflow spreads out in all directions, cutting off the supply of warm moist air, and the storm cell dissipates rapidly. However, in strong shear (i.e. $|\mathbf{S}| \simeq 10^{-2}\ \mathrm{s}^{-1}$ over the lowest several kilometers), two factors act to extend the longevity of the storm: storm-relative low-level inflow from the east prevents the cold air from moving out ahead of the storm (Wilhelmson & Klemp 1978, Thorpe & Miller 1978), and lifting pressure gradients reinforce new updraft growth on the southern and northern flanks of the central updraft (Schlesinger 1980, Rotunno & Klemp 1982). In response to these forcing influences, the updraft splits gradually into two cells that move laterally apart, as illustrated in Figure 3*b*.

The lifting vertical pressure gradients appear to be the fundamentally important factor in splitting the cloud into two cells that then move apart. Although the cold low-level downdraft outflow increases the convergence along the updraft flanks, numerical simulations demonstrate that updraft splitting occurs even if this central downdraft is prevented from forming (Rotunno & Klemp 1982, 1985). This is accomplished in the models by not allowing any precipitation to fall relative to the surrounding air. Without this falling precipitation, the negative buoyancy required to generate the downdraft is greatly reduced.

The origin of these lifting forces can be elucidated through a detailed analysis of the terms in the vertical momentum equation, as generated by three-dimensional storm-simulation models. By dividing the pressure into components $\pi = \pi_{\mathrm{dn}} + \pi_{\mathrm{b}}$, produced by dynamic interactions π_{dn} and through buoyancy effects π_{b}, the vertical component of (1) can be written in the form

$$\frac{\partial w}{\partial t} = -\underbrace{\mathbf{v}\cdot\nabla w}_{\text{advection}} - \underbrace{\frac{\partial \pi_{\mathrm{dn}}}{\partial z}}_{\substack{\text{dynamics}\\ \text{forcing}}} - \underbrace{\left(\frac{\partial \pi_{\mathrm{b}}}{\partial z} - B\right)}_{\substack{\text{buoyancy}\\ \text{forcing}}} + \underbrace{F_w}_{\text{mixing}}, \tag{6}$$

and expressions for the dynamics and buoyancy components of the pressure are obtained from the divergence of (1):

$$\nabla\cdot(\bar{\rho}\nabla\pi_{\mathrm{dn}}) = -\nabla\cdot(\bar{\rho}\mathbf{v}\cdot\nabla\mathbf{v}) + \nabla\cdot\bar{\rho}\mathbf{F}, \tag{7}$$

$$\nabla\cdot(\bar{\rho}\nabla\pi_{\mathrm{b}}) = \frac{\partial}{\partial z}(\bar{\rho}B). \tag{8}$$

The dynamics forcing term in (6) thus represents the contribution to vertical accelerations from the portion of the vertical pressure gradient created by dynamic interactions, while the buoyancy forcing term includes the pressure gradient arising through buoyancy effects. Since a large portion of the buoyancy pressure gradient may be balanced by the buoyancy term itself, the buoyancy forcing, as defined in (6), represents the net influence of buoyancy on vertical accelerations. The divergence of the turbulent mixing term $\nabla \cdot \bar{\rho}\mathbf{F}$ has a generally minor influence on the pressure, since these terms largely cancel through the continuity equation (2); thus, they are omitted in further discussions (Klemp & Rotunno 1983). Pressure decompositions similar to this have been used to analyze numerous aspects of storms simulated with three-dimensional models (Wilhelmson 1974, Schlesinger 1980, Rotunno & Klemp 1982, 1985, Klemp & Rotunno 1983).

Schlesinger (1980) first noted that the lifting pressure gradients on the flanks of a splitting updraft are dynamic in origin. Rotunno & Klemp (1982) proposed that these lifting pressure gradients on the updraft flanks are induced by the midlevel rotation. To assess the influence of rotation on the pressure we expand the right-hand side of (7) in the form

$$\nabla \cdot (\bar{\rho}\nabla\pi_{\mathrm{dn}}) = -\bar{\rho}\underbrace{\left[\left(\frac{\partial u}{\partial x}\right)^2 + \left(\frac{\partial v}{\partial y}\right)^2 + \left(\frac{\partial w}{\partial z}\right)^2 - \frac{d^2 \ln \bar{\rho}}{dz^2} w^2\right]}_{\text{fluid extension terms}}$$

$$-2\bar{\rho}\underbrace{\left[\frac{\partial v}{\partial x}\frac{\partial u}{\partial y} + \frac{\partial u}{\partial z}\frac{\partial w}{\partial x} + \frac{\partial v}{\partial z}\frac{\partial w}{\partial y}\right]}_{\text{fluid shear terms}}. \tag{9}$$

Consider first the fluid shear term $\partial v/\partial x \cdot \partial u/\partial y$. For a wind field in pure rotation, this term is simply $-\frac{1}{4}\zeta^2$. For purposes of qualitative analysis, note that in the interior of a flow, the Laplacian of a variable is roughly proportional to the negative of the variable itself (i.e. $\nabla^2\pi \sim -\pi$). Combining these relations then yields

$$\pi_{\mathrm{dn}} \sim -\zeta^2, \tag{10}$$

which suggests that the strong midlevel rotation on the updraft flank (see Figure 3*b*) acts to lower the pressure and thereby induces updraft growth on these flanks. The second and third fluid shear terms in (9) are related similarly to the x- and y-components of vorticity, respectively. These terms are amplified by the vortex ring that forms around the updraft because of the horizontal buoyancy gradients on its flanks [i.e. through the baroclinic generation term in Equation (4)]. Therefore, these two rotational terms also contribute to lowering the pressure on the flanks of the midlevel updraft, although (unlike the first term) they are not specific to the flanks that are transverse to the shear. Rotunno & Klemp (1982) found that the fluid extension terms in (9) do not contribute to the lifting pressure gradients on the updraft flanks. Consequently, strong rotation about the vertical axis appears to be the special feature that promotes the splitting within evolving supercell thunderstorms.

As the splitting progresses and the two updraft centers move apart laterally, the downdraft dividing the two cells tilts the vortex lines downward, producing two vortex-pair circulations as shown in Figure 3*b* (Rotunno 1981). As each updraft begins to propagate transverse to the mean wind shear, the direction of storm-relative inflow turns as indicated by the dashed cylindrical arrows in Figure 3*b*. In this configuration, the low-level inflow contains a streamwise component of the horizontal shear vorticity, as well as the transverse component that produced the original vortex pair. Browning & Landry (1963) and Barnes (1970) proposed that the tilting of the streamwise vorticity in the environmental shear was the important contributor to the rotation in supercell storms. Rotunno (1981), Lilly (1982), and Davies-Jones (1984) also emphasized the importance of the streamwise vorticity; using simplified linear models, they demonstrated that if the environmental vorticity vector is parallel to the storm-relative inflow, the vertical vorticity generated through tilting will tend to be in phase with the vertical velocity (as opposed to the transverse component that produces vertical vorticity maxima on the updraft flanks, as shown in Figure 3*a*).

Following Lilly (1986a), we consider a steady updraft propagating transverse to the environmental wind shear (i.e. to the south at velocity c_y in the situation shown in Figure 3). In a coordinate framework relative to the moving updraft, flow approaches from the south with a velocity $-c_y$. The linear vertical vorticity equation (3) then becomes (ignoring mixing)

$$-c_y \frac{\partial \zeta}{\partial y} = \frac{dU}{dz} \frac{\partial w}{\partial y}, \tag{11}$$

which can be integrated immediately, with the result being

$$\zeta = \frac{dU}{dz} \frac{w}{(-c_y)}. \tag{12}$$

Thus, the vertical vorticity is coincident with the vertical velocity in this simplified situation.

Davies-Jones (1984) and Rotunno & Klemp (1985) extended the linear theory by providing interpretations of vertical vorticity generation within storms that remain valid even in fully nonlinear flow. Using the equations of motion as written in (1) and (2), we can show through Ertel's theorem (see Dutton 1976, p. 382) that for a conserved scalar variable Θ depending only on the thermodynamic properties of the flow, it follows that

$$\frac{d}{dt}\left\{\frac{\boldsymbol{\omega}\cdot\nabla\Theta}{\bar{\rho}}\right\} = 0. \qquad (13)$$

Here, we identify Θ as the entropy of the moist fluid flow. Since $\boldsymbol{\omega}\cdot\nabla\Theta = 0$ in the initial, undisturbed state, it will remain zero throughout the evolving storm. This result means that vortex lines must lie along isentropic surfaces. Thus, as isentropic surfaces are drawn up into the storm, their embedded vortex lines are similarly deformed.

This effect was illustrated by Rotunno & Klemp (1985) within a numerically simulated storm evolving in a unidirectional (west-to-east) shear (see Figure 4). In the initial state, the isentropic surfaces are horizontal and the vortex lines in the environmental shear are oriented south to north. As low-level flow rises into the updraft ($t = 10$ min) the vortex lines follow the deforming isentropic surfaces, producing cyclonic vertical vorticity on the southern flank and anticyclonic vorticity along the northern flank (not shown). After the splitting updraft begins moving toward the south ($t = 40$ min), vortex lines are continuously turned into the vertical along the southern flank as the isentropic surfaces are lifted into the updraft. As the evaporatively cooled downdraft (centered on the symmetry plane) brings midlevel air to the ground, a depression appears in the low-level isentropic surface ($t = 40$, 60 min). Vortex lines turned upward along the southern updraft flank turn back downward and produce anticyclonic vorticity on the northern flank ($t = 60$ min). At $t = 40$ and 60 min, the vortex lines shown in the foreground curl underneath along the boundary of the cold downdraft and disappear from view. This redirection of vortex lines is discussed further in Section 6.

4. PREFERENTIAL ENHANCEMENT OF CYCLONICALLY ROTATING STORMS

The splitting process described above leads naturally to a pair of rotating storms—a cyclonic one moving to the right of the mean winds and an anticyclonic one moving to the left. However, right-moving, cyclonically rotating supercell storms are, in fact, observed far more commonly. Davies-Jones (1985) notes that out of 143 storms in which radar detected strong rotation, only 3 rotated anticyclonically. Although it is tempting to attribute this bias to the ambient rotation in the atmosphere caused by the Coriolis force (which is cyclonic in the Northern Hemisphere), scale analyses suggest that this effect should be small (Morton 1966). Numerical storm simulations indicate that the Coriolis force does enhance the cyclonic rotation in the right-moving storm but does not selectively suppress the anticyclonically rotating storm (Klemp & Wilhelmson 1978).

By simulating supercell storms in a variety of wind-shear conditions, Klemp & Wilhelmson (1978) demonstrated that a clockwise turning with

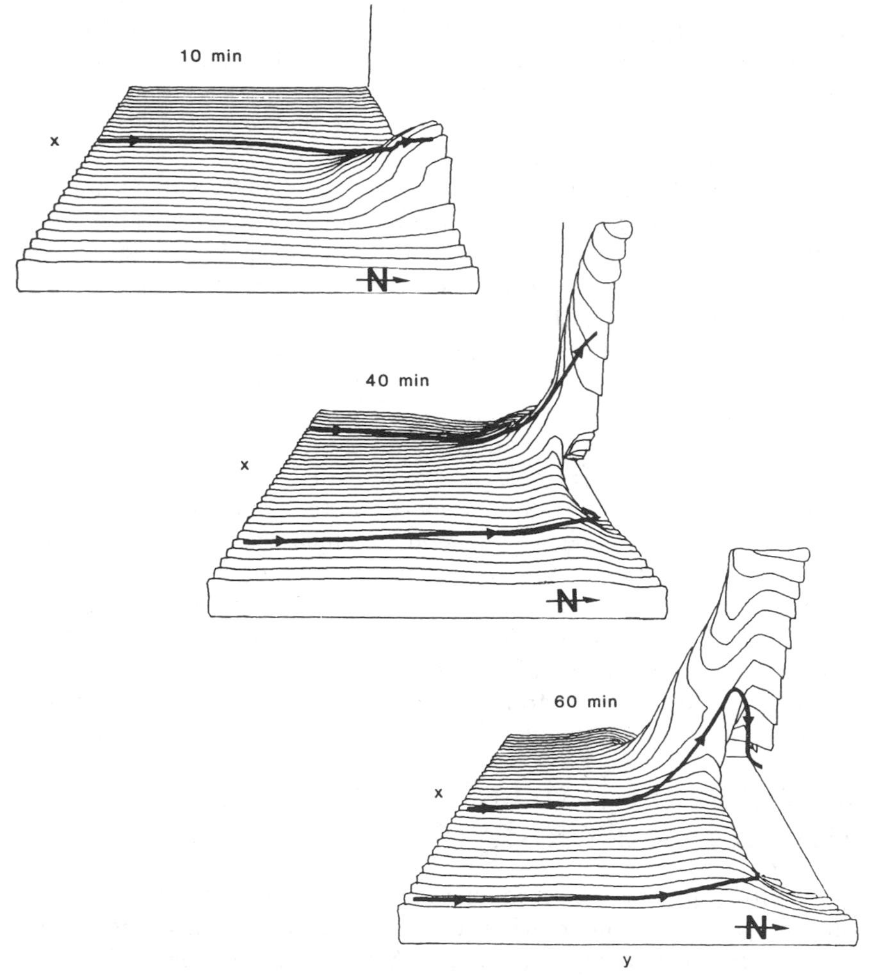

Figure 4 Numerically simulated three-dimensional perspective (viewed from the east) of a low-level isentropic surface that is being drawn up into the storm updraft at $t = 10$, 40, and 60 min. The environmental wind shear is directed from west to east, and the flow is symmetric about the vertical plane along the northern border. Vortex lines (heavy solid lines) lie approximately on the isentropic surface. The base plane spans 20 km in the south-north direction, and the vertical scale is exaggerated by a factor of two. (From Rotunno & Klemp 1985.)

height of the environmental wind-shear vector is the key factor that selectively promotes the cyclonic, right-moving storm. This turning is apparent in the composite wind sounding (Figure 5) compiled by Maddox (1976) for a large number of tornadic storms. Notice that if we superimpose x, y-axes over the U, V-axes in Figure 5, the wind-shear vector **S** is locally tangent to the curve $V(U)$ at all levels in the sounding. In this composite, the wind-shear vector turns clockwise with height, from the north-northeast near the surface to the east-southeast at 700 mbar (about 3 km above the ground).

Figure 6 illustrates the evolving storm structures for two wind profiles—one having a unidirectional environmental wind-shear vector (*top*) and the other having a shear vector that turns clockwise with height through the lower levels (*bottom*). With the winds below 2.5 km given by the dashed line in the accompanying wind plot, the initial cloud splits into two storms after 40 min that move apart as they propagate to the northeast. In the absence of surface drag and Coriolis effects, they evolve into identical, mirror-image right- and left-moving storms. In this simulation the line of symmetry is oriented east-west and moves toward the north at 12.7 m s^{-1}. When the low-level shear vector turns cyclonically with height, the initial convective cell evolves into an intense, cyclonically rotating right-moving storm, while on the left flank, only weaker short-lived cells form along the gust front. Notice that the significant factor here is the turning of the wind-shear vector, *not* the turning of the wind vector itself. (In either wind profile, the ground-relative winds turn clockwise with height.)

Rotunno & Klemp (1982) found that the basis for the selective enhancement of either the right- or left-moving storm is contained within linear theory. As an initially axisymmetric updraft interacts with a mean wind

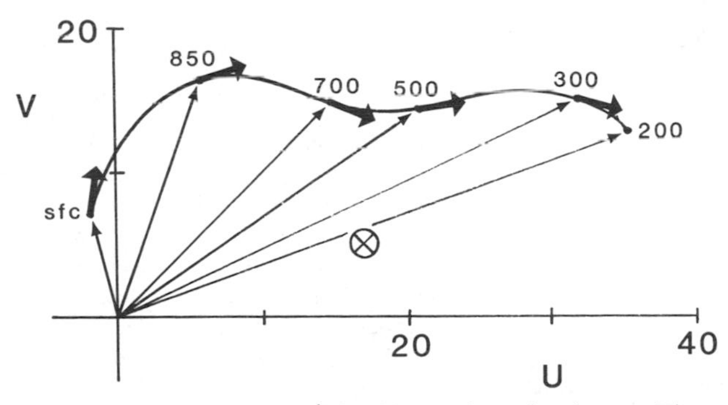

Figure 5 Mean wind sounding (in m s^{-1}) for 62 tornado outbreak cases. The soundings are composited by computing the winds at each level relative to the estimated storm motion. Heavy arrows indicate the direction of the shear vector at each level (labeled in mbar). The estimated mean storm motion is denoted by ⊗. (Adapted from Maddox 1976.)

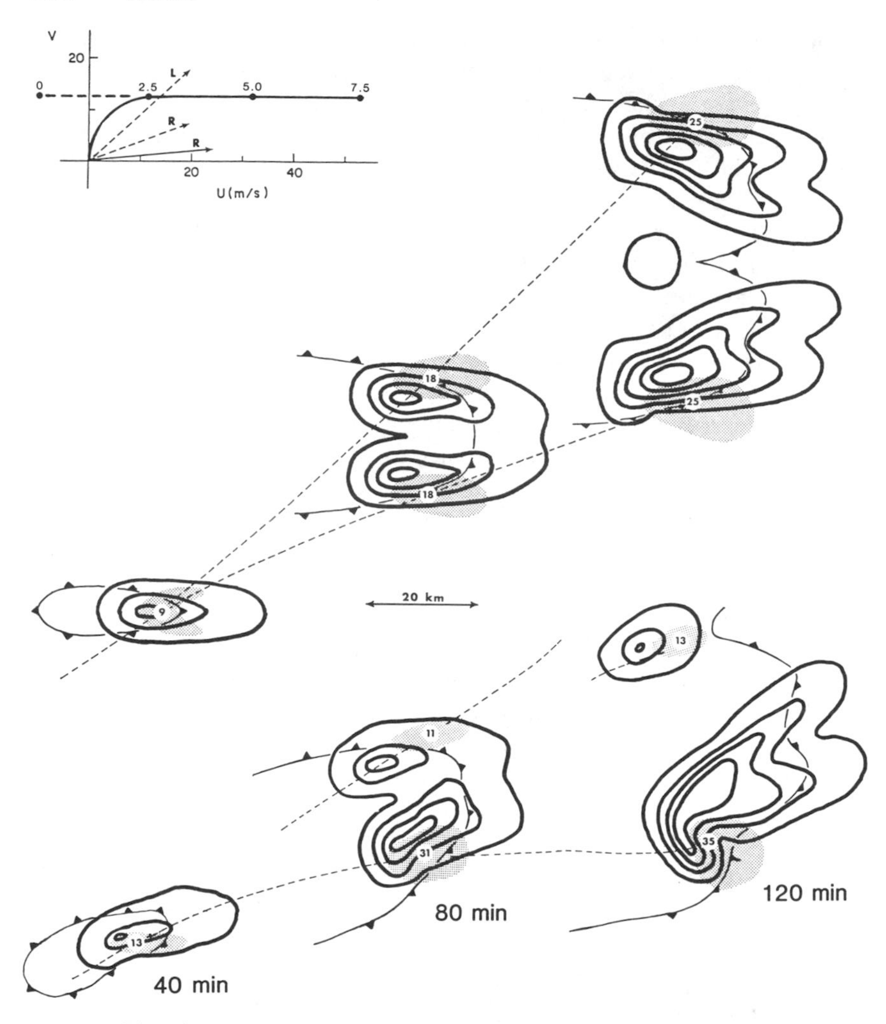

Figure 6 Plan views of numerically simulated thunderstorm structures at 40, 80, and 120 min for two environmental wind profiles (displayed at upper left) having wind shear between the surface and 7.5 km. The storm system in the lower portion of the figure evolves in response to the wind profile, in which **S** turns clockwise with height between the ground and 2.5 km (heavy solid line in wind plot), while the upper system develops when **S** is unidirectional (same wind profile except following the heavy dashed line below 2.5 km). The plan views depict the low-level (1.8 km) rainwater field (similar to radar reflectivity) contoured at 2 g kg^{-1} intervals, the midlevel (4.6 km) updraft (shaded regions), and the location of the surface cold-air outflow boundary (barbed lines). The maximum updraft velocity is labeled (in m s^{-1}) within each updraft at each time. The dashed lines track the path of each updraft center. Arrows in the wind plot indicate the supercell propagation velocities for the unidirectional (dashed) and turning (solid) wind-shear profiles. (Adapted from Klemp & Weisman 1983.)

shear $\mathbf{S}$ that turns clockwise with height, favorable vertical pressure gradients are dynamically induced on the right flank while unfavorable gradients arise on the left. To demonstrate this effect, consider an updraft perturbation in a homogeneous fluid ($B = 0$). The linearized inviscid vertical momentum and pressure equations (6)–(8) are then simply

$$\frac{dw}{dt} = -\frac{\partial \pi}{\partial z}, \tag{14}$$

$$\nabla^2 \pi = -2\mathbf{S} \cdot \nabla_{\mathrm{h}} w. \tag{15}$$

We can evaluate the qualitative behavior of (15) by again using the approximation $\nabla^2 \pi \sim -\pi$, which yields

$$\pi \sim \mathbf{S} \cdot \nabla_{\mathrm{h}} w. \tag{16}$$

Thus, linear theory suggests that as an updraft interacts with the shear flow, a high-to-low pressure gradient develops across the updraft in the direction of the local shear vector at each level. For a constant shear magnitude, this pressure effect increases in amplitude with height beneath the level of the maximum updraft velocity.

To visualize the influences of these shear-induced pressure variations, consider first the unidirectional shear profile represented in Figure 3. Here the shear vector points from west to east at all levels, producing high pressure on the upshear (west) side of the updraft and low pressure on the downshear (east) side, as depicted in Figure 7*a*. As the updraft intensity increases above the ground, these pressure perturbations promote low-level lifting on the downshear side (and descent on the upshear side) that reinforces the storm inflow. However, these vertical pressure gradients do not contribute to a preferential growth on either of the flanks that are transverse to the shear or to storm splitting (which, as discussed in Section 3, is an inherently nonlinear effect).

Expressing the linearized vertical vorticity equation (3) in terms of $\mathbf{S}$, we have

$$\frac{d\zeta}{dt} = \mathbf{k} \cdot (\mathbf{S} \times \nabla_{\mathrm{h}} w), \tag{17}$$

which indicates that the vortex pair generated through tilting is oriented at right angles to the shear vector. For the unidirectional shear in Figure 7*a*, Equation (17) is identical to Equation (5), producing cyclonic and anticyclonic vorticity on the southern and northern flanks, respectively.

Figure 7*b* illustrates the corresponding situation when $\mathbf{S}$ turns clockwise with height through the lower levels of the atmosphere. Here, the winds

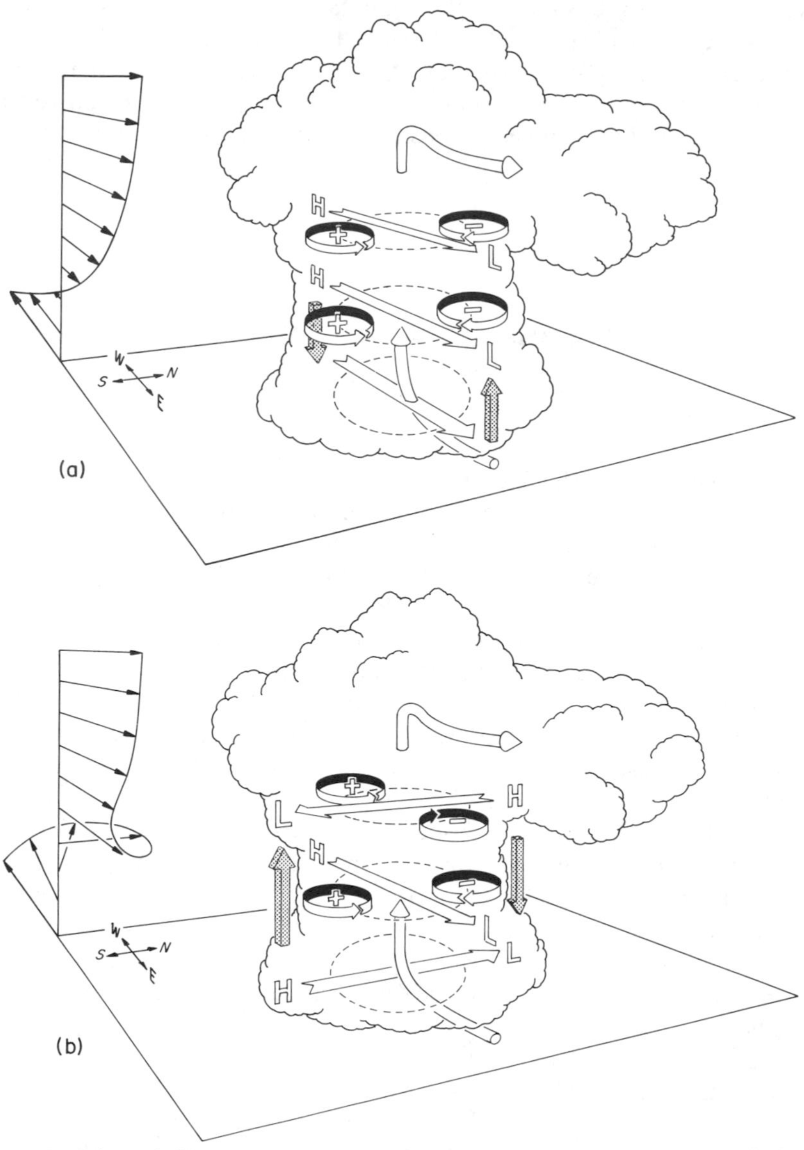

Figure 7 Schematic illustrating the pressure and vertical vorticity perturbations arising as an updraft interacts with an environmental wind shear that (*a*) does not change direction with height and (*b*) turns clockwise with height. The high (H) to low (L) horizontal pressure gradients parallel to the shear vectors (flat arrows) are labeled along with the preferred location of cyclonic (+) and anticyclonic (−) vorticity. The shaded arrows depict the orientation of the resulting vertical pressure gradients. (Adapted from Rotunno & Klemp 1982.)

change with height from easterly, to southerly, and finally to westerly, such that **S** turns through 180° (pointing toward the north at the ground, and turning to the east and then to the south at higher levels). In this situation, the turning shear vector produces vertical pressure gradients that favor ascent on the southern flank (designated the right flank by facing in the direction of the mean-shear vector) and descent on the northern (left) flank. This influence, in conjunction with the nonlinear processes that promote splitting, enhances the development of the right-moving storm and inhibits the growth of the left-moving storm. When there is significant turning of the shear vector, growth on the left flank may be suppressed to the extent that there would be no apparent splitting at all; the initial storm just begins moving to the right of the mean winds at a certain stage in its development (as occurs in the storm simulation with turning shear in Figure 6). As the vortex pair forms perpendicularly to the shear vector, the production of cyclonic vorticity is also on the right flank, where the favorable pressure gradients are promoting new updraft growth. For a wind profile in which **S** turns counterclockwise with height, the situation would be reversed and the left-moving, anticyclonically rotating member of the split pair would be selectively enhanced.

5. STORM PROPAGATION

As storms begin to propagate with a component transverse to the mean winds, those that will become supercell thunderstorms generally continue to intensify and evolve toward a mature structure such as that shown schematically in Figure 2*b*. Their ability to persist in a quasi-steady-state configuration for a period of hours is a truly remarkable feature of these storms. Browning & Foote (1976) analyzed the radar observations of a storm in northeastern Colorado that maintained a supercellular structure for more than 5 hours and produced a swath of damaging hail (some as large as baseballs) over a 300-km-long path.

Comprehensive analyses of tornadic thunderstorms using multiple-Doppler radar observations have provided detailed documentation of the important features of these storms (cf. Ray et al. 1975, 1981, Brandes 1977, 1978, Eagleman & Lin 1977). Numerical simulations have also provided an informative view of supercell thunderstorms, particularly those conducted in idealized environmental conditions, where various environmental influences can be selectively evaluated (cf. Weisman & Klemp 1982, 1984). Figure 8 depicts the model-equivalent of the visible cloud for a right-moving supercell storm during its mature phase for a case of unidirectional wind shear in the absence of Coriolis effects. Even under these idealized conditions, there are strong qualitative similarities

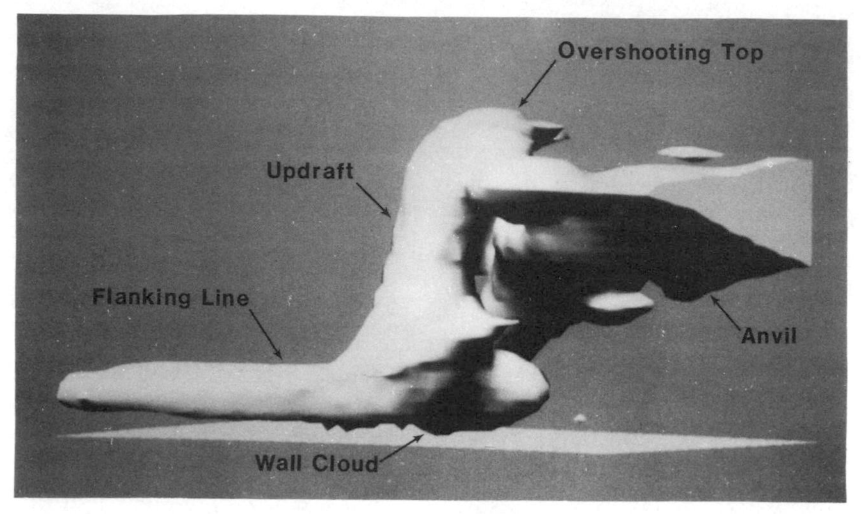

Figure 8 Numerically simulated cloud depicting a supercell storm in its mature phase. The unidirectional environmental wind-shear vector points toward the east (i.e. increasing westerlies with height), and the storm is viewed from the southeast. The shaded surface coincides with the 0.1 g kg^{-1} contour in the model cloud-water field. The storm simulation is described in detail by Rotunno & Klemp (1985).

with the illustration in Figure 1. In this situation, a mirror-image, left-moving storm also develops, which is accommodated using appropriate symmetry conditions along the northern boundary of the computational domain.

The flow structure within this quasi-steady storm (see Figure 9) exhibits features that are characteristic of supercell storms. At low levels, there is a zone of convergence (gust front) where the warm, moist low-level inflow collides with the cold downdraft outflow. Strong cyclonic rotation is visible within the midlevel (4 km) updraft, and this flow turns downstream into the anvil outflow at higher levels. The hooklike feature on the southern side of the rainwater field at low and midlevels corresponds to the hook echo that is frequently observed by radar in tornadic storms. Although the environmental winds blow only in the east-west direction, this storm is propagating toward the south at about 5 m s^{-1}.

The mechanism that causes the transverse propagation of supercell storms has remained an intriguing although illusive issue over the years. Researchers have proposed a variety of theories that appeal to a diverse spectrum of physical processes. Several of these interesting approaches are summarized below.

Analogies With Obstacle Flow

Even in the early storm observations, researchers recognized that the updrafts in large storms remain erect, in defiance of the strong environmental wind shear in which they are embedded. Substantial portions of environmental air therefore diverge and flow around the storm in a manner that resembles flow past a bluff obstacle at high Reynolds number, with separation occurring behind the obstacle (cf. Newton & Newton 1959, Newton 1963, and also the flow at $z = 8$ km in Figure 9). From controlled laboratory experiments it is well known that flow past a cylinder produces high pressure near the forward stagnation point and low pressure along the flanks and in the separated region at the rear. If the cylinder is rotating,

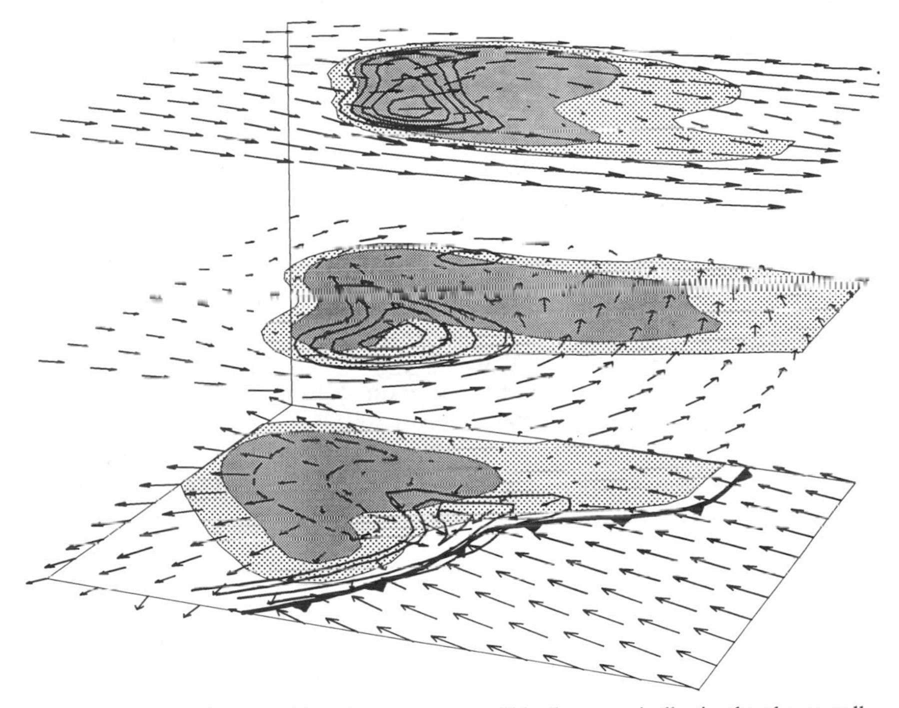

Figure 9 Perspective view of the flow structure within the numerically simulated supercell thunderstorm shown in Figure 8, displayed along horizontal cross sections located at 0, 4, and 8 km above the ground. The horizontal flow vectors are spaced 2 km apart (every second grid point) and are scaled such that the distance separating adjacent vectors corresponds to 25 m s^{-1}. Vertical velocity is contoured in 5 m s^{-1} intervals with the zero line removed except at the lowest level, where w is contoured at $z = 0.25$ km in 1 m s^{-1} increments. Regions of precipitation exceeding 0.5 and 2 g kg^{-1} are differentially shaded, and the barbed line denotes the location of the cold-air boundary at the surface.

a net pressure gradient (lift) arises that is transverse to the mean-flow direction; this is called the Magnus effect (Prandtl & Tietjens 1934a). Byers (1942) and later Fujita & Grandoso (1968) proposed that cyclonically rotating storms deviate to the right of the mean winds as a result of the horizontal pressure gradients produced by this effect.

A more detailed examination of the flow in and around storms reveals significant complications in the obstacle-flow analogy. Updrafts are entraining and/or detraining substantial amounts of air over most of their vertical height. As the updraft is embedded within a strongly sheared flow, both the magnitude and direction of the storm-relative flow may vary significantly with height. In fact, at the lower levels, the storm-relative flow direction is reversed (see the low-level environmental flow in Figures 2*b* and 9), which would cause the Magnus effect to work in the wrong direction. In numerical storm simulations, Rotunno & Klemp (1982) found that the pressure gradient across the updraft aligns more consistently with the direction of the shear vector at each level as suggested by linear theory [see Equation (16)] than with the direction of the storm-relative flow. Finally, the updraft is not a tangible entity that can be deflected laterally. Since the updraft is continually regenerated from below, horizontal forces acting on the rising air parcels do not necessarily promote updraft propagation.

Variations on the obstacle-flow analogy were proposed by Newton & Newton (1959) and Alberty (1969). They argued that low pressure on the flanks (Alberty) and downwind (Newton & Newton) of the storm, produced by obstacle-flow effects, cause vertical pressure gradients that induce lifting on the flanks and, thereby, transverse propagation. I believe this approach correctly focuses attention on the influence of vertical pressure gradients on propagation, although the inference of pressure distributions from obstacle-flow analogies suffers from the complications mentioned above.

Rotationally Induced Propagation

Based on analyses of numerically simulated supercell storms, Rotunno & Klemp (1985) proposed that storm propagation transverse to the environmental wind shear is dynamically induced by the strong midlevel rotation that develops along the flank of the storm. This mechanism is basically the same as the one responsible for storm splitting (described in Section 3); the strong midlevel rotation on the updraft flank lowers the pressure locally, which promotes lifting pressure gradients and thus new updraft growth that displaces the updraft laterally.

The factors influencing propagation were evaluated by examining again the forcing terms in the vertical momentum equation (6). Rotunno &

Klemp found the dynamically induced vertical pressure gradient to be the only term in (6) that causes new growth on the flank of the updraft that is transverse to the wind shear. Although the buoyancy forcing within the updraft is substantial, it is nearly coincident with the updraft and therefore does not contribute to propagation. The nature of the dynamics forcing can be further clarified by considering the terms in (9) that govern the dynamics pressure π_{dn}. By defining $\pi_{dn} = \pi_s + \pi_e$, they computed the contribution to π_{dn} from the fluid shear terms (π_s) and the fluid extension terms (π_e) and established that the shear terms were responsible for the propagation. The fluid extension terms have their strongest influence, as expected, along the centerline of the updraft.

To further isolate the dynamically forced pressure gradients, Rotunno & Klemp (1985) recomputed the storm simulation shown in Figures 8 and 9 but now prohibited the formation of precipitating water drops. This was accomplished by allowing water vapor to condense and release latent heat, but not allowing liquid water to fall relative to the air. With the physics altered in this manner, the initial storm still splits and evolves into a mirror-image pair of storms that exhibit strong rotation at midlevels and propagate apart, transverse to the mean wind shear. The downdrafts are much weaker, and no cold outflow occurs beneath the storm (essentially eliminating the downdraft branch of the circulation depicted in Figure 2*b*). Thus, even without the precipitation-driven downdrafts, the forces promoting transverse propagation continue to operate.

The relationship between the updraft and the shear-induced low pressure is depicted in a three-dimensional perspective (viewed from the east) in Figure 10. The low-π_s region corresponds closely in location and amplitude to the full perturbation pressure π; it induces new lifting on the southern flank of the updraft and thereby propagation of the entire storm toward the south. Recalling the qualitative relationship $\nabla^2\pi_s \sim -\pi_s$, one would expect low pressure to arise where the fluid shear terms are negative [see Equation (9)]. For a fluid in pure rotation these terms are negative, whereas for a purely shearing flow they are positive or zero (see Prandtl & Tietjens 1934b, p. 82). Thus, Rotunno & Klemp (1985) attributed the lowered pressure to the rotational component of the flow structure.

This study suggests that the gust front may be less important in governing supercell storm propagation than previously believed. Beneath the storm, strong lifting is usually maintained by the convergence of the cold-downdraft outflow and the warm moist inflow (clearly visible in the low-level cross section in Figure 9). Weaver & Nelson (1982) and others have documented that new convection forming along this gust front–induced convergence line frequently governs storm propagation. However, in this simulation without precipitation (and thus without a low-level gust front),

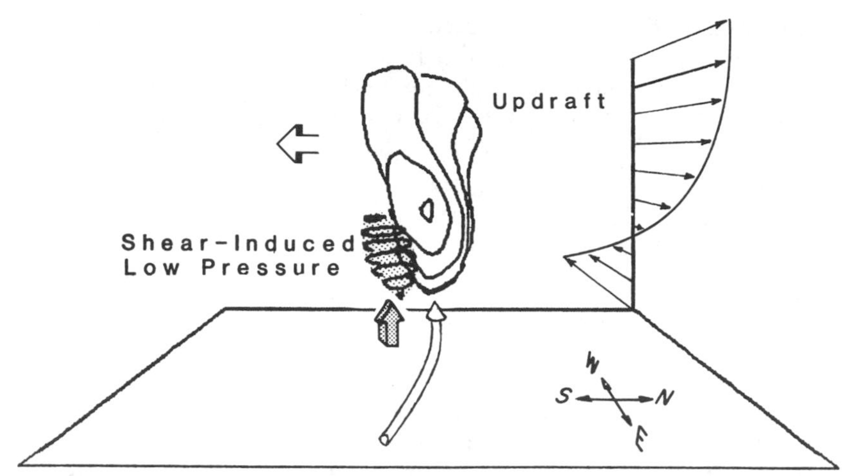

Figure 10 Three-dimensional perspective (looking west) of the updraft (>20 m s^{-1}) together with the shear-induced pressure (<-3.6 mbar) for a simulation with no precipitation in an environment with unidirectional (westerly) wind shear. The cylindrical arrow indicates the direction of storm-relative inflow, the shaded arrow denotes the rotationally induced forcing, and the open arrow shows the southerly direction of propagation. The displayed surface plane is 20 × 20 km, and the vertical scale is exaggerated by a factor of two. (Adapted from Rotunno & Klemp 1985.)

transverse propagation occurs in a similar fashion to that in the full simulation. This result indicates that the cold downdraft outflow may play a secondary role in supercell storm propagation.

In a related study, Weisman & Klemp (1984) computed the contributions of the dynamics and buoyancy forcing terms in (6) to the overall strength of the updraft for both supercellular and nonsupercellular storms. By integrating these terms along updraft trajectories, they found that the dynamics forcing contributed about 60% of the updraft intensity in the supercell storm but only about 35% of the updraft intensity in the nonsupercellular storm. These results suggest that the dynamically induced vertical pressure gradients are fundamentally important in driving a supercell's updraft circulation, as well as in displacing it laterally.

Forced Propagating Gravity Waves

Although the updrafts within convective storms may be highly unstable, they generate substantial perturbations in the surrounding stable environment. An approach developed by Lindzen (1974) and Raymond (1975, 1976) proposes that the motion of convective storms is controlled by propagating gravity waves, forced by the moist convection. The convective heating is parameterized, typically by relating it to the lifting at the level of cloud base caused by the gravity-wave motion. The gravity waves and convection can thus interact symbiotically to generate instabilities that select the preferred modes—termed *wave-CISK* by Lindzen (1974). This mechanism is a variation on CISK (Conditional Instability of the Second Kind), which was introduced by Charney & Eliassen (1964) and Ooyama (1964) to explain the role of small-scale convection in intensifying hurricanes.

In applying wave-CISK, one solves the linear wave equations subject to an appropriate forcing term that represents the nonlinear aspects of the system, most particularly the moist convection. To illustrate this approach, we consider the linear inviscid equations, ignoring vertical variations in the mean density. The thermodynamic equation, expressed in terms of the buoyancy B, can be written as

$$\frac{\partial B}{\partial t}+\mathbf{V}\cdot\nabla_{\mathrm{h}}B+N^2 w = Q, \tag{18}$$

where N is the buoyancy frequency (also called the Brunt-Väisälä frequency), and Q is the latent heating term, which can be approximated by the vertical mass flux through cloud base at $z = b$ multiplied by the average gradient of nonadiabatic heating between cloud base and cloud top (see Raymond 1975, Lilly 1979):

$$Q = \begin{cases} N^2 b \left.\dfrac{\partial w}{\partial z}\right|_{z=0}, & \text{if } \left.\dfrac{\partial w}{\partial z}\right|_{z=0} > 0; \\ 0, & \text{otherwise.} \end{cases} \tag{19}$$

Combining (18) with the linearizations of (1) and (2) and considering disturbances of the form $w = \hat{w}(z)\exp[i(\mathbf{k}\cdot\mathbf{x}-ct)]$ yields the Taylor-Goldstein equation (cf. Drazin & Reid 1981) with a forcing term:

$$\frac{\partial^2 \hat{w}}{\partial z^2}+\left\{\frac{N^2}{(c-V_k)^2}+\frac{1}{(c-V_k)}\frac{d^2 V_k}{dz^2}-k^2\right\}\hat{w} = \frac{\varepsilon\hat{Q}}{(c-V_k)^2}, \tag{20}$$

where $\mathbf{k}$ is the horizontal vector wave number, c is the phase speed, $V_k = \mathbf{V}\cdot\mathbf{k}/k$ is the component of $\mathbf{V}$ in the direction of $\mathbf{k}$, and $k = |\mathbf{k}|$. The factor ε arises in the Fourier modes from the constraint $Q \geq 0$ imposed in (19), and Lindzen (1974) showed that $\varepsilon = 1/2$ is an appropriate estimate. Solving for the eigenmodes of (20), one seeks the most unstable mode to characterize the nature of the propagating disturbance.

Raymond (1975, 1976) solved this equation for observed wind profiles associated with several supercell and splitting-storm cases. The resulting estimates for the motion of the most unstable modes were generally in good agreement with the observed storm propagation if he overestimated the heating using a value of $\varepsilon = 3/2$. However, to generate splitting and transverse propagating modes, the wave-CISK model required directional turning of the environmental wind-shear vector. This requirement is contrary to the results of three-dimensional numerical cloud models that consistently produce splitting storms (such as that shown in Figures 8 and 9) in strong unidirectional wind shear. Also, the neglect of the nonlinear terms in the momentum equation may be a crucial deficiency when applied to supercell storms developing in strong wind shear.

Raymond (1983) reformulated the wave-CISK equation in terms of a parameterized mass flux and increased the realism of the model by allowing for lagged feedback influences in both updrafts and downdrafts. This model uses the more reasonable value of $\varepsilon = 1/2$ and produces good agreement with observations for simulations of midlatitude and tropical squall lines (Raymond 1984). Currently, however, it does not seem that wave-CISK is a viable candidate for explaining supercell propagation.

Optimization of Helicity

Recently, Lilly (1986b) demonstrated that supercell storms have a highly helical circulation and proposed that this helicity promotes the longevity and transverse propagation of these storms. The helicity H is defined as the inner product of vorticity and velocity, $H \equiv \boldsymbol{\omega} \cdot \mathbf{v}$. A more useful dimensionless parameter, relative helicity RH, results from normalizing the helicity by $|\boldsymbol{\omega}|\,|\mathbf{v}|$ such that RH lies in the range of ± 1. Lilly points to recent research that has established the importance of helicity in stabilizing turbulent flows. Highly helical eddies are found to decay much more slowly than those with low helicity and thus eventually become dominant in the flow. He proposes that high helicity may similarly reduce the turbulent dissipation in supercell storms and thereby enhance their longevity. Consequently, Lilly suggests that as a supercell develops, it should evolve naturally into a structure that optimizes its helicity.

To illustrate the highly helical nature of supercells, Lilly computed RH throughout a numerically simulated storm and showed that below a height of about 10 km, the average value of RH was about 0.5. As an example of purely helical flow, Lilly considered Beltrami flows in which, by definition, the vorticity is everywhere parallel to the velocity, such that $\boldsymbol{\omega} = \kappa \mathbf{v}$. Since $-\nabla^2 \mathbf{v} = \nabla \times \boldsymbol{\omega} = \kappa \boldsymbol{\omega} = \kappa^2 \mathbf{v}$, the proportionality constant κ equals the magnitude of the three-dimensional wave number $|\mathbf{k}|$. If both the mean and perturbation flows are helical, the combined flow is helical only if the Beltrami coefficient for each flow is the same. For a unidirectional (westerly) wind shear, Lilly estimated the transverse propagation speed c_y by inserting κ for the disturbance (storm) into the mean-flow Beltrami equation, with the result

$$\kappa^2 = 2\left(\frac{\pi}{W}\right)^2 + \left(\frac{\pi}{h}\right)^2 = \left(\frac{dU/dz}{-c_y}\right)^2, \tag{21}$$

where the half-wavelengths h and W correspond to the height and width of the updraft, respectively. Solving for the propagation speed yields

$$c_y = \pm\gamma h dU/dz, \quad \text{where} \quad \gamma = (1+2h^2/W^2)^{-1/2}\pi^{-1}. \tag{22}$$

Although this expression is derived from highly idealized assumptions and requires a knowledge of the scale of the storm, it seems to provide qualitatively reasonable estimates of the transverse propagation speed. For example, setting $h = W = 10$ km and $dU/dz = 0.005\ \mathrm{s}^{-1}$ ($5\ \mathrm{m\ s}^{-1}\ \mathrm{km}^{-1}$), we have $c_y \simeq 9\ \mathrm{m\ s}^{-1}$.

The role of helicity in supercell thunderstorm dynamics is an intriguing issue that will surely be further analyzed in future research. However, the concept of optimizing helicity is difficult to test; it implies that a whole range of structures could exist (each with a different overall helicity) and that the stabilizing influence of helicity selects the optimal state. Since in nature and in numerical simulations we see only the structure that does evolve, it is hard to judge whether or not this structure is optimal. Also, I feel the importance of minimizing dissipation in supercell storms must be studied further. Supercells are continuously forced by both buoyant and dynamical processes, and the residence time of air parcels within such a storm is short compared with the storm's lifetime. Therefore, it is not yet clear if minimizing dissipation is a key factor influencing the behavior of the storm.

6. TRANSITION TO THE TORNADIC PHASE

Although not all supercell storms produce tornadoes, most of the intense tornadoes are generated by them. In a radar study of Oklahoma storms during 1971–75, for example, Burgess (1976) found that 62% of the 37 storms that exhibited strong storm-scale rotation developed tornadoes, whereas none occurred in storms that did not rotate.

When a storm does move into its tornadic phase, significant alteration of the storm-scale structure occurs that disrupts the nearly steady configuration illustrated in Figures 2*b* and 9. Lemon & Doswell (1979) provide an excellent description of storm features that are consistently observed during this transition. These features include a rapid increase in low-level rotation, a decrease in updraft intensity, a small-scale downdraft forming behind the updraft, and a flow at low levels in which cold-outflow and warm-inflow air spiral around the center of circulation. This low-level flow is depicted schematically in Figure 11; as the downdraft (labeled RFD) adjacent to the updraft intensifies, downdraft outflow progresses cyclonically around the center of rotation (marked by the northern encircled T), which is the likely location for tornado formation. As this outflow pushes into the path of the oncoming moist inflow, a new updraft and center of rotation may also develop tornadic intensity (denoted by the southern encircled T in Figure 11). At the same time, the spreading downdraft outflow cuts off the supply of warm moist air to the original circulation center (called *occlusion*), causing the original updraft to weaken.

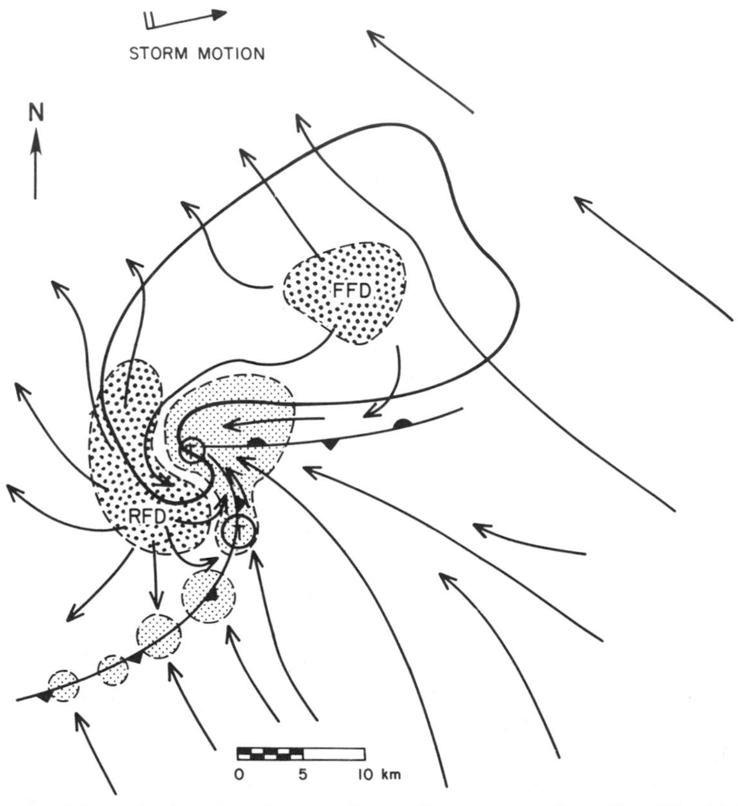

Figure 11 Schematic plan view of a tornadic thunderstorm near the surface. The thick line encompasses the radar echo. The barbed line denotes the boundary between the warm inflow and cold outflow and illustrates the occluding gust front. Low-level position of the updraft is finely stippled, while the forward-flank (FFD) and rear-flank (RFD) downdrafts are coarsely stippled. Storm-relative surface flow is shown along with the likely location of tornadoes (encircled T's). (From Lemon & Doswell 1979, as adapted by Davies-Jones 1985.)

Although a supercell storm may persist in a nearly steady configuration for up to several hours, the transition to the tornadic phase illustrated in Figure 11 may take place in less than about 10 min. Barnes (1978) and Lemon & Doswell (1979) have suggested that this transition is initiated by the rear-flank downdraft (see Figure 11), which forms at midlevels, descends to the surface, and then intensifies the low-level rotation by producing strong shear (Barnes) or temperature gradients (Lemon & Doswell) between this downdraft and the updraft.

More recent numerical storm simulations (Klemp & Rotunno 1983, Rotunno & Klemp 1985) and observational studies (Brandes 1984a,b) indicate a reverse sequence of events; the low-level rotation intensifies, followed by formation of the rear-flank downdraft. Figure 12 illustrates schematically the flow structure within a numerically simulated supercell evolving in a unidirectional wind shear (as in Figures 8 and 9) at a time when the low-level rotation is intensifying rapidly, but prior to the formation of the occluded gust front shown in Figure 11.

Klemp & Rotunno (1983) proposed that the rear-flank downdraft that promotes this occlusion is, in fact, dynamically induced as strong low-level

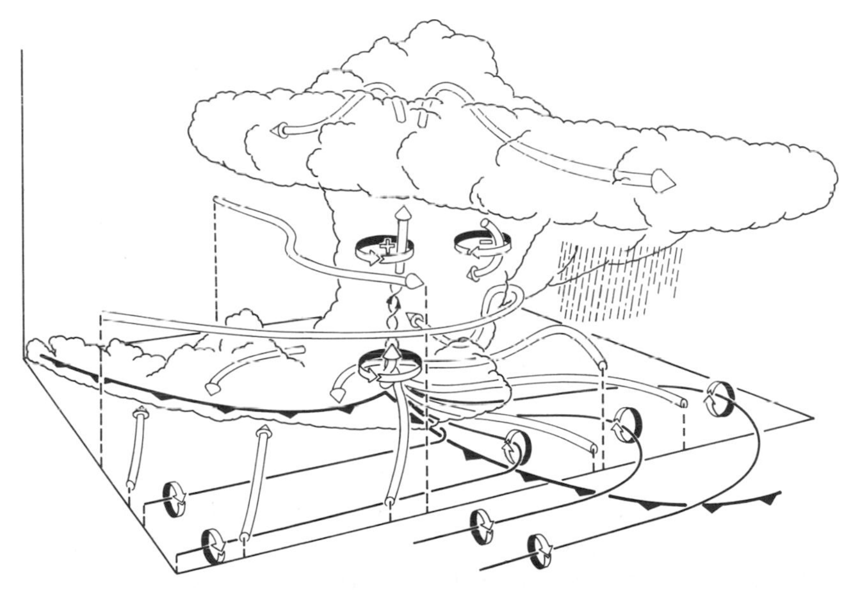

Figure 12 Three-dimensional schematic view of a numerically simulated supercell thunderstorm at a stage when the low-level rotation is intensifying. The storm is evolving in westerly environmental wind shear and is viewed from the southeast. The cylindrical arrows depict the flow in and around the storm. The thick lines show the low-level vortex lines, with the sense of rotation indicated by the circular-ribbon arrows. The heavy barbed line marks the boundary of the cold air beneath the storm.

rotation lowers the pressure locally and draws down air from above. By decomposing the pressure field at a time when the simulated rear-flank downdraft was intensifying, they demonstrated that the fluid shear term $\partial v/\partial x \cdot \partial u/\partial y$ was responsible for virtually the entire adverse vertical pressure gradient near the ground. As discussed earlier, a region in pure rotation satisfies $\partial v/\partial x \cdot \partial u/\partial y = -\frac{1}{4}\zeta^2$, and thus qualitatively we have $\pi \sim -\zeta^2$. Since the intensifying ζ is largest near the ground, a downward-directed pressure gradient results, which in turn promotes the downdraft. The retarding influence of rotation has been recognized in a variety of fluid flows (cf. Binnie & Hookings 1948) and has been called the *vortex valve* effect; Lemon et al. (1975) suggested that as the rotation increases within the storm, this effect may be responsible for its collapse.

An expanded view of the low-level flow in Figure 12 is displayed in Figure 13*a* and indicates the location of the rotation-induced low pressure. As the rear-flank downdraft intensifies, the downdraft outflow spreads out near the ground and, as shown in Figure 13*b*, initiates a new center of convergence and rotation farther east along the gust front (as also depicted in Figure 11). The descending air in the rear-flank downdraft evaporates the cloud water and produces a region of cloud-free air (called the *clear slot*) immediately behind the convergence line, as indicated in Figure 13*b*.

In the numerical simulation just described, the maximum low-level vertical vorticity remained less than one half the maximum at midlevels for over an hour, and then in less than 10 min it intensified to double the midlevel maximum. What factors are responsible for this rapid amplification of the low-level rotation? Analyses of the storm simulations demonstrate clearly that the intensification is stimulated by the baroclinic generation of strong horizontal vorticity [see Equation (4)] along the low-level boundary of the cold air pool forming beneath the storm (Klemp & Rotunno 1983, Rotunno & Klemp 1985). This horizontal vorticity is then tilted into the vertical and strongly stretched as the inflow enters the low-level updraft. To see how this situation arises, notice that in the evolving storm, precipitation is swept around to the northern side of the cyclonically rotating storm. As it falls to the north and northeast of the updraft, evaporation cools the low-level air (see Figure 9). With time, this cold pool of air advances progressively into the path of the low-level inflow to the storm. By the time shown in Figure 9, a significant portion of the inflow is approaching along the boundary of this cold air pool. The horizontal temperature gradients thus baroclinically generate horizontal vorticity that is nearly parallel to the inflowing streamlines. This process generates horizontal vorticity that is several times the magnitude of the mean shear vorticity and that is more favorably oriented to be tilted into vertical cyclonic vorticity. This same mechanism may also be responsible for tornadoes that form occasionally in nonsupercellular storms. If a storm encounters a preexisting cold front or an outflow boundary from another storm, strong horizontal vorticity (baroclinically generated along that boundary) may be swept into the storm and amplified.

The low-level vortex lines depicted in Figures 12 and 13 further illustrate the baroclinic vorticity generation mechanism. Since the environmental shear is westerly, the horizontal vortex lines embedded in the shear are oriented south-north with the sense of rotation as indicated in the undis-

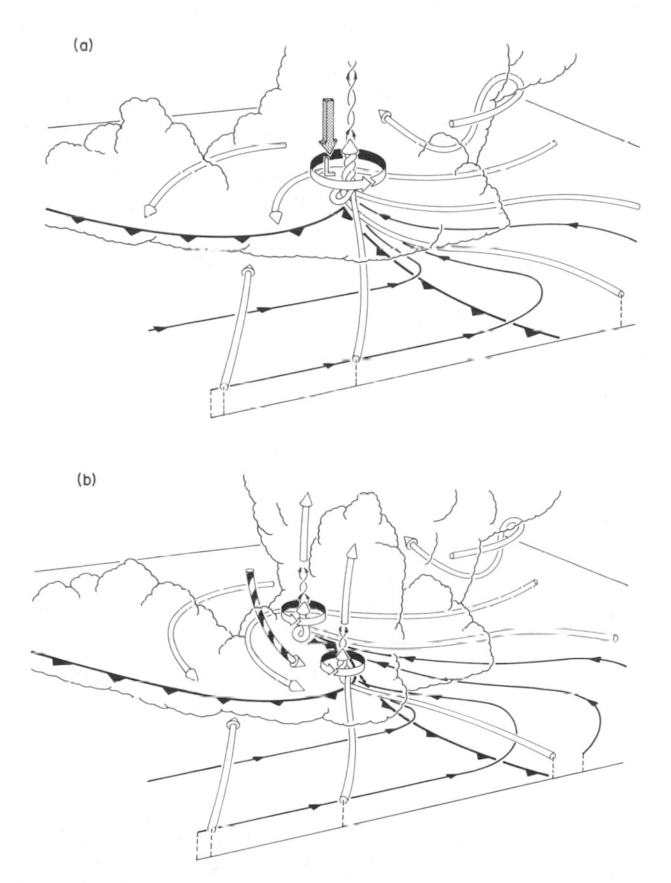

Figure 13 Expanded three-dimensional perspective, viewed from the southeast, of the low-level flow (*a*) at the time depicted in Figure 12, and (*b*) about 10 min later after the rear-flank downdraft has intensified. Features are drawn as described in Figure 12, except that the vector direction of vortex lines are indicated by arrows along the lines. The shaded arrow in (*a*) represents the rotationally induced vertical pressure gradient, and the striped arrow in (*b*) denotes the rear-flank downdraft.

turbed region southeast of the storm in Figure 12. As these vortex lines penetrate the low-level pool of cold air, they turn rapidly toward the center of convergence and are swept into the updraft. At this stage, the low-level updraft is located along the boundary between the warm and cold air, and it intertwines the warm and cold flow in the rising air. As the rear-flank downdraft intensifies, this baroclinic generation supports the rapid intensification of rotation in the secondary updraft forming farther to the east along the gust front in Figure 13*b*. After the original updraft is cut off from the warm inflow, it begins to dissipate while the new updraft continues to strengthen. The strong rotation may then cause a new downdraft that spreads out at the surface, cuts off the inflow to this updraft, and promotes yet another convergence center farther east. (In Figure 13, visualize the eastern circulation center in 13*b* becoming the center shown in 13*a*, and then repeating the cycle.) Such redevelopments, accompanied by successive tornadoes, are not uncommon in tornadic storms (Burgess et al. 1982).

In recent years new analysis techniques have been devised to calculate the pressure and buoyancy fields from the three-dimensional wind fields derived from multiple-Doppler radar observations (cf. Gal-Chen 1978, Hane & Scott 1978). Although the individual schemes vary in their approach, they all compute the pressure and buoyancy fields from the momentum and continuity equations (1)–(2) after estimating all of the kinematic terms from the radar data. These thermodynamic retrieval procedures provide the potential for dramatic advances in convective storm research and for more comprehensive intercomparisons between models and observations. Already, researchers have begun to apply these analysis techniques in studying tornadic storms (Pasken & Lin 1982, Lin & Pasken 1982, Brandes 1984a).

Brandes (1984a) used retrieved thermodynamic data to investigate the transition to the tornadic phase in two storms. He also found that the intensifying low-level rotation was colocated with a region of lowered pressure that promoted the rear-flank downdraft. However, while the derived buoyancy field supported the concept of baroclinic vorticity generation in one storm, it did not in the other. Brandes (1984b) suggested that the stretching of vertical vortex tubes within the low-level updraft may be a more important factor in amplifying the vorticity. This stretching certainly plays an important role in spinning up the vertical vorticity in the baroclinic mechanism as well. A more basic question is, What causes this process to begin to amplify suddenly? I believe the procedures used in retrieving thermodynamic variables are particularly sensitive to the storm structure near the ground, where there are strong gradients in both the kinematic and thermodynamic fields but often a lack of data from the radar. Further studies will certainly help to clarify these issues.

7. DISCUSSION

The advancements in thunderstorm research have documented clearly that the strong rotation within supercells is a dominant factor in shaping the very special characteristics of these storms. This rotation is derived primarily from horizontal vorticity embedded in the environmental wind shear that is swept up into the storm and tilted toward a vertical axis. The strong rotation on the flanks of the midlevel updraft promotes splitting of the initial convective cell as well as transverse propagation of the storm. This rotation also contributes to the formation of a long-lived structure in which the precipitating downdrafts support, rather than destroy, continued convection. As evaporatively cooled air moves into the path of the storm's inflow, baroclinic generation of horizontal vorticity along the boundary of the cold air causes the intensification of low-level rotation that may trigger the transition of a storm into its tornadic phase.

An important question that remains unresolved is, How is the tornadic circulation embedded within the storm-scale structure? The tornado itself appears nearly axisymmetric, while the storm structure in the vicinity of the tornado (along the boundary between warm-updraft and cold-downdraft air) is highly asymmetric. In both numerical models and observations, the issue is complicated by the nearly two orders of magnitude difference in the horizontal scale of the tornado and the parent storm. Future research with high-resolution radar and more powerful supercomputers will undoubtedly contribute to the resolution of this outstanding issue.

To date, tornadic thunderstorm research has led to a refined understanding of the environmental conditions that promote these storms and an improved ability to identify the salient features within an evolving storm that indicate a strong likelihood of impending tornadic activity. Other factors, however, such as the storm-initiation processes and the interactions between storms and with the larger-scale environment, greatly complicate the prospects for forecasting the precise time and location of tornadoes with a significant lead time. Further research in these areas will be necessary in order to improve this outlook substantially.

Literature Cited

Achtemeier, G. L. 1969. Some observations of splitting thunderstorms over Iowa on August 25–26, 1965. *Preprints, Conf. Severe Local Storms, 6th*, pp. 89–94. Boston: Am. Meteorol. Soc.

Alberty, R. L. 1969. A proposed mechanism for cumulonimbus persistence in the presence of strong vertical shear. *Mon. Weather Rev.* 97: 590–96

Barnes, S. L. 1970. Some aspects of a severe right-moving thunderstorm deduced from mesonetwork observations. *J. Atmos. Sci.* 27: 634–48

Barnes, S. L. 1978. Oklahoma thunderstorms on 29–30 April 1970. Part I: Morphology of a tornadic storm. *Mon. Weather Rev.* 106: 673–84

Binnie, A. M., Hookings, G. A. 1948. Laboratory experiments on whirlpools. *Proc. R. Soc. London Ser. A* 194: 348–415

Brandes, E. A. 1977. Gust front evolution and tornado genesis as viewed by Doppler radar. *J. Appl. Meteorol.* 16: 333–38

Brandes, E. A. 1978. Mesocyclone evolution and tornadogenesis: some observations. *Mon. Weather Rev.* 106: 995–1011

Brandes, E. A. 1984a. Relationships between radar-derived thermodynamic variables and tornadogenesis. *Mon. Weather Rev.* 112: 1033–52

Brandes, E. A. 1984b. Vertical vorticity generation and mesocyclone sustenance in tornadic thunderstorms: the observational evidence. *Mon. Weather Rev.* 112: 2253–69

Brooks, E. M. 1949. The tornado cyclone. *Weatherwise* 2: 32–33

Browning, K. A. 1964. Airflow and precipitation trajectories within severe local storms which travel to the right of the winds. *J. Atmos. Sci.* 21: 634–39

Browning, K. A. 1968. The organization of severe local storms. *Weather* 23: 429–34

Browning, K. A., Donaldson, R. J. Jr. 1963. Airflow structure of a tornadic storm. *J. Atmos Sci.* 20: 533–45

Browning, K. A., Foote, G. B. 1976. Airflow and hailgrowth in supercell storms and some implications for hail suppression. *Q. J. R. Meteorol. Soc.* 102: 499–533

Browning, K. A., Landry, C. R. 1963. Airflow within a tornadic storm. *Preprints, Weather Radar Conf., 10th*, pp. 116–22. Boston: Am. Meteorol. Soc.

Browning, K. A., Ludlam, F. H. 1962. Airflow in convective storms. *Q. J. R. Meteorol. Soc.* 88: 117–35

Burgess, D. W. 1976. Single Doppler radar vortex recognition. Part 1: mesocyclone signatures. *Preprints, Conf. Radar Meteorol., 17th*, pp. 97–103. Boston: Am. Meteorol. Soc.

Burgess, D. W., Wood, V. T., Brown, R. A. 1982. Mesocyclone evolution statistics. *Preprints, Conf. Severe Local Storms, 12th*, pp. 422–24. Boston: Am. Meteorol. Soc.

Byers, H. R. 1942. Nonfrontal thunderstorms. *Misc. Rep. No. 3*. Chicago: Univ. Chicago Press. 26 pp.

Byers, H. R., Braham, R. R. Jr. 1949. *The Thunderstorm.* Washington DC: Gov. Print. Off. 287 pp.

Charba, J., Sasaki, Y. 1971. Structure and movement of the severe thunderstorms of 3 April 1964 as revealed from radar and surface mesonetwork data analysis. *J. Meteorol. Soc. Jpn.* 49: 191–213

Charney, J. G., Eliassen, A. 1964. On the growth of the hurricane depression. *J. Atmos. Sci.* 21: 68–75

Davies-Jones, R. P. 1984. Streamwise vorticity: the origin of updraft rotation in supercell storms. *J. Atmos. Sci.* 41: 2991–3006

Davies-Jones, R. P. 1985. Tornado dynamics. See Kessler 1985, pp. 197–236

Drazin, P. G., Reid, W. H. 1981. *Hydrodynamic Stability*. Cambridge: Cambridge Univ. Press. 525 pp.

Dutton, J. A. 1976. *The Ceaseless Wind.* New York: McGraw-Hill. 579 pp.

Eagleman, J. R., Lin, W. C. 1977. Severe thunderstorm internal structure from dual-Doppler radar measurements. *J. Appl. Meteorol.* 16: 1036–48

Fankhauser, J. C. 1971. Thunderstorm-environment interactions determined from aircraft and radar observations. *Mon. Weather Rev.* 99: 171–92

Fujita, T., Grandoso, H. 1968. Split of a thunderstorm into anticyclonic and cyclonic storms and their motion as determined from numerical model experiments. *J. Atmos. Sci.* 25: 416–39

Gal-Chen, T. 1978. A method for the initialization of the anelastic equations: implications for matching models with observations. *Mon. Weather Rev.* 106: 587–606

Hammond, G. R. 1967. Study of a left moving thunderstorm of 23 April 1964. *ESSA Tech. Memo. IERTM-NSSL 31*, Natl. Severe Storms Lab., Norman, Okla. 75 pp.

Hane, C. E., Scott, B. C. 1978. Temperature and pressure perturbations within convective clouds derived from detailed air motion. Preliminary testing. *Mon. Weather Rev.* 106: 654–61

Hitschfeld, W. 1960. The motion and erosion of convective storms in severe vertical wind shear. *J. Meteorol.* 17: 270–82

Houze, R. A. Jr., Hobbs, P. V. 1982. Organization and structure of precipitating cloud systems. *Adv. Geophys.* 24: 225–315

Kessler, E., ed. 1985. *Thunderstorm Morphology and Dynamics.* Norman: Univ. Okla. Press. 411 pp. 2nd ed.

Klemp, J. B., Rotunno, R. 1983. A study of the tornadic region within a supercell thunderstorm. *J. Atmos. Sci.* 40: 359–77

Klemp, J. B., Weisman, M. L. 1983. The dependence of convective precipitation patterns on vertical wind shear. *Preprints, Conf. Radar Meteorol., 21st*, pp. 44–49. Boston: Am. Meteorol. Soc.

Klemp, J. B., Wilhelmson, R. B. 1978. Simulations of right- and left-moving storms produced through storm splitting. *J. Atmos. Sci.* 35: 1097–1110

Klemp, J. B., Wilhelmson, R. B., Ray, P. S.

1981. Observed and numerically simulated structure of a mature supercell thunderstorm. *J. Atmos. Sci.* 38 : 1558–80

Kropfli, R. A., Miller, L. J. 1976. Kinematic structure and flux quantities in a convective storm from dual-Doppler radar observations. *J. Atmos. Sci.* 33 : 520–29

Lemon, L. R., Doswell, C. A. III. 1979. Severe thunderstorm evolution and mesocyclone structure as related to tornadogenesis. *Mon. Weather Rev.* 107 : 1184–97

Lemon, L. R., Burgess, D. W., Brown, R. A. 1975. Tornado production and storm sustenance. *Preprints, Conf. Severe Local Storms, 9th*, pp. 100–4. Boston: Am. Meteorol. Soc.

Lilly, D. K. 1979. The dynamical structure and evolution of thunderstorms and squall lines. *Ann. Rev. Earth Planet. Sci.* 7 : 117–61

Lilly, D. K. 1982. The development and maintenance of rotation in convective storms. In *Intense Atmospheric Vortices*, ed. L. Bengtsson, J. Lighthill, pp. 149–60. Berlin: Springer-Verlag

Lilly, D. K. 1986a. The structure, energetics and propagation of rotating convective storms. Part I: Energy exchange with the mean flow. *J. Atmos. Sci.* 43 : 113–25

Lilly, D. K. 1986b. The structure, energetics and propagation of rotating convective storms. Part II: Helicity and storm stabilization. *J. Atmos. Sci.* 43 : 126–40

Lin, Y. J., Pasken, R. 1982. Thermodynamic structure of a tornadic storm as revealed by dual-Doppler data. *Preprints, Conf. Severe Local Storms, 12th*, pp. 405–8. Boston: Am. Meteorol. Soc.

Lindzen, R. S. 1974. Wave-CISK in the tropics. *J. Atmos. Sci.* 31 : 156–79

Maddox, R. A. 1976. An evaluation of tornado proximity wind and stability data. *Mon. Weather Rev.* 104 : 133–42

Miller, L. J. 1975. Internal airflow of a convective storm from dual-Doppler radar measurements. *Pure Appl. Geophys.* 113 : 765–85

Miller, M. J., Pearce, R. 1974. A three-dimensional primitive equation model of cumulus convection. *Q. J. R. Meteorol. Soc.* 100 : 133–54

Morton, R. B. 1966. Geophysical vortices. In *Progress in Aeronautical Sciences*, ed. D. Küchmann, 7 : 145–94. Oxford/London: Pergamon. 220 pp.

Newton, C. W. 1963. Dynamics of severe convective storms. In *Meteorological Monographs*, ed. D. Atlas et al., 5 : 33–58. Boston: Am. Meteorol. Soc. 247 pp.

Newton, C. W., Fankhauser, J. C. 1964. On the movements of convective storms, with emphasis on size discrimination in relation to water-budget requirements. *J. Appl. Meteorol.* 3 : 651–68

Newton, C. W., Katz, S. 1958. Movement of large convective rain storms in relation to winds aloft. *Bull. Am. Meteorol. Soc.* 32 : 129–36

Newton, C. W., Newton, H. R. 1959. Dynamical interactions between large convective clouds and environment with vertical shear. *J. Meteorol.* 16 : 483–96

Ogura, Y., Phillips, N. A. 1962. Scale analysis of deep and shallow convection in the atmosphere. *J. Atmos. Sci.* 19 : 173–79

Ooyama, K. 1964. A dynamical model for the study of tropical cyclone development. *Geofis. Int.* 4 : 187–98

Pasken, R., Lin, Y. J. 1982. Pressure perturbations within a tornadic storm derived from dual-Doppler wind data. *Preprints, Conf. Severe Local Storms, 12th*, pp. 257–60. Boston: Am. Meteorol. Soc.

Prandtl, L., Tietjens, O. G. 1934a. *Applied Hydro- and Aero-Mechanics*. New York: Dover. 311 pp. 1957 ed.

Prandtl, L., Tietjens, O. G. 1934b. *Fundamentals of Hydro- and Aero-Mechanics*. New York: Dover. 270 pp. 1957 ed.

Ray, P. S., Doviak, R. J., Walker, G. B., Sirmans, D., Carter, J., Bumgarner, B. 1975. Dual-Doppler observation of a tornadic storm. *J. Appl. Meteorol.* 14 : 1521–30

Ray, P. S., Johnson, B. C., Johnson, K. W., Bradberry, J. S., Stephens, J. J., et al. 1981. The morphology of several tornadic storms on 20 May 1977. *J. Atmos. Sci.* 38 : 1643–63

Raymond, D. J. 1975. A model for predicting the movement of continuously propagating convective storms. *J. Atmos. Sci.* 32 : 1308–17

Raymond, D. J. 1976. Wave-CISK and convective mesosystems. *J. Atmos. Sci.* 33 : 2392–98

Raymond, D. J. 1983. Wave-CISK in mass flux form. *J. Atmos. Sci.* 40 : 2561–72

Raymond, D. J. 1984. A wave-CISK model of squall lines. *J. Atmos. Sci.* 41 : 1946–58

Rotunno, R. 1981. On the evolution of thunderstorm rotation. *Mon. Weather Rev.* 109 : 171–80

Rotunno, R., Klemp, J. B. 1982. The influence of the shear-induced pressure gradient on thunderstorm motion. *Mon. Weather Rev.* 110 : 136–51

Rotunno, R., Klemp, J. B. 1985. On the rotation and propagation of simulated supercell thunderstorms. *J. Atmos. Sci.* 42 : 271–92

Schlesinger, R. E. 1975. A three-dimensional numerical model of an isolated deep convective cloud: preliminary results. *J. Atmos. Sci.* 32 : 934–57

Schlesinger, R. E. 1980. A three-dimensional numerical model of an isolated deep thunderstorm. Part II: Dynamics of updraft splitting and mesovortex couplet evolution. *J. Atmos. Sci.* 37: 395–420

Thorpe, A. J., Miller, M. J. 1978. Numerical simulations showing the role of the downdraught in cumulonimbus motion and splitting. *Q. J. R. Meteorol. Soc.* 104: 873–93

Weaver, J. F., Nelson, S. P. 1982. Multiscale aspects of thunderstorm gust fronts and their effects on subsequent storm development. *Mon. Weather Rev.* 110: 707–18

Weisman, M. L., Klemp, J. B. 1982. The dependence of numerically simulated convective storms on vertical wind shear and buoyancy. *Mon. Weather Rev.* 110: 504–20

Weisman, M. L., Klemp, J. B. 1984. The structure and classification of numerically simulated convective storms in directionally varying wind shears. *Mon. Weather Rev.* 112: 2479–98

Wilhelmson, R. B. 1974. The life cycle of a thunderstorm in three dimensions. *J. Atmos. Sci.* 31: 1629–51

Wilhelmson, R. B., Klemp, J. B. 1978. A three-dimensional numerical simulation of splitting that leads to long-lived storms. *J. Atmos. Sci.* 35: 1037–63

Wilhelmson, R. B., Klemp, J. B. 1981. A three-dimensional numerical simulation of splitting severe storms on 3 April 1964. *J. Atmos. Sci.* 38: 1581–1600

Ann. Rev. Fluid Mech. 1987. 19: 403–35

THERMOCAPILLARY INSTABILITIES

Stephen H. Davis

Department of Engineering Sciences and Applied Mathematics, Northwestern University, Evanston, Illinois 60201

1. INTRODUCTION

An interface $\mathcal{S}$ between two immiscible fluids possesses localized properties, the most prominent of which is the interfacial (or surface) tension σ. This represents the magnitude of the force per unit length normal to a cut in the interface (see Levich 1962).

The surface tension σ usually depends on the scalar fields in the system (e.g. the electrical field, the temperature field), as well as on the concentration of foreign materials on the interface (Levich 1962). In the present article we focus on a single such field—the temperature T. Thus, we consider only thermocapillarity and pose an equation of state

$$\sigma = \sigma(T). \tag{1.1}$$

Surface tension enters the description of the dynamics of the system through the force balance at the interface $\mathcal{S}$.

On the one hand, the jump in normal stress at $\mathcal{S}$ balances surface tension times twice the mean curvature H of $\mathcal{S}$. In the absence of viscosity, this is the Laplace relation, which states that the pressure is larger on the concave side of $\mathcal{S}$ by an amount $|2H\sigma(T)|$. Thus, thermocapillarity can alter the capillary pressure jump or give it variations from point to point, depending on T.

On the other hand, there is a jump in shear stress at $\mathcal{S}$ balanced by the surface-tension gradient. If we represent the equation of state (1.1) by a linear law

$$\sigma = \sigma_0 - \gamma(T - T_0), \tag{1.2}$$

then the surface-tension gradients on $\mathcal{S}$ are proportional to γ; temperature

0066–4189/87/0115–0403$02.00

gradients along $\mathscr{S}$ induce shear stresses on $\mathscr{S}$ that result in fluid motion. This is shown in Figure 1, where a temperature gradient is *imposed* along $\mathscr{S}$. For common liquids, we have $\gamma = -d\sigma/dT > 0$, so that there is surface flow from the hot end toward the cold end. Since the bulk fluids are viscous, they are dragged along; bulk-fluid motion results from interfacial temperature gradients. This is called the *thermocapillary effect* (Levich 1962).

Thermocapillary effects can dominate the dynamics in the containerless processing of crystals, the behavior of weld pools, the rupture of thin films, the movement of contact lines, and the propagation of flames over liquid fuels. In many situations in nature as well as in science and industry, the transport of heat across interfaces can be dramatically increased through the presence of additional mixing processes triggered by instabilities. The control of these instabilities can be attained only through the understanding of them.

Clearly, the presence of an interface with variable surface tension can modify already known instabilities. However, surface tension is a localized force at the interface, and thermocapillarity can drive its own instabilities. In this article we concentrate on these processes; we aim to describe their mechanisms in detail and define highly simplified systems that highlight these. After having done this, we then turn to more complicated cases.

The simplified cases we examine involve a simple geometry—the *one-layer system*—in which there is a liquid layer whose lower boundary is a rigid plate and whose upper boundary is an interface with a passive gas (having negligible viscosity and density), as shown in Figure 2. In practice an interface may be subject to a temperature gradient $[\nabla\bar{T}]_{\text{EXT}}$ with arbitrary orientation. We let $(\mathbf{e}_1, \mathbf{e}_2, \mathbf{e}_3)$ be unit vectors in the (x, y, z) directions.

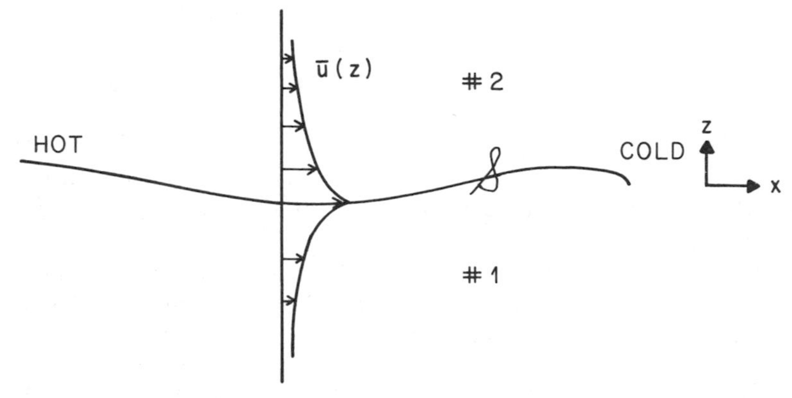

Figure 1 Sketch of two-fluid system in which a temperature gradient is imposed along the interface $\mathscr{S}$ between two immiscible fluids #1 and #2.

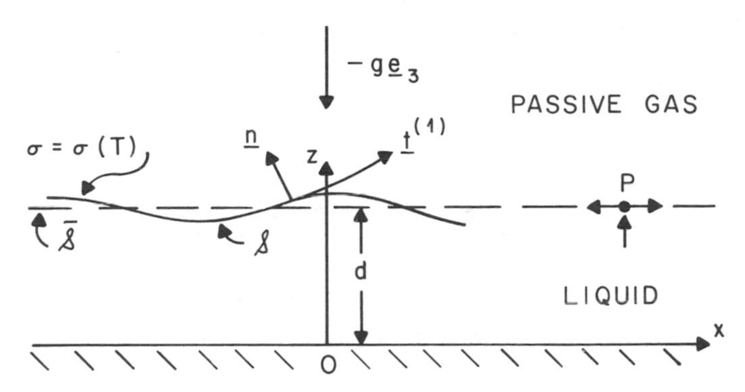

Figure 2 Sketch of the one-layer system.

If

$$\mathbf{e}_1 \cdot [\nabla \bar{T}]_{\mathrm{EXT}} = \mathbf{e}_2 \cdot [\nabla \bar{T}]_{\mathrm{EXT}} = 0, \tag{1.3}$$

then $[\nabla \bar{T}]_{\mathrm{EXT}}$ is imposed normal to $\bar{\mathscr{S}}$, the mean position (z is constant) of the interface, and there exists a purely static basic state. Instabilities of this state lead first to steady Marangoni convection, as identified by Pearson (1958). If

$$\mathbf{e}_3 \cdot [\nabla \bar{T}]_{\mathrm{EXT}} = 0, \tag{1.4}$$

then $[\nabla \bar{T}]_{\mathrm{EXT}}$ is imposed along $\bar{\mathscr{S}}$, and no static states exist. Thermocapillarity can drive a steady shear flow whose instabilities are often time-periodic hydrothermal waves, as identified by Smith & Davis (1983a,b). We further discuss in Section 5 systems in which (*a*) $[\nabla \bar{T}]_{\mathrm{EXT}}$ has arbitrary orientation to $\bar{\mathscr{S}}$ and (*b*) $\bar{\mathscr{S}}$ is not planar.

2. FORMULATION

Consider the *one-layer model* consisting of a liquid layer bounded below by a rigid plane and above by a passive gas (having negligible density and viscosity). The plane lies at $z = 0$ and the mean position of the interface lies at $z = d$, as shown in Figure 2. The liquid is a Newtonian fluid with constant values of the viscosity μ, the reference density ρ_0, the specific heat c_{p}, the thermal conductivity k, and the volume expansion coefficient α; $\kappa = k/\rho_0 c_{\mathrm{p}}$ is the thermal diffusivity, and $\nu = \mu/\rho_0$ is the kinematic viscosity. The surface tension σ of the interface varies with temperature T of the liquid, as given in Equation (1.2). Gravity of magnitude g acts downward.

The system is subject to an *imposed temperature gradient* $[\nabla \bar{T}]_{\mathrm{EXT}}$,

$$[\nabla \bar{T}]_{\mathrm{EXT}} = -b_{\parallel}\mathbf{e}_1 - b_{\perp}\mathbf{e}_3 \tag{2.1}$$

where $\mathbf{e}_i$ are the unit vectors in the (x, y, z) directions. Thus $|b_\parallel|$ and $|b_\perp|$ are the magnitudes of horizontal and vertical temperature gradients, respectively. Let us define

$$b = (b_\parallel^2 + b_\perp^2)^{1/2}. \tag{2.2}$$

We consider the fully three-dimensional system and scale all distances on the average liquid depth d. The velocity vector $\mathbf{v} = (u, v, w)$, pressure p, temperature difference $T - T_0$, surface tension σ, and time t are referred to scales $V_M = \gamma bd/\mu$, $\mu V_M/d = \gamma b$, bd, σ_0, and $\mu/\gamma b$, respectively. As a result, there arise the following dimensionless groups:

$$R = \frac{\rho_0 \gamma b d^2}{\mu^2}, \qquad P = \frac{\nu}{\kappa}, \qquad \mathrm{Ra} = \frac{\alpha b g d^4}{\kappa \nu}, \qquad S = \frac{\rho_0 d \sigma_0}{\mu^2}. \tag{2.3a,b,c,d}$$

Here R is the Reynolds number, P is the Prandtl number, Ra is the Rayleigh number, and S is the nondimensional surface-tension number. Another useful dimensionless group is the Marangoni number, defined as

$$M = RP. \tag{2.3e}$$

In most of what follows, either $b_\parallel = 0$ or $b_\perp = 0$, so that the ratio $b_\parallel/b_\perp$ is not an additional parameter.

The governing equations for the liquid layer are the Navier-Stokes, the energy, and the continuity equations, subject to the Boussinesq approximation:

$$R\left(\frac{\partial \mathbf{v}}{\partial t} + \mathbf{v}\cdot\nabla\mathbf{v}\right) = -\nabla p + \nabla^2\mathbf{v} + \mathrm{Ra}\; T\mathbf{e}_3, \tag{2.4a}$$

$$M\left(\frac{\partial T}{\partial t} + \mathbf{v}\cdot\nabla T\right) = \nabla^2 T, \tag{2.4b}$$

$$\nabla\cdot\mathbf{v} = 0. \tag{2.4c}$$

The interface $\mathscr{S}$ lies at $z = 1 + \eta(x, y, t)$. The kinematic condition is

$$w = \eta_t + u\eta_x + v\eta_y \quad \text{on} \quad z = 1 + \eta(x, y, t), \tag{2.5a}$$

and the stress conditions are

$$\mathbf{n}\cdot\mathbf{T}\cdot\mathbf{n} = 2H(SR^{-1} - T) \quad \text{on} \quad z = 1 + \eta(x, y, t), \tag{2.5b}$$

and for $\alpha = 1, 2$,

$$\mathbf{t}^{(\alpha)}\cdot\mathbf{T}\cdot\mathbf{n} = -\mathbf{t}^{(\alpha)}\cdot\nabla T \quad \text{on} \quad z = 1 + \eta(x, y, t). \tag{2.5c}$$

Here $\mathbf{T}$ is the stress tensor of the liquid, and H is the mean curvature of the interface

$$2H = [\eta_{xx}(1+\eta_y^2) - 2\eta_x\eta_y\eta_{xy} + \eta_{yy}(1+\eta_x^2)]\,(1+\eta_x^2+\eta_y^2)^{-3/2}. \tag{2.5d}$$

Subscripts denote partial differentiation. In Equation (2.5b), $SR^{-1} - T$ is the surface tension of the interface, and the last term in (2.5c) is the surface-tension gradient along the interface. Here $\mathbf{n}$ is the unit normal to $\mathscr{S}$ pointing out of the liquid, and $\mathbf{t}^{(\alpha)}$, $\alpha = 1, 2$, are orthonormal tangent vectors to $\mathscr{S}$.

The velocity scale V_M, the "Marangoni velocity scale," represents a balance between surface-tension gradients on the interface and the shear stresses generated by them, as captured by Equation (2.5c). It is convenient to define the capillary (or crispation) number C as

$$C = \frac{\mu\kappa}{d\sigma_0} = RS^{-1}M^{-1} = P^{-1}S^{-1}. \tag{2.6}$$

3. $[\nabla\bar{T}]_{\mathrm{EXT}} \perp \bar{\mathscr{S}}$: MARANGONI INSTABILITY

3.1 *Basic State*

We consider the situation in which, from Equation (2.1), $b_\parallel = 0$ and $b_\perp > 0$, so that the liquid layer is heated from below normal to the mean position of the interface.

Suppose that the rigid plate of Figure 2 is a perfect heat conductor fixed at temperature T_B, $\hat{T} = T_B$ at $z = 0$, while the interface is cooled by air currents according to the law $-k\nabla\hat{T}\cdot\mathbf{n} = h(\hat{T} - T_{\mathrm{AIR}}) + Q_0$, where h is the unit thermal surface conductance, Q_0 is an imposed heat flux to the environment, and carets denote dimensional quantities. If we non-dimensionalize as in Section 2, identifying T_0 with T_T (to be defined in a moment), then we have the two thermal boundary conditions as follows:

$$T = \frac{T_B - T_T}{b_\perp d} \quad \text{at} \quad z = 0 \tag{3.1}$$

and

$$\nabla T\cdot\mathbf{n} + B\left[T + \left(\frac{T_T - T_{\mathrm{AIR}}}{b_\perp d}\right)\right] + Q = 0 \quad \text{at} \quad z = 1 + \eta(x, y, t), \tag{3.2}$$

where the surface Biot number B is given by

$$B = \frac{hd}{k} \tag{3.3a}$$

and

$$Q = \frac{Q_0}{b_\perp k}. \tag{3.3b}$$

We seek a *basic state* in which the interface is flat,

$$\bar{\eta} = 0, \tag{3.4a}$$

the fluid is static,

$$\bar{\mathbf{v}} = 0, \tag{3.4b}$$

the heat transfer is purely conductive,

$$\bar{T} = 1 - z, \tag{3.4c}$$

and the pressure is hydrostatic,

$$\bar{p} = -\mathrm{Ra}(1-z)^2. \tag{3.4d}$$

Thus, we select T_T to be the interface temperature in the basic state, and we have that

$$T_\mathrm{B} - T_\mathrm{T} = b_\perp d, \qquad T_\mathrm{T} - T_\mathrm{AIR} = (b_\perp d)B^{-1}, \qquad Q = 0. \tag{3.5}$$

The stability problem is obtained by disturbing all quantities $\mathbf{v}$, T, p, and η. The governing system for this nonlinear problem has the following nondimensional groups:

$$M^{(\perp)}, P, C, B, \mathrm{Ra}^{(\perp)}, \varepsilon, \text{ and } (k_1, k_2). \tag{3.6}$$

Here $R^{(\perp)}$ is determined from Equation (2.3e), S is determined from Equation (2.6), and we have inserted a superscript "$\perp$" to remind us that the imposed temperature gradient is normal to $\bar{\mathcal{S}}$ (in the basic state), so that $b = b_\perp$ in the definition of $M^{(\perp)}$. In addition, ε measures the amplitude of a disturbance (at some initial time), and (k_1, k_2) are the wave numbers (or, more generally, the horizontal scales) of the disturbance.

3.2 *Linear Stability Theory*

In the limit $\varepsilon \to 0$, the disturbance equations and boundary conditions can be linearized about the basic state. This theory was first examined and explained by Pearson (1958). He seeks neutral conditions, which correspond to the onset of steady *Marangoni convection.* His curves are sketched in Figure 3. The result, as shown by the solid curves in Figure 3, is that pure conduction is unstable if $M^{(\perp)} > M_\mathrm{L}^{(\perp)}$ when the lower plate is a fixed-temperature boundary, where

$$M_\mathrm{L}^{(\perp)} \approx 79.6, \qquad k_\mathrm{L} \approx 1.99 \quad \text{for} \quad B = 0. \tag{3.7}$$

Figure 3 also shows Pearson's result for the case when the lower plate is a fixed heat-flux boundary. Here the dashed curves apply, and we have

$$M_L^{(\perp)} \approx 48, \qquad k_L \to 0 \quad \text{for} \quad B = 0. \tag{3.8}$$

Notice that an increase in B results in a stabilization of the basic state. In the above, k is the overall wave number,

$$k = (k_1^2 + k_2^2)^{1/2}. \tag{3.9}$$

Pearson's theory is subject to the restrictions that Ra = 0 and that the interface is nondeformable, an approximation that is formally attained in the limit $C \to 0$ ($S \to \infty$). This can be seen from Equations (2.5b) and (2.6), since in this limit $H \to 0$ and hence the interface remains planar. Since this convective instability is time independent, the above results are independent of P.

Gravity is present in experiments on Earth, so the above stability criteria may be modified. Nield (1964) considers the coupled thermocapillary/buoyancy instability problem in which the interface is nondeformable ($C \to 0$).

Scriven & Sternling (1964) consider a two-layer model in which each layer has infinite depth and examine the local behavior of the system near the interface. They consider only the solutocapillary case with mass transfer in either direction; here surface-tension variations are caused by surface-active materials rather than by heat. They allow $C \neq 0$, so that interfacial deflection is permissible, and find profound differences in behavior from the Pearson model. In fact they find that $M_L^{(\perp)} = 0$, so that the system is always unstable. This zero critical value of $M^{(\perp)}$ occurs for $k \to 0$, i.e. for very long waves. Smith (1966) rationalizes this dilemma by

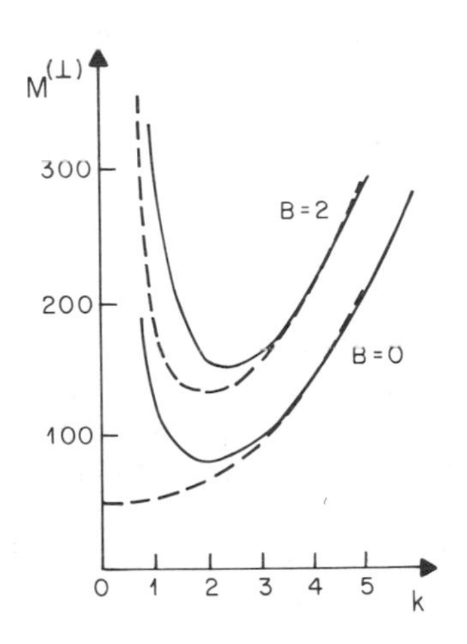

Figure 3 Critical Marangoni number versus overall wave number for Marangoni convection, as calculated by Pearson (1958) for the lower plate as a fixed-temperature surface (solid curves) or as a fixed heat-flux surface (dashed curves).

reconsidering the one-layer model but allowing surface deflection and gravity. He neglects buoyancy, but since the basic state has a hydrostatic pressure field through which the interface deflects, gravity waves are generated. These gravity waves stabilize the long-wave instabilities found by Scriven & Sternling, and in most practical situations they regain the Pearson result.

The question of the influence of surface deflection at long wavelengths can be clarified by direct examination of long waves. Davis (1983) discusses a nonlinear evolution equation for the interfacial height $z = h$ appropriate to slow variations along the layer compared with those normal to the layer. When this equation is expressed in the present notation, it takes the form

$$M^{(\perp)}C\frac{\partial h}{\partial t}+\nabla\cdot\left\{\left(\frac{1}{2}M^{(\perp)}Ch^2-\frac{1}{3}Gh^3\right)\nabla h\right\}+\nabla\cdot\left(\frac{1}{3}h^3\nabla\nabla^2 h\right)=0, \quad (3.10)$$

where G is the Bond number

$$G=\frac{\rho_0 g d^2}{\sigma_0}. \quad (3.11)$$

If we linearize this about $h = 1$ and use normal modes

$$\eta=\eta_0 e^{\sigma t+i(k_1 x+k_2 y)}, \quad (3.12)$$

we obtain the following characteristic equation:

$$\sigma=[M^{(\perp)}]^{-1}C^{-1}\left\{\left(\frac{1}{2}M^{(\perp)}C-\frac{1}{3}G\right)-\frac{1}{3}k^2\right\}k^2. \quad (3.13)$$

The equivalent of the Scriven & Sternling result ($G = 0$) is that if $k^2 < \frac{3}{2}M^{(\perp)}C$, the system is unconditionally unstable. The equivalent of the Smith result is that if

$$M^{(\perp)}<\frac{2}{3}\frac{G}{C}, \quad (3.14)$$

then all long-wave small-amplitude disturbances decay. The Pearson model is regained if we first let $C \to 0$ and then let $G \to 0$. A 1-mm layer of silicone oil having $\rho_0 \approx 1$ g cm^{-3}, $\sigma_0 \approx 21$ dyne cm^{-1}, and $\gamma \approx 0.07$ dyne cm^{-1} °C^{-1} subject to a temperature gradient $b_\perp \approx 20$ °C cm^{-1} has $G/CM^{(\perp)} \approx 70$. Clearly, $G/CM^{(\perp)}$ is proportional to d, and thus the long-wave instabilities can always be made significant if the layer is made thin enough.

Davis & Homsy (1980) examine directly the effects of interface deflection on the Nield model; they examine the one-layer model with buoyancy for

$B = 0$ and use perturbation theory in powers of C to find corrections to the critical conditions (3.7). They find that surface deflection is stabilizing for $M^{(\perp)} < M_0 \approx 67$ but destabilizing for $M^{(\perp)} > M_0$. Thus, the system dominated by thermocapillary effects is destabilized, while that dominated by buoyancy is stabilized. In fact the Marangoni-number correction (S. Rosenblat, G. M. Homsy & S. H. Davis, unpublished) at $O(C)$ is

$$\eta_1 \sim \int \mathbf{e}_3 \cdot \mathbf{T}_0 \cdot \mathbf{e}_3, \tag{3.15}$$

where $\mathbf{T}_0$ is the order-unity stress tensor, and the integral is taken at $z = 1$ over one period in x and y. Thus, the stabilization for $M < M_0$ also means that there is a surface elevation above a rising convective current, consistent with the result of Jeffreys (1951) for pure Bénard convection. Likewise, the destabilization for $M > M_0$ means that there is a surface depression above a rising current, consistent with the result of Davis & Segel (1963) and Scriven & Sternling (1964) for pure Marangoni convection.

In experiments it is convenient to impose a temperature across the gap between two horizontal plates that enclose the liquid layer bounded by a thin air gap present to help control the thermal environment of the interface. Thus, the two-layer problem is of practical importance. This problem has been considered by Zeren & Reynolds (1972) and then corrected and extended by Ferm & Wollkind (1982).

3.3 *Physical Mechanisms*

The instability identified by linear theory can be explained following Pearson (1958). If the plate is heated, as shown in Figure 2, a conductive temperature profile is achieved in which $b_\perp > 0$.

Marangoni instability Assume that a disturbance creates a hot spot (compared with its neighbors) at a point P on $\mathscr{S}$. If the surface tension decreases with temperature ($\gamma > 0$), there is a net surface traction away from P. Since the fluid is viscous, subsurface fluid is dragged away from P. By conservation of mass, an upflow beneath P is created. This rising fluid, since it comes from below, has been warmed by the conductive profile $\bar{T}$; it can maintain the heat excess at P if $\gamma b_\perp d$ is large enough, i.e. if either the temperature gradient is steep or if the surface tension is highly temperature sensitive. The fluid is recycled by cooling along the interface and falling at a distance away from P. Thus, if $M^{(\perp)}$ is large enough, as we have seen, a steady, finite-amplitude convection can be maintained.

The above described convection is most easily generated if $B = 0$. When B is raised from zero, the fluid rising from below P loses some of its heat to the atmosphere, leaving less available to generate surface-tension gradients. As shown in Figure 3, increasing B gives rise to larger values of $M_L^{(\perp)}$ necessary to maintain the convection. From Equation (3.2) in the limit $B \to \infty$ the interface is isothermal, surface-tension gradients are absent, and so $M_L^{(\perp)} \to \infty$. Marangoni convection is absent, though buoyancy-driven convection is still possible.

The result that $M < M_0$ ($M > M_0$) corresponds to both (*a*) elevations (depressions) above rising currents and (*b*) a stabilization (destabilization) due to weak surface deflections is in accord with the calculations of Scriven & Sternling (1964) and Smith (1966) for the case of pure Marangoni convection. Scriven & Sternling conjectured (incorrectly) that surface deflections will be *in general* destabilizing in the expectation that increasing the number of degrees of freedom of a mechanical system (by allowing deflection) will, if anything, lower the stability limits. For $M < M_0$, the convection is buoyancy dominated and the free surface is elevated above a rising current. Such a bulging of the interface augments the cooling to the atmosphere. The opposite situation occurs for $M > M_0$, where thermocapillarity is dominant.

3.4 *Energy Stability Theory*

Energy theory is a variational formulation whose output is a criterion for stability against disturbances of arbitrary amplitude ε.

Davis (1969) formulates the theory for the Nield model—specifically, for the limiting case $C \to 0$ in which surface deflection is absent. He finds a critical value $M_E^{(\perp)}$ of $M^{(\perp)}$ for which $M^{(\perp)} < M_E^{(\perp)}$ is a sufficient condition for stability. $M_E^{(\perp)}$ is a function of B but is independent of P. Below the energy-theory value, stability is guaranteed; above the linear-theory value, instability is guaranteed (Nield 1964). Between the values, the basic state may be stable for ε small but unstable for ε large enough. This region of possible subcritical instabilities vanishes when $M^{(\perp)} = 0$, which is the pure Bénard (gravitational) instability limit, consistent with the result of Joseph (1965). The theory has been extended to negative values of $R^{(\perp)}$ or $M^{(\perp)}$ by Lebon & Cloot (1982). Where or whether subcritical instabilities exist in the range $M_E^{(\perp)} < M < M_L^{(\perp)}$ is a question that can only be answered by bifurcation theory or by direct numerical simulation.

Davis & Homsy (1980) reformulate the energy theory to include effects of surface deformation. They then seek perturbations for small C and find that these perturbations tend to increase $M_E^{(\perp)}$ with C for small C. The numerical values obtained by Davis & Homsy for the degree of stabilization may be inaccurate; Velarde & Castillo (1981) recompute these.

3.5 *Bifurcation Theory*

The postinstability behavior of the system can be examined by bifurcation theory using perturbations in powers of ε near $M^{(\perp)} = M_L^{(\perp)}$.

Scanlon & Segel (1967) pose the one-layer model in the absence of gravity. For simplicity they let $P \to \infty$ and allow the layer to be infinitely deep, an assertion that makes the surface deflection zero at each order of ε. They consider the interaction of two disturbances that allow a competition between (two-dimensional) rolls and (three-dimensional) hexagonal convection and obtain a pair of amplitude equations governing the selection. These have the form

$$\frac{dA}{dt} = \sigma A - \beta AB - A(c_1 A^2 + c_2 B^2), \tag{3.16a}$$

$$\frac{dB}{dt} = \sigma B - \frac{1}{4}\beta A^2 - B\left[\frac{1}{2}c_2 A^2 + (4c_1 - c_2)B^2\right], \tag{3.16b}$$

where σ is the linear-theory growth rate proportional to $M^{(\perp)} - M_{\mathrm{L}}^{(\perp)}$, and the c_i and β are positive (computable) coefficients. The A and B are the time-dependent amplitudes in w_1, the linearized vertical component of velocity,

$$w_1 = \left\{A(t) \cos \frac{1}{2}\sqrt{3}\,kx \cos \frac{1}{2}ky + B(t) \cos ky\right\} f(z), \tag{3.17}$$

and $f(z)$ is a (computable) eigenfunction of the linearized theory. Here, if $A = 0$, $B \neq 0$, there is roll-cell convection; if $A = \pm 2B$, then hexagonal convection is present.

Scanlon & Segel (1967) find that in the range near $M^{(\perp)} = M_{\mathrm{L}}^{(\perp)}$, where validity is expected, the only stable convective state has hexagonal planform; this convection is characterized by upflow in cell centers. As $M^{(\perp)}$ is increased from zero, the conductive state is stable until $\delta M = 0$, where

$$\delta M = (M^{(\perp)} - M_{\mathrm{L}}^{(\perp)})/M_{\mathrm{L}}^{(\perp)}. \tag{3.18}$$

However, hexagonal convection is stable from $\delta M = -0.023$ onward. Thus, as $M^{(\perp)}$ is increased, conduction will remain stable until the system jumps to convection for $-0.023 < \delta M < 0$. This stable subcritical convection creates a dynamic hysteresis (Scanlon & Segel 1967) behavior in the response of the system.

The magnitudes of A and B are formally $O(\varepsilon)$. If β were zero, the equilibria of system (3.16) would have $A, B = O(\sigma^{1/2})$, so that we would have $\varepsilon = O(\sigma^{1/2})$. Now, if $\beta \neq 0$, then strictly speaking one should retain both the quadratic and cubic terms in system (3.16) only if they are comparable, i.e. if $\beta\varepsilon^2 \sim \varepsilon^3$. Thus, only if $\beta = O(\varepsilon)$ can higher powers in A and B be neglected. Since Scanlon & Segel find for β a numerical value independent of ε, their results may be only suggestive.

Why can one consider only the two disturbances defined in Equation

3.17? Any horizontal structure $\exp [ik_1x + ik_2y]$ with $k_1^2 + k_2^2 = k^2$ is allowable. Kraska & Sani (1979) attempt to use such a wider class, which includes the pair of Equation (3.17). They fail to obtain closure; the more disturbances that they include, the more that are required. However, their analysis may be in error, given that the adjoint operator that they present is incorrect.

Cloot & Lebon (1984) consider the nonlinear Nield model in which $C \to 0$ but $B = 0$ or 1, $P = 7$, 70, or 500, and $\text{Ra}^{(\perp)}$ is arbitrary. The full range of horizontal planforms is included, and power-series representations in ε are examined. They find that hexagons are preferred near $\text{M}_\text{L}^{(\perp)}$ and extend to subcritical values of $M^{(\perp)}$. For $\text{Ra}^{(\perp)}$ and $P = 7$, they begin at $\delta M = -0.0003$, again showing the narrow range of subcritical convection. Furthermore, Cloot & Lebon (1984) examine supercritical convection and determine when such steady states are stable against disturbances of various wavelengths and planforms. When $M^{(\perp)}$ is close to zero, so that the state is nearly pure Bénard convection, stable hexagons appear only for $k > k_\text{L}$. As $\text{Ra}^{(\perp)}$ is decreased (i.e. $M^{(\perp)}$ is increased), the stable range shifts downward and falls to $k < k_\text{L}$ for small enough $R^{(\perp)}$. This analysis is exhaustive in that the whole $\text{Ra}^{(\perp)}$-range is covered. However, the comment made above about the analysis of Scanlon & Segel (1967) applies here as well. Formally, both quadratic and cubic terms can only be retained simultaneously if $M^{(\perp)} = O(\varepsilon)$.

In all of the above, the convecting layer is unbounded laterally, so that the horizontal wave vector (k_1, k_2) forms a continuous spectrum. The linear theory determines the wave-vector magnitude $k = k_\text{L}$ at $M^{(\perp)} = M_\text{L}^{(\perp)}$ but not its phase. Thus, the planform is undetermined. Further, at $M^{(\perp)} > M_\text{L}^{(\perp)}$, a whole range of k corresponds to modes that grow according to linear theory. The above theories aim at obtaining the preferred planform (hexagons) and wave number through nonlinear selection. An alternative approach is to focus on a modified problem, i.e. a convection layer with lateral boundaries. These boundaries break the continuous spectrum and allow analyses of one or a few competing modes, presumably present when the container width a (scaled on layer depth) is not too large.

Rosenblat et al. (1982a,b) consider such situations when the container is circular and rectangular. In each case they simplify the problem by posing "slippery sidewalls," thus allowing solution of the linear stability problem by separation of variables. Further they take $C \to 0$ and, for their main analysis, $B = \text{Ra}^{(\perp)} = 0$ as well. P is arbitrary, and now a is present as well.

We focus here on the cylindrical case, where the linearized problem has a vertical velocity component that can be represented as follows:

$$w_1 = A_m(t) f(z) \cos m\theta g_m(r), \tag{3.19}$$

where $f(z)$ is a similar structure function as that in Equation (3.17), except that the present layer has finite depth, g_m is an appropriate Bessel function, and m is an integer giving the azimuthal variations. Figure 4 shows the neutral curve for the problem; note that except for a small range of radii a ($1.7 < a < 2.0$), the most "dangerous" mode is nonaxisymmetric.

For most aspect ratios the weakly nonlinear theory focuses on a single mode and gives an amplitude equation of the form

$$\frac{dA_m}{dt} = \sigma_m A_m - c_m A_m^3, \qquad m \neq 0, \tag{3.20}$$

$c_m > 0$, corresponding to supercritical, pure mode m convection. Here σ_m is the linear-theory growth rate, and c_m is a computable Landau constant. In the range $1.7 < a < 2.0$, there is pure axisymmetric convection with

$$\frac{dA_0}{dt} = \sigma_0 A_0 - \beta_0 A_0^2 - c_0 A_0^3. \tag{3.21}$$

Both upflow and downflow at the container center are possible in each case. Liang et al. (1969) find similar behavior in pure Bénard convection in a slippery-walled cylinder.

When the aspect ratios have special values as shown in Figure 4, linear theory is ambiguous, since $M_L^{(\perp)}$ is equal for two modes and a nonlinear theory must be used to make a pattern selection. Thus, when a is fixed

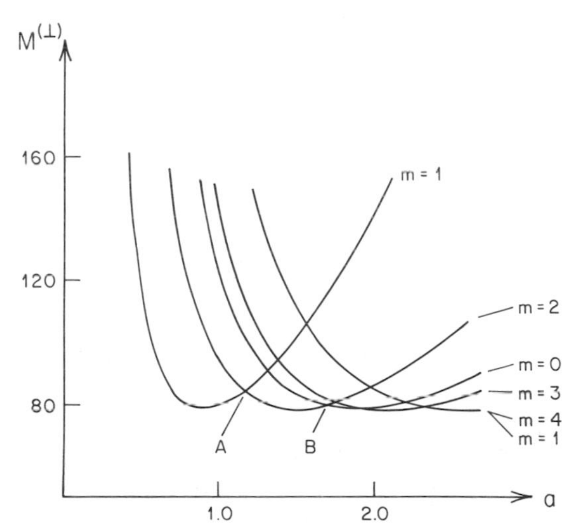

Figure 4 Critical curves of Marangoni number versus box radius for Marangoni instability in a circular cylinder with "slippery" walls ($B = 0$; from Rosenblat et al. 1982a). The envelope of minima is the neutral curve. The points A and B indicate double eigenvalues where modes of two azimuthal wave numbers m have the same $M_L^{(\perp)}$.

near one of the crossovers and $M^{(\perp)}$ is increased through $M_L^{(\perp)}$, there can be a sequence of transitions by primary, then secondary or tertiary, bifurcations. Besides giving information on such sequences at these special values of a, the sequences can mirror the transitions far from these special points, which would be widely separated in $M^{(\perp)}$ and otherwise inaccessible to analysis. Rosenblat et al. (1982a) analyze the double eigenvalues in neighborhoods of points A and B of Figure 4 using an asymptotic simplification of a series-truncation representation as an aid to the calculation of the coefficients.

Near point A, modes $m = 1$ and $m = 2$ interact. If the linearized vertical velocity component w_1 is represented as

$$w_1 = \{A_1(t)g_1(r)\cos\theta + A_2(t)g_2(r)\cos 2\theta\}f(z), \tag{3.22}$$

then the amplitudes near $a_A \approx 1.20$ satisfy

$$\frac{dA_1}{dt} = (\sigma_1 + \Delta)A_1 + c_1A_1A_2 - A_1(c_2A_1^2 + c_3A_2^2), \tag{3.23a}$$

$$\frac{dA_2}{dt} = \sigma_2A_2 - c_4A_1^2 - A_2(c_5A_1^2 + c_6A_2^2). \tag{3.23b}$$

Here the c_i are (computable) coefficients, σ_i are the growth rates of the pure linear-theory modes, both of which vanish at $a = a_A$, $M^{(\perp)} = M_L^{(\perp)}$, and Δ is a splitting parameter whose sign shifts consideration to either $a < a_A$ or $a > a_A$. The two individual bifurcation points are thus at $\delta M = 0$ and $\delta M = \Delta$.

For $a < a_A$, as $M^{(\perp)}$ is increased through $M_L^{(\perp)}$, a mixed $m = 1$–2 mode bifurcates, followed by a transition to a pure $m = 2$ mode at $\delta M > \Delta$, beyond the second critical value of linear theory. In contrast, linear theory predicts that there would be only an $m = 1$ mode at this value of a. For $a > a_A$, as $M^{(\perp)}$ is increased through $M_L^{(\perp)}$, a pure $m = 2$ mode bifurcates. This can persist, or alternatively a *time-periodic* (Hopf) secondary bifurcation can occur, leading to stable oscillatory convection at $\delta M < \Delta$.

When $a \approx a_B$, the modes $m = 2$ and $m = 0$ compete. Here various dynamic hysteresis and jump phenomena are predicted, and the predicted behavior is quite complex. Clearly, if such predicted phenomena do occur in experiment, there is a very rich catalogue of behavior left to be examined.

3.6 *Experiments*

There have only been a handful of quantitative experiments on Marangoni convection.

Bénard (1900) observes convection in a layer of spermaceti just beyond critical conditions, measures the wave number of the convection, observes that the convection is hexagonal, and notes that the flow is upward under surface depressions and downward under surface elevations. The wave number corresponding to Bénard's experiments is approximately 2.2, a value consistent with the linear theory of Marangoni convection. The sense of the surface deflection is in agreement with theory due to Davis & Segel (1963) and Scriven & Sternling (1964).

Koschmieder (1967) conducts experiments in thin layers of silicone oil under a thin air gap, the layers being so thin that buoyancy-driven convection is negligible in both phases. He observes that the initial transient pattern in a dish is a set of concentric rings, but that these quickly break up into hexagons. Experiments in the same apparatus but with the liquid in contact with the upper rigid lid, thus eliminating Marangoni convection, always yield steady convection patterns in the form of concentric rings. Koschmieder measures the critical Marangoni number approximately and the wave number quite accurately. He finds fair agreement with the theory of Nield for the former quantity, and he obtains $\alpha \approx 2.0$.

Hoard et al. (1970) conduct experiments in silicone oil similar to those of Koschmieder and make visual observations of the form and wave number of the convective patterns, as well as estimates of the critical Marangoni and Rayleigh numbers. They also find the characteristic hexagonal pattern, together with good agreement between the measured and predicted values of the wave number and critical parameters. They further discuss the effect on the convection that occurs when there is relatively poor insulation of the vertical lateral boundaries of their apparatus.

Palmer & Berg (1971) and Pantaloni et al. (1979) report series of experiments in which both the fluid properties and the depth of a silicone-oil layer are varied over the range from buoyancy-driven to thermocapillary-driven convection. Good agreement with the theory is seen, especially in the experiments of the latter authors.

Ferm & Wollkind (1982) have made perhaps the most careful comparisons between the linear theory of two-layer Marangoni-Bénard instability and the experiments of Koschmieder (1967) and Palmer & Berg (1971); their results show that the agreement is quite good for several features though only qualitative for others.

Finally, Koschmieder & Biggerstaff (1986) reexamine the two-layer experiment using silicone oil and air and find for liquid layers thinner than 2 mm an anomalous convection structure at $M^{(\perp)}$ an order of magnitude lower than is predicted by Nield (1964). This pattern is replaced by the expected hexagonal structure when $M^{(\perp)}$ is near $M_L^{(\perp)}$. No convincing explanation of this behavior is given.

3.7 *Further Work*

Clearly, linear stability and energy theories on the laterally unbounded layer are in hand; the addition of further complicating effects can be introduced at the price of added complexity. Nonlinear theories using leading-order bifurcation methods are seemingly exhausted by the work of Cloot & Lebon (1984). However, the necessity of including higher-order terms when $M^{(\perp)} > O(\varepsilon)$ needs to be examined.

Marangoni convection in laterally bounded systems has barely been touched. The work of Rosenblat et al. (1982a,b) examines the case of "slippery" sidewalls. It would be extremely valuable to do the linearized problem for no-slip walls, even for $B = 0$, $C \to 0$, as long as non-axisymmetric modes are allowed. This determines the shapes of the neutral curves and in particular the sequence of interweavings analogous to those of Figure 4. Further, the bifurcation theories near double eigenvalues in such cases would determine whether the predictions of the above theories are reliable. In particular, the prediction of secondary bifurcation to time-periodic convection is untested.

All experiments done so far address only the first instability at $M_L^{(\perp)}$. There is a dire need for experiments at higher $M^{(\perp)}$, exploring further transitions even in a qualitative way. The explosion of interest in recent years in Bénard convection stems from phenomena uncovered in high-Ra experiments unexplained by the theory of the moment. The range of phenomena in Marangoni convection should be rich, since it includes not only those of the nonlinear dynamics of the bulk fluid but also those of an interface that might become contorted and even disjointed.

4. $[\nabla \bar{T}]_{\mathrm{EXT}} \parallel \mathscr{S}$: HYDROTHERMAL INSTABILITIES

4.1 *Basic State*

We consider the situation in which, from Equation (2.1), $b_\parallel \neq 0$ and $b_\perp = 0$, so that the liquid layer is heated horizontally, tangent to the mean position of the interface. Gravity is ignored, so that $\mathrm{Ra}^{(\parallel)} = G = 0$.

Suppose that the rigid plate of Figure 2 is a zero-heat-flux surface. The interface is cooled by air currents according to law (3.2), where we exert an axial temperature gradient by imposing the following condition on the air temperature (in dimensional form):

$$T_{\mathrm{AIR}}(x) = -b_\parallel x. \tag{4.1}$$

The imposed axial temperature field induces an axial surface-tension gradient that drives motion in the layer. We consider two basic states, as defined by Smith & Davis (1983a).

First, there is the *linear-flow basic state* in which the interface is flat,

$$\bar{\eta} = 0, \tag{4.2a}$$

$$\bar{\mathbf{v}} = (\bar{u}(z), 0, 0), \qquad \bar{u}(z) = z, \tag{4.2b}$$

the temperature is

$$\bar{T} = -x + \bar{\theta}(z), \qquad \bar{\theta}(z) = \frac{1}{6} M^{(\parallel)}(1 - z^3), \tag{4.2c}$$

the pressure gradient is

$$\bar{p}_x = 0, \tag{4.2d}$$

and

$$\bar{Q} = \frac{1}{2} M^{(\parallel)}. \tag{4.2e}$$

Here the velocity profile is linear, and the temperature field consists of the impressed axial nondimensional profile $(-x)$ plus the *flow-induced vertical structure* $\bar{\theta}(z)$, a balance between horizontal convection and vertical conduction; these are shown in Figure 5*a*. The linear-flow basic state is an *exact solution* of the governing system.

Second, there is the *return-flow basic state*, in which the interface is flat,

$$\bar{\eta} = 0, \tag{4.3a}$$

$$\bar{\mathbf{v}} = (\bar{u}(z), 0, 0), \qquad \bar{u}(z) = \frac{3}{4} z^2 - \frac{1}{2} z, \tag{4.3b}$$

the temperature is

$$\bar{T} = -x + \bar{\theta}(z), \qquad \bar{\theta}(z) = -\frac{1}{48} M^{(\parallel)}(3z^4 - 4z^3 + 1), \tag{4.3c}$$

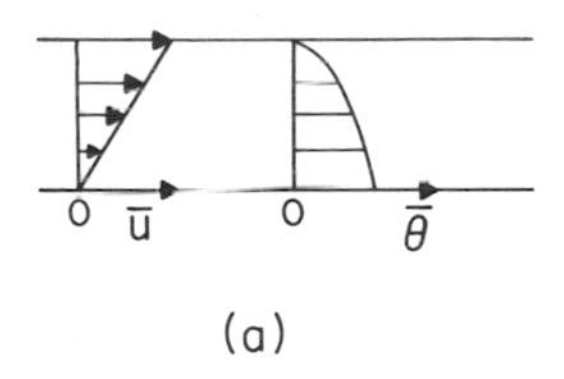

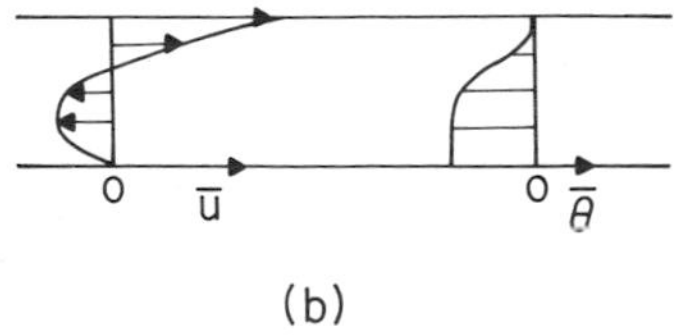

Figure 5 Sketches of the velocity and flow-induced temperature profiles for the (*a*) linear-flow basic state and (*b*) return-flow basic state.

the pressure gradient is

$$\bar{p}_x = \frac{3}{2}, \tag{4.3d}$$

and

$$\bar{Q} = 0. \tag{4.3e}$$

Here the velocity profile is quadratic, and the temperature field consists of the impressed axial profile $(-x)$ plus the *flow-induced vertical structure* $\bar{\theta}(z)$; these are shown in Figure 5*b*. The return-flow profile basic state is an *exact solution* for $C \to 0$.

The linear-flow state simulates an *open system*. For example, if the plate at $z = 0$ had finite extent, then near the center of the plate the solution (4.2) might hold while the fluid would exit the channel (at large positive x), recirculate below the plate, say, and then return through the open end (at large negative x).

The return-flow state has a nonzero pressure gradient, obtained by setting to zero the flow rate across any vertical section; it simulates the flow in a *closed system* (say, a long slot). Here the endwalls create the pressure gradient causing the near-surface flow to return near the bottom of the layer. The solution (4.3) represents the "core flow" away from the ends, where the flows turn. Sen & Davis (1982) use matched asymptotic expansions to obtain this core flow, the endwall-region corrections, and the matching of these. The approximation they use there is that $C = O(A^4)$, $R^{(\|)} = O(1)$, and $M^{(\|)} = O(1)$ as $A \to 0$. Here A is the aspect ratio of the slot. Note that large numerical values of $R^{(\|)}$ or $M^{(\|)}$ are allowed here.

Smith & Davis (1983a,b) have identified two distinct classes of instabilities associated with the above-defined basic states. There are convective instabilities that are driven by mechanisms within the bulk of the layer and that do not depend strongly on the deflection of the interface. There are surface-wave instabilities that depend intrinsically on the interaction of the flow and surface deflection. We discuss these separately.

4.2 *Linear Theory: Convective Instabilities–Linear Flow*

In this case we follow Smith & Davis (1983a), setting $C = 0$ and examining the basic state for instability using normal modes for any quantity $\phi(x, y, z, t)$ of the form

$$\Phi'(z) \exp\{i(k_1 x + k_2 y) + \sigma t\}, \tag{4.4a}$$

where

$$\sigma = v + i\omega \tag{4.4b}$$

is the complex growth rate.

In the present case we use the following parameters to characterize the system: $M^{(\parallel)}$, P, B, k_1, and k_2. Since these are linear theories, it follows that $\varepsilon \to 0$. Smith & Davis (1983a) find that increasing B delays the instability, as in the case of Marangoni convection. Hence, we discuss here only the limit $B \to 0$.

The critical conditions $M_L^{(\parallel)}$ versus P are obtained by minimization over (k_1, k_2). Figure 6 shows that the neutral curve consists of three parts: when $P > 1.60$, there is stationary convection in the form of longitudinal rolls $(k_1 = 0)$; when $0.60 < P < 1.60$, there are two-dimensional $(k_2 = 0)$ waves that travel downstream; when $P < 0.60$ there are oblique waves that are nearly longitudinal rolls $(k_1 \approx 0)$ traveling nearly cross stream, but with a component of phase speed directed opposite to the surface flow. The largest value of $M_L^{(\parallel)} = 21.3$ occurs at $P = 0.60$. Therefore, for $B = 0$ the linear flow is unstable for any P when $M^{(\parallel)} > 21.3$.

4.2.1 FLOW-INDUCED MARANGONI CONVECTION The static layer of Section 3 has symmetry for every (k_1, k_2) pair, and so the planform in Marangoni convection is undetermined by linear theory. The addition of shear in the basic state breaks this symmetry in favor of stationary longitudinal rolls when $P > 1.60$ and $B = 0$. Smith & Davis (1983a) find that $k_1 = 0$ is the only possible stationary mode.

The interpretation of this instability stems from the flow-induced vertical temperature distribution $\bar{\theta}(z)$ given in Equation (4.2c). This form makes the layer "heated from below." Let us write the linearized disturbance

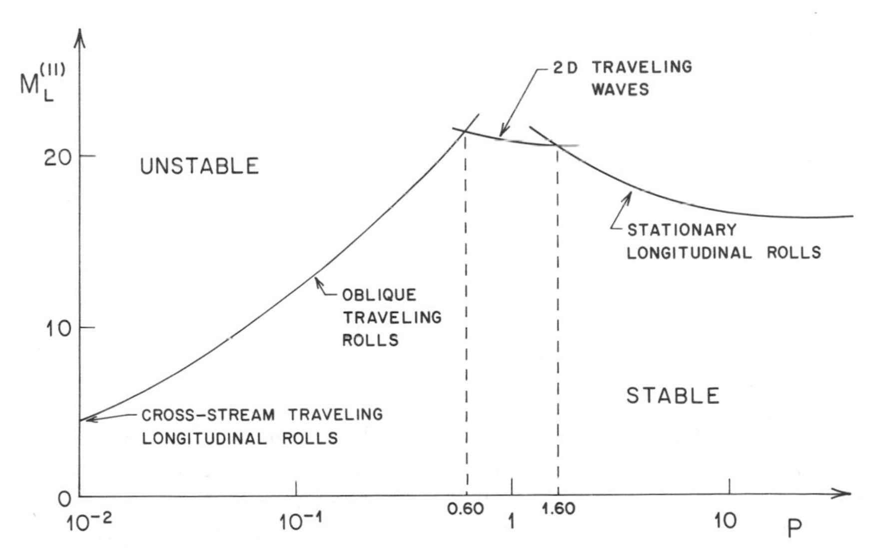

Figure 6 Critical curves ($B = 0$) for the linear-flow basic state (from Smith & Davis 1983a).

equations for longitudinal rolls in normal-mode form:

$$(L-\sigma M^{(\parallel)}P^{-1})U' = M^{(\parallel)}P^{-1}\frac{d\bar{u}}{dz}W', \tag{4.5a}$$

$$(L-\sigma M^{(\parallel)}P^{-1})LW' = 0, \tag{4.5b}$$

$$(L-M^{(\parallel)}\sigma)T' = M^{(\parallel)}\bar{T}_x U' + M^{(\parallel)}\bar{T}_z W', \tag{4.5c}$$

where

$$L = \frac{d^2}{dz^2} - k_2^2. \tag{4.5d}$$

For $P \to \infty$, $U' \to 0$, Equations (4.5) reduce to the equivalent of those of Pearson (1958), except that in nondimensional form the magnitude of the vertical temperature gradient in Equation (4.5c) is not unity. In the Pearson case (Figure 3 above), we have the equivalent of $M_L^{(\parallel)} \approx 48$. If we estimate that the profile $\frac{1}{6}M^{(\parallel)}(1-z^3)$ has the effective average gradient $\frac{1}{6}M^{(\parallel)}$, then Pearson's result, applied in the present case, should give

$$\frac{1}{6}[M_L^{(\parallel)}]^2 \approx 48, \tag{4.6a}$$

or

$$M_L^{(\parallel)} \approx 17. \tag{4.6b}$$

This is in excellent agreement with the computed value for $P \to \infty$ of Smith & Davis (1983a), namely $M_L^{(\parallel)} \approx 15.5$.

Now, when P is finite but $P \gg 1$, we see from Equation (4.5a) that

$$U' \sim -M^{(\parallel)}P^{-1}\frac{d\bar{u}}{dz}W', \tag{4.7}$$

where $d\bar{u}/dz \equiv 1$. When $W' > 0$, there is upflow beneath a surface hot line (along the x-direction). From relation (4.7), it follows that $U' < 0$ and $U' = O(P^{-1})$. This in turn produces a convective cooling, as seen from Equation (4.5c). Here we have $\bar{T}_x = -1$, $\bar{T}_z < 0$, so that the destabilizing term $M^{(\parallel)}\bar{T}_z W'$ is opposed by the stabilizing term $M^{(\parallel)}\bar{T}_x U'$. Thus, as shown in Figure 6, $M_L^{(\parallel)}$ increases as P is decreased from infinity. Smith & Davis (1983a) argue similarly that the stabilization as P decreases is accompanied by an increase of k_{2_L}.

Flow-induced Marangoni instability Our discussion on the mechanism for this instability paraphrases that of Smith & Davis (1983a). Assume that a disturbance creates a hot line L (compared with its neighbors) in the flow direction on $\mathscr{S}$. If surface tension decreases with temperature

($\gamma > 0$), there is a net surface traction away from L in the cross-stream directions. Since the fluid is viscous, subsurface fluid is dragged away from L. By conservation of mass, an upflow beneath L is created. This rising fluid, since it comes from below, has been warmed by the flow-induced vertical profile and reinforces the heat excess of the hot line, promoting sustained Marangoni convection. When $P \to \infty$ the fluid rises with zero downstream velocity perturbation, but when $P < \infty$ this perturbation is nonzero. The rising fluid moves into a region where the basic-state speed is higher; it moves the particle upstream from a cooler downstream location, cooling the hot line and opposing Marangoni convection. As P decreases from infinity, it strengthens its opposition, leading to larger $M_L^{(\|)}$.

4.2.2 HYDROTHERMAL INSTABILITY For $P \to 0$ the preferred modes are longitudinal rolls ($k_1 = 0$) that propagate in the cross-flow directions and have $M_L^{(\|)} \sim P^{1/2}$. As P is increased to $P = 0.60$, the axes of the rolls rotate slightly into oblique waves that propagate in directions pointing against the flow up to 7.5° to the cross-stream direction. Thus, for $P < 0.60$, they are almost transversely propagating. They arise from an instability that involves a transfer of energy from the imposed horizontal (axial) temperature gradient to the disturbances through perturbations in the horizontal velocity field.

Smith (1986) shows for small P that the inertia dominates the viscous forces. Equation (4.5a) shows that $U' < 0$ and is of magnitude $M^{(\|)}$ when $W' > 0$; the perturbation induced is an upstream component of velocity almost oppositely phased with T'. In Equation (4.5c) the term $M^{(\|)}\bar{T}_x U'$ dominates the term $M^{(\|)}\bar{T}_z W'$.

Hydrothermal instability at small P Physically this mechanism can be described as follows (Smith 1986). Assume that a disturbance creates a hot line L (compared with its neighbors) in the flow direction on $\mathscr{S}$. If the surface tension decreases with temperature ($\gamma > 0$), there is a net surface traction away from L in the cross-stream directions. Since the fluid is viscous, subsurface fluid is dragged away from L. By conservation of mass, an (almost in-phase) upflow beneath L is created. Since the fluid rises into a region where the basic-state speed is higher, it has an inertially driven (out-of-phase) negative downstream velocity component. This upstream moving particle has been cooled by the imposed axial temperature gradient and so lowers the heat excess in the hot line. In turn the cooling weakens the upflow, but the upstream velocity continues to increase, since there is still upflow. The fluid inertia causes the decreasing temperature at L to overshoot and become negative. The hot line L is now a cool line, and the process reverses through the thermocapillary effect. An increase in $M^{(\|)}$ strengthens the downstream velocity perturbation, leading to a maintained convective state through the extraction of energy from the imposed axial temperature gradient. The flow-induced vertical temperature gradient makes the layer "heated from below" on a static basis, but it is the mean horizontal velocity distribution that has the major effect on $M_L^{(\|)}$. As P is increased from zero, viscous effects increase and so a larger $M^{(\|)}$ is required to maintain the heat excess of the hot line. This is associated with an increased k_L.

For $0.60 < P < 1.60$, two-dimensional hydrothermal waves become the preferred mode. This instability is related to the *Hydrothermal instability at large P* discussed below.

4.3 *Linear Theory: Convective Instabilities–Return Flow*

The critical conditions $M_L^{(\|)}$ versus P are obtained by minimization over (k_1, k_2). Figure 7 shows a smooth neutral curve that for $P \to 0$ has $M_L^{(\|)} \sim P^{1/2}$ and $k_1 \to 0$; it is a longitudinal roll propagating cross stream. As P is increased, $M_L^{(\|)}$ increases and the rolls have axes that rotate into oblique waves; for $P \to \infty$ the waves are nearly two-dimensional ($k_2 \ll 1$), propagating in the directions 7.90° from the upstream direction, and we have $M_L^{(\|)} \to 398$.

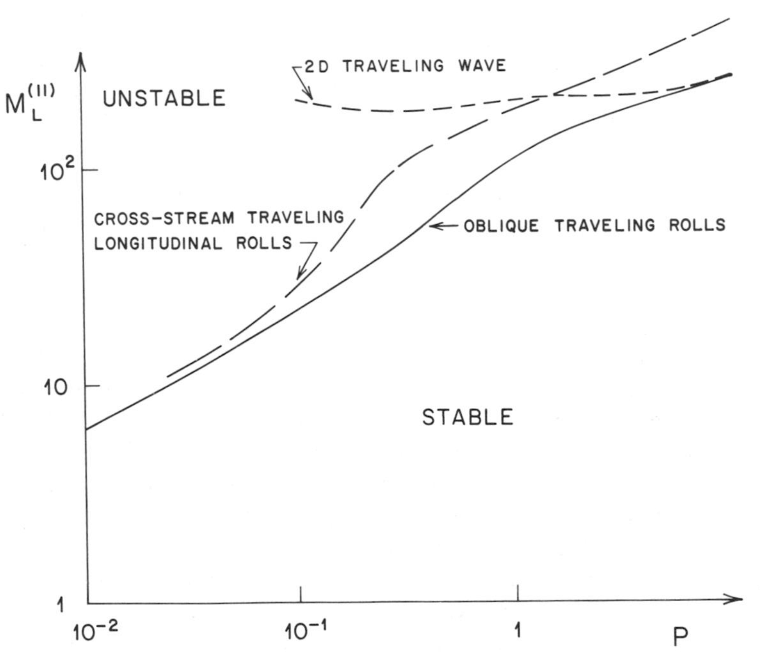

Figure 7 Critical curves ($B = 0$) for the return-flow basic state (from Smith & Davis 1983a).

4.3.1 FLOW-INDUCED MARANGONI CONVECTION The flow-induced vertical distribution $\bar{\theta}(z)$ given in Equation (4.3c) corresponds to the layer being "cooled from below." Thus, no stationary convective instability is present.

4.3.2 HYDROTHERMAL INSTABILITY We saw earlier for the linear flow for $P \ll 1$ that instability arises from a transfer of energy to the disturbances from the imposed horizontal temperature field through a horizontal-convection mechanism. For small P, the return flow is susceptible to the same instability. Now, however, the flow-induced vertical gradient opposes this, so that a slightly larger value of $M^{(\|)}$ is required here to sustain the instability than was the case for the linear-flow profile. More importantly, the instability is retarded here by the alteration of the velocity profile (Smith 1986), since $|d\bar{u}/dz|$ is smaller here than it was in the linear-flow case.

For large values of P, energy transfer from the vertical flow-induced temperature distribution to the disturbances through vertical convection becomes the dominant mechanism of instability. For $P \to \infty$, let us consider a hot line L oriented cross stream to the flow along the y-axis. Figure 8 shows the instantaneous flow lines; these are centered near the interface. The hot line induces thermocapillary stresses that cause flow upstream and downstream as shown. A vertical upflow beneath L is induced by conservation of mass. This upflow cools L as a result of the flow-induced vertical profile shown. The warm fluid leaves L on $\mathscr{S}$, moves upstream to point a, and descends toward point b. As is shown in Equation (4.5c), this flow is most effective in heating a point b near the maximum of $\bar{\theta}_z$. Calculations (Smith 1986) show that the temperature at point b can be nearly 20 times that at L. Vertical conduction then strongly heats point a, and the hot spot effectively moves upstream, consistent with the results of Smith & Davis (1983a). We can summarize this argument, taken from Smith (1986), as follows:

Hydrothermal instability at large P Assume that a disturbance creates a hot line L (compared with its neighbors) cross stream to the flow direction

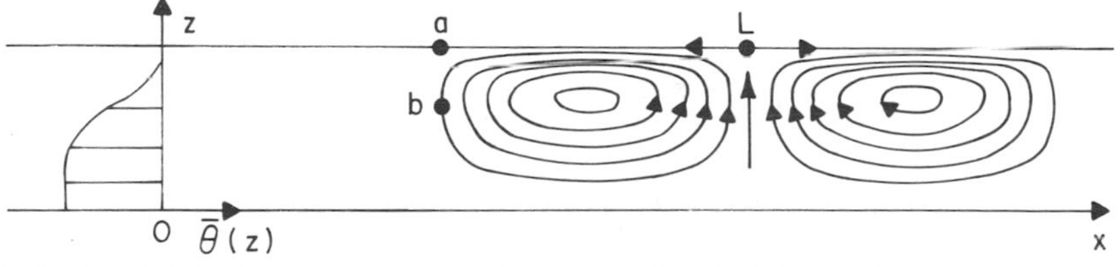

Figure 8 Sketch of the streamlines and flow-induced vertical temperature profiles for the one-layer system with the return-flow basic state.

on $\mathscr{S}$. If the surface tension decreases with temperature ($\gamma > 0$), there is a net surface traction away from L in the upstream and downstream directions. Since the fluid is viscous, subsurface fluid is dragged away from L. By conservation of mass, an upflow beneath L is created. If the original disturbance has closed streamlines, the downward flow leaving $\mathscr{S}$ upstream of L causes intense subsurface heating near the maximum of the gradient of the flow-induced vertical temperature profile. This elevated temperature internal to the layer produces a large conductive heating of $\mathscr{S}$ upstream of L. Thus, L effectively moves upstream. The heating upstream of L is proportional to $M^{(\|)}$, so that instability is maintained for large enough $M^{(\|)}$. The energy for the instability comes from the flow-induced vertical temperature field, which on a static basis is stabilizing.

By the above mechanism, $\bar{T}_z$ determines the value of $M_{\mathrm{L}}^{(\|)}$ and the distance upstream that the hot line travels. Both of these should decrease as $\bar{T}_z \to 0$; thus, in the linear-flow case in which $\bar{T}_z < 0$, $M_{\mathrm{L}}^{(\|)}$ is quite small and the hot line is induced to move downstream, consistent with the preferred mode for $0.60 < P < 1.60$.

Xu & Davis (1984) consider a return-flow profile for a circular-cylindrical geometry in which an axial temperature gradient is imposed. For $C \to 0$ they find convective instabilities both qualitatively and quantitatively similar to those discussed above.

4.4 *Linear Theory: Surface-Wave Instabilities–Linear Flow*

In this case we follow Smith & Davis (1983b) and examine the linearized problem ($\varepsilon \to 0$) with $C \neq 0$. Since the instability is hydrodynamic (and not thermal) in nature, we use the parameters $R^{(\|)}$, P, B, S, and (k_1, k_2).

Since surface deflection is important, we retain this effect and list the two-dimensional linearized interfacial boundary conditions applicable on $z = 1$:

$$w' = \eta'_t + \bar{u}(1)\eta'_x, \tag{4.8a}$$

$$\mathbf{e}_3 \cdot \mathbf{T} \cdot \mathbf{e}_3 = [S(R^{(\|)})^{-1} - \bar{T}(x,1)]\eta'_{xx} + 2\frac{d\bar{u}}{dz}\eta'_x, \tag{4.8b}$$

$$\mathbf{e}_1 \cdot \mathbf{T} \cdot \mathbf{e}_3 = -\frac{d^2\bar{u}}{dz^2}(1)\eta' + \bar{T}_z(x,1)\eta'_x + T'_x, \tag{4.8c}$$

$$T'_z + B[\bar{T}_z(x,1)\eta' + T'] + \bar{T}_{zz}(x,1)\eta' - \bar{T}_x(x,1)\eta'_x = 0. \tag{4.8d}$$

The key limit here is $B \to \infty$, $P \to 0$, which from condition (4.8d), relationship (2.3e), and Equation (2.4b) shows that the thermal field decouples from the hydrodynamic field, i.e. $T' = 0$; this is the *isothermal limit* of the system. The basic-state shear flow is set up by thermocapillarity, but the instability is a purely isothermal one, directly analogous to that identified by Miles (1960) and corrected and extended by Smith & Davis (1982). In this limit, a "wind stress," equal to the thermocapillary gradient, sets up a flow. A variant of Squire's theorem focuses attention on two-dimensional traveling waves that lead to a critical value $R_L^{(\|)}$ of the Reynolds number. This limiting case typifies the results for a wide range of parameters. For example, if $S \geqq 10^4$, the change in $R_L^{(\|)}$ with B is less than 2% for any P, and for $B = 0$, $R_L^{(\|)}$ changes by less than 7.5% for $P \leqq 10$. When $P \leqq 10^{-2}$, $R_L^{(\|)}$ is constant to within 2% of any S. Figure 9 shows a typical case with the nonisothermal and isothermal cases compared.

Note two restrictions. Firstly, aside from the limiting case $B \to \infty$, $P \to 0$, the system is nonisothermal, and Squire's theorem does not hold, but Smith & Davis (1983b) still consider only two-dimensional disturbances. Secondly, the term in the normal-stress boundary condition $S[R^{(\|)}]^{-1} - \bar{T}(x, 1) = S[R^{(\|)}]^{-1} + x$ precludes the use of normal modes. Smith & Davis (1983b) replace this term by $S[R^{(\|)}]^{-1}$ and argue that the resulting normal-mode predictions are valid when "the surface tension does not vary much over a characteristic wavelength of a disturbance." Thus, the resulting wavelength λ must be much shorter than $S[R^{(\|)}]^{-1}$, i.e.

$$\lambda \ll S[R^{(\|)}]^{-1}. \tag{4.9}$$

4.5 *Linear Theory: Surface-Wave Instabilities–Return Flow*

The return-flow basic state, unlike the linear-flow state, is susceptible to long-wave instabilities. In fact for $k_1 \to 0$, $B = O(1)$, we have

$$R_L^{(\|)} \sim (k_1^2 S)^{1/2}. \tag{4.10}$$

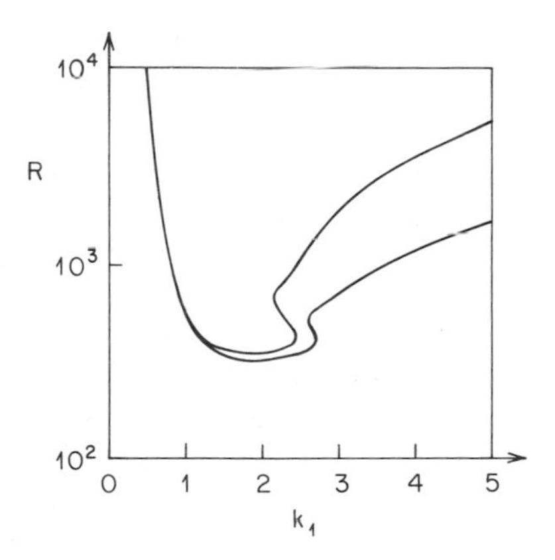

Figure 9 Neutral curves for the linear flow with $S = 10^4$. The upper curve is for the isothermal problem of Smith & Davis (1982), and the lower curve is for the thermocapillary layer with $P = 0$ and $B = 0$ (from Smith & Davis 1983b).

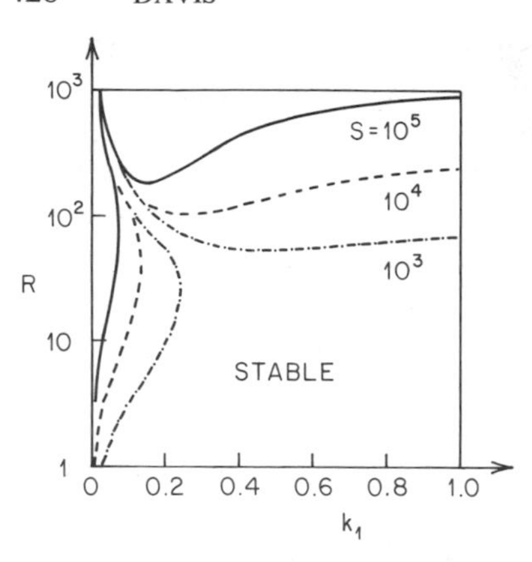

Figure 10 Neutral curves for the return flow with $P = 1$, $B = 1$, and various values of S (from Smith & Davis 1983b).

Thus, when $S = O(1)$, it follows that $R_L^{(\|)} \sim 0$; and if S is so large that $k_1^2 S = O(1)$, there is a nonzero critical value.

When the neutral curves are extended by numerically solving the disturbance system, an extremely complicated structure is found. Figure 10 shows the two-branched structure typically obtained. Recall that restriction (4.9) excludes the small-k_1 portion of this figure from having physical validity.

Xu & Davis (1985) consider a return-flow profile for a circular-cylindrical geometry in which an axial temperature gradient is imposed. For $C \ll 1$ but $C \neq 0$ they find that the axial flow suppresses the capillary instabilities (that would lead to break-up), so that coherent jets can be longer than is possible in the isothermal case. These jets are susceptible to surface-wave instabilities.

4.6 *Nonlinear Theory*

There has been no energy-theory analysis for systems having $[\nabla \bar{T}]_{\text{EXT}} \| \bar{\mathscr{S}}$.

The only bifurcation theory is that of Smith (1985). He considers convective instabilities of the return-flow basic state. These hydrothermal instabilities are studied with $C = B = 0$ and $10^{-3} < P < 10$. By the linear theory of Section 4.3, there are always two oblique waves that propagate with components in the direction opposite to the surface flow; one propagates left (cross stream) and one propagates right. If we write the linear-theory component of velocity w_1 as

$$w_1 \sim A_L(t) f_L(z) e^{i(k_1 x + k_2 y)} + A_R(t) f_R(z) e^{i(k_1 x - k_2 y)}, \tag{4.11}$$

then $(k_1, \pm k_2)$ are the critical values of the wave numbers, and the $f(z)$ are the (computable) eigenfunctions. Smith (1985) finds that

$$\frac{dA_L}{dt} = \sigma A_L - A_L(c_1 A_L^2 + c_2 A_R^2), \tag{4.12a}$$

$$\frac{dA_R}{dt} = \sigma A_R - A_R(c_2 A_L^2 + c_1 A_R^2), \tag{4.12b}$$

where the c_i are (computable) complex coefficients, and σ is the linear-theory growth rate of the modes. By analyzing system (4.12), Smith finds that (*a*) the instabilities are always *supercritical*, and (*b*) the only stable supercritical states are the pure modes $A_L = 0$, $A_R \neq 0$ or $A_L \neq 0$, $A_R = 0$. All mixed modes, including the cross-stream standing-wave case, are unstable.

Thus, the nonlinear theory fixes the relative phases of the linearized modes by determining the selected linear combination of the left- and right-propagating modes. In the Smith (1985) analysis, only pure modes are stable in the laterally unbounded layer. If lateral boundaries were present, it might happen that the preferred mode would be waves standing in the cross-stream direction but propagating against the surface flow.

4.7 *Discussion*

When $[\nabla \bar{T}]_{EXT} \| \bar{\mathscr{S}}$, surface-tension gradients drive shear flows. These shear flows induce vertical temperature profiles whose structure controls the instability characteristics of the system.

The flow-induced temperature distribution for the linear-flow state is "heated from below," giving rise to flow-induced steady Marangoni convection at large P in the form of longitudinal rolls. This instability is absent in the return-flow state, since there the flow-induced temperature distribution is "cooled from below."

The imposed tangential temperature gradient is the source of energy for hydrothermal instabilities when P is small. For $P \to 0$ these are longitudinal waves that propagate cross stream. As P is increased, the axes of the rolls rotate and then have a phase velocity with a component directed opposite to the free-surface flow. In the return-flow case, the axes rotate nearly 90° as $P \to \infty$, and these waves derive their energy from the "statically stable" vertical distribution of temperature. In the linear-flow case, the axes rotate only slightly and give way to two-dimensional waves at P near unity; these waves derive their energy from the vertical thermal structure. At high P, the flow-induced Marangoni convection preempts the hydrothermal waves.

The flow-induced vertical temperature field also serves as the gross measure of instability. For example, in the "statically unstable" linear-flow case we have $M_L^{(\|)} \approx 15.5$ for $P \to \infty$, whereas in the "statically stable" return-flow case we have $M_L^{(\|)} \approx 398$ for $P \to \infty$.

The surface-wave instability is most prominent at low P. It takes the form of, presumably, two-dimensional traveling waves.

The surface-wave instabilities are hydrodynamic in nature and thus are measured by $R^{(\parallel)}$, not $M^{(\parallel)}$. The convective instabilities are measured by $M^{(\parallel)}$, not $R^{(\parallel)}$. Except for the cases of "statically unstable" flow-induced vertical temperature profiles at large P, the instabilities are all time periodic, not steady. Except for narrow ranges and for, presumably, surface-wave instabilities, the instabilities are three dimensional.

Clearly, there is a wealth of new areas to pursue. How sensitive are these instabilities to changes in flow conditions, to heating conditions, or to gravity effects? There are two values of P shown in Figure 6 where the linear-flow state has two coexisting instabilities. If these interact nonlinearly, the selection process may shed light on further transitions at large $M^{(\parallel)}$. The nonlinear theory is nearly unexplored.

There are no present quantitative experiments for the case of $[\nabla \bar{T}_{\mathrm{EXT}}] \parallel \bar{\mathscr{S}}$, neither for the initial instability nor for higher transitions.

5. GENERALIZATIONS

In this section we wish to generalize on the systems discussed earlier.

5.1 $[\nabla \bar{T}]_{\mathrm{EXT}}$ *Oblique to* $\bar{\mathscr{S}}$

Consider now the one-layer model subject to an *arbitrary* external gradient in the form (2.1). If $b_{\parallel}$ and $b_{\perp}$ are arbitrary, then the corresponding Marangoni numbers $M^{(\parallel)}$ and $M^{(\perp)}$ are independent parameters.

In this case the basic state we seek has a planar interface,

$$\bar{\eta} = 0. \tag{5.1a}$$

The tangential heating will induce an axial flow of the form

$$\bar{\mathbf{v}} = (\bar{u}(z), 0, 0), \tag{5.1b}$$

where $\bar{u}(z)$ depends on the boundary conditions and so may have the form (4.2b) or (4.3b). There is a temperature field of the form (4.2c) or (4.3c),

$$\bar{T} = -x + \bar{\theta}_{\parallel}(z), \tag{5.1c}$$

where $\bar{\theta}_{\parallel}$ represents the flow-induced vertical temperature field.

In addition the normal heating will induce a vertical temperature profile of the form (3.4c),

$$\bar{T} = \bar{\theta}_{\perp}(z), \tag{5.1d}$$

which, because of linearity, does not interact with that of Equation (5.1c); the full temperature distribution is

$$\bar{T} = -x + \bar{\theta}_{\perp}(z) + \bar{\theta}_{\parallel}(z). \tag{5.1e}$$

Likewise, the vertical pressure gradients superpose, and the velocity profiles are those of the tangential heating.

The stability characteristics of this basic state can in part be presumed by examining the vertical temperature profile $\bar{\theta}_{\perp}(z)+\bar{\theta}_{\parallel}(z) \equiv \bar{\theta}(z)$. If $d\bar{\theta}/dz < 0$, then the fluid is "heated from below," and flow-induced Marangoni convection should appear for large enough P. In fact, for $P \to \infty$, we saw in Section 4.2.1 how the Pearson (1958) result can be used for the case $M^{(\perp)} = 0$. We can here generalize this result for the linear flow with additional ($M^{(\perp)} > 0$) heating from below to the form

$$\frac{1}{6}[M_{\mathrm{L}}^{(\parallel)}]^2 + M_{\mathrm{L}}^{(\perp)} \approx 48, \tag{5.2}$$

where we assume that the lower plate is a prescribed heat-flux surface. Figure 11 shows the estimated regions of stability. Smith (1986, private communication) finds for $M^{(\perp)} = 0$ that

$$M_{\mathrm{L}}^{(\parallel)} \approx 15.5(1+\alpha P^{\beta}). \tag{5.3}$$

When $\alpha = 0.455$, $\beta = -0.808$, this expression correlates well with the results of Smith & Davis (1983a) for $P \geqq 1$. Since the pure Marangoni instability is independent of P, then we might have the law

$$\frac{1}{6}[M_{\mathrm{L}}^{(\parallel)}]^2 + M_{\mathrm{L}}^{(\perp)} \approx 48(1+\alpha P^{\beta})^2 \tag{5.4}$$

governing the combined cases. Note that formulas (5.2) and (5.4) are obtained here only for $M_{\mathrm{L}}^{(\perp)} > 0$. However, they should be extendable to the range $M_{\mathrm{L}}^{(\perp)} < 0$, shown in Figure 11, since the extra vertical temperature field merely retards the "unstable" stratification.

The hydrothermal instabilities at large P draw their energy from the vertical temperature distribution. If one adjusts $M^{(\perp)}$ to give strong stability against flow-induced Marangoni convection, then it promotes hydrothermal instabilities. However, as shown in Section 4, these instabilities occur at values of $M^{(\parallel)}$ an order of magnitude higher than the values at which steady convection would occur.

The hydrothermal instabilities at small P draw their energy from the tangential temperature distribution; this profile is nearly unaltered by an extra heating from below, so these instabilities should not be too sensitive to such alteration.

The surface-wave instabilities are, to a large measure, unaffected by the

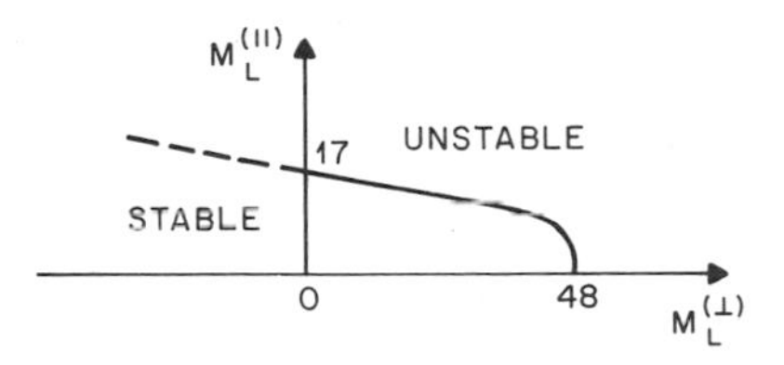

Figure 11 Critical vertical Marangoni–horizontal Marangoni number curves estimated from the linear theories.

thermal field for $P \to 0$. They would not change much as $M^{(\perp)}$ is increased. However the effect of larger P could be appreciable; there is not enough information available to know.

5.2 *Nonplanar* $\bar{\mathscr{S}}$

Consider a cylindrical container of radius a heated from below (Figure 12*a*). If the vessel is a rigid boundary and the contact angle is 90°, then there is a static basic state having a planar interface.

When the contact angle is, say, less than 90°, there is a meniscus at a wall (Figure 12*b*) of dimension $G^{-1/2}a$, where G is the Bond number given in Equation (3.11). If the walls are insulators, say, this curved interface will have nonuniform temperature and hence nonuniform surface tension. Thermocapillarity will induce a meniscus convection, as shown. Clearly, when $G \gg 1$, this is a relatively small, local effect. When $M^{(\perp)}$ is small, its only gross dynamical consequence would be the creation of a very small enhancement of heat transport over that of pure conduction. It can, however, round the instability curves so as to produce an imperfect bifurcation. With no meniscus, the case of axisymmetric convection would be a steady solution of an equation of the form (3.21), as shown on the skewed parabola of Figure 13. When meniscus convection is present, an altered curve would be expected. Now if $G \leqq 1$, as might be possible in outer space and as shown in Figure 12*c*, the meniscus would fill the whole container, and the imperfection would be very large. The relation between this problem and that of Figure 12*a* would be tenuous. In a certain sense this problem is closer to one in which the interface is heated along its boundary.

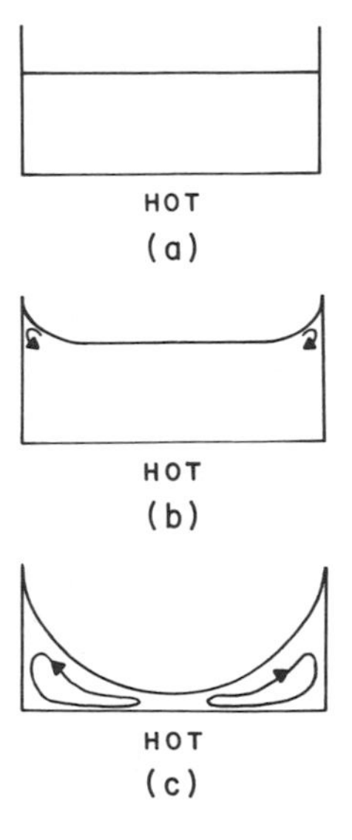

Figure 12 Sketch of a partially filled cylinder heated from below in which (*a*) the contact angle is 90°, (*b*) the contact angle is less than 90° and the Bond number is large, and (*c*) the contact angle is less than 90° and the Bond number is small.

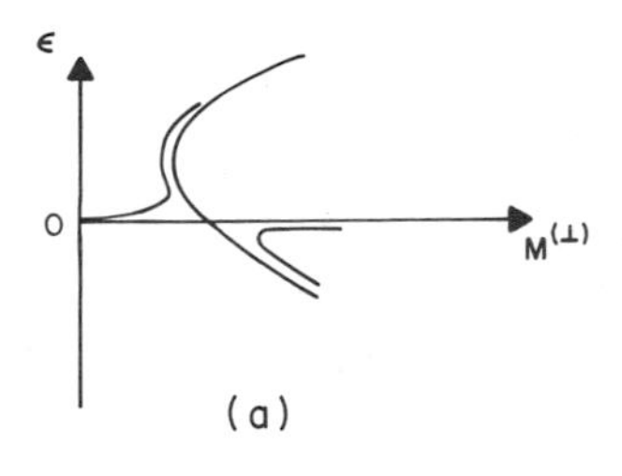

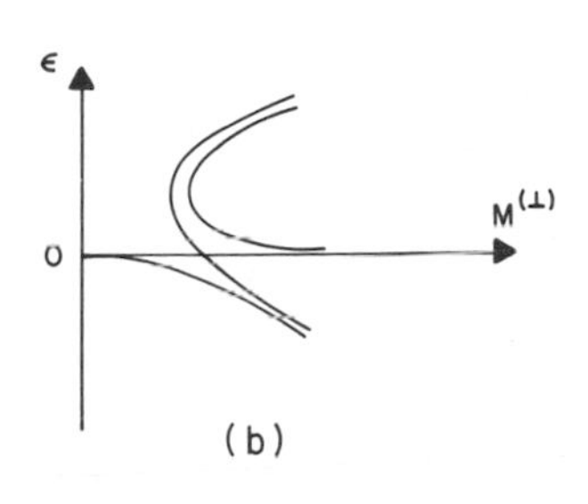

Figure 13 Sketch of the bifurcation to axisymmetric convection when imperfections are absent (the skewed parabolas) and when they are present (the disconnected curves). Figures (*a*) and (*b*) indicate the difference between thermal imperfections of different signs.

In fact, away from the walls the radial velocity would have a return-flow character and, perhaps, be susceptible to hydrothermal instabilities.

Notice that in the above discussion we have addressed only the axisymmetric convection state. On general grounds (see Tavantzis et al. 1978) one would expect imperfections only when the meniscus flow and the principal eigenfunction of the linearized Marangoni problem (for a planar interface) are not orthogonal. Orthogonality would occur for all azimuthal wave numbers $m \neq 0$ when the meniscus is axisymmetric.

6. CONCLUSIONS

In this article we have focused narrowly on thermocapillary instabilities in planar layers. What we learn of the mechanics and behavior here is applicable to more complicated systems, as discussed somewhat in the previous section.

There are a number of excellent sources of information on other features of variable surface-tension effects. Among these are Sternling & Scriven (1959), Levich (1962), Levich & Krylov (1969), Kenning (1968), Sørensen (1979), Velarde & Castillo (1981), and Normand et al. (1977).

Acknowledgments

I would like to express my appreciation to my colleague Prof. M. K. Smith. His keen insights have enlivened the subject, and his critical abilities have enhanced this article.

This work was supported by the National Science Foundation, Fluid Mechanics Program.

Literature Cited

Bénard, H. 1900. Les tourbillons cellulaires dans une nappe liquide. *Rev. Gén. Sci. Pures Appl.* 11: 1261–68

Cloot, A., Lebon, G. 1984. A nonlinear stability analysis of the Bénard-Marangoni problem. *J. Fluid Mech.* 145: 447–69

Davis, S. H. 1969. Buoyancy–surface tension instability by the method of energy. *J. Fluid Mech.* 39: 347–59

Davis, S. H. 1983. Rupture of thin liquid films. In *Waves on Fluid Interfaces*, ed. R. E. Meyer, pp. 291–302. New York: Academic

Davis, S. H., Homsy, G. M. 1980. Energy stability theory for free surface problems: buoyancy-thermocapillary layers. *J. Fluid Mech.* 98: 527–53

Davis, S. H., Segel, L. A. 1963. Surface elevation in Bénard cells. *Am. Math. Soc. Notices* 10: 496

Ferm, E. N., Wollkind, D. J. 1982. Onset of Rayleigh-Bénard-Marangoni instability: comparison between theory and experiment. *J. Non-Equilib. Thermodyn.* 7: 169–90

Hoard, C. Q., Robertson, C. R., Acrivos, A. 1970. Experiments on the cellular structure in Bénard convection. *Int. J. Heat Trans.* 13: 849–56

Jeffreys, H. 1951. The surface elevation on cellular convection. *Q. J. Mech. Appl. Math.* 4: 283–88

Joseph, D. D. 1965. On the stability of the Boussinesq equations. *Arch. Ration. Mech. Anal.* 20: 59–71

Kenning, D. B. R. 1968. Two-phase flow with nonuniform surface tension. *Appl. Mech. Rev.* 21: 1101–11

Koschmieder, E. L. 1967. On convection under an air surface. *J. Fluid Mech.* 30: 9–15

Koschmieder, E. L., Biggerstaff, M. I. 1986. Onset of surface-tension-driven Bénard convection. *J. Fluid Mech.* 167: 49–64

Kraska, J. R., Sani, R. L. 1979. Finite amplitude Bénard-Rayleigh convection. *Int. J. Heat Mass Transfer* 22: 535–46

Lebon, G., Cloot, A. 1982. Buoyancy and surface-tension driven instabilities in presence of negative Rayleigh and Marangoni numbers. *Acta Mech.* 43: 141–58

Levich, V. G. 1962. *Physicochemical Hydrodynamics*. Englewood Cliffs, NJ: Prentice-Hall

Levich, V. G., Krylov, V. S. 1969. Surface-tension-driven phenomena. *Ann. Rev. Fluid Mech.* 1: 293–316

Liang, S. F., Vidal, A., Acrivos, A. 1969. Buoyancy-driven convection in cylindrical geometries. *J. Fluid Mech.* 36: 239–56

Miles, J. W. 1960. The hydrodynamic stability of a thin film of liquid in uniform shearing motion. *J. Fluid Mech.* 8: 593–610

Nield, D. A. 1964. Surface tension and buoyancy effects in cellular convection. *J. Fluid Mech.* 19: 341–52

Normand, C., Pomeau, V., Velarde, M. G. 1977. Convective instability: a physicist's approach. *Rev. Mod. Phys.* 49: 581–624

Palmer, H. J., Berg, J. C. 1971. Convective instability in liquid pools heated from below. *J. Fluid Mech.* 47: 779–87

Pantaloni, J., Bailleux, R., Salan, J., Velarde, M. G. 1979. Rayleigh-Bénard instability: new experimental results. *J. Non-Equilib. Thermodyn.* 4: 201–18

Pearson, J. R. A. 1958. On convection cells induced by surface tension. *J. Fluid Mech.* 4: 489–500

Rosenblat, S., Davis, S. H., Homsy, G. M. 1982a. Nonlinear Marangoni convection in bounded layers. Part 1. Circular cylindrical containers. *J. Fluid Mech.* 120: 91–122

Rosenblat, S., Homsy, G. M., Davis, S. H. 1982b. Nonlinear Marangoni convection in bounded layers. Part 2. Rectangular cylindrical containers. *J. Fluid Mech.* 120: 123–38

Scanlon, J. W., Segel, L. A. 1967. Finite amplitude cellular convection induced by surface tension. *J. Fluid Mech.* 30: 149–62

Scriven, L. E., Sternling, C. V. 1964. On cellular convection driven by surface tension gradients: effect of mean surface tension and viscosity. *J. Fluid Mech.* 19: 321–40

Sen, A. K., Davis, S. H. 1982. Steady thermocapillary flows in two dimensional slots. *J. Fluid Mech.* 121: 163–84

Smith, K. A. 1966. On convective instability induced by surface tension gradients. *J. Fluid Mech.* 24: 401–14

Smith, M. K. 1985. The nonlinear stability of thermocapillary shear layers. *Bull. Am. Phys. Soc.* 30: 1732

Smith, M. K. 1986. Instability mechanisms in dynamic thermocapillary liquid layers. *Phys. Fluids* 29: 3182–86

Smith, M. K., Davis, S. H. 1982. The instability of sheared liquid layers. *J. Fluid Mech.* 121: 187–206

Smith, M. K., Davis, S. H. 1983a. Instabilities of dynamic thermocapillary liquid layers. Part 1. Convective instabilities. *J. Fluid Mech.* 132: 119–44

Smith, M. K., Davis, S. H. 1983b. Instabilities of dynamic thermocapillary liquid layers. Part 2. Surface-wave instabilities. *J. Fluid Mech.* 132: 145–62

Sørensen, T. S. 1979. *Dynamics and Instability of Fluid Interfaces, Lecture Notes in Physics*, Vol. 105. Berlin: Springer-Verlag

Sternling, C. V., Scriven, L. E. 1959. Interfacial turbulence: hydrodynamic instability and the Marangoni effect. *AIChE J.* 5: 514–23

Tavantzis, J., Reiss, E. L., Matkowsky, B. J. 1978. On the smooth transition to convection. *SIAM J. Appl. Math.* 34: 322–37

Velarde, M. G., Castillo, J. L. 1981. *Convection Transport and Instability Phenomena*, eds. J. Zierep, H. Oertel Jr. Karlsruhe: Braun Verlag

Xu, J.-J., Davis, S. H. 1984. Convective thermocapillary instabilities in liquid bridges. *Phys. Fluids* 27: 1102–7

Xu, J.-J., Davis, S. H. 1985. Instability of capillary jets with thermocapillarity. *J. Fluid Mech.* 161: 1–26

Zeren, R. W., Reynolds, W. C. 1972. Thermal instabilities in two-fluid horizontal layers. *J. Fluid Mech.* 53: 305–27

Ann. Rev. Fluid Mech. 1987. 19: 437–63

MAGNETIC FLUIDS

Ronald E. Rosensweig

Corporate Research Science Laboratories, Exxon Research and Engineering Company, Clinton Township, Annandale, New Jersey 08801

INTRODUCTION

A colloidal magnetic fluid, or ferrofluid, consists typically of a suspension of monodomain ferromagnetic particles such as magnetite (size about 100 Å) in a nonmagnetic carrier fluid. A surfactant covering the particles prevents particle-to-particle agglomeration, and Brownian motion prevents particle sedimentation in gravitational or magnetic fields. The number of ferromagnetic particles in 1 cm^3 may be as great as 10^{18}; by comparison, 1 cm^3 of air under standard conditions contains 2.7×10^{19} molecules. The ferrofluids are attracted strongly by magnetic fields with forces that easily overcome gravity (see Figure 1).

The continuum description of magnetic-fluid flow, termed ferrohydrodynamics (FHD), has developed since its introduction over 20 years ago (Neuringer & Rosensweig 1964) as a field of study similar to magnetohydrodynamics and electrohydrodynamics (Shaposhnikov & Shliomis 1975, Rosensweig 1985). However, the field forces arise not from current flow or the presence of free charge but from magnetically polarizable matter subjected to applied magnetic field. A bibliography lists 822 papers and patents appearing in the period 1980–83 (Charles & Rosensweig 1983).

MAGNETIC-FLUID BEHAVIOR

Theory admits the possibility of ferromagnetism in a homogeneous liquid, but no substance is known whose Curie temperature exceeds its melting point. As a result of this situation, a size-reduction process for production of colloidal ferrofluid was devised (Papell 1965) and then further refined (Rosensweig et al. 1965, Kaiser & Rosensweig 1967). Khalafalla & Reimers (1973b, 1974) developed chemical precipitation syntheses suited for rapid production. Proprietary ferrofluids are produced in the United States by

0066–4189/87/0115–0437$02.00

Ferrofluidics Corporation, Nashua, New Hampshire. English, French, and Dutch activities in ferrofluids have been reviewed by Charles & Popplewell (1980), Martinet (1983), and Scholten (1983), respectively.

Magnetization of Ferrofluid

The particles in a colloidal ferrofluid, each with its embedded magnetic moment, are analogous to molecules of a paramagnetic gas. In the absence of an applied field, the particles are randomly oriented, and the fluid has no net magnetization. However, for ordinary field strengths the tendency of the dipole moments to align with the applied field is partially overcome by thermal agitation. Langevin's classical theory has been adapted to yield the superparamagnetic relationship between the applied field and the resultant magnetization of the particle collection. For a colloidal ferrofluid composed of particles of one size, we have

$$\frac{M}{\phi M_{\mathrm{d}}} = \coth \alpha - \frac{1}{\alpha} \equiv L(\alpha), \tag{1}$$

Figure 1 Response of a pool of magnetic fluid to the magnetic field of a steady electrical current flowing through a vertical, straight conductor. (*a*) Current off; (*b*) current on. (Photos by the author; originally appeared in *International Science and Technology*, July 1966).

$$\alpha = \frac{\pi}{6} \frac{\mu_0 M_d H d^3}{kT}, \tag{2}$$

where $L(\alpha)$ is the Langevin function, d is particle diameter (m), H is applied magnetic field (A m^{-1}), ϕ is volume fraction of magnetic solids, M is the magnetization of the ferrofluid (A m^{-1}), M_d is the domain magnetization of the particles (A m^{-1}), k is Boltzmann's constant (1.38×10^{-23} N m K^{-1}), T is the absolute temperature (in degrees Kelvin), and μ_0 is the permeability of free space ($4\pi \times 10^{-7}$ H m^{-1}). The domain magnetization of magnetite (the most typical magnetic particle) is 4.46×10^5 A m^{-1}, and that of iron is 17.3×10^5 A m^{-1}.

For large values of α we have $L(\alpha) = 1$, and the ferrofluid exhibits saturation magnetization, a condition often approached in practice. For small values of the parameter α we have $L(\alpha) = \alpha/3$, and thus the initial susceptibility $\chi_i = M/H$ is given by

$$\chi_i = \frac{\pi}{18} \phi \mu_0 \frac{M_d^2 d^3}{kT}. \tag{3}$$

Typical values of χ_i range from 1 to 5, although a value of 40 is achieved in aqueous products of phase separation. When the initial permeability is appreciable, it is no longer permissible to neglect the interaction between the magnetic moments of the particles, and the right side of Equation (3) is multiplied by the factor $3(\chi_i + 1)/(2\chi_i + 3)$. Thus, the initial susceptibility is theoretically increased by up to 50% when χ_i is much larger than unity.

Aqueous solutions and melts of certain paramagnetic salts such as manganese dichloride, ferric chloride, and holmium nitrate display appreciable magnetic response in intense magnetic fields that would saturate the colloidal ferrofluids.

Magnetic Relaxation

There are two mechanisms by which the magnetization of a colloidal ferrofluid can relax after the applied field has been changed (Shliomis 1974a). In the first mechanism, the relaxation occurs by particle rotation in the liquid. In the second, the relaxation is due to rotation of the magnetic moment within the particle. If a ferrofluid is solidified—by freezing, for example—only the second mechanism is operative. The particle-rotation mechanism is characterized by a Brownian rotational diffusion time τ_B having hydrodynamic origin. The intrinsic rotational process is known as the Néel mechanism, characterized by the time constant τ_N.

Néel relaxation predominates and *intrinsic superparamagnetism* attains when

$$\frac{\tau_N}{\tau_B} \ll 1, \tag{4}$$

whereas the Brownian mechanism predominates and the material exhibits *extrinsic superparamagnetism* when

$$\frac{\tau_N}{\tau_B} \gg 1. \tag{5}$$

When, as in ferrohydrostatics, the magnetization vector **M** is collinear with the magnetic-field vector **H**, then the body-torque density $\mu_0\mathbf{M} \times \mathbf{H}$ is absent and the state of stress is symmetric. However, the state of stress becomes asymmetric in dynamic flow, and the rate of relaxation can become an important variable. Denoting τ_F as a characteristic flow time, asymmetric stress is present if

$$\frac{\tau_B}{\tau_F} \gg 1 \quad \text{and} \quad \frac{\tau_N}{\tau_F} \gg 1, \tag{6}$$

and the stress state is symmetric if

$$\frac{\tau_B}{\tau_F} \ll 1 \quad \text{or} \quad \frac{\tau_N}{\tau_F} \ll 1. \tag{7}$$

Figure 2 presents a map of the behavior defined by these relationships. In quasi-equilibrium ferrohydrodynamics, **M** is sensibly collinear with **H**, as in ferrohydrostatics. The viscous stress tensor is then symmetric, and it is immaterial whether relaxation is dominated by the Néel or the Brownian relaxation mechanism. When **M** does not align with **H**, there exists a nonequilibrium state of magnetization and a concomitant state of asymmetric stress.

Brownian relaxation is faster than Néel relaxation for particles of sufficiently large size. The crossover occurs at about 10 nm for magnetite, 8.5 nm for iron, and 4 nm for cobalt.

Continuum Description

A very general and exact set of continuum equations applicable to polar, dipolar, or nonpolar fluids or solids may be written (Dahler & Scriven 1961, Rosensweig 1985) as

$$\frac{\partial \rho}{\partial t} + \nabla \cdot (\rho \mathbf{v}) = 0, \tag{8}$$

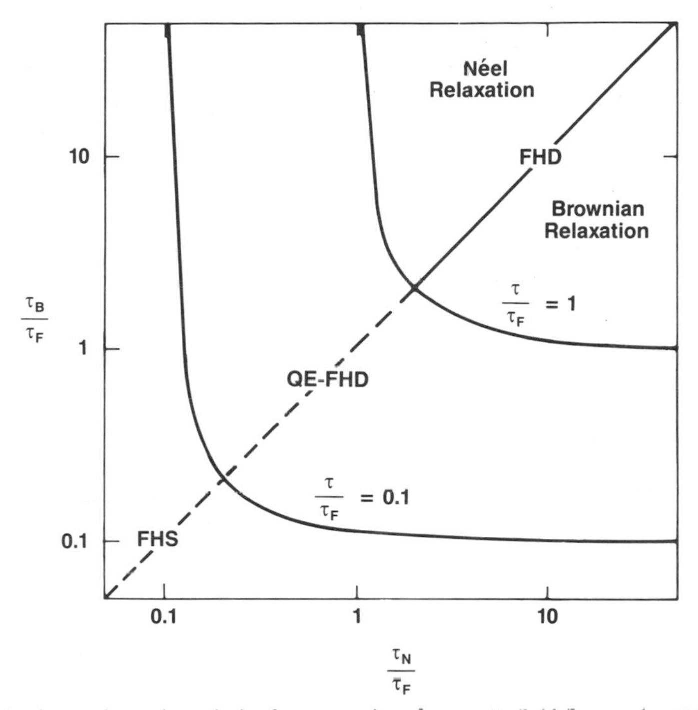

Figure 2 Approximate boundaries for categories of magnetic-fluid flow as determined by the magnetization relaxation processes of two types. FHD denotes ferrohydrodynamics, QE-FHD quasi-equilibrium ferrohydrodynamics, and FHS ferrohydrostatics. Stress is asymmetric within the FHD region. $\tau^{-1} = \tau_N^{-1} + \tau_B^{-1}$.

$$\frac{\partial \rho \mathbf{v}}{\partial t} + \nabla \cdot (\rho \mathbf{v}\mathbf{v}) = \nabla \cdot \mathbf{T} + \rho \mathbf{F}, \tag{9}$$

$$\frac{\partial \rho \mathbf{s}}{\partial t} + \nabla \cdot (\rho \mathbf{v}\mathbf{s}) = \nabla \cdot \mathbf{C} + \mathbf{A} + \rho \mathbf{G}. \tag{10}$$

An analogous theory has developed independently from interest in microelastic materials (Eringen 1966, Tanahashi et al. 1983). The equations express the conservation of mass, the balance of momentum, and the balance of internal angular momentum, respectively. Here ρ is the mass density, $\mathbf{v}$ the velocity, $\mathbf{s}$ the spin angular momentum per unit mass, $\mathbf{T}$ the surface stress tensor, $\mathbf{F}$ the external force per unit mass (gravitational constant $\mathbf{g}$ in what follows), $\mathbf{G}$ the body couple per unit mass, and $\mathbf{A} = \boldsymbol{\varepsilon} : \mathbf{T}$ the conversion rate per unit volume of external or orbital angular momentum to spin angular momentum.

Equations (8) and (9) have the same form as in ordinary fluid mechanics. T contains magnetic, pressure, and viscous stresses; however, the viscous part of the stress tensor is unsymmetric. The spin field is given by $\mathbf{s} = I\boldsymbol{\omega}$, where $\boldsymbol{\omega}$ is the angular velocity of microscale rotation of particles, and I is the average moment of inertia per unit volume. In the balance equation for $\mathbf{s}$ the tensor C accounts for surface couple due to diffusion of spin angular momentum.

Constitutive Equations

Both T and $\mathbf{G}$ will have a dependence on $\mathbf{M}$. A relaxation equation for $\mathbf{M}$ proposed by Shliomis (1974b) takes account of rotational Brownian motion and has the form

$$\frac{D\mathbf{M}}{Dt} = \boldsymbol{\omega} \times \mathbf{M} - \frac{1}{\tau}(\mathbf{M} - \mathbf{M}_0), \tag{11}$$

where $\mathbf{M}_0 = M_0\mathbf{H}/H$, $D/Dt = \partial/\partial t + \mathbf{v}\cdot\nabla$, and τ is the Brownian rotational relaxation time. Magnetization $\mathbf{M}$ is defined by

$$\mathbf{B} = \mu_0(\mathbf{H} + \mathbf{M}), \tag{12}$$

where μ_0 is the magnetic permeability in vacuum. The magnetic field $\mathbf{H}$ and magnetic induction $\mathbf{B}$ are described by the magnetostatic limit of Maxwell's equation in the absence of electric currents:

$$\nabla \times \mathbf{H} = \mathbf{0}, \qquad \nabla\cdot\mathbf{B} = 0. \tag{13a,b}$$

A more elaborate fluid-mechanical description is needed if Néel relaxation is included. The 16 governing equations[1] above contain the 41 unknowns ρ, $\mathbf{v}$, T, $\mathbf{s}$ (or $\boldsymbol{\omega}$), C, $\mathbf{A}$, $\mathbf{G}$, $\mathbf{H}$, $\mathbf{B}$, $\mathbf{M}$, and M_0. The unknowns exceed the number of equations, so it is necessary to augment the equation set with other constitutive equations relating the unknowns to each other. These have been formulated as follows (Condiff & Dahler 1964, Cowley & Rosensweig 1967) to give 27 additional equations with 2 additional unknowns, p and p^*:

$$\mathsf{T} = \lambda(\mathrm{tr}\,\mathsf{D})\mathsf{I} + 2\eta\mathsf{D} - \mathsf{I}\times\mathbf{A} - [(p^* + H^2/2)\mathsf{I} + \mathbf{HB}], \tag{14}$$

$$\mathsf{C} = \lambda'(\mathrm{tr}\,\mathsf{D}')\mathsf{I} + 2\eta'\mathsf{D}', \tag{15}$$

$$\mathbf{A} = \zeta(\boldsymbol{\Omega} - 2\boldsymbol{\omega}), \quad \text{where} \quad \boldsymbol{\Omega} \equiv \nabla \times \mathbf{v}, \tag{16}$$

$$\mathbf{G} = \mu_0\mathbf{M} \times \mathbf{H}, \tag{17}$$

[1] By a theorem of Helmholtz, to uniquely specify a vector, both its curl and divergence must be specified. Thus (13a,b) is counted as a total of three equations.

$$p^* = p - \mu_0 \int_0^H \rho^2 \left[\frac{\partial (M/\rho)}{\partial \rho} \right]_{H,T} dH, \tag{18}$$

$$p = p(\rho, T), \tag{19}$$

$$M_0 = M_0(H, \rho, T), \tag{20}$$

where $\mathbf{D} = \nabla \mathbf{v}$, $\mathbf{D}' = \nabla \boldsymbol{\omega}$, and $\boldsymbol{\Omega}$ is the vorticity. For incompressible fluid, we have tr $\mathbf{D} = \nabla \cdot \mathbf{v} = 0$. Here λ is the bulk coefficient of viscosity, λ' the bulk coefficient of spin viscosity, η the shear coefficient of viscosity, and η' the shear coefficient of spin viscosity. The integral term in (18) includes the effect of magnetostriction. The thermodynamic pressure p is related to mass density and temperature T with an equation of state (19), equilibrium magnetization is specified through a magnetic equation of state indicated by (20), and ζ is termed the vortex viscosity.

The constitutive relations together with the governing equations yield a determinate system when used with appropriate boundary and initial conditions. Boundary conditions satisfied by the magnetic field are

$$[\mathbf{B} \cdot \mathbf{n}] = 0 \quad \text{and} \quad [\mathbf{H} \cdot \mathbf{t}] = 0, \tag{21a,b}$$

where $\mathbf{n}$ is the unit normal vector at an interface, $\mathbf{t}$ is a unit tangential vector, and brackets indicate difference across the interface.

Although this set of ferrohydrodynamic equations is quite complex, physically interesting analytical solutions are known for special cases. It should be mentioned also that global energy conservation furnishes a natural technique at times for solving problems (Tsebers & Maiorov 1980a, Rosensweig et al. 1983b, Bacri & Salin 1984).

EQUILIBRIUM FLOWS

In flows with slowly shifting orientation of magnetic field relative to translating and rotating fluid elements, the following simplifications apply:

$$\mathbf{M} = \mathbf{M}_0 \quad \text{or} \quad \mathbf{M} \times \mathbf{H} = \mathbf{0}, \tag{22}$$

$$\mathbf{A} = \mathbf{0}, \tag{23}$$

$$\mathbf{C} = \mathbf{0}. \tag{24}$$

Thus, magnetization is parallel to the field, and antisymmetric stresses and couples disappear. The flow is in a state of quasi-equilibrium. For incompressible fluid the equation of continuity (8) and the constituted momentum balance (9) reduce to

$$\nabla \cdot \mathbf{v} = 0, \tag{25}$$

$$\frac{\partial \mathbf{v}}{\partial t}+\mathbf{v}\cdot\nabla\mathbf{v} = -\nabla p^*+\mu_0 M\nabla H+\eta\nabla^2\mathbf{v}+\rho\mathbf{g}. \tag{26}$$

The term $\mu_0 M\nabla H$, where M and H are magnitudes, has the form of Kelvin's body force density (i.e. the force density on an isolated magnetized body) and hence is suggestive of a magnetic body force distribution in the medium. Combinations of this term with components of the ∇p^* term yield alternative, equivalent formulations. For flow that is inviscid ($\eta = 0$), irrotational ($\mathbf{\Omega} = \mathbf{0}$), isothermal ($T =$ constant), and steady ($\partial/\partial t = 0$), (26) admits the integral representing a generalization of Bernoulli's equation (Rosensweig 1966a),

$$p^*+\rho\frac{v^2}{2}+\rho gh-\mu_0\bar{M}H = \text{constant}, \tag{27}$$

where $\bar{M}$ is field-averaged magnetization,

$$\bar{M} \equiv \frac{1}{H}\int_0^H M\,dH, \tag{28}$$

v is the magnitude of $\mathbf{v}$, and h is the elevation in the gravitational field. Very often it is desired to apply (27) between two points along a stream tube in the interior of a fluid configuration having a free surface, and a boundary condition is required to relate p^* to the environmental pressure p_0. The condition is obtained by equating the normal-stress difference $[\mathbf{n}\cdot\mathbf{T}]$ from (14), evaluated at the interface, to the capillary pressure p_c due to surface tension σ and mean interfacial curvature $\mathscr{H} = \nabla\cdot\mathbf{n}/2$. The result is the boundary condition to accompany (27):

$$p^*+\frac{\mu_0}{2}M_n^2 = p_0+2\mathscr{H}\sigma. \tag{29}$$

Compared with ordinary fluid mechanics, the unique feature of (29) is the appearance of the magnetic normal surface stress $\mu_0 M_n^2/2$, in which $M_n = \mathbf{n}\cdot\mathbf{M}$ is the normal component of magnetization evaluated at the interface.

One of the simplest posed problems in ferrohydrodynamics is the response of a spherical droplet to a uniform magnetic field. The field $\mathbf{H}$ within the particle is spatially uniform, $\mathbf{v} = \mathbf{0}$, and h is sensibly constant, so from (27) p^* is spatially uniform within the droplet. At the equator of the droplet, we have $M_n = 0$, which increases to a maximum value at the poles. From (29) it is apparent that surface curvature cannot remain constant, with the result that the droplet elongates along the field direction.

The Conical Meniscus

Numerous responses of ferrofluids are defined, and conveniently analyzed, using the generalized Bernoulli equation (27) and its accompanying boundary condition (29). For example, the photographs of Figure 1 illustrate the ferrohydrostatic response of a pool of ferrofluid to a steady electric current I passed through a vertical rod running through the center of the pool. The current produces an azimuthal field external to the wire with magnitude $H = I/2\pi r$. Applying (27) between a point in the fluid in the distant interface (where the field is negligible and the pressure is p_0) and another point in the interface closer to the current-carrying rod (where the elevation is Δh above the distant interface) gives

$$p^* + \rho g \Delta h - \mu_0 \bar{M} H = p_0. \tag{30}$$

Because $\mathbf{M}$ is collinear with $\mathbf{H}$, we have $M_n = 0$ at the free surface. If we neglect capillary pressure, Equation (29) then gives $p^* = p_0$. Hence, it follows that

$$\Delta h = \frac{\mu_0 \bar{M} H}{\rho g}. \tag{31}$$

For linearly magnetizable fluid we have $M = \chi H$, where the susceptibility χ is constant and $\bar{M} = \chi H/2$ from (28). Thus the elevation at the conductor surface, where $r = R$, is

$$\Delta h = \frac{\mu_0 \chi I^2}{8\pi^2 \rho g R^2}. \tag{32}$$

This phenomenon helps validate the deduction that force in magnetic fluid results from the gradient of field magnitude and is independent of field direction.

The remainder of this section is concerned with a number of important ferrohydrostatic problems.

Rotary-Shaft Seal

The components of a staged magnetic-fluid rotary-shaft seal are shown in Figure 3 (Rosensweig 1971). The seals are in widespread use in computer peripheral equipment and semiconduuctor manufacturing processes as pressure-, vacuum-, and exclusion-sealing devices (Moskowitz 1975). A ring magnet forms part of a magnetic circuit in which an intense magnetic field is established in the gaps between the teeth on a magnetically permeable shaft and the surface of an opposing pole block. Ferrofluid intro-

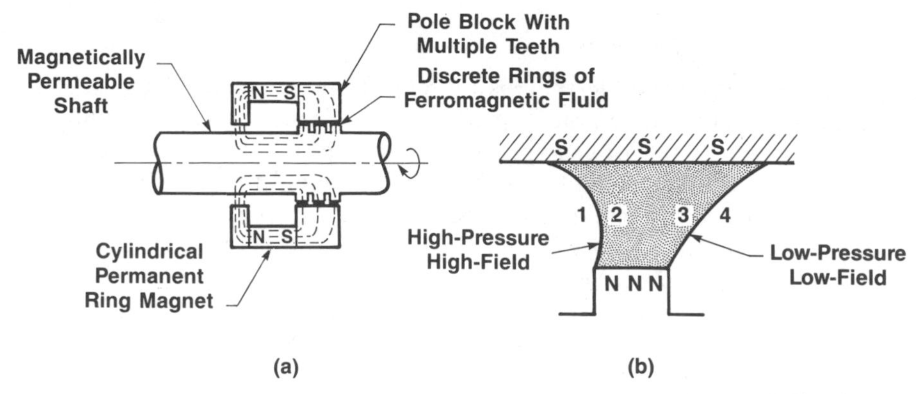

Figure 3 (*a*) Components of a multistage, magnetic-fluid rotary-shaft seal. (*b*) Sketch to analyze the pressure supported across one stage of the seal.

duced into the gaps forms discrete liquid rings capable of supporting a pressure difference while maintaining zero leakage. The seals operate without wear as the shaft rotates because the mechanical moving parts do not touch. Applying (27) between horizontal points 3 and 2 in the sketch of Figure 3*b* when the shaft is stationary, and recognizing that the normal component of magnetization is zero at both interfaces, yields

$$\Delta p = p_4 - p_1 = \mu_0 \bar{M} H \tag{33}$$

for the case when the field is negligible at station 2. Substituting (SI units) $\mu_0 = 4\pi \times 10^{-7}$ H m^{-1}, $\mu_0 H = 1.8$ T (18,000 G), and $\mu_0 M = 0.07$ T (700 G) gives $\Delta p = 10^5$ N m^{-2}, or about 1 atm. Perry & Jones (1976) measured static burst pressures of fully filled gaps and found good agreement with values calculated from Equation (33).

The typical density of pole teeth is 10 per centimeter of shaft length. When loaded with ferrofluid and then initially pressurized, the individual rings of fluid bubble gas through to the interstage zones and then reseal. The author devised an experimental rotary seal that withstood over 60 atm pressure differential; more typical of commercial practice is 1 to 2 atm differential for vacuum seals and 0.1 atm differential in exclusion seals.

Buoyancy, Levitation, and Hydrostatic Bearings

The net force acting on a magnetic or nonmagnetic body immersed in a magnetic fluid in the presence of a magnetic field is formulated as

$$\mathbf{F} = \int_S \mathbf{n} \cdot \mathbf{T} \, dS + \int_V \rho' \mathbf{g} \, dV, \tag{34}$$

where the stress tensor $\mathbf{T}$ is given by (14), and ρ' is the mass density of the ferrofluid. Using (14), (27), and the divergence theorem, we can transform the force to

$$\mathbf{F} = \int_S \left[\left(H_n B_n - \int_0^H B \, dH \right) \mathbf{n} + H_t B_n \mathbf{t} \right] dS + \int_V (\rho - \rho') g \mathbf{k} \, dV, \tag{35}$$

where subscript n denotes the normal component and t the tangential component of the field vector evaluated at the body-fluid interface, and $\mathbf{n}$ and $\mathbf{t}$ are the corresponding unit vectors. The last term in (35) represents the usual Archimedean buoyancy. The first integral on the right permits the total magnetically induced force to be computed from knowledge of the magnetic-field distribution. For an immersed body that is nonmagnetic, using the magnetic-field boundary conditions $[B_n] = 0$ and $[H_t] = 0$, we can show that (35) reduces to

$$\mathbf{F} = -\int_S \left(\frac{1}{2} \mu_0 M_n^2 + \mu_0 \bar{M} H \right) \mathbf{n} \, dS + \int_V (\rho - \rho') g \mathbf{k} \, dV. \tag{36}$$

In levitation we have $\mathbf{F} = \mathbf{0}$, and for stable levitation of a neutrally buoyant body such that a positive restoring force accompanies any small displacement from equilibrium, it is required that $\delta H > 0$. Thus the magnetic field must possess a local minimum of field magnitude. This is possible with opposed magnets, in which case ferrofluid circumvents Earnshaw's theorem and permits the stable levitation of magnetic bodies in static magnetic fields (Rosensweig 1966a).

Levitation of a magnetic proof mass (Figure 4) furnishes a ferrohydrostatic bearing support that is the basis of sensitive but rugged accelerometers. A similarly magnetized cylinder levitated inside a cylindrical case is used for viscous damping of rotary motion. Unlike in hydrodynamic or fluid film bearings, the support force in a ferrohydrostatic bearing is passive and requires no motion (Rosensweig 1966b).

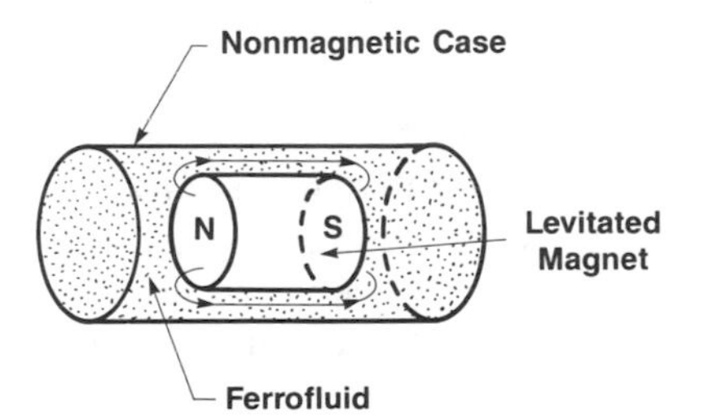

Figure 4 Self-levitation of a magnet in a ferrofluid. The supported magnet is free to translate or rotate along the axis.

Sink-Float Separation

Sink-float processes utilize the high apparent density imparted to a pool of ferrofluid subjected to a gradient magnetic field in order to separate mixtures of materials on the basis of their mass density. Rosensweig (1969) disclosed methods for continuous processing, Andres (1976) extensively employed paramagnetic salt solutions, and Fay & Quets (1980) developed the use of magnetized grates in which the ferrofluid inventory is small and the magnetic power is minimized. An informative, approximate expression for the apparent density can be developed from Equation (36) assuming a strong field intensity, such that

$$\frac{M_{\mathrm{n}}^2}{\bar{M}H} \ll 1. \tag{37}$$

Then, applying the divergence theorem to the surface integral of (36) and assuming constant densities and uniform field gradient, we obtain for the force on the submerged object

$$\mathbf{F} = [-\mu_0 M \nabla H + (\rho - \rho')g\mathbf{k}]V. \tag{38}$$

With the field gradient oriented in the vertical direction, recognizing that the force $\mathbf{F} = \mathbf{0}$ at the point of neutral buoyancy, we find that the object's density ρ' equals the ferrofluid's apparent density ρ_{a} given by

$$\rho_{\mathrm{a}} = \rho - \frac{\mu_0 M}{g}\frac{dH}{dz}. \tag{39}$$

The minus sign on the last term indicates that the magnetic force transmitted to the immersed body is equal and opposite to the magnetic force on an equal volume of magnetic fluid. Using available ferrofluids and iron-yoke electromagnets, we may float any known nonmagnetic solid (e.g. lead or tungsten). The technique has been demonstrated on a semiworks scale in the separation of metal mixtures (Khalafalla & Reimers 1973a, Shimoiizaka et al. 1980) and in laboratory use as a means for high-resolution mass-density determination (Hughes & Birnie 1980).

Skjeltorp (1985) has shown that a monolayer of uniform-size polystyrene spheres in a ferrofluid will crystallize on a triangular lattice in the presence of a magnetic field applied perpendicular to the layer. An effective magnetic dipole moment is associated with each sphere as a result of the polarization of the surrounding ferrofluid by the field. The dipole holes, oriented antiparallel to the external fluid, repel each other and produce a crystallization (see Figure 5). The system provides a directly observable model for viewing solid-state phase-change phenomena.

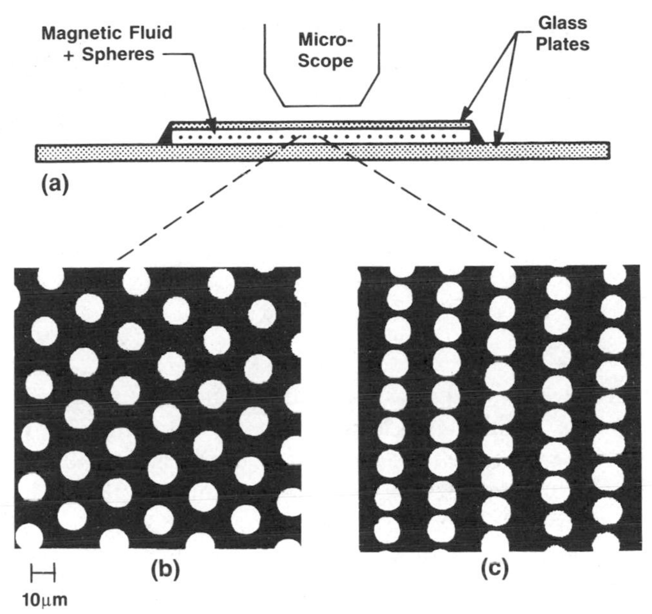

Figure 5 (*a*) Experimental apparatus to study pattern formation of 10-μm inert spheres in a layer of ferrofluid. (*b*) Field oriented normal to the layer produces a triangular lattice of mutually repelling holes. (*c*) In-plane magnetization of the fluid induces chaining of the holes (Skjeltorp 1985).

In a metal-finishing process, abrasive grains are suspended in a magnetic fluid and subjected to the buoyant force generated in an applied magnetic field (Tani & Kawata 1985). The magnetic repulsion produces a polishing pressure, said to yield a high-stock-removal rate and a smooth surface finish.

Hydrodynamic Bearings

Bearing spindles supported on a film of ferrofluid have been developed recently as an improvement over traditional ball-bearing spindles. The ferrofluid lubricant, present as a film with typical thickness 10^{-3} cm, achieves a higher rotational stability and accuracy such that upcoming generations of magnetic and optical disk drives might store an order of magnitude more information on the same disk (P. Stahl, personal communication, 1985). Acoustically quiet and free of running wear, the magnetic-fluid spindle operates under the laws governing ordinary hydrodynamic bearings with the additional feature that the lubricant is retained magnetically (see Figure 6). The magnetic containment is different than in pressure seals and more economical in the mass of magnets employed for the volume of ferrofluid that is captured.

To better appreciate the mechanism of containment, consider applying the generalized Bernoulli equation between two points in the ferrofluid located at opposite ends of a spindle having its axis oriented at angle θ to the vertical direction. It follows that the upper interface (located at elevation h') must self-position to a region of field intensity H' that *exceeds* the field H at the lower interface (located at elevation h) by an amount satisfying the relationship

$$(\mu_0 \bar{M} H)' - (\mu_0 \bar{M} H) = \rho g(h' - h)$$
$$= \rho g L \cos \theta, \qquad (40)$$

where L is the length of the magnetic-fluid region. As an amusing consequence, when a horizontal bearing ($\theta = \pi/2$) is rotated to the vertical orientation ($\theta = 0$), the magnetic fluid is displaced slightly, sometimes upward. As another consequence this "magnetic bottle" is inoperative unless it is completely filled. [Choose $h' > h$ at a point where H' is close to zero, in which case Equation (40) fails to be satisfied.]

Peak pressure occurs near the center in the running spindle and is elevated about 1 to 2 atm above ambient pressure, a level greater than a single seal stage can sustain. Recirculation of the ferrofluid through a longitudinal passage in the shaft reduces the end pressure that the seals must sustain to a manageably small value.

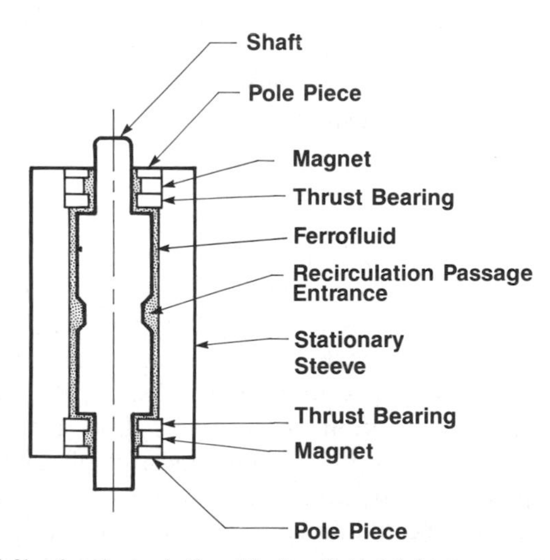

Figure 6 Fluid-film bearing spindle with ferrofluid lubricant captured in a "magnetic bottle." The bulk of the captured fluid is contained in a region of negligible magnetic field.

Impactless Printing

A high-speed, silent, plain-paper recording process has been developed in which spatial undulations of the magnetic-fluid surface are established in the field of a comb of magnetized pins (Maruno et al. 1983). Printing occurs on command, in which electric pulses under computer control cause ejection of droplets of the ferrofluid, which serves as the ink, from surface peaks.

Assuming magnetization is normal to the interface and neglecting capillary pressure gives for the surface elevation

$$\Delta h = \frac{1}{\rho g}\left(\mu_0 \bar{M} + \mu_0 \frac{M^2}{2}\right). \tag{41}$$

Samples of the printed output exhibit a resolution exceeding that of a dot-matrix printer.

FLOW INSTABILITIES

Flow instability at the free surface of a ferrofluid relates to the state of stress upon the interface. Eliminating p^* between Equations (27) and (29) yields a magnetically augmented form of the Young-Laplace equation describing stress equilibrium at the interface between a magnetic and a nonmagnetic fluid. When both phases are magnetizable, but immiscible, the equation can be derived in a more general form:

$$\begin{aligned} &\frac{1}{2}(\rho_2 v_2^2 - \rho_1 v_1^2) - g z_0(\rho_1 - \rho_2) + \mu_0[(\bar{M}H)_1 - (\bar{M}H)_2] \\ &+ \frac{1}{2}\mu_0(M_{1n}^2 - M_{2n}^2) - 2\mathscr{H}\sigma = \text{constant}, \end{aligned} \tag{42}$$

where subscript 1 refers to fluid below the interface, 2 to fluid above, and z_0 is interfacial deflection. When the potentials for the velocity and magnetic fields are determined in both regions, Equation (42) conveniently couples the solutions together.

Normal-Field Instability

One very characteristic instability occurs when a uniform magnetic field exceeding a critical intensity is applied perpendicular to the interface of magnetic fluid, producing spontaneously a hexagonal pattern of peaks and valleys on the interface (Cowley & Rosensweig 1967). If we represent the interfacial deflection as $z_0 = \hat{z}_0 \operatorname{Re}\{\exp[i(\omega t - \mathbf{k}\cdot\mathbf{x})]\}$, the following dispersion equation is determined from a linearization of Equation (42) relating frequency ω to wavelength k when one phase is nonmagnetic:

$$\rho\omega^2 = \rho g k + \sigma k^3 - \frac{k^2 \mu_0 M_0^2}{1 + \mu_0/\mu}. \tag{43}$$

Incipient instability occurs when both the following conditions are met:

$$\omega^2 = 0, \qquad \frac{\partial \omega^2}{\partial k} = 0. \tag{44a,b}$$

The critical spacing $\lambda_c = 2\pi/k_c$ and magnetization M_c then are given by

$$\lambda_c = 2\pi\left(\frac{\sigma}{g\rho}\right)^{1/2}, \qquad M_c^2 = \frac{2}{\mu_0}\left(1 + \frac{\mu_0}{\mu}\right)(\rho g \sigma)^{1/2}. \tag{45a,b}$$

Tests by Cowley & Rosensweig (1967) supported these theoretical relationships.

Subsequently, several authors have determined nonlinear features in the evolution of two-dimensional disturbances (roll cells), but the results do not apply to real systems that are three-dimensional. Incisive analytical results for three-dimensional disturbances (Gailitis 1977) indicate that peaks of finite height develop discontinuously on the flat surface, with each tall peak surrounded by six shallow troughs. The first-order nature of the transition on a wide pool of ferrofluid was confirmed recently (Boudouvis et al. 1986). When field strength increased, the interface remained almost flat until the critical field intensity was reached (see Figure 7). A first-order transition to peaks of about 4 mm height was observed that increased further with field strength, following the curve shown in the figure. Peak heights were determined from laser-beam reflection. When the field was decreased the same curve was followed back down, but the peaks persisted down beyond the original critical field strengths. Then the level configuration was once again attained through a hard transition. The measurements are in satisfactory agreement with computation (solid curve in Figure 7) based on the Galerkin/finite-element method. Regular states were followed by first-order continuation in parameters, whereas singular states (bifurcation and turning points) were circumvented with an adaptive continuation technique.

Bacri & Salin (1984) earlier reported measurements of the hysteresis in a closely related system with ferrofluid confined between narrowly spaced parallel plates such that peaks are distributed along a single row. A related problem is the observed jump and hysteresis in elongated droplet shape in increasing and decreasing applied magnetic field (Bacri & Salin 1983).

When the ferrofluid layer is thin and a normal field is applied, a topological instability can occur in which the fluid continuity is ruptured, producing individual drops in hexagonal array. When the time required to apply the magnetic field is much less than the time associated with viscous flow in the thin film, and field in excess of the critical intensity is applied, selection of the characteristic length in the pattern is determined by the fastest growing mode, resulting in smaller drop size and spacing (Bashtovoi et al. 1985). An analogy has been drawn in which the ferrofluid drops represent a two-dimensional crystal exhibiting solid-state features of dislocations, disclinations, and melting (Skjeltorp 1983); the magnetic-field intensity plays the role of temperature. Transition of a type II superconductor from the normal to the superconducting state produces a hexagonal pattern of magnetization (Essmann & Trauble 1967) having a plausible mathematical analogy to the normal-field instability.

Malik & Singh (1984, 1985) treat weakly nonlinear aspects of wave

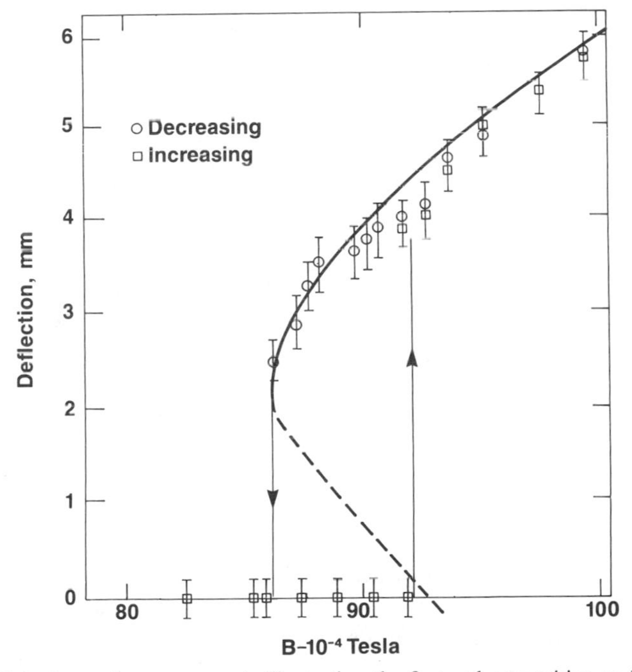

Figure 7 Experimental measurements illustrating the first-order transition and hysteresis in surface deflection for the magnetic-fluid normal-field instability. The solid curve represents computed behavior (Boudouvis et al. 1986).

propagation and growth at magnetic-fluid planar interfaces using the method of multiple scales. The interfacial disturbance amplitude is governed by the nonlinear Klein-Gordon equation when the magnetic field is normal to the interface. Solutions are found for wave profiles having the form of bell-shaped solitons (hyperbolic-secant shape) able to survive collisions with each other, and kinks and antikinks (positive and negative hyperbolic-tangent shape) unable to survive collisions and thus known as solitary waves. If we apply the same technique with tangential orientation of magnetic field, a nonlinear Schrödinger equation arises with similar solutions. Experiments are needed in the study of these waves; a likely complication in the case with normal field is the known metastability of the two-dimensional disturbances.

Labyrinthine Instability

Because a normal field destabilizes the interface and a tangential field is stabilizing, it is initially surprising that the labyrinthine instability arises at all. The instability occurs in a layer of magnetic fluid having a free surface contained between closely spaced flat surfaces when a field is applied normal to the containing surfaces (Romankiw et al. 1975, Tsebers & Maiorov 1980a). A form of the labyrinthine phenomenon is established when ferrofluid is introduced from a point source into the cell containing the nonmagnetic liquid (see Figure 8). Owing to the uncrowded geometry, a double structure develops in which the fingers outline a pattern of bigger fingers; the structure is reminiscent of the tertiary structure of proteins in living cells.

Analogous to the formation of labyrinthine magnetic domains in garnets (Druyvesteyn & Dorleijn 1971), and transition in type I superconductors (Faber 1958), the labyrinthine patterns are suggestive of brain structure, brain coral, insect cuticle and the color pattern in cuttlefish (H. F. Nijhout, personal communication, 1983),[2] and a feedback pattern when a video camera looks at its video monitor (Crutchfield 1984). Several interesting theorems have been deduced concerning onset and development of the patterns in ferrofluids (Tsebers & Maiorov 1980b), with the analyses based on global energy conservation.

A discrete, cellular model using stochastic choices of energetically favored moves generates simulations of labyrinthine patterns (Rosensweig 1986). Total energy is computed as the sum of surface energy, dipole-field interaction, and dipole-dipole repulsion under the assumption of nearest-

[2] See cuticle photography of *rhodnius prolixus* in P. A. Lawrence, "Computers show how cells communicate," *New Sci.*, Vol. 53, No. 785 (1972), pp. 475–77; the cuttlefish (Sepia Officinala) is illustrated in W. D. Russell-Hunter, *A Life of Invertebrates* (New York: Macmillan, 1979).

Figure 8 Labyrinthine pattern exhibiting a doubly bifurcated structure results from point-source introduction of hydrocarbon-base ferrofluid into a flat round cell containing an aqueous solution. The cell spacing is 1.65 mm, the magnetic field intensity is 0.011 T, and the magnetic-fluid susceptibility is 1.6 (R. E. Rosensweig, unpublished).

neighbor interactions. The labyrinthine pattern can be induced experimentally in a thin layer of dielectric fluid subjected to a uniform applied electric field (Rosensweig et al. 1983b).

Convective Instability

As illustrated in Figure 9, a convective instability arises from and is driven by heat conduction through a magnetized fluid layer having a temperature-dependent magnetization (Curtis 1971, Shliomis 1974a, Berkovsky & Bashtovoi 1980). The effect is strongest when the applied field gradient is oriented in the direction of the temperature gradient causing the heat to flow. Convective motions of the fluid set in at a critical value of the magnetic Rayleigh number, similar to the mechanism of convective instability of thermally expansive fluid in a gravitational field (cf. Bénard instability). The Rayleigh number N_{Ra} in either case represents the dimensionless ratio

$$N_{\mathrm{Ra}} = \frac{\text{Heat convection}}{\text{Heat conduction}} \cdot \frac{\text{Driving force}}{\text{Viscous force}}, \tag{46}$$

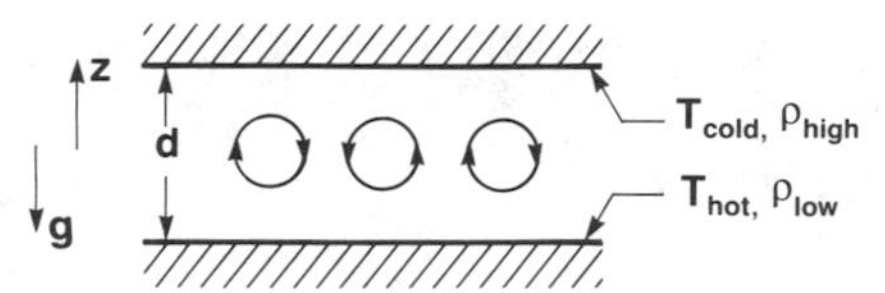

(a) Unmagnetized Fluid Heated From Below in the Gravitational Field.

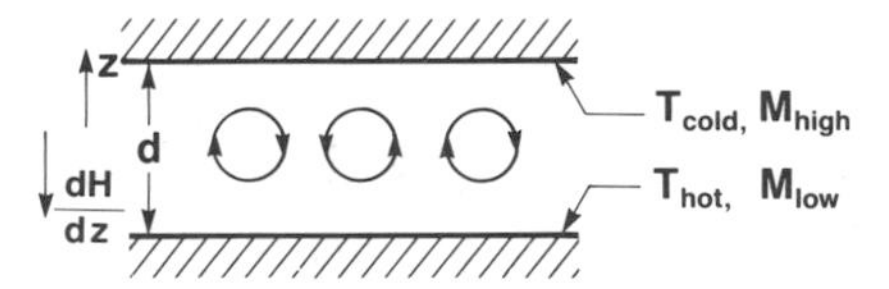

(b) Magnetic Fluid Heated in the Presence of an Applied Magnetic Field Gradient.

Figure 9 Convective instability in heated layers.

with heat convection $\rho v c_0 \Delta T$, heat conduction $k\Delta T/d$, and viscous force $\eta v/d$ [where c_0 is the specific heat per unit mass, k the thermal conductivity, v a characteristic velocity (cancels in the ratio), ΔT the temperature difference, and d the wall spacing]. The driving force in (46) is given by

$$\text{Driving force} = \begin{cases} gd\Delta\rho & \text{(buoyancy)} \\ \mu_0 KMG\Delta T & \text{(magnetism).} \end{cases} \tag{47}$$

Here the pyromagnetic coefficient is $K = (\partial M/\partial T)/M$, and G is the magnetic-field gradient. Convection sets in when N_{Ra} exceeds 1708 in a horizontal layer and 1558 in a vertical layer. The convective flow represents an example of the magnetocaloric process in which thermal energy is converted to the mechanical energy of motion (Resler & Rosensweig 1964), and it offers opportunities in the passive cooling of electrical circuits, machinery, and processes (Matsuki et al. 1977).

FLOW STABILIZATION

The interaction of a magnetic field with a magnetic fluid preserves the equilibrium of certain flows that otherwise become unstable. Limited use has been made of flow stabilization, perhaps because the effects are little known or appreciated.

Tangential-Field Mechanism

The simplest stabilizing condition is illustrated with a uniform magnetic field applied tangentially to the horizontal, undisturbed, planar surface of a pool of magnetic liquid in a gravitational field. With the field oriented along the y-direction and the fluid free of mean motion, the dispersion relation found by Zelazo & Melcher (1969) is

$$\rho\omega^2 = \rho g k + \sigma k^3 + k_y^2 \frac{\mu_0 M_0^2}{1+\dfrac{\mu}{\mu_0}}. \tag{48}$$

While a normal field is destabilizing, the tangential orientation stabilizes a ferrofluid interface in one plane. The magnetization remains stabilizing if the fluid layer is inverted so that its surface faces downward; the gravitational term in (48) then reads $-\rho g k$ and represents the destabilizing influence of Rayleigh-Taylor problems.

A variation of the tangential magnetic-field concept permits stabilization of fluid cylinders such as fluid jets. The equation of the deformed surface of the stream is written in the form

$$r = r_0 + \zeta(z, t), \qquad \zeta = a_0 e^{i(\omega t - kz)}, \tag{49a,b}$$

and r_0 is the radius of the unperturbed fluid cylinder. The corresponding dispersion equation (Taktarov 1975) is

$$\omega^2 = \frac{k^2 H_0^2(\mu-\mu_0)^2 I_0(kr_0)\,K_0(kr_0)}{\rho[\mu_1 I_1(kr_0)K_0(kr_0)+\mu_0 I_0(kr_0)K_1(kr_0)]} - \frac{\sigma k I_1(kr_0)}{\rho r_0^2 I_0(kr_0)}\,(1-k^2 r_0^2). \tag{50}$$

From (49), we see that instability results when ω^2 is negative. With a field absent, the first term on the right side of (50) vanishes, and what remains is the classical Rayleigh result: Onset occurs for $k_0 r_0 < 1$ ($\lambda > 2\pi r_0$), i.e. when the wavelength of the disturbance exceeds the perimeter of the jet. The magnetic term shifts the critical wavelength to larger values and decreases the maximum growth rate of unstable modes. Thus, in a stream of magnetic fluid stabilized by a longitudinal magnetic field, the intact length of the jet increases, and the droplet size when breakup finally occurs is greater. Magnetic stabilization of a magnetic-fluid jet has been observed experimentally (Bashtovoi & Krakov 1978).

It has been demonstrated experimentally and theoretically that the fingering or Saffman-Taylor instability occurring in the displacement of a less viscous fluid by a more viscous fluid in a porous medium is similarly stabilizable using a tangential, uniform magnetic field (Rosensweig et al. 1978).

Drag Reduction

The separation of flow from the wall of a pipeline, such as occurs at bends or in the presence of changes in the cross section, increases hydraulic drag and reduces the flow rate. Accordingly, the development of methods for controlling flow separation is important scientifically and could be technically significant. The use of magnetic fluid, coating a wall and held to it by an inhomogeneous magnetic field, shows some promise in this direction.

A convenient analytical model for studying the conditions for development of separation in two-dimensional viscous flow specifies flow in a channel with equally spaced sinusoidal walls (Medvedev & Krakov 1984). A constant-thickness layer of magnetic fluid is held in position at both walls by a system of periodically arranged permanent magnets or current-carrying conductors. The magnetic forces tend to stabilize the interface between magnetic and nonmagnetic liquids, and in the analysis the interface is assumed to be fixed. Figure 10*a* illustrates the results of a regular perturbation analysis for dependence of the critical parameter $A\varepsilon^2$ on the relative thickness of the magnetic fluid layer $(1-h)$, where $A = a/L$, $\varepsilon = 2\pi L/\lambda$, and $h = l/L$. Here L is the channel half-width, l the layer thickness, and λ the wavelength of the sinusoid having amplitude a. Flow separation is dependent only on geometrical parameters and not on the viscosity of the fluid. With increased thickness of the magnetic layer the separation of the flow in the center region gets more difficult, and at a relative thickness of 0.835 or greater no separation is possible. Analysis of higher-order terms shows that this solution is valid for small Reynolds number and not too large a/L. Figure 10*b* illustrates a double-eddy pattern of streamlines predicted by the model for a flow with separation.

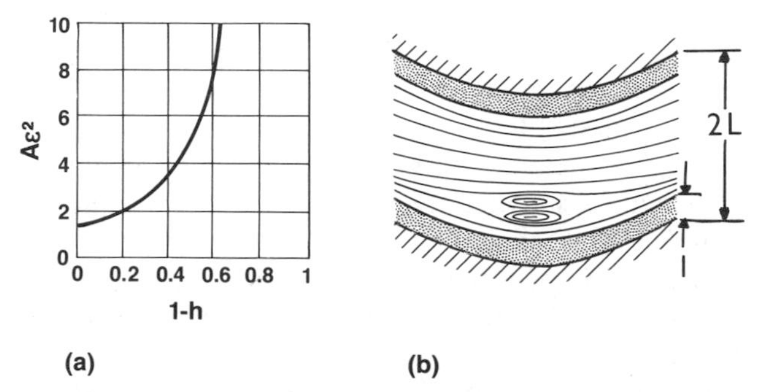

Figure 10 Drag control in channel flow using a surface layer of magnetic fluid positioned against the wall. (*a*) Stability envelope as function of geometrical parameters. (*b*) Streamlines of the separated flow (after Medvedev & Krakov 1984).

FLOWS WITH ASYMMETRIC STRESS

A problem that has yielded to analysis, giving a solution in good agreement with experiments, concerns the effective viscosity of magnetized fluid in plane shear (Shliomis 1972). An additional resistance to that of the usual viscous resistance is established, since the magnetically-oriented particles are impeded from rotating freely. The problem is fairly complex, since Brownian motion and hydrodynamic forces exert a disorienting effect on the particles. Earlier treatments neglecting Brownian rotation predicted saturation of viscosity increase in an applied field that is low by one to two orders of magnitude. The viscosity increase when accounting for Brownian motion is given for most cases by the asymptotic expression

$$\frac{\Delta\eta}{\zeta} = \frac{\mu_0 M_0 H\tau/4\zeta}{1+\mu_0 M_0 H\tau/4\zeta}\sin^2\beta, \tag{51}$$

where β is the angle between $\mathbf{H}$ and $\mathbf{\Omega}$. When $\mathbf{H}$ is parallel to $\mathbf{\Omega}$, the viscosity is independent of the presence of the field. (The particles are free to rotate with the angular velocity $\mathbf{\Omega}/2$ of the fluid.)

Careful measurements show that while the pipe-flow friction coefficient of a ferrofluid increases with magnetic field strength in the laminar-flow regime, there is no magnetic-field effect in a turbulent-flow regime (Kamiyama et al. 1983).

Considerable study has been devoted to the phenomenon of swirl flow generated by a rotating magnetic field; such a flow was observed by Moskowitz & Rosensweig (1967). For example, a steady-state solution of (9) and (10) for the swirl flow in a cylindrical container of radius R using boundary conditions $v(R) = 0$, $\omega(R) = 0$ was developed by Zaitsev & Shliomis (1969) in terms of modified Bessel functions. The velocity field is that of solid-body rotation in the core with a thin layer having steep gradients near the wall.

Observations of torque and of rotation of nonmagnetic concentric cylinders separated by magnetic fluid offer a means for concomitant measurement of normal and spin viscosities (Berkovsky et al. 1984).

The influence of microrotational motion on heat transfer in a particle suspension might be appreciable. Expressions for the apparent thermal conductivity $\boldsymbol{\lambda}$, which now is a tensor quantity, and the total heat-flux vector $\mathbf{q}$ in a ferrofluid with internal rotations, given by Berkovsky et al. (1980) are

$$\boldsymbol{\lambda} = (\lambda_0+\lambda_1)\mathbf{I}-\lambda_1\mathbf{rr}+\lambda_2\boldsymbol{\varepsilon}\cdot\mathbf{r}, \tag{52}$$

$$\mathbf{q} = -\boldsymbol{\lambda}\cdot\nabla T = -\lambda_0\nabla T-\lambda_1[\nabla T-\mathbf{r}(\mathbf{r}\cdot\nabla T)]+\lambda_2\mathbf{r}\times\nabla T, \tag{53}$$

where

$$\mathbf{r} = \mathbf{R}/R, \qquad \mathbf{R} = 1/2\mathbf{\Omega} - \boldsymbol{\omega}, \qquad \lambda_1 = \lambda_1(R), \qquad \lambda_2 = \lambda_2(R), \tag{54}$$

and λ_0 is the thermal conductivity in the absence of an applied field. The last term in the equation for $\mathbf{q}$ implies that heat transfer is possible in the direction normal to the temperature gradient, a thermal analogue of the Hall effect. An estimate of the various coefficients indicates that a rather large particle radius (~ 1000 Å) and high angular velocity $|\mathbf{R}| = 10^4\ \mathrm{s}^{-1}$, for example, are required to make λ_1/λ_0 and λ_2/λ_0 of order unity.

The enhancement of heat-transfer rate by a factor of up to 30 in an asymmetric stress field between concentric cylinders has been reported in systematic measurements by Shulman et al. (1977). Because the particles of the ferrosuspension in this study were larger than colloidal in size, there is a need to study the effect with actual ferrofluid.

In an analytical study, heat transfer in an infinitely extended square array of rotating, thermally conducting circular cylinders with fluid-filled interstices was computed for large rotary Peclet number using singular perturbation analysis (Nadim & Brenner 1985). The study is of interest in representing an accurate solution of a model system having features of an asymmetrically stressed ferrofluid.

MAGNETIC TWO-PHASE FLOW

Multiphase flows arise in connection with fluidized beds, sedimentation of particles and drops, boiling of liquids, foams, and other systems. The ferrohydrodynamics of these systems is in its infancy. Using an averaging process to describe such flows, we can derive a theory of interpenetrating continua as an alternative to postulating a set of continuum equations at the outset (Jackson 1985, Rosensweig 1985). A merit of this approach is that uncertainty is concentrated into terms that arise naturally and that can then be constituted empirically.

A recent development is the use of ferrohydrodynamic principles to study a magnetically stabilized operating range of fluidized solids in which turbulence is prevented (Rosensweig 1979, Rosensweig et al. 1983a). The size of the fluidized particles (50–1000 μm) is large compared with the size of particles in a colloidal ferrofluid. Translational motion of fluid relative to the magnetizable particles must be incorporated into the analytical description, but Brownian motion of the particles is negligible.

In the magnetized fluidized system, instability (when it appears) develops in the bulk. While transition to turbulence in a one-phase flow of an incompressible fluid is a complex three-dimensional problem, transition in the seemingly complex two-phase fluidized system reduces to a mathematically tractable, one-dimensional problem.

Stably fluidized moving beds are very well suited for the countercurrent contacting of fluids with particles in reactors and separation processes, the essential operations in nearly all chemical-processing networks (Lucchesi et al. 1979). Sedimentation and other problems have also been studied (Gogosov et al. 1979).

Literature Cited

Andres, U. 1976. *Magnetohydrodynamic and Magnetohydrostatic Methods of Mineral Separation*. New York: Wiley

Bacri, J. C., Salin, D. 1983. Dynamics of the shape transition of a magnetic ferrofluid drop. *J. Phys. Lett.* 44: 415–20

Bacri, J. C., Salin, D. 1984. First order transition in the instability of a magnetic fluid interface. *J. Phys. Lett.* 45(11): 558–64

Bashtovoi, V. G., Krakov, M. S. 1978. Surface instability in the nonisothermal layers of magnetized fluids. *Magnetohydrodynamics* 14(3): 285–90

Bashtovoi, V. G., Krakov, M. S., Reks, A.G. 1985. Instability of a flat layer of magnetic liquid for supercritical magnetic fields. *Magnetohydrodynamics* 21(1): 19–24

Berkovsky, B. M., Bashtovoi, V. 1980. Instabilities of magnetic fluids leading to a rupture of continuity. *IEEE Trans. Magn.* 16(2): 288–97

Berkovsky, B. M., Vislovich, A. N., Kashevsky, B. E. 1980. Magnetic fluid as a continuum with internal degrees of freedom. *IEEE Trans. Magn.* 16(2): 329–42

Berkovsky, B. M., Ivanova, N. I., Kashevskii, B. E. 1984. A viscometric method for magnetic liquids. *Magnetohydrodynamics* 20(2): 3–10

Boudouvis, A. G., Puchalla, J. L., Scriven, L. E., Rosensweig, R. E. 1986. Normal field instability and patterns in pools of ferrofluid. *J. Magnetism Magn. Mater.* In press

Charles, S. W., Popplewell, J. 1980. Ferromagnetic liquids. In *Ferromagnetic Materials*, ed. E. P. Wohlfarth, 2: 509–59. Amsterdam: North-Holland

Charles, S. W., Rosensweig, R. E. 1983. Magnetic fluids bibliography. *J. Magnetism Magn. Mater.* 39(1,2): 192–220

Condiff, D. W., Dahler, J. S. 1964. Fluid-mechanical aspects of antisymmetric stress. *Phys. Fluids* 7(6): 842–54

Cowley, M. D., Rosensweig, R. E. 1967. The interfacial stability of a ferromagnetic fluid. *J. Fluid Mech.* 30(4): 671–88

Crutchfield, J. P. 1984. Space-time dynamics in video feedback. *Physica* 10D: 229–45

Curtis, R. A. 1971. Flows and wave propagation in ferrofluids. *Phys. Fluids* 14(10): 2096–2102

Dahler, J. S., Scriven, L. E. 1961. Angular momentum of continua. *Nature* 192: 36–37

Druyvesteyn, W. F., Dorleijn, J. W. F. 1971. Calculations on some periodic magnetic domain structures; consequences for bubble devices. *Philips Res. Rep.* 26: 11–28

Eringen, A. C. 1966. Theory of micropolar fluids. *J. Math. Mech.* 16(1): 1–18

Essmann, U., Trauble, H. 1967. The direct observation of individual flux lines in type II superconductors. *Phys. Lett. A* 24(10). 526–27

Faber, T. E. 1958. The intermediate state in superconducting plates. *Proc. R. Soc. London Ser. A* 248: 460–81

Fay, H., Quets, J. M. 1980. Density separation of solids in ferrofluids with magnetic grids. *Sep. Sci. Technol.* 15(3): 339–69

Gailitis, A. 1977. Formation of the hexagonal pattern on the surface of ferromagnetic fluid in an applied magnetic field. *J. Fluid Mech.* 82(3): 401–13

Gogosov, V. V., Naletova, V. A., Shaposhnikova, G. A. 1979. Models of multiphase polarized and magnetized media. *Fluid Mech. Sov. Res.* 8: 96–113

Hughes, J. M., Birnie, R. W. 1980. Density determination of microcrystals in magnetic fluids. *Am. Mineral.* 65(3–4): 396–401

Jackson, R. 1985. Hydrodynamic stability of fluid-particle systems. In *Fluidization*, ed. J. F. Davidson, R. Cliff, D. Harrison, pp. 47–72. New York: Academic. 2nd ed.

Kaiser, R., Rosensweig, R. E. 1967. Study of ferromagnetic liquid. *CFSTI Rep. NASA-CR-91684*, Avco Corp., Wilmington, Mass. 238 pp.

Kamiyama, S., Koike, K., Oyama, T. 1983.

Pipe flow resistance of magnetic fluid in a nonuniform transverse magnetic field. *J. Magnetism Magn. Mater.* 39(1,2): 23–26
Khalafalla, S. E., Reimers, G. W. 1973a. Separating non-ferrous fluid metals in incinerator residue using magnetic fluids. *Sep. Sci.* 8: 161–78
Khalafalla, S. E., Reimers, G. W. 1973b. Magnetofluids and their manufacture. *US Patent No. 3,764,540*
Khalafalla, S. E., Reimers, G. W. 1974. Production of magnetic fluids by peptization techniques. *US Patent No. 3,843,540*
Lucchesi, P. J., Hatch, W. H., Mayer, F. X., Rosensweig, R. E. 1979. Magnetically stabilized beds—new gas solids contacting technology. *Proc. World Pet. Congr., 10th*, Pap. SP-4, 4: 419–25 (discussion). Philadelphia: Heyden
Malik, S. K., Singh, M. 1984. Nonlinear dispersive instabilities in magnetic fluids. *Q. Appl. Math.* 42(3): 359–71
Malik, S. K., Singh, M. 1985. Nonlinear Kelvin Helmholtz instability in hydromagnetics. *Astrophys. Space Sci.* 109(2): 231–39
Martinet, A. 1983. The case of ferrofluids. In *Aggregation Processes in Solution*, Ch. 18, pp. 1–41. New York: Elsevier
Maruno, S., Yubakami, K., Soga, S. 1983. Plain paper recording process using magnetic fluids. *J. Magnetism Magn. Mater.* 39(1,2): 187–89
Matsuki, H., Yamasawa, K., Murakami, K. 1977. Experimental considerations on a new automatic cooling device using temperature-sensitive magnetic fluid. *IEEE Trans. Magn.* 13(5): 1143–45
Medvedev, V. F., Krakov, M. S. 1984. Effect of magnetic liquid on separation of flow in sinusoidal channel. *Magnetohydrodynamics* 20(2): 15–20
Moskowitz, R. 1975. Dynamic sealing with magnetic fluids. *ASLE Trans.* 18(2): 135–43
Moskowitz, R., Rosensweig, R. E. 1967. Nonmechanical torque-driven flow of a ferromagnetic fluid by an electromagnetic field. *Appl. Phys. Lett.* 11(10): 301–3
Nadim, A., Brenner, H. 1985. *Taylor dispersion in concentrated suspensions of rotating cylinders*. Presented at AIChE Ann. Meet., Chicago (Pap. No. 22h)
Neuringer, J. L., Rosensweig, R. E. 1964. Ferrohydrodynamics. *Phys. Fluids* 7(12): 1927–37
Papell, S. S. 1965. Low viscosity magnetic fluid obtained by the colloidal suspension of magnetic particles. *US Patent No. 3,215,572*
Perry, M. P., Jones, T. B. 1976. Dynamic loading of a single-stage ferromagnetic liquid seal. *J. Appl. Phys.* 49(4): 2334–38
Resler, E. L. Jr., Rosensweig, R. E. 1964. Magnetocaloric power. *AIAA J.* 2(8): 1418–22
Romankiw, L. T., Slusarczak, M., Thompson, D. A. 1975. Liquid magnetic bubbles. *IEEE Trans. Magn.* 11(1): 25–28
Rosensweig, R. E. 1966a. Fluidmagnetic buoyancy. *AIAA J.* 4(10): 1751–58
Rosensweig, R. E. 1966b. Buoyancy and stable levitation of a magnetic body immersed in a magnetizable fluid. *Nature* 210(5036): 613–14
Rosensweig, R. E. 1969. Material separation using ferromagnetic liquid techniques. *US Patent No. 3,700,595*
Rosensweig, R. E. 1971. Magnetic fluid seals. *US Patent No. 3,620,584*
Rosensweig, R. E. 1979. Magnetic stabilization of the state of uniform fluidization. *Ind. Eng. Chem. Fundam.* 18(3): 260–69
Rosensweig, R. E. 1985. *Ferrohydrodynamics*. New York: Cambridge Univ. Press. 344 pp.
Rosensweig, R. E. 1986. Lattice model of the magnetic fluid labyrinth. In *Physics of Complex and Supermolecular Fluids*, ed. S. A. Safran, N. A. Clark. New York: Wiley. In press
Rosensweig, R. E., Nestor, J. W., Timmins, R. S. 1965. Ferrohydrodynamic fluids for direct conversion of heat energy. *Mater. Assoc. Direct Energy Convers. Proc. Symp. AIChE Chem. Eng. Ser.*, 5: 104–18, 133–37 (discussion)
Rosensweig, R. E., Zahn, M., Lee, W. K., Hagan, P. S. 1983a. Theory and experiments in the mechanics of magnetically stabilized fluidized solids. In *Theory of Dispersed Multiphase Flow*, ed. R. E. Meyer, pp. 359–84. New York: Academic
Rosensweig, R. E., Zahn, M., Shumovich, R. 1983b. Labyrinthine instability in magnetic and dielectric fluids. *J. Magnetism Magn. Mater.* 39(1,2): 127–32
Rosensweig, R. E., Zahn, M., Vogler, T. 1978. Stabilization of fluid penetration through a porous medium using a magnetizable fluid. In *Thermomechanics of Magnetic Fluids*, ed. B. Berkovsky, pp. 195–211. Washington, DC: Hemisphere
Scholten, P. C. 1983. How magnetic can a magnetic fluid be? *J. Magnetism Magn. Mater.* 39(1,2): 99–105
Shaposhnikov, I. G., Shliomis, M. I. 1975. Hydrodynamics of magnetizable media. *Magnetohydrodynamics* 11(1): 37–46
Shimoiizaka, J., Nakatsuka, K., Fujita, T., Kounosu, A. 1980. Sink float separators using permanent magnets and water based magnetic fluid. *IEEE Trans. Magn.* 16(2): 368–71

Shliomis, M. I. 1972. Effective viscosity of magnetic suspensions. *Sov. Phys. JETP* 34(6): 1291–94

Shliomis, M. I. 1974a. Magnetic fluids. *Sov. Phys. Usp.* 17(2): 153–69

Shliomis, M. I. 1974b. Certain gyromagnetic effect in a liquid paramagnetic. *Sov. Phys. JETP* 39(4): 701–4

Shulman, Z. P., Kordonskii, V., Demchuk, S. A. 1977. Effect of a heterogeneous rotating magnetic field on the flow and heat exchange in ferrosuspensions. *Magnetohydrodynamics* 13(4): 406–9

Skjeltorp, A. T. 1983. Studies of two-dimensional lattices using ferrofluid. *J. Magnetism Magn. Mater.* 37: 253–56

Skjeltorp, A. T. 1985. Ordering phenomena of particles dispersed in magnetic fluids. *J. Appl. Phys.* 57(1): 3285–90

Taktarov, N. G. 1975. Breakup of magnetic liquid jets. *Magnetohydrodynamics* 11(2): 156–58

Tanahashi, T., Sawada, T., Ando, T., Motoicha, I., Torii, H. 1983. A note on the analytical treatment in ferrohydrodynamics. *Bull. JSME* 26: 1509–19

Tani, Y., Kawata, K. 1985. Development of high-efficient fine finishing process using magnetic fluid. *J. Jpn. Soc. Lubr. Eng.* 30(7): 472–76

Tsebers, A. O., Maiorov, M. M. 1980a. Magnetostatic instabilities in plane layers of magnetizable liquids. *Magnetohydrodynamics* 16(1): 21–28

Tsebers, A. O., Maiorov, M. M. 1980b. Structure of interface of a bubble and magnetic fluid in a field. *Magnetohydrodynamics* 16(3): 15–20

Zaitsev, V. M., Shliomis, M. I. 1969. Entrainment of ferromagnetic suspension by a rotating field. *J. Appl. Mech. Tech. Phys.* 10(5): 696–700

Zelazo, R. E., Melcher, J. R. 1969. Dynamics and stability of ferrofluids: surface interactions. *J. Fluid Mech.* 39(1): 1–24

Ann. Rev. Fluid Mech. 1987. 19: 465–91

VON KÁRMÁN SWIRLING FLOWS

P. J. Zandbergen and D. Dijkstra

Department of Mathematics, Twente University of Technology,
7500 AE Enschede, The Netherlands

INTRODUCTION

Swirling flows have many interesting features and occur frequently both in nature and in technology. Large rotating-flow systems are present in the atmosphere, in the oceans, and around the famous Great Red Spot of the planet Jupiter. It is apparent that the damage that can be caused by tornadoes is sufficient reason in itself to study this subject, but it is certainly not the only one. Technical applications can be found in the fabrication of computer memories by crystal-growth processes (Langlois 1985), in viscometry, in lubrication, and in the area of rotating-flow machinery such as centrifuges and turbines.

Two special cases of interesting and intriguing rotating-flow problems are the flow above an infinite rotating disk and between two infinite coaxial rotating disks. The first publication on these problems appeared 65 years ago and was written by one of the giants of fluid dynamics: Theodore von Kármán (1921). He considered the problem of the flow induced by an infinite rotating disk where the fluid far from the disk is at rest. By using a similarity principle, he was able to reduce the full system of Navier-Stokes equations to a pair of nonlinear ordinary differential equations in the axial coordinate.

The von Kármán similarity principle still applies when the problem is generalized (Batchelor 1951) to include the case where the fluid itself, far from the disk, is rotating as a solid body. This introduces the parameter s, which is the ratio of the angular velocity of the fluid at infinity to the angular velocity of the disk. At the disk surface, suction or injection can also be admitted (Stuart 1954, Kuiken 1971). Another generalization is to consider the flow between two infinite coaxial rotating disks. This intro-

0066–4189/87/0115–0465$02.00

duces a second parameter—the Reynolds number based on the distance separating the two disks. Note that the problem of one disk does not contain a characteristic length and hence no Reynolds number.

For reasons of later reference and also to show more clearly the relation between the two problems described above, we give the equations at this stage. If we use a cylindrical coordinate system (r, ϕ, z) with the z-axis coinciding with the axis of rotation, the corresponding velocity components (u, v, w) according to the von Kármán similarity principle are given by

$$u = r\Omega f'(x), \qquad v = r\Omega g(x), \qquad w = -2(\nu\Omega)^{1/2} f(x). \tag{1}$$

Here Ω is a suitable characteristic angular velocity, the prime denotes differentiation with respect to the nondimensional axial coordinate $x = z(\Omega/\nu)^{1/2}$, and ν is the kinematic viscosity.

The incompressible, laminar, stationary Navier-Stokes equations for the one-disk case now reduce to the von Kármán similarity equations

$$f''' + 2ff'' = f'^2 + s_1^2 - g^2, \tag{2}$$

$$g'' + 2fg' = 2f'g, \tag{3}$$

with boundary conditions

$$x = 0, \qquad f = f_0, \qquad f' = 0, \qquad g = s_0, \tag{4}$$

$$x \to \infty, \qquad f' \to 0, \qquad g \to s_1. \tag{5}$$

For the two-disk problem, the equations can be given as

$$f'''' + 2ff''' = -2gg', \tag{6}$$

$$g'' + 2fg' = 2f'g. \tag{7}$$

The boundary conditions now are

$$x = 0, \qquad f = f_0, \qquad f' = 0, \qquad g = s_0, \tag{8}$$

$$x = \mathrm{Re}^{1/2}, \qquad f = f_1, \qquad f' = 0, \qquad g = s_1. \tag{9}$$

Fundamental parameters governing the problem are the Reynolds number Re and the ratio parameter s (occasionally replaced by σ), given by

$$\mathrm{Re} = H^2\Omega/\nu, \qquad s = s_1/s_0, \qquad \sigma = s_0/s_1, \tag{10}$$

where H is the dimensional distance separating the disks. This distance, which appears as $\mathrm{Re}^{1/2}$ in (9), may be rescaled to 1, as many authors do. Over 10 different notations are encountered in the literature. For the connection between these notations, see Holodniok et al. (1981).

Clearly, Equation (6) can be integrated to give

$$f''' + 2ff'' = f'^2 + \kappa - g^2, \tag{11}$$

where κ is a constant of integration associated with the pressure. For the two-disk problem κ is unknown, but in the single-disk case it is known to be s_1^2, so that (11) and (2) are identical in this case.

The original von Kármán problem is governed by Equations (2)–(5) with $s_0 = 1$, $f_0 = s_1 = 0$. The first adequate numerical solution was obtained by Cochran (1934), and his results show that f, $f' > 0$ while $g' < 0$. Hence the fluid is drawn in axially and thrown out radially, in agreement with physical intuition (centrifugal fan).

Related to the original von Kármán problem is the case where the disk is at rest while the fluid at infinity is in solid-body rotation ($s_0 = 0$, $s_1 = 1$). This problem was considered by Bödewadt (1940), and his solution exhibited oscillations in the velocity components, in contrast with the monotonic behavior in the von Kármán case.

Further significant progress was made in the early 1950s by Batchelor (1951) and Stewartson (1953). Based on qualitative arguments, Batchelor predicted that in the case of the two-disk problem with only one disk rotating, the fluid in the core would rotate with constant angular velocity and boundary layers would be present at each disk. He also considered the case where both disks rotate at the same speed but in opposite senses ($s = -1$). Batchelor argued that in this case the main body of the fluid would be counter-rotating in two parts, with a transition layer in between and, of course, with boundary layers at the disks.

Stewartson (1953), on the other hand, claimed that in both cases considered above the fluid in the core would not rotate at all. He predicted these results on the basis of a low-Reynolds-number expansion and an experiment with disks in open air. Since the results calculated by Bödewadt (1940) did not fit his predictions, Stewartson claimed that the Bödewadt solution was inaccurate. This rejection is unjustified, since it is now known that the Bödewadt solution is essentially correct.

The views developed by Stewartson and Batchelor were so conflicting, but forwarded by two of the great men of fluid dynamics, that this in itself would be reason enough to raise considerable interest in the problem. Since then, especially in the last 25 years, there has been a continuous flow of papers both on the one-disk and the two-disk problem. Both problems have been attacked from many sides. There is a whole body of literature on rigorous mathematical questions about existence and (non)uniqueness of solutions, as well as many papers on the numerical calculations of solutions. These problems have been attacked by asymptotic expansions to clarify the behavior at infinity for the one-disk case or to study in detail various peculiarities of the solutions. Also, use has been made of parameter power-series expansions.

For the two-disk problem there are a multitude of solutions at sufficiently high Reynolds number, as was first exhibited by Mellor et al. (1968). Many of these solutions are closely connected to the one-disk problem, but others are not. There are only a few papers in the literature where these connections have been recognized (Zandbergen 1980).

For the one-disk problem it appears that solutions do not exist in all cases (Evans 1969), and further that there are families of infinitely many solutions for certain ranges of the parameter $s = s_1/s_0$ (Zandbergen & Dijkstra 1977).

We believe that there are at least two reasons for the continuous interest in the present subject. The first reason is that the system of governing equations considered here is a system of ordinary differential equations that still is an exact representation of the Navier-Stokes equations. The second is that the solution appears to be nonunique.

We have chosen to review the literature by showing the main characteristics of any solution. We do this by first stating a number of general facts about certain features of a solution. Then we turn to a survey of the one-disk problem, followed by a discussion of the two-disk problem. In each of the two cases the more fundamental mathematical aspects are considered first, with numerical solutions and expansion methods discussed subsequently.

In order to demonstrate more clearly the relation between solutions to the one-disk and the two-disk problem, we have devoted a separate section to this question.

Another very interesting problem is the stability of the stationary solutions to the one-disk and two-disk problems. Here, only a few results have been obtained, and these stability results are also considered in a separate section.

Finally, we consider the case where the radii of the disks are finite. The question of whether and to what extent the Kármán swirling flow occurs in a finite radial geometry is of course quite interesting and of fundamental importance, particularly when experimental results are considered.

It should be remarked that reviews by other authors on certain aspects of the present problem have been used intensively (McLeod 1975, Parter 1982, van Wijngaarden 1985, Brady & Durlofsky 1986).

BASIC FACTS AND BASIC INVISCID SOLUTIONS

There are some basic facts with regard to a possible solution that play a fundamental role in understanding the behavior of any solution of the equations.

McLeod (1971, 1975) has observed that the quantity

$$\phi = f''^2 + g'^2 \tag{12}$$

is governed by the differential equation

$$\phi'' + 2f\phi' = 2(f'''^2 + g''^2). \tag{13}$$

Since the right-hand side is always nonnegative, simple manipulation leads to the following lemma:

Lemma 1 For any solution of Equations (2, 3) or (6, 7), ϕ' is either identically zero or has at most one zero (say x_0). In the latter case, we have sign (ϕ') = sign $(x - x_0)$.

Now it is easy to derive a similar lemma for

$$\psi = f'f''' + g'^2, \tag{14}$$

since this quantity satisfies

$$\psi' + 2f\psi = \phi'/2. \tag{15}$$

Lemma 2 For any solution of the equations, ψ is either identically zero or has at most two zeros. In the latter case, ψ is positive before the smallest zero.

It seems that this last lemma has not been observed in the literature. The two lemmas are extremely useful when deriving rigorous results, as is outlined in the next section.

It is evident from many papers on multiple solutions that sequences of "large humps" or "hills" almost always occur that are amenable to matched asymptotic expansions. Hence, many papers on these aspects are available (Tam 1969, Rasmussen 1971, Kuiken 1971, Ockendon 1972, Watts 1974, Bodonyi 1975, Matkowsky & Siegmann 1976, Dijkstra 1980, Kreiss & Parter 1983). Dijkstra considered the possibility of chains of "large-hill" solutions and derived asymptotic results for such chains. In effect, these are large regions of inviscid flow separated from each other by viscous transition layers. In the inviscid regimes the highest order derivatives in the equations may be neglected, with the result that

$$f(x) = -\lambda \sin^2 \mu(x - x_0) - C/(4\lambda\mu^2), \qquad g(x) = \pm 2\mu f(x). \tag{16}$$

With $\lambda/\mu \gg 1$ this *basic inviscid solution* formally satisfies the full equations to leading order, and the solution extends over the range

$$0 < x - x_0 < \pi/\mu. \tag{17}$$

At the end points, singularities develop in higher-order terms (Kuiken

1971), but the solution can be continued through viscous layers at each end.

The constant C in (16) can be identified with κ in (11) for the two-disk case, and it matches s_1^2 in Equation (2) for the case of one disk.

In the literature, at least two fundamental limits of (16) are encountered, namely

$$\lambda \to \infty, \qquad \mu \to 0, \qquad \lambda\mu^2 \to \text{constant}, \tag{18}$$

$$\lambda \to \infty, \qquad \mu \to 0, \qquad \lambda\mu^5 \to \text{constant}. \tag{19}$$

The limit (18) has been considered, for example, by Dijkstra (1980), while examples of the stronger inviscid case (19) can be found in papers by Watts (1974) and Kreiss & Parter (1983). In the last paper, a rigorous proof of the validity of (16) for the two-disk problem is given.

In a number of papers the term in (16) containing C is ignored, since it is of minor importance for the description of the basic inviscid solution itself. This term becomes important, however, in the limit (18) as the solution approaches the viscous regions at the end points of the range (17). With regard to such a viscous region, we now summarize the basic results as obtained in many papers. With $x_0 = 0$, we have to leading order in the viscous sublayer

$$f(x) = -(Ax^2 + C/A)/2, \qquad A = 2\lambda\mu^2 > 0, \tag{20}$$

while g satisfies

$$g'' - (Ax^2 + C/A)g' + 2Axg = 0. \tag{21}$$

It is precisely this linear differential equation and asymptotic properties of its solution that are fundamental in investigations where solutions of type (16) are considered.

The Case $C = 0$

A special case of Equation (21) that frequently occurs in the literature is the situation where $C = 0$ or $A \to \infty$. In this case the equation can be reduced to Kummer's equation, and in terms of standard confluent hypergeometric functions the general solution is

$$g(x) = \alpha U(-2/3, 2/3, \xi) + \beta e^{\xi} U(4/3, 2/3, -\xi), \qquad \xi = Ax^3/3, \tag{22}$$

where α and β are arbitrary constants. The asymptotic behavior is given by

$$U(-2/3, 2/3, Ax^3/3) \sim \tfrac{1}{2}(3 \text{ sign } x - 1)\,(A/3)^{2/3}x^2, \qquad A|x|^3 \to \infty, \tag{23}$$

while the function multiplying β increases exponentially as $x \to \infty$ and becomes exponentially small in the other direction.

A direct consequence of (23) is that with $\beta = 0$ and $\alpha \neq 0$, the solution (22) has the property

$$g''(-\infty)/g''(+\infty) = -2. \tag{24}$$

Moreover, g changes sign exactly once.

Now, consider two adjacent solutions of basic inviscid type (16) with parameters interconnected through a viscous sublayer at $x_0 = 0$. The parameters of the solution for $x > 0$ are given by (λ_1, μ_1), and those for $x < 0$ by (λ_2, μ_2). With the $\pm$ sign in (16) absorbed in g, a matching argument yields to leading order

$$\lambda_2\mu_2^2/(\lambda_1\mu_1^2) = 1, \qquad \mu_2/\mu_1 = 2. \tag{25}$$

Hence, the amplitude λ in (16) becomes four times as large, whereas the range given in (17) grows by a factor of two as the solution passes through the viscous layer in the positive direction, provided that $C = 0$. A chain of these solutions can be constructed in the same way. Second-order results have been given by Dijkstra (1980).

Another possibility for continuing a solution (16) in the positive direction is the direct match through a viscous region with the asymptotic boundary conditions (5) of the one-disk problem for $s_1 = 0$ (Kuiken 1971, Ockendon 1972). The argument hinges on the fact that Equation (2) for f to leading order reduces to

$$f''' + 2ff'' = f'^2. \tag{26}$$

Hence, the system (2, 3) decouples as above. Apparently, such a simple argument does not apply to the case where s_1, and hence C, can no longer be neglected.

General Values of C

As far as we know, the case where $C < 0$ in Equation (21) has not been explored in the literature. Dijkstra (1980) considered the case where $C \geq 0$. The solution of (21) then essentially depends on

$$a = CA^{-4/3}, \qquad A = 2\lambda\mu^2, \tag{27}$$

and the value -2 in (24) must be replaced by $\rho(a)$. Consequently, the ratio μ_2/μ_1 in (25) becomes $|\rho|$. Some particular values of ρ are the following:

a	0	0.2	0.4	1.00	1.5	5.00	∞	
ρ	-2	-1.48	-1.07	-0.31	0.05	0.75	1.	(28)

It follows that chains of basic inviscid solutions (16) are possible with increasing or decreasing ratios of the amplitude λ, depending on the value of a.

If a has the value 0 [i.e. $C = 0$ or $A \to \infty$ in Equation (27)], then $\rho = -2$, which is the case discussed in the preceding subsection and the case that almost exclusively appears in the literature cited earlier in this section. For the general one-disk problem, the value of s_1 in Equation (5), and hence C, will be nonzero. This means that in a sense there is too much emphasis in the literature on the special case $C = 0$.

The exploration of second-order effects for $C \neq 0$ has not been encountered in the literature.

The direct continuation of a solution (16) into the asymptotic behavior (5) of the one-disk solution was considered by Zandbergen (1980). A crucial result in his investigation is the so-called *circulation shock*, where the positive half-space is close to rigid rotation ($f \sim 0$, $g \sim 1$), while in the other half-space there is no rotation at all ($g \to 0$ exponentially as $x \to -\infty$) and f is given by (20). This is equivalent to the solution (16) in the limit given in (18). In general, a small perturbation in the positive half-space of this circulation shock will produce a long chain of basic inviscid solutions in the negative half-space.

THE ONE-DISK PROBLEM

The mathematical definition of the one-disk problem has been given in the Introduction [Equations (2)–(5)]. The fundamental parameter is the ratio s (or σ) given by Equation (10).

Existence and Uniqueness

The early rigorous results for the one-disk case were obtained by von Kármán & Lin (1961) and Howarth (1961). In their papers an existence proof is given for the case $s = 0$ with suction f_0 sufficiently large.

A large number of papers on existence appeared in the years around 1970. McLeod (1969a) removed the condition on f_0, and he was the first to prove existence of a solution for $s = 0$ and arbitrary values of f_0, which includes the original case $f_0 = 0$ of von Kármán (1921). The solution appears to have the property that both f' and g' do not change sign. This implies monotonicity of the axial and angular velocity, in agreement with the numerical solution obtained by Cochran (1934). A more general class of differential equations was considered by Lan (1971), whose existence proof for $s = 0$ was different from the one by McLeod.

For general values of s there is no monotonicity of the velocities, and the problem is much harder to tackle. Near rigid rotation ($s \sim 1$) or for sufficiently large suction, existence results have been obtained by Watson (1966), Hartman (1971), and Bushell (1972). Hartman showed that the generalized problem with $f'(\infty) \neq 0$ has a solution only when $s = 0$.

Again, McLeod (1971) was the first to prove a most complete theorem that guarantees the existence of a solution for $s \geq 0$ and arbitrary values of f_0. This solution appears to have an angular velocity with constant sign, which does not necessarily imply that any solution has this property. Uniqueness of the solution considered by McLeod has not been proven, although numerical results hint at this. Of course, there is sufficient numerical evidence (for instance, for $s = 0$, $f_0 = 0$) to know that the problem has infinitely many other solutions, but in these others g is no longer of constant sign (Zandbergen & Dijkstra 1977, Lentini & Keller 1980, Dijkstra 1980).

A shorter proof of McLeod's theorem was given by Hartman (1972). McLeod's proof hinges on Lemma 1 as formulated in the previous section. This lemma implies that the quantity ϕ defined by (12) is either identically equal to zero or decreases monotonically to zero from above as $x \to \infty$, for any solution of the problem (McLeod 1970). In addition, we have $-2f_0\phi(0)+f''(0)\,(s_1^2-s_0^2) < 0$, and this means that for $s = 1$ the only possibility is the trivial solution $f = f_0$, $g = s_0$, whereas the case $s = -1$, $f_0 \leq 0$ does not possess a solution. It should be noted that for $s = -1$ and f_0 sufficiently large, there does exist a solution (Watson 1966).

The existence results reviewed above as well as other papers have been summarized by McLeod (1975).

The full consequences of Lemma 2 in the preceding section have not been explored, but an immediate result is that $\psi > 0$ for the one-disk case and hence that $|f''(0)| < |g'(0)|$, provided that $f > 0$. A similar result holds when $f < 0$.

A final rigorous result concerns the asymptotic behavior at infinity. McLeod (1969a,b) showed that if there is a solution of the problem, then it must have a specific asymptotic behavior as $x \to \infty$. This behavior may be obtained in the obvious way by linearizing the equations about the state at infinity.

Expansions

Rogers & Lance (1960) have performed the procedure of linearizing about the state at infinity (i.e. $f = f_\infty$, $g = s_1$) for general values of s_1 in condition (5). In complex notation this gives

$$f'(x)+ig(x) \sim is_1 + c\exp(\gamma x), \qquad x \to \infty, \tag{29}$$

where c is an arbitrary complex constant, while γ has a negative real part p and satisfies

$$\gamma^2+2f_\infty\gamma-2is_1 = 0, \qquad \gamma = p+iq. \tag{30}$$

If the boundary conditions at the origin are ignored and s_1 is prescribed, then the behavior (29) essentially depends on three real parameters, namely

f_∞ and the complex constant c. It should be remarked that the decay (29) is oscillatory unless $s_1 = 0$, which gives $q = 0$. For this case Cochran (1934) extended (29) to four terms, while Bödewadt (1940) used six terms for the case $s_1 = \infty$, i.e. $\sigma = 0$. The Cochran approximation was later extended by Benton (1966) and Ackroyd (1978) to give more accurate results for the original von Kármán problem.

The full asymptotic expansion contains only exponentials and has been analyzed by Dijkstra (1978). He showed that sufficiently close to rigid rotation, the expansion converges over the full range $x \geq 0$ and satisfies the boundary conditions at the origin for a nonporous disk with appropriate values of f_∞ and c. The numerical details have been worked out by P. Gragert & D. Dijkstra (unpublished) up to 30 terms.

Another method of expansion is to linearize the flow over the full range $x \geq 0$ about the state of rigid rotation (Rogers & Lance 1960). For the case $f_0 = 0$ this method produces a power series in powers of $s-1$ or, if required, $\sigma - 1$. The series consists of exponentials multiplied by polynomials and has been computed by Van Hulzen (1980) up to 12 terms using formula manipulation with exact rational coefficients. An estimate of the radius of convergence was found to be $|s-1| \leq 1.16$, with a singularity developing at $s = -0.16$. In an attempt to elucidate the structure of this singularity, Weidman & Redekopp (1976) linearized the equations about the solution at $s = 0$ and computed eight terms of the expansion in powers of s. They located the singularity at $s = -0.154$, with a power 4/3 appearing there. Later, Zandbergen & Dijkstra (1977) showed that both the location of the singularity and the power 4/3 are incorrect.

Numerical Solutions

The first reliable numerical solution of a one-disk problem was given by Cochran (1934), who considered the original von Kármán case $s = 0$, $f_0 = 0$. Bödewadt (1940) calculated results for a disk at rest in a rigidly rotating fluid. His numerical solution has been subject to criticism (Stewartson 1953), but this criticism is unjustified.

A real step forward was the paper by Rogers & Lance (1960), which is outstanding in many ways. They were the first to give adequate numerical results for general nonnegative values of s with no suction, and they also considered some cases with $f_0 \neq 0$. Moreover, they presented the asymptotic behavior (29) of the solution at infinity and calculated some terms of the perturbation expansion about $s = 1$. They used a shooting method from the disk toward infinity, which is rather cumbersome. The reason is that Equation (30) not only has a solution with $p < 0$ but also one with $p > 0$, and this may easily lead to divergence of the numerical solution when integrating in the direction of increasing x. Rogers & Lance also considered some values of $s < 0$ and concluded that below -0.2 no solution could be found unless suction was admitted.

The influence of suction has been considered in detail by Evans (1969). Just like Rogers & Lance, he replaces infinity by $x_m = 12$, but he uses a shooting method in two directions starting at 0 and 12 and proceeding inward to the midpoint of the range of integration. Evans found that in the case of $f_0 = 0$, no numerical solution could be obtained in the range

$$-1.35 < s < -0.161, \tag{31}$$

and further that at $s = -1$ a numerical solution of the form (16) exists provided that $f_0 \geq 0.1$. Note that according to the rigorous papers cited above, the problem with $s = -1$ and $f_0 = 0$ has no solution.

Zandbergen & Dijkstra (1977) designed a highly accurate method with the aim of clarifying the behavior of the no-suction solution as s approaches -0.16 from above. They extended (29) with an extra term to second order and used the result at $x = x_m$ to provide starting values for a shooting technique toward the disk. This incorporation of the asymptotic behavior means that the method essentially shoots from infinity down to $x = 0$. The boundary conditions at the disk were iteratively satisfied by exploiting the three real parameters on which the behavior (29) depends. In this way they could reach a precision of at least 10 decimal places. Near the critical point $s_{cr} = -0.16$, the quantity f_∞ was used instead of s as a prescribed parameter, and it was found that the solution had the form

$$f(x) = f_1(x) \pm (s - s_{cr})^{1/2} f_2(x) + O(|s - s_{cr}|), \qquad s_{cr} = -0.16053876, \tag{32}$$

which disproves the predictions by Weidman & Redekopp (1976). It is clear that a second solution branch merges with the first one at $s = s_{cr}$, and Zandbergen & Dijkstra continued the second branch up to $s = +0.07452563$, where a third branch appeared. In a subsequent paper, Dijkstra & Zandbergen (1978) indicated an infinity of solution branches oscillating around $s = 0$ (Figure 1). All the higher branches are marked by the occurrence of the basic inviscid solution (16) or a chain of such solutions. Qualitatively, it can be stated that each time the solution branch folds, another basic inviscid solution is added onto the already existing chain (Zandbergen 1980, Dijkstra 1980).

Similar solutions have been obtained by Lentini & Keller (1980) and White (1978). The first authors used a finite-difference technique and arclength continuation to proceed from one solution branch to the next. Like Zandbergen & Dijkstra (1977), they exploited the asymptotic behavior (29) to reduce the x-range of the calculation. However, they applied two asymptotic relations with first-order accuracy, whereas the third relation employed was only of order zero. This is probably the reason why their results do not possess the full accuracy claimed.

We now turn to the left end point of the range (31), where the structure of the solution has been elucidated by Ockendon (1972) and Bodonyi (1975). In her pioneering paper Ockendon showed by matched asymptotic expansions that a basic inviscid solution (16) can be constructed, which, through viscous sublayers near the end points (17), satisfies the boundary conditions at the disk and at infinity. She was the first to point out that a chain of basic inviscid solutions was also possible where the last specimen of the chain merges with the behavior of the solution at infinity. This means that many nonunique solutions, later on presented by several authors, can be traced back to the paper by Ockendon, and this also refers to the case of two disks.

A key role in the theory appears to be played by the two values $s = s^* = \pm 1.43553098$, or

$$s^{*^2} = \kappa^* = 2.06074919. \tag{33}$$

It is illuminating to summarize the argument leading to (33). For simplicity we do so by means of the theory of Bodonyi so that we can avoid the small suction used by Ockendon.

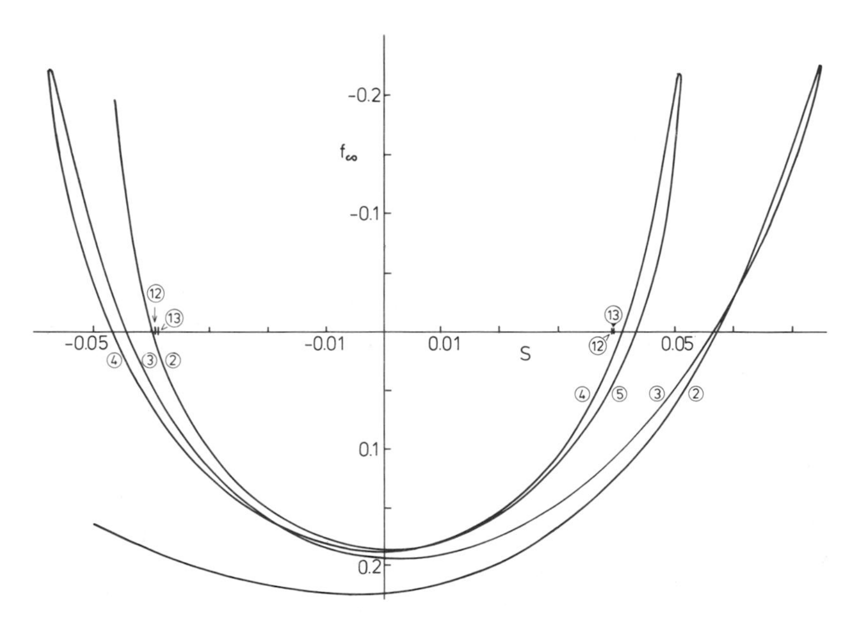

Figure 1 Values of f_∞ for small s on the higher solution branches. Basic branch and part of branch 2 have been omitted.

With suction $f_0 = 0$ and $g(0) = s_0 = 1$ we consider in terms of (16) and (20) the limit $\mu \to 0$, $A \to \infty$, so that λ also tends to infinity. To leading order, the term C/A in (20) is zero and f satisfies the conditions at the disk. Near the disk, g is given by (22) with constant $\beta = 0$ to avoid an exponential explosion as x increases. The constant α follows from the condition $g(0) = 1$, which completes the leading order. With the known solution g, one returns to the equation for f; it can be shown that to prevent exponential growth at this stage, the constant s_1^2 in (2) must have a particular value, namely (33).

The theory by Bodonyi appears to be in good agreement with his numerical solution as $s \to -1.4355$ from below. However, he obtained only one solution, whereas the paper by Ockendon implies that a whole family of solutions should exist there and also at $s = +1.4355$. Both these families have been constructed by Zandbergen (1980) in an apparently rather unknown paper. The qualitative behavior of the solution branches are depicted in Figures 2 and 3. The approach by Zandbergen arises from an argument applied from $x = \infty$ down to $-\infty$ and may produce a solution for any point on any branch. Moreover, it reveals the intimate connection between the branches and the families.

This approach departs from Equation (29), which in general contains

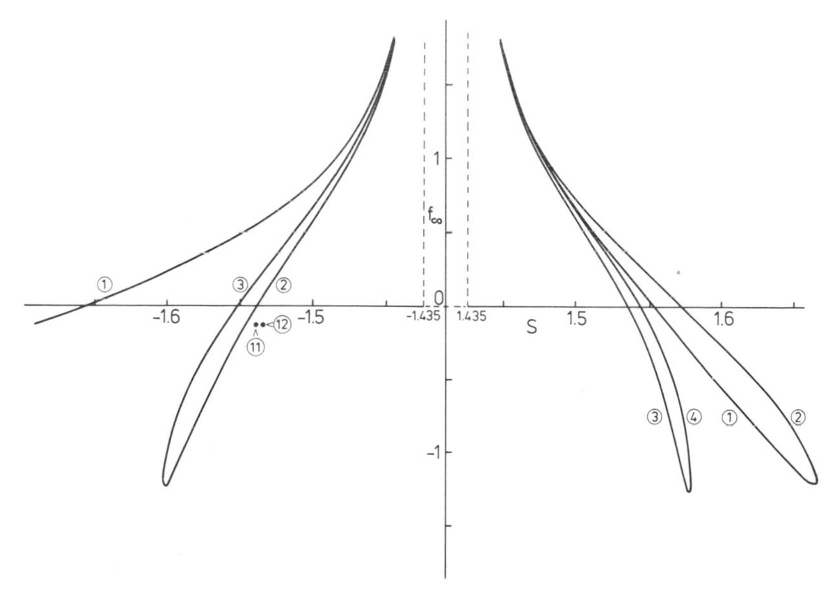

Figure 2 Behavior of f_∞ for values of s near the critical values $\pm s^*$ [Equation (33)]. The connection of branches takes place at infinity.

an infinite number of points $x = x_n$ where f' vanishes. At these potential disk locations the value $g(x_n)$ will be different from s_1, but the difference tends to zero as $x_n \to \infty$. With this interpretation, the asymptotic solution (29) is in fact the state of almost-rigid rotation, so that one may equally as well start from this state. When an almost-rigid rotation solution is continued beyond the disk at $x = 0$ in the negative direction and the suction f_0 is varied, it is possible to construct a circulation shock, as outlined above. As $x \to -\infty$ the solution f then is the parabola (20), while g becomes exponentially small, similar to the result (22) with $\alpha = 0$. If such a circulation shock is perturbed, e.g. by a perturbation δf_0 of the suction f_0, then the constant α no longer vanishes; as a consequence, g will start to behave like a multiple of x^2 [Equation (23)]. Hence, a basic inviscid solution (16) develops where μ will be small with the perturbation δf_0. Beyond this basic inviscid solution a chain of such solutions can be constructed with global behavior dictated by (28). The number of specimens in the chain will depend on the magnitude of μ, i.e. on the value of δf_0. In the viscous transition layer between two adjacent basic solutions, there exists a point x_k, say, where f' vanishes and f is small. In fact $f(x_k)$ can be made to vanish simultaneously with f' for a suitable perturbation δf_0. If the original disk at $x = 0$ is removed to position x_k, a zero-suction solution to the problem is constructed and the value $g(x_k)$ determines the ultimate value of s, which almost always appears to be close to either 0 or $\pm s^*$, the square root of κ^* [Equation (33)].

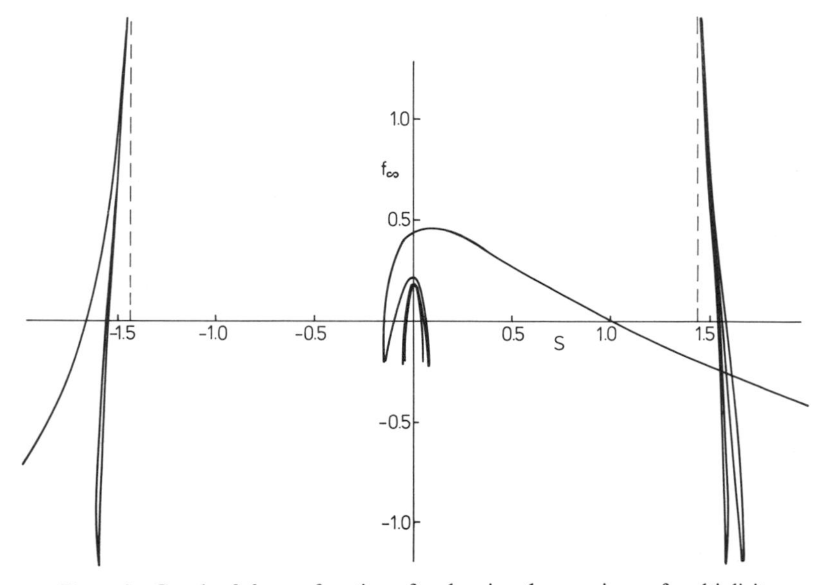

Figure 3 Graph of f_∞ as a function of s, showing three regimes of multiplicity.

It should be noted that the theory above does not apply when $s = 0$, but adequate asymptotic results are available to fill this gap (Kuiken 1971, Dijkstra 1980).

The argument can be repeated for other suitable locations x_k, which means that the solution families appearing in different s-regimes are in fact intimately connected and may be traced back to one root, namely a circulation shock.

It should be observed that for the exact state of a circulation shock, the quantity ϕ [Equation (12)] has a finite value at $x = -\infty$. In general, this does not hold for a perturbation of the shock, since in that case the solution runs into a singularity before it reaches $x = -\infty$. With a singularity located at $x = x^*$ and $\xi = x - x^*$, the behavior is given by

$$f \sim 2/\xi + D^2 (\cos 2\theta - 5\sqrt{7} \sin 2\theta - 22)/352 + E\sqrt{\xi} \cos \Phi,$$

$$g \sim D \sin \theta/\xi^{3/2}.$$

where $\theta = \frac{1}{2}\sqrt{7} \log \xi + \theta_0$, $\Phi = \frac{1}{2}\sqrt{15} \log \xi + \Phi_0$.

In a sense this result obtained by Zandbergen (1980) may be interpreted as the general solution of the governing equations, since it contains five arbitrary constants, namely x^*, D, E, θ_0, and Φ_0. It is remarkable that this singular solution has not been exploited in the literature.

As we conclude this section on the one-disk problem, we note that if the existence of a circulation shock could be proven with rigor, then techniques employed by Kreiss & Parter (1983) for the two-disk configuration could be used to prove the existence of any of the solutions discussed above.

THE TWO-DISK PROBLEM

The definition of the two-disk problem was given in Equations (6)–(11). The suction or blowing quantities f_0 and f_1 are zero.

On the one hand, the two-disk problem might look easier than the case of one disk, since now the range is finite; on the other hand, we have here the added complexity of the Reynolds number Re.

Nonunique solution structures open up as Re grows above ~ 55. However, the question of nonuniqueness goes back further than 3 April 1974, the date when Watts published his value of 55; the question about uniqueness was first discussed in the early 1950s and can be phrased concisely as follows: Does the core rotate (Batchelor) or not (Stewartson)?

Existence and Uniqueness

The first worker to give an existence theorem for the flow between two disks was Hastings (1970). By means of a fixed-point theorem, he proved existence for $\text{Re} = 1$ and both s_0 and s_1 sufficiently small. In a frequently cited paper, Elcrat (1975) claims to have proven existence for a much larger region of values (s_0, s_1) than Hastings. Elcrat essentially assumes that f'' remains bounded and finds that there exists a solution provided that

$$0 \leq H^2 s_0 < B \quad \text{and} \quad -s_0 \leq s_1 \leq 0$$

or

$$H^2(s_0^2 - s_1^2) < B^2 \quad \text{and} \quad 0 < s_1 \leq s_0.$$

His results heavily depend on the property that the function g in these cases should be monotonic. However, the lemma in which he claims to have proven this property is false in the last case where the disks are rotating in the same sense. This means that existence has not been proven for this case.

In contrast with Elcrat, McLeod & Parter (1977) proved that g will not be monotonic if $0 < s_1 < s_0$ provided that Re is sufficiently large, depending on s_0 and s_1.

In an earlier paper McLeod & Parter (1974) considered the case $s = -1$, i.e. the disks are counterrotating at the same speed. The existence of an odd solution was proven for all Re > 0. Here, "odd" means that both f and g are odd functions with respect to the midpoint between the disks. Moreover, McLeod & Parter gave an asymptotic analysis for Re $\rightarrow \infty$ and concluded that their solution consists of a boundary layer at each disk, while in the main body f is linear and g vanishes exponentially when leaving the neighborhood of each disk. In both the boundary layers the solution is virtually the same as that obtained by Cochran (1934) for the original von Kármán problem. Hence, McLeod & Parter have shown the existence of a Stewartson-type solution.

An interesting and impressively rigorous analysis for the two-disk configuration was given by Kreiss & Parter (1983). They proved the existence and nonuniqueness of solutions at sufficiently large Re. The structures considered are of the form (16) or are a chain of such expressions. In fact, they give a rigorous analysis of a case originally studied by Watts (1974), who computed a solution of the type (16) at Re = 57. Following the work of Kuiken (1971) and Ockendon (1972), Watts also presented a matched asymptotic analysis and derived explicit leading-order results. In terms of the present notation, his solution is given by (16) with $x_0 = 0$ and (19) the relevant limit. With normalization to $s_0 = 1$, the parameters are

$$\mu = \pi/\sqrt{\text{Re}}, \qquad \lambda = \frac{3}{2}\left[\frac{\Gamma(2/3)}{\Gamma(1/3)}\right]^3 \mu^{-5}. \tag{34}$$

For this value of μ and arbitrary λ, the solution (16) with $C = 0$ satisfies the homogeneous boundary conditions on f at both disks, so that we may call it an eigenfunction. The amplitude λ, however, follows from the condition $g(0) = 1$ and a matching argument identical to the reasoning presented after Equation (33).

Although Watts does not give a value for the constant κ in (11), it is not surprising that κ will approach κ^* [Equation (33)] in the limit considered (Habers 1984). On the other hand, it is surprising that the solution is virtually independent of s_1: The boundary condition on g at the second disk can always be satisfied (Watts 1974, Habers 1984). The reason is that in this boundary layer, the solution g is again (22), where now x measures distance from the second disk and α follows from a matching condition while β can be chosen so as to satisfy the boundary condition on g. The function multiplying β decreases exponentially in this case and does not influence the global solution in the main body between the disks.

The Watts solution consists of one single basic inviscid component (16), but this is not the only possibility. For Re sufficiently large, Kreiss & Parter (1983) have proven the existence of solutions consisting of a chain of n basic inviscid components, and they derive formulas governing the global behavior of such a chain. It seems that they were unaware of the work by Dijkstra (1980), whose results for a chain have been given in the section on basic facts. For the present case this leads to (25), which is exact in the limit considered and an immediate consequence of the standard solution (22). This means that there is no need for the analysis contained in the appendix of the paper by Kreiss & Parter or in the survey by Parter (1982).

Prior to these papers, Kreiss & Parter (1981) investigated the asymptotic properties of certain solutions, but for a summary of fundamental aspects we refer to Parter (1982). He raises a number of open questions from which it may be concluded that the Batchelor-type solutions are apparently hard to tackle rigorously.

Numerically, however, much more is known about this type of solution, but let us first consider some low-Reynolds-number expansions.

Expansions

If the two-disk problem is normalized such that the range (9) becomes $[0, 1]$ and s_0 is taken to be unity, then it is found for small Re that an expansion proceeding in powers of Re^2 can be constructed where each term is a polynomial in the normalized coordinate $x/\mathrm{Re}^{1/2}$ and $s = s_1/s_0$.

Stewartson (1953) considered this expansion, but only the first term was presented. Hoffman (1974) computed eight terms of the expansion and found that convergence was limited due to square-root behavior in the complex Re-plane for negative Re^2. By means of rational approximation techniques, the Reynolds-number regime of validity could be extended considerably.

Comparable results have been obtained by Hulshof (1983) with formula manipulation.

Numerical Solutions

The first important paper of a large series of papers devoted to numerical solution of the two-disk problem was again due to Lance & Rogers (1962). They used a Runge-Kutta shooting technique to solve the equations for several values of s in the range $|s| \leq 1$ and Re up to 1000. For larger Reynolds numbers their results agreed with the view of Batchelor, e.g. for $s = 0$ and $s = -0.3$. These solutions show a definite plateau in g, that is, a definite core rotation. For the particular case $s = -1$ Lance & Rogers took advantage of the odd character of the solution by integrating over half of the range, and a definite Stewartson solution without core rotation was the result at Re = 1023.

Further support for the Batchelor view was advanced by Pearson (1965) in a time-dependent finite-difference calculation. His Batchelor-type solution for $s = 0$ and Re = 1000 demonstrates that this flow is stable in time. For the case $s = -1$, however, the steady-state motion was completely different from the Lance & Rogers result. In fact, the flow computed by Pearson at $s = -1$ shows the first numerical evidence for the basic inviscid solution (16), which differs from the predictions of both Batchelor and Stewartson for the counterrotating case (Tam 1969).

When the results discussed above are combined, it will be clear that for $s = -1$ and Re = 1000 there was already nonuniqueness in 1965, but the first demonstration of this feature was the milestone paper by Mellor et al. (1968). They considered the flow between a rotating and a stationary disk (that is, $s = 0$) and found numerical solutions of both Batchelor and Stewartson type for Re above ~ 200. In addition, they computed solutions consisting of two or three "cells" that are of the basic inviscid type (16). Experimentally, Mellor et al. found a Batchelor solution at Re = 100. This paper was a stimulus for others to extend the solution space or to confirm the results discussed above experimentally or more accurately (Nguyen et al. 1975, Roberts & Shipman 1976, Pesch & Rentrop 1978, Wilson & Schryer 1978).

More extensive calculations have been performed by Szeto (1978) and Keller & Szeto (1980) where s ranges from -1 to 1 and Re up to 1000. It emerges that the solution is unique for all s in this range if Re is below 55. This is in accordance with the results obtained by a student of the present authors (Habers 1984). Near Re = 55 the strong inviscid solution of Watts (1974) enters into the solution space (see the subsection in this section on existence). As Re grows, the solution structure has an increasingly complex behavior, which can also be inferred from the papers by Holodniok et al. (1977, 1981). Their method uses finite differencing and Newton iteration, which is fast and robust. For higher Re they report parasitic solutions that disappear as the number of gridpoints is increased. It should be remarked that finite-difference methods usually have limited accuracy, but to enhance precision one can resort to multiple shooting techniques. An example for $s = -1$ and Re = 2000 is the Stewartson solution computed by Pesch & Rentrop (1978).

To illustrate the complexity of the solution space, we return to the investigation by Holodniok et al. (1981). At Re = 625 they calculated a multiplicity of solutions for s ranging between -1 and 1. From this paper we present here Figure 4. It will be clear from this figure that a number of

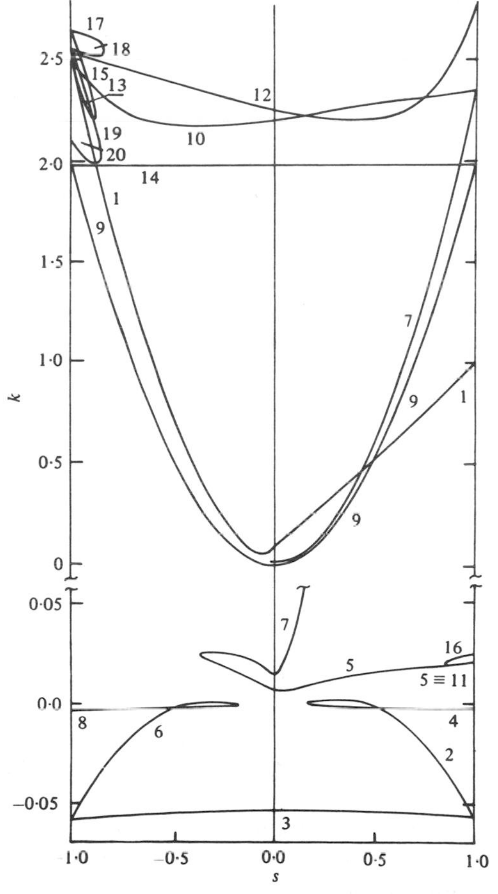

Figure 4 The integration constant κ for the two-disk problem [Equation (11)] as function of s at Re = 625 (Holodniok et al. 1981). Branch 1 contains rigid rotation; note branch 14 with $\kappa = 2.0$ [Equation (33)]. Reproduced with permission of the *Journal of Fluid Mechanics*.

critical values of s occur in the same way as in the one-disk case. These dangerous bends require special numerical techniques; otherwise the solution procedure may suddenly fail as s arrives at such a critical point.

The connection of the branches takes place at the end points $s = \pm 1$, which physically corresponds to the interchange of the disks. Mathematically, this means that if $f(x)$, $g(x)$ is a solution of the problem for $s = \pm 1$, then $\pm f(\mathrm{Re}^{1/2} - x)$, $\pm g(\mathrm{Re}^{1/2} - x)$ is also a solution for the same value of κ.

From additional graphs presented by Holodniok et al. or from various other papers it can be deduced that the inviscid solution (16) plays a fundamental role with regard to multiplicity, in the same way as in the case of one disk. This connection and other relations between the two problems have been only occasionally recognized in the literature, and we shall now try to say something more about this feature.

CONNECTION BETWEEN THE TWO PROBLEMS

We consider only nonporous disks, and we use the angular velocity Ω of the disk at $x = 0$ as a reference velocity. This means that $f_0 = f_1 = 0$, $s_0 = 1$, $s_1 = s$.

It is remarkable that little attention has been paid in the literature to possible relations between the one-disk and two-disk problems. Nevertheless, the famous conjectures by Batchelor (1951) and Stewartson (1953) can be traced back to different perceptions of the way in which two one-disk solutions can be connected to give a two-disk result (Rasmussen 1971).

We first consider the problem of two disks with one disk at rest ($s = 0$). Following Batchelor and Rasmussen we assume a bulk rotation between the two disks with $g = s'$ and boundary layers at the disks. As $\mathrm{Re} \to \infty$ the solution in the boundary layer at the rotating disk tends to a one-disk solution with $g(0) = 1$ and $g(\infty) = s'$. This solution will have an associated axial velocity at infinity, represented by $f(\infty)$, that depends on s'. At the other disk, which is at rest, a Bödewadt (1940) solution develops with an axial outflow that far from that disk has the dimensional form

$$w_\infty = 1.349\ (\nu\Omega s')^{1/2}.$$

(Incidentally, we note that a more accurate value of the coefficient is 1.349426085.)

The value of s' assumed above should be such that the two axial velocities match. The result is $s' = 0.313$, and the dimensional axial inflow to the rotating disk becomes $0.755(\nu\Omega)^{1/2}$ as compared with $0.884(\nu\Omega)^{1/2}$ for the original von Kármán case of a single disk. It should be noted that the value stated for s' was already obtained by Grohne (1956), who gave 0.314.

The prediction put forward by Stewartson for the case considered results in a bulk flow given by

$$g = 0, \qquad f = 0.442\ (1 - x/\mathrm{Re}^{1/2})^2,$$

with only one boundary layer whose solution is given by the Cochran (1934) result for the original von Kármán problem.

Next we consider the case of two counterrotating disks, i.e. the case $s = -1$. According to Stewartson the angular velocity in the bulk vanishes again, and the solution in the two boundary layers is the same as in the boundary layer above. In this case there is a radial bulk velocity that is negative in order to replace the mass thrown out by each disk. It should be remarked that the generalized one-disk problem with $f'(\infty) \neq 0$ only has a solution if $g(\infty) = 0$ (Hartman 1971). It should also be noted that the Batchelor prediction for the case $s = -1$, as described in the introduction, has not shown up in the literature.

In the constructions above, use has been made of the basic one-disk solution branch (that is, the branch containing the rigid-rotation solution). It is clear that the construction may be repeated for the higher one-disk solution branches. Then the basic inviscid solution (16) or a chain of such solutions will appear over a part of the range between the disks. Because of the range required by the inviscid components, the Reynolds number should be sufficiently large. With this key it is again possible to interpret a number of two-disk solutions as composites of two appropriate one-disk solutions.

In the cases considered so far, the relation between the two problems essentially leads to the matching concept that "infinity" for each disk separately is located somewhere in the range between the two disks. It should be remarked that this concept does not apply to the strong inviscid solution (16) with parameters (34) due to Watts (1974). Therefore, this solution deserves the name "eigenfunction of the two-disk problem."

There is also a completely different relation between the two problems, and this connection has been put forward and explored by Zandbergen (1980). He considers a one-disk solution with $f(\infty) = 0$. Note that there are infinitely many such one-disk solutions (Figure 1). For each individual solution there exists an infinite sequence of points x_n with $f'(x_n) = 0$ [see Equation (29)]. Since $f(\infty) = 0$, we have $f(x_n) \to 0$ as $x_n \to \infty$, and by means of a suitable small perturbation of this solution it is possible to satisfy $f(x_n) = 0$. The result is a two-disk solution with $\mathrm{Re} = x_n^2$. The number of two-disk solutions constructed in this way on the basis of the infinity of one-disk solutions becomes ∞^2 if Re is unlimited.

CONNECTION WITH FINITE DISKS

This is a subject in itself, especially since all technical applications are in this area. Obviously, experimental verification of the validity of von Kármán's similarity approach can only be obtained with finite disks.

It is not our purpose here to give a comprehensive review of the literature on this subject, but it may be worthwhile to identify the phenomena that might be expected.

Early experiments were performed by Schultz-Grunow (1935) and Stewartson (1953). The latter found experimental support for his theoretical predictions by means of a cardboard-disk configuration in open air. Picha & Eckert (1958) conducted experiments with both shrouded and unshrouded disks and found that the presence of a housing seriously affected the flow between the disks as compared with the case of an open end. In the last case the experiments revealed some agreement with Stewartson's predictions, whereas the closed configuration with one disk at rest showed a definite core rotation, as conjectured by Batchelor. This core rotation was later confirmed by Bien & Penner (1970).

More recent and extensive experiments have been conducted by Szeri et al. (1983) and Dijkstra & Van Heijst (1983). In both papers numerical solutions of the partial differential equations were also presented, and good agreement with the experimental results was established. Dijkstra & Van Heijst used a closed configuration with an aspect ratio H/R of 7%. For the case $s = 0$ and Re = 1000, good agreement with the Batchelor-type similarity solution was found over the full axial domain and over at least 50% of the radial domain. For negative values of s below -0.15, it was discovered that the flow field in the meridional plane consisted of two cells leading to a stagnation point at the slower rotating disk. This phenomenon can be visualized by means of particles of suitable density that will crowd together at the stagnation point. In the three-dimensional configuration a ring will then appear. The position of this ring is found to be strongly dependent on s.

Further numerical investigations for finite shrouded disks have been performed by Pao (1970, 1972) and Lugt & Haussling (1973).

Recently, the relevance of similarity solutions to the flow induced by disks of large but finite radius has been extensively investigated by Brady & Durlofsky (1986). In the limit that they consider, the radial part of the Laplacian drops out, and this causes some difficulties with the radial boundary conditions. In this recommendable paper these authors clearly point out the difference between open- and closed-end flows.

NONSTATIONARY FLOWS AND STABILITY

One question that seems paramount concerns the stability of solutions, particularly because of the multitude of stationary solutions.

There are two completely different approaches. The first is to consider truly three-dimensional perturbations of the basic Kármán flow. The second is to maintain the assumptions (1) but with f and g also time dependent. In fact, the flows described by the latter approach can be described as nonstationary von Kármán flows, but it should be kept in mind that stability considerations based on this approach may be of limited value. We first describe results obtained with this nonstationary approach.

An early example is the steady-state solution for $s = 0$ and $\mathrm{Re} = 1000$, as calculated by Pearson (1965) with the time-dependent equations. His results reveal the temporal stability of Batchelor's prediction for this case.

Bodonyi & Stewartson (1977) considered the transient problem for one disk with $s = -1$. This problem has no stationary solution, which might explain the complete breakdown of their solution in a finite time, namely $\Omega t - 2.36$.

Bodonyi (1978) solved the nonstationary equations for one disk near the Bödewadt solution. In terms of the ratio parameter σ [Equation (10)], he recovers the Bödewadt result at $\sigma = 0$, but for $\sigma = -0.1$ his numerical results indicate a limit cycle. Beyond -0.15 the calculation diverges.

In the same year Szeto confirmed the conclusion given above from Pearson's paper (Szeto 1978). In addition, he found that the prediction by Stewartson for the two-disk case with $s = 0$ is unstable.

Recently, Bodonyi & Ng (1984) returned to the subject and considered the spectrum for temporal perturbations of a possible one-disk solution. They obtained a continuum spectrum of stable modes and a discrete spectrum with unstable eigenvalues, and they concluded that the one-disk problem loses stability for $\sigma < -0.03$. Moreover, it was inferred that all higher branches near $s = 0$ are unstable.

The first approach, which can be considered as directly investigating stability, has seen many important contributions during the last few years, both theoretically and experimentally. Wilkinson & Malik (1985) showed that steady wave patterns emanate from point sources on the disk. Measurements indicate that the critical Reynolds number, defined by $R(\Omega/\nu)^{1/2}$, is around 280. As is shown by Malik et al. (1981), it is important to take account of curvature effects in the linear-stability analysis. More refined calculations have been made by Mack (1985). Second-harmonic interactions have been considered by Malik (1986) and Itoh (1985).

It should be noted that the first approach as well as the second one require much more research to solve all the remaining problems. The relations between the first and the second approach have so far not been explored.

CONCLUDING REMARKS

We have tried to trace the development of research on the problem of infinite disks from its birth in 1921 until the present.

A number of questions remain unanswered. From the rigorous point of view, Kreiss & Parter (1983) have proven the existence and nonuniqueness of a special class of solutions for the two-disk case, but there are other solutions that do not fit this concept. In addition, the existence of non-unique solutions to the one-disk problem has not been proven (as far as we know; see also the remark at the end of the one-disk section). Also, Parter (1982) concludes his review with a number of open problems.

The question of the validity of von Kármán's approach in a finite geometry is of major importance and has been considered in detail only very recently (Dijkstra & Van Heijst 1983, Brady & Durlofsky 1986). We believe that in this area more papers are to be expected.

Stability is also a subject where further work is needed.

In this review we have also tried to offer some insight into the multiplicity of the solution structure. This feature shows, be it rigorously by matched asymptotics or numerically, the rich mathematical structure of a special case of a remarkable set of full partial differential equations—the Navier-Stokes equations.

ACKNOWLEDGMENT

We would like to thank Dr. H. K. Kuiken for bringing a number of papers to our attention.

Literature Cited

Ackroyd, J. A. D. 1978. On the steady flow produced by a rotating disc with either surface suction or injection. *J. Eng. Math.* 12: 207–20

Batchelor, G. K. 1951. Note on a class of solutions of the Navier-Stokes equations representing steady rotationally-symmetric flow. *Q. J. Mech. Appl. Math.* 4: 29–41

Benton, E. R. 1966. On the flow due to a rotating disk. *J. Fluid Mech.* 24: 781–800

Bien, F., Penner, S. S. 1970. Velocity profiles in steady and unsteady rotating flows for a finite cylindrical geometry. *Phys. Fluids* 13: 1665–71

Bödewadt, U. T. 1940. Die Drehströmung über festem Grunde. *ZAMM* 20: 241–53

Bodonyi, R. J. 1975. On rotationally symmetric flow above an infinite rotating disk. *J. Fluid Mech.* 67: 657–66

Bodonyi, R. J. 1978. On the unsteady similarity equations for the flow above a rotating disc in a rotating fluid. *Q. J. Mech. Appl. Math.* 31: 461–72

Bodonyi, R. J., Ng, B. S. 1984. On the stability of the similarity solutions for swirling flow above an infinite rotating disk. *J. Fluid Mech.* 144: 311–28

Bodonyi, R. J., Stewartson, K. 1977. The unsteady boundary layer on a rotating disk in a counter-rotating fluid. *J. Fluid Mech.* 79: 669–88

Brady, J. F., Durlofsky, L. 1986. On rotating disk flow: similarity solution versus finite disks. *J. Fluid Mech.* In press

Bushell, P. J. 1972. On von Kármán's equations of swirling flow. *J. London Math. Soc.* 4(2): 701–10

Cochran, W. G. 1934. The flow due to a rotating disc. *Proc. Cambridge Philos. Soc.* 30: 365–75

Dijkstra, D. 1978. *TW-Memo. 205*, Dep. Math., T.H.T. Enschede, Neth.

Dijkstra, D. 1980. On the relation between adjacent inviscid cell type solutions to the rotating-disk equations. *J. Eng. Math.* 14: 133–54

Dijkstra, D., Van Heijst, G. J. F. 1983. The flow between two finite rotating disks enclosed by a cylinder. *J. Fluid Mech.* 128: 123–54

Dijkstra, D., Zandbergen, P. J. 1978. Some further investigations on nonunique solutions of the Navier-Stokes equations for the Kármán swirling flow. *Arch. Mech. Stosow.* 30: 411–19

Elcrat, A. R. 1975. On the swirling flow between rotating coaxial disks. *J. Differ. Equat.* 18: 423–30

Evans, D. J. 1969. The rotationally symmetric flow of a viscous fluid in the presence of an infinite rotating disc with uniform suction. *Q. J. Mech. Appl. Math.* 22: 467–85

Grohne, D. 1956. Zur laminaren Strömung in einer kreiszylindrischen Dose mit rotierendem Deckel. *ZAMM*, pp. 17–20 (Sonderheft)

Habers, J. H. A. 1984. *Niet eenduidige oplossingen van de Navier-Stokes vergelijkingen voor een stroming tussen 2 draaiende platen bij lagere Reynoldsgetallen.* MS thesis. Dep. Math., T.H.T. Enschede, Neth. 44 pp. (In Dutch)

Hartman, P. 1971. The swirling flow problem in boundary layer theory. *Arch. Ration. Mech. Anal.* 42: 137–56

Hartman, P. 1972. On the swirling flow problem. *Indiana Univ. Math. J.* 21: 849–55

Hastings, S. P. 1970. An existence theorem for some problems from boundary layer theory. *Arch. Ration. Mech. Anal.* 38: 308–16

Hoffman, G. H. 1974. Extension of perturbation series by computer: viscous flow between two infinite rotating disks. *J. Comput. Phys.* 16: 240–58

Holodniok, M., Kubicek, M., Hlavacek, V. 1977. Computation of the flow between two rotating coaxial disks. *J. Fluid Mech.* 81: 689–99

Holodniok, M., Kubicek, M., Hlavacek, V. 1981. Computation of the flow between two rotating coaxial disks: multiplicity of steady-state solutions. *J. Fluid Mech.* 108: 227–40

Howarth, L. N. 1961. A note on the existence of certain viscous flows. *J. Math. Phys.* 40: 172–76

Hulshof, B. J. A. 1983. *Automatic error cumulation control, illustrated by a problem in fluid mechanics.* MS thesis. Dep. Math., T.H.T. Enschede, Neth. 70 pp.

Itoh, N. 1985. Stability calculations of the three-dimensional boundary layer flow on a rotating disk. In *Laminar-Turbulent Transition*, ed. V. V. Kozlov, pp. 463–70. New York: Springer-Verlag

Keller, H. B., Szeto, R. K. H. 1980. Calculation of flows between rotating disks. In *Computing Methods in Applied Sciences and Engineering*, ed. R. Glowinski, J. L. Lions, pp. 51–61. Amsterdam: North-Holland

Kreiss, H. O., Parter, S. V. 1981. On the swirling flow between rotating coaxial disks, asymptotic behavior 1, 2. *Proc. R. Soc. Edinburgh Sect. A* 90: 293–346

Kreiss, H. O., Parter, S. V. 1983. On the swirling flow between rotating coaxial disks: existence and nonuniqueness. *Commun. Pure Appl. Math.* 36: 55–84

Kuiken, H. K. 1971. The effect of normal blowing on the flow near a rotating disk of infinite extent. *J. Fluid Mech.* 47: 789–98

Lan, C. C. 1971. On functional-differential equations and some laminar boundary layer problems. *Arch. Ration. Mech. Anal.* 42: 24–39

Lance, G. N., Rogers, M. H. 1962. The axially symmetric flow of a viscous fluid between two infinite rotating disks. *Proc. R. Soc. London Ser. A* 266: 109–21

Langlois, W. E. 1985. Buoyancy-driven flows in crystal-growth melts. *Ann. Rev. Fluid Mech.* 17: 191–215

Lentini, M., Keller, H. B. 1980. The von Kármán swirling flows. *SIAM J. Appl. Math.* 38: 52–64

Lugt, H. J., Haussling, H. J. 1973. Devel-

opment of flow circulation in a rotating tank. *Acta Mech.* 18: 255–72
Mack, L. M. 1985. The wave pattern produced by point source on a rotating disk. *AIAA Pap. No. 85-0490*
Malik, M. R. 1986. Wave interactions in three-dimensional boundary layers. *AIAA Pap. No. 86-1129*
Malik, M. R., Wilkinson, S. P., Orszag, S. A. 1981. Instability and transition in rotating disk flow. *AIAA J.* 19: 1131–37
Matkowsky, B. J., Siegmann, W. L. 1976. The flow between counter-rotating disks at high Reynolds numbers. *SIAM J. Appl. Math.* 30: 720–27
McLeod, J. B. 1969a. Von Kármán's swirling flow problem. *Arch. Ration. Mech. Anal.* 33: 91–102
McLeod, J. B. 1969b. The asymptotic form of solutions of von Kármán's swirling flow problem. *Q. J. Math. Oxford* 20: 483–96
McLeod, J. B. 1970. A note on rotationally symmetric flow above an infinite rotating disc. *Mathematika* 17: 243–49
McLeod, J. B. 1971. The existence of axially symmetric flow above a rotating disk. *Proc. R. Soc. London Ser. A* 324: 391–414
McLeod, J. B. 1975. Swirling flow. In *Lecture Notes in Mathematics*, 448: 242–55. Berlin: Springer-Verlag
McLeod, J. B., Parter, S. V. 1974. On the flow between two counter-rotating infinite plane disks. *Arch. Ration. Mech. Anal.* 54: 301–27
McLeod, J. B., Parter, S. V. 1977. The non-monotonicity of solutions in swirling flow. *Proc. R. Soc. Edinburgh Sect. A* 76: 161–82
Mellor, G. L., Chapple, P. J., Stokes, V. K. 1968. On the flow between a rotating and a stationary disk. *J. Fluid Mech.* 31: 95–112
Nguyen, N. D., Ribault, J. P., Florent, P. 1975. Multiple solutions for flow between coaxial disks. *J. Fluid Mech.* 68: 369–88
Ockendon, H. 1972. An asymptotic solution for steady flow above an infinite rotating disc with suction. *Q. J. Mech. Appl. Math.* 25: 291–301
Pao, H. P. 1970. A numerical computation of a confined rotating flow. *J. Appl. Mech.* 37: 480–87
Pao, H. P. 1972. Numerical solution of the Navier-Stokes equations for flows in the disk-cylinder system. *Phys. Fluids* 15: 4–11
Parter, S. V. 1982. On the swirling flow between rotating coaxial disks: a survey. In *Lecture Notes in Mathematics*, 942: 258–80. Berlin: Springer-Verlag
Pearson, C. E. 1965. Numerical solutions for the time-dependent viscous flow between two rotating coaxial disks. *J. Fluid Mech.* 21: 623–33
Pesch, H. J., Rentrop, P. 1978. Numerical solution of the flow between two counter-rotating infinite plane disks by multiple shooting. *ZAMM* 58: 23–28
Picha, K. G., Eckert, E. R. G. 1958. Study of the air flow between coaxial disks rotating with arbitrary velocities in an open or enclosed space. *Proc. US Natl. Congr. Appl. Mech., 3rd*, pp. 791–98
Rasmussen, H. 1971. High Reynolds number flow between two infinite rotating disks. *J. Aust. Math. Soc.* 12: 483–501
Roberts, S. M., Shipman, J. S. 1976. Computation of the flow between a rotating and a stationary disk. *J. Fluid Mech.* 73: 53–63
Rogers, M. H., Lance, G. N. 1960. The rotationally symmetric flow of a viscous fluid in the presence of an infinite rotating disk. *J. Fluid Mech.* 7: 617–31
Schultz-Grunow, F. 1935. Der Reibungswiderstand rotierender Scheiben in Gehäusen. *ZAMM* 14: 191–204
Stewartson, K. 1953. On the flow between two rotating coaxial disks. *Proc. Cambridge Philos. Soc.* 49: 333–41
Stuart, J. T. 1954. On the effects of uniform suction on the steady flow due to a rotating disk. *Q. J. Mech. Appl. Math.* 7: 446–57
Szeri, A. Z., Schneider, S. J., Labbe, F., Kaufman, H. N. 1983. Flow between rotating disks, Part 1, Basic flow. *J. Fluid Mech.* 134: 103–31
Szeto, R. K. H. 1978. *The flow between rotating coaxial disks.* PhD thesis. Calif. Inst. Technol., Pasadena
Tam, K. K. 1969. A note on the asymptotic solution of the flow between two oppositely rotating infinite plane disks. *SIAM. J. Appl. Math.* 17: 1305–10
Van Hulzen, J. A. 1980. Computational problems in producing Taylor coefficients for the rotating disk problem. *Sigsam Bull. ACM* 14: 36–49
van Wijngaarden, L. 1985. On multiple solutions and other phenomena in rotating fluids. *Fluid Dyn. Trans.* 12: 157–79
von Kármán, T. 1921. Uber laminare und turbulente Reibung. *ZAMM* 1: 233–52
von Kármán, T., Lin, C. C. 1961. On the existence of an exact solution of the equations of Navier-Stokes. *Commun. Pure Appl. Math.* 14: 645–55
Watson, J. 1966. On the existence of solutions for a class of rotating disc flows and the convergence of a successive approximation scheme. *J. Inst. Math. Its. Appl.* 1: 348–71

Watts, A. M. 1974. Preprint 74, Dep. Math., Univ. Queensland, Brisbane, Aust.
Weidman, P. D., Redekopp, L. G. 1976. On the motion of a rotating fluid in the presence of an infinite rotating disk. *Arch. Mech. Stosow.* 28: 1011–24
White, A. B. 1978. *Rep. CNA 132*, Cent. Numer. Anal., Univ. Tex., Austin
Wilkinson, S. P., Malik, M. R. 1985. Stability experiments in the flow over a rotating disk. *AIAA J.* 23: 588–95
Wilson, L. O., Schryer, N. L. 1978. Flow between a stationary and a rotating disk with suction. *J. Fluid Mech.* 85: 479–96
Zandbergen, P. J. 1980. New solutions of the Kármán problem for rotating flows. In *Lecture Notes in Mathematics*, 771: 563–81. Berlin: Springer-Verlag
Zandbergen, P. J., Dijkstra, D. 1977. Non-unique solutions of the Navier-Stokes equations for the Kármán swirling flow. *J. Eng. Math.* 11: 167–88

Ann. Rev. Fluid Mech. 1987. 19 : 493–530

ISOLATED EDDY MODELS IN GEOPHYSICS

G. R. Flierl

Department of Earth, Atmospheric, and Planetary Sciences, Massachusetts Institute of Technology, Cambridge, Massachusetts 02139

INTRODUCTION

Geophysical fluid flows often appear to be dominated by a strong, but localized, vortical structure that lasts for many circulation times even when relatively turbulent flows are impinging upon it. A prime example is Jupiter's Great Red Spot, still swirling strongly some 300 years after our first observations. The *Voyager* photographs offer a vivid impression of the robustness of the Red Spot despite the shearing flows surrounding it and the small vortices impinging upon it. In the Earth's ocean, we find similar long-lived structures—e.g. Gulf Stream rings—that can progress over thousands of kilometers. Satellite imagery shows that the rings exist in a highly variable eddy field, yet they retain their distinctive chemical and biological water characteristics over hundreds of revolutions. In the atmosphere, too, synoptic meteorologists often tend to characterize a "blocking" event as an isolated phenomenon. In these situations, a strong, nearly stationary disturbance diverts and weakens, or "blocks," the normal westerly flows and alters the storm tracks, often for several weeks. (Unfortunately, this phenomenon, in the fluid medium we are able to observe best, is the least clearly isolated!) Other synoptic-scale atmospheric vortices, such as hurricanes, are likewise localized but, because of the extremely active role of the thermodynamic processes, may require quite different explanations.

From such observations, we can extract a set of dynamical properties characterizing these flows:

1. Synoptic-scale dynamics—flows nearly in geostrophic balance.

0066–4189/87/0115–0493$02.00

2. Spatially isolated structures—eddy flows vanish as $|x|$, $|y|$ become large.
3. Order-one contributions from the nonlinearity.
4. Significant gradients of potential vorticity (*PV*, the dynamical conserved quantity) in the background fields.

This list of qualities has encouraged theoreticians to develop local models for strong geophysical flows. The models fall into three basic classes, as summarized by Malanotte-Rizzoli (1982). The first class, KdV solitons, is well covered in her review and her previous work (Malanotte-Rizzoli & Hendershott 1980, Malanotte-Rizzoli 1980). A second class is characterized by analytic relationships between the potential vorticity and the stream function even at large amplitude with closed streamlines. They can be found over topography, in weak shear, or when the scale is on the order of the geometric mean of the deformation radius and the Earth's radius—the so-called intermediate scale. Recently, M. Swenson (private communication) has suggested that the expansion procedures used in deriving these solutions are problematical, and thus caution is urged.

In this review, we examine the least well-synthesized class of isolated eddy models, the "modons," a term coined by Stern (1975), who found an example arising from a variational principle and regarded it as a nonlinear mode of the oceanic system. We share Stern's belief that modon models can indeed provide important insights into the roles of dispersion and strong nonlinearity in isolated geophysical flows. As we see, modons do manifest the properties required by the observations. In addition, modons also show the following properties:

5. They transport a significant volume of fluid at the phase speed of the motion, or, in the case of a stationary feature, have closed streamlines and trapped fluid.
6. They have anomalous potential vorticity in the core: *PV* is not an analytic functional of Ψ but is multivalued and is discontinuous at some order.
7. They translate at a speed depending on both size and amplitude, a common property for nonlinear disturbances; c also generally lies outside the range of speeds obtained by linear, sinusoidal waves.

Thus, the modon model is an appealing choice for explaining the isolated, strong-eddy features common within geophysical flows.

Since the scales of these strong geophysical eddies are typically synoptic or larger, this review uses primarily the quasi-geostrophic system of equations (QG; cf. Pedlosky 1979), which filter out all wave modes except for

Rossby waves:

$$\frac{\partial}{\partial t}\nabla^2\psi + J(\psi, \nabla^2\psi) + \beta\frac{\partial}{\partial x}\psi - \frac{f_0}{\bar{\rho}}\frac{\partial}{\partial z}(\bar{\rho}w) = 0,$$

$$\frac{\partial}{\partial t}\frac{\partial\psi}{\partial z} + J\left(\psi, \frac{\partial\psi}{\partial z}\right) + \frac{N^2}{f_0}w = 0. \tag{1}$$

Here ψ is the geostrophic stream function, w is the vertical velocity, N is the Brunt-Väisälä frequency, $f_0+\beta y$ is the Coriolis parameter, and $J(A, B)$ is the Jacobian $A_x B_y - A_y B_x$. These can also be written in the form

$$\frac{\partial}{\partial t}PV + J(\psi, PV) = 0,$$

with the potential vorticity

$$PV = \nabla^2\psi + \frac{1}{\bar{\rho}}\frac{\partial}{\partial z}\bar{\rho}\frac{f_0{}^2}{N^2}\frac{\partial}{\partial z}\psi + \beta y \equiv (\nabla^2 + L_z)\psi + \beta y. \tag{2}$$

We consider motions embedded in a zonal flow $\bar{u}(y, z) = -(\partial/\partial y)\bar{\psi}(y, z)$ and split the stream function as follows:

$$\psi = \bar{\psi}(y, z) + \left(\frac{N}{\bar{\rho}f_0}\right)^{1/2}\phi(x - ct, y, \zeta), \quad \text{where} \quad \zeta = \int^z \frac{N}{f_0}.$$

The stretched vertical coordinate and the factor multiplying ϕ transform the $\nabla^2 + L_z$ operator into a three-dimensional Laplacian $\mathbb{V}^2 = \nabla^2 + \partial^2/\partial\xi^2$. A translating reference frame, moving with speed c, has also been introduced. With these modifications the QG equations become

$$\frac{\partial}{\partial t}\mathbb{V}^2\phi - \Gamma\phi + J(\bar{\psi} + cy, (\mathbb{V}^2 - v^2)\phi) + \left(\frac{N}{\bar{\rho}f_0}\right)^{1/2} J(\phi, \mathbb{V}^2\phi) = 0. \tag{3}$$

The term

$$v^2 = \frac{\beta - \dfrac{\partial^2}{\partial y^2}\bar{u} - L_z\bar{u}}{c - \bar{u}} - \left(\frac{\bar{\rho}f_0}{N}\right)^{1/2} L_z\left(\frac{N}{\bar{\rho}f_0}\right)^{1/2} \equiv \bar{\mathrm{P}}_\psi - \Gamma \tag{4}$$

is a generalization of the "potential well" of Malguzzi & Malanotte-Rizzoli (1984); its relationship to wave propagation was first pointed out by Charney & Drazin (1961) (see also Charney & Flierl 1981), who called

$n = i\nu$ the "index of refraction." For boundary conditions, we consider a lower boundary with topographic elevation $b(y)$, which gives

$$\frac{\partial}{\partial t} D\phi + J(\bar{\psi} + cy, (D - \nu_{\mathrm{b}})\phi) + \left(\frac{N}{\bar{\rho} f_0}\right)^{1/2} J(\phi, D\phi) = 0 \quad \text{at} \quad z = 0, \tag{5}$$

where

$$D\phi = \frac{N}{f_0} \frac{\partial}{\partial \xi} \left(\sqrt{\frac{N}{\bar{\rho} f_0}} \phi \right), \qquad \nu_{\mathrm{b}} = \frac{\bar{u}_z(0) - \dfrac{N^2(0)}{f_0} b_y}{\bar{u}(0) - c}.$$

The QG equations (3, 5) are used throughout most of this review, although comments upon the similarities and differences with the primitive equations (cf. Holton 1979) are made where appropriate.

There are other dynamical equations with modonlike solutions, most notably in plasma-physics problems, and some very important work on these structures is underway by researchers interested in vortices in plasmas. The Hasegawa-Mima equations for electrostatic drift waves in a magnetic field, which are isomorphic to the quasi-geostrophic model, of course have modon solutions (Makino et al. 1981, Meiss & Horton 1983), but analogous solutions to other systems of equations [Alfvén vortices (Shukla et al. 1985), ballooning modes (Shukla 1985), flute vortices (Yu et al. 1985)] have also been found. The Vlasov equations also show nonlinear coherent structures: In unstable circumstances, rolled-up vortices form much like those found in the shear layer (Berk et al. 1970), while in the stable case, nonlinear "holes" form. The latter appear to be key elements to understanding the plasma dynamics (Berman et al. 1985). We do not comment in any more detail upon these nongeophysical models but reiterate that they are a potential source of important insights.

In this review we present general conditions concerning the existence of isolated eddy solutions, derive several examples of modon structures, and discuss (more speculatively) the potential role of modon models in flows that are only partially isolated, so that the coherent structures couple to radiating Rossby waves. The approach is to synthesize and generalize much of the previous work; many of our results and discussions, while implied by and building upon the literature, have not appeared in the forms we advance. We hope that the reader will gain an understanding of both the power and the limitations of modon models of geophysical flows.

GENERAL PROPERTIES OF ISOLATED EDDY MODELS

Far-Field Structure

It is simplest to begin a discussion of isolated features by looking in the far field: Here the motion associated with the feature should be weak, and (3) can be linearized. If we presume that the disturbance propagates without change in shape $\partial/\partial t = 0$, we have

$$(\nabla^2 - \nu^2)\phi = 0. \tag{6}$$

Similarly, boundary conditions lead to

$$D\phi - \nu_b\phi = 0. \tag{7}$$

The search for QG isolated solutions, then, involves examining (6, 7) to find ϕ's that decay at large distances; nonlinear terms will be introduced later to make the solution well behaved near the origin.

For solutions that are decaying in x, y, and z, it is necessary that ν^2 be positive. The last term in the expression (4) for ν^2 tends to have this sign, but the other term

$$\bar{q}_\psi \equiv \frac{\beta - \bar{u}_{yy} - L_z\bar{u}}{c - \bar{u}} \tag{8}$$

could have either sign. Our criterion becomes

$$\frac{\beta - \bar{u}_{yy} - L_z\bar{u}}{c - \bar{u}} > \left(\frac{\bar{\rho} f_0}{N}\right)^{1/2} \frac{1}{\bar{\rho}} \frac{\partial}{\partial z} \frac{\bar{\rho} f_0^2}{N^2} \frac{\partial}{\partial z} \left(\frac{N}{\bar{\rho} f_0}\right)^{1/2}.$$

In the case of an isothermal atmosphere [$N = \text{constant}$, $\bar{\rho} = \rho_0 \exp(-z/H)$], the right-hand side simplifies to $-f_0^2/4N^2H^2$; for an exponentially stratified ocean [$\bar{\rho} = \text{constant}$, $N = N_0 \exp(z/h)$], it becomes $-3f_0^2/4N^2h^2$.

The equation above serves to restrict the possible phase speeds of isolated disturbances. As an example, consider an isothermal atmosphere with no zonal flows; the isolated, steady-translating disturbances must have $\beta/c > -0.25/R_d^2$ so that $c > 0$ or $c < -4\beta R_d^2$, where the deformation radius is $R_d = NH/f$. Like KdV solitons, the isolated disturbances have phase speeds disjoint from those of wavelike solutions; although this is usually perceived as a result of nonlinearity (and indeed nonlinear interactions are required to push the disturbance at these speeds), this argument demonstrates that the linear dispersion relationship, but with an imaginary wave number, still plays an essential role. The imaginary wave number, of course, corresponds to the decaying fields at large distances.

The eastward phase speeds, in the context of the linear dynamics applying on the edges of the feature, may seem counterintuitive to those accustomed to Rossby et al.'s (1939) argument that β (or northward increases in background potential vorticity) is responsible for westward propagation. It is still true that clockwise flow, for example, tends to increase the disturbance PV on the eastern side and to decrease it on the western side. The distinction lies in the relationship between the vorticity and the flow direction: For cyclonic isolated features with the velocity decaying outward, the vorticity is positive, whereas for a wavelike cyclonic structure the vorticity is negative. The vorticity changes (+ on east, − on west) described above thus correspond to an eastward shift for isolated structures.

It has not been previously noted that the relationship between the horizontal and vertical scales of an isolated feature is opposite to the familiar result of Charney & Drazin (1961). If the horizontal scale is L, the vertical structure equation

$$\frac{\partial^2}{\partial \xi^2}\phi = (v^2 - \nabla^2)\phi \sim \left(v^2 - \frac{1}{L^2}\right)\phi$$

leads to trapped solutions only for *long* waves when the v^2 term can overcome the $1/L^2$ term; it is difficult to construct small-scale features that are trapped vertically. One caveat on these conclusions is that c, which enters through the v^2 term, is not necessarily the same for different scale structures. We also remark that the meridional and vertical structures of the eddy and of $u - c$ are solutions to the same operator, but the forcing terms ($-\partial^2/\partial x^2$ and β) are such that the eddy scales will tend to be smaller.

It is also useful to consider solutions that are not completely isolated in the three-dimensional sense applied previously. Propagation in the vertical or horizontal direction could be accepted in the presence of boundaries that reflect the energy back to form a normal mode. For example, let us add top and bottom boundaries and suppose that v^2 is such that (6) can be separated into vertical and horizontal structure equations, $v^2 = v^2_{(y)} + v^2_{(z)}$, and $\phi = \Phi(x, y)F(z)$ with a separation constant λ^2; then the vertical equation is

$$F_{\xi\xi} = v^2_{(z)}F - \lambda^2 F,$$

where $\lambda^2 > 0$, corresponding to a normal mode in z. The horizontal structure satisfies

$$\nabla^2\Phi = v^2_{(y)}\Phi + \lambda^2\Phi,$$

and we can see that trapping will be rather more likely, since the λ^2 term contributes in the proper sense to enhance horizontal decay. In other cases, trapping may be only partial, with some regions in y or z having $v^2 < 0$; these cases are discussed further below.

Point-Vortex Models

Helmholtz (1858) was the first to realize that flows with vorticity could usefully be modeled by replacing the centers of high or low vorticity with singularities having the appropriate circulation—point vortices. Von Kármán's (1911) model for the vortex street, including a derivation of the translational characteristics and the stability of various forms, was a striking example of the power of this idealization. Since that time, a number of exact solutions have been discovered [some of which, like the cluster of vortices exploding into dipoles (Aref 1982), are of singular geometric beauty]. For the two-dimensional equations (with $\beta = 0$, $\partial/\partial z = 0$), a Hamiltonian form for the motions of the point vortices can be derived (Kirchhoff 1876), and chaotic solutions exist when more than three interacting vortices are considered.

We can find a large set of isolated solutions that are the singular analogues of modons by examining point-vortex solutions to (3). Consider the case in which v^2 is constant; this occurs when the mean flow is restricted to solutions of

$$\left[\frac{\partial^2}{\partial y^2} + L_z - \left(\frac{\bar{\rho} f_0}{N}\right)^{1/2} L_z \left(\frac{N}{\bar{\rho} f_0}\right)^{1/2} - v^2\right](\bar{u} - c) = \beta. \tag{9}$$

The equation for the disturbance becomes

$$\frac{\partial}{\partial t}\nabla^2\phi - \Gamma\phi + J\left(\bar{\psi} + cy + \left(\frac{N}{\bar{\rho} f_0}\right)^{1/2}\phi, \nabla^2\phi - v^2\phi\right) = 0.$$

The boundary conditions, by similar manipulation, are

$$\frac{\partial}{\partial t} D\phi + J\left(\bar{\psi} + cy + \left(\frac{N}{\bar{\rho} f_0}\right)^{1/2}\phi, D\phi - v_b\phi\right) = 0,$$

assuming that v_b is constant.

When $v^2 = \Gamma$—the potential vorticity of the background state is uniform—we can use standard point-vortex theory [cf. Aref (1983) for a review]. Helpful insights may be gained from analytical solutions with only a small number of vortices, or the dynamical equations may be integrated numerically with computational effort less than that required to solve discretized versions of (1). There are computational advantages to point-vortex models because (*a*) potential vorticity will be exactly conserved, (*b*) resolution can be concentrated in regions where it is required, and (*c*) boundary conditions at infinity can be exactly represented.

The conserved quantity, potential vorticity, is represented as a constant plus a set of delta functions with coefficients for the strength of each vortex. The inviscid advection equation for potential vorticity then simplifies because the constant term does not contribute and the singularities translate with the flow. The next requirement is that the flows must be derivable from the potential-vorticity distribution; for flows determined by a stream function, this involves knowledge of the Green's function [cf. Hoskins et al. (1985), who also described an approximate procedure for the case of an isentropic but not quasi-geostrophic system]. The boundary conditions on rigid horizontal walls can be satisfied if there are no temperature gradients along the boundary and the eddy has only point sources of $D\phi$. Given the positions of all the sources, the flow at each position can be calculated, and the evolution of the flow is found by solving a set of ordinary differential equations.

For geophysical flows, demanding that $\bar{\mathrm{P}}_\psi$ be zero is highly restrictive. Essentially, one has eliminated Rossby-wave processes, both in the context originally thought of by Rossby and in the more general sense where waves in shear flows are considered as Rossby waves supported by the gradient of mean potential vorticity. While point-vortex models on the β-plane have been proposed (cf. Zabusky & McWilliams 1982), a large number of vortices must be introduced even far from the dominant structure in order to represent the background gradient in the potential-vorticity field properly. [We note that the mapping of Weinstein (1983) of the β-plane equations into the f-plane equations in reality introduces a free-surface slope that cancels β and forces $\bar{\mathrm{P}}_\psi = 0$; it is not able to represent easily vortices in a fluid at rest at infinity.]

Because creation of vorticity by advection of the mean gradient is a fundamental process in geophysical flows, point-vortex models will not in general be such useful idealizations. However, there is one circumstance in which point vortices do provide exact solutions that can lend considerable insight. It is true that we cannot write a general time-dependent solution when the PV gradient terms v^2 or v_b are nonzero constants, but we can examine steadily translating states corresponding to a moving set of vortices. These satisfy

$$(\nabla^2 - v^2)\phi = \sum_i s_i \delta(x-x_i)\delta(y-y_i)\delta(z-z_i) \tag{10}$$

under the boundary conditions

$$(D - v_b)\phi = \sum_i s_{b_i} \delta(x-x_i)\delta(y-y_i). \tag{11}$$

In addition, however, we must impose a highly restrictive condition that

$$\dot{x}_i = 0 \Rightarrow c = \bar{u} - \left(\frac{N}{\bar{\rho} f_0}\right)\phi_y, \qquad \dot{y}_i = 0 \Rightarrow \frac{N}{\bar{\rho} f_0}\phi_x = 0 \tag{12}$$

at each point where a vortex resides: The vortices translate at the local flow speed, and, for steadily translating structures, the translation speeds must all be $(c, 0)$.

MONOPOLAR SOLUTIONS These only exist when (*a*) the mean *PV* gradients are zero or (*b*) the vortex resides at a critical layer where $\bar{q}_y = 0$. We can see this by attempting to satisfy (10–12) with a single vortex. In that case, we require $c = \bar{u}(y_1)$ by (12), which implies that the vortex resides at the critical layer. However, to avoid a singularity in v^2 or, alternatively, since we are presuming v^2 is a finite constant, we must then require that $\bar{q}_y = 0$ at y_1 also.

As an example, consider a model for Jupiter's Great Red Spot like that of Ingersoll & Cuong (1981). The mean flow is taken to be

$$\bar{u}(y) = u_0 \sin ky.$$

To make v^2 constant (here we assume constant N and $\bar{\rho}$), select $c = -\beta/k^2$ so that $v^2 = -k^2$. We take a normal-mode structure in z [so that the vorticity singularity is actually of the form $s\delta(x)\delta(y-y_1)F(z)$] to find

$$\phi = -\frac{s}{2\pi} K_0[\sqrt{(\lambda_1^2 - k^2)(x^2 + (y-y_1)^2)}]F(z),$$

with the vortex residing at the point

$$\sin ky_1 = -\beta/u_0 k^2.$$

(This expression agrees fairly well with Ingersoll & Cuong's much more complex and nonsingular solution.) The streamlines are sketched in Figure 1; one can see why this kind of solution is an attractive model for representing the Red Spot.

DIPOLE SOLUTIONS The constraint (12) places severe demands upon the symmetry of the vortex configuration; dipoles and vortex streets represent the simplest structures that might allow all the induced fluid velocities to be zonal. The dipole solutions are of interest here, since they will be isolated in space.

When the flow is barotropic (singularities at all levels in z at each of the two x-points) with no mean flow and vortices of strength s and $-s$ at $y = a/2$ and $-a/2$, respectively, we find that the stream function is

(Zabusky & McWilliams 1982)

$$\psi = -\frac{s}{2\pi}\left[K_0\left(\sqrt{\frac{\beta}{c}}\,r_-\right) - K_0\left(\sqrt{\frac{\beta}{c}}\,r_+\right)\right], \qquad r_\pm^2 = x^2 + \left(y \pm \frac{a}{2}\right)^2,$$

and the translational speed is

$$c = \frac{s}{2\pi}\sqrt{\frac{\beta}{c}}\,K_1\left(\sqrt{\frac{\beta}{c}}\,a\right).$$

This nonlinear dispersion relationship, $c = c(a, s)$ (plotted in Figure 2), shares many characteristics with the nonsingular modons. In particular, the speed is disjoint from that of linear waves: c for barotropic motions must be positive (eastward traveling). It becomes proportional to vortex strength and independent of β at large amplitude when vortex-vortex interactions dominate and β-induced flows are negligible. For smaller s, however, the eastward speed of the vortex becomes considerably slower than the f-plane model would predict; the vorticity changes induced by north-south motion are no longer negligible compared with the circulation

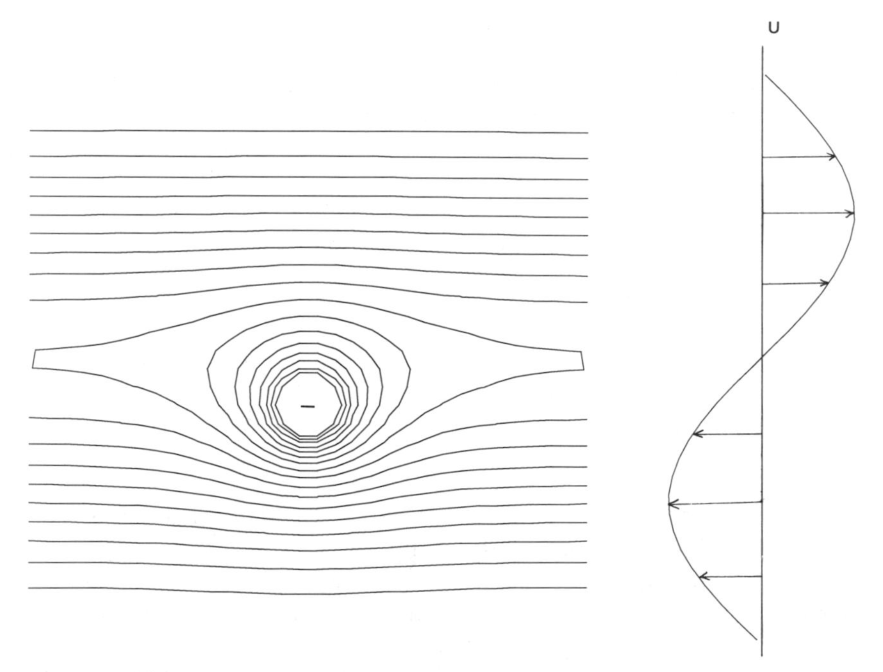

Figure 1 An anticyclonic baroclinic point vortex embedded in a flow $\bar{u}(y) = U_0 \sin(ky)$. Parameters are $U_0 = 1$, $k = 1$, $\beta = 0.5$, and $s = -4\pi$.

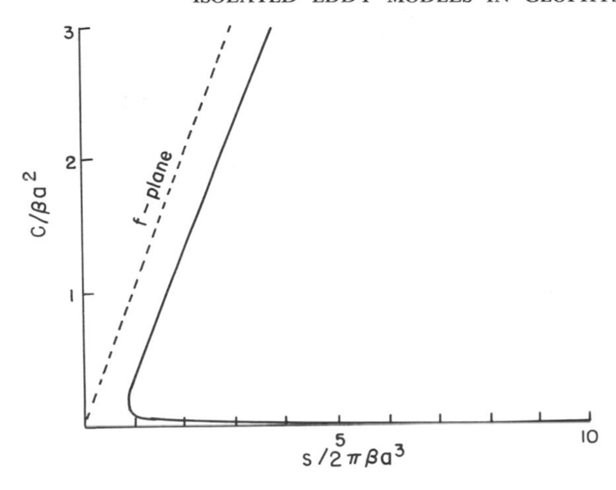

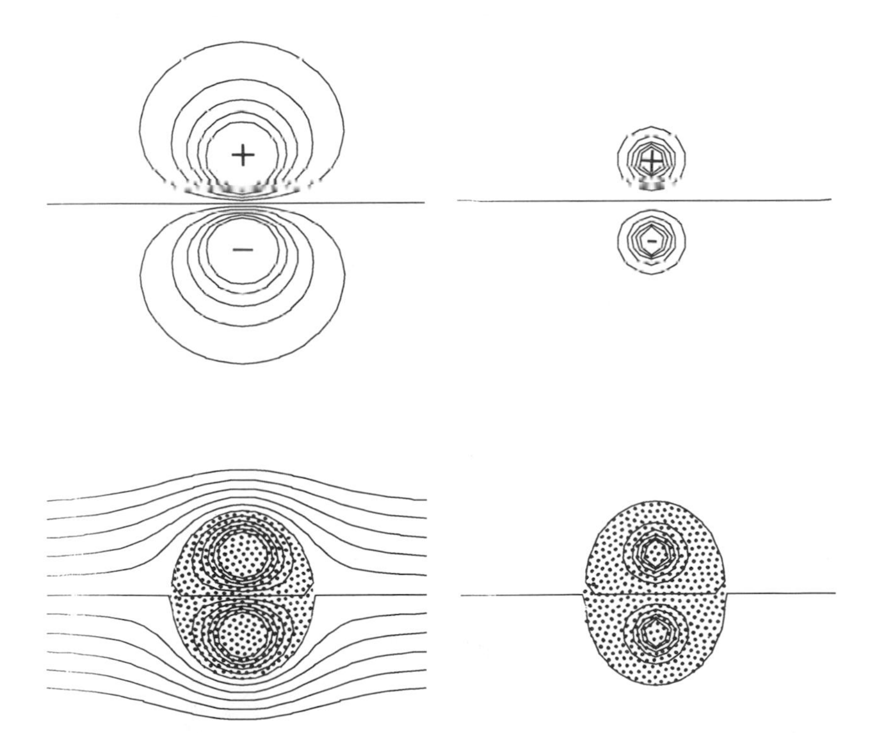

Figure 2 Barotropic dipole on the β-plane. (*top*) Dispersion relationship $c/\beta a^2$ vs. $s/2\pi\beta a^3$; (*middle*) ψ and $\psi + cy$ for $\beta = 1$, $a = 1$, $s/2\pi\beta a^3 = 1.53$, $c = 0.879$; (*bottom*) ψ and $\psi + cy$ for $\beta = 1$, $a = 1$, $s/2\pi\beta a^3 = 1.53$, $c = 0.05$.

of the point vortex. The vortices must be sufficiently strong ($s > 1.7\pi\beta a^2$) for a steady solution to exist at all.

The stream-function patterns in the inertial and comoving reference frames (shown in Figure 2) reveal one important feature of dipole structures: A significant volume of fluid (shaded) is carried along with the structure as it propagates. The requirement for trapped fluid is just that $\psi + cy$ has closed contours; since irrotational flows cannot have interior extrema, trapped fluid (transported at the phase speed of the disturbance) occurs only in motions with vorticity.

A peculiar and unnoticed feature of these solutions is the existence of two different patterns with the same vortex strengths and spacings but different propagation speeds and, of course, quite different stream functions (Figure 2). The multiple states again exemplify the importance of the flow structure in the far field.

If we now pass to the stratified case, we find dipolar structures related to the "hetons" of Hogg & Stommel (1985). In a uniformly stratified, infinitely deep fluid with vortices of strength $\pm s$ at $(0, \pm a/2, \pm h/2)$, the stream function is

$$\psi = -\frac{Ns}{4\pi f}\left[\frac{\exp\left(-\sqrt{\frac{\beta}{c}}r_-\right)}{r_-} - \frac{\exp\left(-\sqrt{\frac{\beta}{c}}r_+\right)}{r_+}\right],$$

$$r_\pm^2 = x^2 + \left(y \pm \frac{a}{2}\right)^2 + \left(z \pm \frac{h}{2}\right)^2 \frac{N^2}{f_0^2},$$

and the translational speed is

$$c = \frac{Nsa}{4\pi f_0}\left[\sqrt{\frac{\beta}{c}} + \frac{1}{\sqrt{a^2 + \frac{N^2h^2}{f_0^2}}}\right]\left(a^2 + \frac{N^2h^2}{f_0^2}\right)^{-3/2}$$

$$\times \exp\left(-\sqrt{\frac{\beta}{c}\left(a^2 + \frac{N^2h^2}{f_0^2}\right)}\right).$$

Stratification does not make qualitative differences: The translational speeds are reduced as the vortices are separated vertically, since for a given horizontal separation, the vertical influence of the singularities decays over a scale fa/N. Other features are similar to the barotropic solution—for example, the existence of a minimum strength in the presence of β and the approach to an f-plane limit for strong vortices. Sketches of various fields are shown in Figure 3.

When the system has upper and lower boundaries, we can derive solutions by adding image vortices. In the limit where the vortices approach the upper and lower boundaries, we recover the stratified equivalent of Stommel & Hogg's "hetons": The temperature perturbation becomes single signed, and the warm fluid is transported with the phase speed of the motion. Note that "hetons" are very similar to the barotropic modon with a baroclinic "rider" solution of Flierl et al. (1980); indeed, Mied & Lindemann (1982) explicitly point out the resemblance of a baroclinic eddy with a tilted axis to the modon solutions, and they explore the motion and evolution of such features.

If we use an isothermal atmospheric model, we find that v^2 = constant for stationary disturbances when

$$\bar{u} = \beta R_{\mathrm{d}}^2 \left[\frac{1}{\frac{1}{4} - v^2 R^2} + A_+ \exp \frac{z}{H}\left(\frac{1}{2} + vR\right) + A_- \exp \frac{z}{H}\left(\frac{1}{2} - vR\right) \right].$$

Let two point vortices be placed one scale height up in the atmosphere ($z = H$) so that $\xi = NH/f_0 - R$, and let images be added to satisfy the lower boundary condition (in this case $\phi = 0$, since $\bar{u}$ will be chosen to be zero at $z = 0$). The solution is

$$\begin{aligned}\phi = -\frac{3s}{2\pi^2} \Bigg\{ & k_0\left(v\sqrt{x^2 + \left(y - \frac{a}{2}\right)^2 + (\xi - R)^2}\right) \\ & - k_0\left(v\sqrt{x^2 + \left(y + \frac{a}{2}\right)^2 + (\xi - R)^2}\right) \\ & - k_0\left(v\sqrt{x^2 + \left(y - \frac{a}{2}\right)^2 + (\xi + R)^2}\right) \\ & + k_0\left(v\sqrt{x^2 + \left(y + \frac{a}{2}\right)^2 + (\xi + R)^2}\right),\end{aligned}$$

where k_0 is the modified spherical Bessel function $k_0(z) = (\pi/2z)\exp(-z)$. Given the separation a of the vortices, we can then find the vortex strength s required so that the induced flow $u = -(\bar{\rho}f_0/N)^{1/2}\phi_y(x \pm a/2, R)$ cancels $\bar{u}(z = H)$; typical flow fields are given in Figure 4.

It should be noted that non-quasi-geostrophic equivalents of the stratified point vortices are quite different. In the single-layer, f-plane shallow-

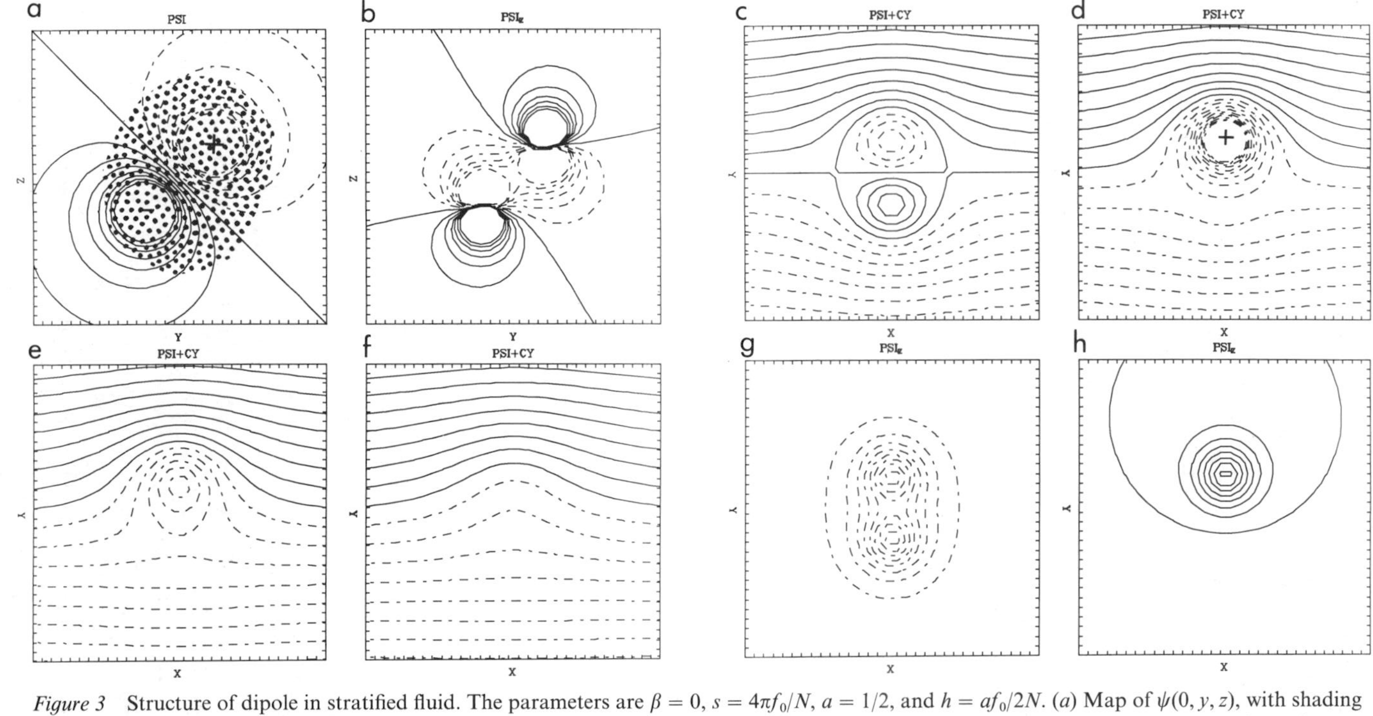

Figure 3 Structure of dipole in stratified fluid. The parameters are $\beta = 0$, $s = 4\pi f_0/N$, $a = 1/2$, and $h = af_0/2N$. (*a*) Map of $\psi(0, y, z)$, with shading indicating the extent of the trapped fluid; (*b*) similar figure for ψ_z (or temperature); (*c–f*) four maps of the stream function in the moving reference frame at $z = 0, 0.5, 1.0$, and 1.5; (*g, h*) maps of ψ_z versus x and y at the levels $z = 0, 0.5$.

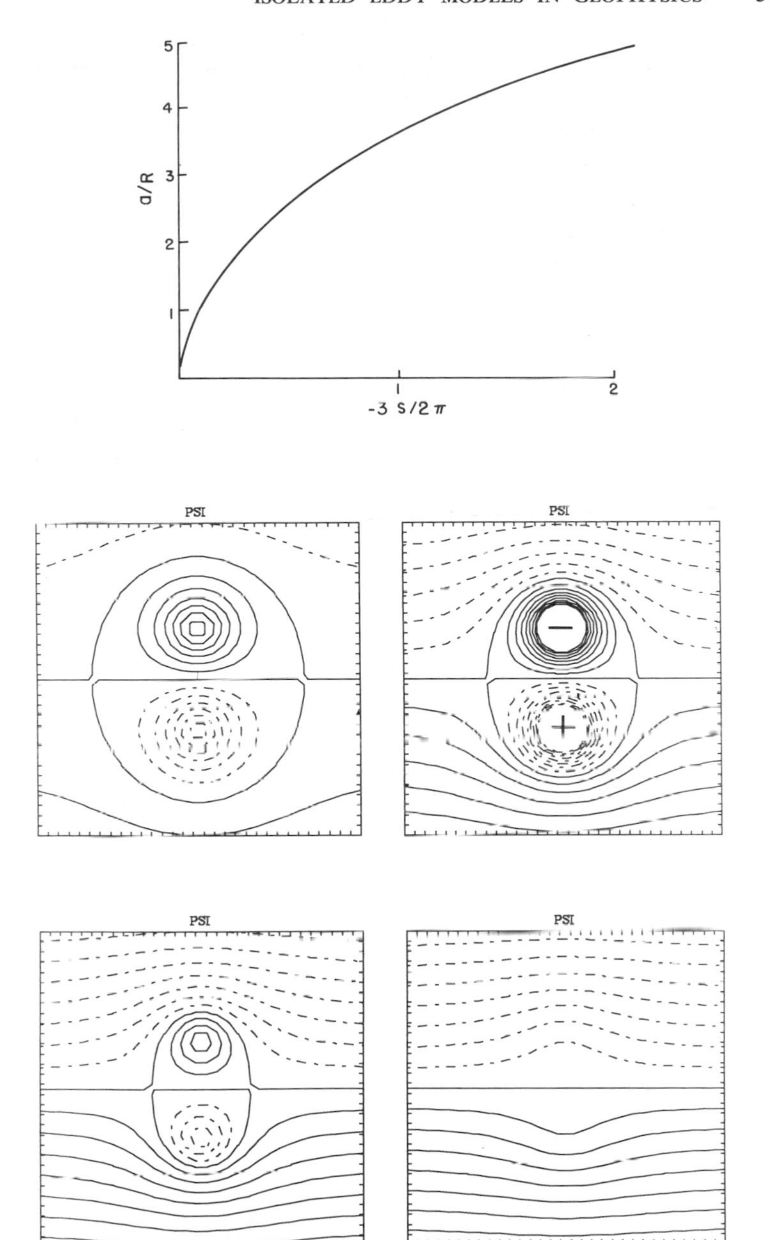

Figure 4 Dipole at $z = H$ in an isothermal atmosphere. Upper panel shows the size-strength relationship a/R vs. $-3s/2\pi$ necessary to maintain the eddy stationary in the case $\nu R = 1/4$, $A_+ = 3$, $A_- = -25/3$. In the middle and lower panels, stream-function patterns $\psi/\beta R^3$ are shown at various levels $z = 0.5H$, H, $1.5H$, and $2H$ for the case $a = 2R$, $-3s/2\pi = 0.40$.

water model, for example, a radially symmetric vortex having the same structure in the far field as the solution to the equivalent barotropic equations can be constructed. However, Flierl (1979) found that the fluid thickness then vanishes at a finite radius; unlike the QG or two-dimensional model, one cannot make this radius smaller by increasing the vorticity in the core. If we wish to calculate the evolution of the fields in a singular shallow-water model, we are therefore forced to track curves of singularity, rather than points, and the computations become analytically intractable. In addition, there is no simple form of a Green's function, so contour-dynamics methods, which have proved so successful for two-dimensional flows (cf. Zabusky & Overman 1983), will not succeed. Furthermore, the anticyclonic constant-PV eddies have a region near the center in cyclostrophic balance with the centrifugal terms opposing and dominating the Coriolis forces; this sort of balance is certainly not common in synoptic-scale geophysical flows. However, we believe that the insights gained from singular solutions are of sufficient value to accept the use of the more approximate dynamics of the QG system.

Integral Relationships

The fact that a single point vortex satisfies the equations of motion on the f-plane but has no equivalent on the β-plane is not accidental. Rather, the integral theorem of Flierl et al. (1983) demonstrates that an isolated structure in the β-plane cannot be a simple, monopolar vortex. Here we present a slightly different version of this theorem to emphasize more strongly the role of external contributions to the potential vorticity (β, topography) as opposed to internal contributions associated with mean flows.

Consider the Boussinesq primitive equations, with a zonal flow $\bar{u}(y,z)$ explicitly factored out from the eddy flow $\mathbf{u}(x,y,z,t)$, $w(x,y,z,t)$,

$$\left(\frac{\partial}{\partial t}+\bar{u}\frac{\partial}{\partial x}\right)\mathbf{u}+\mathbf{u}\cdot\nabla\bar{u}+w\bar{u}_z+\mathbf{u}\cdot\nabla\mathbf{u}+w\mathbf{u}_z+f\hat{k}\times\mathbf{u}=-\nabla p,$$

$$0=-p_z+B,$$

$$\nabla\cdot\mathbf{u}+wz=0,$$

$$\left(\frac{\partial}{\partial t}+\bar{u}\frac{\partial}{\partial x}\right)B+\mathbf{u}\cdot\nabla\bar{B}+w\bar{B}_z+\mathbf{u}\cdot\nabla B+wB_z=0,$$

where ∇ and $\mathbf{u}$ include only x- and y-dimensions, and B is the buoyancy. We now integrate these equations vertically between two material surfaces

$z = s_0(x, y, t)$ and $z = s_1(x, y, t)$. Using

$$w(x, y, s_i, t) = \left(\frac{\partial}{\partial t} + \bar{u}\frac{\partial}{\partial x} + \mathbf{u}\cdot\nabla\right)s_i\bigg|_{z=s_i}$$

and Leibniz' rule as appropriate, we find

$$\frac{\partial}{\partial t}(s_1 - s_0) + \frac{\partial}{\partial x}\int u + \frac{\partial}{\partial y}\int v = 0, \tag{13}$$

$$\frac{\partial}{\partial t}\int u + \frac{\partial}{\partial x}\int (\bar{u}+u)^2 + \frac{\partial}{\partial y}\int (\bar{u}+u)v - f\int v = -\frac{\partial}{\partial x}\int p + p_1 s_{1x} - p_0 s_{0x},$$

$$\frac{\partial}{\partial t}\int v + \frac{\partial}{\partial x}\int (\bar{u}+u)v + \frac{\partial}{\partial y}\int v^2 + f\int u = -\frac{\partial}{\partial y}\int p + p_1 s_{1y} - p_0 s_{0y}.$$

We shall work with the continuity equation (13) and the overall momentum balances derived from integrating the last two equations. We assume that the flow is isolated at all depths $s_1 > z > s_0$, and thus that boundary contributions to the integrals vanish. [The form of this discussion was much influenced by work of Killworth (1986).]

$$\frac{\partial}{\partial t}\iiint u - f_0 \iiint v - \beta \iint y \int v = \iint (p_1 s_{1x} - p_0 s_{0x}), \tag{14}$$

$$\frac{\partial}{\partial t}\iiint v + f_0 \iiint u + \beta \iint y \int u = \iint (p_1 s_{1y} - p_0 s_{0y}). \tag{15}$$

We now consider a number of special cases in which these balances (13–15) simplify to give significant information about the flow.

FLAT UPPER AND LOWER BOUNDARIES, FLOW ISOLATED AT ALL DEPTHS In this case (Flierl et al. 1983) we take the material surfaces to be the upper and lower boundaries so that $\dot{s}_0 = \dot{s}_1 = 0$. The continuity equation implies the existence of a mass-transport stream function

$$\int u = -\Psi_y, \qquad \int v = \Psi_x,$$

and the momentum equations simplify to

$$0 = \iint p_1 s_{1x} - p_0 s_{0x}, \tag{16}$$

$$\beta \iint \Psi = \iint p_1 s_{1y} - p_0 s_{0y}. \tag{17}$$

When the boundaries are flat ($\nabla s_0 = \nabla s_1 = 0$), we derive the "no net angular momentum" theorem:

$$\iint \Psi = 0 \Rightarrow \iiint \hat{k} \cdot \mathbf{r} \times \mathbf{u} = 0.$$

The basic physics here is simple: Due to the increase in the Coriolis parameter northward, there is a net north/south force on a cyclonic circulating flow. For this flow configuration the net westward transport on this northern side must balance the net eastward transport on the southern side; yet the southward Coriolis force on the latter is not as strong as the northward Coriolis force on the former. In the case considered here, there is no other overall force that can balance this "Rossby force"; isolated features therefore cannot exist and remain isolated unless they have no net angular momentum.

What happens when the initial condition is isolated but has net angular momentum? Flierl et al. (1983) argued that long barotropic Rossby waves are generated and cause boundary contributions (particularly $\oint \Psi_t \hat{x} \cdot \hat{n}$ in the y-momentum equation) that alter the momentum balance. We demonstrate this process by solving numerically for the evolution of a barotropic vortex $\Psi = \exp(-0.5r^2)$ ($\iint \Psi = 2\pi$ at $t = 0$) using the barotropic vorticity equation. The results, shown in Figure 5, are familiar from the work of Firing & Beardsley (1976) and McWilliams & Flierl (1979): The vortex breaks apart rapidly, and waves propagate quickly to the boundaries of the domain. In contrast, Figure 5 also shows the evolution of a state $\Psi = x \exp(-0.5r^2)$ ($\iint \Psi = 0$ at $t = 0$) with no net angular momentum; note the much weaker radiation and longer persistence of the structure.

SLOPING LOWER BOUNDARY If the lower boundary slopes as $b = dx + ey$, we find the constraints

$$d \iint p_0 = 0,$$

$$e \iint p_0 + \beta \iint \Psi = 0.$$

If the topographic contours are not latitude lines, both the average perturbation in the bottom pressure and the angular momentum must be zero. The former condition (which also holds for $\beta = 0$ and $d = 0$) is equivalent to the result of M. Mory & M. E. Stern (private communication; cf. Mory 1983). They demonstrated that significant pressure perturbations exist in the fluid above a blob of heavy fluid on a slope, where one might have expected that the upper layer could be only weakly perturbed. This statement follows directly from the hydrostatic equation

$$p_{\text{upper}} = p_0 - g\frac{\Delta\rho}{\rho}h - gey,$$

so that

$$\iint p_{\text{upper}} = -g\frac{\Delta\rho}{\rho}\iint h.$$

Thus if the motion is isolated in the upper layer, as well as (of course) in the blob, there must be strong currents in the upper layer.

Finally, if $d = 0$ but $e, \beta \neq 0$, there may be net angular momentum if there is sufficient flow near the bottom to provide a torque that can balance the Rossby force. In the quasi-geostrophic limit, the bottom currents must be on the order of $-\beta H/df_0$ times the interior flows.

STEADILY TRANSLATING SOLUTIONS To conserve potential vorticity, any translating structure must move only east or west (and if it has bottom

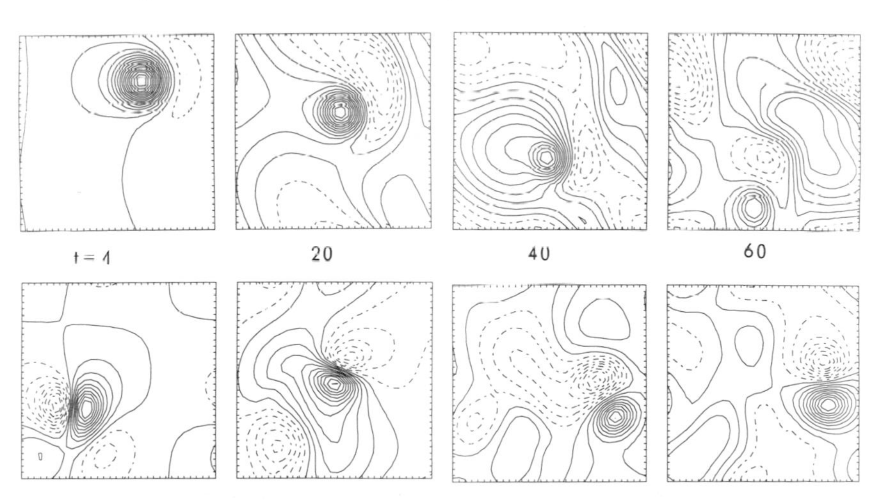

Figure 5 Evolution of barotropic flows with and without net angular momentum. The initial conditions are described in the text. Here $\beta = 0.1$ and the contouring grid spacing is 0.4. A pseudo-spectral code was used (W. K. Dewar & B. Klinger, private communication). The upper panels show the case with initial net angular momentum, and the lower four panels the case without.

signals, the topography must slope north-south). We can replace $\partial/\partial t$ in the continuity equation (13) by $-c(\partial/\partial x)$. In addition, the thickness between the two vertical surfaces will be written in terms of a mean thickness $\bar{h}(y)$ and an anomaly $h(x, y)$. We then find

$$\frac{\partial}{\partial x}\left(\int u - ch\right) + \frac{\partial}{\partial y}\int v = 0,$$

or

$$\int u = ch - \Psi_y, \qquad \int v = \Psi_x.$$

Using these in the horizontally integrated momentum equations gives

$$0 = \iint p_1 s_{1x} - p_0 s_{0x}$$

$$c \iint fh + \beta \iint \Psi = \iint p_1 s_{1y} - p_0 s_{0y}.$$

(The origin is chosen to be at the center of mass of the h-field.)

The results of Nof (1981, 1983, 1985) and Killworth (1983) are obtained by postulating that the integrals involving $p_1 s_{1y}$ and $p_0 s_{0y}$ vanish; an expression for the translational speed can thereby be derived. This might apply when the fluid below $z = s_0$ is motionless and the isopycnals are level; then the deep hydrostatic relation gives

$$p_0 = \int_{z_{\text{ref}}}^{s_0} B(z)\, dz, \tag{18}$$

so that

$$\iint -p_0 s_{0y} = \iint s_0 p_{0y} = \iint s_0 s_{0y} B(s_0) = 0.$$

We come back later to the question of when this assumption can work and the implications of its breakdown; for now, we simply remark that the original justification was in terms of the two-layer model with a lower layer that was infinitely deep and motionless. The upper interface is the (rigid) lid, $\nabla s_1 = 0$, and the translation speed is given by

$$c = -\beta \iint \Psi \Big/ \iint fh, \tag{19}$$

which represents a balance between the Rossby force and the net Coriolis force on the propagating anomaly of mass.

In applying this theory to the oceanic data, we find a number of difficulties: As an example, we consider the motion of a Gulf Stream ring (82B). The azimuthal velocity field is shown in Figure 6, superimposed on the isopycnals. The volume anomaly can be calculated fairly easily, although the data do not show an asymptotic limit to $s(r)$ very clearly; thus we are forced to assume that the outermost station is representative of the waters around the ring. With this assumption, $\iint h$ can be estimated as a function of σ_θ (potential density) on the lower isopycnal (Figure 6). The integral of Ψ cannot be calculated so accurately, since $v(r)$ does not fall off rapidly in this data set. If we assume that v vanishes beyond 90 km, we find the $\iint \Psi$ values as shown. Finally, we plot the speed of motion from (19). A second difficulty then arises: What material surface represents the bottom of the eddy, since the isopycnals clearly do not level off as required? It is encouraging to find that there is a whole range of isopycnals in the thermocline that yield rather similar estimates of the translational speed c (-0.6 to -0.8 cm s^{-1}), although these speeds are rather small compared with the observed values of about -4 cm s^{-1} (Evans et al. 1985). The remainder may be due to the general mean southwestward drift in the Slope Water.

Consider now the consistency of the "region of no motion" assumption (18). First, it is quite clear that this cannot be exactly correct for a finite-depth fluid. The proof is simple: If the deeper fluid is at rest, the motion in the deep region is also isolated, and the Flierl et al. (1983) integral result holds when applied from the surface to a level beneath the deepest extent of s_0. Then $\iint \Psi$ must be zero, which implies that $\iint h = 0$ and c is indeterminate. Clearly, (19) is only an approximate result, and the motion will not be isolated at all depths. The consequences of this are discussed in later sections.

MODONS

Exact Solutions

One approach to producing more realistic representations of isolated vortex structures is to "desingularize" the point-vortex solutions by spreading the anomaly in potential vorticity over a finite area $\pi\varepsilon^2$. The problem becomes one of finding the free boundary of the constant-*PV* fluid; a perturbation calculation with ε small compared with the separation shows that it becomes elliptical. Pierrehumbert (1980) and Wu et al. (1984) have solved the $\beta = 0$ case numerically, finding c and r for the whole range of ε. The vortex takes the form of a flattened ellipse and speeds up as ε

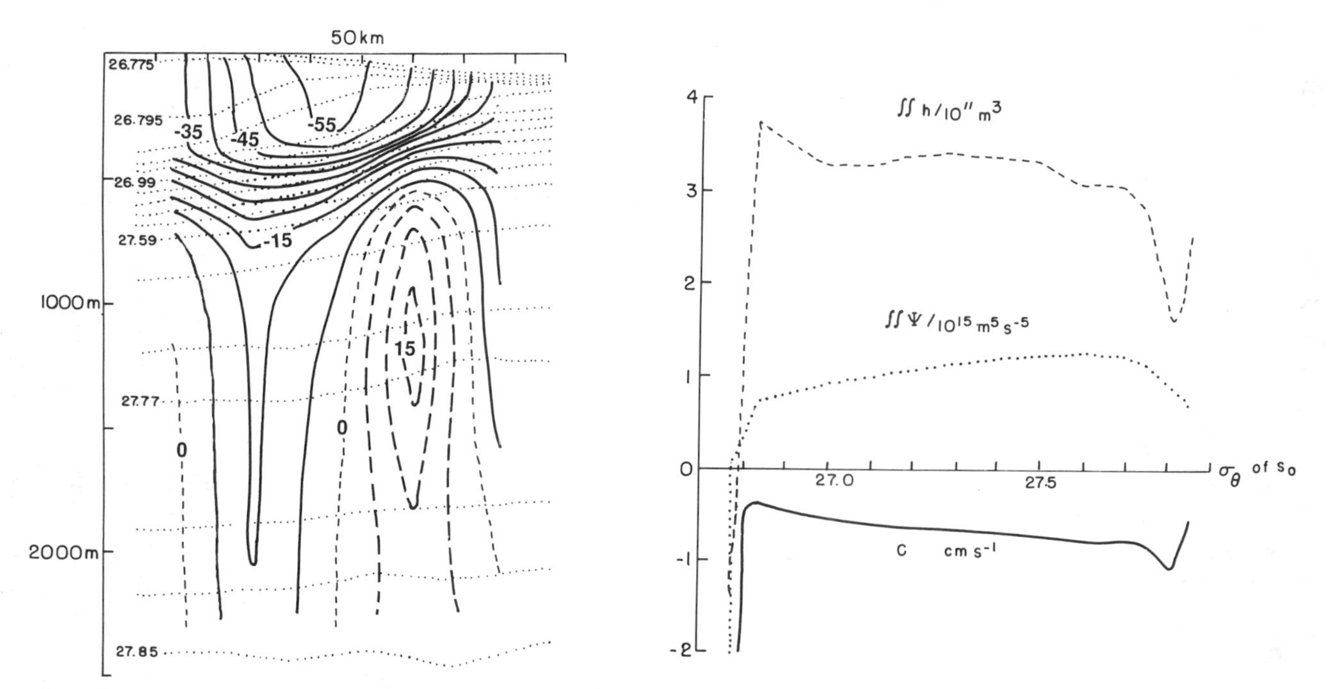

Figure 6 (*left*) Azimuthally averaged azimuthal velocity for Ring 82B in June, 1983. These preliminary data were provided by T. Joyce and will appear in final form in T. Joyce, T. McDougall, & G. Flierl (in preparation). The velocities are in centimeters per second, with dashed lines indicating clockwise flow. The light dots indicate σ_θ surfaces. (*right*) Integrated properties from the surface to the σ_θ surface indicated.

increases. This problem has yet to be solved for β-plane vortices, although similar techniques could be applied.

Like the desingularized examples considered above, the modon solutions, following the work of Larichev & Reznik (1976), build upon an isolated exterior field connected smoothly to an interior region with a different relationship between the vorticity and the stream function. The only differences are (*a*) the region of anomalous potential vorticity is assumed to coincide with the trapped region, (*b*) for analytic simplicity the boundary is taken to be circular, and (*c*) there is a linear relationship between Ψ and PV in the interior (rather than constant PV). (The latter is probably only important in that it makes the fields smoother and allows expression of the solutions in closed form.) The solution procedure consists of writing linear exterior and interior equations for the total stream function, including the cy term:

$$(\nabla^2 + L_z)\Psi + \beta y = A + B\Psi = P(\Psi). \tag{20}$$

In the exterior region where $\Psi \to \bar{\psi} + cy$, we have $A = 0$ and $B = \bar{q}_\psi$ [from (8)], and we need $\bar{q}_\psi$ to be constant (or at most a function of z). The mean flow, then, will be in the class satisfying (9); we note that this restrictive assumption is applied only for the sake of making analytical progress. Presumably, solutions with other mean profiles can be obtained numerically. The exterior solution is just $\Psi = \bar{\psi} + cy + (N/\bar{\rho}f_0)^{1/2}\phi_e(r, \theta, z)$ with $\nabla^2\phi_e = \gamma^2\phi_e$. In the interior, we take $A = A(z)$ and $B = -k^2$, and the solution is

$$\Psi = \mathscr{A}(z) - \frac{\beta y}{k^2} + \left(\frac{N}{\bar{\rho}f_0}\right)^{1/2} \phi_i(r, \theta, \xi),$$

where

$$(L_z + k^2)\mathscr{A} = A(z)$$

and

$$\nabla^2\phi_i = -k^2\phi_i.$$

The Helmholtz equations for the ϕ's are deceptively simple; the difficulties come in determining the location of the free boundary

$$r = a(\theta, z),$$

where $\Psi(a(\theta, z), \theta, z) = \psi_0(z)$ and $\Psi_r(a(\theta, z), \theta, z)$ is continuous. Table 1 lists published modon solutions; here we discuss only two solutions to illustrate both their character and some of the difficulties of the analysis involved in finding steady states.

Table 1 Examples of modon solutions

Reference	Dynamics	Background PV field	Interior PV form	Comments
Lamb (1879)	2D	None	Linear	As a variant of Hill's spherical vortex
Batchelor (1967)	2D	None	Linear	Same solution as above
Pierrehumbert (1980)	2D	None	Constant	—
Wu et al. (1984)	2D	None	Constant	Somewhat different approach
Stern (1975)	Barotropic	β	Linear	Only $c = 0$ case
Larichev & Reznik (1976)	Barotropic and 1.5-layer	β	Linear	Propagating solutions
Flierl (1979)	Stratified	β	Linear + $F(z)$	Stationary with rider
Flierl et al. (1980)	Barotropic, 1.5- and 2-layer	β	Linear, linear + constant	Full discussion of all cases with circular boundary
Berestov (1979)	Constant N	β	Linear	Continuously stratified
Berestov (1981)	Constant N	β	Linear	Continuously stratified with rider
Boyd (1985)	Equatorial	β	Linear	—
Dewar (1986)	Non-QG, 2-layer	β	Constant PV	Bridges gap to Nof's (1981) work
Moore & Saffman (1971)	2D	Constant shear	Constant	Found two forms, one unstable, elliptical
Ingersoll & Cuong (1981)	1.5-layer	$U/k \cos(ky) + \beta y$	Linear	Great Red Spot model
Swenson (1982)	Barotropic	β and shear	Linear & weakly nonlinear	Various solutions with weak shear or weak β
Flierl et al. (1983)	2D	Remote vortex	Linear	Propagates in circle
Tribbia (1984)	Barotropic and 1.5-layer	β on sphere	Linear	—
Verkley (1984)	Barotropic	β on sphere	Linear	—

The first example is Berestov's (1979) stratified eddy solution, chosen because it not only is characteristic of modon solutions (and indeed is the simplest) but also is less well known. Consider a uniformly stratified ocean ($N = \bar{\rho} =$ constant), infinitely deep, with the surface at $z = 0$ and with no mean flows. Suppose the boundary of the trapped fluid is the half-sphere (in stretched z-space) $r = a$, where r is the spherical radius $r = [x^2 + y^2 + (Nz/h)^2]^{1/2}$, and let a, ψ_0, and A be constant. Then the solutions ϕ_e and ϕ_i to the exterior and interior Helmholtz equations must satisfy

$$\phi_e(a, \theta, \varphi) = \psi_0 - ca \sin\theta \cos\varphi = \psi_0 - \frac{\beta}{\nu^2} a \sin\theta \cos\varphi,$$

$$\phi_i(a, \theta, \varphi) = \psi_0 + \frac{\beta}{k^2} a \sin\theta \cos\varphi - \frac{A}{k^2},$$

where θ, φ are the angles from the east and horizontal directions, respectively. The solutions are

$$\phi_e = \psi_0 \frac{k_0(\nu r)}{k_0(\nu a)} - \frac{\beta a}{\nu^2} \frac{k_1(\nu r)}{k_1(\nu a)} \sin\theta \cos\varphi,$$

$$\phi_i = \left(\psi_0 - \frac{A}{k^2}\right) \frac{j_0(kr)}{j_0(ka)} + \frac{\beta a}{k^2} \frac{j_1(kr)}{j_1(ka)} \sin\theta \cos\varphi$$

[ignoring the constant $(N/\bar{\rho} f_0)$ factor], with j_i and k_i the ordinary and modified spherical Bessel functions. Matching the tangential velocity gives the nonlinear dispersion relation

$$c \sin\theta \cos\varphi + \phi_{e_r}(a, \theta, \varphi) = -\beta/k^2 \sin\theta \cos\varphi + \phi_{i_r}(a, \theta, \varphi),$$

$$c\left[1 - \frac{\nu a k_1'(\nu a)}{k_1(\nu a)}\right] = -\beta/k^2\left[1 - \frac{ka j_1'(ka)}{j_1(ka)}\right],$$

or

$$\frac{k_2(\nu a)}{\nu a k_1(\nu a)} = -\frac{j_2(ka)}{ka j_1(ka)} \tag{21}$$

(using $\nu^2 = \beta/c$), as shown in Figure 7. This is the typical form of a modon relationship: it relates the speed (inherent in ν), the size a, and the amplitude of the modon. The latter is essentially a function of k, since the interior structure and quantities such as maximum ψ or the peak velocity

$$u(0,0,0) = \beta a^2 \frac{1}{(ka)^2}\left[1 - \frac{1}{3}\frac{ka}{j_1(ka)}\right]$$

depend on ka.

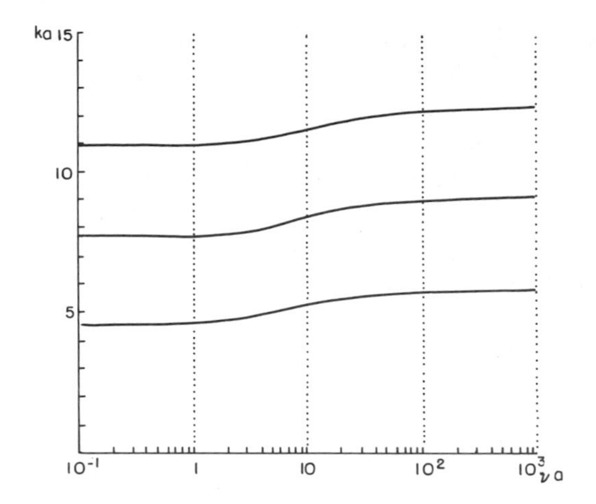

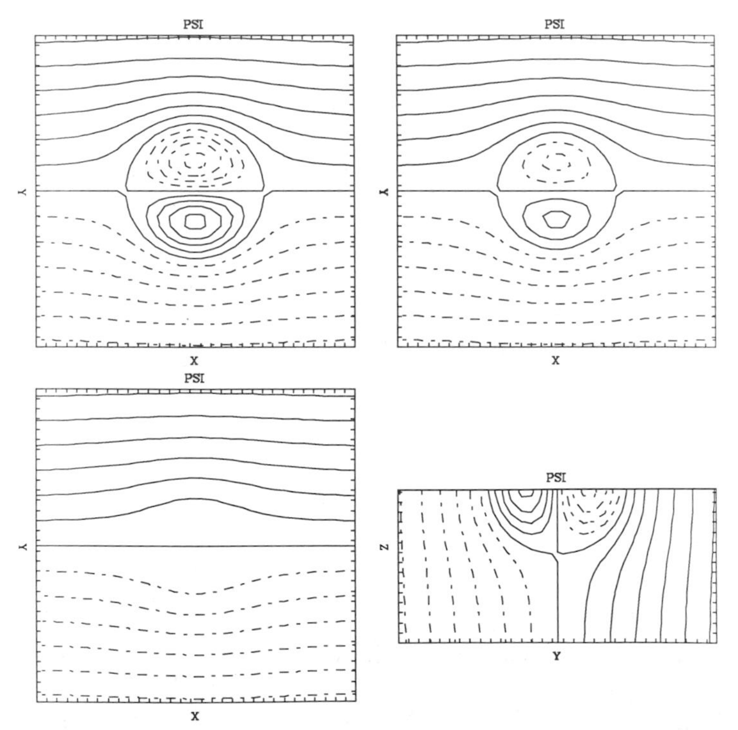

Figure 7 Solution for Berestov's modon. (*top*) The relationship between ka and νa; (*middle* and *lower left*) $\psi(x, y)$ plots at $z = 0$, $-1.0f/aN$, and $-2.0fa/N$; (*bottom right*) $\psi(y, z)$ at $x = 0$.

Note that there is one free parameter (ψ_0) determining the amplitude of the radial symmetric (or "rider") field. Flierl et al. (1980) give various examples of the changes in shape that can be produced by rider fields. We only show the simplest ($\psi_0 = 0$) structure using the smallest ka choice in Figure 7. [In the plasma applications, the modon with the lowest ka is known as the "ground state." There are arguments in Berestov (1981) that the higher ka values do not lead to valid solutions. These appear to be based on the fact that P in (20) is a single-valued functional in the northern part of the domain and attempting to rule out multiple-valued P's. However, examination of the ground-state modon reveals that P is not single valued when the whole plane is considered; in fact, the only proper restriction is that the boundary between different forms of the potential-vorticity functional must be a streamline. The higher-ka solutions are correct solutions, although their stability and their physical importance are certainly open to question.]

As a second example, we sketch the solution for a two-layer stationary modon embedded in a *baroclinic* flow (Figure 8):

$$\nabla^2\Psi_1+\lambda^2(\Psi_2-\Psi_1)+\beta y = P_1(\Psi_1),$$

$$\nabla^2\Psi_2+\lambda^2(\Psi_1-\Psi_2)+\beta y = P_2(\Psi_2),$$

where $\lambda^2 = f_0^2/g(\Delta\rho/\rho)H_1 = 1/2R_d^2$ (R_d is the deformation radius) and $\Psi_1 \to -\bar{u}_1 y$, $\Psi_2 \to -\bar{u}_2 y$ in the far field. Thus we have

$$P_1(\Psi_1) = -\frac{\beta+\lambda^2(\bar{u}_1-\bar{u}_2)}{\bar{u}_1}\Psi_1,$$

$$P_2(\Psi_2) = -\frac{\beta+\lambda^2(\bar{u}_2-\bar{u}_1)}{\bar{u}_2}\Psi_2,$$

in the exterior of the modon. The exterior solution is

$$\Psi_1 = -\bar{u}_1 y+AK_1(qr)\sin\theta,$$

$$\Psi_2 = -\bar{u}_2 y+\alpha AK_1(qr)\sin\theta,$$

with

$$\left[q^2+\frac{\beta-\lambda^2\bar{u}_2}{\bar{u}_1}\right]\left[q^2+\frac{\beta-\lambda^2\bar{u}_1}{\bar{u}_2}\right] = \lambda^4. \tag{22}$$

Analysis of this equation reveals that there is one positive root for q only when $\bar{u}_1+\bar{u}_2 > \beta/\lambda^2$ (in the case that $\bar{u}_1$, $\bar{u}_2 > 0$)—the barotropic component of the flow must exceed the long-wave speed. The coefficient

$$\alpha = -\frac{1}{\lambda^2}\left(q^2 + \frac{\beta - \lambda^2 \bar{u}_2}{\bar{u}_1}\right)$$

can be shown to be negative; if the streamlines diverge at one level, they will converge at the other. Thus the closed streamline will occur in only one layer; let that be layer 1, so that

$$AK_1(qa) = \bar{u}_1 a,$$

where a is the modon radius.

The interior solution takes the form

$$\psi_i = d_i y + [e_i J_1(k_1 r) + f_i J_1(k_2 r)] \sin\theta,$$

where the k_i's satisfy

$$\left[k_1^2 + \frac{\beta - \lambda^2 \bar{u}_2}{\bar{u}_1}\right]\left[k_2^2 + \frac{\beta - \lambda^2 \bar{u}_2}{\bar{u}_1}\right] = -\lambda^4. \tag{23}$$

[The long derivation is quite similar to that in Flierl et al. (1980, Sect. 6) and uses the fact that $P_2(\Psi_2)$ is the same inside the modon as outside.] Finally, we also find from the matching conditions (ψ_1, ψ_{1r}, ψ_2, ψ_{2r}, and P_2 all continuous) that

$$k_j a \frac{J_1(k_j a)}{J_2(k_j a)} = -\frac{qaK_1(qa)}{K_2(qa)} \tag{24}$$

must hold for each k_j. This dispersion relationship is like that sketched in Figure 8 and is tabulated in Flierl et al. (1980). The solution procedure involves choosing q and a; Equation (24) then gives k_1 and k_2, Equation (23) yields $(\beta - \bar{u}_2/2R_d^2)/\bar{u}_1$, and Equation (22) gives $(\beta - \bar{u}_1/2R_d^2)/\bar{u}_2$. From these, we can calculate $\bar{u}_1$ and $\bar{u}_2$.

For a model with the two equal-thickness layers (as in Flierl et al. 1980), we find no solutions with $\bar{u}_1, \bar{u}_2 > 0$ if we use the lowest two roots for k_1 and k_2. We do find such solutions with the first and third root, as sketched in Figure 8. Note that the modons have two unpleasant features: the convergent streamlines in layer 2, and the extremely complex and presumably delicate structure within the trapped region. Thus the two-layer model prediction is quite different from the equivalent barotropic structure of McWilliams (1980). The former problem can be viewed as arising because the basic flow state is not favorable to trapping stationary eddies, and the confinement in x, y is achieved by using a modal structure in z. The complex structure is probably related to the linearity of the PV/Ψ relationship and the choice of circular boundaries.

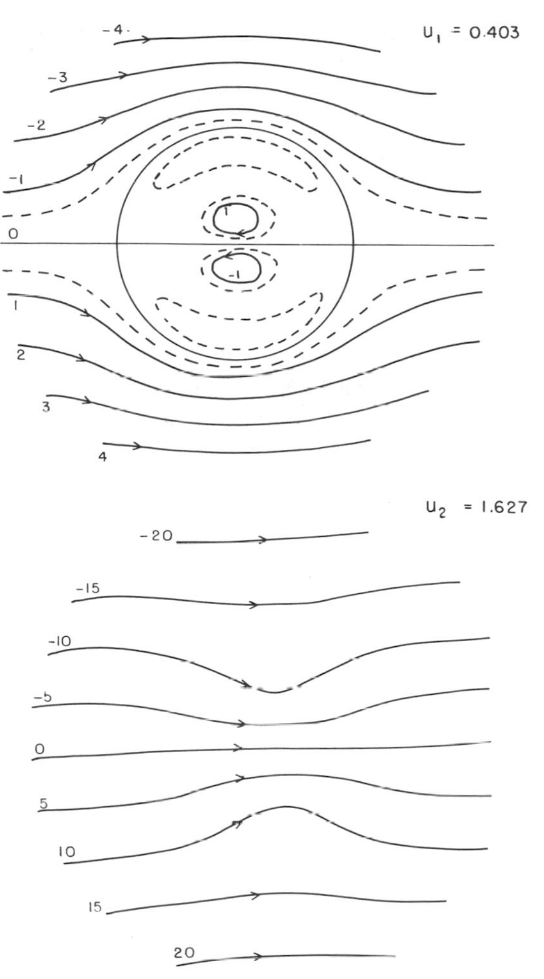

Figure 8 Two-layer modons with baroclinic shear. The closed streamline is in layer 1 with the weaker flow; the open streamlines are in the layer with stronger flow. The units are βR_d^2 for U's and $a = 6.5R_d$.

Radiating Modons

There are two difficulties with the formalism presented previously that make its application to geophysical situations difficult. First, the requirement that $v^2 > 0$ in the far field will rarely hold over the whole region of interest. Use of a normal-mode structure in the vertical may be able to offset negative v^2 regions (positive index of refraction zones), but it is in general more likely that the y, z modes will be "leaky," with some regions of negative ν^2. Second, the integral theorem requiring no net angular momentum is quite severe. While structures such as Gulf Stream rings or the Great Red Spot may have corresponding countercurrents below the main feature, it seems unlikely that these are able to compensate completely for the upper region's angular momentum. (In the case of rings, the observations are still sketchy, with some data indicating counterflow and other data not.) Understanding of the coupling between nonlinear coherent structures and waves seems both an important and a fruitful direction for research.

Some understanding of this coupling has been obtained by using a two-layer model with the coherent structure in one layer and the waves confined to the other (Flierl 1984). As an example, we consider again the two-layer QG equations but now with unequal layer depths $H_1 = \delta H_2$ with $\delta \ll 1$. In the frame of reference moving at speed c, we have

$$\left[\delta\frac{\partial}{\partial t}+(\bar{u}_1-c)\frac{\partial}{\partial x}+J(\psi_1,\cdot)\right][\nabla^2\psi_1+\lambda^2(\delta\psi_2-\psi_1)]$$
$$+[\beta+\lambda^2(\bar{u}_1-\bar{u}_2)]\psi_{1x}=0,$$
$$\left[\delta\frac{\partial}{\partial t}+(\bar{u}_2-c)\frac{\partial}{\partial x}+\delta J(\psi_2,\cdot)\right][\nabla^2\psi_2+\lambda^2(\psi_1-\delta\psi_2)]$$
$$+[\beta-\delta\lambda^2(\bar{u}_1-\bar{u}_2)]\psi_{2x}=0,$$

where we have also assumed that changes other than east-west propagation occur on a slow time scale, and that $|\psi_2| \sim \delta|\psi_1|$. The latter assumption implies that the flow is dominantly baroclinic. The lowest order problem in layer 1 yields

$$\nabla^2\psi_1-\lambda^2\psi_1+[\beta+\lambda^2(\bar{u}_1-\bar{u}_2)]y = P_1((c-\bar{u}_1)y+\psi_1), \tag{25}$$

which has a modon solution

$$\psi_1 = \begin{cases} -(c-\bar{u}_1)a\dfrac{K_1(\nu r)}{K_1(\nu a)}\sin\theta, & r>a, \\ (c-\bar{u}_1)a\dfrac{\nu^2}{k^2}\left[\dfrac{J_1(kr)}{J_1(ka)}-\dfrac{r}{a}\left(1+\dfrac{k^2}{\nu^2}\right)\right]\sin\theta, & r<a, \end{cases} \tag{26}$$

as long as

$$\nu^2 = \lambda^2+\frac{\beta+\lambda^2(\bar{u}_1-\bar{u}_2)}{c-\bar{u}_1} \geq 0.$$

In an oceanic application, where layer 1 represents the water above the thermocline, we might take $\bar{u}_j = 0$ so that we need $c \leq -\beta/\lambda^2$. (The case with the equal sign has special significance, as discussed below.) For an atmospheric application, layer 1 would be the lower layer; if we take $\beta - \lambda^2\bar{u}_2$ to be negative, we could have stationary solutions.

Consider now the layer-2 equation at lowest order:

$$\nabla^2\psi_2 + \frac{\beta}{\bar{u}_2 - c}\psi_2 = -\lambda^2\psi_1.$$

Since $\beta/(\bar{u}_2 - c)$ can be positive (and is for both examples above), corresponding to $v^2 < 0$ in this layer, the solution to this forced problem will characteristically have a Rossby-wave field far from the eddy. The next order in layer 1 shows the interaction between the structure and the wave field. The simplest expression for this is the energy balance for the layer:

$$\frac{\partial}{\partial t}\frac{1}{2}\iint |\nabla\psi_1|^2 + \lambda^2\psi_1^2 = \lambda^2(\bar{u}_1 - c)\iint \psi_1 \frac{\partial}{\partial x}\psi_2 + \ldots.$$

The correlation between ψ_1 and ψ_{2x} is negative, and $\bar{u}_1 - c$ is positive for the problems considered; thus energy drains out of the feature at a rate that depends on the exact shape of the feature and the value of $\beta/(\bar{u}_2 - c)$. The decay rate scales by $(H_1/H_2)\beta R$, but in the case considered by Flierl (1984), the proportionality constant was also quite small. We anticipate, then, that even nonisolated structures can be long lived if the nonlinear advection is sufficiently strong and properly organized.

Meiss & Horton (1983) provide another approach to a radiating problem: They assume that v^2 varies slowly, becoming negative far from the dipole structure. The radiation occurs in the same horizontal strata as the modon. The problem is solved by expanding the linear waves near the turning point and matching to the far-field solution of the modon. They find that the energy loss is exponentially small unless the speed of the modon approaches the long linear wave limit.

We next consider the case of Equation (25) corresponding to $v^2 = 0$. We can now find a rider component of the flow that does not integrate to zero:

$$\psi = \begin{cases} -(c - \bar{u}_1)\dfrac{a}{r}\sin\theta, & r > a, \\[2ex] -(c - \bar{u}_1)a\left[\dfrac{r}{a} - \dfrac{2J_1(kr)}{kaJ_0(ka)}\right]\sin\theta & \\[2ex] +A_0[J_0(kr) - J_0(ka)], & r < a, \end{cases} \tag{27}$$

$$J_1(ka) = 0.$$

The assumption here is that the $A(z)$ term in Equation (20) is nonzero. The solution is *not* isolated in the Flierl et al. (1983) sense, since $\psi = O(1/r)$; however, the nearby isolated solutions $\nu^2 > 0$ have the undesirable property that the rider field has a zero spatial average, as some manipulation of the formulas in Flierl et al. (1980) demonstrates. In fact, the QG form of (19),

$$(c+\beta/\lambda^2)\iint \psi_1 = 0,$$

clearly requires that the $\nu^2 = 0$ case be the only one with isolated monopolar eddies; it is therefore of considerable interest for modeling structures such as Gulf Stream rings.

The solution (27) is the QG equivalent of the isolated eddy model of Nof (1981, 1983, 1985) and Killworth (1983). It captures several features of the nongeostrophic eddy solution—the westward propagation consistent with (19), the weaker dipolar structure of flows with characteristic strength βa^2 (Killworth 1983, Flierl 1984), and the stronger azimuthally symmetric circulation (if we make A large). Others—the possibility of the surfacing of the thermocline, the fact that anticyclones/cyclones travel faster/slower than the long-wave speed—cannot be reproduced by the QG model but require terms of higher order in Rossby number. The difference in propagation rate can be potentially significant, since the loss of energy to baroclinic modes in the far field may depend on whether the eddy moves faster or slower than the fastest linear sinusoidal disturbance.

The contribution from the lower layer, of course, will lead to loss of energy into barotropic Rossby waves; however, the expected lifetime of the feature is still long (several hundred rotation times), and it seems worthwhile to regard it as essentially an equilibrium structure.

Stability of Modons

The question of the stability of the modon structures has recently excited the interest of a number of people, especially those working on the application of Arnol'd's theorem to fluid dynamics. This is understandable, not only because the problem by its two-dimensional nature is resistant to ordinary analysis, but also because the modons are exact solutions to the QG equations with nontrivial contributions from the $J(\psi, \nabla^2\psi)$ term. This work has still not led to definite conclusions; here we only indicate some of the directions being taken.

The earliest results came from simply using modons as initial conditions in numerical models and observing the evolution (McWilliams et al. 1981, McWilliams & Zabusky 1982, Makino et al. 1981). These experiments certainly indicated that these structures were robust solutions, even in the presence of considerable numerical noise, and were not subject to any strong exponential instability. Additional experiments on eastward-propagating modons with imposed noise indicated that there was a threshold perturbation amplitude that depended on the scale content, with smaller scales being less effective in destroying the feature for the same rms perturbation vorticity. It is not clear what the influence of even larger scales might be, although Swenson (1982) argues that weak shear across a modon will distort but not destroy it. The numerical experiments also suggest that adding a radially symmetric rider field leads to instability (McWilliams & Flierl 1979, M. Swenson, private communication).

If we follow linear-instability theory, we find that exponentially growing perturbations $[\varphi \sim \exp(\sigma t)]$ upon equivalent barotropic modons (the brand that seems to have excited the most interest) satisfy

$$\sigma(\nabla^2-\lambda^2)\varphi+J(\Psi,(\nabla^2-\lambda^2-P')\varphi)=0, \tag{28}$$

where Ψ is the total stream function (including the cy-term) from Equation (26) and

$$v^2=\begin{cases}\dfrac{\beta}{c}+\lambda^2, & r>a\\ -k^2, & r<a\end{cases}, \qquad P'=\begin{cases}\dfrac{\beta}{c}, & r>a,\\ -(k^2+\lambda^2), & r<a.\end{cases}$$

Matching conditions at the perturbed interface imply that φ and φ_r are continuous (since Ψ, Ψ_r and Ψ_{rr} are). If we follow the arguments of Arnol'd (1965, 1969), Blumen (1968), or Charney & Flierl (1981), we find that

$$\operatorname{Re}\sigma\left\{\iint|\nabla\varphi|^2+\lambda^2|\varphi|^2-\iint_{r<a}\frac{[(\nabla^2-\lambda^2)\varphi]^2}{k^2+\lambda^2}+\iint_{r>a}\frac{[(\nabla^2-\lambda^2)\varphi]^2}{\beta/c}\right\} \tag{29}$$

$$=\operatorname{Re}\sigma\,\mathscr{L}=0.$$

A proof of linear stability is obtained if it is demonstrated that $\mathscr{L}\neq 0$. We can synthesize and simplify the results of Pierini (1985) and Swaters (1986) using an approach similar to that of the latter author: If a generalized perturbation wave number η is defined by

$$\lambda^2+\eta^2=\frac{\iint[(\nabla^2-\lambda^2)\varphi]^2}{\iint|\nabla\varphi|^2+\lambda^2|\varphi|^2}, \qquad \eta^2>0,$$

we can rewrite (29) as

$$\mathscr{L} = \frac{k^2-\eta^2}{k^2+\lambda^2} \iint |\nabla\varphi|^2+\lambda^2|\varphi|^2+\left(\frac{c}{\beta}+\frac{1}{k^2+\lambda^2}\right) \iint_{r>a} [(\nabla^2-\lambda^2)\varphi]^2.$$

If $c > 0$ and $\eta^2 < k^2$, then $\mathscr{L} > 0$; thus eastward-traveling modons are stable to disturbances with large enough scale. Likewise, if $c < -(\beta/\lambda^2)$ and $\eta^2 > k^2$, then $[(c/\beta)+(k^2+\lambda^2)^{-1}] < 0$ and $\mathscr{L} < 0$; accordingly, westward-traveling modons are stable to small-scale disturbances. These arguments (which resemble the Fjortoft restriction that energy must be transferred to both larger and smaller scales) are not conclusive, since they do not demonstrate stability to all perturbations; on the other hand, one can argue that the perturbations to blocking events, for example, are too small to disrupt a modon (Swaters 1986).

The arguments of Laedke & Spatschek (1985) are more subtle; they have attempted to demonstrate that westward-traveling modons are stable to all disturbances (other than trivial translational modes) by applying other constraints on $\mathscr{L}$ that, in effect, force η^2 to be greater than k^2 by eliminating the gravest modes of the operator formed by a variation of (29). The approach is unusual and thought provoking; however, we feel that further exploration of these techniques is still required.

CONCLUDING REMARKS

The discussions in this review suggest a number of open questions that must be addressed before any final assessment of the usefulness of isolated eddy models for geophysical flows can be made. The lifetime of modon structures in the presence of dissipation (Swaters 1985) or dispersion [i.e. radiation into propagating regions (Meiss & Horton 1983)] has been explored to some degree, but with little of the complexity of geophysical flows. The generation mechanism is barely understood. One fascinating study by Pierrehumbert & Malguzzi (1984) suggests that localized weak forcing may lead to local multiple equilibria, with the high-amplitude state being a modon maintained against dissipation by the forcing. The forcing may be provided by air-sea temperature contrasts or by the alteration in storm tracks associated with the split flow. Their study, as well as a similar effort by Baines (1983), indicates one approach to the generation of isolated structures. Another is suggested by McWilliams (1983), who demonstrated that the independent and radiating vortices may bond together to form an isolated dipole; in a turbulent simulation, McWilliams (1984) also found evidence of fairly long-lived coherent vortices (the β-effect was absent, and thus wave processes were relatively weak). Finally, Aref & Siggia (1980) showed theoretically that a wake or jet flow would break up into dipolar structures; the soap-film experiments of Basdevant et al. (1984) give striking examples of this phenomenon. In the ocean, observations of dipolar structures emerging from fronts suggest that coherent features may arise naturally from instabilities.

Advances in mathematical and computational techniques are also required. Modon models have been criticized as being simplistic, under the impression that such structures could not be found or would not persist in the complex horizontal and vertical shears found in real flows. We have argued that modons are not fundamentally so limited and believe that the simplicity of known solutions is related more to our limited powers of analysis. Computational methods such as contour dynamics have allowed considerable extension of isolated eddy models in nonrotating two-dimensional flows (cf. Zabusky & Overman 1983) but are not yet prevalent for geophysical modeling. Replacing QG theory with shallow-water or density-coordinate models has also been a strong undercurrent of isolated eddy modeling. Again, this issue was only touched upon in this review; we believe that allowing order-one thickness changes is certainly fruitful for increasing the realism of our models. Of course, there is a natural mechanism for isolating the eddy in such models if the interface surfaces! We believe that as our techniques develop, the family of isolated or partially isolated solutions will grow rapidly, and commensurately our understanding will deepen.

Likewise, application of isolated eddy models to oceanic, atmospheric, or planetary data is only beginning. As we remarked, many phenomena evoke such interpretations but we, as yet, have difficulty making comparisons more quantitative. Many of the features distinguishing a "modon" from other types of disturbances are subtle and difficult to measure unambiguously. Our analysis methods are pivoted on more linear ideas—Fourier methods, for example—and may hide the nonlinear and isolated nature of the phenomenon.

The point-vortex and modon models seem to offer a valuable paradigm for explaining isolated, nonlinear atmospheric, oceanic, and planetary phenomena in a simple way. When this understanding is coupled with an appreciation for the complex interactions between the coherent structures and the turbulence and wave fields surrounding them, we believe that the role of isolated eddies will become as apparent in the theory of geophysical flows as the structures themselves are in the observations.

ACKNOWLEDGMENTS

The support of the National Science Foundation (grant #OCE-8504080) and the Office of Naval Research (grant #N00014-80-C-0273) is gratefully acknowledged. Paola Malanotte-Rizzoli and Gordon Swaters most graciously read drafts on short notice and offered many helpful comments. I thank especially Norma Kroll for her careful reading and critique of the language and style of this review, and Ellen Silverburg for typing the manuscript.

Literature Cited

Aref, H. 1982. Point vortex motions with a center of symmetry. *Phys. Fluids* 25 : 2183–87

Aref, H. 1983. Integrable, chaotic, and turbulent vortex motions in two-dimensional flows. *Ann. Rev. Fluid Mech.* 15 : 345–89

Aref, H., Siggia, E. D. 1980. Vortex dynamics of the two-dimensional turbulent shear layer. *J. Fluid Mech.* 100 : 705–37

Arnol'd, V. I. 1965. Conditions for nonlinear stability of stationary plane curvilinear flows of an ideal fluid. *Sov. Math.* 6 : 773–77

Arnol'd, V. I. 1969. On an a-priori estimate in the theory of hydrodynamic stability. *Am. Math. Soc. Trans.* 79 : 267–69

Baines, P. G. 1983. A survey of blocking mechanisms with application to the Australian region. *Aust. Meteorol. Mag.* 31 : 27–36

Basdevant, C., Couder, Y., Sadourny, R. 1984. Vortices and vortex-couples in two-dimensional turbulence, or long-lived couples are Batchelor's couples. *Lect. Notes Phys.* 230 : 327–46

Batchelor, G. K. 1967. *An Introduction to Fluid Dynamics*. Cambridge : Cambridge Univ. Press. 615 pp. (pp. 534–35)

Berestov, A. L. 1979. Solitary Rossby waves. *Izv. Acad. Sci. USSR Atmos. Oceanic Phys.* 15 : 443–47

Berestov, A. L. 1981. Some new solutions for the Rossby solitons. *Izv. Acad. Sci. USSR Atmos. Oceanic Phys.* 17 : 60–64

Berk, H. L., Neilsen, C. E., Roberts, K. V. 1970. Phase space hydrodynamics of equivalent nonlinear systems: experimental and computational observations. *Phys. Fluids* 13 : 980–95

Berman, R. H., Tetreault, D. J., Dupree, T. H. 1985. Simulation of phase space hole growth and the development of intermittent plasma turbulence. *Phys. Fluids* 28 : 155–76

Blumen, W. 1968. On the stability of quasi-geostrophic flow. *J. Atmos. Sci.* 25 : 929–31

Boyd, J. P. 1985. Equatorial solitary waves. Part 3: Westward-travelling modons. *J. Phys. Oceanogr.* 15 : 46–54

Charney, J. G., Drazin, P. G. 1961. Propagation of planetary scale disturbances from the lower into the upper atmosphere. *J. Geophys. Res.* 66 : 83–109

Charney, J. G., Flierl, G. R. 1981. Oceanic analogues of large-scale atmospheric motions. In *Evolution of Physical Oceanography (Scientific Surveys in Honor of Henry Stommel)*, ed. B. A. Warren, C. Wunsch, pp. 504–48. Cambridge, Mass: MIT Press

Dewar, W. K. 1986. On the structure and propagation of isolated, thick, warm eddies. *J. Phys. Oceanogr.* In press

Evans, R. H., Baker, K. S., Brown, O. B., Smith, R. C. 1985. Chronology of warm-core ring 82B. *J. Geophys. Res.* 90(C5): 8803–11

Firing, E., Beardsley, R. C. 1976. The behavior of a barotropic eddy on the beta plane. *J. Phys. Oceanogr.* 6 : 57–65

Flierl, G. R. 1979. A simple model for the structure of warm and cold core rings. *J. Geophys. Res.* 84(C2) : 781–85

Flierl, G. R. 1984. Rossby wave radiation from a strongly nonlinear warm eddy. *J. Phys. Oceanogr.* 14 : 47–58

Flierl, G. R., Larichev, V. D., McWilliams, J. C., Reznik, G. M. 1980. The dynamics of baroclinic and barotropic solitary eddies. *Dyn. Atmos. Oceans* 5 : 1–41

Flierl, G. R., Stern, M. E., Whitehead, J. A. Jr. 1983. The physical significance of modons: laboratory experiments and general physical constraints. *Dyn. Atmos. Oceans* 7 : 233–63

Helmholtz, H. 1958. On integrals of the hydrodynamical equations which express vortex motion. Transl. P. G. Tait, 1867, in *Philos. Mag.* 33 : 485–512

Hogg, N. G., Stommel, H. M. 1985. The heton, an elementary interaction between discrete baroclinic geostrophic vortices, and its implications concerning heat flow. *Proc. R. Soc. London Ser. A* 397 : 1–20

Holton, J. R. 1979. *An Introduction to Dynamic Meteorology*. New York : Academic. 391 pp.

Hoskins, B. J., McIntyre, M. E., Robertson, A. W. 1985. On the use and significance of isentropic potential vorticity maps. *Q. J. R. Meteorol. Soc.* 111 : 877–946

Ingersoll, A. P., Cuong, P. G. 1981. Numerical model of long-lived Jovian vortices. *J. Atmos. Sci.* 38: 2067–76
Killworth, P. D. 1983. On the motion of isolated lenses on a beta-plane. *J. Phys. Oceanogr.* 13: 368–76
Killworth, P. D. 1986. On the propagation of isolated multilayer and continuously stratified eddies. *J. Phys. Oceanogr.* 16: 709–16
Kirchhoff, G. R. 1876. *Vorlesungen über Matematische Physik*, Vol. 1. Leipzig: Teubner. 466 pp.
Laedke, E. W., Spatschek, K. H. 1985. Dynamical properties of drift vortices. *Phys. Fluids* 28: 1008–10. See also *Phys. Fluids* 29: 133–42
Lamb, H. 1879. *Hydrodynamics*. Cambridge: Cambridge Univ. Press. 738 pp. (pp. 244–45)
Larichev, V. D., Reznik, G. M. 1976. Two-dimensional Rossby soliton: an exact solution. *Rep. USSR Acad. Sci.* 231: 1077–79
Makino, M., Kamimura, T., Taniuti, T. 1981. Dynamics of two-dimensional solitary vortices in a low-beta plasma with convective motion. *J. Phys. Soc. Jpn.* 50: 980–89
Malanotte-Rizzoli, P. 1980. Solitary Rossby waves over variable relief and their stability. Part II. Numerical experiments. *Dyn. Atmos. Oceans* 4: 261–94
Malanotte-Rizzoli, P. 1982. Planetary solitary waves in geophysical flows. *Adv. Geophys.* 24: 147–224
Malanotte-Rizzoli, P., Hendershott, M. C. 1980. Solitary Rossby waves over variable relief and their stability. Part I. The analytical theory. *Dyn. Atmos. Oceans* 4: 247–60
Malguzzi, P., Malanotte-Rizzoli, P. 1984. Nonlinear stationary Rossby waves on non-uniform zonal winds and atmospheric blocking. Part I: the analytical theory. *J. Atmos. Sci.* 41: 2620–28
McWilliams, J. C. 1980. An application of equivalent modons to atmospheric blocking. *Dyn. Atmos. Oceans* 5: 43–66
McWilliams, J. C. 1983. Interactions of isolated vortices. II. Modon generation by monopole collision. *Geophys. Astrophys. Fluid Dyn.* 24: 1–22
McWilliams, J. C. 1984. The emergence of isolated, coherent vortices in turbulent flow. *J. Fluid Mech.* 146: 21–43
McWilliams, J. C., Flierl, G. R. 1979. On the evolution of isolated, nonlinear vortices. *J. Phys. Oceanogr.* 9: 1155–82
McWilliams, J. C., Flierl, G. R., Larichev, V. D., Reznik, G. M. 1981. Numerical studies of barotropic modons. *Dyn. Atmos. Oceans* 5: 219–38
McWilliams, J. C., Zabusky, N. J. 1982. Interactions of isolated vortices. I: Modons colliding with modons. *Geophys. Astrophys. Fluid Dyn.* 19: 207–27
Meiss, J. D., Horton, W. 1983. Solitary drift waves in the presence of magnetic shear. *Phys. Fluids* 26: 990–97
Mied, R. P., Lindemann, G. J. 1982. The birth and evolution of eastward propagating modons. *J. Phys. Oceanogr.* 12: 213–30
Moore, D. W., Saffman, P. G. 1971. Structure of a line vortex in an imposed strain. In *Aircraft Wake Turbulence*, ed. J. H. Olsen, A. Goldburg, M. Rogers, pp. 339–53. New York: Plenum
Mory, M. 1983. Theory and experiment of isolated baroclinic vortices. In *1983 Summer Study Program in Geophysical Fluid Dynamics, Woods Hole Oceanogr. Inst. Tech. Rep. WHOI-83-41*, pp. 114–32
Nof, D. 1981. On the beta-induced movements of isolated baroclinic eddies. *J. Phys. Oceanogr.* 11: 1662–72
Nof, D. 1983. On the migration of isolated eddies with application to Gulf Stream rings. *J. Mar. Res.* 41: 399–425
Nof, D. 1985. Joint vortices, eastward propagating eddies and migratory Taylor columns. *J. Phys. Oceanogr.* 15: 1114–37
Pedlosky, J. 1979. *Geophysical Fluid Dynamics*. Berlin: Springer-Verlag. 629 pp.
Pierini, S. 1985. On the stability of equivalent modons. *Dyn. Atmos. Oceans* 9: 273–80
Pierrehumbert, R. T. 1980. A family of steady, translating vortex pairs with distributed vorticity. *J. Fluid Mech.* 99: 129–44
Pierrehumbert, R. T., Malguzzi, P. 1984. Forced coherent structures and local multiple equilibria in a barotropic atmosphere. *J. Atmos. Sci.* 41: 246–57
Rossby, C.-G., and collaborators. 1939. Relation between variations in the intensity of the zonal circulation of the atmosphere and the displacements of the semi-permanent centers of action. *J. Mar. Res.* 2: 38–55
Shukla, P. K. 1985. Nonlinear drift-ballooning modes. *Phys. Rev. A* 32: 1858–61
Shukla, P. K., Yu, M. Y., Varma, R. K. 1985. Drift-Alfvén vortices. *Phys. Fluids* 28: 1719–21
Stern, M. E. 1975. Minimal properties of planetary eddies. *J. Mar. Res.* 40: 57–74
Swaters, G. E. 1985. Ekman layer dissipation in an eastward-travelling modon. *J. Phys. Oceanogr.* 15: 1212–16
Swaters, G. E. 1986. Stability conditions and a-priori estimates for equivalent barotropic modons. *Phys. Fluids* 29: 1419–32
Swenson, M. 1982. Isolated 2D vortices in

the presence of shear. In *1982 Summer Study Program in Geophysical Fluid Dynamics, Woods Hole Oceanogr. Inst. Tech. Rep. No. WHOI-82-45*, pp. 324–36

Tribbia, J. J. 1984. Modons in spherical geometry. *Geophys. Astrophys. Fluid Dyn.* 30: 131–68

Verkley, W. T. M. 1984. The construction of barotropic modons on a sphere. *J. Atmos. Sci.* 41: 2492–2504

von Kármán, T. 1911. Über den Mechanismus der Widerstendes, den bewegter Körper in einer Flüssigkeit Erfährt. *Nachr. Ges. Wiss. Göttingen Math.-Phys. Kl.* 1911: 509–17 (Part 2 in 1912; see also von Kármán, Th., Rubach, H. 1912. *Phys. Z.* 13: 49–59)

Weinstein, A. 1983. Hamiltonian structure for drift waves and geostrophic flow. *Phys. Fluids* 26: 388–90

Wu, H. M., Overman, E. A., Zabusky, N. J. 1984. Steady-state solutions of the Euler equations in two dimensions: rotating and translating V-states with limiting cases. I. Numerical algorithms and results. *J. Comput. Phys.* 53: 42–71

Yu, M. Y., Shukla, P. K., Varma, R. K. 1985. Flute vortices in nonuniform magnetic fields. *Phys. Fluids* 28: 2925–27

Zabusky, N. J., McWilliams, J. C. 1982. A modulated point vortex model for geostrophic beta-plane dynamics. *Phys. Fluids* 25: 2175–82

Zabusky, N. J., Overman, E. A. 1983. Regularization of contour dynamical algorithms. I. Tangential regularization. *J. Comput. Phys.* 52: 351–73

Ann. Rev. Fluid Mech. 1987. 19 : 531–75

RECENT DEVELOPMENTS IN RAPID-DISTORTION THEORY

A. M. Savill

Fluid Dynamics Section, Cavendish Laboratory, University of Cambridge, Cambridge CB3 0HE, England

INTRODUCTION

The aim of this review is to outline some of the recent developments that have occurred in the use of Rapid-Distortion Theory (hereafter RDT) for calculating the response of turbulence to applied strains. These developments include the application of the theory to a wider range of flows and also changes in the manner it is applied. In particular, the intention is to indicate how RDT may be of practical use by identifying where exact results of this linear theory have proved valuable and by explaining how the strict "rapid" conditions have been relaxed to consider the rather slow strain rates of most flows of interest. The ways in which RDT can be an aid to the closure of transport-equation models, notably through the provision of stress-intensity ratio predictions and the potential improvements to the current approximations for pressure-strain correlations, are also discussed. To provide a background and framework for these latest results, it is first necessary to briefly review the development of RDT itself and earlier applications of the theory. However, it is not possible to do this full justice here, and for a more detailed appreciation, the reader is referred to the excellent earlier review of Hunt (1978); both editions of Townsend's monograph *The Structure of Turbulent Shear Flow* (1956, 1976); the PhD theses of Maxey (1978), Durbin (1978), and Savill (1979); and previous articles by Deissler (1961, 1970, 1975), Mathieu (1971), Hunt (1973), Moffatt (1981), Maxey (1982), and Savill (1982a). It is also impractical in such a short article to make reference to all of the work in this field, and so the selection of material for inclusion has necessarily been based on the author's own, inevitably biased opinion of the most significant recent research results. At the same time, in discussing the connections

0066–4189/87/0115–0531$02.00

with and implications for calculation schemes based on approximations to stress transport equations, it has been necessary to digress into other areas of turbulence modeling. However, in regarding "recent" to mean the period since Hunt's (1978) review, a conscious effort has been made to bring this assessment right up to date; this is reflected in the number of references to unpublished results and private communications, while some of the author's own analysis is published here for the first time. Some suggestions are also made about the directions in which further progress may perhaps be made, both through applying the theory to support experimental work in new areas and by using it in conjunction with Reynolds-stress transport (hereafter RST) equations to increase the structural input, and hence range of application, of existing turbulence models at several levels of closure.

Two areas of recent work that have deliberately been omitted concern the use of RDT to model diffusion and the flow around bluff bodies; these are the subjects of another very recent review by Hunt (1984a). A more general discussion of diffusion has also appeared in this series (Hunt 1985b). These omissions aside, it is hoped that this article will provide a critical appraisal of the present state-of-the-art and a guide to possible future developments.

BASIC RAPID-DISTORTION THEORY

The basic theory as discussed by Batchelor & Proudman (1954) makes no assumption about eddy structure but simply considers that turbulence contains a range of scales and associates these with ranges of wave numbers. It is assumed that the turbulence is weak, in the sense that the initial root-mean-square (rms) value of the u component of the fluctuations is much less than a characteristic mean-velocity scale, $u_{\text{rms}} \ll U_0$, so that the turbulence interacts strongly with the mean flow but only weakly with itself. This allows one to linearize the equations of motion in the strictly rapid limit where the distortion time is far less than the Lagrangian time scale of the turbulence, $t_{\text{d}} \ll t_L \equiv L/u_{\text{rms}}$ (where L is an integral scale).

In the extension of the theory adopted by the Cavendish group (e.g. Townsend 1970, 1976, 1980, Elliott 1976, Savill 1979), which is singled out for discussion here because it is of the greatest interest in analyzing shear flows, the turbulent motion is considered to be made up of many individual flow structures that do not overlap one another. Each eddy is seen as an organized swirling motion of finite size and simple form that is identifiable for times long compared with local time scales. The large-scale, low-wave-number eddies interact with the mean flow and gain energy from it. At the same time these eddies are distorted, and the stresses set up in them are effectively the Reynolds stresses. They pass energy on to small scales, where it is eventually dissipated by viscosity.

With the additional assumption of local homogeneity (over the scale of the large eddies, an important limitation that is not always true), and in the limit of high-turbulence Reynolds number, $u_{rms}L/\nu \gg 1$, Fourier series for the velocity and pressure field can be defined and substituted into the general transport equation for velocity fluctuations to derive the rapid strain equations for Fourier component amplitude $a_i(k)$ and wave number k:

$$\frac{\partial a_i}{\partial t}(k) = -a_l(k)\frac{\partial U_i}{\partial x_l} + \frac{2k_i k_j}{k^2}\frac{\partial U_j}{\partial x_l}a_l(k), \tag{1}$$

$$\frac{\partial k_i}{\partial t} = -k_j\frac{\partial U_j}{\partial x_i}. \tag{2}$$

The initial and distorted a, k are then linearly related by distortion matrices that have been derived for several simple distortions by Savill (1979), Townsend (1980), and Maxey (1982). The Reynolds-stress intensities may be recovered by defining a distorted spectrum function $F_{ij}(k)_d$ such that $\langle u_i u_j \rangle = F_{ij}(k)_d$ and $F_{ij}(k)_d = A_{ip}A_{jq}F_{pq}(k)$, where this initial spectrum is usually assumed to be isotropic. Exceptions are provided by the work of Deissler (1961), Sreenivasan & Narasimha (1978), and Maxey (1982) in which the initial spectrum tensor was chosen to be axisymmetric.

An alternative approach is to form linearized equations for the two-point velocity correlations and to solve directly for the spectrum tensor, neglecting third-order and higher moments (Pearson 1959, Deissler 1961, 1970, Fox 1964, Loiseau 1973, Zimont & Sabel'nikov 1975, V. W. Sumner, unpublished results).

Much early work made use of explicit results for the irrotational straining of homogeneous isotropic (Batchelor & Proudman 1954), and later axisymmetric (Sreenivasan & Narasimha 1978), turbulence to compare the theory with experiment for several distorting duct configurations (Townsend 1954, Tucker & Reynolds 1968, Reynolds & Tucker 1975, Tan-atichat et al. 1980, Tan-atichat & Harandi 1986).

However, following the derivation of results for uniform shearing (Moffatt 1967, Townsend 1970), the emphasis changed to consider the response of turbulence to such rotational straining and, more recently, the effect of additional strain rates on sheared turbulence (Keffer et al. 1978, Townsend 1980, Sreenivasan 1985). Deissler (1975) and Loiseau (1973), in comparing RDT with the homogeneous shear-flow experiments of Champagne et al. (1970), Harris et al. (1977), and Mulhearn (1975), assumed that the equilibrium shear-flow structure would be attained in the limit of large total shear. However, Mathieu (1971) and later Maxey (1982) have pointed out that the effective shear α experienced by the flow structure on average is limited by the finite distortion time, such that the transport of α may be modeled by

$$\frac{\partial \alpha}{\partial t} = \frac{\partial U_1}{\partial x_3} - \frac{\alpha}{t_d}, \tag{3}$$

which is consistent with the later data of Tavoularis & Corrsin (1981).

For inhomogeneous shear flows, Townsend (1970) assumed that the total effective shear experienced by an average eddy is the integral of the rate of strain along its path since entrainment. He thus estimated values of α for the various plane shear flows from a similar transport equation that incorporated a term to account for diffusion by the turbulence, and he found that these values covered a small range (despite the fact that they were overestimates due to the omission of a Maxey-type relaxation term). This result suggests that the structure is always relatively new and likely to be similar for all these flows. Using the values for α, Townsend was able to predict quite accurately eight out of the nine principal correlation functions measured by Grant (1958), a remarkable result.

It is perhaps equally surprising that although RDT should strictly only be applicable to some of the very rapid axisymmetric contractions considered by Tan-atichat et al. (and then, as they point out, only when the turbulence is of sufficiently small scale relative to the characteristic length scale of the contraction), the theory qualitatively reproduces experimental trends in different Reynolds-stress components for the majority of the irrotational strain distortions studied. This is particularly true when the calculations are corrected for the normal development (in many cases just decay) of the turbulence in the following way:

If one relaxes the rapid condition to consider distortions acting over times less than but of order t_L, there is then time for the large eddies to pass energy down a cascade to smaller eddies during the distortion. However turbulence/turbulence interactions between the large eddies themselves (similar wave numbers) may still be ignored if the turbulence remains weak, and so the only part of the nonlinear term that needs to be modeled is the transfer between disparate wave numbers. This will be roughly uniform over the extent of a large eddy and so may simply be described by a form of eddy viscosity (Pearson 1959, Townsend 1970). Such an approximation has the advantage that both viscous and nonlinear effects can be encompassed in a single additional term of the form $-\nu_T k^2 a_i$.

Furthermore, when this term is added to (1), the new equation has an integrating factor, so that the original solutions may be corrected simply by dividing through by this factor. Hunt (1978) has shown how ν_T may be estimated on the assumption that most distortions act to pile up vortex lines, while Townsend (1980) calculated this quantity from a balance between the mean-flow energy entering the large eddies as a result of the shear and the transfer of energy out of these structures to smaller scales. The same analysis leads to a more general result:

$$\nu_T = 0.12\alpha. \tag{4}$$

The disadvantage of including such an eddy transfer term is that the calculations are no longer independent of the exact form of the initial spectrum tensor.

A simpler, but less accurate, approach due to Ribner & Tucker (1953) is to assume the turbulence develops as if no distortion occurred and then subtract this development from the calculated results (Tucker & Reynolds 1968, Prabhu et al. 1974, Tan-atichat et al. 1980, Sreenivasan 1985). However, it is important to note that even after such corrections have been applied, RDT cannot be used to calculate the magnitudes of the stresses because the linearization introduces an arbitrary velocity scale.

Townsend's (1970) successful use of RDT to predict the correlation contours for simple shear flows has been taken as evidence that the energy-containing eddies are produced from newly entrained or "recycled" essentially isotropic material by the action of the mean shear in precisely the manner described by RDT. In order to check this idea it is first important to know what these eddies look like. Their actual shape has been deduced from his experimental correlation functions by Grant (1958) using a mixture of inference and guesswork, and by Payne (1966) using a computational approach based on eigenfunctions of these two-point correlations due to Lumley (1965). Both arrived at a similar picture of inclined, double-roller eddies that has subsequently been refined by the pattern-matching work of Mumford (1982, 1983). [In fact, Payne had to make assumptions about some of the eigenfunctions to obtain solutions, since full data were not available. Recently, Moin (1984) has published eigenfunction decompositions for a boundary layer obtained from numerical simulations, and M. R. Maxey & L. Sirovich (private communication) are currently studying the interpretation of such eigenfunctions in the light of RDT. One approach they are following is to examine various inviscid rapid distortion problems, with rigid free-slip boundaries imposed, for which correlations and eigenfunctions can be evaluated to see if these correspond to recognizable flow structures.]

Consideration of the equation for turbulent vorticity in the presence of mean shear then indicates that this has two effects on a vortex element: A mean vortex line can be stretched along its length, generating vorticity fluctuations, while the vertical component of the turbulent vorticity is rotated and stretched by the mean motion, generating a streamwise component. A combination of lifting, shearing, and stretching thus produces a horseshoe vortex loop, exactly the type of structure observed in the boundary layer by Head & Bandyopadhyay (1981), Wallace et al. (1983; see also Wallace 1982), and Smith (1984), and revealed by the numerical simulations of Moin & Kim (1985), although rather larger in scale, and also suspected in other flows. If the double-roller eddies are regarded as just one part of these structures (the "legs"), then it would appear RDT can indeed describe the formation of such structures.

There are two problems with this interpretation. First, it relies on the existence of a continual source of isotropic material provided by an entrainment process that, it could be argued, would order newly entrained fluid. Second, in the strict rapid limit, mean vortex lines can only be rotated through small angles, so that RDT should only describe the initial perturbation and not the subsequent generation and distortion of large eddies. This is why many people feel it is largely fortuitous that the theory provides such a useful and adequate description of sheared turbulence.

However, whether entrainment is "active," involving large-scale overturning and engulfment of external fluid, or "quiescent," with just small-scale "nibbling" at the interface (Townsend 1966), it is possible that little ordering actually occurs. In the first case, it appears that material is rapidly swirled into the flow and only rather more slowly has vorticity diffused into it, so that most of the ordering occurs before the flow becomes turbulent. In the latter case the process may act over too small a region for this to be significant. Equally, when one considers the effect of other processes such as bursting, vortex tangling, and decay, it does not seem unreasonable to assume that a continual source of unorganized, essentially isotropic material is available for new generation of larger-scale eddies.

I believe that the reason RDT can describe these eddies so well is that by predicting the initial linear perturbation of mean vortex lines, it effectively sets a "blueprint" for the type of structure to be produced, and that during the subsequent nonlinear development of the horseshoe vortex as lift-up and stretching occur, it maintains this basic form. (It is worth noting that such a structure once formed is essentially self-propelling, driven by self-induction.) Individual dimensions may be greatly altered, but it is not perturbed into a different type of structure by interactions with its neighbors, whatever their scale. Maxey (1982) has expressed a similar sentiment, suggesting that RDT describes the initial development of the structure and that nonlinear processes then tend to limit this development rather than radically alter the structure. If this is so, one can justify using the theory to model the plane shear flows and confidently proceed in applying it to their subsequent distortion. It would be convenient if the transformation matrices for more complex distortion histories were the product of simple distortions that together produce the total distortion, but unfortunately this commutative property holds only for the wave-number matrices and is not valid for any sequence including rotation. Where possible, the transformation matrix for the coefficients should be obtained by integration of the following equation:

$$\frac{dA_{ij}}{dt} = \left(2\frac{k_i k_l}{k^2}\frac{\partial U_l}{\partial x_m} - \frac{\partial U_i}{\partial x_m}\right)A_{mj}. \tag{5}$$

This was the procedure followed by Hunt (1973) and V. W. Sumner (unpublished results), but it is limited to the few flows where the mean flow can be solved explicitly. For other situations, Townsend (1980) has devised a numerical scheme in which the integration is done by "quadrature" such that the complex distortion is split into a sequence of simple, small distortions. For economy and accuracy these should be chosen so that the total distortion during, as well as at the end of, each step is close to the actual continuous distortion and so that no step has the effect of partially cancelling the influence of the previous step. The adequacy of results can be tested by reducing step size. The final Fourier coefficients are then calculated from the original ones with a transformation matrix that is the product of all the matrices for each step. This technique has been used to model a complex duct distortion of a wake (Elliott & Townsend 1981) and a curved mixing layer (Castro & Bradshaw 1976) by Elliott (1976), V. W. Sumner (unpublished results), and Townsend (1980). More recently, A. A. Townsend (private communication) has successfully applied a similar procedure to the turbulent Couette flow between concentric cylinders studied by Smith & Townsend (1982). He argues that the high levels of $\overline{uv}/\overline{q^2}$ and $\overline{v^2}/\overline{q^2}$ found in the center of this flow result from the distortion of parcels of boundary-layer material convected away from the wall by the toroidal eddies that span the flow. This effect is modeled by applying a shear to isotropic turbulence to represent the boundary layer and then adding on an amount of the constant-circulation irrotational straining, which characterizes Couette flow, equal to the rate of this strain in the flow multiplied by the time of flight determined from the toroid rotation rate (i.e. it is assumed that the toroid acts entirely passively instead of applying a distortion to the boundary-layer fluid itself).

Sreenivasan (1985) has avoided the problems of describing complex strain fields when applying RDT to his experimental investigation of the effect of two different contractions on a homogeneous shear flow; this was done by comparing calculations for irrotational strain and shear separately. The ratio of irrotational-strain rate to shear in the two cases was 0.12 and 1. Corrections for natural development of the shear flow were particularly important for the irrotational straining in the first case, and none of the calculations closely predicted the data, as one might expect. Nevertheless, in both cases the irrotational-strain-only RDT calculations reproduced the data better than the shear-only calculations, which suggests that the turbulence structure responds more readily to the additional strain rate than to subsequent shearing. This result is consistent with other observations (as discussed in the excellent earlier article in this series by Smits & Wood 1985) that extra strain rates e have a more dramatic (often a factor of 10 greater) effect than their magnitude alone would suggest. In fact, as the total effective strain experienced by the flow structure is limited by the eddy lifetime rather than by the total strain actually applied, it is clear that the addition of a new strain rate will always have a greater effect than the further application of one to which the structure has already responded and may even have reached equilibrium state. These apparently anomalous responses are therefore modeled quite naturally by RDT, which emphasizes the value of the theory in this context. Indeed, by modeling the eddy structure and its response to distortion, RDT has the potential to form the basis of a truly predictive model. In particular, it could handle situations where one strain rate acts to reorient the eddy structure so that it is then preferentially amplified by a second strain rate. It would seem that transport-equation models must fail for such "nonlinear" responses to a combination of strain rates. By comparison, an RDT-based scheme, provided it were applicable in the semirapid limit of most experiments and accounted for all energy-containing motions in the flow, should only break down if there were major changes in the type of eddy structure. Even this may not be a restriction, as indicated in the next section, which describes a first attempt at devising such a model.

It is apparent from the earlier consideration of the vorticity equation in the presence of mean shear that thinking in terms of the distortion of turbulent vorticity provides a better physical description of the flow structure and the manner in which it responds to straining than considering the effects on the velocity fluctuations. In a recent fascinating review, Hunt (1985a) has emphasized the value of this "vortex-dynamics" approach to complex turbulent flows and provided more examples of its application than can be mentioned here in detail. In particular, however, he shows how the changes in vorticity and velocity fluctuations may be estimated from considerations of the stretching and rotation of line elements in the flow.

The main problem of this approach is in calculating the velocity field induced by the changes in vorticity, which requires solving $\nabla \wedge \mathbf{u} = \boldsymbol{\omega}$ subject to $\nabla \cdot \mathbf{u} = 0$, a task that normally requires the solution of four partial differential equations. However, Goldstein (1978) has rediscovered a little known, less cumbersome means of solving these equations due to Serrin (1959), and Goldstein & Durbin (1980) have used this method in practice to calculate changes in turbulence of arbitrary scale in an inlet contraction. [Hunt (1985a) has produced a new geometrical derivation of this solution, while Durbin (1981) has shown that for irrotational distortions, the vorticity-distortion matrix is simply the inverse of the transformation matrix for wave numbers.]

Goldstein (1979) has also considered the turbulence generated by the interaction of random-entropy fluctuations with nonuniform mean flows, which is of importance to studies of the generation of core or combustion noise, as well as to studies of the fatigue life of turbine blades and of jet engines. In a previous volume in this series, Goldstein (1984) has considered the use of RDT calculations for determining the generation of sound by high-Reynolds-number turbulent shear flows as an alternative to the acoustic-analogy approach, which has also been discussed in detail in an earlier volume of this series by Ffowcs Williams (1977).

This requires the inclusion of compressibility effects in RDT. Hunt (1978) has previously applied the theory in compressible mean flow when considering the amplification of turbulence observed in the compression phase of an internal-combustion engine. More recently, Jayaram et al. (1985) have applied RDT concepts to modeled Reynolds-stress equations expressed in terms of Favre-averaged variables and used the resulting scheme to successfully reproduce the experimental data of Jayaram & Smits (1985) for a supersonic turbulent boundary layer experiencing the combined effects of bulk compression and streamline curvature, $\delta/R = 0.1$, which satisfied the RDT conditions. Their second flow, with $\delta/R = 0.02$ (which did not satisfy the rapid condition), could not be modeled in this way. RDT has also been used to model the evolution of Reynolds stresses in response to bulk dilatation (Dussage & Gaviglio 1981), and both shock-boundary layer (Debieve et al. 1982) and shock wave free turbulence interactions (Debieve & Lacharme 1985).

Among other recent uses of RDT, Stretch & Britter (1985a,b) and Stretch (1986) have adapted the theory to consider density fluctuations in order to investigate the influence of stable stratification on the structure of locally homogeneous turbulent shear flows. They point out that their comparison of RDT calculations with experimental data for stably stratified inhomogeneous boundary layers, and hence dense contaminant dispersion, requires some caution. However, the theoretical results do provide some valuable physical insight into some of the mechanisms that operate in such flows.

Gartshore et al. (1983) have used RDT to show how external turbulence induces irrotational fluctuations in an initially turbulence-free shear layer; these fluctuations then interact with the shear to produce rotational velocity fluctuations and mean Reynolds stresses. These RDT calculations of such "interaction-at-a-distance" effects (Gartshore & Savill 1982, Savill & Zhou 1983) were shown to be in good agreement with experimental data.

Finally, Balakumar (1983), Balakumar & Widnall (1986), and H. Atassi (unpublished results) have made use of RDT approximations to investigate the modifications to the normal velocity fluctuations that occur as a result of the distortion imposed by mounting manipulator plates [otherwise known as large-eddy breakup devices (LEBUS)] in a turbulent boundary layer (Mumford & Savill 1984, Savill 1984, Anders et al 1984, Plesniak & Nagib 1985, Guezennec & Nagib 1985). Following a similar line to that of Goldstein & Atassi (1976) for the analysis of airfoils in gusts, they find that v' is significantly reduced behind such devices; that this suppression increases with chord length, but that for tandem plates the suppression is the square of that for a single element; and that the presence of the ground plane decreases the effectiveness of the manipulator. All of these results appear to be consistent with experimental findings. At the same time, T. B. Gatski (private communication) has shown that the observed reduction and subsequent recovery of skin friction downstream of these devices may be explained, at least qualitatively, in terms of an RDT analysis applied to the momentum-transport equations.

DISTORTED-STRUCTURE MODELING

The success of RDT in modeling plane turbulent shear flows suggests that these flows are all dominated by the same, single group of energy-containing eddies. However, since the work of Grant (1958), it has become increasingly apparent (e.g. Townsend 1976, 1979, Savill 1979, 1982a,b, Mumford et al. 1980, Mumford 1982, 1983, Keffer 1982, Roshko 1982, and others) that we can identify three principal types of eddy that appear to be part of a regenerative cycle of motions. These are the following:

1. Inhomogeneous, large anisotropic swirling motions [termed "lasmos" by Savill (1979), equivalent to Grant's "jets"] that have most of their circulation in the plane of the mean shear and are assumed here to be essentially two dimensional, although their cross-stream scale may be only two times their streamwise extent. They appear to be responsible for the engulfment of external fluid during periods of active entrainment (Townsend 1970, 1979).

2. Roller eddies. Typically inclined at angles of 45° to the flow direction, these are also anisotropic and appear in different groupings and with different degrees of long-range order in each plane flow (Mumford 1982, 1983, Savill 1982b).
3. Quasi-isotropic small scales of the inertial subrange and dissipative regions that are believed to be the decay products of the larger eddies and may collectively take the form of either rods or ribbons within the flow (Kuo & Corrsin 1972).

Savill (1978, 1979, 1982a) has incorporated this "three-eddy" picture in an RDT calculation scheme based on that of Townsend (1980) to produce a "Distorted-Structure Model" (hereafter DSM); this model considers how each group of eddies responds to the mean-flow rates of strain, accounts for their finite lifetime and the nonlinear interactions that occur between them, and incorporates a mechanism for the generation of longitudinal eddies in response to certain extra strain rates. It is possible that the lasmos and roller eddies may be physically connected so that they form parts of larger, horseshoe structures, but this is not essential to the current modeling.

RDT effectively assumes that neither the lasmos nor the small scales (which together might be regarded as the "non-RDT" eddies) make significant contributions to the stress intensities. The easiest way to make an allowance for such contributions is to estimate their inherent anisotropy and then assign a percentage of the total turbulence intensity, q_s and q_l, to both classes of structure. In DSM the small scales are assumed isotropic [Townsend (1970), Elliott (1976), and Keffer et al. (1978) have made similar allowances for their isotropizing influence], while the anisotropy of the lasmos has been estimated from a experiment conducted by Keffer (1965) on a wake passing through a distorting duct, where the sense of strain was such that eventually the lasmos were essentially the only structures remaining in the flow. There are thus only three parameters that need to be specified in order to model the various undisturbed plane shear flows: α [and hence ν_T from (4)], q_s and q_l. The effective shear strain can be based on local estimates (which can be accurate provided a time representative of the average eddy lifetime is adopted for t_d) or calculated from Townsend's effective strain equation with the addition of Maxey's damping factor; in both cases the results are similar. The experimental data of Townsend (1980) and Mumford et al. (1980) suggest that small scales typically contain approximately 10% of the total turbulence energy in the plane wake flows, while the energy holding of the lasmos is around 20%. With variations of these estimates, Savill (1979) found that it was possible to reproduce all the stress-intensity ratios, $\overline{u_i u_j}/\overline{q^2}$, for a plane mixing layer, wake, and the outer region of a boundary layer.[1] This result provides support for the notion that these flows all have the same "universal" three-eddy structure, and that the differences between them can be explained in terms of differences in the relative strengths and average strain histories of these basic elements.

At any given instant a particular shear flow will contain a combination of lasmos, small scales, and a whole range of roller eddies with shapes, sizes, and inclinations dependent on their age. Some eddies will have only recently arisen within the flow and so will have encountered little strain, whereas others will be old, highly strained, and on the point of breaking up. (In DSM it is tacitly assumed that an individual eddy may sustain a maximum strain of 3α.) Ideally, perhaps one should identify several groups of structures, assign each a representative α, follow how each group responds to distortion, and then combine solutions. In practice it is possible to characterize each of the simple shear flows by a single value of α (typically 4), but it is clear that this is an average over all the individually strained eddies in the flow at any instant and represents a balance between the generation and decay of these structures. In equilibrium flows this balance is maintained at a particular level and α is constant, but when the flow is subjected to applied strains the balance is upset. Therefore, in order to apply DSM to distorted shear layers it is necessary to consider how the distortion affects the average strain history of the "RDT" rollers as well as how each of the "non-RDT" eddy types will respond to the additional strain rates (which may be characterized by parameters defined purely in terms of the imposed distortion geometry) and the continued shearing (modeled by an α_s that needs to be determined).

The small scales may simply result from the breakup of larger eddies or they may perhaps be produced continually at the edge of the rollers, where material might be torn away from the "surface" of the eddy as a result of the high local strain rates acting there (A. A. Townsend, private communication). One could imagine initial vortex sheets produced in this way being shredded into the forms seen by Kuo & Corrsin (1972). Because

[1] RDT is not applicable to the inner layer because the appreciable curvature of the mean velocity profile results in growing wave modes that are not accounted for in the analysis (see discussion by Maxey 1978). In the outer layer it is natural to identify the RDT rollers with the legs of hairpin vortices, and their tops [which in cross section appear as transverse vortices, have been termed "typical eddies" by Falco (1977), and seem to be responsible for quiescent entrainment] with the lasmos. The problem is that the scales of these eddies are far smaller than those predicted by the RDT correlations, so for the purposes of DSM it is assumed that RDT simply provides a useful representation or animation of the boundary-layer structure rather than an exact simulation. It is further assumed that anisotropy of the hairpin tops is similar to but less pronounced than the wake lasmos.

these fine-structure regions typically have dimensions 15 to 30 times the sizes of the scales they contain, some interaction with the mean flow may occur. If so, this would be accounted for automatically in the model but overpredicted because the effective shear experienced by such eddies would be considerably less than that experienced by the rollers as a result of the shorter lifetimes of the small scales. [This is the essence of J. C. R. Hunt & A. A. Townsend's recent (unpublished) attempts to extend RDT modeling to consider smaller scales by assuming an effective shear for the low-wave-number part of the energy spectrum, but reducing this by a factor $(1+\varepsilon^{2/3}/k^{1/3})$ for the high-wave-number region.] However, experiments in plane shear flows indicate that little shear stress is set up in the subrange, so this possibility is ignored in DSM. The effect of distortions that lead to the breakup (or stabilization) of larger scales or increased (reduced) rates of transfer of energy are modeled by increases (decreases) in q_s.

The response of the lasmos to distortion has to be considered simultaneously with that of the "RDT" rollers, not because they may be parts of the same structure, but because their principal planes of circulation are essentially perpendicular to one another (Townsend 1979) and so any strain that acts to amplify one will tend to suppress the other. If the additional strain tends to put energy into the roller elements and at the same time suppresses the lasmos (thus reducing entrainment), then one would expect the rollers to survive longer and experience a larger total shear, which would increase the total effective shear $\alpha+\alpha_s$. However, if these highly strained eddies are not continually replaced, $\alpha+\alpha_s$ must decrease eventually to account for the effective "destraining" of the flow structure on average.

In the opposite case, where the lasmos are amplified, the associated increase in entrainment and mixing could be expected to result in a reduction in the lifetime of the rollers, so that rapid destraining might occur ($\alpha+\alpha_s \rightarrow 0$) and the lasmos could become completely dominant, as in Keffer's (1965) flow. Such changes have to be reflected in alterations to q_s and q_l.

A further consideration is that if, as a result of a distortion, the turbulence energy increases, then nonlinear interactions will become more significant. Comparison of DSM calculations with experiments suggests that if the turbulence intensity is less than 10%, turbulence/turbulence interactions can still be ignored and the enhanced energy cascade from large to small scales can be modeled by increasing the energy-transfer coefficient ν_T by a factor equal to the rise in $\overline{q^2}$. It is clear that in general this "constant" must be a function of $\overline{q^2}$ as well as α.

In higher-intensity flows it is no longer possible to disregard the greatly increased frequency of turbulence/turbulence interactions, and it is then necessary to account explicitly for the extra energy exchanges between the large eddies themselves. The curved mixing-layer experiment of Castro & Bradshaw (1976) suggests that these interactions act mainly to redistribute energy among the normal-stress components, with the result that these tend to become "locked-on" to variations in $\overline{q^2}$. Some progress may therefore be made by including an empirical redistribution scheme triggered when q exceeds 10% [see Tan-atichat et al. (1980) for an alternative discussion].

It is important to realize that neither DSM nor RDT as used by Townsend (1980) can be used as a forward-marching scheme. Instead, for each station through a distortion the calculations start from the virtual origin at isotropy, apply the appropriate shear, and then step through the appropriate strain history in order to reach the appropriate stage. The reasons for this are the finite lifetime of the eddies and the fact that one is always having to consider an averaged description of the flow at each instant. These are fundamental points. At a first station near the start of the distortion, one effectively takes a "snapshot" of the flow structure and then represents it by some average eddy, but the more highly strained eddies that are present at this stage will not reach a second station further on, whereas new, relatively unstrained structures that replace them were not present at the first station. Thus, in order to move directly from station to station it would be necessary to devise some way of "forgetting" part of the spectrum function calculated at the first station and at the same time introduce a correction for the newly entrained material. It is far easier to consider the flow at the two different positions as resulting from the evolution of two different sets of eddies with different strain histories since their assumed isotropic starting conditions. Thus, the model parameters must be set specifically for each distortion step, with continuity provided by their variation with applied strain.

An outcome of this is that if an impulsive distortion or successive distortions are applied, the calculations correctly mimic the flow structure's memory of the earlier strain history. If the distortion persists, memory of the initial conditions will eventually be lost, and it will be possible for a new "moving" equilibrium to be set up in which the eddies continually experience the same average distortion strain history. Alternatively, if the distortion ends, the recovery is automatically initiated as the memory of the extra strain rates dies away, and the dominating influence of the mean shear is reestablished. Full recovery can be attained if the intensities of the lasmos and small scales simultaneously return to their original levels. One would expect this process to be completed in the time it takes for the whole of the flow structure present at the end of distortion to be replaced by new structures that have experienced only shear, and a simple estimate of this "turnover time" suggests a recovery length in boundary layers of order 40δ, close to the value found in many cases. Furthermore, the delayed recovery that occurs following distortions that lead to the formation of turbulent longitudinal eddies can be explained if it is assumed these very stable structures are formed from highly strained roller eddies. By persisting into the recovery region these structures then retain some memory of the distortion within the flow, and complete recovery is delayed until they decay.

Regarding the distortion imposed on boundary layers by drag-reducing outer-layer manipulator devices, it has been observed (Mumford & Savill 1984) that one effect of these is to introduce vortices into the flow of opposite sign to the mean-flow vorticity. At the same time the imposed momentum deficit reduces the mean shear near the wall. There is thus an effective deshearing of the flow for which the model predicts an increase in the average inclination of the eddy structure to the stream direction and a reduction in $\overline{uv}/q^2$ (the farther the structures move away from the 45° maximum straining angle, the less efficient they are at extracting energy from the mean flow), both of which have been observed.

Savill (1982a) has successfully applied DSM to several curved flows, including recovery, that satisfy the relaxed-RDT conditions, $u_{\mathrm{rms}}/U_0 < 1$, $t_d/t_L \leq 1$, for which the model should be valid. (Figure 1 represents a comparison of DSM and other model predictions for the principal shear-stress ratio in Castro & Bradshaw's experiment.) In each case the curvature was approximated using the simplest sequence of simple distortion matrices close to the true geometry: subsequent shear (parameter α_s), constant-vorticity rotation (parameter ϕ), and (where necessary) secondary shear; here ϕ is the flow rotation angle θ, and $\alpha_s = -2\theta + Q\theta$, where Q is the experimental ratio of shear to rotation, which sets the initial value of α_s. The results are perhaps better than one might have expected for test cases in which even these less stringent conditions were barely satisfied. This probably reflects the weakness of choosing u_{rms} and L as the appropriate velocity and length scales. In practice, eddies seem to have longer lifetimes and to be more resilient to distortion than these scales suggest. If more appropriate estimates of their strength (based on their total turbulence intensity) and dimensions (taken from pattern matching and flow visualization) are used, the conditions are less restrictive and thus more completely satisfied in most experimental flows.

An important feature of DSM is that the inclusion of all the necessary parameters, and to some extent their magnitudes, may be justified on physical grounds. Admittedly there is still an initial learning process involved as the model is applied to different shear flows subjected to new distortions, since it is based on assumptions that only provide a guide to the levels required of the necessary parameters, and their values need to be refined slightly before data are exactly reproduced. However, a predictive potential is apparent even from the curved-flow results. A. M. Savill (unpublished results) has subsequently applied the model to a wider range of test cases and found that although the parameters α, q_s, and q_l have to be specified separately for each of the various shear flows, their subsequent variation for a given distortion is largely independent of flow type.

Of course, DSM is not in itself a complete model because it is founded on the linearized RDT equations, and so the reference turbulence intensity is artificial. Comparison with experimental results must therefore be made through stress-intensity ratios, normally restricted to the maximum shear-stress positions in free-shear flows or to an average over the outer region of a wall layer [although, in principle, values at any position in the flow could be evaluated, and A. M. Savill (unpublished results) has shown that the stress-intensity ratios along the centerline of a distorted wake may be evaluated by assuming that the eddies on either side of this region make equal and, in the case of $\overline{uv}$, opposite contributions to the stresses that are set up there and then combining independent DSM calculations for the two sets of structures]. The model is thus of the greatest practical value

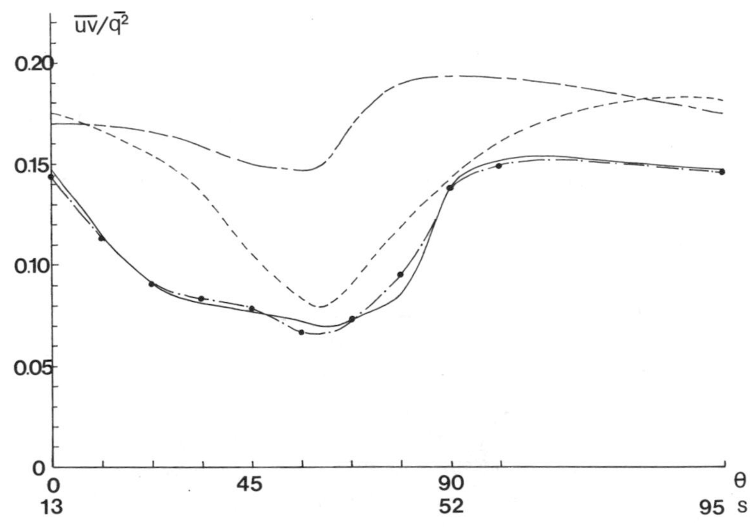

Figure 1 Comparison of model performance for the curved mixing layer of Castro & Bradshaw (1976). ——: curve through experimental data; — — —: k-ε; - - - - - -: RST (Gibson & Rodi 1981); —●—: DSM.

in situations where layer-averaged or representative structure parameter predictions can usefully be allied to estimates of the true turbulence intensity.

In this context DSM might be regarded as midway between an algebraic stress-transport model and a full Reynolds-stress model in the hierarchy of closure schemes. However, it has close connections with some of the more advanced approaches, notably the hairpin modeling being developed by Perry & Chong (1982), Perry et al. (1985), and H. McDonald and colleagues (Liu et al. 1985). Both models can benefit from the computational "data" being produced by Large-Eddy Simulations (LES) and Full Navier-Stokes Simulations (FNS), especially from the work of Moin & Kim (1985) and Kim (1985). Indeed, V. W. Sumner (unpublished results) has suggested that RDT could be considered as a type of large, energy-containing eddy simulation with an abrupt cutoff, in which case one might regard DSM as adding in a subgrid model. RDT has also been used to calibrate two-point closure spectral models by Cambon et al. (1981) and Bertoglio (1981).

RDT AS AN AID TO TRANSPORT-EQUATION MODELING

Comparison of RDT and RST Modeling

The differences between applying RDT and RST models to calculate the development of turbulent flows may be seen by considering the associated approximations to the full Reynolds-stress equations:

$$\frac{D\overline{u_i u_j}}{Dt} = -\underbrace{\left(\overline{u_i u_k}\frac{\partial U_j}{\partial x_k} + \overline{u_j u_k}\frac{\partial U_i}{\partial x_k}\right)}_{\text{Production}}$$

$$+\underbrace{\frac{\partial}{\partial x_k}\left[\underbrace{(\overline{u_i u_j u_k})}_{\text{Velocity}} + \underbrace{\overline{p(\delta_{jk}u_i + \delta_{ik}u_j)}}_{\text{Pressure}} - \underbrace{\nu\frac{\partial \overline{u_i u_j}}{\partial x_k}}_{\text{Viscous}}\right]}_{\text{Diffusion}}$$

$$\underbrace{-\overline{p\left(\frac{\partial u_i}{\partial x_j} + \frac{\partial u_j}{\partial x_i}\right)}}_{\text{Pressure–Strain}} - \underbrace{2\nu\overline{\frac{\partial u_i}{\partial x_k}\frac{\partial u_j}{\partial x_k}}}_{\text{Dissipation}}. \tag{6}$$

In Reynolds-stress models of the type described by Launder et al. (1975), four principal assumptions are made:

(*a*) The unknown third-order correlations representing velocity diffusion have to be approximated in terms of calculable second-order correlations. (*b*) The dissipation is assumed to be isotropic and is either modeled as $\overline{u_i u_i}^{3/2}$ divided by a prescribed length scale l_ε or derived from another, empirical transport equation. (*c*) The pressure strain is normally divided into two parts, ϕ_{ij1} and ϕ_{ij2}, although this separation is not unique (Lumley 1975). The first, representing the nonlinear, third-order influence of turbulence/turbulence interactions that promote an exchange of energy among stress components, is usually referred to as the "return-to-isotropy" part and modeled in terms of second-order correlations related to the turbulence anisotropy. The second, "rapid" part (so-called because during rapid distortion of, for example, isotropic turbulence, this is the only part acting, but equally appropriate because these terms are then instantly anisotropic, whereas all others only gradually develop anisotropy) represents the linear influence of the mean strain on the turbulence and is approximated by a product of the local strain rates and Reynolds stresses. (*d*) The pressure diffusion is normally assumed to be zero, although this is known to be incorrect (and to adversely affect model performance) in shock–boundary-layer interactions (Vandromme & Ha Minh 1985) and adverse pressure gradients (Orlandi 1981a).

By comparison, in RDT the nonlinear third-order correlation terms $\overline{u_i u_j u_k}$ and ϕ_{ij1}, and often the dissipation, are set equal to zero rather than approximated. However, the rapid pressure strain and pressure diffusion are calculated directly. It is found that the RDT result for ϕ_{ij2} is only identical to the Launder et al. (1975) prescription in the limit of weak, homogeneous distortion of initially isotropic turbulence. In general, as Hunt (1978) has pointed out, the model constants must be a function of strain history. Maxey (1982) notes that the use of RDT to calculate stress-intensity ratios is consistent with including a Rayleigh damping term in Equation (6) to replace ϕ_{ij1} and ε, and that this is equivalent to using the normal Rotta (1951) model for ϕ_{ij1} employed by Launder et al. (1975) with $C_1 = 2.0$. Since the recommended value is about 3.0, this implies that the influence of this term will be underestimated using RDT.

Hunt (1978) also argues that because RDT is based on rotational deductions from and approximations to the equations of motion, then under conditions for which the approximations hold it should be preferable and more physically revealing than methods based on the RST equations. However, both models have their limitations.

RST models should be valid in the limit $t_d \gg t_L$ when local equilibrium holds, but precisely because the values of the model constants are based on experimental data for simple equilibrium flows, when they are used to predict more complex strain fields they may respond too slowly (e.g. Brown et al. 1985) and may tend to recover too rapidly (e.g. Rodi & Scheuerer 1981, U. R. Müller, unpublished results).[2]

RDT, on the other hand, provides a model that is not restricted by local approximations and describes the continuing influence of strain history, but that is only valid in the limit $t_d \ll t_L$, which is not satisfied in most turbulent flows of interest. This is precisely why DSM was developed, but even if this were a complete model, it would not be obviously superior to an RST model. Instead, these two different approaches to the closure problem are essentially complementary and should be used in combination. The following sections discuss how RDT-based calculations may be used, either directly or through the mathematical and physical insight they provide, to suggest improvements to transport-model approximations at the Reynolds-stress and lower levels of closure.

Eddy-Viscosity Models

PROVISION OF SHEAR-STRESS TRANSPORT Various modifications have been suggested to the basic eddy-viscosity approach of Boussinesq (1877),

$$-\overline{u_i u_j} = \nu_e\left(\frac{\partial U_i}{\partial x_j} + \frac{\partial U_j}{\partial x_i}\right) - \frac{2}{3}\delta_{ij}k, \tag{7}$$

and of Kolmogorov (1942) and Prandtl (1945),

$$\nu_e = C_\mu k^{1/2} L. \tag{8}$$

In particular, Pope (1975) has tried to overcome some of the limitations that this approach imposes (isotropic eddy viscosity, alignment of Reynolds stresses with their generating rates of strain, no dependence on the rotational invariant, similar scaling of L and l_ε, constancy of C_μ) in his effective eddy-viscosity hypothesis:

$$-\overline{u_i u_j} = \underbrace{C_\mu k^2/\varepsilon}_{\nu_e}\left(\frac{\partial U_i}{\partial x_j} + \frac{\partial U_j}{\partial x_i}\right) + k/\varepsilon \nu_e (C_1 + P/\varepsilon - 1)^{-1}\left(\frac{7C_2+1}{11}\right)\left(\frac{\partial U_l}{\partial x_i}\frac{\partial U_l}{\partial x_j} - \frac{\partial U_i}{\partial x_l}\frac{\partial U_j}{\partial x_l}\right), \tag{9}$$

[2] B. E. Launder (private communication) has pointed out that the question of whether model calculations predate and so predict, or postdate and hence "postdict," experimental results is of less importance than whether the model used empirical input from the same type of flow as it is being used to calculate. Since the second-order closures employed by his group at the University of Manchester and by others elsewhere are based on input solely from simple flows, the responses that these suggest occur in more complex flows can reasonably be called predictions. The author agrees with this sentiment and for reasons of clarity has used the terms "postdicted" or "reproduced" to indicate when models have made use of flow-specific modifications.

where

$$C_\mu = \frac{4}{15}(C_1 + P/\varepsilon - 1)\left(1 - 2\mathbf{r}^2\left(\frac{7C_2+1}{11}\right)^2 - \frac{2}{3}\left(\frac{5-9C_2}{11}\right)^2 \mathbf{s}^2\right)^{-1}, \tag{10}$$

with $\mathbf{r}$ and $\mathbf{s}$ the rotational and shear invariants, respectively. Here the coefficients are determined from the Launder et al. (1975) modeled-RST equations with standard constants C_1 and C_2. [The predictions are equivalent to those obtained using the Algebraic Stress Model (ASM) proposed by Rodi (1976).]

However, even with the inclusions of such modifications, an eddy-viscosity formulation implies that the stresses are determined locally, although transport of both k and l_ε may be accounted for, whereas the exact RST equations show that these may be convected by both the mean and fluctuating velocity fields. For any simple ν_e assumption to be valid, these terms must be negligible. In practice, as we have seen, this is not so, and the stresses depend on the strain history of the flow. Another particularly clear example of stresses lagging behind local strain influence is provided by the curved-wake experiments of Nakayama (unpublished results).

One way of dealing with such memory "effects" is to follow the line of reasoning of Builtjes (1977), who developed a more general expression by considering turbulence as a non-Newtonian fluid:

$$-\overline{u_i u_j} = \nu_{e_0}\delta_{ij}\overline{q^2} + \nu_e\left(\frac{\partial U_i}{\partial x_j} + \frac{\partial U_j}{\partial x_i}\right) + \nu_{e_m} L_m\, \partial/\partial x_l \left(\frac{\partial U_i}{\partial x_j} + \frac{\partial U_j}{\partial x_i}\right), \tag{11}$$

where $\nu_{e_0} = -1/3$, L_m is a memory length, and ν_{e_m} is chosen so that $\int_0^t \overline{u_i u_j u_k}$ is modeled by $\nu_{e_m} L_m$.

An alternative is to drop the eddy-viscosity concept and work instead in terms of structure parameters a_{ij}, equivalent to the stress-intensity ratios defined in the section on distorted structure modeling. This was effectively the approach adopted by Bradshaw et al. (1967), who assumed constancy of a_{12} in developing their shear-stress transport model for boundary layers from the transport equation for turbulence energy. In an extension to this, Harsha (1974) put a_{12} equal to $|\overline{uv}|/k$ to allow for the change of sign across ducts, jets, and wakes. A similar line of approach has recently been taken by Johnson & King (1984, 1985), resulting in a "hybrid eddy-viscosity/Reynolds-shear-stress" formulation that includes a modeled ordinary differential equation for $\overline{uv}_{\max}$, again with a_{12} constant. To account for the influence of strain history on $\overline{uv}$ directly, Mathieu (1971) proposed that

$$\frac{\overline{uv}}{k} = f\left(\int_0^x \frac{\partial U}{\partial y}\frac{dx}{U}\right), \tag{12}$$

where f is a function of the accumulated strain along a streamline. Jeandel et al. (1978) constructed an appropriate transport equation incorporating an empirical relation between a_{12} and the distortion ratio determined from the data of Champagne et al. (1970). In an extension to this work, Maxey (1978, 1982) and Townsend (1970, 1980) have calculated the structure parameters a_{12}, a_{11}, a_{22}, a_{33} using linear RDT, with values of the effective shear strain estimated either by fitting to experimental data or from a model equation describing the transport of this quantity; they combined these parameters with an equation for turbulence intensity in order to calculate Reynolds-stress distributions in oscillating pipe flow and in flow over water waves.

More recently, A. M. Savill (unpublished work) has taken this approach a stage further by using DSM predictively to estimate all the stress-intensity ratios around both sides of a jet-engine rotor blade, with model coefficients determined from mean-flow strain-rate data provided by an inviscid calculation and the previous successful applications of the model to curved, accelerating, and distorting-duct flows. An allowance for the influence of free-stream turbulence on the eddy structure was made on the basis of flow-visualization studies, with the simplifying assumption that both boundary layers were essentially fully developed. The predicted a_{12}, a_{11}, and a_{22} were then read into a heat-transfer program incorporating a McDonald & Fish (1973) boundary-layer-averaged k-l scheme that normally assumes constant values of 0.15, 0.5, and 0.2, respectively, for these quantities but also includes scaling for transition and wall cooling. The usual corrections for curvature and acceleration were switched off.

Unfortunately, although the inclusion of the DSM parameters improved the pressure-surface predictions, a compatibility problem arose because both boundary layers were in reality transitional and the McDonald & Fish scheme ascribes all of the associated structure-parameter scaling to low-Reynolds-number effects, whereas it seems likely that some of the scaling will be connected with the development of the flow structure that (like the response to subsequent straining) is implicit in the DSM calculations. In an attempt to avoid such "double accounting," the predictions were repeated, with the transition process simulated by allowing the model coefficients describing the flow structure to develop from an initially isotropic starting condition at the leading edge [Hart (1985) has now shown that this is a poor approximation] toward their fully developed values at the end of transition; this position is indicated by the original heat-transfer program output. Using the resulting "developing" structure parameters without further scaling produced additional improvements in the heat-transfer predictions over both the pressure and suction surfaces of the blades, as indicated by Figure 2 (A. E. Forest, unpublished results).

ZONAL MODELING In discussing the concepts of zonal and universal modeling, Savill (1981a) has suggested that current models for complex, three-dimensional flows such as blade cooling, mixers, and combustors might be improved, in a way similar to the two-dimensional boundary-layer heat-transfer calculations. Such improvements could be made by using the mean-velocity and shear-stress output of the models to divide the flow up into regions dominated by different types of distorted shear layers, then applying DSM to calculate representative structure parameters for each of these regions, and finally using these in place of the stress ratios determined from the eddy-viscosity model. This process of "region identification" would result in a simple form of a "universal" zonal model, one in which the same turbulence model is used in each flow zone.

Such a technique has subsequently been applied to the case of a single cooling jet injected normally into a turbine-blade boundary layer (A. M. Savill & A. J. White, unpublished results), where the flow naturally divides up into a *boundary layer* approaching and interacting with a *curved jet* forming a *wake* dominated by longitudinal vortices. Identification of individual shear layers, their centerlines and sense of curvature, etc., was successfully achieved on the basis of mean-strain and stress-intensity output from a three-dimensional Navier-Stokes computation incorporating a k-ε model (White 1980), supported by the developing shear-flow data of Elliott (1976) and Gibson & Younis (1983) and by the flow visualization conducted by Foss (1980). The DSM stress-intensity ratio predictions for several xy- and xz-sections through the flow are compared with the more

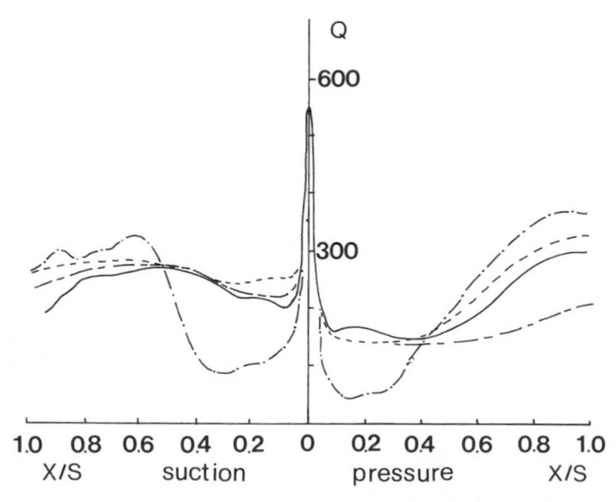

Figure 2 Heat-transfer predictions around the pressure and suction surfaces of a turbine blade. ——: experimental data; – —— –: McDonald & Fish (1973) k-l model; · —— ·: fully developed DSM calculations with low-Re scaling; ------: DSM calculations for developing flow.

recent experimental data of Andreopoulos & Rodi (1984) in Figure 3. The $\overline{uv}/\overline{q^2}$ data indicate that the original PACE code predicted the jet region too high above the wall. Since the DSM calculations were based on these predictions, they too are geometrically incorrect, but the magnitude and shape of the profiles are close to experimental values. One might expect the secondary shear-stress ratios to be a more sensitive guide to performance, and, after vertical scaling on the basis of the $\overline{uv}$ comparison, the DSM predictions closely follow the experimental results, whereas the PACE results are clearly in error. In other regions the unmodified predictions were not believable ($a_{ii} > 0.7$, $a_{ij} > 0.3$) and locally even unphysical (a_{ii} negative). However, it is interesting to note that the DSM calculations indicate that the presence of several strain rates acting simultaneously tends to prevent the formation of extreme stress levels. In such situations it seems unlikely that a large fraction of the total turbulence intensity will be directed into any one normal-stress component, and equally opposing strains limit the levels attained by the shear stresses. This effective damping of anisotropy may explain why k-ε models have often been found to perform surprisingly well in such complex flows (a point made by W. Rodi, B. E. Launder, and others at the 1985 Refined Turbulence Modeling Meeting at Hydraulics Research Ltd.).

The advantages of such "one-iteration" PACE/DSM or McDonald & Fish/DSM schemes are of course rather limited, and because it is unlikely DSM could be directly incorporated as a subroutine in a truly iterative scheme, it may be necessary instead to follow the formulation of Mathieu and devise functional fits to the various strain/shear-layer combinations that could be used in this way. An alternative approach would be to use the RDT-founded calculations to make appropriate modifications to the basic eddy-viscosity approximation.

Generalizing (7) to allow for anisotropy and nonalignment of stress and strain, we have

$$-\overline{u_i u_j} = C_{ij}\nu_e\left(\frac{\partial U_i}{\partial x_j} + B_{lm}\frac{\partial U_l}{\partial x_m}\right). \tag{13}$$

The DSM predictions allow one to identify the dominant strain rates for each Reynolds-stress component. Thus for the cooling jet flow, one finds $\overline{uw}$ depends on $\partial U/\partial z$ as expected, but also on $\partial W/\partial y$, so that

$$\overline{uw} = -C_{13}\nu_e(\partial U/\partial z + B_{13}\partial W/\partial y). \tag{14}$$

[Andreopoulos & Rodi (1984) arrived at a similar conclusion from a study of the various terms contributing to the production of $\overline{uw}$ in the transport equation for this shear stress.] Furthermore, A. J. White (private com-

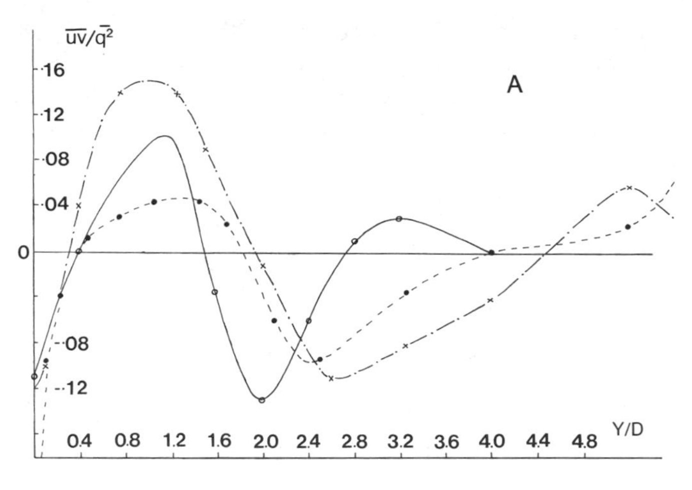
uv/q²
·16
·12
·08
·04
0
-·08
-·12
A
0.4 0.8 1.2 1.6 2.0 2.4 2.8 3.2 3.6 4.0 4.4 4.8
Y/D

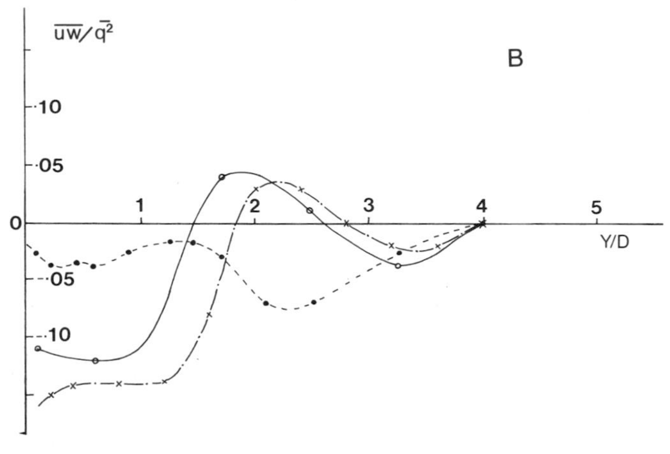
uw/q²
·10
·05
0
-·05
-·10
B
1 2 3 4 5
Y/D

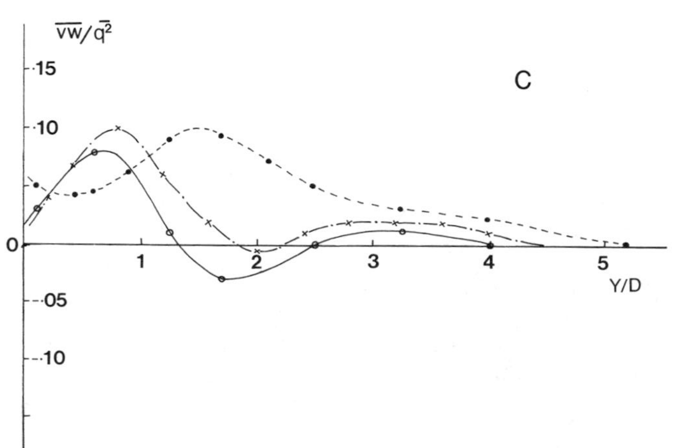
vw/q²
·15
·10
0
-·05
-·10
C
1 2 3 4 5
Y/D

munication) has suggested that B_{ij} could be obtained from Pope's effective eddy-viscosity relationship and then the C_{ij} coefficients determined from a fit to the DSM output.

COUNTERGRADIENT TRANSPORT The type of modification expressed by (13) is necessarily a compromise because this equation remains a local approximation even though the coefficients are determined by the strain history. This limitation is evident in a two-dimensional flow (zone) containing asymmetry, such as a wall jet, where the shear stress $\overline{uv}$ does not fall to zero at the associated velocity extrema, resulting in a negative generation term in the turbulence energy equation [although for the reasons outlined by Hinze (1970), this need not imply a transfer of energy back to the mean flow]. Such "negative production" regions may be successfully modeled in a simplistic manner using DSM in the same way as for the wake centerline discussed earlier. Following the suggestion of Bradshaw (1975), we may consider a wall jet as a jet plus boundary layer and assume that the stress intensities in the negative-production region are due to contributions, proportional to turbulence intensity, from the jet on one side and an inner boundary layer on the other. The same procedure was adopted to determine the a_{ij} values at the boundary between distorted shear layers in the more complex three-dimensional flows. Similarly, Savill & Zhou (1983) have suggested that the merging region of a wake/boundary-layer interaction may be modeled by adding Reynolds-stress contributions from the wake eddies and outer boundary-layer structures. By modeling the strain history of two different sets of eddies and then effectively mixing them together in this way, it is possible to account at least partially for both local, gradient diffusion by the smaller scales and nonlocal, bulk convection or countergradient transport by the larger scales, despite the fact that DSM is based on linearized RDT equations, where diffusion by the velocity field is ignored.

The wake/boundary-layer situation may be further complicated by the movement of large-scale wake vortices, carrying shear stress, across the maximum-velocity line into the boundary-layer region, where they may become more ordered; this ordering results in a genuine reverse transfer of energy to more coherent motion (Zhou & Squire 1985) as well as a particularly large negative-production region (Bario et al. 1982). This situation might perhaps be modeled by allowing wake lasmos to partially cancel the stress contribution from the boundary-layer eddies. However, again the value of RDT appears limited, and the alternative option of a

---

Figure 3 Structure-parameter modeling of a normal jet in a cross-flow with velocity ratio 2. —○—: experimental data of Andreopoulos & Rodi (1984); --●--: PACE k-ε model; ·— × —·: DSM. (*A*) $\overline{uv}/\overline{q^2}$; (*B*) $\overline{uw}/\overline{q^2}$; (*C*) $\overline{vw}/\overline{q^2}$.

further modification to the eddy-viscosity hypothesis—the addition of an extra term to account explicitly for the nonlocal influence of large scales—seems more practical.

Following Hinze (1970) and others, Zhou & Squire (1982, 1985) and Savill & Zhou (1983) have shown that adding a term of the form

$$\frac{1}{\sqrt{2\pi}} \frac{\partial(\overline{v^2})^{1/2}}{\partial y} [(\overline{v^2})^{1/2}]^3 t^3 \frac{\partial^2 U}{\partial y^2}, \tag{15}$$

[where $(\overline{v^2})^{1/2}$ and t are the velocity and time scales representative of the large eddies] to (7) provides a close fit of calculations to experimental data when the main part of the eddy-viscosity expression is derived assuming local equilibrium. Such an additional term can be justified physically to represent the contribution of large-scale motions and is similar in form to the memory term in Builtjes' lag equation.

The Dissipation Transport Equation

This equation may be written as

$$\underset{\text{Advection}}{\frac{D\varepsilon}{Dt}} = \underset{\text{Production}}{-C_{\varepsilon_1} \overline{u_i u_j} \frac{\partial U_i}{\partial x_j}} - \underset{\text{Dissipation}}{C_{\varepsilon_2} \frac{\varepsilon^2}{k}} + \underset{\text{Diffusion}}{D\varepsilon} . \tag{16}$$

The 1980–81 Stanford Conference on Complex Turbulent Flows showed that the ε-equation is the weakest link in most of the k-ε and algebraic or Reynolds-stress models that make use of it, no matter what level of closure is used. In particular, it fails to reproduce the universal length-scale distribution observed in nonequilibrium flows, instead predicting too large a length scale near walls, notably under adverse pressure-gradient conditions. [The pseudo-vorticity equation of Wilcox & Rubesin (1980) suffers from a similar limitation.] Despite this deficiency there appears to be considerable reticence among computors to develop an alternative length-scale-producing equation, largely as a result of the considerable effort that would have to put into verification against experimental results and comparison with existing codes for a wide range of test cases, but also simply because the ε limitations are well known and so may be countered by appropriately "tweaking constants." (The results of the 1985 EUROMECH 202 Colloquium on Measurement Techniques in Low-Speed Flows indicated that modelers are perfectly entitled to vary their constants, particularly in approximating third-order correlations, by up to 15%.) Such a procedure can certainly be justified if the necessary adjustments to model constants can be made to occur automatically in response to the changes in flow conditions that require them. Launder et al. (1981) appear to have made considerable progress toward this end. While following the very promising alternative approach of considering different parts of the energy spectrum separately and developing a transport equation for the rate of energy transfer out of the large, energy-containing eddies [strictly the quantity one needs to model, and not equal to dissipation except under local equilibrium conditions (Launder & Schiestel 1978)], Hanjalic & Launder (1978, 1980) found that this transfer seems to be promoted by irrotational straining. Arguing that the transferred energy must eventually be dissipated, they suggested the addition of an appropriate term to the production of dissipation. Their extended ASM scheme was successfully applied to a range of adverse pressure-gradient boundary layers (see also Rodi & Scheuerer 1986) and to plane and round jets. Their later use of a tensor-invariant curvature transformation leads to a similar increase in dissipation when streamlines start to curve. However, the work of Newley (1985) indicates that there may be other difficulties with models that are not based on the Launder et al. (1975) approach, and he suggests that a revision of the closure may be required to model the quantitatively different effects of concave and convex curvature. The implication is that this should be tied more closely to the strain history, as indicated by the success of RDT-based calculations applied to these types of distortion.

Pope (1978) independently introduced a dissipation term, proportional to the stretching of mean vorticity, to explain the difference in spreading rate between plane and round jets on the grounds that stretching of the large eddies increases their rate of energy transfer to smaller scales, and hence the dissipation rate, and that these large eddies tend to be aligned with the mean vorticity. This interpretation is supported by the large-scale simulations of Kerr (1983), and also by the computations of Kida & Hunt (1984), who have attempted to investigate the energy transfer between large and small scales using RDT. [Their work is complemented to an extent by the spectral modeling of Orlandi & Crocco (1985) and the experiments of Veeravalli & Warhaft (1985).] Kida & Hunt considered the large scales to act as a random straining on the weaker small scales, and in the limit of short distortion times they found that irrotational straining does indeed enhance the transfer rate, but that rotational straining only helps to reduce the anisotropy. Lumley (1978) has pointed out that Pope's term is incorrectly independent of turbulence anisotropy, and with his colleagues (Shih et al. 1985a,b) has made a simple correction for this. They also allow C_{ε_2} to be a function of anisotropy (Lumley & Newman 1977).

The experimental investigation of a wake passing through a constant-area distorting duct performed by Elliott (1976), together with the further analysis and application of linearized RDT by Townsend (1980), Elliott & Townsend (1981), and the subsequent improved DSM calculations of

Savill (1979), provided further evidence for such an increase in dissipation due to irrotational straining. However, in this case it appears that the dissipation occurring at small scales was enhanced more than the energy transfer out of large scales, with the result that the inertial subrange disappeared toward the end of the distortion, an observation that was allowed for in the DSM postdictions. This suggests that in some instances the large eddies may lose energy directly to viscosity rather than through intermediate smaller scales via inertial forces. M. M. Gibson (private communication) has suggested that this effect of sustained straining might explain why Reynolds-stress models sometimes predict $\overline{uv}$ too large when they produce the correct levels for normal stresses. The implication is that a strain-dependent dissipation term should be added to the shear-stress transport equation, which would take the modeled equations outside the normally assumed framework of isotropic dissipation.

In contrast, the influence of uniform rotation appears to be a reduction in the rate of dissipation. The LES and FNS computations conducted by Bardina et al. (1985) suggest that such rotation tends to align vortex tubes with the axis of rotation and so prevent the tangling and mutual interactions that lead to an energy cascade. As a result they have added an additional damping term to the dissipation equation that is a function of the rotational tensor. Audpoix et al. (1983) have also studied uniform rotation with LES and agree that such a term is needed, but they suggest that the production of dissipation must be a function of rotation as well. Since simple shear can be decomposed into plane strain plus rotation, it may be that such modifications will help predict the differences between these and strained flows. This was a notable deficiency of all the models tested at Stanford in 1981, with the exception of RDT, which was also the only model that could account for the influence of rotation (Savill 1981b).

For several reasons then, it would seem that RDT-based calculations may have a role to play in qualifying and quantifying additional strain- and anisotropy-related terms to extend the range of applicability of current ε closure models and to avoid the complexity of multiscale schemes.

A simpler solution is suggested by the work of Hunt et al. (1986) and Newley (1985). Building on Hunt's (1984a) RDT analysis of the influence of strain rates on line elements (in this case, in the flow over two-dimensional hills), they have replaced the dissipation equation in the Launder et al. (1975) scheme by an algebraic expression for the dissipation length scale:

$$1/l_\varepsilon = (1/0.09\delta + 1/\kappa y) + \beta\left(\frac{\partial U/\partial z}{k^{1/2}} - \left\langle\frac{\partial U/\partial z}{k^{1/2}}\right\rangle\right) + 1/L_0, \tag{17}$$

where ⟨⟩ denotes a streamwise average. The first part of this equation relates l_ε to the mixing length via a smoothed version of the normal ramp function, whereas the second part represents the influence of the mean-strain-rate development on l_ε, and the third accounts for any imposed length scale. Newley has found that this simple model captures the essential features of the experimentally observed length-scale variation and provides accurate predictions when implemented in the Launder et al. scheme. The implication is that the use of such a shear-dependent length scale, proportional to both the distance above the surface and also the magnitude and history of the mean strain rate, might in some cases obviate the need for a transport-equation prescription.

Pressure-Strain Closure

A limitation of current ASM and RST models is that they use local approximations to the pressure-strain terms, whereas in principle a global prescription is required. Nevertheless, these models have produced acceptable results for many flows, which suggests that the approximations are worth persevering with, and in recent years considerable effort has been expended to ensure that they, and other modeled terms, are at least consistent with certain basic principles. In particular, they should satisfy realizability (Schumann 1977, Lumley 1978), which for the Reynolds stresses implies $\langle u_i u_j \rangle > 0$, and they should be consistent with RDT for high strain rates (Reynolds 1982). Material indifference (Speziale 1979, 1980) does not appear to be necessary (Gence & Mathieu 1980, Lumley 1983), but a further requirement that the approximate terms in the modeled equations be of the same form as those that they replace in the exact equations, resulting in what Donaldson (1968) termed "invariant modeling," is often imposed. However, several computors have successfully used noninvariant approximations to improve predictions for specific strain rates (the various additions to the ε-equation discussed previously are examples), and Bradshaw (1981) has concluded that such modifications should be regarded as acceptable. This author's view is that although tensor invariance would necessarily be required of a strictly universal model, this need not be the case for a model in which strain-dependent parameters are automatically used only when the flow enters a region where such strain rates are acting. Also, Shih & Lumley (1985) have pointed out that coefficients determined by the imposition of realizability may not be suitable for general turbulence. Instead, comparison with experiment is needed to scale the coefficients in such a way that they approach these values when the simulated turbulence approaches a realizability condition.

Most second-order models, including that of Launder et al. (1975), make use of a version of the linear approximation to ϕ_{ij1} due to Rotta (1951),

$$\phi_{ij1} = C_1 \frac{\varepsilon}{\overline{q^2}} \left(\overline{u_i u_j} - \frac{1}{3} \delta_{ij} \overline{q^2} \right) = C_1 \varepsilon b_{ij}, \quad \begin{cases} C_1 \sim 3.0 (2 \leq C_1 \leq 5), \\ b_{ij}: \text{anisotropy factor}, \end{cases} \tag{18}$$

(note that some researchers use k in place of q^2, in which case C_1 is halved), and either of the two Launder et al. (1975) (LRR) models for ϕ_{ij2}:

$$\begin{aligned} \text{LRR1}: \phi_{ij2} = & -\left(\frac{C_2+8}{11}\right)\left(P_{ij} - \frac{2}{3}\delta_{ij}P\right) \\ & -\left(\frac{30C_2-2}{55}\right)\frac{\overline{q^2}}{2}\left(\frac{\partial U_i}{\partial x_j} + \frac{\partial U_j}{\partial x_i}\right) \\ & -\left(\frac{8C_2-2}{11}\right)\left(D_{ij} - \frac{2}{3}\delta_{ij}P\right), \end{aligned} \tag{19}$$

where P is the production of turbulence energy, $C_2 = 0.4$ (if $C_1 = 3.0$) in order to fit homogeneous-flow data, and (19) reduces exactly to the RST expression for weak homogeneous distortion of initially isotropic turbulence,

$$\phi_{ij2} = 0.2\overline{q^2}\left(\frac{\partial U_i}{\partial x_j} + \frac{\partial U_j}{\partial x_i}\right); \tag{20}$$

or to a reduced form, because the first term dominates (19):

$$\text{LRR2}: \phi_{ij2} = -C_2'\left(P_{ij} - \frac{2}{3}\delta_{ij}P\right). \tag{21}$$

Here $C_2' = 0.6$ in order to satisfy (20), as used by Gibson & Launder (1978), Gibson & Rodi (1981), and Gibson & Younis (1981), who noted that C_1 must then be 3.6 to fit plane homogeneous shear flows. Chung & Adrian (1979) also suggested that $C_2' = 0.6$, but that C_1 is related to the anisotropy ratios $\overline{u^2}/\overline{v^2}$ and $\overline{u^2}/\overline{w^2}$ and turbulence Reynolds number Re_L.

The work of Lumley (1978) and Speziale (1980) indicates that a more general form of (18) is required to satisfy tensor invariance:

$$\phi_{ij1} = C_1' b_{ij} + \gamma(b_{ij}^2 + 2\mathrm{II}\ \delta_{ij}/3), \tag{22}$$

where II is the second invariant of b_{ij}, and C_1 and γ should be functions of turbulence anisotropy, specified in terms of the second- and third-order invariant III, and Re_L, although the work of Lumley & Newman (1977) indicated that $\gamma = 0$ if high-Reynolds-number flows with weak anisotropy are to return to isotropy. The FNS simulations of Rogallo (1981) for a range of homogeneous flows suggest that the linear, Rotta part of (22) alone does indeed provide a fairly good fit to the computed "data" for ϕ_{ij1}, and similarly LRR1 provides a fairly good fit to ϕ_{ij2} if C_1 and C_2 vary with anisotropy and Re_L in accordance with Lumley's (1978) predictions. The later simulations of Feiereisen et al. (1981a,b) indicated that the model developed by Lumley & Newman (1977) based on (22) was overcomplex, but they also implied that $C_1 = 0.7$, assuming the linear term alone is sufficient to model ϕ_{ij1}, although C_1 must be at least 2 to give a return to isotropy. However, Feiereisen et al. found that anisotropy of dissipation (which also acts as a redistribution term but is assumed to be zero in the Launder et al. scheme) was as large as 0.85 $b_{ij}/\overline{q^2}$; if this was assumed to be included in ϕ_{ij1}, they found that C_1 became a function of i and j with a best-fit "constant" value of about 2.7. Vasquez-Malebran & Boysan (1985) have used analytical solutions to transport equations for the invariants II and III to show that the Rotta model gives satisfactory agreement with various experiments on the decay of axisymmetric turbulence, but that it is insufficient for nonaxisymmetric flows when a nonlinear term dependent on II and III is required.

The more recent experiments of Choi (1983) support these findings in that they show that all components return toward isotropy at the same rate in axisymmetric flows but at different rates in plane flows. Furthermore, Choi finds that the rate of return varies, being slow for low-Reynolds-number turbulence and following axisymmetric contraction, faster for a plane flow, but fastest after axisymmetric expansion, although there was no evidence for a final period of decay right to isotropy. On the basis of his results, he has constructed an expression for ϕ_{ij1} that satisfies realizability. This also suggests that $\gamma = 0$, but that C_1 becomes a function of both II and III. Shih & Lumley (1985) have incorporated this in their general Reynolds-stress model, satisfying realizability (which requires the addition of a higher-order term in b_{ij} to the model for ϕ_{ij2} also, and similar functional fits to experiment need to be found for the coefficients in this). Such a model was first developed by Domingos (1981), who stressed the need for data with III > 0.

Weinstock (1981, 1982) has calculated C_1 from first principles and confirmed that it varies not only with anisotropy ratios $\overline{u^2}/\overline{v^2}$ and $\overline{u^2}/\overline{w^2}$, but also with i and j, and it may even be negative under certain conditions, as implied by the experimental results of Wyngaard (1980). Weinstock & Burk (1985) have calculated the theoretical values for C_{1ij} for the weak shear examined by Champagne et al. (1970) and the stronger shear of

Harris et al (1977). As indicated in Table 1, the normal-component values are all rather small. Indeed, they found that C_{111} is less than 2 over a wide range of anisotropy ratio $\overline{u^2}/\overline{v^2}$. This suggests that ϕ_{ij1} does not cause a return to isotropy, but instead merely resists large anisotropy, a weaker effect in line with Choi's results. [Lumley (1978) and others have arrived at a similar conclusion that the Rotta term overpredicts the tendency to isotropy.] They also found a strong dependence of C_{1ij} on dissipation anisotropy and estimated values of this that were as large as those values reported by Feiereisen et al. (1981a). Consequently, their C_{1ij} values were consistent with those from the eddy-simulation data.

Kida & Hunt (1984) have used RDT to calculate the pressure-strain correlation produced by their random large-scale/small-scale turbulence interaction that effectively provides an estimate of ϕ_{ij1} for this type of nonlinear interaction. Again, a weaker tendency to isotropy is found than that predicted by the Rotta model. More significantly, these RDT calculations also show that the evolution of stress anisotropy cannot in general be described solely in terms of the velocity anisotropy tensor but rather must also be a function of the wave-number anisotropy. It may be that this extra structural effect is indirectly modeled to some extent by the nonlinear terms deduced by Choi, but no specific dependence has yet been incorporated in any model. Both effects were evident in large-eddy simulations presented at EUROMECH 199 (1985).

At the same time, J. C. R. Hunt & A. A. Townsend (private communications) have pointed out that there is in fact no need for a final period of more rapid decay, since anisotropy tends to increase ultimately because the large eddies take longer to decay. This is another structural effect, and it may be enhanced in flows that have been subjected to axisymmetric straining, since such straining can lead to the formation of very

Table 1 Comparison of pressure-strain model coefficients deduced for plane shear flows[a]

					RST		RDT	
					LRR1	LRR2	AMS	GAM
Component	WB	HCG	WB	CHC	$C_2 = 0.4$	$C_1 = 3.6$	$C_2 = C_{2ij}$	$C_2 = 1.13$
(ij)	C_1	C_2'	C_1	C_2'	C_2'	C_2'	C_2'	C_2'
11	1.96	0.73	2.04	0.725	0.71	0.6	0.85	0.68
22	2.34	0.50	2.64	0.53	0.55	0.6	−0.35	−0.30
33	0.30	0.96	1.04	0.92	0.87	0.6	0.90	0.98
12	3.6	0.64	3.6	0.55	0.6	0.6	0.6	0.6

[a] Acronyms as follows: RST, Reynolds-Stress Transport; RDT, Rapid-Distortion Theory; WB, Weinstock & Burk (1985); HCG, Harris et al. (1977); CHC, Champagne et al. (1970); LRR1 and LRR2, equations of Launder et al. (1975); AMS, A. M. Savill (unpublished results); GAM, Gence et al. (1978).

persistent longitudinal vortices. The implication is that memory effects may play an important role under certain conditions, and these are also not accounted for explicitly in existing model schemes. Again, however, such effects may be partially accounted for by the higher-order invariant terms, since Gence & Mathieu (1980) have found that when the sign of III is positive (which implies that one large component is losing energy to two smaller ones), the recovery is slower than when III is negative (when two larger components transfer energy to one smaller one). Furthermore, Le Penven et al. (1984) point out that instead there may even be a return to axisymmetry when III is positive. The calculations of Vasquez-Malebran & Boysan (1985) support these findings, but they also indicate that the response will in general depend on the initial anisotropy as well, so that the redistribution of energy is in general determined by the magnitude and the direction of individual energy transfers.

Regarding the modeling of Launder et al. (1975), LRR2 is now more widely used than LRR1 because the increased complexity introduced by the latter has not seemed to be justified by improvements in model performance. However, the simulations of Rogallo (1981) indicated that fitting the Rotta model plus LRR1 simultaneously reproduces the Launder et al. recommended constants, while the simulations carried out by Feiereisen et al. for plane homogeneous shear showed that the reduced LRR2 scheme was not as good as LRR1, but that C_2' should also be a function of i and j.

Maxey (1982) has compared linearized RDT calculations of the rapid pressure-strain correlations with LRR1 ($C_2 = 0.4$) using RDT predictions for $\overline{u_i u_j}$ in a uniform shear flow and initial anisotropy ratios between 1 and 2. He found that although the two prescriptions agree for very small strains, differences quickly appear; in particular, up to a total strain of 1, ϕ_{332} is approximately half the RDT value, and although ϕ_{222} is of the correct magnitude, it is opposite in sign, which suggests that C_2 is also a function of i and j. Gence et al. (1978) have similarly compared RDT calculations for both shear and pure straining distortions with the experiments of Gence & Mathieu (1979, 1980) and have arrived at an optimum compromise constant value of $C_2 = 1.13$. Recently, Weinstock & Burk (1985) obtained estimates for ϕ_{ij2} for the Champagne et al. (1970) and Harris et al. (1977) flows by subtracting their theoretical values for ϕ_{ij1} from the experimentally determined ϕ_{ij}. They have then compared these with LRR1, which they note reduces to

$$\phi_{ij2} = C'_{2ij}\left(P_{ij} - \frac{2}{3}P\delta_{ij}\right), \tag{23}$$

the same form as the abridged LRR2 expression, where $C'_{211} = -(9-3g)/11$, $C'_{222} = (12-15g)/11$, $C'_{233} = (6+9g)/11$, and $C'_{212} = 0.6$. They find good agreement if $g = C_2 = 0.4$ (within 5% for $C_2 = 0.42$; see Table 1), which suggests that the discrepancies pointed out by Leslie (1980) were largely due to the deficiencies of the Rotta model. A. M. Savill (unpublished results) has reanalyzed Maxey's results by modifying C_{2ij} to obtain a better fit between LRR1 and the RDT calculations for each component; then, by taking $g = C_{2ij}$, he found the C'_{2ij} values implied by these and compared them with those determined using the Gence et al. (1978) result and with the Weinstock & Burk (1985) findings. The magnitudes are similar, although RDT predicts that C'_{222} should be of opposite sign. This provides further support for the contention that the LRR1 model can be a reasonable prescription, at least in the limit of homogeneous flows subjected to simple shearing. Stretch (1986) has extended the comparison between linearized RDT calculations and the LRR1 prescription for ϕ_{ij2} to buoyant flows. He finds that the discrepancy over the sign of ϕ_{222} disappears at Richardson numbers above 0.25.

Further progress has always been limited by the lack of experimental data, or computer-simulation "data" for more complex inhomogeneous flows. However, R. J. Adrian and colleagues have now developed a probe with which it is possible to make direct measurements of pressure-velocity and pressure–rate-of-strain correlations and have produced the first set of detailed measurements of these quantities in the mixing layer of a round jet (Nithianandan 1980, Chang et al. 1985). Nithianandan has used his data for $\overline{p\partial u_i/\partial x_i}$ to estimate C_{211} for different values of C_1 (assuming LRR1) along the centerline of the mixing layer (see Figure 4). His results are particularly valuable because the flow passed through several stages of development in the region studied. From an eddy-structure point of view, there were significant changes in the intensity and hence the influence of the different scales of eddies, with associated variations in the nonlinear interactions between large scales and energy transfer to small scales. However, this means that to obtain an estimate of the true distribution of C_{211} it is necessary to make an allowance for the effect of these changes on C_1. A. M. Savill (unpublished results) has made a preliminary attempt at this, using arguments similar to those on which the nonlinear approximations embodied in DSM are founded to estimate the likely variation of C_1, and obtained a smooth reduction in C_{211} from an initial value of 1.1 toward 0.6 in the fully developed region shown in Figure 4. A more rigorous result could be obtained by calculating C_{1ij} from the theoretical expressions of Weinstock (1981, 1982) and then using C_{111} to derive C_{211}. Chang et al. (1985) have provided some further data, but information on other components in this and other flows has yet to appear.

It is difficult to constructively sum up these many and varied findings, but perhaps a further reference to the work of Sreenivasan (1985) will help to illustrate the current position regarding pressure-strain modeling. He compared data obtained by balance for the shear flow passing through both of his contractions with the Rotta model. For the first duct (irrotational strain to shear ratio of 0.12) he found that C_1 had to be different for different components, but that the Launder et al. (1975) combination of the Rotta model and LRR2 together gave acceptable results with just constant values for C_1 and C_2' of 3.3 and 0.53, respectively. These findings are typical, and the values of C_1 and C_2' are close to those expected, which implies that the Launder et al. scheme is satisfactory for both plane shear flows and those subjected to weak additional straining. Indeed, Newley (1985) has subsequently shown that the Launder et al. model exactly reproduces the stress-intensity changes predicted by RDT for such distortions, irrespective of the initial anisotropy.

However, the values of $C_{111} = 6$, $C_{122} = 3.6$, and $C_{133} = 17.1$ that Sreenivasan obtained were very high. They support the other observations that irrotational straining increases the rates of energy transfer, but at the same time they point to the difficulty of finding a single model that can handle even a small range of simple strains and their combination. The obstacles to achieving this are further highlighted by the results for the second duct, with a strain ratio of about one. Here the simple Rotta model broke down, and it was also not possible to fit the overall Rotta plus LRR2 model, no matter what values of C_1 and C_2' were used. In fact, it appears

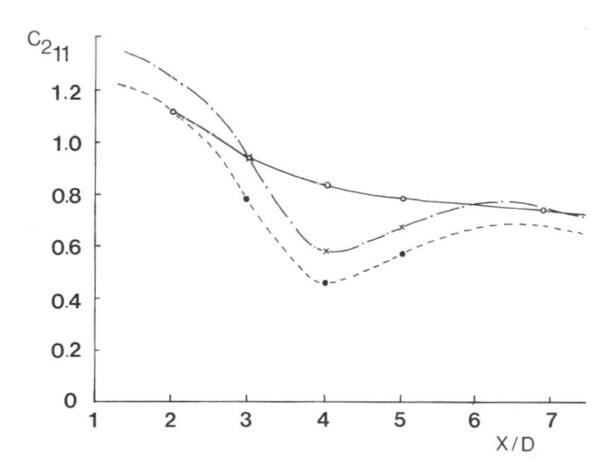

Figure 4 Values of C_{211} along the centerline of the initial mixing layer of a round jet (Nithianandan 1980). --●--: for $C_1 = 2.5$; ·—×—·: for $C_1 = 3.0$; —○—: for an assumed variation of C_1 (A. M. Savill, unpublished results).

that the redistribution of energy among components was extremely complicated in this case, with $\overline{u^2}$ starting out as a "donor" and $\overline{w^2}$ a "receiver," but eventually their roles were reversed, while $\overline{v^2}$ acted as a receiver throughout with the result that there was a tendency away from isotropy in one region. This is an extreme example of the influence of strain history and initial anisotropy on the redistribution of energy among components that was indicated by earlier results, and it shows the necessity for more complex models, for ϕ_{ij1} at least, once one moves away from simple flows responding to weak additional strain rates. At present these more general closures are based on higher-order invariants of the anisotropy tensor that need to be fitted to a wider range of experiments. However, if the predetermined signs and magnitudes of II and III can be related more directly to the influence of wave-number anisotropy and strain/structure memory effects predicted by RDT, it may be possible to achieve greater generality. If not, it would seem that alternative nonlinear terms need to be devised that do model these effects. If ϕ_{ij1} is accurately determined it may be that LRR1 is adequate for some cases, but in general more complex invariant expressions seem to be required. Lecointe et al. (1985) have shown that RDT can also be usefully employed in this context to calibrate relationships derived from isotropic functional theory.

SOME SPECULATION ON FUTURE DEVELOPMENTS

It seems likely that RDT will soon find further practical applications in high-speed flows such as supersonic boundary layers and wakes passing through cascades, where the strain rates are sufficiently high for the linear theory to be valid. Models, such as DSM, where the rapid condition is relaxed and some account is taken of both nonlinear effects and contributions from a wider range of eddy scales, are already being considered for smeared shock flows; such flows are typified by the supersonic boundary layer subjected to compression and weak curvature that was studied by Jayaram & Smits (1985). Extending these models to stratified flows would be a logical progression, and later they could perhaps be used to provide structure-parameter input to the type of depth-averaged calculation schemes envisaged for some environmental flows by D. B. Spalding and W. Rodi (unpublished, 1985). There may also be scope for further calculations with both these approximate, nonlinear and linear RDT models of the "manipulative" types of distortion that lead to drag reductions and that are of such academic and industrial interest at present.

More important advances may come from using the theory to establish flow eddy structure/strain-dependent closure for existing turbulence models, particularly for pressure strain at the RST level. The sensible starting point for this process would be homogeneous flows, since even simple RDT performs well on these and there are several Stanford 1980–81 Conference test cases available for a range of applied strains. At that meeting, no RST model could provide satisfactory predictions for all of these with the same set of constants, and they all predicted $\overline{uv}/\overline{q^2}$ increases with total strain, contrary to the experimental findings. RDT predicted the correct trend for this parameter, and as stated previously it was the only model that could cope with solid-body rotation. At the same time, most of the available pressure-strain data are for these flows, albeit by balance or from FNS and LES calculations. Ideally, an assessment of RDT modeling of all these test cases would lead to the derivation of an algebraic expression describing the rapid-pressure-strain redistribution for small total strains, but it is more likely that one would have to construct a functional fit to a range of strain parameters. This could then be implemented in an RST scheme for testing against the same set of flows. Provided improvements and a degree of generality were achieved, the next step would be to apply the modified RST scheme to some of the curved flows, which have already been modeled using RDT and RST separately, to check performance and make further modifications if justified. At this stage it would be convenient if one could have sufficient confidence in the new pressure closure to ascribe the remaining discrepancies with experimental results to weaknesses in other areas, particularly turbulent diffusion and dissipation, and then work on these.

Unfortunately, two things are likely to prevent this. Firstly, replacing just the current model approximations to the rapid part of the pressure strain is unsatisfactory because one could always make compensating changes to the coefficients in the model used for the return-to-isotropy part if this was retained. A much more satisfactory solution might be to replace the whole pressure-strain influence, and pressure diffusion, by an RDT-based prescription. At present, the information on the turbulence/turbulence part of ϕ_{ij} is discarded when one linearizes the rapid-strain equations, but perhaps in the future it will be possible to derive an additional functional fit to account for this by using the Kida & Hunt (1984) technique. An alternative might be to consider modeling these nonlinear terms within the rapid-strain equations themselves.

Secondly, as Bradshaw (1981) has pointed out, RDT strictly only provides a description of pressure strain in the limit of very high strain rates (relative to $e/\partial u/\partial y$) and so should be regarded as just a qualitative guide to the much weaker distortions of most experiments. Quantitative improvements to model closure would thus require that modifications be made on the basis of DSM-type calculations in which the RDT approximations have been modified to encompass such conditions. This has not previously been possible because there was insufficient information on which to base the required corrections to the pressure field for the influence of lasmos, small scales, and the associated nonlinear energy transfers. However, the direct pressure-velocity data emerging from the work of R. J. Adrian and his colleagues is allowing for a start to be made on this problem. By comparing RDT and DSM calculations with their pressure-strain data for a mixing layer, where there is already some understanding of the strengths of large-scale eddies and turbulence/turbulence interactions at different stages of development, one may begin to estimate the necessary correction factors. It seems reasonable to assume that small scales will act only as a slight damping on the pressure-velocity correlations, since Mulhearn (1975) and others have shown that pressure fluctuations are more highly correlated across shear layers than are velocity fluctuations.

Close to walls there is a third difficulty, since a correction must be made to the pressure-strain terms to account for the influence of the solid boundary. Several damping functions have been proposed, but for the sake of completeness it would be better if this effect could also be described within the RDT-based closure. This may not be possible, but the indications from the work of Hunt & Graham (1978), Maxey (1978), Durbin (1978), and Hunt (1984a), who have looked at the effect of eddy impingement on RDT calculations, are that there is some chance that this final correction might be constructed. Their RDT modeling has already indicated that current approximations that assume the same near-wall variation for all components of pressure-strain correlation are incorrect, and it has led to modifications aimed at remedying this weakness (e.g. Prud'homme & Eligobashi 1983). A more rigorous method of including contributions to both pressure and velocity correlations from large- and small-scale eddies, and one that would immeasurably improve the DSM approach itself, would be to calculate their influence directly using RDT. The effect of the lasmos might perhaps be modeled through an extension of Kida & Hunt's ideas for modeling or by considering these large-scale motions as simple two-dimensional vortices with circulation confined to transverse, longitudinal or inclined planes, constructing distortion matrices to describe them, and then allowing these to act on the smaller-scale anisotropic and very small-scale isotropic eddies (A. A. Townsend, private communication).

At present, such refinements are a long way off, but, continuing on the same extremely optimistic note, it has been suggested that RDT/DSM might make an even more valuable contribution to turbulence modeling if it could be converted into a prescription for length scale to replace the ε-equation. Although the theory can already be used to accurately predict the correlation length scales in all three coordinate directions of the roller-eddy components of strained turbulent-shear-layer structure, and could be modified to include the contribution of the lasmos and small scales to these, the problem would be to relate these scales to the mixing and dissipation length scales required by calculation schemes. This may well prove an insuperable difficulty, but it merits investigation, since it might allow the development of a shear-stress transport model based solely on a turbulence energy equation with RDT-based strain/structure responsive closure that avoids gradient diffusion, ε, and the eddy-viscosity hypothesis. Whether this is a desirable goal is open to debate, but any progress in this direction should be valuable. It seems likely that future turbulence models will be more physical than mathematical, and the main aim of turbulence modeling at present must surely be to get more structural information into the closure at all levels. RDT seems to offer one way of achieving, or at least aiding, this process. Furthermore, if we concentrate on the two weakest links in RST modeling, namely the pressure strain and dissipation, there is the possibility of cascading any improvements down through models employing a lower level of closure, even perhaps to one-equation models including pressure diffusion (Orlandi 1981b), in line with the recommendations of the Stanford Computors Report (see Kline et al. 1981, 2: 957–60).

ACKNOWLEDGMENTS

I would like to thank Dr. A. A. Townsend for the many discussions we have had on the subject of Rapid Distortion Theory and for his interest in, and guidance of, the ideas that led to the development of Distorted Structure Modeling and provided the framework for this review. Special thanks are due to Dr. J. C. R. Hunt for stimulating my awareness of the wider aspects of the theory and for providing much relevant information. I would also like to express my thanks to the many people who have helped with the collection of research material, especially those who have made available material either not widely distributed or at present unpublished. The private communications of Prof. R. J. Adrian, Dr. K.-S. Choi, Dr. A. E. Forest, Dr. M. M. Gibson, Prof. B. E. Launder, Prof. J. L. Lumley, and Dr. A. J. White have been particularly valuable. Finally, thanks are due to Dr. R. B. Price for encouraging the development of this work and preparation of the review as part of the research carried out by the author under Rolls-Royce University Research Project No. 239.

Literature Cited

Anders, J. B., Hefner, J. N., Bushnell, D. M. 1984. Performance of large-eddy breakup devices at post-transitional Reynolds numbers. *AIAA Pap. No. 84-0345*

Andreopoulos, J., Rodi, W. 1984. Experimental investigation of jets in a cross flow. *J. Fluid Mech.* 138: 93–128

Audpoix, B., Cousteix, J., Liandrat, J. 1983. Effects of rotation on isotropic turbulence. *Proc. Symp. Turbul. Shear Flows, Karlsruhe, 4th*, pp. 9-7–12

Balakumar, P. 1983. Application of unsteady aerodynamics to large-eddy breakup devices. *Fluid Dyn. Res. Lab. Rep. 83-5*, Dep. Aeronaut. Astronaut., Mass. Inst. Technol., Cambridge

Balakumar, P., Widnall, S. E. 1986. Application of unsteady aerodynamics to large-eddy breakup devices in turbulent flow. *Phys. Fluids* 29: 1779–87

Bardina, J., Ferziger, J. H., Rogallo, R. S. 1985. Effect of rotation on isotropic turbulence: computation and modelling. *J. Fluid Mech.* 154: 321–36

Bario, F., Charnay, G., Papailou, K. D. 1982. An experiment concerning the confluence of a wake and a boundary layer. *Trans. ASME J. Fluids Eng.* 104: 18–28

Batchelor, G. K., Proudman, I. 1954. The effect of rapid distortion of a fluid in turbulent motion. *Q. J. Mech. Appl. Math.* 7: 83–103

Bertoglio, J.-P. 1981. A model of three-dimensional transfer in non-isotropic homogeneous turbulence. *Proc. Symp. Turbul Shear Flows, 3rd, Davis*, pp. 17.1–6

Boussinesq, J. 1877. Theorie de l'écoulement tourbillant. *Mem. Pre. par. div. Sav.*, Vol. 23, Paris

Bradshaw, P. 1975. In *Project Squid Report on Turbulence in Internal Flows*, ed. S. N. B. Murthy, pp. 243–62. New York: Plenum

Bradshaw, P. 1981. Complex strain fields. See Kline et al. 1981, 2: 700–12

Bradshaw, P., Ferriss, D. H., Atwell, N. P. 1967. Calculation of boundary layer development using the turbulent energy equation. *J. Fluid Mech.* 28: 593–616

Brown, J. L., Kussoy, M. I., Coakley, T. J. 1985. Turbulent properties of axisymmetric shock-wave/boundary-layer interaction flows. *Proc. IUTAM Symp. Turbul. Shear-Layer/Shock-Wave Interactions, Palaiseau*, pp. 21–23 (Abstr.)

Builtjes, P. J. H. 1977. *Memory effects in turbulent flows*. PhD thesis. Lab. Aero-Hydrodyn., Univ. Delft, Neth.

Cambon, C., Jeandel, D., Mathieu, J. 1981. Spectral modelling of homogeneous non-isotropic turbulence. *J. Fluid Mech.* 104: 247–62

Castro, I. P., Bradshaw, P. 1976. The turbulence structure of a highly curved mixing layer. *J. Fluid Mech.* 73: 265–304

Champagne, F. H., Harris, V. G., Corrsin, S. 1970. Experiments on nearly homogeneous turbulent shear flow. *J. Fluid Mech.* 41: 81–140

Chang, P., Adrian, R. J., Jones, B. G. 1985. Fluctuating pressure and velocity fields in the near field of a round jet. *T&AM Rep. 475*, Univ. Ill., Urbana

Choi, K.-S. 1983. *A study of the return to isotropy of homogeneous turbulence*. PhD thesis. Cornell Univ., Ithaca, N.Y.

Chung, M. K., Adrian, R. J. 1979. Evaluation of variable coefficients in second order turbulence models. *Proc. Symp. Turbul. Shear Flows, 2nd, London*, pp. 10.43–48

Debieve, J.-F., Lacharme, J. P. 1985. A shock wave-free turbulence interaction. *Proc. IUTAM Conf. Turbul. Shear-Layer/Shock-Wave Interaction, Palaiseau*, pp. 52–53 (Abstr.)

Debieve, J.-F., Gouin, H., Gaviglio, J. 1982. Evolution of the Reynolds stress tensor in a shock wave–turbulence interaction. *Indian J. Technol.* 20: 90–97

Deissler, R. G. 1961. Effects of inhomogeneity and of shear flow in weak turbulent fields. *Phys. Fluids* 4: 1187–98

Deissler, R. G. 1970. Effect of initial condition on weak homogeneous turbulence with uniform shear. *Phys. Fluids* 13: 1868–69

Deissler, R. G. 1975. Comparison of theory and experiment for homogeneous turbulence with shear. *Phys. Fluids* 18: 1237–40

Domingos, J. J. D. 1981. Pressure strain: exact results and models. *Proc. Symp. Turbul. Shear Flows, 3rd, Davis*, pp. 19.26–30

Donaldson, C. 1968. See Kline et al. 1968, 1: 114–18

Durbin, P. A. 1978. *Rapid distortion theory of turbulent flows*. PhD thesis. Univ. Cambridge, Engl.

Durbin, P. A. 1981. Distorted turbulence in axisymmetric flow. *Q. J. Mech. Appl. Math.* 34: 489–500

Dussage, J. P., Gaviglio, J. 1981. Bulk dilatation effects on Reynolds stresses in the rapid expansion of a turbulent boundary layer at supersonic speeds. *Proc. Symp. Turbul. Shear Flows, 3rd, Davis*, pp. 2.33–38

Elliott, C. J. 1976. *Eddy structure in turbulent flow*. PhD thesis. Univ. Cambridge, Engl.

Elliott, C. J., Townsend, A. A. 1981. The development of a turbulent wake in a distorting duct. *J. Fluid Mech.* 113: 433–67

Falco, R. E. 1977. Coherent motions in the outer region of turbulent boundary layers. *Phys. Fluids* 20: 124–32 (Suppl.)

Feiereisen, W. J., Reynolds, W. C., Ferziger, J. H. 1981a. Numerical simulation of a compressible homogeneous turbulent shear flow. *Rep. No. TF-13*, Dep. Mech. Eng., Stanford Univ., Calif.

Feiereisen, W. J., Shirani, E., Ferziger, J. H., Reynolds, W. C. 1981b. Direct simulation of homogeneous turbulent shear flows on the Illiac IV computer: applications to compressible and incompressible modelling. *Proc. Symp. Turbul. Shear Flows, 3rd, Davis*, pp. 19.31–37

Ffowcs Williams, J. E. 1977. Aeroacoustics. *Ann. Rev. Fluid Mech.* 9: 447–68

Foss, J. F. 1980. Interaction region phenomena for the jet in a cross flow problem. *Rep. SFB 80/E/161*, Univ. Karlsruhe, W. Germ.

Fox, J. 1964. Velocity correlations in weak turbulent shear flows. *Phys. Fluids* 7: 562–64

Gartshore, I. S., Savill, A. M. 1982. Some effects of free stream turbulence on the flow around bluff bodies. *Proc. EUROMECH 160, Berlin*

Gartshore, I. S., Durbin, P. A., Hunt, J. C. R. 1983. The production of turbulent stress in a shear flow by irrotational fluctuations. *J. Fluid Mech.* 137: 307–29

Gence, J. N., Mathieu, J. 1979. On the application of successive plane strains to grid-generated turbulence. *J. Fluid Mech.* 93: 501–14

Gence, J. N., Mathieu, J. 1980. The return to isotropy of an homogeneous turbulence having been submitted to two successive plane strains. *J. Fluid Mech.* 101: 555–66

Gence, J. N., Angel, Y., Mathieu, J. 1978. Partie linéaire des corrélations mettant en jeu la pression dans une turbulence homogène associée à un cisaillement et à un effet de gravité. *J. Méc.* 17: 329–57

Gibson, M. M., Launder, B. E. 1978. Ground effects on pressure fluctuations in the atmospheric boundary layer. *J. Fluid Mech.* 86: 491–511

Gibson, M. M., Rodi, W. 1981. A Reynolds-stress closure model of turbulence applied to the calculation of a highly curved mixing layer. *J. Fluid Mech.* 103: 161–82

Gibson, M. M., Younis, B. A. 1981. Calculations of a turbulent wall jet on a curved wall with a Reynolds stress model of turbulence. *Proc. Symp. Turbul. Shear Flows, 3rd, Davis*, pp. 4.1–6

Gibson, M. M., Younis, B. A. 1983. Turbulent measurements in a developing mixing layer with mild destabilizing curvature. *Exp. Fluids* 1: 23–30

Goldstein, M. E. 1978. Unsteady vortical and entropic distortion of potential flow around arbitrary obstacles. *J. Fluid Mech.* 89: 433–68

Goldstein, M. E. 1979. Turbulence generated by the interaction of entropy fluctuations with non-uniform mean flow. *J. Fluid Mech.* 93: 209–24

Goldstein, M. E. 1984. Aeroacoustics of turbulent shear flows. *Ann. Rev. Fluid Mech.* 16: 263–85

Goldstein, M. E., Atassi, H. 1976. A complete second-order theory for the unsteady flow about an airfoil due to a periodic gust. *J. Fluid Mech.* 74: 741–65

Goldstein, M. E., Durbin, P. A. 1980. The effect of finite turbulence spatial scale on the amplification of turbulence by a contracting stream. *J. Fluid Mech.* 98: 473–508

Grant, H. L. 1958. The large eddies of turbulent motion. *J. Fluid Mech.* 4: 149–90

Guezennec, Y. G., Nagib, H. M. 1985. Documentation of the mechanisms leading to net drag reduction in manipulated turbulent boundary layers. *AIAA Pap. No. 85-0519*

Hanjalic, K., Launder, B. E. 1978. Turbulent transport modelling of separating and reattaching shear flows. I. Basic multiscale development and its application to thin shear flows. *Rep. TF/78/9*, Dep. Mech. Eng., Univ. Calif., Davis

Hanjalic, K., Launder, B. E. 1980. Sensitizing the dissipation equation to irrotational strains. *Trans. ASME J. Fluids Eng.* 102: 34–40

Harris, V. G., Graham, J. A. H., Corrsin, S. 1977. Further experiments in nearly homogeneous shear flow. *J. Fluid Mech.* 81: 657–88

Harsha, P. T. 1974. A general analysis of free turbulent mixing. *Rep. AEDC-TR-73-177*, Arnold Eng. Dev. Center, Arnold AFB, Tenn.

Hart, M. 1985. *Boundary layers on turbine blades*. PhD thesis. Univ. Cambridge, Engl.

Head, M. R., Bandyopadhyay, P. 1981. New aspects of turbulent boundary layer structure. *J. Fluid Mech.* 107: 297–338

Hinze, J. O. 1970. Turbulent flow regions with shear stress and mean velocity gradient of opposite sign. *Appl. Sci. Res.* 22: 163–75

Hunt, J. C. R. 1973. A theory of turbulent flow around two-dimensional bluff bodies. *J. Fluid Mech.* 61: 625–706

Hunt, J. C. R. 1978. A review of the theory

of rapidly distorted turbulent flow and its applications. *Fluid Dyn. Trans.* 9: 121–52
Hunt, J. C. R. 1984a. Flow round bluff obstacles. *VKI Lect. Ser.*
Hunt, J. C. R. 1984b. Turbulence structure in thermal convection and shear-free boundary layers. *J. Fluid Mech.* 138: 161–84
Hunt, J. C. R. 1985a. Vorticity and vortex dynamics in complex turbulent flows. *Proc. CANCAM '85*, pp. 1–21
Hunt, J. C. R. 1985b. Turbulent diffusion from sources in complex flows. *Ann. Rev. Fluid Mech.* 17: 447–85
Hunt, J. C. R., Graham, J. M. R. 1978. Free-stream turbulence near plane boundaries. *J. Fluid Mech.* 84: 209–36
Hunt, J. C. R., Newley, T. M. J., Stretch, D. D. 1986. *Proc. IMA Conf. Stratified Flow.* In press
Jayaram, M., Smits, A. J. 1985. The distortion of a supersonic turbulent boundary layer by bulk compression and surface curvature. *AIAA Pap. No. 85-0299*
Jayaram, M., Dussage, J.-P., Smits, A. J. 1985. Analysis of a rapidly distorted, supersonic, turbulent boundary layer. *Proc. Symp. Turbul. Shear Flows, 5th, Ithaca*
Jeandel, D., Brison, J. F., Mathieu, J. 1978. Modelling methods in physical and spectral space. *Phys. Fluids* 21: 169–82
Johnson, D. A., King, L. S. 1984. A new turbulence closure model for boundary layer flows with strong adverse pressure gradients and separation. *AIAA Pap. No. 84-0175*
Johnson, D. A., King, L. S. 1985. Transonic separated flow predictions based on a mathematically simple, nonequilibrium turbulence closure model. *Proc. IUTAM Symp. Turbul. Shear-Layer/Shock-Wave Interactions, Palaiseau*, pp. 4–5 (Abstr.)
Keffer, J. F. 1965. The uniform distortion of a turbulent wake. *J. Fluid Mech.* 22: 135–59
Keffer, J. F. 1982. General discussion on free shear flows. *Proc. IUTAM Conf. Struct. Complex Turbul. Flow, Marseilles*, pp. 273–78. Berlin: Springer-Verlag
Keffer, J. F., Kawall, J. G., Hunt, J. C. R., Maxey, M. R. 1978. The uniform distortion of thermal and velocity mixing layers. *J. Fluid Mech.* 86: 465–90
Kerr, R. M. 1983. Higher order derivative correlations of velocity and temperature in isotropic and sheared numerical turbulence. *Proc. Symp. Turbul. Shear Flows, 4th, Karlsruhe*, pp. 14.9–12
Kida, S., Hunt, J. C. R. 1984. Interaction between turbulence of different scales over short times. *Proc. IUTAM Int. Congr. Theor. Appl. Mech., 26th, Lyngby*, Pap. No. 453
Kim, J. 1985. Evolution of a vortical structure associated with the bursting event in a channel flow. *Proc. Symp. Turbul. Shear Flows, 5th, Ithaca*, pp. 9.23–28
Kline, S. J., Morkovin, M. J., Sovran, G., Cockrell, J. J., eds. 1968. *Proc. AFOSR-IFP-Stanford Conf. Comput. Turbul. Boundary Layers, Stanford*
Kline, S. J., Cantwell, B. J., Lilley, G. M., eds. 1981. *Proceedings of 1980–81 AFOSR-HTTM-Stanford Conference on Complex Turbulent Flows, Stanford*
Kolmogorov, A. N. 1942. Equation of turbulent motion of an incompressible fluid. *Izv. Akad. Nauk SSSR Ser. Fiz.* 6: 56–58
Kuo, A. Y., Corrsin, S. 1972. Experiments on the geometry of the fine-structure regions in fully turbulent fluid. *J. Fluid Mech.* 56: 447–79
Launder, B. E., Schiestel, R. 1978. Sur l'utilisation d'échelles temporelles multiples en modélisation des écoulements turbulents. *C. R. Acad. Sci. Paris.* 286A: 709
Launder, B. E., Reece, G. J., Rodi, W. 1975. Progress in the development of a Reynolds stress turbulence closure. *J. Fluid Mech.* 68: 537–66
Launder, B. E. (Chairman), Cousteix, J., Hanjalic, K., Rodi, W., Savill, A. M., et al. 1981. Ad hoc Committee of Computors Report: what have we achieved, where do we go. See Kline et al. 1981, 2: 957–62
Lecointe, Y., Piquet, J., Visonneau, M. 1985. Rapid term modelling of Reynolds stress closures with the help of rapid distortion theory. *Proc. Symp. Turbul. Shear Flows, 5th, Ithaca*, pp. 12.7–12
Le Penven, L., Gence, J. N., Compte-Bellot, G. 1984. On the approach to isotropy of homogeneous turbulence: effect of the partition of the kinetic energy among the velocity components. *Proc. Fundam. Fluid Mech. Conf.*
Leslie, D. C. 1980. Analysis of a strongly sheared, nearly homogeneous turbulent shear flow. *J. Fluid Mech.* 98: 435–48
Liu, N.-S., Shamroth, S. J., McDonald, H. 1985. On hairpin vortices as model of wall turbulence structure. *Proc. Symp. Turbulent Shear Flows, 5th, Ithaca*, pp. 2.1–6
Loiseau, M. 1973. Doct. Ing. thesis. Univ. Lyon, Fr.
Lumley, J. L. 1965. *Proc. Int. Colloq. Atmos. Turbul. and Radio Wave Propag., Moscow*
Lumley, J. L. 1975. Pressure-strain correlation. *Phys. Fluids* 18: 750
Lumley, J. L. 1978. Computational modelling of turbulent flows. *Adv. Appl. Mech.* 18: 124–76
Lumley, J. L. 1983. Turbulence modelling. *J. Appl. Mech.* 105: 1097–1103
Lumley, J. L., Newman, G. R. 1977. The

return to isotropy of homogeneous turbulence. *J. Fluid Mech.* 82: 161–78
Mathieu, J. 1971. *V.K.I. Lect. Ser.*, Vol. 36
Maxey, M. R. 1978. *Aspects of unsteady turbulent shear flow, turbulent diffusion and tidal dispersion*. PhD thesis. Univ. Cambridge, Engl.
Maxey, M. R. 1982. Distortion of turbulence in flows with parallel streamlines. *J. Fluid Mech.* 124: 261–82
McDonald, H., Fish, R. W. 1973. Practical calculations of transitional boundary layers. *Int. J. Heat Mass Transfer* 16: 1729–44
Moffatt, H. K. 1967. The interaction of turbulence with strong wind shear. *Proc. Int. Colloq. Atmos. Turbul. and Radio Wave Propag., Moscow*, ed. A. M. Yaglom, V. I. Tatarsky, pp. 139–54. Moscow: Nauka
Moffatt, H. K. 1981. Some developments in the theory of turbulence. *J. Fluid Mech.* 106: 27–48
Moin, P. 1984. Probing turbulence via large eddy simulation. *AIAA Pap. No. 84-0174*
Moin, P., Kim, J. 1985. The structure of the vorticity field in turbulent channel flow. Part 1. Analysis of instantaneous fields and statistical correlations. *J. Fluid Mech.* 155: 441–64
Mulhearn, P. J. 1975. On the structure of pressure fluctuations in turbulent shear flow. *J. Fluid Mech.* 71: 801–12
Mulhearn, P. J., Luxton, R. E. 1975. The development of turbulence structure in a uniform shear flow. *J. Fluid Mech.* 68: 577–90
Mumford, J. C. 1982. The structure of the large eddies in fully developed turbulent shear flows. Part 1. The plane jet. *J. Fluid Mech.* 118: 241–68
Mumford, J. C. 1983. The structure of the large scale eddies in fully developed turbulent shear flows. Part 2. The plane wake. *J. Fluid Mech.* 137: 447–56
Mumford, J. C., Savill, A. M. 1984. Parametric studies of flat plate, turbulence manipulators including direct drag results and laser flow visualisation. *Proc. ASME Symp. Laminar Turbul. Boundary Layers, New Orleans, Fluids Eng. Div.*, 11: 41–51
Mumford, J. C., Savill, A. M., Townsend, A. A. 1980. Identification of flow patterns in turbulent flows. *Proc. EUROMECH 132, Lyon*, pp. IV1–1c
Newley, T. M. J. 1985. *Turbulent air flow over hills*. PhD thesis. Univ. Cambridge, Engl.
Nithianandan, C. K. 1980. *Fluctuating velocity-pressure field structure in a round jet turbulent mixing region*. PhD thesis. Univ. Ill., Urbana
Orlandi, P. 1981a. Model of low Reynolds number wall turbulence for equilibrium layers. *Trans. ASME J. Fluids Eng.*
Orlandi, P. 1981b. Summary report on a one-equation turbulence model including the low Reynolds number wall effect. See Kline et al. 1981, 3: 1472–78
Orlandi, P., Crocco, L. 1985. Interaction between isotropic turbulent fields of different scales. *Proc. Symp. Turbul. Shear Flows, 5th, Ithaca*, pp. 15.1–5
Payne, F. R. 1966. Large eddy structure of a turbulent wake. *Rep.*, Dep. Aerosp. Eng., Pa. State Univ., University Park
Pearson, J. R. A. 1959. The effect of uniform distortion on weak homogeneous turbulence. *J. Fluid Mech.* 5: 274–88
Perry, A. E., Lim, K. L., Henbest, S. M. 1985. A spectral analysis of smooth flat-plate boundary layers. *Proc. Symp. Turbul. Shear Flows, 5th, Ithaca*, pp. 9.29–34
Perry, A. E., Chong, M. S. 1982. On the mechanism of wall turbulence. *J. Fluid Mech.* 119: 173–217
Plesniak, M. W., Nagib, H. M. 1985. Net drag reduction in turbulent boundary layers resulting from optimized manipulation. *AIAA Pap. No. 85-0518*
Pope, S. B. 1975. A more general effective-eddy hypothesis. *J. Fluid Mech.* 72: 331–40
Pope, S. B. 1978. An explanation of the turbulent round-jet/plane-jet anomaly. *AIAA J.* 16: 279–81
Prabhu, A., Narasimha, R., Sreenivasan, K. R. 1974. *Adv. Geophys.* 18B: 317
Prandtl, L. 1945. Über ein neues Formelsystem für die ausgebildete Turbulenz. *Nachr. Akad. Wiss. Göttingen* 1945: 6–19
Prud'homme, M., Eligobashi, S. 1983. Prediction of wall-bounded turbulent flows with an improved version of a Reynolds stress model. *Proc. Symp. Turbul. Shear Flows, 4th, Karlsruhe*, pp. 1.7–12
Reynolds, A. J., Tucker, H. J. 1975. The distortion of turbulence by general uniform irrotational strain. *J. Fluid Mech.* 68: 673–93
Reynolds, W. C. 1982. Physical and analytical foundation concepts and new directions in turbulence modelling and simulation. In *Turbulence Models and Their Applications*, 2: 56
Ribner, H. S., Tucker, M. 1953. The spectrum of turbulence in a contracting stream. *NACA Rep. No. 1113*
Rodi, W. 1976. A new algebraic relation for calculating the Reynolds stresses. *ZAMM* 56: T219–21
Rodi, W., Scheuerer, G. 1986. Scrutinizing the k-ε model under adverse pressure gradient conditions. *Trans. ASME J. Fluids Eng.* 108: 174–79
Rogallo, R. S. 1981. Numerical experiments

in homogeneous turbulence. *NASA TM 81315*
Roshko, A. 1982. *Proc. IUTAM Conf. Struct. Turbul. Flow, Marseilles*, pp. 274–75. Berlin: Springer-Verlag
Rotta, J. C. 1951. Statistiche Theorie nichthomogenen Turbulenz. *Z. Phys.* 129: 547–72
Savill, A. M. 1978. Distorted structure modelling—a more physical approach to rapid distortion theory. *CEGB RD/B/N4691*, Cent. Electr. Generat. Board
Savill, A. M. 1979. *Effects on turbulence of curved or distorting mean flow*. PhD thesis. Univ. Cambridge, Engl.
Savill, A. M. 1981a. Zonal modelling? See Kline et al. 1981, 2: 999–1004
Savill, A. M. 1981b. Results of the 1980–81 Stanford Conference—a personal assessment. Part 2. Homogeneous flows. *Rep. TSF 81/4*, Fluid Dyn. Sect., Cavendish Lab., Univ. Cambridge, Engl.
Savill, A. M. 1982a. A new structural model of turbulent shear flows and its application to a selection of curved flows including recovery. *Proc. IAHR Symp. Refined Modelling Flows, Paris*, 1: 219–37
Savill, A. M. 1982b. The turbulence structure of a highly curved wake. *Proc. IUTAM Conf. Struct. Complex Turbul. Flow, Marseilles*, pp. 185–97. See also Keffer 1982, pp. 275–76
Savill, A. M. 1984. The drag reduction mechanisms of flat plate, turbulence manipulators. *Proc. EUROMECH 181, Saltsjöbaden*
Savill, A. M., Zhou, M. D. 1983. Wake/boundary layer and wake/wake interactions—smoke flow visualisation and modelling. *Proc. Asian Congr. Fluid Mech., 2nd, Beijing*, pp. 743–52. Beijing: Sci. Press
Schumann, U. 1977. Realizability of Reynolds stress turbulence models. *Phys. Fluids* 20: 721–25
Serrin, J. 1959. Mathematical principles of classical fluid mechanics. In *Handbuch der Physik*, 9: 123–52. Berlin: Springer-Verlag
Shih, T.-H., Lumley, J. L. 1985. Modelling of pressure correlation terms in Reynolds stress and scalar flux equations. *Rep. FDA-85-3*, Cornell Univ., Ithaca, N.Y.
Shih, T.-H., Lumley, J. L., Chen, J. Y. 1985a. Second order modelling of a passive scalar in a turbulent shear flow. *Rep. FDA-85-15*, Cornell Univ., Ithaca, N.Y.
Shih, T.-H., Chen, J. Y., Lumley, J. L. 1985b. Second order modelling of boundary-free turbulent shear flows with a new model form of pressure correlation. *FDA-85-07*, Cornell Univ., Ithaca, N.Y.
Smith, C. R. 1984. A synthesized model of the near wall behavior in turbulent boundary layers. *Proc. Symp. Turbul., 8th, Rolla, Mo.*, pp. 1–27
Smith, G. P., Townsend, A. A. 1982. Turbulent Couette flow between concentric cylinders at large Taylor numbers. *J. Fluid Mech.* 123: 187–217
Smits, A. J., Wood, D. H. 1985. The response of turbulent boundary layers to sudden perturbations. *Ann. Rev. Fluid Mech.* 17: 321–58
Speziale, C. G. 1979. Invariance of turbulent closure models. *Phys. Fluids* 22: 1033–37
Speziale, C. G. 1980. Closure relations for the pressure-strain correlation of turbulence. *Phys. Fluids* 23: 459–63
Sreenivasan, K. R. 1985. The effect of contraction on a homogeneous turbulent shear flow. *J. Fluid Mech.* 154: 187–213
Sreenivasan, K. R., Narasimha, R. 1978. Rapid distortion of axisymmetric turbulence. *J. Fluid Mech.* 84: 497–516
Stretch, D. 1986. *The dispersion of slightly dense contaminants in a turbulent boundary layer*. PhD thesis. Univ. Cambridge, Engl.
Stretch, D., Britter, R. 1985a. Thoughts, calculations and experiments on the decay of stratified grid turbulence. *Proc. IUTAM Symp. Mixing Stratified Fluids, Univ. W. Aust.*
Stretch, D., Britter, R. 1985b. Rapid distortion calculations on stably-stratified turbulent shear flows. *Symp. Turbul. Diffusion, 7th, Boulder*, pp. 184–87 (Abstr.)
Tan-atichat, J., Harandi, S. 1986. Effects of acoustic disturbances on measured flow characteristics through a contraction. *AIAA Pap. No. 86-0766-CP*
Tan-atichat, J., Nagib, H. M., Drubka, R. E. 1980. Effects of axisymmetric contractions on turbulence of various scales. *NASA CR 165136*
Tavoularis, S., Corrsin, S. 1981. Experiments in nearly homogeneous turbulent shear flow with a uniform mean temperature gradient. Parts 1 and 2. The fine structure. *J. Fluid Mech.* 104: 311–48, 349–68
Townsend, A. A. 1954. The uniform distortion of homogeneous turbulence. *Q. J. Mech. Appl. Math.* 7: 104–27
Townsend, A. A. 1966. The mechanism of entrainment in free turbulent flows. *J. Fluid Mech.* 26: 689–715
Townsend, A. A. 1970. Entrainment and the structure of turbulent flow. *J. Fluid Mech.* 41: 13–46
Townsend, A. A. 1956, 1976. *The Structure of Turbulent Shear Flow*. Cambridge: Cambridge Univ. Press. 1st, 2nd eds.
Townsend, A. A. 1979. Flow patterns of large eddies in a wake and in a boundary layer. *J. Fluid Mech.* 95: 515–37
Townsend, A. A. 1980. The response of

sheared turbulence to additional distortion. *J. Fluid Mech.* 98: 171–91
Tucker, H. J., Reynolds, A. J. 1968. The distortion of turbulence by irrotational plane strain. *J. Fluid Mech.* 32: 657–73
Vandromme, D., Ha Minh, H. 1985. Physical analysis of turbulent boundary-layer/shock-wave interactions using second order closure predictions. *Proc. IUTAM Symp. Turbul. Shear-Layer/Shock-Wave Interactions, Palaiseau*, pp. 19–21 (Abstr.)
Vasquez-Malebran, S. A., Boysan, F. 1985. On the modelling of the return to isotropy of homogeneous non-isotropic turbulence. *Proc. Symp. Turbul. Shear Flows, 5th, Ithaca*, pp. 12.25–30
Veeravalli, S., Warhaft, Z. 1985. The interaction of two distinct turbulent velocity scales in the absence of mean shear. *Proc. Symp. Turbul. Shear Flows, 5th, Ithaca*, pp. 15.11–17
Wallace, J. M. 1982. On the structure of bounded turbulent shear flow: a personal view. *Dev. Theor. Appl. Mech.* 11: 509 21
Wallace, J. M., Balint, J.-L., Mariaux, J.-L., Morel, R. 1983. Observations on the nature and mechanisms of the structure of turbulent boundary layers: a survey and new results. *ASCE Spec. Conf., West Lafayette, Ind.*
Weinstock, J. 1981. Theory of pressure–strain-rate correlation for Reynolds-stress turbulence closures. Part 1. Off-diagonal elements. *J. Fluid Mech.* 105: 369–96
Weinstock, J. 1982. Theory of pressure-strain rate. Part 2. Diagonal elements. *J. Fluid Mech.* 116: 1–30
Weinstock, J., Burk, S. 1985. Theoretical pressure strain term: resistance to large anisotropies of stress and dissipation. *Proc. Symp. Turbul. Shear Flows, 5th, Ithaca*, pp. 12.13–18
White, A. J. 1980. The prediction of the flow and heat transfer in the vicinity of a jet in a cross-flow. *ASME-80-WA/HT-26*
Wilcox, D. C., Rubesin, M. W. 1980. Progress in turbulence modeling for complex flow fields including effects of compressibility. *NASA TP 1517*
Wyngaard, J. C. 1980. In *Turbulent Shear Flows 2*, pp. 352–65. Berlin: Springer-Verlag
Zhou, M. D., Squire, L. C. 1982. The interaction of a wake with a boundary layer. *Proc. IUTAM Conf. Struct. Complex Turbul. Flow, Marseilles*, pp. 376–87. Berlin: Springer-Verlag
Zhou, M. D., Squire, L. C. 1985. The interaction of a wake with a turbulent boundary layer. *Aeronaut. J.*, Pap. No. 1256, pp. 72–81
Zimont, V. L., Sabel'nikov, V. A. 1975. Distortion of homogeneous turbulence in channels with variable cross-section area. Transl. from *Izv. Akad. Nauk SSSR, Mekh. Zhidk. Gaza No. 10*

Ann. Rev. Fluid Mech. 1987. 19: 577–600

RAREFACTION WAVES IN LIQUID AND GAS-LIQUID MEDIA

S. S. Kutateladze,† V. E. Nakoryakov, and A. A. Borisov

Institute of Thermophysics, Siberian Branch of the USSR Academy of Sciences, 630090 Novosibirsk, USSR

Introduction

In recent years the thermodynamics of a substance having parameters close to critical values has attracted considerable attention from physicists, physicochemists, and thermal physicists. One can single out two basic reasons stimulating the increased interest in this problem:

1. Since the mid-1950s, the findings of experimental studies have gone beyond the scope of the existing concepts based on the van der Waals equation for real gas.
2. The processes occurring in the apparatuses of interior ballistics, nuclear power, and chemical technology lie in wide ranges of temperature and pressure, including the critical state of a working body.

The thermodynamics of a substance in the critical state has been extensively studied, and a strong dependence of the thermodynamic parameters on the temperature and pressure near the critical point has been revealed. However, processes such as heat exchange and finite-amplitude waves (shock waves) have not been as thoroughly investigated in the region of the critical point. In fact, the question of the dynamics of finite-amplitude perturbations of pressure, density, and temperature in the critical region has not been considered at all. One should mention just one study devoted to shock waves near the critical point, namely that by Zel'dovich (1946), who analyzed the entropy condition for shock-wave stability and theoretically the possible existence of rarefaction shock waves. The effect of the sign of $(\partial^2 P/\partial V^2)_S$ on the structure of compression and rarefaction waves was discussed by Thompson & Lambrakis (1973).

† Author now deceased.

0066–4189/87/0115–0577$02.00

In practice, there are no theoretical or experimental works dealing with the propagation of finite-amplitude waves near the critical point. This situation may be explained by the difficulties involved in solving the thermodynamic equations when the singular behavior of the thermodynamic parameters (heat capacity, compressibility, etc.) at the critical point is taken into account. Possibly, experimental work on the propagation of finite-amplitude waves near the critical point has been complicated by the difficulties encountered by the experimentalist in obtaining the critical state in a measurement cell having a considerable length (several meters).

The peculiarity of the critical point lies in the fact that even small perturbations of the parameters of a medium lead to a qualitative change in the character of the anomalies of the thermodynamic parameters because of the infinite "susceptibility" near the critical point. The anomalies observed in that vicinity—opalescence, complete light absorption, dramatic growth of sound absorption, infinite increase of heat capacity—are of great interest for a researcher. The nonanalytic character of the change in the physical properties near the critical point, and the universality in the behavior of both the equilibrium and kinetic quantities of substances different in nature, make the critical point an object of great concern for those fields of physics where cooperative effects determine the character of a phenomenon (nuclear physics, elementary-particle physics, biophysics).

Before we discuss the propagation processes of finite-amplitude waves in the critical region, we first consider briefly the properties of a substance in this region, since it is clear that the final solution will depend on the parameters of the unperturbed state, critical or near-critical in our case.

Thermodynamic Properties of a Substance Near the Critical Point

In the 1960s, Widom (1965), Kadanoff (1966), and Patashinsky & Pokrovsky (1966) suggested the thermodynamic law of corresponding states near the critical point. According to the statistical hypothesis of similarity, the nonanalytic (singular) part of the free energy is assumed to be a generalized homogeneous function of its argument. It may, therefore, be represented as a function of one variable that is a combination of temperature and density. In this case, the asymptotic behavior is described by simple power laws. The hypothesis of similarity results in functional relationships between the power indices.

L. Kadanoff, A. Z. Patashinsky, and V. L. Pokrovsky have advanced the hypothesis of similarity of critical fluctuations, linking the law of corresponding states in thermodynamics and between the critical indices to the behavior of correlation functions in the critical region. They have assumed the correlation functions to be homogeneous in their arguments near the critical point. The similarity hypothesis states that the singular dependence of the physical quantities on the temperature and density is a consequence of the divergence of the correlation radius, since this radius is the only significant scale for a nonanalytic change. The physics of critical phenomena is explained by large regions having the same density rather than by details in the behavior of the density on large scales. The similarity hypothesis was discovered and generalized by Migdal (1968) and Polyakov (1968) using the methods of quantum field theory. Wilson (1971) succeeded in obtaining the formalism allowing an explicit calculation of the critical indices, which was referred to as the renormalization-group method. One consequence of the equations of the renormalization group, in addition to the similarity laws, is the universality principle.

For kinetic coefficients, the divergence was accounted for by Kawasaki (1966) by means of the dynamic hypothesis of similarity and the theory of interacting modes. Indeed, near the critical point the density fluctuations are very intensive, and the velocity gradient created by viscous shear forces at the fluid boundary leads easily to the homogeneity in density. When the fluid returns to a homogeneous condition, energy is dissipated. This may be interpreted as a consequence of interaction between the viscous and acoustic modes, and an abnormally large viscosity occurs as a result.

Near the critical point the isochoric heat capacity $C_V/T \sim \partial^2 P/\partial T^2$ is temperature dependent as $|T-T_c|^{-\alpha}$. The compressibility is given by $\partial\rho/\partial\mu = \rho\partial\rho/\partial P = |T-T_c|^{-\gamma}$, where the most probable values for the exponents are $\alpha = 0.12$ and $\gamma = 1.23$, according to Anisimov (1974).

VELOCITY AND ABSORPTION OF SOUND WAVES In what follows we discuss the propagation of finite-amplitude compression waves of arbitrary duration. In accordance with its Fourier expansion, such a wave may be represented as a set of harmonics of different small-amplitude frequencies, i.e. as a set of sound waves. The regularities of the propagation of these waves near the critical point are therefore of interest.

The experiments performed by Schneider (1951) are likely to be among the fundamental studies on the measurement of ultrasound at the critical point. Schneider investigated the velocity and absorption of ultrasonic waves in overheated and saturated vapors of SF_6 and discovered the minimum ultrasonic velocity at this point. In this work he also measured the damping per wavelength at a frequency of 0.6 MHz and found an abnormally high value in the vicinity of the critical point.

As Botch & Fixman (1965) assumed, the observed damping is due to the dynamic heat capacity connected with large-scale density fluctuations. For numerical calculations they used Ornstein-Zernike's equation for the correlation function and the Debye form of a dependence of correlation length on temperature. The damping per wavelength $\alpha_\lambda = \alpha\lambda = \alpha c/f$, found from the imaginary part of the complex sound velocity, has the form $\alpha_\lambda \sim \omega^{-0.25} \cdot I_2(t, \omega)$, where the obtained integral I_2 is a rapidly varying function of the temperature t and is very weakly dependent on the frequency ω.

The expression for the classical damping, complemented by the bulk viscosity ξ, is of the form

$$\alpha = \alpha_{\text{clas}} + \frac{2\pi^2 f^2 \xi}{\rho c^3} = 2\pi^2 f^2 [4\eta/3 + \xi + \kappa(c_V^{-1} - c_p^{-1})]/\rho c^3.$$

The main contribution to sound absorption is determined by the term $\frac{4}{3}\eta + \xi$, since in the frequency band we are concerned with the role of the term $\kappa(c_V^{-1} - c_p^{-1})$ seems to be insignificant. Kadanoff & Martin (1963) have derived an expression for the low-frequency and long-wave limits of this quantity:

$$\lim_{\omega \to 0} \lim_{q \to 0} \omega^2 q^{-4} \operatorname{Im}[c(q, \omega)] = \frac{4}{3}\eta + \xi.$$

Here $\operatorname{Im}[c(q, \omega)]$ is the imaginary part of the double Fourier transform of the dynamic correlation function for the density fluctuation

$$\langle [\rho(r, t) - \langle \rho(r, t) \rangle][\rho(0, 0) - \langle \rho(0, 0) \rangle] \rangle.$$

Kadanoff & Swift (1968) and Kawasaki (1966) have calculated the functional form of $\frac{4}{3}\eta + \xi$ using the schemes of interaction between modes, including the thermal and sound modes as intermediate states.

The situation is complicated by the fact that there are at least three isolated frequency regions. In region I (the lowest frequencies $\omega \leq \omega_1 = \kappa/\rho c_p \xi^2$) the sound wave splits into two thermal modes, and as a result, damping is characterized by a very strong divergence (approximately as t^{-2}) and a quadratic dependence on the frequency. In regions II and III ($\omega_1 \ll \omega \ll c/\xi$) Kadanoff & Swift (1968) predict that $\alpha \sim \omega^2 t^{-\nu-\alpha} \sim \omega^2 t^{-2/3}$. Kawasaki (1966), however, indicates a different behavior in these two regions: In region II ($\omega_1 \ll \omega \ll \omega_2$), damping of the sound wave is still connected with the contribution of thermal modes to the bulk viscosity, but $\alpha \sim t^2$ and thus is not dependent on the frequency ω; in region III ($\omega_2 \ll \omega \ll c/\xi$), sound waves are a dominant intermediate state and $\alpha \sim \omega^2 t^{-2/3}$, just as in Kadanoff's study. The characteristic frequencies ω_1, ω_2, and ω_3 are temperature dependent, and, according to Kawasaki's approximate estimation, they change, respectively, as $a_1 t^2$, $a_2 t^{4/3}$, and $a_3 t^{2/3}$, where a_1, a_2, and a_3 are roughly equal to 10 MHz. Hence, region I usually lies below the ultrasonic frequencies, and the upper limit of c/ξ is obviously above these frequencies for all the interesting and accessible values of ω. Summing up the above, we obtain

Region I $\omega < \omega_1$: $\alpha \sim \omega^2 t^{-2}$,

Region II $\omega_1 \ll \omega \ll \omega_2$: $\alpha \sim \omega^0 t^2$,

Region III $\omega_2 \ll \omega \leq c/\xi$: $\alpha \sim \omega^2 t^{-2/3}$.

We should note that as the critical point is approached, ω_2 rapidly decreases and the experimental frequency can shift from region II to region III. This means that when both terms have comparable magnitudes, one can expect a complex dependence of the form $\alpha \sim A\omega^2 t^{-2/3} + Bt^2$. We should also mention that the expression for absorption holds only when $\omega\tau_{\text{therm}} \gg 1$, where $\tau_{\text{therm}} = \rho c_V/\kappa q^2$ is the characteristic time of heat transfer, κ is the coefficient of heat conduction, and q is the wave number. The condition $\omega\tau_{\text{therm}} \gg 1$ is closely connected with the condition of the adiabatic nature of a sound wave.

FINITE-AMPLITUDE WAVES NEAR THE CRITICAL POINT Fisher (1957) and Kamensky & Pokrovsky (1969) have treated theoretically questions dealing with the peculiarity of the propagation of finite-amplitude perturbations of pressure in a substance near the thermodynamic critical point. We are not aware of any experimental studies on this subject in the literature. Considering the behavior of the speed of sound in the critical region, Fisher (1957) analyzed the critical point and the critical adiabat. The shock waves in a supercritical state were qualitatively shown to have no peculiarities in their propagation. In a rarefaction wave the system will split into two phases, followed by a change in the state along the coexistence curve, and the behavior of the wave must depend considerably on the amplitude. A rarefaction wave of very small amplitude (sound) cannot propagate at the critical point, since $(\partial P/\partial \rho)_S = 0$. If, therefore, one attempts to excite sound using a harmonic source, the system in a critical state will serve as an "acoustic diode": All the compression half-waves will be transmitted, whereas all the rarefaction waves will be "cut." With sound of small but finite amplitude, the propagation of rarefaction waves seems to be possible, but their velocity will be very low and will be considerably different from that of compression waves. The signals from the harmonic source will, therefore, be generated as a very complex nonlinear wave, with breaks in it from the very beginning. Kamensky & Pokrovsky (1969) have shown that if a phase transition from a one-phase to a two-phase state occurs in the sound wave, then the problem becomes nonlinear, even for small amplitudes, owing to a jump in the speed of sound on the phase coexistence curve. As a result, the "acoustic-diode" effect has been calculated quantitatively.

In Jouguet's formula the subscripts 1 and 2 designate the parameters of a substance prior to and after the wave, respectively. The conjectural closeness of the Poisson abiabat to the Hugoniot shock adiabat for a weak wave permits the following two conditions to be regarded as satisfied: (*a*) the thermodynamic shock-wave possibility condition $S_2 > S_1$, and (*b*) the mechanical shock-wave stability condition $c_1 + u_1 < D < c_2 + u_2$, where c, u, and D are the speed of sound, the wave velocity, and the medium velocity, respectively. The minimum value of heat capacity at which the existence of rarefaction shock waves is possible is equal to 80 J mol^{-1} K^{-1}. In Figure 1 the regions are shown in coordinates P, V, where $(\partial^2 P/\partial V^2)_S < 0$. According to Borisov & Khabakhpashev (1982), increasing the molar heat capacity increases the rarefaction shock wave existence region.

Novikov (1948) analyzed one more possibility for the existence of rarefaction shock waves, which is referred to the region of two-phase states of a substance, namely to the flow of wet steam. During the heat-insulated reversible flow of wet steam, its state changes according to the adiabat

$$dP/dV = -\gamma P/V.$$

In this equation, γ and V are the adiabatic exponent and the specific volume for wet steam, respectively. Differentiating with respect to P at

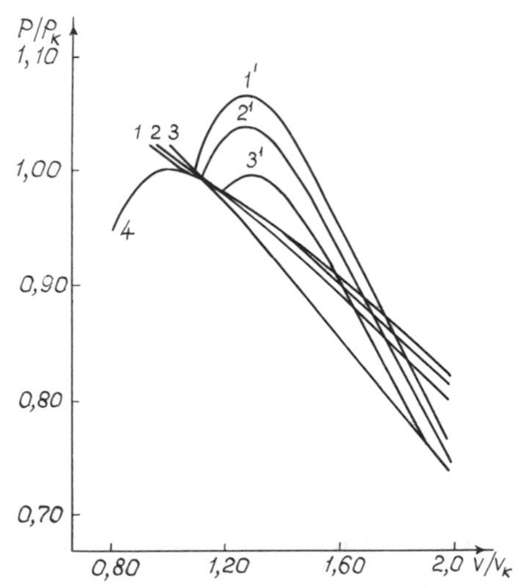

Figure 1 Regions of abnormal thermodynamic properties. 1': $-c_V = 40$, 2': $-c_V = 32$, 3': $-c_V = 24$; curves 1, 2, and 3 are the adiabats of van der Waals gas that correspond to these heat capacities; curve 4 denotes the boundary of the two-phase region.

$S = \text{constant}$, we get

$$(\partial^2 P/\partial V^2)_S = V[(\gamma+1)/P + (\partial\gamma/\partial P)_S]/\gamma^2 P.$$

An analysis of the experimental data for water vapor shows that for pressures from $P_c = 22.5$ MPa to $P = 21.5 \pm 0.21$ MPa, we have $\partial\gamma/\partial P < -(\gamma+1)/P$ and thus $(\partial^2 P/\partial V^2)_S < 0$; thus the rarefaction shock wave can exist.

Fortov & Krasnikov (1970) calculated the shock adiabats of cesium at the parameters typical for shock tubes with heating. In the range of experimentally accessible parameters, an adiabat has an inflection caused by the cesium excitation and ionization processes.

Kahl & Mylin (1969) considered the question of the possible existence of rarefaction shock waves in a van der Waals–Maxwell fluid and concluded that their existence in any phase or in a two-phase state is impossible. However, they pointed out that a rarefaction shock wave can exist in the range of parameters very close to the critical ones at $c_V^* \geq 16.65$ and at $c_V^* \geq 86$ in the two-phase medium. Existence of a rarefaction shock wave depends on the equation of state and on the specific heat at constant volume c_V. The form of the specific heat used here takes into account the translational, rotational, and vibrational energy of the molecules. The disadvantage of this work is that the abnormal increase of c_V in the vicinity of the critical point is neglected.

One should also mention Thompson's (1971) studies. He has analyzed the structure of the steady isentropic gas flow in a nozzle and of Prandtl-Meyer flow with $(\partial^2 P/\partial V^2)_S < 0$. Based on calculations using tabular data, Lambrakis & Thompson (1972) identified a number of high-molecular substances (hydrocarbons and fluorocarbonic compounds) whose range of abnormal thermodynamic properties includes the critical isotherm. Using the first integral of the Navier-Stokes one-dimensional equation, Thompson & Lambrakis (1973) found the structure of a plane stationary rarefaction wave in general form.

For media with arbitrary equations of state, Galin (1958, 1959) showed that if $(\partial^2 P/\partial V^2)_S$ changes sign, it is impossible to answer unambiguously the question of the possible existence of compression and rarefaction shock waves for the whole region. Shock transitions were shown to be possible only when the shock adiabat does not lie in the (V, P) plane to the right of a vector drawn from the initial to the final state. For the general case, the wave adiabat was constructed by Sidorenko (1968). He showed that if the wave adiabat has two points of inflection, then there exist two shock waves linked by a continuous wave or two continuous waves separated by a shock wave. This conclusion was based on theorems that he proved later (Sidorenko 1982).

Thus, an analysis of the experimental and theoretical studies devoted to the behavior of a substance in the region of the thermodynamic critical point enables the following conclusions to be drawn:

1. In the range of temperatures and densities of a substance close to the critical values ($T \lesssim T_c, \rho \lesssim \rho_c$), the behavior of the substance is well described by the van der Waals equation of state, and by the scaling equation of state in the immediate vicinity of the critical point ($T \approx T_c, \rho = \rho_c$). The anomalies of the thermodynamic quantities in the asymptotic region ($T \to T_c, \rho \to \rho_c$) are described by simple power laws. The critical exponents are universal for different substances.
2. Experiments on sound speed and absorption in the critical region indicate the availability of a minimum of the speed of sound at $T = T_c$. In the vicinity of the critical point the speed of sound slows down as the pressure decreases, while the absorption of sound waves depends on both the frequency and the temperature difference.

A theoretical and experimental investigation of the dynamics of finite-amplitude waves near the critical liquid-vapor point would be of extreme interest. The evolution of the initial signal should be strongly dependent on the values of the thermophysical parameters, which show unusual behavior near the critical point.

Experimental Study of Finite-Amplitude Waves Near the Critical Point

Compression waves in media near the critical point can be of considerable length (up to a few tens of centimeters). Thus for an experimental study of the dynamics of compression and rarefaction waves, the length of a test section—the shock tube—must constitute a few meters.

The maximum sizes of the cells that have been used so far for the measurement of the thermodynamic properties of substances close to the critical point have been no more than a few centimeters, and hence the mass of the substances studied has been no more than a few tens of grams.

In the experiments under discussion, it was important to ensure highly accurate temperature control of the test section. For this purpose, the shock tube was contained in a closed hydrodynamic system. The experimental set-up was described in detail by Borisov et al. (1983).

The experiments demonstrated for the first time the existence of non-broadening rarefaction waves in a substance at near-critical conditions. Figure 2 illustrates the structure of rarefaction waves in Freon-13. The oscilloscope traces in Figure 2*A* were triggered simultaneously. In Figure

2*B* the triggering of trace 4 with respect to trace 3 was done with a time delay of 3.8×10^{-2} s. From the oscillograms in Figure 2, the velocity of the rarefaction shock wave is calculated to be 50 m s^{-1} under the given initial conditions. The width of the rarefaction shock wave is 33.7×10^{-3} m. This value of the width is in agreement with the theoretical estimates given below. The initial parameters of the substance in front of the rarefaction wave were chosen near the thermodynamic critical point of Freon-13. From the Figure 2 oscillograms it can be seen that the rarefaction wave propagates as a surface of sudden, very sharp change in the state of the substance (pressure) followed by a gradual change.

In the theory of nonlinear waves (Whitham 1974), the steady state of a

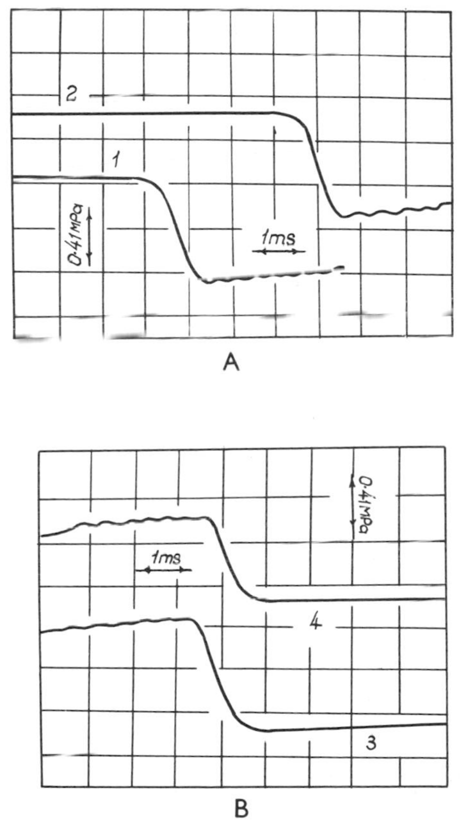

Figure 2 The evolution of a rarefaction shock wave in Freon-13 near the critical point. Curves 1, 2, 3, and 4 correspond to the transducers placed at a distance of 0.15, 0.30, 0.45, and 2.35 m from the diaphragm, respectively.

shock wave is characterized by the time $\tau = \Delta P/(dP/dt)_{max}$, where ΔP is the amplitude difference and $(dP/dt)_{max}$ is the maximum steepness of the wave profile. This time must be short compared with the characteristic propagation time t_p of a wave in a shock tube. In our case the ratio τ/t_p is quite small (1.35×10^{-2}). The values of ΔP and $(dP/dt)_{max}$ for the wave profiles shown in Figure 2 were determined by numerical differentiation of the curves $P(t)$.

In the wave-propagation process, the time τ remained unchanged and was 6.75×10^{-4} s. In addition, during the experiments it was also established that the steepness of the rarefaction wave had not changed after traveling through the high-pressure chamber. In these experiments both the initial state of the substance and the state of the substance in the wave itself were always in a one-phase region near the critical point. Thus our experiments clearly demonstrated that the observed phenomenon is the rarefaction shock wave, defined by Ya. B. Zel'dovich as a surface of sudden, very sharp change in the state of a substance that propagates relative to the unperturbed substance.

To confirm that negative shock waves exist only in the critical region, experiments were performed on the evolution of rarefaction waves far from the critical point in Freon-13 and in nitrogen. We present here some of the results. The initial parameters ahead of the wave were chosen far from the critical point, as $T/T_c = 0.98$, $P/P_c = 0.83$, $\rho/\rho_c = 0.465$ for Freon-13 (Figure 3). As expected, in none of the experiments did rarefaction shock waves develop. During its evolution, the rarefaction wave

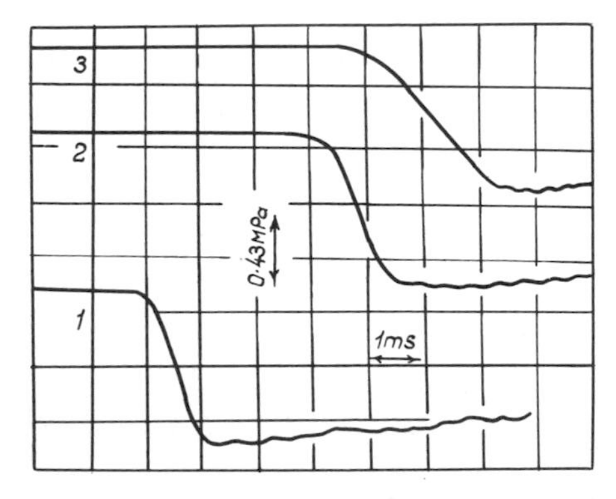

Figure 3 The structure of a rarefaction wave in Freon-13 outside the abnormal region. Here $P_0 = 3.30$ MPa, $\rho_0 = 270$ kg m^{-3}, $T_0 = 295.15$ K. Curves 1, 2, and 3 correspond to transducers at distances of 0.15, 0.45, and 2.35 m from the diaphragm, respectively.

spreads in such a way that over the same distances as in Figure 2, its steepness is reduced by a factor of five or more.

Experiments performed at different times gave good agreement for all oscillograms.

Theory of the Evolution of Finite Perturbations Close to the Critical Liquid-Vapor Point

It is known that near the thermodynamic critical point, compressibility, heat capacity, and kinetic coefficients become abnormally high, and they tend to infinity when the thermodynamic variables are exactly at the critical point. Under conditions near the critical state, the medium becomes very sensitive to external thermal or dynamic perturbations, and the anomalies in turn have a strong effect on the structure of these perturbations.

Consider the propagation of finite perturbations of pressure, density, or particle velocity in a medium whose initial state is close to the critical point. The perturbations are assumed to be long wave, i.e. the correlation radius is much smaller than the characteristic length of the perturbations. In this case, the contribution from relaxation processes is significantly smaller than that of dissipation.

The complete set of equations describing the motion of the medium near the critical point has, in the hydrodynamic approximation, the form

$$\rho[v_t+(v\nabla)v] = -\nabla P+\eta\Delta v+\left(\xi+\frac{\eta}{3}\right)\text{grad div } v,$$

$$\rho T[S_t+(v\nabla)S] = \kappa\Delta T+\xi\ (\text{div } v)^2+\frac{\eta}{2}\left(\frac{\partial v_i}{\partial x_k}+\frac{\partial v_k}{\partial x_i}-\frac{2}{3}\delta_{ik}\frac{\partial v_c}{\partial x_c}\right)^2,$$

$$P = P(\rho, S),$$

$$\rho_t+\text{div } \rho v = 0,$$

where S is the entropy, P the pressure, ρ the density, and v the velocity of the medium in laboratory coordinates (x, t). Here and in what follows, suffixes t and x indicate partial derivatives.

DERIVATION OF AN EQUATION FOR PERTURBATIONS PROPAGATING IN ONE DIRECTION We assume that the relative deviations $(P-P_0)/P_0$, $(\rho-\rho_0)/\rho_0$, and $(v-v_0)/c_0$ from the equilibrium values P_0, ρ_0, and v_0, which are caused by the wave, are quantities of the first order of smallness μ. In addition, we assume the dissipative effects to also be small, i.e. ξ, η, $\kappa \sim \mu$. Neglecting in the original set of equations the terms exceeding the second order of smallness, we reduce this system to the form

$$(\rho' + \rho_0)\frac{\partial v'}{\partial t} = -\nabla P' + \left(\xi + \frac{\eta}{3}\right)\nabla(\nabla v'),$$

$$\rho_0 T_0 \frac{\partial S'}{\partial t} = \kappa \nabla^2 T',$$

$$\frac{\partial \rho'}{\partial t} + \rho_0 \nabla v' = 0,$$

$$P' = \left(\frac{\partial P}{\partial \rho}\right)_S \rho' + \frac{1}{2}\left(\frac{\partial^2 P}{\partial \rho^2}\right)_S \rho'^2 + \left(\frac{\partial P}{\partial S}\right)_\rho S'.$$

Expressing T' via pressure

$$T' = P'\left(\frac{\partial T}{\partial P}\right)_S$$

and substituting it into the Laplacian of the equation of heat conduction, we obtain

$$\rho_0 T_0 \frac{\partial S'}{\partial t} = \kappa \left(\frac{\partial T}{\partial P}\right)_S \nabla^2 P'.$$

From the equation of motion, we have

$$\rho_0 \frac{\partial}{\partial t}\nabla v' = -\nabla(\nabla P') + \frac{\partial}{\partial t} O(\mu^2).$$

Taking into account that $\nabla(\nabla P') = \nabla^2 P'$, we obtain

$$\rho_0 T_0 \frac{\partial S'}{\partial t} = -\kappa \left(\frac{\partial T}{\partial P}\right)_S \rho_0 \left[\frac{\partial}{\partial t}\nabla v' + \frac{\partial}{\partial t} O(\mu^2)\right].$$

Since $\kappa O(\mu^2) = O(\mu^3)$, the latter equation may be integrated as follows:

$$S' = -\frac{\kappa}{T_0}\left(\frac{\partial T}{\partial P}\right)_S \nabla v'.$$

Then the equation of state takes the form

$$P' = c_0^2 \rho' - \frac{\kappa}{T_0}\left(\frac{\partial P}{\partial S}\right)_\rho \left(\frac{\partial T}{\partial P}\right)_S \nabla v' + \frac{1}{2}\left(\frac{\partial^2 P}{\partial \rho^2}\right)_S \rho'^2, \tag{1}$$

where

$$c_0^2 = \left(\frac{\partial P}{\partial \rho}\right)_S; \qquad P' = P - P_0, \qquad \rho' = \rho - \rho_0,$$

$$v' = v - v_0, \qquad v_0 = 0.$$

Further we consider only the propagation of plane waves, since in the shock-tube experiment this case has the most applicability. However, we note that the assumptions will be valid for cylindrically and spherically symmetrical waves. In these cases also, a single equation may be obtained for the perturbed values.

Thus, retaining terms of the second order of smallness in the equation of motion and continuity and replacing the pressure using the equation of state, we obtain the following set of equations for plane waves:

$$(\rho_0 + \rho')v'_t + \rho_0 v' v'_x = -c_0^2 \rho'_x - \left(\frac{\partial^2 P}{\partial \rho^2}\right)_S \rho' \rho'_x + b v'_{xx}, \tag{2}$$

$$\rho'_t + (\rho_0 + \rho')v'_x + v\rho'_x = 0, \tag{3}$$

where

$$b = \xi + \frac{4}{3}\eta + \frac{\kappa}{T_0}\left(\frac{\partial P}{\partial S}\right)_\rho \cdot \left(\frac{\partial T}{\partial P}\right)_S.$$

By differentiating (2) with respect to x and (3) with respect to t, then eliminating v'_{xt} from the resulting equation using the substitutions $v' = c_0\rho'/\rho_0$, $\rho'_t = -\rho_0 v'_x$ in the terms of second order of smallness, we obtain a hyperbolic wave equation for ρ':

$$c_0^2 \rho'_{xx} + \frac{2c_0^2}{\rho_0}\left[1 + \frac{\rho_0}{2c_0^2} \cdot \left(\frac{\partial^2 P}{\partial \rho^2}\right)_S\right](\rho' \rho'_x)_x = \rho'_{tt} + \frac{b c_0}{\rho_0}\rho'_{xxx}.$$

This equation describes the propagation of nonlinear waves in both directions and enables us to solve problems dealing with wave interactions. However, in what follows we analyze only waves that propagate in one direction. In the experiment, this corresponds to a rarefaction wave forming after the bursting of the diaphragm separating the low- and high-pressure chambers.

Considering perturbations that propagate in one direction $x > 0$, we search for the solution to the last set of equations, again to second order, using the method of Khokhlov (1961). Because of the weak nonlinearity of the processes under consideration, the perturbations of velocity and density, v' and ρ', can be represented as slowly varying functions of the coordinates, so that v' and $\rho' = F(\mu x, t - x/c_0)$. This means that the perturbations change their shape only slowly as they evolve in the x-direction.

In this case, Equations (2) and (3) in the variables $z = \mu x$, $\tau = t - x/c_0$ take the form

$$\mu c_0^2 \frac{\partial \rho'}{\partial x} - c_0 \frac{\partial \rho'}{\partial \tau} + (\rho_0 + \rho') \frac{\partial v'}{\partial \tau} - \frac{\rho_0}{c_0} v' \frac{\partial v'}{\partial \tau} - \left(\frac{\partial^2 P}{\partial \rho^2}\right)_S \frac{\rho'}{c_0^2} \frac{\partial \rho'}{\partial \tau} = -\frac{b}{c_0^2} \frac{\partial^2 v'}{\partial \tau^2},$$

$$\mu \rho_0 \frac{\partial v'}{\partial x} + \frac{\partial \rho'}{\partial \tau}\left(1 - \frac{v'}{c_0}\right) - \frac{1}{c_0}(\rho_0 + \rho') \frac{\partial v'}{\partial \tau} = 0.$$

The equations describing the evolution of v' and ρ' must be of similar form, since they are characteristic of the same wave process. In this case, the nonlinear relations of the type $v' = \rho' c_0/\rho_0$ must be complemented by terms of the second order of smallness and by some derivatives with unknown coefficients a and d such that $a \sim 1$, $d \sim O(\mu)$, i.e.

$$\rho' = \frac{\rho_0}{c_0} v' + \frac{a\rho_0}{c_0} v'^2 + \frac{d\rho_0}{c_0} \frac{\partial v'}{\partial \tau},$$

$$v' = \frac{c_0}{\rho_0} \rho' - \frac{ac_0}{\rho_0^2} \rho'^2 - \frac{dc_0}{\rho_0} \frac{\partial \rho'}{\partial \tau}.$$

By substituting now the perturbation of the velocity v' as a function of density, i.e. $v = f_1(\rho')$ into the equation of motion, and $\rho' = f_2(v')$ into the equation of continuity, we get

$$\mu \frac{c_0}{\rho_0} \frac{\partial \rho'}{\partial x} - 2 \frac{\rho'}{\rho_0^2} \frac{\partial \rho'}{\partial \tau} \left[ac_0 + \left(\frac{\partial^2 P}{\partial \rho^2}\right)_S \frac{\rho_0}{2c_0^2}\right] = \left[\frac{b}{c_0^2 \rho_0} + d\right] \frac{1}{\rho_0} \frac{d^2 \rho'}{\partial \tau^2},$$

$$\mu \frac{\partial v'}{\partial x} - 2 \frac{v'}{c_0^2} \frac{\partial v'}{\partial \tau} [1 - ac_0] = -\frac{d}{c_0} \frac{\partial^2 v'}{\partial \tau^2}.$$

From a comparison of coefficients of the nonlinear and dissipative terms, we find

$$a = \frac{1}{2c_0} - \left(\frac{\partial^2 P}{\partial \rho^2}\right) \frac{\rho_0}{4c_0^3}; \qquad d = -\frac{b}{2\rho_0 c_0^2}.$$

Substituting a and d into the equation for the velocity perturbation, we obtain

$$\mu v'_x - \left[1 + \frac{\rho_0}{2c_0^2} \left(\frac{\partial^2 P}{\partial \rho^2}\right)_S\right] \frac{v'}{c_0^2} v'_\tau = \frac{b}{2\rho_0 c_0^3} v'_{\tau\tau}. \tag{4}$$

We have now obtained the Burgers equation, written in a coordinate system moving with velocity c_0. The equations for the perturbations P' and ρ' are obtained from (4) by replacing v' by $(P'/\rho_0 c_0)v'$ and $c_0\rho'/\rho_0$, respectively. For ideal gases the expression in square brackets is simply $(\gamma+1)/2$. For real gases, however, the second derivative near the critical point has the abnormal form $(\partial^2 P/\partial \rho^2)_S < 0$, and the overall expression in the brackets may become negative. In this case, Equation (4) describes the evolution of a rarefaction shock wave. In fact, in the laboratory system of coordinates, Equation (4) for the pressure can be rewritten as

$$P_t + c_0 P_x + \frac{1}{2\rho_0^3 c_0^2}\left(\frac{\partial^2 P}{\partial V^2}\right)_S PP_x/\rho_0 c_0 = \frac{P_{xx}}{2\rho_0}\left[\frac{4}{3}\eta + \xi + \kappa(c_V^{-1} - c_P^{-1})\right], \quad (5)$$

where V is the specific volume.

ANALYSIS OF THE EQUATION DERIVED The factor in front of the nonlinear term in (5) can be written as

$$\alpha = (\partial^2 P/\partial V^2)_S/2\rho_0^3 c_0^2 = 1 + \frac{\rho_0}{2c_0^2}\left(\frac{\partial^2 P}{\partial \rho^2}\right)_S = 1 + \rho_0 c_0 \left(\frac{\partial c_0}{\partial P}\right)_S,$$

from which it follows that a change in sign of α is associated with an abnormal decrease in the speed of sound when the pressure increases near the critical point (Figure 4). From (5), the velocity of a pressure jump with amplitude P' is as follows:

$$D(P') = c_0 + \frac{P'}{2}\left[\left(\frac{\partial c_0}{\partial P}\right)_S + \frac{1}{\rho_0 c_0}\right]. \quad (6)$$

It can be seen from Figure 4 that in a rarefaction wave, the derivative $(\partial c_0/\partial P)_S$ is negative near the critical point and quickly attains that value at which the expression in the square brackets in (6) also becomes negative. Hence we have $P'[(\partial c_0/\partial P)_S + 1/\rho_0 c_0] > 0$, and therefore the velocity at each point of the wave profile increases when the modulus of the pressure amplitude increases. This leads to the formation of a discontinuity in the rarefaction wave, i.e. to the formation of a rarefaction shock wave. A similar effect for the temperature jumps in superconducting fluids was reported by Khalatnikov (1971). Furthermore, the jump velocity satisfies the condition for mechanical stability ($c_0 < D < c_1 - v$). In fact, if we substitute the values of the jump velocity from (6) for the speed of sound in the perturbed state c_1 and the particle velocity v, we get

$$c_0 < D < c_0 + P'[(\partial c_0/\partial P)_S + 1/\rho_0 c_0].$$

In this case the entropy condition

$$S_1 - S_0 = (\partial^2 P/\partial V^2)_S (V_0 - V_1)^3/12T_0$$

is satisfied automatically, since $(\partial^2 P/\partial V^2)_S < 0$ for the same range of parameters as the coefficient $(\partial c_0/\partial P)_S + 1/\rho_0 c_0$ in the condition for mechanical stability.

The factor in the square brackets in (5) takes abnormally high values close to the critical point, and it may be evaluated by one of Kawasaki's (1966) models. Let us estimate the width of the transition zone, i.e. the "thickness" δ of a rarefaction shock wave. Its value is determined by

$$\delta = \frac{bc_0}{P'(\partial^2 P/\partial V^2)_S/2\rho_0^3 c_0^2}. \tag{7}$$

By means of the theory of interacting modes, Kadanoff & Swift (1968) determined the damping factor β via the unknown b:

$$\beta = b\pi\omega/c_0^2\rho,$$

which near the critical point behaves as $\beta = \omega\varepsilon^{-2}$, where $\varepsilon = |T - T_c|/T_c$.

Such a formula for β holds for values $\omega\varepsilon^{-2} < 5 \times 10^{-3}$. In our case this inequality is fulfilled. We now estimate the nonlinearity coefficient using Bethe's formula (Lambrakis & Thompson 1972):

$$\left(\frac{\partial^2 P}{\partial V^2}\right)_S = \left(\frac{\partial^2 P}{\partial V^2}\right)_T - \frac{3T}{c_V}\left(\frac{\partial P}{\partial T}\right)_V \frac{\partial^2 P}{\partial V \partial T} + \frac{3T}{c_V^2}\left(\frac{\partial P}{\partial T}\right)_V \left(\frac{\partial^2 P}{\partial T^2}\right)_V$$
$$+ \frac{T}{c_V^2}\left(\frac{\partial P}{\partial T}\right)_V^3 [1 - T(\partial c_V/\partial T)_V/c_V].$$

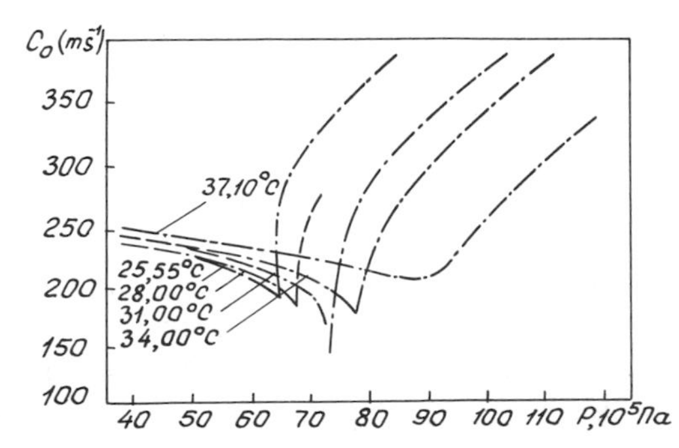

Figure 4 The dependence on pressure of the speed of sound in carbonic acid near the critical point.

It can be shown that when the critical point is approached, the first term on the right-hand side goes to zero according to the definition, the second term also goes to zero, and the third term tends to zero as $\varepsilon^{2\alpha_1}$, where $\alpha_1 \simeq 0.1$ is the critical exponent of the heat capacity $c_V \sim |T-T_c|^{-\alpha_1}$ (T_c is the critical temperature). The last term increases as $\varepsilon^{-1+2\alpha_1}$ when $T \to T_c$. Thus we have

$$\alpha = V^3(\partial^2 P/\partial V^2)_S/2c_0^2 = \varepsilon^{-1+2\alpha_1}\varepsilon^{-\alpha_1} = \varepsilon^{-1+\alpha_1}$$

and hence

$$\delta = \delta_0\{|T-T_c|/T_c\}^{-1-\alpha_1/2}.$$

Here δ_0 is a value independent of the temperature difference $T-T_c$ and is therefore the transition-zone width of a compression shock wave far from the critical point. The value of δ_0 is 0.88×10^{-7} m, i.e. of the order of several mean free paths. For the case in our experiment, where $T-T_c = 10^{-3}$, we obtained $|T-T_c|/T_c = 3 \times 10^{-6}$, $\delta/\delta_0 = 629{,}540$, from which it follows that $\delta = 0.88 \times 10^{-5} \times 629{,}540 = 5.6$ cm. Thus the transition-zone width of the rarefaction shock wave near the critical point is determined not by the mean free path of molecules but by the other scale, which is nearly one million times greater than the mean free path. Close to the critical point, indeed, a new scale appears, a correlation radius, which also increases infinitely as $T \to T_c$. The correlation radius here seems to be a determining scale for the width of the rarefaction shock wave.

According to Rudenko & Soluyan (1974), Equation (5) has a solution describing a stationary shock wave:

$$P = P_0 + \Delta P/[1+\exp(\varepsilon\alpha\Delta P/bc)],$$

where

$$\Delta P = P_{\xi=-\infty} - P_{\xi=+\infty}, \qquad \xi = x - Dt, \qquad \partial P/\partial\xi|_{\xi=\pm\infty} = 0.$$

It is seen from the solution that if $\alpha < 0$, there exists a steady solution in the form of a rarefaction shock wave. In the case $\alpha > 0$, no steady solutions exist in the form of a rarefaction shock wave; rarefaction shock waves decay away in the process of evolution.

THE STRUCTURE OF ARBITRARY-AMPLITUDE COMPRESSION AND SHOCK WAVES IN VAN DER WAALS GAS WITH CONSTANT HEAT CAPACITY In the previous sections we have considered the behavior of finite-amplitude waves near the critical point. In what follows we do not impose restrictions on the wave amplitude and instead consider the solutions for arbitrary amplitudes. For this purpose, we take advantage of the following. In a recent paper, Zel'dovich (1981) points out that thermodynamic equilibrium in the critical region is established slowly because of strong fluctuations. Thus, immediately after a rapid change in the state, equilibrium will occur only in first order (neighbor molecules). Therefore, the assumption has been made that such a change may be described by classical theories that do not take into account the fluctuations and that "just in this sense a retoration of the van der Waals critical point occurs in rapid processes." Owing to this situation, in what follows we use the van der Waals equation of state, and the heat capacity is assumed to be large but constant. The van der Waals equation of state is usually written as follows: $(P+a/V^2)(V-b) = R_G T$, where a and b are van der Waals constants, $R_G = R/\mu$, R is the universal gas constant, and μ is the molecular mass.

Let the specific isochoric heat c_V be constant. This assumption is valid for the same range of parameters where the van der Waals equation is applicable. Then for the specific internal energy, we have $E = c_V T - a/V$.

In addition, consider the equations for the isentropic and shock adiabats. The dimensionless equation of the isentropic adiabat for a van der Waals gas with constant heat capacity can be written in a fairly simple form:

$$P^* = (P_0 + 3/v_0^2)\,[(v_0 - 1/3)/(v - 1/3)]^{\gamma} - 3/v^2, \tag{8}$$

where $P^* = P/P_c$, $v = V/V_c$, $\gamma = 1 + 1/c_V$, and $c_v = c_V/R_G$; the subscripts "c" and "0" indicate the critical thermodynamic values and the initial values, respectively.

We also write down the dimensionless expression for $(\partial^2 P/\partial V^2)_S$:

$$(\partial^2 P^*/\partial v^2)_S = \gamma(\gamma+1)\,(P^* + 3/v^2)/(v - 1/3)^2 - 18/v^4.$$

The dimensionless equation of the shock adiabat for a van der Waals gas with constant heat capacity can be written as

$$P^* = [2P_0 c_V(v_0 - 1/3) - P_0(v - v_0) - 6c_V(v - 1/3)/v^2 + 6c_V(v_0 - 1/3)v_0^2 + 6(1/v - 1/v_0)]/[2c_V(v - 1/3) + (v - v_0)]. \tag{9}$$

It can be seen from a comparison of formulas (8) and (9) that the equation of the shock adiabat is of a more complicated form than the equation of the isentropic adiabat. However, in the ranges of pressure, volume, and heat capacity under consideration, these adiabats essentially coincide. This fact is further used below to substantiate the difference scheme.

ALGORITHM OF NUMERICAL SOLUTION When solving numerically the problem of the decay of an arbitrary breakup it is convenient to write the mass, momentum, and energy conservation laws for an arbitrary mobile element of the medium with a boundary through which there is no flux of substance.

In the one-dimensional case, we have

$$\oint \rho \, dx = 0, \qquad \oint \rho u \, dx - p \, dt = 0,$$

$$\oint \rho(E + u^2/2) \, dx - pu \, dt = 0,$$

where ρ is the density of the medium and u is its velocity. The dissipative terms are neglected in the equations, since with the grid steps Δx and Δt appropriately chosen, the "viscosity" and "heat conduction" inherent in the difference scheme itself are of the same order as the real viscosity and heat conduction.

Consider an ideal model of the shock tube: a cylindrical tube with closed ends that consists of two sections, the high- and low-pressure chambers, separated by a partition. At a certain moment of time the partition is removed and the compression wave travels to the low-pressure chamber, while the rarefaction wave propagates toward the high-pressure chamber. To consider this process, we use the method suggested by Godunov (1959).

The difference analogues of the equations for the ith cell can be written as follows:

$$\rho_i' \Delta x_i' = \rho_i \Delta x_i, \qquad (\Delta x_i = x_{i+1} - x_i, x_i' = x_i + \bar{u}_i \Delta t),$$

$$\rho_i \Delta x_i (u_i' - u_i) = (\bar{P}_i - \bar{P}_{i+1}) \Delta t,$$

$$\rho_i \Delta x_i (E_i' - E_i) = P_i (\bar{u}_i - \bar{u}_{i+1}) \Delta t,$$

where x_i and x_{i+1} are the coordinates of the cell boundaries (contact discontinuities), $\bar{u}_i$ and $\bar{u}_{i+1}$ the velocities of the cell boundaries, and $\bar{P}_i$ and $\bar{P}_{i+1}$ the pressures at the boundaries of the cell; all primed symbols indicate the quantities relating to the moment of time $t' = t + \Delta t$. For weak shock and simple waves, good calculational accuracy of $\bar{u}_i$ and $\bar{P}_i$ can be achieved using a "sound approximation":

$$u_i = (p_{i-1} - p_i + a_{i-1} u_{i-1} + a_i u_i)/(a_{i-1} + a_i),$$

$$P_i = [a_{i-1} a_i (u_{i-1} - u_i) + a_{i-1} P_i + a_i P_{i-1}]/(a_{i-1} + a_i),$$

where $a_i = \rho_i c_i$ and c is the adiabatic speed of sound. The application here of the two last equations is justified, in addition, by the coincidence mentioned above of the shock and isentropic adiabats.

Thus the set of equations is mathematically closed and well founded from a physical point of view. In a numerical solution, the time step Δt

has been chosen so as to ensure that the difference scheme is stable and has good convergence. The approximation and stability of a similar difference scheme were considered in detail by Godunov (1976).

The calculations are presented in Figures 5–8. The dimensionless coordinate is $\xi = x/\Delta x_0$, where Δx_0 is the initial size of the calculational cell. The dimensionless time is given by $\tau = t/t_*$, where t_* is the characteristic time of the process (the time for which the stationary profile of a shock wave is established). The dimensionless perturbation of pressure is $\Delta P_L^* = P^* - P_L^*$ in the compression wave and $\Delta P_H^* = P_H^* - P^*$ in the rarefaction wave. (P_L and P_H are the initial pressures in the low- and high-pressure chambers, respectively.)

Figure 5 demonstrates the evolution of a compression wave at $c_V = 30$. Curves 1–4 correspond to the moments of time $\tau = 1$, 10, 20, and 30. In this case, the perturbed and unperturbed states were in the range of abnormal thermodynamic properties. (The initial parameters on the discontinuity were the following: $P_L^* = 0.912$, $v_L = 1.54$, $P_H^* = 1.040$, and $v_H = 1.11$.) In Figure 6, the unperturbed state was outside this range ($P_L^* = 0.846$, $v_L = 1.85$, $P_H^* = 1.160$, and $v_H = 1.00$). It can be seen from Figure 5 that the width of the compression wave front increases with time, just as is the case with the rarefaction wave in an ideal gas. Figure 6 shows that the forepart of the wave front, corresponding to the range of normal thermodynamic properties, remains steep, and the back part of the wave front, corresponding to the range of abnormal thermodynamic properties, spreads. So far rarefaction shock waves with the spreaded back part of the front have been observed only in media with relaxation (so-called partially dispersed waves) (Becker 1974). However, in those experiments the width of the spread part was constant and was determined by the relaxation time, whereas in the case under discussion, the width of this part of the front is proportional to the path passed by the wave.

The evolution of a rarefaction wave at $c_V = 20$ is illustrated in Figures 7 and 8. Curves 1–4 correspond to the same moments of time as in Figure

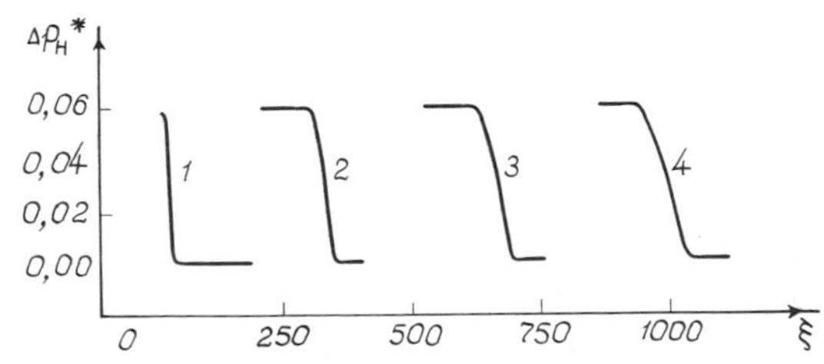

Figure 5 The evolution of a compression wave whose parameters are inside the abnormal region.

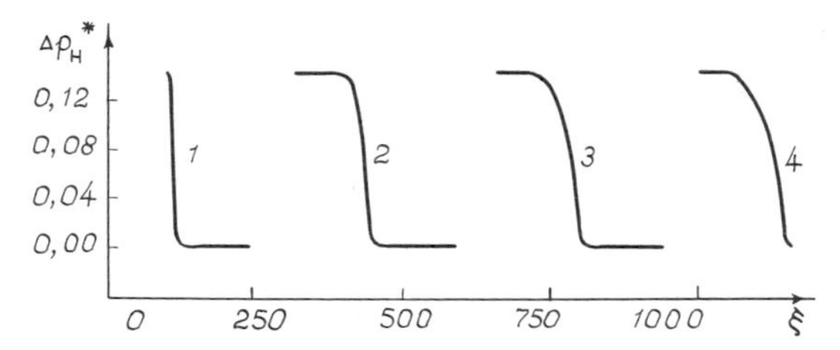

Figure 6 The evolution of a compression wave whose parameters are outside the abnormal region.

5, and curve 5 corresponds to $\tau = 40$. In Figure 7, the perturbed and unperturbed states were in the region of abnormal thermodynamic properties (the initial parameters at the discontinuity were $P_L^* = 0.834$, $v_L = 2.00$, $P_H^* = 0.960$, $v_H = 1.25$), and in the case of Figure 8 the perturbed state was outside this region ($P_L^* = 0.610$, $v_L = 3.33$, $P_H^* = 0.960$, $v_H = 1.25$). It can be seen from Figure 7 that the width of the rarefaction wave front does not increase with time, i.e. the rarefaction shock wave does occur. Figure 8 shows also that the forepart of the wave front, which corresponds to the range of abnormal thermodynamical properties, remains steep, and the back part of the front, which corresponds to the range of normal thermodynamic properties, spreads. Both of these forms of the structure of rarefaction waves have been predicted theoretically and observed experimentally.

If we take into account the abnormal behavior of heat capacity near the liquid-vapor critical point, then a considerable extension of the class of substances in which rarefaction shock waves can propagate should result. Since the behavior of a substance near the critical point is universal for all simple fluids, the phenomenon of rarefaction shock waves will be observed also for all substances that are close to their liquid-vapor critical points. In addition, the results should not vary qualitatively if the scaling equation of state is used instead of the van der Waals equation of state.

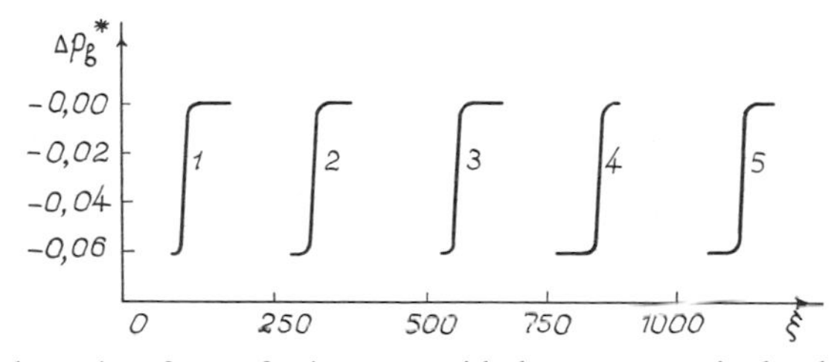

Figure 7 The dynamics of a rarefaction wave with the parameters in the abnormal region.

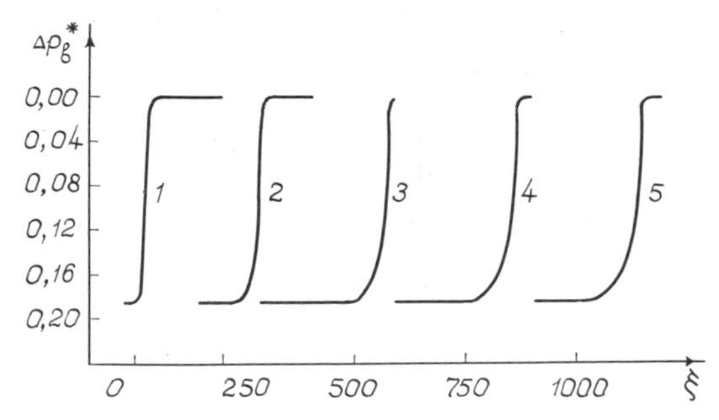

Figure 8 The dynamics of a rarefaction wave whose final state is outside the abnormal region.

Conclusions

The experimental and theoretical results presented above enable one to conclude that the existence of a rarefaction shock wave near the liquid-vapor critical point is due to the abnormal decrease in the speed of sound when the pressure increases. We have made here no quantitative comparison between the experimental and calculated results because to do this, it is necessary to know accurate power dependencies of the heat capacity and compressibility, as well as the coefficients of viscosity and heat conduction of Freon-13, on the temperature and density. Unfortunately, no similar studies have been made so far, neither for Freon-13 nor for any other substance. Most commonly, the temperature dependencies of the thermodynamic and kinetic properties are measured.

Many results are, nevertheless, in good qualitative agreement. One can therefore argue that the theoretical models considered above are capable not only of accounting for the phenomena observed in the experiments but also of predicting new, interesting effects.

Compare the results obtained with the data on wave propagation in solids after their polymorphic transformations. In the works performed by Erkman (1961) and Ivanov & Novikov (1961), after explosive loading of iron and steel samples, spalls with smooth surfaces were observed. This phenomenon has been attributed to an interaction of rarefaction shock waves. The existence of a rarefaction jump in solids subjected to polymorphic transformations is associated with the fact that the phase transition leads to the appearance of a convex-upward section on the adiabat, i.e. the section on which the average value of the second-order pressure derivative with respect to the volume is less than zero (at constant entropy).

As for compression waves, they always propagate as shock waves (one or two) in solids that have undergone polymorphic transformations. The compression waves, which consist of the jump and the spreading back part of the front and which are possible near the liquid-vapor critical point, cannot be observed in solids. This is because $(\partial^2 P/\partial V^2)_S > 0$ at any point in a solid where the adiabat has no singularities.

Thus the similarity between the processes occurring in solids, subjected to polymorphic transformations, and those near the liquid-vapor critical point is associated with the abnormal behavior of the adiabat, whereas their difference is due to the physical nature of this anomaly. If in solids this anomaly is the phase transition, then close to the critical point there is a strong change in the thermodynamic properties (compressibility, etc.) of a single-phase substance.

ACKNOWLEDGMENT

We are indebted to Al. A. Borisov and G. A. Khabakhpashev for their reading of our manuscript and helpful comments.

Literature Cited

Anisimov, M. A. 1974. Investigation of the critical phenomena in liquids. *Usp. Fiz. Nauk* 114: 249–94

Becker, E. 1974. Relaxation effects in gas flow. *Fluid Dyn. Trans.* 7: 9–30

Borisov, A. A., Khabakhpashev, G. A. 1982. Structure of the compression and rarefaction waves in van der Waals gas with constant specific heat. *J. Appl. Mech. Tech. Phys. (USSR)* 23: 115–20

Borisov, A. A., Borisov, Al. A., Kutateladze, S. S., Nakoryakov, V. E. 1983. Rarefaction shock wave near the critical liquid-vapour point. *J. Fluid Mech.* 126: 59–73

Botch, W., Fixman, M. 1975. Sound absorption in gases in the critical region. *J. Chem. Phys.* 42: 199–204

Erkman, J. O. 1961. Smooth spalls and the polymorphism in iron. *J. Appl. Phys.* 32: 939–44

Fisher, J. Z. 1957. Sound propagation in the critical point. *Acoust. J.* 3: 208

Fortov, V. E., Krasnikov, Yu. G. 1970. On setting up a thermodynamically complete equation of state of imperfect plasma on the basis of dynamic experiments. *Zh. Eksp. Teor. Fiz.* 59: 1645–56

Galin, G. Ya. 1958. Shock waves in media described by an arbitrary equation of state. *Dokl. Akad. Nauk SSSR* 119: 1106–9

Galin, G. Ya. 1959. On theory of shock waves. *Dokl. Akad. Nauk SSSR* 127: 55–58

Godunov, S. K. 1959. Finite-difference method for numerical calculation of breaking solutions of hydrodynamic equations. *Math. J.* 19: 271–306

Godunov, S. K. 1976. *Numerical Solution of Multi-Dimensional Problems of Gas Dynamics*. Moscow: Nauka. 400 pp.

Ivanov, A. G., Novikov, S. A. 1961. Shock rarefaction waves in iron and steel. *Zh. Eksp. Teor. Fiz.* 40: 1880–82

Kadanoff, L. P. 1966. Scaling laws for Ising models near T_c. *Physics* 2: 263–72

Kadanoff, L. P., Martin, P. G. 1963. Hydrodynamic equations and correlation functions. *Ann. Phys. (NY)* 24: 419–69

Kadanoff, L. P., Swift, J. 1968. Transport coefficients near the liquid-gas critical point. *Phys. Rev.* 166: 89–101

Kahl, G. D., Mylin, D. C. 1969. Rarefaction shock possibility in a van der Waals–Maxwell fluid. *Phys. Fluids* 12: 2283–91

Kamensky, V. G., Pokrovsky, V. L. 1969. Finite amplitude sound near the critical point. *Zh. Eksp. Teor. Fiz.* 56: 2148–54

Kawasaki, K. 1966. Correlation-function approach to the transport coefficients near the critical point. *Phys. Rev.* 150: 291–306

Khalatnikov, I. M. 1971. *Theory of Superfluids*. Moscow: Nauka. 320 pp.

Khokhlov, R. V. 1961. Theory of shock radio waves in nonlinear lines. *Radiotekh. Electron.* 6: 917–25

Lambrakis, K. C., Thompson, P. A. 1972. Existence of real fluids with negative fundamental derivative. *Phys. Fluids* 15: 933–35

Migdal, A. A. 1968. Diagram technique near the Curie point and phase transition of the second kind in Bose-fluids. *Zh. Eksp. Teor. Fiz.* 55: 1964–79

Novikov, I. I. 1948. Existence of shock waves rarefaction. *Dokl. Akad. Nauk SSSR* 59: 1545–46

Patashinsky, A. Z., Pokrovsky, V. L. 1966. On the behaviour of the ordering system near the phase transition point. *Zh. Eksp. Teor. Fiz.* 50: 439–47

Polyakov, A. M. 1968. Microscopic description of critical phenomena. *Zh. Eksp. Teor. Fiz.* 55: 1026–38

Rudenko, O. V., Soluyan, S. I. 1974. *Theoretical Foundations of Nonlinear Acoustics.* Moscow: Nauka. 288 pp. Transl., 1977, by Consultants Bureau (New York)

Schneider, W. G. 1951. Sound velocity and sound absorption in the critical temperature region. *Can. J. Chem.* 29: 243–52

Sidorenko, A. D. 1968. Wave adiabats for media with arbitrary state equation. *Dokl. Akad. Nauk SSSR* 178: 818–21

Sidorenko, A. D. 1982. Wave adiabats for media with an arbitrary equation of state. *Appl. Math. Mech.* 46: 241–47

Thompson, P. A. 1971. A fundamental derivative in gasdynamics. *Phys. Fluids* 14: 1843–49

Thompson, P. A., Lambrakis, K. C. 1973. Negative shock waves. *J. Fluid Mech.* 60: 187–208

Whitham, G. B. 1974. *Linear and Nonlinear Waves.* New York: Wiley. 622 pp.

Widom, B. 1965. Equation of state in the neighborhood of the critical point. *J. Chem. Phys.* 43: 3898–3905

Wilson, K. G. 1971. Renormalization group and critical phenomena. *Phys. Rev. B* 4: 3174–3205

Zel'dovich, Ya. B. 1946. The possibility of shock rarefaction waves. *Zh. Eksp. Teor. Fiz.* 16: 363–64

Zel'dovich, Ya. B. 1981. Establishment of the van der Waals critical point in fast processes. *Zh. Eksp. Teor. Fiz.* 80: 2111–12

SUBJECT INDEX

W

Y

Z

CUMULATIVE INDEXES

CONTRIBUTING AUTHORS, VOLUMES 1–19

CHAPTER TITLES, VOLUMES 1–19

Annual Reviews Inc.

ORDER FORM

A NONPROFIT SCIENTIFIC PUBLISHER

4139 El Camino Way
P.O. Box 10139
Palo Alto, CA 94303-0897 • USA

Now you can order
TOLL FREE
1-800-523-8635
(except California)

Annual Reviews Inc. publications may be ordered directly from our office by mail or use our Toll Free Telephone line (for orders paid by credit card or purchase order, and customer service calls only); through booksellers and subscription agents, worldwide; and through participating professional societies. Prices subject to change without notice. ARI Federal I.D. #94-1156476

- **Individuals:** Prepayment required on new accounts by check or money order (in U.S. dollars, check drawn on U.S. bank) or charge to credit card — American Express, VISA, MasterCard.
- **Institutional buyers:** Please include purchase order number.
- **Students:** $10.00 discount from retail price, per volume. Prepayment required. Proof of student status must be provided (photocopy of student I.D. or signature of department secretary is acceptable). Students must send orders direct to Annual Reviews. Orders received through bookstores and institutions requesting student rates will be returned.
- **Professional Society Members:** Members of professional societies that have a contractual arrangement with Annual Reviews may order books through their society at a reduced rate. Check with your society for information.
- **Toll Free Telephone orders:** Call 1-800-523-8635 (except from California) for orders paid by credit card or purchase order and customer service calls only. California customers and all other business calls use 415-493-4400 (not toll free). Hours: 8:00 AM to 4:00 PM, Monday-Friday, Pacific Time.

Regular orders: Please list the volumes you wish to order by volume number.
Standing orders: New volume in the series will be sent to you automatically each year upon publication. Cancellation may be made at any time. Please indicate volume number to begin standing order.
Prepublication orders: Volumes not yet published will be shipped in month and year indicated.
California orders: Add applicable sales tax.
Postage paid (4th class bookrate/surface mail) **by Annual Reviews Inc.** Airmail postage or UPS, extra.

ANNUAL REVIEWS SERIES		Prices Postpaid per volume USA/elsewhere	Regular Order Please send: Vol. number	Standing Order Begin with: Vol. number
Annual Review of ANTHROPOLOGY				
Vols. 1-14	(1972-1985)	**$27.00/$30.00**		
Vol. 15	(1986)	**$31.00/$34.00**		
Vol. 16	(avail. Oct. 1987)	**$31.00/$34.00**	Vol(s). ________	Vol. ________
Annual Review of ASTRONOMY AND ASTROPHYSICS				
Vols. 1-2, 4-20	(1963-1964; 1966-1982)	**$27.00/$30.00**		
Vols. 21-24	(1983-1986)	**$44.00/$47.00**		
Vol. 25	(avail. Sept. 1987)	**$44.00/$47.00**	Vol(s). ________	Vol. ________
Annual Review of BIOCHEMISTRY				
Vols. 30-34, 36-54	(1961-1965; 1967-1985)	**$29.00/$32.00**		
Vol. 55	(1986)	**$33.00/$36.00**		
Vol. 56	(avail. July 1987)	**$33.00/$36.00**	Vol(s). ________	Vol. ________
Annual Review of BIOPHYSICS AND BIOPHYSICAL CHEMISTRY				
Vols. 1-11	(1972-1982)	**$27.00/$30.00**		
Vols. 12-15	(1983-1986)	**$47.00/$50.00**		
Vol. 16	(avail. June 1987)	**$47.00/$50.00**	Vol(s). ________	Vol. ________
Annual Review of CELL BIOLOGY				
Vol. 1	(1985)	**$27.00/$30.00**		
Vol. 2	(1986)	**$31.00/$34.00**		
Vol. 3	(avail. Nov. 1987)	**$31.00/$34.00**	Vol(s). ________	Vol. ________

ANNUAL REVIEWS SERIES		Prices Postpaid per volume USA/elsewhere	Regular Order Please send:	Standing Order Begin with:
			Vol. number	Vol. number
Annual Review of COMPUTER SCIENCE				
Vol. 1	(1986)	**$39.00/$42.00**		
Vol. 2	(avail. Nov. 1987)	**$39.00/$42.00**	Vol(s). ______	Vol. ______
Annual Review of EARTH AND PLANETARY SCIENCES				
Vols. 1-10	(1973-1982)	**$27.00/$30.00**		
Vols. 11-14	(1983-1986)	**$44.00/$47.00**		
Vol. 15	(avail. May 1987)	**$44.00/$47.00**	Vol(s). ______	Vol. ______
Annual Review of ECOLOGY AND SYSTEMATICS				
Vols. 1-16	(1970-1985)	**$27.00/$30.00**		
Vol. 17	(1986)	**$31.00/$34.00**		
Vol. 18	(avail. Nov. 1987)	**$31.00/$34.00**	Vol(s). ______	Vol. ______
Annual Review of ENERGY				
Vols. 1-7	(1976-1982)	**$27.00/$30.00**		
Vols. 8-11	(1983-1986)	**$56.00/$59.00**		
Vol. 12	(avail. Oct. 1987)	**$56.00/$59.00**	Vol(s). ______	Vol. ______
Annual Review of ENTOMOLOGY				
Vols. 10-16, 18-30	(1965-1971, 1973-1985)	**$27.00/$30.00**		
Vol. 31	(1986)	**$31.00/$34.00**		
Vol. 32	(avail. Jan. 1987)	**$31.00/$34.00**	Vol(s). ______	Vol. ______
Annual Review of FLUID MECHANICS				
Vols. 1-4, 7-17	(1969-1972, 1975-1985)	**$28.00/$31.00**		
Vol. 18	(1986)	**$32.00/$35.00**		
Vol. 19	(avail. Jan. 1987)	**$32.00/$35.00**	Vol(s). ______	Vol. ______
Annual Review of GENETICS				
Vols. 1-19	(1967-1985)	**$27.00/$30.00**		
Vol. 20	(1986)	**$31.00/$34.00**		
Vol. 21	(avail. Dec. 1987)	**$31.00/$34.00**	Vol(s). ______	Vol. ______
Annual Review of IMMUNOLOGY				
Vols. 1-3	(1983-1985)	**$27.00/$30.00**		
Vol. 4	(1986)	**$31.00/$34.00**		
Vol. 5	(avail. April 1987)	**$31.00/$34.00**	Vol(s). ______	Vol. ______
Annual Review of MATERIALS SCIENCE				
Vols. 1, 3-12	(1971, 1973-1982)	**$27.00/$30.00**		
Vols. 13-16	(1983-1986)	**$64.00/$67.00**		
Vol. 17	(avail. August 1987)	**$64.00/$67.00**	Vol(s). ______	Vol. ______
Annual Review of MEDICINE				
Vols. 1-3, 6, 8-9	(1950-1952, 1955, 1957-1958)			
11-15, 17-36	(1960-1964, 1966-1985)	**$27.00/$30.00**		
Vol. 37	(1986)	**$31.00/$34.00**		
Vol. 38	(avail. April 1987)	**$31.00/$34.00**	Vol(s). ______	Vol. ______
Annual Review of MICROBIOLOGY				
Vols. 18-39	(1964-1985)	**$27.00/$30.00**		
Vol. 40	(1986)	**$31.00/$34.00**		
Vol. 41	(avail. Oct. 1987)	**$31.00/$34.00**	Vol(s). ______	Vol. ______

FROM

NAME ____________________

ADDRESS ____________________

____________ ZIP CODE ________

PLACE
STAMP
HERE

ANNUAL REVIEWS SERIES		Prices Postpaid per volume USA/elsewhere	Regular Order Please send:	Standing Order Begin with:
			Vol. number	Vol. number
Annual Review of SOCIOLOGY				
Vols. 1-11	(1975-1985)	**$27.00/$30.00**		
Vol. 12	(1986)	**$31.00/$34.00**		
Vol. 13	(avail. Aug. 1987)	**$31.00/$34.00**	Vol(s). ______	Vol. ______

Note: Volumes not listed are out of print.

SPECIAL PUBLICATIONS		Prices Postpaid per volume USA/elsewhere	Regular Order Please Send:
Annual Reviews Reprints: **Cell Membranes, 1975-1977**			
(published 1978)	Softcover	**$12.00/$12.50**	______ Copy(ies).
Annual Reviews Reprints: **Immunology, 1977-1979**			
(published 1980)	Softcover	**$12.00/$12.50**	______ Copy(ies).
Intelligence and Affectivity: Their Relationship During Child Development, by Jean Piaget			
(published 1981)	Hardcover	**$8.00/$9.00**	______ Copy(ies).
Telescopes for the 1980s			
(published 1982)	Hardcover	**$27.00/$28.00**	______ Copy(ies).
The Excitement and Fascination of Science, Volume 1			
(published 1965)	Clothbound	**$6.50/$7.00**	______ Copy(ies).
The Excitement and Fascination of Science, Volume 2			
(published 1978)	Hardcover	**$12.00/$12.50**	
	Softcover	**$10.00/$10.50**	______ Copy(ies).

TO: **ANNUAL REVIEWS INC.,** a nonprofit scientific publisher
4139 El Camino Way
P.O. Box 10139
Palo Alto, CALIFORNIA 94303-0897

Please enter my order for the publications listed above. California orders, add sales tax. Prices subject to change without notice.

Institutional purchase order No. ______

Amount of remittance enclosed $ ______

Charge my account ☐ VISA ☐ MasterCard ☐ American Express

INDIVIDUALS: Prepayment required in U.S. funds or charge to bank card below. Include card number, expiration date, and signature.

Acct. No. ______

Exp. Date ______ ______
Signature

Name ______
(Please print)

Address ______
(Please print)

______ Zip Code ______

______ Send free copy of current ***Prospectus*** ☐

Area(s) of Interest

Federal I.D.#94-1156476